D0079125

Sustainable Development of the Biosphere

The International Institute for Applied Systems Analysis

The International Institute for Applied Systems Analysis (IIASA) is a nongovernmental research institution, bringing together scientists from around the world to work on problems of common concern. Situated in Laxenburg, Austria, IIASA was founded in October 1972 by the academies of science and equivalent organizations of twelve countries. Its founders gave IIASA a unique position outside national, disciplinary, and institutional boundaries so that it might take the broadest possible view in pursuing its objectives, listed opposite.

The national member organizations of the Institute are listed below.

To promote international cooperation in solving problems arising from social, economic, technological, and environmental change.

To create a network of institutions in the national member organization countries and elsewhere for joint scientific research.

To develop and formalize systems analysis and the sciences contributing to it, and promote the use of analytical techniques needed to evaluate and address complex problems.

To inform policy advisors and decision makers about the potential application of the Institute's work to such problems.

Austria
The Austrian Academy of Sciences

Bulgaria
The National Committee for Applied Systems Analysis and Management

Canada
The Canadian Committee for IIASA

Czechoslovakia
The Committee for IIASA of the Czechoslovak Socialist Republic

Finland
The Finnish Committee for IIASA

France
The French Association for the Development of Systems Analysis

German Democratic Republic
The Academy of Sciences of the German Democratic Republic

Germany, Federal Republic of
Association for the Advancement of IIASA

Hungary
The Hungarian Committee for Applied Systems Analysis

Italy
The National Research Council

Japan
The Japan Committee for IIASA

Netherlands
The Foundation IIASA–Netherlands

Poland
The Polish Academy of Sciences

Sweden
The Swedish Council for Planning and Coordination of Research

Union of Soviet Socialist Republics
The Academy of Sciences of the Union of Soviet Socialist Republics

United States of America
The American Academy of Arts and Sciences

Sustainable Development of the Biosphere

Editors

William C. Clark and R. E. Munn

International Institute for Applied Systems Analysis, Laxenburg, Austria

Published on behalf of the International Institute for Applied Systems Analysis by Cambridge University Press with support of the Government of Canada and the German Marshall Fund of the United States.

CAMBRIDGE UNIVERSITY PRESS
Cambridge
London New York New Rochelle
Melbourne Sydney

Published by the Press Syndicate of the University of Cambridge.
The Pitt Building, Trumpington Street, Cambridge CB2 1RP.
32 East 57th Street, New York, NY 1002, USA.
10 Slamford Road, Oakleigh, Melbourne 3166, Australia.

First published 1986

Typeset by J. W. Arrowsmith Ltd, Bristol, UK.
Printed in Austria by Gistel Druck, Vienna

British Library cataloguing in publication data
Sustainable development of the biosphere.
1. Ecology 2. Economic development
I. Clark, W. C. II. Munn, R. E.
304.2′8 QH541

Library of Congress cataloging in publication data
Sustainable development of the biosphere.
Includes index.
1. Economic development – Environmental aspects.
2. Environmental policy, I. Clark, W. C. (William C.)
II. Munn, R. E.
HD75.6.S87 1986 363.7 86-12996

ISBN 0 521 32369 X hard cover
ISBN 0 521 31185 3 paperback

Preface

Mankind has often degraded the local and even the regional environment. Only recently, however, has the possibility been envisaged that local actions could lead to irreversible global changes. The most awful example, of course, is the *nuclear war scenario.* But consider also the worldwide rise in atmospheric concentrations of CO_2, methane, and chlorofluorocarbons.

There are few historical analogies to guide us on the consequences of these global changes, and the predictions of simulation models are rather uncertain. How then is society to respond? This is a typical high-impact, low-probability risk management problem for which the stakes are very high indeed.

Within this context, the IIASA Feasibility Study on Sustainable Development of the Biosphere was formulated in early 1983. A Prospectus outlining the Study was prepared by Dr. W. C. Clark and circulated widely in the summer of 1983. Subsequent meetings in Ottawa, Cambridge (MA), Mainz, Toronto, Moscow, and Leningrad helped to clarify the framework for the Study, which has the following points:

(1) To synthesize in policy terms our understanding of global ecological and geophysical systems as they are linked with industrial and resource development activities.
(2) To characterize the issues of global environmental change in terms of their ability to inhibit or promote regional development.
(3) To explore institutional and organizational designs for more effective international research, policymaking, and management, concerning interactions between environment and regional development.

These ideas were elaborated at a Task Force Meeting held at IIASA, 27–31 August, 1985. The consensus of that Meeting was that it was timely and, indeed, urgent to begin a long-term research investigation of the sustainable development of the biosphere – in the context of resource management.

The overview papers presented at the Task Force Meeting were subjected to very wide ranging reviews at the Meeting and subsequently. These papers are contained in this volume, together with a number of commentaries written later by persons not necessarily present at the Meeting. We believe that the authors have provided a new perspective on resource development in a world of ecological uncertainty and surprise.

Subsequent to the Workshop, a number of important developments have taken place with respect to the Feasibility Study. These include a successful Symposium and Planning meeting, Moscow, 14–16 March, 1985, when:

(1) The name of the study was changed to *Ecologically Sustainable Development of the Biosphere* to emphasize the ecological aspects of development.
(2) A memorandum assuring collaboration between Soviet and IIASA scientists was signed by Academician Yuri Izrael (Chairman of the USSR State Committee for Hydrometeorology and Control of the Natural Environment) and Professor Thomas Lee (Director of IIASA).

Another important development in 1985 was the approval by the IIASA Council (June 1985) of the Biosphere Project as one of IIASA's major activities during the next several years. Several national members of IIASA have indicated their intention to support the Project, and the Canadians have already launched a counterpart study.

We are indebted to the Government of Canada for substantial financial support provided to the organization of the Task Force Meeting and to subsequent editing and publication of this book. In

fact, it is fair to say that the Meeting might not have happened without the intellectual and financial commitment made by Canada.

Additional support for the Task Force Meeting was provided by the German Marshall Fund of the United States.

William C. Clark and R. E. Munn
Editors

Contents

PART ONE

OVERVIEW

Introduction

Contributor

About the Contributor

William C. Clark leads the studies on *Sustainable Development of the Biosphere* at the International Institute for Applied Systems Analysis (IIASA), in Laxenburg, Austria. Educated at Yale University and the University of British Columbia, Clark joined IIASA from the Institute for Energy Analysis in Oak Ridge, Tennessee (USA). His research has included basic ecological studies on the stability and resilience of ecological systems, policy analysis for resource and environmental management, work on understanding societal risk-taking behavior, and a retrospective assessment of human development strategies being pursued in the Third World. He is a member of the US National Academy of Science's Board on Atmospheric Sciences and Climate and its Committee on the Applications of Ecological Theory to Environmental Problems.

Clark is a coeditor of *Environment* magazine. He is also coauthor of *Redesigning Rural Development: A Strategic Perspective* (Johns Hopkins, 1982) and *Adaptive Environmental Assessment and Management* (John Wiley, 1978), and editor of the *Carbon Dioxide Review* (Oxford University Press, 1982). In 1983, Dr. Clark was named a MacArthur Prize Fellow.

Chapter 1 Sustainable development of the biosphere: themes for a research program

W. C. Clark

Introduction

The long sweep of human history has involved a continuing interaction between peoples' efforts to improve their well-being and the environment's ability to sustain those efforts. Environmental constraints have always limited where and how people conduct their lives. Sometimes the constraints have yielded to technological or organizational innovation, allowing further social development. At other times development has yielded to environmental constraints, contributing to social stagnation and human suffering. The productive capacity of individual ecosystems has been sometimes enhanced, sometimes degraded as a result of their encounters with the human species.

Throughout most of history, the interactions between human development and the environment have been relatively simple and local affairs. But the complexity and scale of these interactions are increasing. What were once local incidents of pollution shared throughout a common watershed or air basin now involve multiple nations – witness the concerns for acid deposition in Europe and North America. What were once acute episodes of relatively reversible damage now affect multiple generations – witness the debates over disposal of chemical and radioactive wastes. What were once straightforward questions of ecological preservation versus economic growth now reflect complex linkages – witness the feedbacks among energy and crop production, deforestation and climatic change that are evident in studies of the atmospheric "greenhouse effect".

Economic and environmental interdependence

These linkages are both economic and environmental. Even one hundred years ago the demand for wood and agricultural products in Europe was shaping patterns of deforestation and wetland conversion in lands surrounding the Indian Ocean. More recently, changes in world energy prices, precipitated by events in the Middle East, have increased the utilization of biomass for fuel throughout the world. In the future, the degree to which today's developing countries participate in the international grain markets could well shape the use and degradation of environmentally marginal land in the American Great Plains.

Humanity is thus entering an era of chronic, large-scale, and extremely complex *syndromes* of interdependence between the global economy and the world environment. Relative to the earlier generations of problems, these emerging syndromes are characterized by profound scientific ignorance, enormous decision costs, and time and space scales that transcend those of most social institutions. The difficulties posed by these changes will intensify over the next century as the number of people on Earth, the amount of industrial production, and the demand for agricultural products increase doubly or more.

A major challenge of the coming decades is to learn how long-term, large-scale interactions between environment and development can be better managed to increase the prospects for ecologically sustainable improvements in human well-being. Management is not the same as prediction.

The distinction is an important one, for management can be improved despite the enormous uncertainties and downright ignorance that will continue to make detailed predictions illusory. This point was eloquently argued by the distinguished members of the Bellagio Conference on Science, Technology, and Society:

> The world faces serious problems today, which require concerted effort by all nations for their solution. Much has been written about these problems, and the limitations within which solutions can be found. But the limitations are not those frequently assumed. Nature offers us many opportunities to readjust our technologies to solve problems. Nevertheless, some physical limitations, particularly those imposed by the ecological balance of which man is part, are real and must be respected. We must learn to live with a naturally dynamic ecosystem and not make unreasonable demands for short term stability. Rather, those long term trends that are more likely to determine the survivability of human society must be identified and properly managed. In ecological terms we urge that greater heed be given to resilience rather than stability [1].

Which long-term trends of environmental change could most severely limit societies' development? How can we obtain more usable knowledge about these limits and the options for avoiding or relaxing them? What scientific research, institution building, or technological innovation might be undertaken over the next decade to increase future freedom of action in the quest for a sustainable development of the biosphere?

Intellectual foundations

At the time of the Stockholm Conference on the Environment in 1972, efforts to address such questions were constrained by the virtual absence of relevant data, explanatory hypotheses, and interdisciplinary research experience. Over the last decade, however, advances in a number of fields have begun to lay the foundations for effective research on interactions of the global economy and the world environment. Consider the following specifics:

(1) Studies of the biosphere have been revolutionized by the emergence of an integrative global perspective that emphasizes biogeochemical processes and their connections with climate. Plans for new international research programs on "Global Ecology", "Global Habitability", "Global Change", and the like, stress the needs and opportunities for close collaboration among biologists, atmospheric specialists, oceanographers, geologists, and other earth scientists [2].

(2) Efforts to understand the broad patterns and processes of human development have been comparably stimulated by the rise of the "world system" view of modern history. This has focused on the interplay of institutions, technologies, and resources over what Braudel has called *la longue durée*, thus providing fertile ground for the collaboration of economists, historians, and geographers [3].

(3) Parallel advances have occurred in the formulation of concepts and methods needed to deal with the multiple equilibria, thresholds, and discontinuities that increasingly seem to be the norm rather than the exception in interactions between human activities and the environment. Atmospheric scientists, ecologists, students of sociotechnical systems, and others are converging on a usable understanding of how such complex behaviors arise from nonlinear systems operating across a range of time and space scales [4].

(4) Finally, over the last decade or so a great deal of experience has accumulated in the conduct of interdisciplinary studies that link incomplete science with policy problems. The strengths and limitations of alternative approaches to synthesis and analysis have become better understood through this experience. New hybrid styles are emerging combining the best of the expert committee and formal modeling approaches [5].

The challenge is to capitalize on these independent advances by integrating them in ways that can help to fashion more effective tools for understanding and managing interactions between the global economy and the world environment.

A program on sustainable development

This book concludes the first phase of a new international program, *Ecologically Sustainable Development of the Biosphere*, based at the International

Institute for Applied Systems Analysis (IIASA). It involves a collaborative network of historians, engineers, geographers, environmental scientists, economists, managers, and policy people from around the world. The program began in 1982 when IIASA's then Director, C. S. Holling, asked the present author, then at Oak Ridge Associated Universities (USA), to undertake an assessment of what additional work beyond that already underway would most help to realize the potential for better management of interactions between development and environment [6]. My initial draft report to IIASA was reviewed, critiqued, and built upon through a series of meetings hosted by the Canadian Government in Ottawa and Toronto, the American Academy of Arts and Sciences in Cambridge and Washington, and the Soviet Academy of Sciences in Moscow and Leningrad. Emerging from this process was a conviction that the program would best complement existing work on environment–development interactions by emphasizing four characteristics: a synoptic perspective, a long-term time horizon, a regional to global scale, and a management orientation.

Synoptic perspective

Most studies of interactions between development and environment have concentrated on individual problems like soil erosion, ozone depletion, "greenhouse" effects, deforestation, acid deposition, and the like. In fact, however, individual human activities, such as fossil fuel burning or land-clearance, simultaneously affect many such problems. Conversely, many of the individual problems are simultaneously affected by multiple economic activities and natural processes. Studies of individual causes and effects are inadequate for policymakers who confront the increasingly complex and threatening syndromes of environmental and economic interdependence. The need is rather for a synoptic approach that addresses those interdependencies directly.

Long-term time horizon

Many environmental problems reflect unanticipated, long-term consequences of development activities originally undertaken for their short-term benefits. Most environmental studies, however, focus on the more or less immediate impacts and ameliorative measures. Needed is a complementary program that adopts a sufficiently long-term time horizon to encompass the full interplay between developmental and environmental processes. Analyses of these processes indicate a time horizon one century into the future and, for perspective, at least two centuries into the past.

Regional to global scale

Many of the most intractable environmental problems arise because either the causes or the consequences of development activities extend far beyond the local or even national areas responsible for managing these activities. Most environmental studies, nonetheless, focus on local and national interactions with human development. The need is for complementary work to analyze the relevant processes over spatial scales sufficiently broad to encompass the extensive interactions of contemporary development activities with the world's environment. Analysis of these processes lead us to focus on interactions that become important on scales of large regions (e.g., 1×10^6 km^2 or more) to the globe as a whole. Our effort is an attempt to "think globally while acting locally" – that is, to link our large-scale perspective on consequences and implications to the smaller scales at which specific actions and policy choices are undertaken, and at which their impacts are experienced.

Management orientation

Little is to be gained from just another effort to predict impacts of development on the environment over the broad time and space scales that concern us here. Our challenge is rather to characterize potentially intense interactions between future development efforts and the environment, and to identify specific policies and management actions that could make these interactions more to societies' liking, and less threatening to global life-support systems.

The plan of this chapter

The international meetings and discussions described above outlined a set of review papers that would provide innovative points of departure for the planning of a specific research program. Authors were identified and asked to explore the contributions that their respective areas of interest might make to an assessment of long-term, large-scale interactions between development and environment. Special attention was paid to identifying linkages between traditionally unrelated disciplines and to articulating agendas for collaborative research. The initial essays were reviewed in late 1984 at an international workshop attended by the essay authors, other scholars, and policymakers. The essays were revised, often extensively, to reflect discussions at the workshop. A further critical dimension was provided by the independent *Commentaries* that accompany each chapter in this volume. Taken together, these contributions provide the foundations and initial scaffolding on which our continuing inquiry into a sustainable development of the biosphere is being built.

In the remainder of this introductory chapter, I draw from the chapters a sketch of the central questions that need to be better understood if we are to shape a more usable understanding of long-term, large-scale interactions between environment and development.

An Earth transformed by human action

Today's concerns about a sustainable development of the biosphere are part of a long intellectual tradition. The evolution of this tradition reflects both changing scales of interactions between human activities and the environment and changing visions of what these interactions should be. We cannot expect the Biosphere Program to free itself from its cultural and historical contexts, nor, indeed, should it. But if the Program is to be much more than propaganda in support of one or another preferred social agenda, then it must be built on a self-awareness of these contexts and the biases they entail. I have found the writings of Gilbert White, Lynton Caldwell, Ian Burton, and Robert Kates to be particularly helpful in establishing this self-awareness [7]. Related perspectives in this volume are provided by Williams, Timmerman, and Ravetz.

Malthusian perspectives

Concerns about possible environmental limits to human development extend back about as far as one is willing to look. Plato, in his *Critias*, was already lamenting that agricultural activities had transformed the land of Attica into the "bones of a wasted body . . . the richer and softer parts of the soil having fallen away, and the mere skeleton being left [8]." The terms of the modern debate, however, were defined by Malthus. Writing almost two centuries ago, he characterized human well-being in terms of the ratio of environmental resources to human population. Malthus' assessment was bleak, for he saw in food production a land-limited resource that could not possibly be increased quickly enough to keep pace with a growing human population [9].

Events proved Malthus wrong, at least for most of the world over most of the last two hundred years. But, as Williams and Timmerman point out in their contributions to this volume, the Malthusian vision has nonetheless persisted, shaping the thinking of subsequent generations. *Figure 1.1* shows how both the numerator and the denominator of Malthus' argument have evolved through time [10]. Malthus' worries about agricultural limits to Britain's growth have become today's worries about the limits that food, materials, energy, water, and the biosphere's absorptive capacity might impose on human development of the planet.

Malthus was provoked to write his *Essay on the Principle of Population* by what he saw as the utopian visions of Condorcet. Supporters of Adam Smith's competing vision were immediately provoked to refute Malthus. Ever since, the debate has swung between two poles of thought. As characterized by Burton and Kates:

> In its extreme form, one pole is determinist in its view of nature, Malthusian in its concern with the adequacy of resources, and conservationist in its prescription for policy. The opposite pole is possibilist in its attitude toward nature, optimistic in its view of technological advance and the sufficiency of resources and generally concerned with technical and managerial problems of development [11].

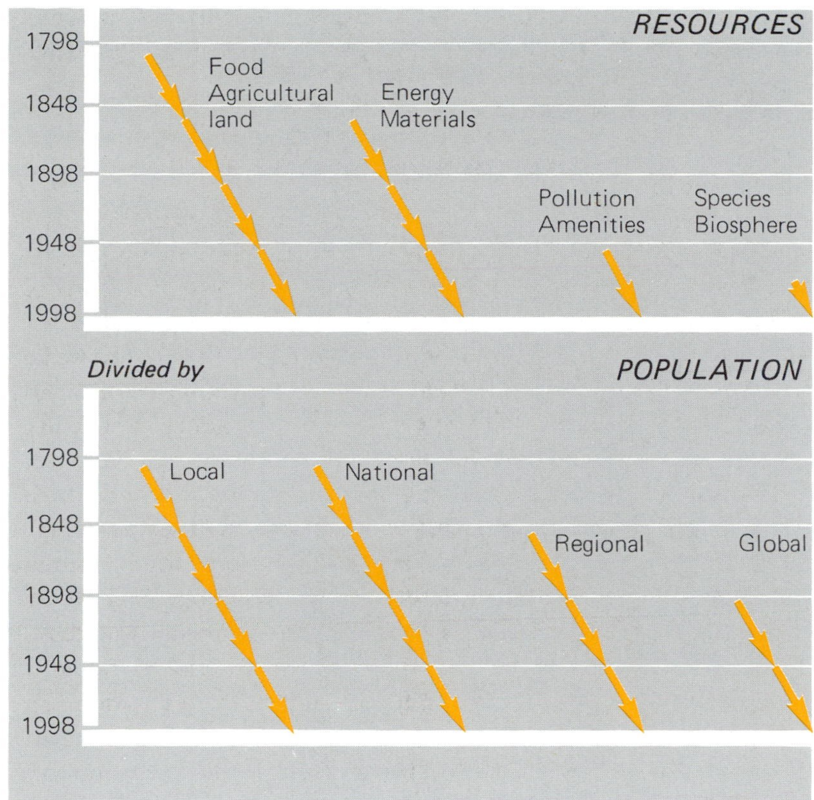

Figure 1.1 Historical changes in the numerator and denominator of the "Malthus Ratio", from Kates and Burton [10].

Both poles have been articulately and extensively defended. The conservationist tradition runs from George Perkins Marsh's *The Earth as Modified by Human Action* (1874), through F. Osborn's *Our Plundered Planet* (1948), Harrison Brown's *The Challenge of Man's Future* (1954), Rachel Carson's *Silent Spring* (1962), and the Meadows' *Limits to Growth* (1972), to the US Government's mammoth report, *Global 2000* (1980). Exponents of the possibilist view have ranged from Reclus' *La Terre et les Hommes* (1876), through numerous publications from Resources for the Future and the Hudson Institute in the 1960s and 1970s, to Julian Simon's *The Resourceful Earth* (1984) and the World Resources Institute's recent celebration of *The Global Possible* [12].

The rhetorical gifts of advocates on both sides of the debate have been so strong as to leave many bystanders, myself included, under the spell of whichever has been most recently encountered. But as Alvin Weinberg has remarked, "one's reaction . . . is as much a reflection of one's own preferences and biases as it is an objective estimate of the correctness of [the] divergent views [13]." My own biases are possibilist, though of the "despairing optimist" variety portrayed by Rene Dubos. As is evident elsewhere in this volume, other participants in our Biosphere Program have entered from the conservationist side of the debate. We have not attempted to eliminate or even to balance the biases these different perspectives entail. Rather, as urged by Thompson (this volume), we are striving for a pluralism that keeps our biases explicit while benefiting from the multiple perspectives they generate.

The increasing scale of interactions

The debate between the conservationists and the possibilists concerns the prospects and implications of interactions between human development and the environment. The swings of this debate, however, should not obscure the increasingly global scale that these interactions have assumed over the last hundred years [14].

The emergence of a global perspective on the impacts of human habitation on the Earth can be traced back to the nineteenth century. A convenient reference point is George Perkins Marsh's *The Earth as Modified by Human Action*, published in 1874 [15]. Drawing on his own extensive travels and reading, Marsh set out "to indicate the character and, approximately, the extent of the changes produced by human action in the physical conditions of the globe we inhabit." Successive chapters "traced the history of man's industry as exerted upon Animal and Vegetable Life, upon the Woods, upon the Waters, and upon the Sands." Though global in the scope of his concerns and examples, most of Marsh's examples described modifications of individual localities or small regions as a result of human development. The possibility of larger modifications to come, however, was raised in a final chapter on "Great projects of physical change accomplished or proposed by man". Marsh's prescient account included discussion of a sea level Panama canal, diversion of the Nile and Colorado Rivers, adjustment of the water level in the Caspian and Aral Seas by diverting the rivers of central Russia, urban waste disposal, intentional weather modification, and the unintended cumulative impact of agricultural and industrial activity. His balanced and graciously written analysis pointed out the human benefits that such developments could entail, but also cautioned that unexpected and deleterious environmental side-effects might occur.

Fifty years later, the world's population had increased by almost 50%. Its industrial activity had grown more than five-fold. The increasing scale and significance of these human developments as agents of global environmental change were forcefully articulated by a remarkable group of scientists. These included the French theologian and paleontologist Pierre Teilhard de Chardin, the Austrian-born American biophysicist Alfred J. Lotka, and, above all, the Russian mineralogist Vladimir Ivanovitch Vernadsky.

The writings of Teilhard de Chardin were, as often, obscure if inspirational. But Lotka could have been contributing to a present-day government report when he wrote, in 1924:

> [W]hatever may be the ultimate course of events, the present is an eminently atypical epoch. Economically, we are living on our capital; biologically we are changing radically the complexion of our share in the carbon cycle by throwing into the atmosphere, from coal fires and metallurgical furnaces, ten times as much carbon dioxide as in the natural biological process of breathing . . . [16].

Vernadsky's perspective was even deeper and more prophetic. Recapitulating in 1945 themes he had first annunciated in lectures and books 20 years earlier, he described the concept of:

> . . . the biosphere, the only terrestrial envelope where life can exist. Basically, man cannot be separated from it . . . He is geologically connected with its material and energetic structure [17].

Echoing the Italian geologist Stoppani, Vernadsky argued that the most significant aspect of man's development was not his technology *per se*, but rather the sense of global knowledge and communication engendered by that technology [18]. He portrayed this "noosphere" or "realm of thought" as

> a new geological phenomenon on our planet. In it for the first time man becomes a large-scale geologic force . . . Chemically, the face of our planet, the biosphere, is being sharply changed by man consciously, and even more so, unconsciously [19].

Vernadsky and his contemporaries had neither the data nor the instruments to convert their insights into useful analytical tools for understanding the global environmental impacts of human development. Over the last 50 years, however, and especially since the International Geophysical Year (IGY) of 1956–1957, the needed data and measurements have begun to be accumulated. Equally important, so have the breathtaking photographs from space that portray the biosphere from a global perspective. From this new knowledge and vision has been built a still incomplete, but nonetheless quantitative and detailed picture of the global environmental consequences of human development.

The major features of this picture were sketched in a recent report of the International Council of Scientific Union's Scientific Committee on problems of the Environment (SCOPE). They are worth quoting at length:

> ● Man's activities on earth today induce fluxes of carbon, nitrogen, phosphorus and sulphur that are of similar magnitude to those associated with the natural global cycles of these elements; in limited areas man's influence dominates the cycles. The

likely increase of man's activities during the remainder of this and during the next century will undoubtedly mean significant disturbances of the global ecosystem.

- The most important ways whereby man is interfering with the global ecosystem are:

  – fossil fuel burning which may a) double the atmospheric CO_2 concentration by the middle of the next century; b) further increase the emissions of oxides of sulphur and nitrogen very significantly;

  – expanding agriculture and forestry and the associated use of fertilizers (nitrogen and phosphorus) significantly alter the natural circulation of these nutrients;

  – increased exploitation of the fresh water system both for irrigation in agriculture and industry and for waste disposal.

- According to our present understanding, the most important impacts of these changes in the long-term perspective are:

  – a gradual change towards a warmer climate, the details and implications of which we know very little about;

  – the concentration of ozone will decrease in the stratosphere, due to the increased release of N_2O and chlorine compounds and increase in the troposphere, due to the increased release of NO_x and hydrocarbons:

  – an increase of the areas affected by lake and stream acidification in mid-latitudes and possibly also in the tropics; the ion balance of the soils may be significantly disturbed, as is now being found with regard to aluminum;

  – a decrease of the extent of tropical forests, which will enhance the rate of increase in atmospheric CO_2 concentration and release other minor constituents to the atmosphere; this may also contribute to soil degradation;

  – due to loss of organic matter and nutrients, soil deterioration will occur and this implies a reduced possibility for the vegetation to return to pristine conditions. . . ;

  – a trend toward the eutrophication of estuarine and coastal marine areas;

  – more frequent development of anoxic conditions in fresh-water and marine systems and sediments.

- The development and continuation of highly productive units in agriculture and forestry means an increasing dependence on technological advances that, to be properly directed, requires profound knowledge about long term modifications of the soil.

- The long-term implications of exploiting the natural resources of the earth are not well understood, nor do we understand what is permissible in order to guarantee that present or future (possibly higher) levels of productivity will not later decline . . . [20].

The modern world described by the SCOPE report supports three times the human population and 100 times the industrial activity that it did a century ago, when Marsh described the Earth "as modified by human action." As the SCOPE report shows, this vast increase in human activity has left the Earth not just modified, but fundamentally transformed.

The Earth as garden

What images are appropriate for thinking about an Earth transformed by human action? As emphasized by Holling, Timmerman, and Thompson in their contributions to this volume, the choice of the images or myths we use to structure our accounts of the world is a fundamental one, which radically constrains the questions we ask and answers we get. Much of what Burton and Kates characterized as conservationist thinking seems to be based on an image of nature in its "original state", perturbed by sporadic human blundering. The concerned possibilists, in contrast, have embraced the image of "spaceship Earth" as a creation of and responsibility for enlightened engineering.

Neither image, however, will quite do. On the one hand, the Earth *has* been transformed by human activity, with hardly a corner that is not now being managed, at least in some limited sense. On the other hand, we have learned just enough about the planet and its workings to see how far we are from having either the blueprints or the operator's manual that would let us turn that diffuse and stumbling management into the confident captaincy implied by the "spaceship" school of thought.

Reflecting on these inadequacies, Harvey Brooks has argued that today's Earth "is more like a garden than a primeval forest, even if the garden is ill-kept [21]." I like the "garden" image, not the least because it emphasizes the human *use* of Earth for productive purposes. I should add, however, that my personal gardening experience is not of trim English hedges, but rather of the tense encounter in the foothills of the American Appalachian Mountains between vines, bugs, beasts, tornados, and the would-be gardener.

If we accept the garden image as a useful one, two questions arise: What kind of garden do we want? What kind of garden can we get?

The first of the questions – "What kind of garden do we want?" – ultimately calls for an expression of values. The values on which we have based this study – the kinds of garden we want – are suggested in our choice of title: *The* Sustainable *Development of the Biosphere.* The commonsense meaning of "sustainable" is a good first approximation of our intended meaning. We seek to distinguish gardening strategies that can be sustained into the indefinite future from those that, however successful in the short run, are likely to leave our children bereft of nature's support.

There is a strong anthropomorphic bias in these "sustainability" values – people worry about nature primarily in terms of what nature means for people's own welfare. If environmental degradation occurs perhaps it can be traced to people's failure to define their welfare in sufficiently long-run terms to include their descendants. Other less anthropomorphic biases are possible, but I suspect that most of us eventually return to our essential humanness.

The second question raised above is one of feasibility: "What kind of garden can we get?" While not divorced from value judgments, this latter question is fundamentally one of knowledge and know-how. Environmental degradation results not only from insufficient value placed on the long run, but also from sheer ignorance of how environment and development interact. What kinds of long-term development pressures are likely to be the sources of the next century's environmental transformations? What are the implications of these transformations for the biosphere's productive capacity? How can the transformations be managed to shape a garden more to our liking, yet still enhance our children's options?

Managing global change: technologies, institutions and research

What kind of research program might actually produce usable knowledge for the management of long-term, large-scale interactions between development and environment?

Our study has set as its ultimate goal the identification and evaluation of strategic interventions through which societies might change the long-term, large-scale interactions between development and environment. In particular, we seek to show how important technological, institutional, and research strategies that might be set in place over the next decade could affect the prospects for a sustainable development of the biosphere. Before describing these goals in more detail, it is useful to emphasize two things our objectives are not.

First, our objective is *not* to produce yet another set of detailed predictions about future environmental impacts. Such predictive efforts are naive on two counts: technical infeasibility and practical irrelevance. They are infeasible because of the complexity of the environmental syndromes arising from interactions with development, the dependence of these interactions on unknowable future patterns of social choice and evolution, and the incomplete state of the sciences required for their assessment. They are irrelevant because there is little serious demand for detailed predictions on the part of experienced politicians, businessmen, or managers. Such action-oriented people have shown themselves to be less interested in dubious predictions of "what's going to happen" than in understanding how they might intervene to make what actually does happen more to their liking and benefit [22].

Second, our objective is *not* to produce yet another lecture on "what society should do". The syndromes arising from the interactions between development and environment are simultaneously global and local. All people on Earth are affected by them, now and in the future; no person and few nations can do much unilaterally to alter them. What people do today will change both the constraints and opportunities that face later generations. The nature and severity of environmental syndromes – the kind of garden that "we" want – will thus differ greatly among people, times, and places. Useful tools for managing the changes of our global garden will be tools that can serve differing (and evolving) local, national, and generational perceptions of the good and proper life.

The design and evaluation of options for intervention

In practice, technologies, institutions, and research are not independent entities. Studies of the biosphere over the last 100 years provide an excellent example of how the questions asked by researchers are influenced by the kinds of measurement technologies available to them. Likewise, as Harvey Brooks points out in Chapter 11, the assessment of any major technology requires an appreciation of the institutionalized "operating" system and social structure within which it is embedded. Finally, it should by now be universally accepted that significant technologies shape the societies that employ them. Nonetheless, it is useful to focus initially on questions regarding the individual technologies, institutions, and researches that we need to explore in order to evaluate their potential as tools for managing long-term, large-scale interactions of development and environment.

Technologies

Many of the most significant changes in the nature and consequences of long-term, large-scale interactions between development and environment over the last several centuries have been intimately associated with technological innovation. For example, Richards (Chapter 2) tells how the Mallee scrublands of Australia resisted clearance until roller dragging technologies were invented. The fuel wood crisis of eighteenth century Europe was resolved through the development of coal burning technologies, which thereby contributed to the reforestation of much previously cleared land.

In these examples, and many others documented later in this volume, the transformation of the environment is essentially a by-product of technological change. Arguments summarized in the next two sections suggest that the most environmentally significant technologies over the long term and large scale tend to be those directly involved in the transformation of land, water, fuel, and other mineral resources. Within each of these "transforming technology" categories, we seek to characterize several strategic options that might be pursued over the next decade in one or more large regions of the globe. One example is the energy system of horizontally integrated fuel sources and zero pollution emissions described by Häfele and his colleagues in Chapter 6. Another might be the major river diversions now being implemented or planned in several parts of the world [23]. A third is the choice between agricultural technologies (including the genetic engineering of crops) that are more or less intensive in their use of land, fuel, and fertilizer inputs. In all cases, our aim is to characterize how such transforming technologies, adopted at specific rates and over specific areas, would alter the environmental consequences of economic and social development.

Institutions

The institutions through which people, economies, and governments decide upon and implement their activities also provide a potential means of intervening in the interactions between environment and development. At one extreme, these include formal international organizations, such as FAO, IAEA, and UNEP; multilateral agreements, such as the *Law of the Sea* or the *Ozone Convention*; the laws and regulations that embody policies of the world's nation states. Somewhat less formal are the market structures and command hierarchies that probably mediate the majority of interactions between development and environment. At the informal extreme, it is important to appreciate the implications of different cultural and religious beliefs for sustainable development practices. (In a recent experiment several Islamic scholars drafted a corpus of environmental law based on the United States' National Environmental Policy Act, but invoking the Koran rather than the State as the primary source of motivation and authority.)

Experience suggests that some kinds of institutions provide better means of dealing with certain kinds of large-scale, long-term environmental syndromes than others [24]. For example, "commons" problems, such as the depletion of stratospheric ozone by halocarbon releases, are particularly resistant to market-mediated resolution. At least when substitutes have been available, however, they have sometimes been mitigated through formal international negotiations or even moral suasion (see Majone, Chapter 12, and Haigh's *Commentary*). On the other hand, intergenerational trade-offs of the sort involved in soil erosion or hazardous waste disposal require quite different institutional responses. Perry has recently reviewed the performance of various international science organizations with a view toward identifying strengths, weaknesses, and functional gaps [25]. But further work on designing useful typologies of institutions and syndromes is urgently required [26].

Research and monitoring

In the very long run, increased knowledge about the ways in which economic activities interact with the environment may prove to be our most powerful tool for managing a sustainable development of the biosphere. Experience suggests that much of the research agenda will take care of itself. For the long term that concerns us here, the most useful returns will almost certainly come from studies on the fundamental processes that link the chemical, physical, geological, and biological components of the Earth system. The broad research programs now being formulated in discussions of "Global Change", "Global Habitability", and the like, if adequately funded, can be expected to provide the foundations of knowledge on which specific management actions can be built. Our challenge is to provide policymakers with rapid access to, and a critical appreciation of, these foundations, so that development decisions can be taken with the best possible understanding of their environmental implications.

There remain, however, areas in which greater attention to the specific scientific understanding required for informed development choices could generate useful additions to the research agenda. As one example, Dickinson (Chapter 9) argues that for the "greenhouse" syndrome of fossil fuel burning, deforestation, and climatic change, more policy and research attention should be devoted to the nature and consequences of low-probability, high magnitude climatic warmings. And Parry (Chapter 14), in his treatment of the managerial implications of long-term climatic changes, shows that more research on the relation between the changing means and the changing frequency of extreme events might be extremely useful. Still required are systematic means of assessing the special research tasks in the natural sciences that are needed to inform long-term, large-scale development choices.

It is in the area of environmental monitoring, however, that I suspect a consideration of future development possibilities could have the greatest impact on the natural science studies. Michael Gwynne, the Director of the Global Environmental Monitoring System (GEMS), notes in his *Commentary* to Chapter 13 that early monitoring efforts were designed with too little attention to the eventual practical application of their results. For example, it is not entirely unfair to say that much of the history of the Landsat satellite program and its relatives has been one of rather too many answers (those beautiful, false-color photos on everyone's walls) in search of a few urgent questions. Recent monitoring initiatives have done somewhat better in this respect (see Regier and Baskerville, Chapter 3). The new generation of monitoring systems currently being designed need to have ways of being even more responsive to future management needs.

This does not mean subordinating basic scientific goals to short-term perceptions of crisis. It does mean recognizing that the next generation of large-scale environmental monitoring systems will be with us for many decades into the future. Over that period, the environment will be transformed significantly by human activities. Both science and management will be better served if the design of monitoring systems reflects some systematic thought on the nature and magnitude of the possible transformations, the practical problems they may create, and the managerial choices they will require. As Thomas Schelling put it at our planning conference, the challenge is to devise an approach that, if it had been in use 20 years ago, would have led us *then* to begin monitoring the health and growth of the forests that *now* suddenly seem to be suffering from their proximity to intensive industrial activities. (In Chapter 13, Izrael and Munn emphasize the fact that acid deposition networks have traditionally not included biological monitoring of the health of the ecosystem being impacted.)

The biosphere: using our knowledge of global change

In order to evaluate how alternative strategies of technological, institutional, and research management can affect the prospects for a sustainable development of the biosphere, it is necessary to shape a synoptic understanding of how human activities transform the environment. The last decade has seen the emergence of an integrated view of life on Earth that emphasizes the linkages

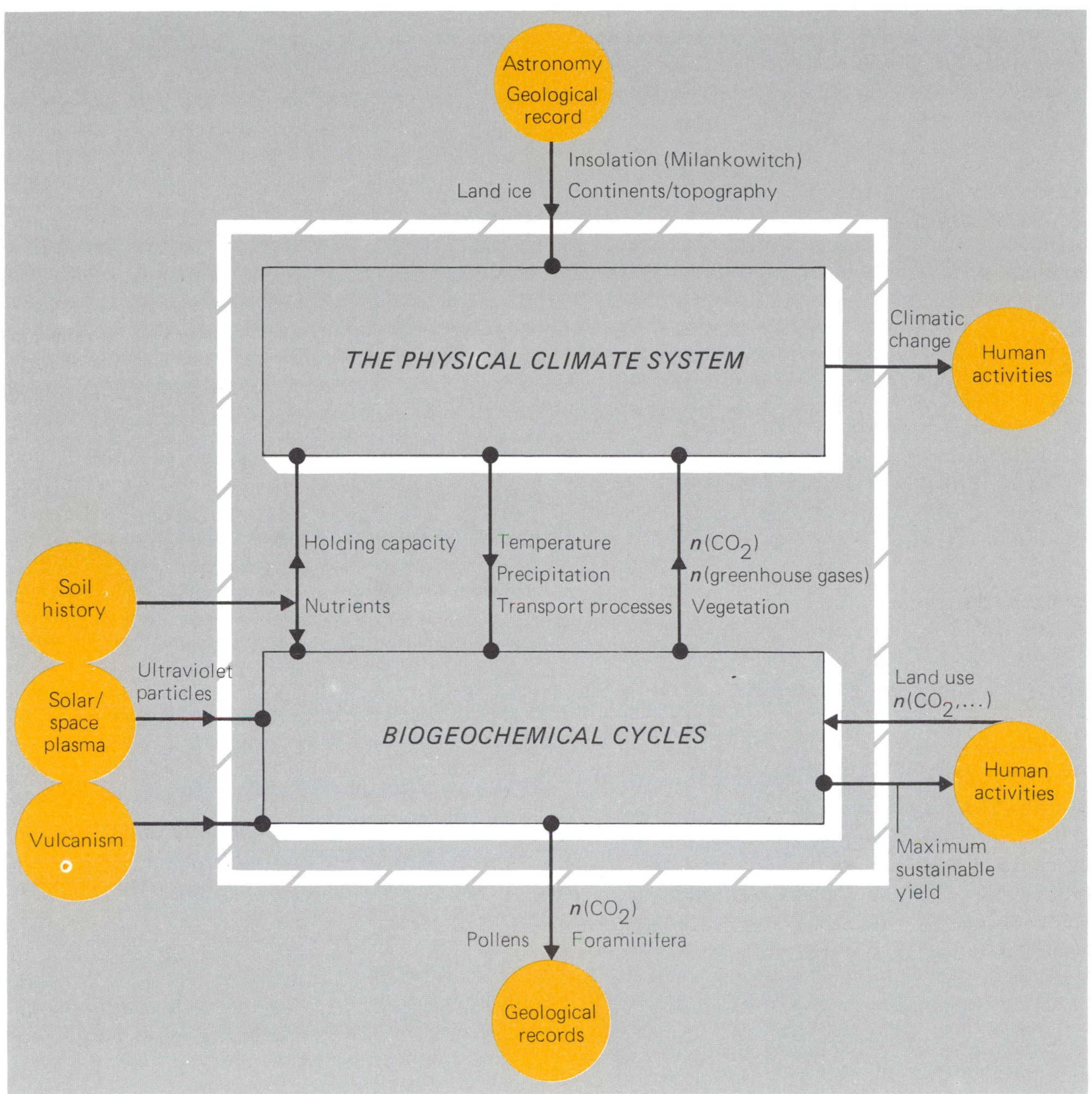

Figure 1.2 The relationships among biological, chemical, and geophysical processes in the biosphere. Redrawn from Bretherton [27].

between the Earth's atmosphere, oceans, soils, and biota (see McElroy, Chapter 7, and Lovelock's *Commentary*). One common thread weaving through these systems is the flow of the major chemical compounds of carbon, nitrogen, oxygen, sulfur, and phosphorus. Another is the flow of energy that drives the atmospheric and oceanic circulation, generates the climate, powers photosynthesis, and surrounds us with light, heat, and ionizing radiation. A third is the ubiquitous influence of life on the interacting flows of materials and energy (*Figure 1.2*). These interactions are complex, often nonlinear, and occur over a great variety of time and space scales. They give the biosphere a propensity for discontinuous change, characterized by threshold responses and multiple domains of stability. As

McElroy and Holling point out in Chapters 7 and 10, the "cycles" and "equilibria" prevalent in earlier writings on the biosphere seem increasingly hard to find.

Earth sciences under the aspect of Gaia

As foreseen by Lotka and Vernadsky, man is indeed one of the living organisms that influences the interactions of the biosphere. But contrary to the conventional wisdom of even a decade ago, the role of nonhuman life-forms in determining the character of the global environment now appears both ubiquitious and powerful [28]. James Lovelock argues that the global environment has remained habitable for life over the past several billion years precisely because of the feedback control that life has exerted on the global environment [29]. Lovelock's "Gaia" or "Mother Earth" hypothesis remains controversial. But whatever the fate of its details, "Gaia" has clearly provided a unifying concept around which chemists, biologists, geophysicists, oceanographers, geographers, and others are now conducting a variety of exciting and cross-disciplinary studies.

This emerging, unified perspective on the sciences of Earth has provided tentative answers to a number of long-standing questions regarding the nature and origins of environmental change. We can now document long-term, large-scale patterns of variation in the biosphere's temperature, chemical composition, and biomass distribution. We have some solid understanding about which natural processes – physical, chemical, and biological – control these pulses. We have the beginnings of an ability to model quantitatively these interactions and to predict future patterns of environmental change [30].

A great deal, nonetheless, remains unknown. For all its accomplishments, the rapid growth in our knowledge of the Earth's biogeochemical and climatic processes has raised at least as many questions as it has answered. We are still unable to make confident predictions regarding such potentially important questions as the climatic effects of the increasing concentration of atmospheric carbon dioxide or the large-scale environmental consequences of massive land-use changes in the tropics. We cannot explain the complex patterns of mortality and morbidity occurring in forests of the industrialized regions. We do not know why the atmospheric concentration of important gases like methane is rising. We understand far too little about how future agricultural practices will affect the crucial role of soil microorganisms in mediating gas exchange between the atmosphere and the biosphere.

Active discussions within the scientific community are exploring how a useful integration of future research among disciplines and nations could best be achieved. The International Council of Scientific Union's recent symposium on "Global Change" outlined some of the priority concerns and likely results of such an integrated program. The "Overview" paper for that symposium concluded:

> It is thus of fundamental importance to develop a knowledge base through a sharply focussed aggregate of research programs sharing a common global view of the Earth system that emphasizes the connectedness of all intervening parts. These programs should ultimately be designed to assess trends and anticipate natural and anthropogenically-caused change over a 50–100 year time-scale. To accomplish this goal, they must focus on the major biogeochemical cycles involving the atmosphere, hydrosphere, lithosphere and biosphere over the full range of time-scales, and they must consider the energy sources and energy transfer processes internal and external to the earth; not merely the major inputs but also those minor contributions which affect man-made systems and may play a triggering role in the causes of global change.... The time interval of approximately one hundred years for which trends in global change should be assessed exceeds by almost an order of magnitude that contemplated in most current or planned programs. Existing programs will thus have to be complemented or expanded accordingly [31].

Toward a synoptic framework

In Chapter 8, Crutzen and Graedel provide an illustration of the kind of synoptic framework this study hopes to create for synthesizing ongoing research of the biosphere in a such a way that it will become usable for managing interactions between development and environment. Crutzen and Graedel focus on the implications of human development activities for the atmosphere. Parallel syntheses are needed for the land and soil, the hydrosphere, and the biota [32].

Valued environmental components

Our basic approach to the creation of a synoptic framework is drawn from experience in environmental impact assessment [33]. We seek to establish the causal relationships between "valued environmental components" and potential sources of environmental change. Valued environmental components, in the sense used here, are attributes of the environment that some party to the assessment believes to be important. Which environmental components are valued in a particular case will depend upon specific social, political, and environmental circumstances, as well as on the level of aggregation appropriate for the intended use. In general, policymakers, interest groups, and scientists may all argue for inclusion of specific components. However the list is eventually derived, it represents an explicit value judgment defining the terms in which users of the assessment are willing to describe the environment aspects of "what kind of garden they want". One of the clearest lessons from assessment experience of the last decade is that unless some definite – and preferably short – list of (most highly) valued environmental components is specified as a focus for assessment, the subsequent "factual" analysis is unlikely to be useful to anyone.

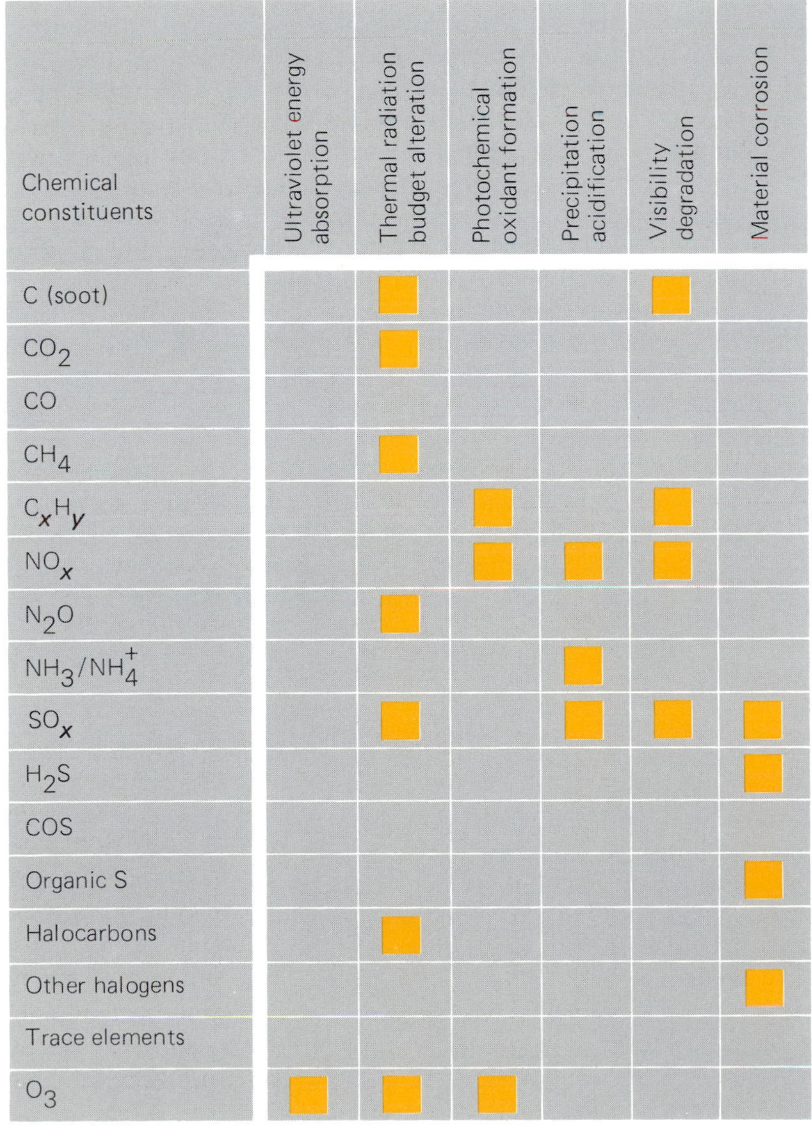

Chemical constituents	Ultraviolet energy absorption	Thermal radiation budget alteration	Photochemical oxidant formation	Precipitation acidification	Visibility degradation	Material corrosion
C (soot)		■			■	
CO_2		■				
CO						
CH_4		■				
C_xH_y			■		■	
NO_x			■	■	■	
N_2O		■				
NH_3/NH_4^+				■		
SO_x		■		■	■	■
H_2S						■
COS						
Organic S						■
Halocarbons		■				
Other halogens						■
Trace elements						
O_3	■	■	■			

Figure 1.3 Major impacts of atmospheric chemistry on valued atmospheric components. The yellow squares indicate that the listed chemical is expected to have a significant *direct* effect on the listed property of the atmosphere. Definitions of the atmospheric properties are given in *Table 1.1*. Data are from Crutzen and Graedel (Chapter 8) and the National Research Council [34], modified as a result of personal communications from R. C. Harris and P. Crutzen.

Table 1.1 Definitions of valued atmospheric components and sources of disturbance.[a]

Valued atmospheric components	*Definition*
Ultraviolet energy absorption	This property reflects the ability of the stratosphere to absorb ultraviolet solar radiation, thus shielding the Earth's surface from its effects. This property is commonly addressed in discussions of "the stratospheric ozone problem"
Thermal radiation budget alteration	This property reflects the complicated relationships through which the atmosphere transmits much of the energy arriving from the sun at visible wavelengths while absorbing much of the energy radiated from Earth at infrared wavelengths. The balance of these forces, interacting with the hydrological cycle, exerts considerable influence on the Earth's temperature. This property is commonly addressed in discussions of "the greenhouse problem"
Photochemical oxidant formation	This property reflects the oxidizing properties of the atmosphere, caused by concentration of a variety of highly reactive gases. The treatment here focuses on local-scale oxidants that are often implicated in problems of "smog", crop damage, and degradation of works of art
Precipitation acidity	This property reflects the acid–base balance of the atmosphere as reflected in rain, snow, and fog. It is commonly addressed in discussions of "acid rain"
Visibility degradation	Visibility is reduced when light of visible wavelengths is scattered by gases or particles in the atmosphere
Material corrosion	This property reflects the ability of the atmosphere to corrode materials exposed to it, often through the chloridation or sulfidation of marble, masonry, iron, aluminum, copper, and materials containing these

Sources of disturbance[b]	*Definition*
Oceans and estuaries	Includes coastal waters and biological activity of the oceans
Vegetation and soils	Does not include wetlands or agricultural systems, for which see below; does include activities of soil microorganisms
Wild animals	Does not include domestic or marine animals, for which see below; does include microbes, except for those of soils, for which see above
Wetlands	An important subcomponent of vegetation and soils; does not include rice, for which see below
Biomass burning	Includes both natural and anthropogenic burning
Crop production	Includes rice, but not forestry; includes fertilization and irrigation
Domestic animals	Includes grazing systems and the microbial flora of the guts of domestic animals
Petroleum combustion	Includes impacts of refining and waste disposal
Coal combustion	Includes impacts of mining, processing, and waste disposal
Industrial processes	Includes cement production and the processing of nonfuel minerals

[a] This table provides definitions of terms used in the text, adapted from Crutzen and Graedel (Chapter 8).
[b] The sources are largely self-explanatory. The notes provided are confined to special considerations important in the text. For more details, see Crutzen and Graedel (Chapter 8).

For purposes of this example, the valued environmental components identified by Crutzen and Graedel and described in *Table 1.1* are an adequate point of departure. The managerial goal is to understand the relationships (if any) between these valued properties of the atmosphere and the natural fluctuations and human activities that might be sources of significant change in them. Recent advances in our understanding of atmospheric chemistry and its interactions with the biosphere raise the practical possibility of specifying such relationships in terms of fundamental biological, chemical, and physical processes [34].

Interactions

Present knowledge regarding the valued atmospheric components affected by changes in specific atmospheric chemicals is given qualitative expression in *Figure 1.3*, in which the convention used indicates only *direct* effects [35]. Note that many components are affected by multiple chemicals. Present knowledge regarding the specific atmospheric chemicals affected by changes in potential sources of disturbance is given qualitative expression in *Figure 1.4*, again using the convention of indicating only direct effects. Note that many

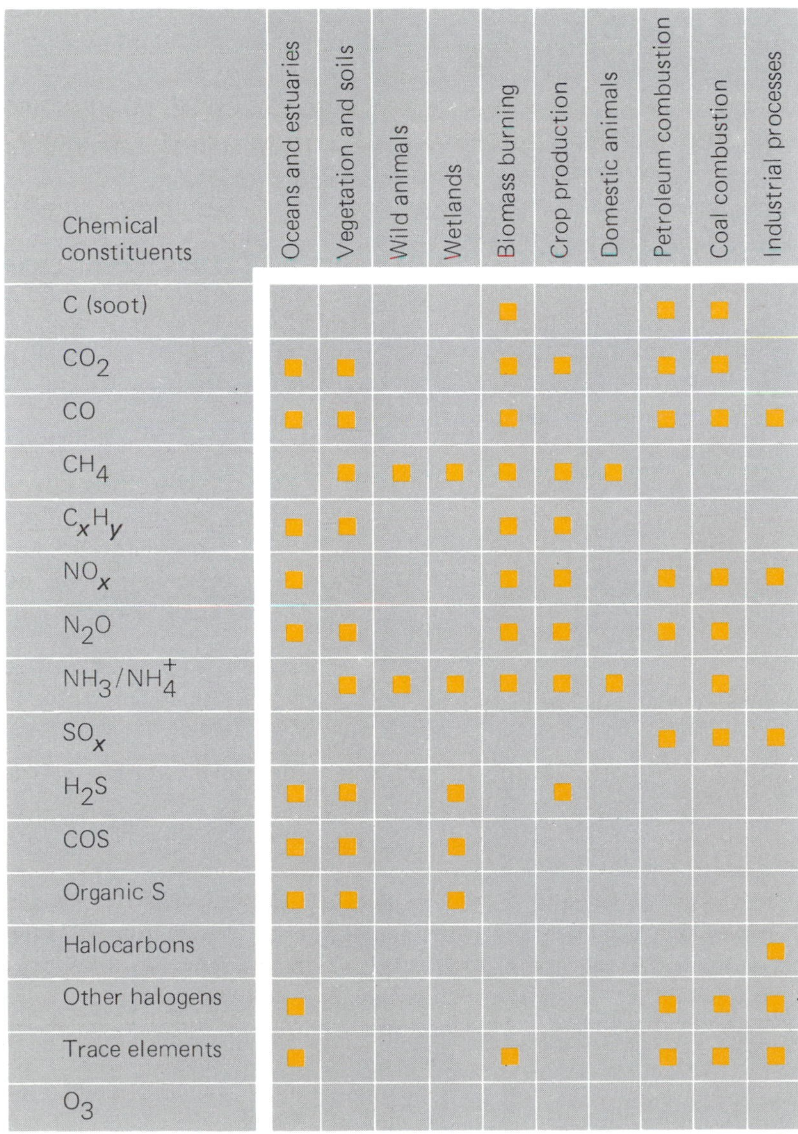

Chemical constituents	Oceans and estuaries	Vegetation and soils	Wild animals	Wetlands	Biomass burning	Crop production	Domestic animals	Petroleum combustion	Coal combustion	Industrial processes
C (soot)					■			■	■	
CO_2	■	■			■	■		■	■	
CO	■	■			■			■	■	■
CH_4		■	■	■	■	■	■			
C_xH_y	■	■			■	■				
NO_x	■				■	■		■	■	■
N_2O	■	■			■	■		■	■	
NH_3/NH_4^+		■	■	■	■	■	■		■	
SO_x								■	■	■
H_2S	■	■		■		■				
COS	■	■		■						
Organic S	■	■		■						
Halocarbons										■
Other halogens	■							■	■	■
Trace elements	■				■			■	■	■
O_3										

Figure 1.4 Sources of major disturbances to atmospheric chemistry. The yellow squares indicate that the listed source is expected to exert a significant *direct* effect on the listed chemical. Definitions of the sources are given in *Table 1.1*. Data sources as for *Figure 1.3*.

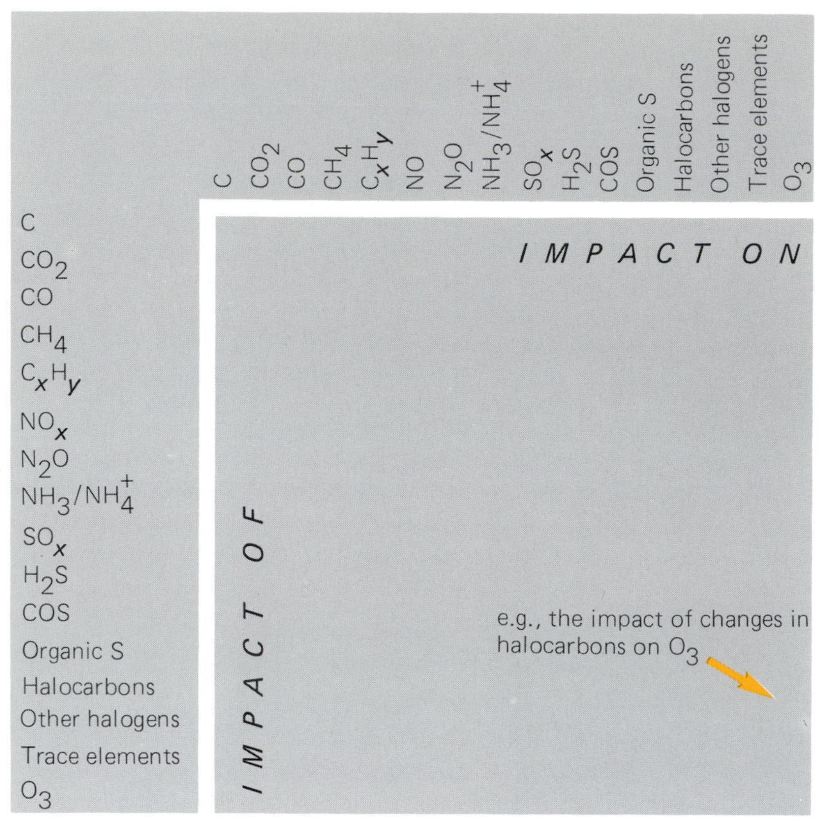

Figure 1.5 Chemical interactions in the atmosphere. A framework for assessing interactions among the chemical compounds listed in *Figures 1.3* and *1.4*.

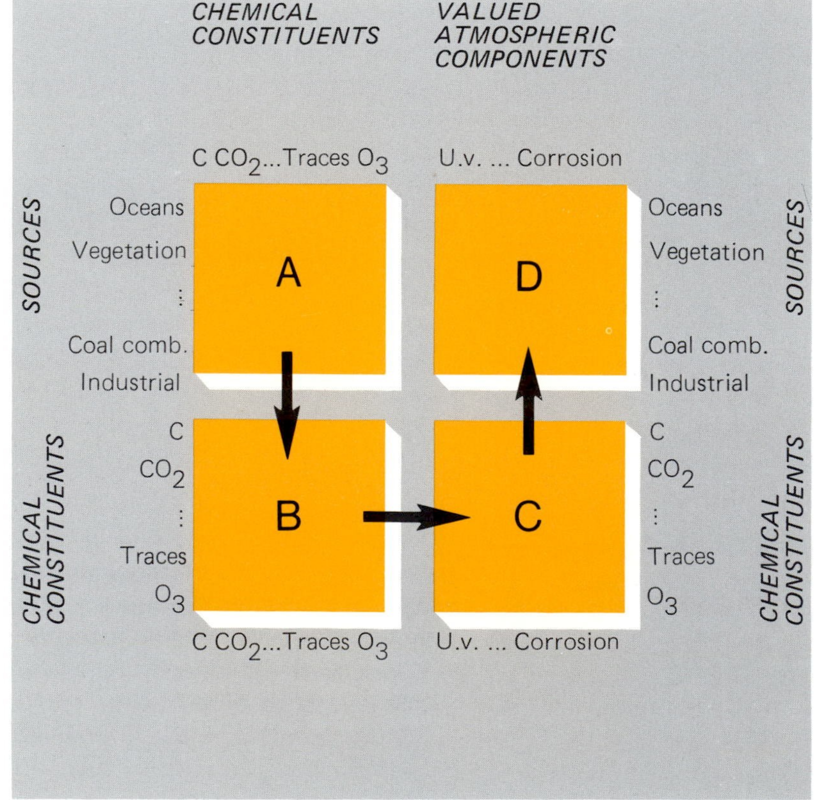

Figure 1.6 The science of impact assessment. An integrating framework for determining the relation between valued atmospheric components and the sources that disturb them (**D**) as a function of the relationships shown in *Figures 1.3* (**C**), *1.4* (**A**), and *1.5* (**B**).

atmospheric chemicals are affected by multiple kinds of sources. The pervasive influence on atmospheric chemistry exerted by changes in the biosphere – in ocean life, plants, soils, and animals – is also evident in the figure.

To complete the chemical connection between sources and valued atmospheric components, it is finally necessary to attend to the matter of indirect effects – the fact that source-induced changes in chemical species *A* may affect a given valued atmospheric component through an intermediate influence on chemical species *B* [36]. Tracking the indirect effects of chemical interactions is one of the central tasks of contemporary atmospheric science. The immense complexity of even the relatively well understood interactions precludes their discussion here [37]. Conceptually, however, the substance of such a discussion can be captured in a matrix constructed along the lines of *Figure 1.5.* The chemical compounds of *Figures 1.3* and *1.4* thus provide the common denominator for an analysis of biogeochemical processes that link sources of changes to consequences of changes in the atmosphere.

The three figures discussed above can be combined to provide a synoptic framework for atmospheric assessment [38]. As suggested in *Figure 1.6,* we can begin with a valued atmospheric component like "precipitation acidification" and its immediate chemical causes (*Figure 1.3*), trace these back through their interactions with other atmospheric chemicals (*Figure 1.5*), and finally identify the sources responsible for initiating those interactions (*Figure 1.4*). The ultimate product is a synoptic matrix that shows the impact of each potential source of atmospheric change on each valued atmospheric component. Crutzen and Graedel's initial effort to construct such a synoptic matrix is shown in *Figure 1.7.* Their assessment is qualitative, as befits the present state of knowledge. It also includes an estimate of the reliability of that knowledge, thus addressing a central goal of our Program as elaborated by Ravetz in Chapter 15.

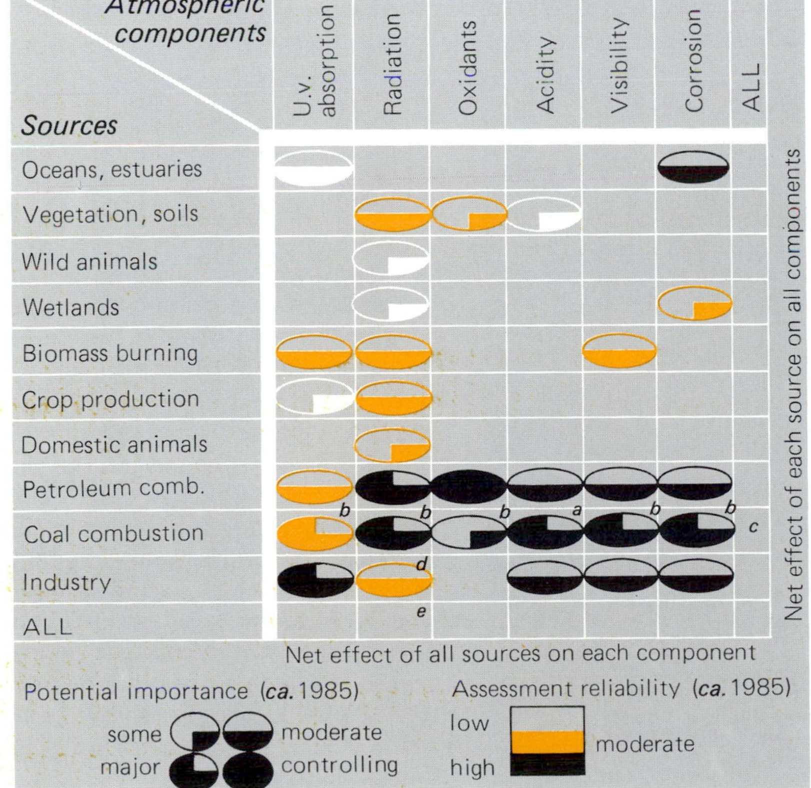

Figure 1.7 A synoptic assessment of impacts on the atmosphere. This figure, adapted from Crutzen and Graedel (Chapter 8) gives a completed version of panel **D** from *Figure 1.6.* The valued atmospheric components defined in *Table 1.1* are listed as the column headings of the matrix. The sources of disturbances to these components are listed as row headings. Cell entries assess the relative impact of each source on each component and the relative scientific certainty of the assessment. "Column totals" would, in principle, represent the net effect of all sources on each valued atmospheric component. "Row totals" would indicate the net effect of each source on all valued atmospheric components. These totals are envisioned as judgmental, qualitative assessments rather than as literal, quantitative summations. The significance of the letters in the cells is described in the text.

Environmental assessments

The simplest atmospheric impact assessments involve only a single cell of the matrix. A typical example is the study of effects of a single source, such as a new coal-fired power station, on a single valued environmental component, such as precipitation acidification (location *a* in *Figure 1.7*). More complex atmospheric assessments have addressed the question of aggregate impacts across different kinds of sources. A contemporary example is the study of the net impact on the Earth's thermal radiation budget caused by chemical perturbations due to fossil fuel combustion, biomass burning, land-use changes, and industrialization (e.g., locations *d* in *Figure 1.7*) [39]. The assessment then becomes a column total (location *e*) in the synoptic framework. Even more useful for the purposes of policy and management are assessments of the impacts of a single source on multiple valued environmental components. The simple study noted above would fall into this category if the impacts of coal combustion were assessed not only on acidification, but also on photochemical oxidant production, materials corrosion, visibility degradation, etc. (e.g., locations *b* in *Figure 1.7*). The impact assessment then becomes a row total in the synoptic matrix (location *c*) [40].

Among the most troublesome interactions between development and environment are those that involve cumulative impacts. In general, cumulative impacts become important when sources of perturbation to the environment are grouped sufficiently closely in space or time that they exceed the natural system's ability to remove or dissipate the resultant disturbance. This is a complicated issue on which systematic work is just beginning [41]. The basic data required to structure such assessments are the characteristic time and space scales of the relevant environmental constituents and development activities. The scales for chemical constituents of the atmosphere plotted in *Figure 1.8* should be accurate to within a factor of 2 or 3 [42]. Data on a comparably defined scale and relevant to a number of human development activities are noted later. Even without the development data,

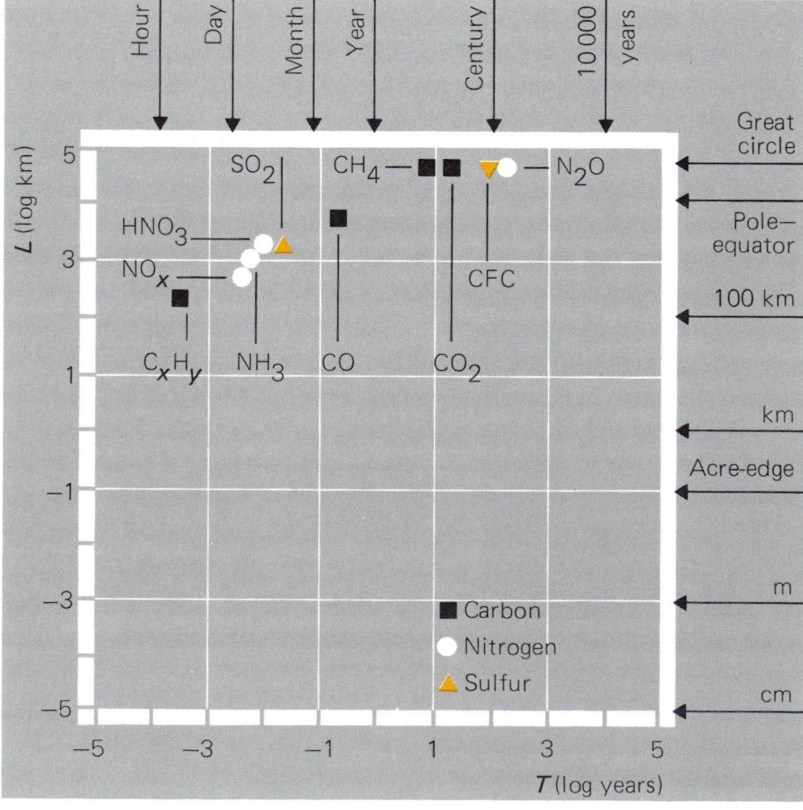

Figure 1.8 Characteristic scales of atmospheric constituents. The figure applies to the clean troposphere above the boundary layer. The abscissa indicates the amount of time required for the concentration of the listed chemicals to be reduced to 30% of their initial values through chemical reactions. The ordinate indicates the mean horizontal displacement (square root of East–West times North–South displacement) likely to occur over that lifetime. Data are from Crutzen [42], modified as a result of personal communications from P. J. Crutzen and R. C. Harris.

however, *Figure 1.8* suggests how scale information can be used to structure cumulative impact assessments.

For example, perturbations to the gases represented in the upper-right corner of *Figure 1.8* accumulate over decades to centuries, and around the world as a whole. Today's perturbations to these gases will still be affecting the environment decades hence; perturbations occuring anywhere in the world will affect the environment everywhere in the world. These gases are all radiatively active (see *Figure 1.3*), thus giving the "greenhouse" syndrome its long-term, global-scale character. At the other extreme, the heavy hydrocarbons and coarse particles represented in the lower-left corner drop out of the atmosphere in a matter of hours, normally traveling a few hundred kilometers or less from their sources. The environmental components of visibility reduction and photochemical oxidant formation associated with these chemicals (*Figure 1.3*) thus take on their acute, relatively local character. The middle of *Figure 1.8* represents a group of chemicals associated with the acidification of precipitation, all with characteristic scales of about a couple of days and a couple of thousand kilometers. Since the world's nation states tend to have spatial dimensions of the same order (*ca.* 1000 km), we should not be surprised to see acidification accumulating across national borders.

Figure 1.8 also reflects the fact that concerns for the accumulation of acidification impacts through time are based not on the accumulation of relevant chemicals in the atmosphere, but rather on the accumulation in other media, such as soil and water, and on the accumulation of sources in societies' growing use of fossil fuels. Expanding the synoptic framework of Crutzen and Graedel to contend with these additional environmental and developmental dimensions is a major part of the Biosphere Program [43].

Extending the framework

The framework developed by Crutzen and Graedel must be expanded before it can provide a truly synoptic environmental framework for assessing interactions between development and the atmospheric environment. A relatively easy addition would be one or more valued environmental components that reflect the role of atmospheric chemicals as direct fertilizers or toxins for plants. Such a modification would allow the integrated treatment of such phenomena as the stimulation of plant growth by carbon dioxide and its inhibition by sulfur oxides – both products of fossil fuel combustion [44]. Somewhat more ambitiously, the approach could be expanded beyond its present chemical focus to include the appropriate physical and biological processes, and the sources of disturbance to them. Dickinson's (Chapter 9) sketch of a comprehensive framework for understanding the impact of human activities on climate shows the potential of such an integrated approach [45]. Ultimately, the need is for a qualitative framework that puts in perspective the impacts of human activities and natural fluctuations, not just on the atmospheric environment, but also on soils, water, and the biosphere as a whole.

Additional extensions to the framework are needed to provide it with spatial and temporal dimensions. Crutzen and Graedel's initial effort focused on present day impacts across a mix of local, regional, and global conditions. They are now preparing separate versions of the framework shown in *Figure 1.7* for the global, regional, and local interactions suggested in *Figure 1.8*. Their assessment will eventually consist of a single, global-scale analysis, plus several analyses for specific, large-scale regions (e.g., Europe), selected to reflect "interesting" interactions of development and environment.

For each of these spatially defined perspectives, the need is for a sequence of figures that show how the relations between sources and valued environmental components change through time. As suggested in *Figure 1.9*, this might consist of a separate version of *Figure 1.7* created to reflect a "slice" through the evolving conditions at 25- or 50-year intervals. In the version we are presently exploring, this sequence will extend several hundred years into the past and a century into the future. The result, it is hoped, will help to put the changing character of interactions between human activities and the environment into a truly synoptic historical and geographical perspective.

This historical dimension, however, focuses attention back on the question of human development patterns. To extend the Crutzen and Graedel framework as proposed, we need to know how

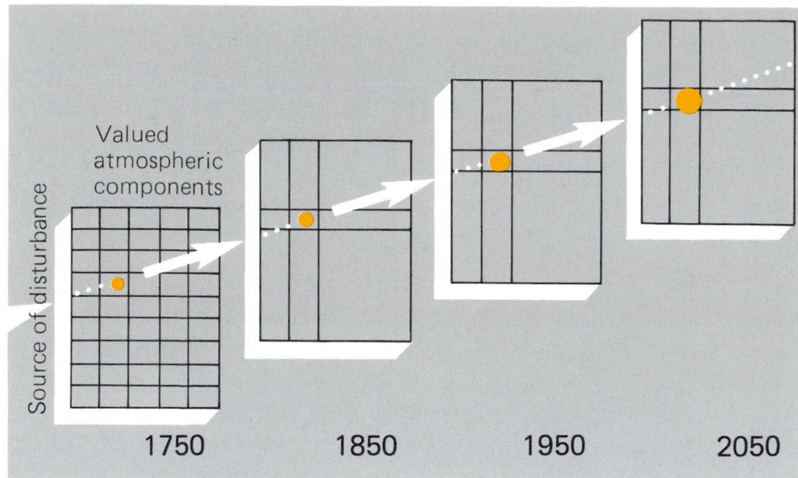

Figure 1.9 A history of disturbances to the atmosphere, as might be expressed through a time series of source–impact matrices, such as that shown in *Figure 1.7.*

much and what kind of relevant human activity was being carried out at specific times in the past, and how much and what kind might be conducted in the future.

Human development: long-term trends and broad-scale patterns

In this section, my focus shifts from the biosphere to human development *per se.* I attempt to sketch some of the major lines of thought that need to be explored for an understanding of the pressures on the environment that arise through people's efforts to improve their well-being. I concentrate here on the long-term, large-scale patterns that reflect the continuities of human development across continents and centuries. I reserve until the next section a consideration of the sources and consequences of the important *dis*continuities – the surprises, innovations, and unique events – that make mere trend extrapolation an inadequate guide to past or future histories of the interactions between human development and the environment.

The modern world system

A dominant feature of human development over the last several centuries has been its increasingly global interdependence. As Richards writes in Chapter 2, "Modern history is only comprehensible when we consider both the unique qualities and experiences of national or local units *and* the processes of world change undergone through time." A growing body of scholarship seeks to analyze the origins, nature, and implications of the "modern world-system" [46]. Some of the most significant of the long-term, large-scale processes discussed in this literature are summarized by Richards on pages 53–54.

(1) Expansion of the European frontier of settlement into the New World, the great Eurasian steppe, and Australasia.
(2) A steady growth in human population from 641 million estimated for 1700 to 4435 million for 1980.
(3) A dramatic growth in the population and spatial extent of the world's cities.
(4) Increased use of fossil fuels and hydroelectric power, which helped to create a revolution in transport, communications, and industrial production.
(5) Development of scientific methods, institutions, and technical means for research and discovery in the biological and physical sciences.
(6) Development of new weaponry with global reach and the capacity for near-global destruction.
(7) Dramatic advances in our ability to cure ill or injured individuals and to control the spread of disease at the collective level.
(8) A steady growth in the scale, efficiency, and stability of large-scale complex organizations (i.e., bureaucracies), in both private and public modes.
(9) The emergence of self-regulating, price-fixing global markets for goods and services.

(10) The emergence of a world division of labor between the developed countries of the North (or core) and the developing countries of the South (or periphery).
(11) The expansion of more intensive sedentary agriculture and the simultaneous compression of tribal peoples engaged in shifting cultivation or in pastoral nomadism.

In our Biosphere Program we seek to illuminate the nature and consequences of interactions between these world development processes and the global environment.

Demographic patterns

Human population growth at the global scale has exhibited perhaps three periods of relatively rapid increase followed by episodes of relative stability. The popular image of an inexorable exponential increase is a distortion produced by the use of short time horizons and inappropriate graphing techniques (*Figure 1.10*).

The most recent period of rapid population increase in the densely settled parts of the world began in the eighteenth century. Malthus wrote just as the momentum of this growth was gathering force. A consensus is now emerging that the momentum will be spent toward the end of the next century, to yield another period of relative stability in human numbers, this time with a world population of between 8 and 11×10^9 people [48].

The last decades of the twentieth century thus occur in the midst of what Kates and Burton have called "the Great Climacteric": a critical period, more persistent than a crisis, where significant change and unusual danger may occur [49]. This latest climacteric extends from roughly 1700 to 2100, during which time the world's population will likely experience about four doublings. The first three have already occurred. The fourth, and probably last in the present climacteric, will near completion within the lives of our grandchildren.

Spatial distributions

To see what these human developments mean in terms of their transformations of the Earth, however, it is essential to understand their spatial

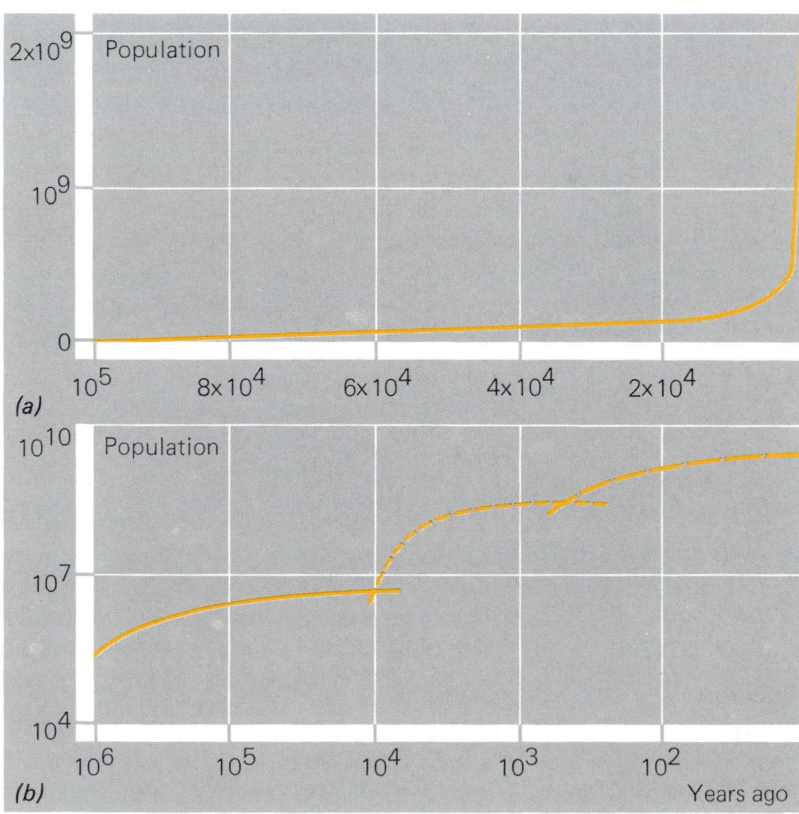

Figure 1.10 Historical growth in human population of the world. (*a*) Typical arithmetic plot of changes over the last 10 000 years. The impression is one of a long, static period followed by a contemporary explosion. (*b*) Logarithmic plot of changes over the last million years. Three population surges appear, reflecting the revolutions of tool-making, of agriculture, and of industry. Figure from Deevey [47].

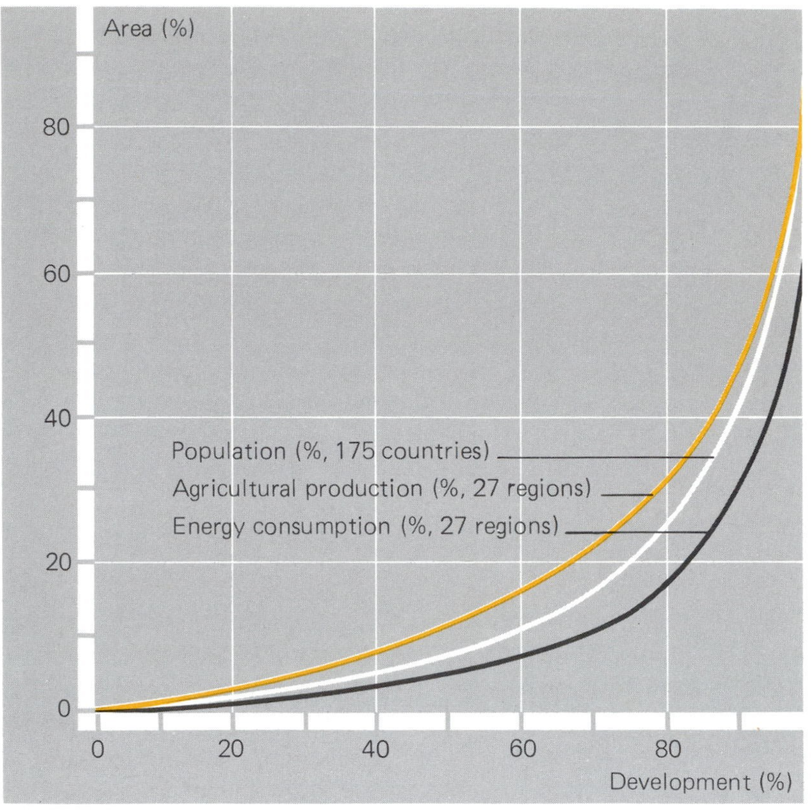

Figure 1.11 Cumulative distribution of human development on the Earth's land. The curves ("Lorenz" curves) are formed by ranking regions of the Earth from most densely to least densely developed, and then plotting the cumulative proportion of the development activity that occurs on increasing portions of the Earth's total land area. Separate curves are plotted for population, agricultural production, and energy consumption. Thus, for example, the middle curve indicates that about 60% of the Earth's population is found on the most crowded 10% of its land area [50].

distribution. One half of the human population now exists on less than 10% of the Earth's land, while three quarters of the population exists on only 20% of the land (see *Figure 1.11*). Similar patterns of concentration hold for economic activity (see below).

The locations of the world's population concentrations are suggested in *Figure 1.12*. The high

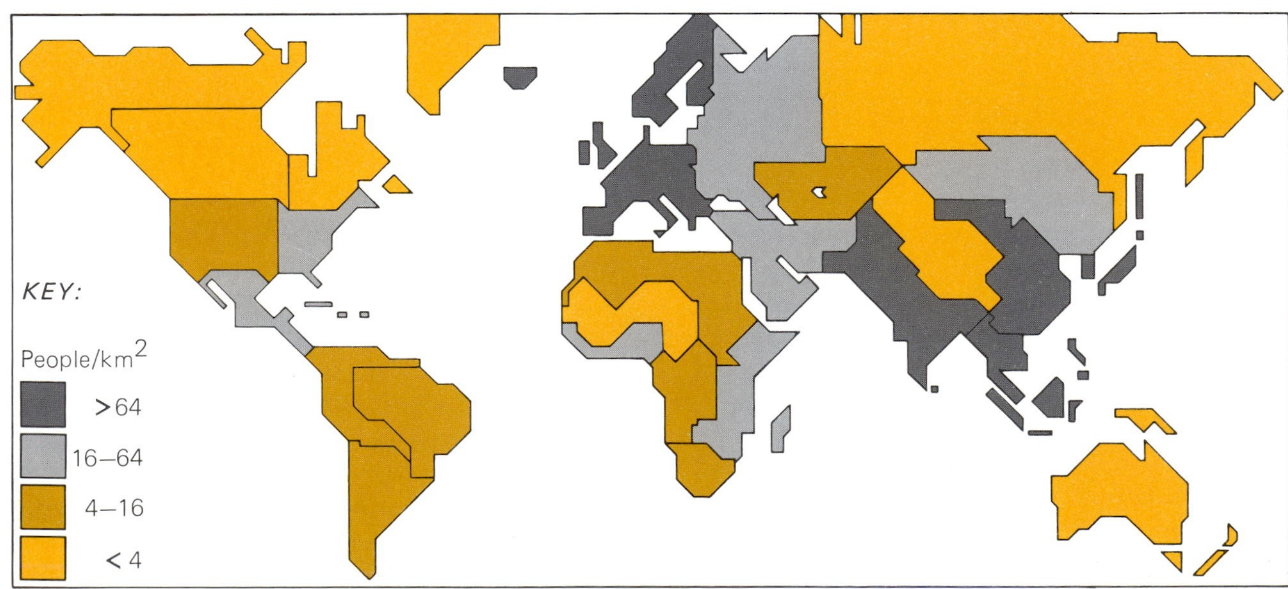

Figure 1.12 Geographical distribution of current population density, mapped for areas of about 5×10^6 km^2. Note that the density scale is geometric [51].

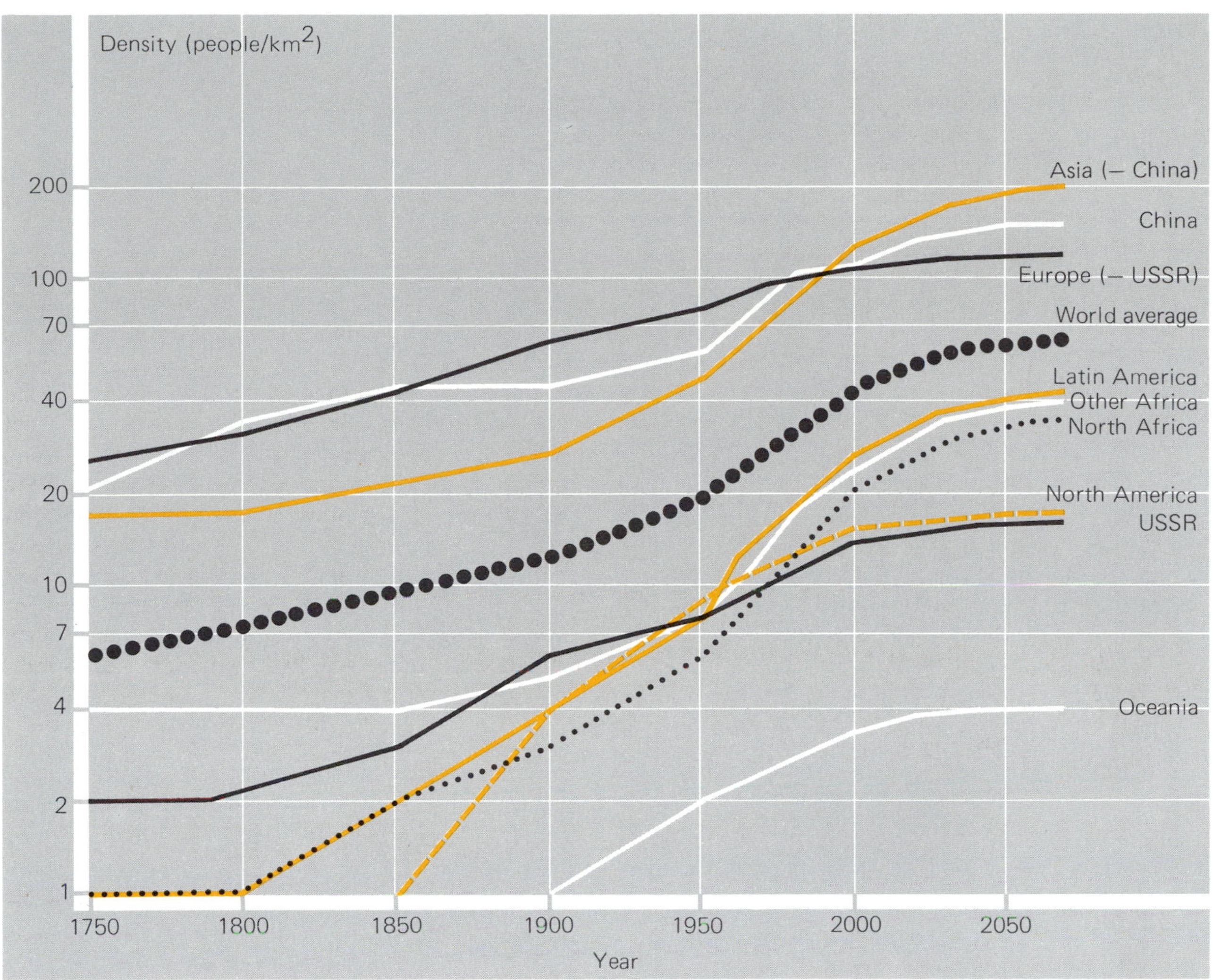

Figure 1.13 History and projections of population densities for the world as a whole and for continental regions (note the logarithmic density scale) [52].

densities of people in Europe and South and East Asia stand out. At the broad regional scale of resolution used here, densities range from more than 200 people per km^2 in China proper, through a world average of about 30, to less than 1 person per km^2 in Alaska and the Canadian Territories.

The patterns of relative crowding reflected in *Figure 1.12* have persisted through much of history and will almost certainly continue into the future. *Figure 1.13* shows the historical growth of population densities on the continental scale since 1750 and typical projections to the end of the next century. These demonstrate that today's relatively high densities in Europe and South and East Asia are essentially permanent structures of human settlement over the three centuries that concern us here. Densities in this relatively crowded 15% of the Earth have remained more than double those of the remaining 85%. Moreover, population densities in Europe and China have remained similar throughout the period. The Americas, Africa, and the USSR have supported remarkably similar densities over most of the last century, diverging only in the more distant past and future. By early in the next century, three major density groupings seem likely to emerge on the continental scale. The highest, as for the last 200 years, will be represented by Asia and Europe. An order of magnitude less dense, on average, will be the lowest group consisting of North America, the USSR, and Oceania. In between will be a medium density group consisting of Latin America and Africa.

Absolute densities

The absolute densities suggested in *Figure 1.13* are also of interest [53]. The density of 20 people per km^2 already reached in Asia and Europe by 1750 is only now being achieved in Africa and Latin America and will not be exceeded, on average, in North America or the USSR even over the next century. More significant is the average density of 100 people per km^2 typical for Asia and Europe today. This is at least double the figure likely to be reached for any other large region of the Earth over the next century. In other words, we already have a century of historical experience in Europe and China with continental population densities of 40 or more people per km^2 – densities as high as, or higher than, those likely to develop in the Americas, Africa, or the USSR over the next 100 years.

The only historically unprecedented densities likely to exist on the continental scale by the end of the next century will probably occur in Asia. Approaching 200 people per km^2, it will reach a density of the order of twice that of contemporary Europe. Even these relatively high future densities, however, are less than those encountered in many individual countries today. Bangladesh, for example, now has over 600 people per km^2, Japan over 300, West Germany almost 250, and the UK more than 200.

The relevance of these population densities for interactions between development and environment will become clearer when we examine what the populations are doing – i.e., their patterns of economic activity.

Economic patterns

Not surprisingly, regions with high population densities generally support high densities of economic activity and impose comparably dense pressures on the environment. But the patterns of economic activity cannot be understood solely in local terms. Indeed, by the beginning of the eighteenth century at the latest, a true "world economy" had emerged. Throughout the entire period that concerns us here, demands for many goods and services arising in one part of the Earth have been met with supplies drawn from half a world away. Prices of many goods and services have tended to be regulated in a global market, with relatively few sustained deviations among regions. Since the beginning of the eighteenth century, the volume of world trade has risen by a factor of 500 or so [54]. International trade now represents about one third of the world's total GNP, a figure that would be even higher if long-distance trade within large countries was included [55].

The expanding world economy consists of a number of interrelated human activities that exert pressures on the total environment. Some of the most important pressures, as we saw earlier, are related to the activities of agricultural production and energy use. Global histories of the environmental consequences of agricultural production and energy consumption would provide invaluable guidance for the assessment of sustainable development options. Unfortunately, as Richards (Chapter 2) and Williams (*Commentary* to Chapter 2) point out, such global reconstructions are only now beginning to be assembled. Richard's own work suggests the potential of such studies. He documents a 300-year wave of global land transformation radiating outward from economically developed core areas into the developing peripheries. Driven primarily by increasing population and by increasingly long-distance trade in agricultural products, this wave has doubled the amount of arable land on Earth since the middle of the nineteenth century. In some regions it has entailed virtually total deforestation, in others the drainage of hundreds of thousands of square kilometers of wetlands. For the world as a whole, it has brought more than 2 million km^2 of land under irrigation.

We still lack synoptic histories of energy-related activities with anything like the depth and resolution of the land transformation studies. Nonetheless, global reconstructions of fossil fuel consumption that extend back to the middle of the last century have been attempted, and reveal the patterns shown in *Figure 1.14.* More detailed analyses reaching back to the early part of this century are summarized by Darmstadter in Chapter 5. The pictures emerging from these studies are incomplete, particularly with regard to the environmental implications of energy

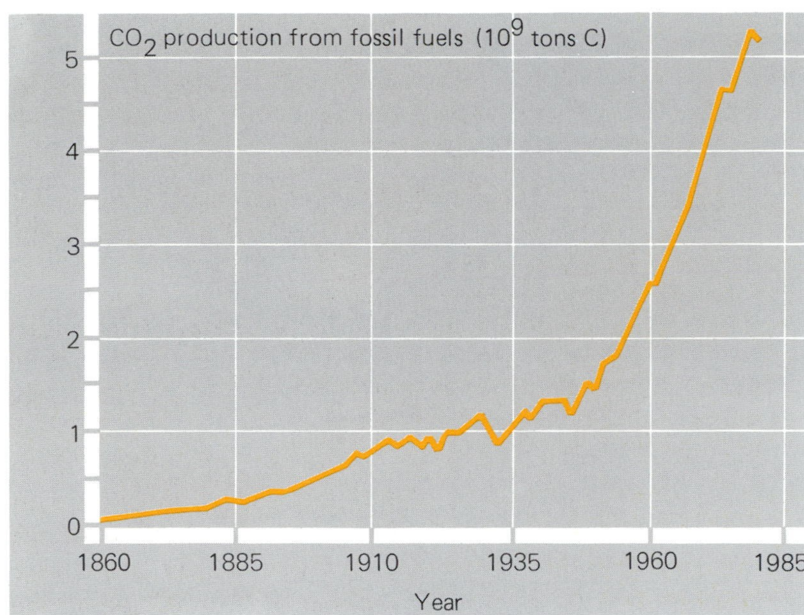

Figure 1.14 History of world fossil fuel energy consumption, as reflected in carbon emissions [56].

consumption based on fuel wood, hydropower, and nuclear fission [57]. But the increasing magnitude, pace, and global inclusiveness of environmental transformations throughout the last centuries are evident.

Present trends

Today's patterns of agricultural and energy activity reflect the patterns in population density already considered. Most agriculture is carried out, and most energy is consumed, on a relatively small amount of the Earth's surface (*Figure 1.11*). Ten percent of the Earth supports nearly one half of the world economy's food production and experiences nearly three quarters of its fuel consumption. The world's present patterns of agriculture and energy densities are shown in *Figures 1.15* and *1.16*. As can be seen in the maps, and even more clearly in the graphical format of *Figure 1.17*,

Figure 1.15 Geographical distribution of current agricultural production density, mapped for areas of about 5×10^6 km^2. Note that the density scale is geometric [58].

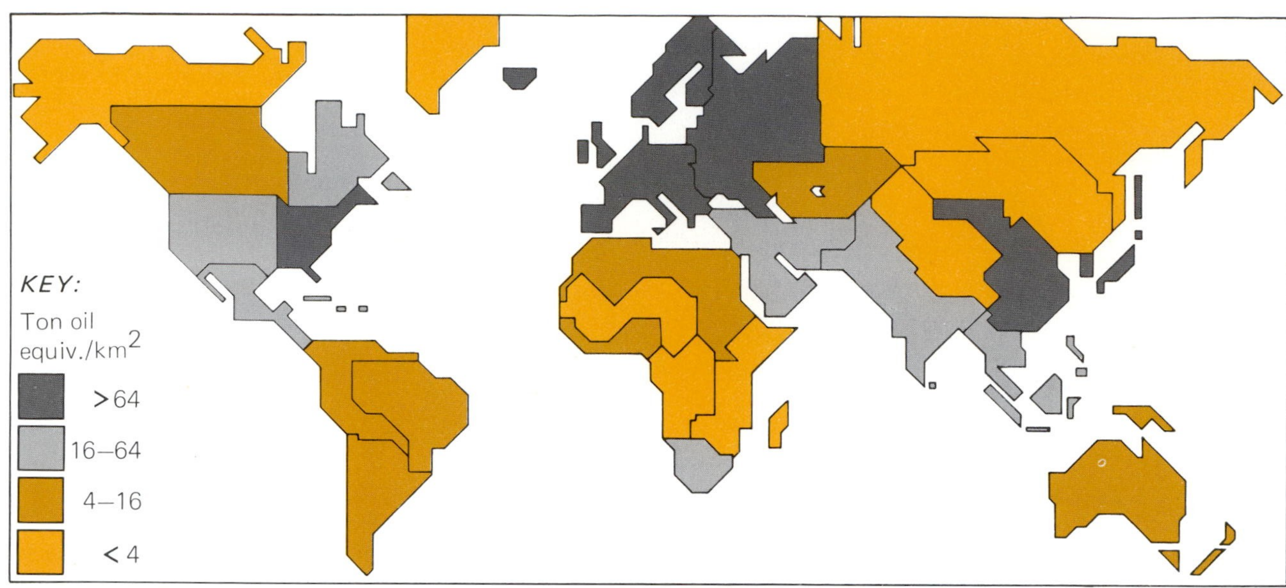

Figure 1.16 Geographical distribution of current energy-consumption density, mapped for areas of about 5×10^6 km^2. Note that the density scale is geometric [59].

densities of energy consumption and agricultural production are correlated. The regions of Earth exposed to the greatest environmental pressures due to crop production also tend to be exposed to the greatest pressures from energy consumption. Partial exceptions occur for India and China, where the ratio of energy use to agricultural production is relatively low, and for Eastern Canada, South Africa, and Oceania, where the ratio is relatively high. Nonetheless, the relationship is surprisingly strong, almost certainly testifying to the fundamental role played by population density in determining the environmental pressures of human development.

Looking to the future

Ultimately, I hope it will be possible to use information like that conveyed in *Figure 1.17* to construct a typology of "kinds" of pressures that different stages and patterns of human development impose on the surrounding environment [61]. A useful effort would be to chart the temporal evolution of various regions through the space of *Figure 1.17*. The objective of such an exercise would be to search for situations in which the past history of development densities and environmental problems in one region could serve as an analog for future histories of environmental problems in another region about to pass through a similar development pattern. I suggest that the long history of relatively high population densities in Europe should be able to teach us something about the future environmental pressures of regions just now reaching the density of population that Europe supported 200 years ago. The power of such analogies should be much stronger as we move from population, the ultimate source of environmental pressures, to the more immediate sources represented by energy consumption and agricultural production activities.

To use the past evolution of regional development densities as a guide to future pressures on the environment will require that something be said of the possible future histories of energy use and agricultural production. The performance of past forecasts of global patterns of development activities is reviewed by Crosson and Darmstadter (Chapters 4 and 5). The record does not inspire much confidence in future predictions [62]. More useful than yet another set of such dubious forecasts would be a carefully selected set of future world development scenarios, constructed with a view toward articulating the environmental constraints that they might entail and the social responses to these constraints that might impove the prospects for a sustainable development of the biosphere.

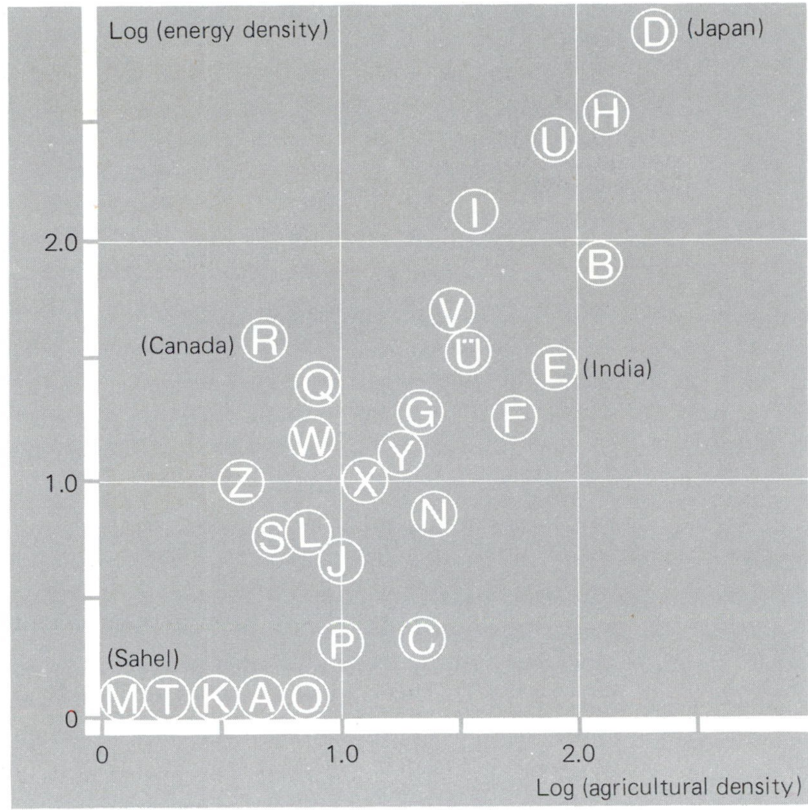

Figure 1.17 Plot of present densities of agricultural production (units are the logarithm of total value added in US \$100 per km^2) versus densities of energy consumption (units are the logarithm of total coal equivalent in 10^3 kg per km^2) [50]. Key: A, China (Turkestan and Tibet); B, China (proper); C, China (Manchuria); D, Japan; E, India; F, Asia (Southeast); G, Near East; H, Europe; I, European USSR; J, USSR (Kazakhstan); K, USSR (Siberia); L, Africa (North); M, Africa (Sahel); N, Africa (West); O, Africa (Mid); P, Africa (East); Q, Africa (South); R, Canada (East); S, Canada (West); T, America (North); U, USA (East); Ü, USA (West); V, America (Middle); X, Brazil; Y, America (Temperate South), Z, Oceania.

Interactions between development and environment: discontinuities and surprise

The intensifying interactions between development and environment have motivated a number of studies over the last several decades. As Brooks notes in Chapter 11, however, many of those studies have been based

> ... on an evolutionary paradigm – the gradual, incremental unfolding of the world system in a manner that can be described by surprise-free models, with parameters derived from a combination of time series and cross sectional analyses of the existing system [63].

This evolutionary approach has its uses, particularly in elucidating the long-term trends and broad-scale patterns discussed in the previous section. But, as C. S. Holling has forcefully argued, experience shows that significant

> ... change can also occur in abrupt, discontinuous bursts. If there is no framework of expectation or understanding then the abrupt changes are perceived as surprises – as crises or opportunity [64].

Surprise-free forecasts are therefore necessary, but insufficient, tools for efforts to improve the management of long-term interactions between development and environment. By leaving out the external shocks, nonlinear responses, and discontinuous behavior so typical of social and natural systems, surprise-free analysis leaves us unprepared to interpret a host of not improbable eventualities. By leaving out the social learning called into play by the resultant crises and climacterics, it also reduces the challenge of designing adaptive management strategies to the mindless, repetitive implementation of past mistakes (see Cantley, *Commentary* to Chapter 11).

A need for concepts and methods

The problem is not that analysts have been unaware of the shortcomings of surprise-free thinking, but rather that they

> ... lack ... usable methodologies to deal with discontinuities and random events. The multiplicity of conceivable surprises is so large and heterogeneous that the analyst despairs of deciding where to begin, and instead proceeds in the hope that in the longer sweep of history surprises and discontinuities will average out, leaving smoother long-term trends that can ... provide a basis for reasonable approximations to the future [65].

But real history turns out to be far from an "averaging out". The distinguished historian W. H. McNeill concluded his remarks at a symposium on "Resources for an Uncertain Future" as follows:

> I believe historians' preoccupation with catastrophe might be useful to economists, if they care to listen. Extreme cases, breakdowns, abrupt interruptions of established market relations – these are not staples of economic theory, and are, I believe, usually dismissed by statistically minded analysts of the norm and its fluctuations. But human societies are a species of equilibrium, and equilibria are liable to catastrophe when, under special limiting conditions, small inputs may produce very large, often unforeseen, and frequently irreversible outputs. I believe there is a branch of mathematics that deals with catastrophe – sudden changes in process; I must say that I, as an historian contemplating the richly catastrophic career of humanity across the centuries, venture to recommend to economists a more attentive consideration of such models – at least when trying to contemplate the deeper past and long-range future [66].

A central challenge of our Biosphere Program is to see how far we can go in developing the methods, models, and concepts necessary to move beyond surprise-free analyses to a more realistic treatment of the interactions between development and environment. The "catastrophe" theories and their relatives referred to by McNeill have played an important role in providing an alternative paradigm to surprise-free analysis [67]. But to move beyond the empty generalities and vague analogies that have so inflated the "catastrophe" literature, it is necessary to assemble a great deal of carefully documented case material and to fashion a family of specific and rigorously tested explanatory models. As described by Holling and Brooks in Chapters 10 and 11, the necessary work has only just begun.

A sampler of potential surprises

To give some specific flavors to the subsequent discussion, let us sketch a sample of the kinds of discontinuous, potentially surprising behaviors likely to be encountered in efforts to manage interactions between development and environment.

Atmosphere–biosphere interactions

Hydroxyl radicals are important "cleaners" of the atmosphere, removing by reaction those gases, such as carbon monoxide, hydrocarbons, and the nitrogen and sulfur compounds, involved in precipitation acidification. An important reaction pathway involves methane, the atmospheric concentration of which is now rising at least in part due to land clearance, biomass burning, and paddy rice production in the tropics. Crutzen [68] argues that where background concentrations of nitrogen oxides in the lower troposphere are below a threshold value of about 100 p.p.t.(v), an increase in methane leads to a decrease in hydroxyl concentration and thus to an increase in the often problematic gases that the hydroxyl radical would otherwise have removed. Above this threshold value an increase in hydroxyl concentration will occur. Where background concentrations of nitrogen oxides are above another threshold value of about 30 p.p.t.(v), an increase in methane leads to strong increases of ozone concentrations. The critical threshold values that separate these reaction pathways are exceeded in regions of heavy industrial development and fossil energy use, but not in less densely developed areas. Thus, as development proceeds and nitrogen oxide emissions increase, the atmosphere experiences a potentially surprising "flip" from a state of low oxidative potential to one of high oxidative potential. Such a "flip" could be implicated in a variety of chemical and biological impacts associated with industrialization.

Resource management interactions

Holling (Chapter 10) and Regier and Baskerville (Chapter 3) describe a variety of renewable resource cases in which abrupt changes intruded on what were supposed to be surprise-free management strategies. I summarize just one of these examples. The regional development of Eastern Canada in the 1950s was heavily dependent on its forest products sector. When further development was threatened by the irruption of a forest pest called the spruce budworm, a successful control strategy was implemented using unprecedentedly large-scale applications of insecticides guided by one of the most sophisticated models of insect population dynamics ever constructed. Protected by the insecticides, both the forest and the forest economy prospered. But their very prosperity made them ever more vulnerable to catastrophic failure in the event that the budworm should escape from the control imposed by the insecticides. Amid continuing institutional denials that the short-term management success had become a long-term liability, a combination of events indeed precipitated a budworm epidemic. The unprecedented extent and intensity of this outbreak left the surprised regional economy scrambling to sell off a sea of dead timber and wondering how to feed all the modern pulpmills that had been designed on the basis of surprise-free forecasts of wood supply.

Technology–economy interactions

Brooks (Chapter 11) presents a menu of examples of technological surprise. Among the most interesting is the case of the USA automobile industry. It began its life in the early part of this century in what Brooks calls a period of "exuberant experimentation", open to new ideas, new firms, and new opportunities presented by the evolving structure of American society. Steam, electric, and diesel power all competed credibly with the internal combustion gasoline engine. By the mid 1920s, however, the lead established by Alfred P. Sloan's managerial innovations gave the gasoline strategy a lead that economies of scale and more rapid progress along the learning curve of manufacturing skills rapidly converted into a total dominance of the market. The domain of feasible experimentation and innovation became progressively more narrow, focusing on incremental cost savings rather than strategic design alternatives. What Brooks calls a "technological monoculture" had emerged. At the same time, however, the very success of the automobile in penetrating throughout the American economy imposed on society increasingly heavy externalities: highway and urban congestion, environmental pollution, a rising toll of traffic accidents. These mounting externalities, plus external shocks like the oil price increases of the 1970s, placed strong pressures for change on the automobile industry, pressures for which its "monoculture" structure gave it little capacity to respond. Unable to adapt to a changing world, a surprised automobile industry saw its customers turning to cheaper, better, and more diversified foreign products.

Toward a general understanding

Careful analyses of many cases such as those noted above have led to the beginnings of a general understanding concerning the nature and origins of at least some kinds of surprise and discontinuous change. In reviewing this understanding, Holling (Chapter 10) emphasizes the critical importance of processes operating on multiple time scales. In particular, he argues that the discontinuous change of one or more "fast" variables (in the cases described above, the oxidant potential of the atmosphere, the density of budworms, and the sales of the USA automobile industry) can often be understood as a consequence of specific alterations in the underlying system structure that result from the continuous change of one or more "slow" variables (i.e., the rate of NO emissions, the density of the forest, the "species diversity" of USA automobile designs). Holling's summary of the key features of "discontinuous change" systems he has studied is quoted below, with my own parenthetical glosses added to provide context [69]. Several of the features are illustrated in *Figure 1.18*.

> (1) There can be a number of locally stable equilibria and stability domains around these equilibria [due to the existence of significant nonlinear processes in the system].
> (2) [Discontinuous j]umps between the stability domains can be triggered by exogenous events . . . the size of these domains is a measure of the sensitivity to such events.
> (3) The stability domains themselves expand, contract, and disappear in response to changes in slow variables . . . and, quite independently of exogenous events, force the system to move between domains.

(4) ... different classes of variability ... emerge from the form of the equilibrium surfaces [which reflect potential stable levels of the fast variables if the slow variables were held constant, and vice versa] and the manner in which they interact [i.e., intersect]. There can be conditions of low equilibrium with little variability.... And there can be dynamic disequilibrium in which there is no global equilibrium condition and the system moves in a catastrophic manner between stability domains.... There also exists the possibility of "chaotic" behavior.

... discontinuous change is [thus] an internal property of each system. For long periods change is gradual and discontinuous behavior is inhibited. Conditions are eventually reached, however, when a [discontinuous] jump event becomes increasingly likely and ultimately inevitable.

The potential of utility of Holling's approach can be illustrated by current debates over the impacts of climatic variability and change on society. Parry (Chapter 14) characterizes the problem as follows:

What, for example, were the real "causes" of the Sahelian crisis... meteorological drought or enhanced vulnerability due to economic and political developments insensitive to an environment that has always been changeable?

He points a way to its resolution, however, with a distinction

... between *contigently necessary* and *contingently sufficient conditions* for an occurrence. It is probable that increased vulnerability was a necessary condition (or precondition) for the Sahelian disaster; without it the economic system might have been more resilient to the meteorological drought. It was not, however, a sufficient condition, because it required a further event... to precipitate the effect [71].

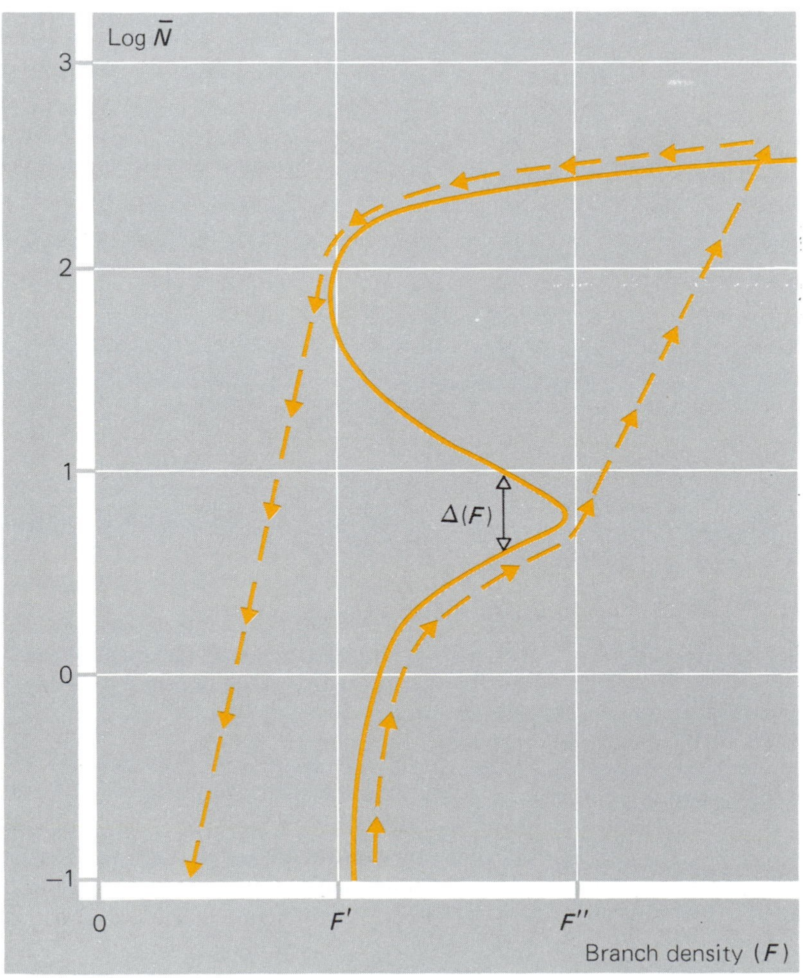

Figure 1.18 Isocline of equilibrium conditions (solid line) and dynamics (dotted line) for a model of interactions between an insect pest and a forest. The insect has a rapid growth rate and behaves as a "fast variable" relative to the "slow variable" of forest growth. For branch densities between F' and F'' there are three equilibrium densities of the insect. The highest and lowest are stable, the middle unstable. As branch density rises from F' to F'', insect densities "track" at equilibrium along the lowest equilibrium surface. As branch densities rise above F'', two of the insect equilibria disappear, and the insect population density rises rapidly and catastrophically to the highest of the equilibrium surfaces. Alternatively, random fluctuations or immigrations of insects of quantity ΔF or greater can "flip" the system from the lowest to the highest equilibrium surface. Once the high insect densities are reached, the forest slowly dies back, with insect densities slowly declining. When forest density drops below F' the two upper equilibria disappear, and the insect population collapses [70].

In the conceptual framework introduced by Holling, the "contingently necessary" economic and political developments would thus seem to play the stage-setting role of slow variables. Meteorological drought is one possible condidate for the exogenous triggering event. And the disaster of ensuing famine is the discontinuous change that society, lacking an adequate framework of understanding, reacts to with surprise.

Whether this perspective on discontinuous change can help to develop tools and understanding that will be useful in managing the interactions between human activities and the environment in places like the Sahel remains to be seen. At a minimum, however, it focuses attention on the need to assess, much more systematically than has normally been the case and across the full range of space and time scales, those processes that might be significantly involved in particular interactions between development and environment. Much more empirical and modeling work needs to be done to explore the potential of the research directions on discontinuous change sketched above. Of particular importance, as pointed out by di Castri in his *Commentary* to Chapter 10, will be efforts to construct a spatial analog to the "fast variable/slow

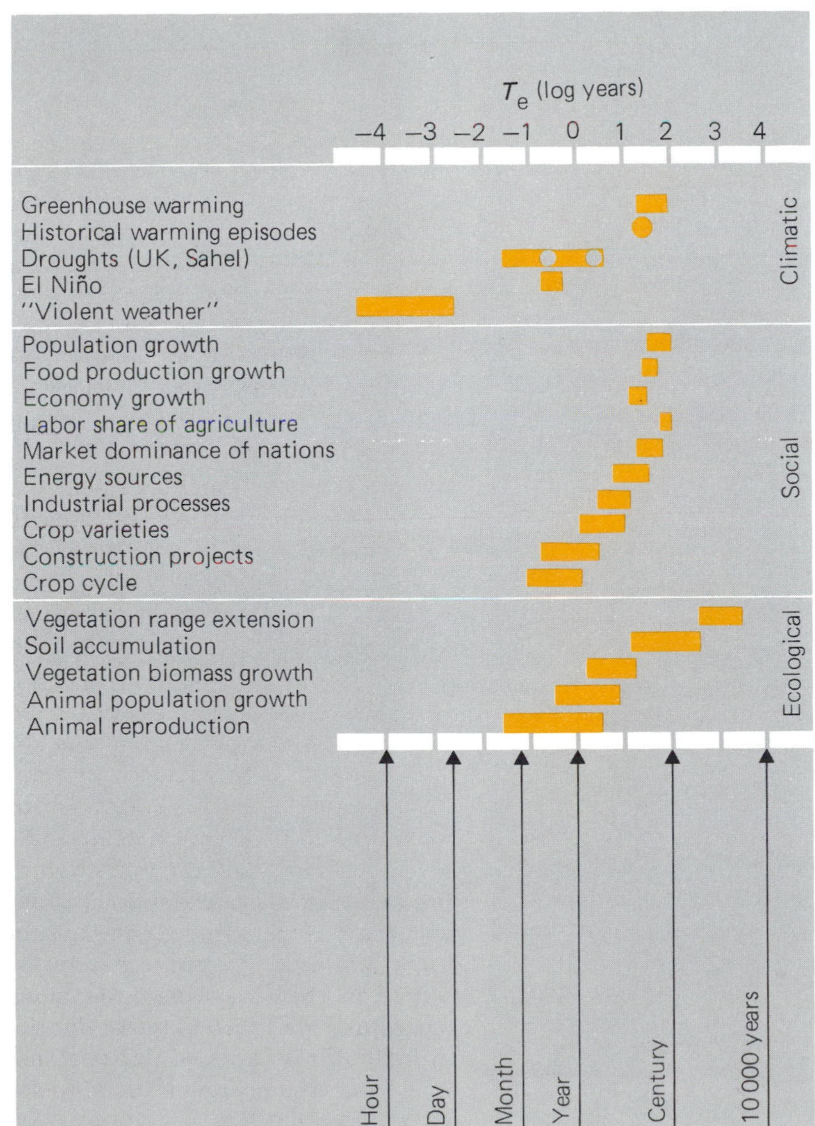

Figure 1.19 Characteristic time scales for climatic, social, and ecological processes. The time plotted is that required for the listed process to change by a factor of *e* or 2.7, plotted in logarithmic units of years. Thus, the time required for an energy source, such as oil, to increase its share of the total world energy market by a factor of 2.7 is about 10 years [73].

variable" distinction that could be used in coupling the

> ... local surprise (of a biological but mostly societal type) and global change (considered chiefly from a biogeochemical point of view) [72].

The initial steps in the Biosphere Program to identify and scale the processes relevent to interactions among climates, ecosystems, and societies have been reported elsewhere and are summarized in *Figures 1.19* to *1.21*.

Next steps

The elements of a general understanding of discontinuous change as outlined above should certainly be pursued through a vigorous program of case studies, modeling, and theoretical analysis. But at

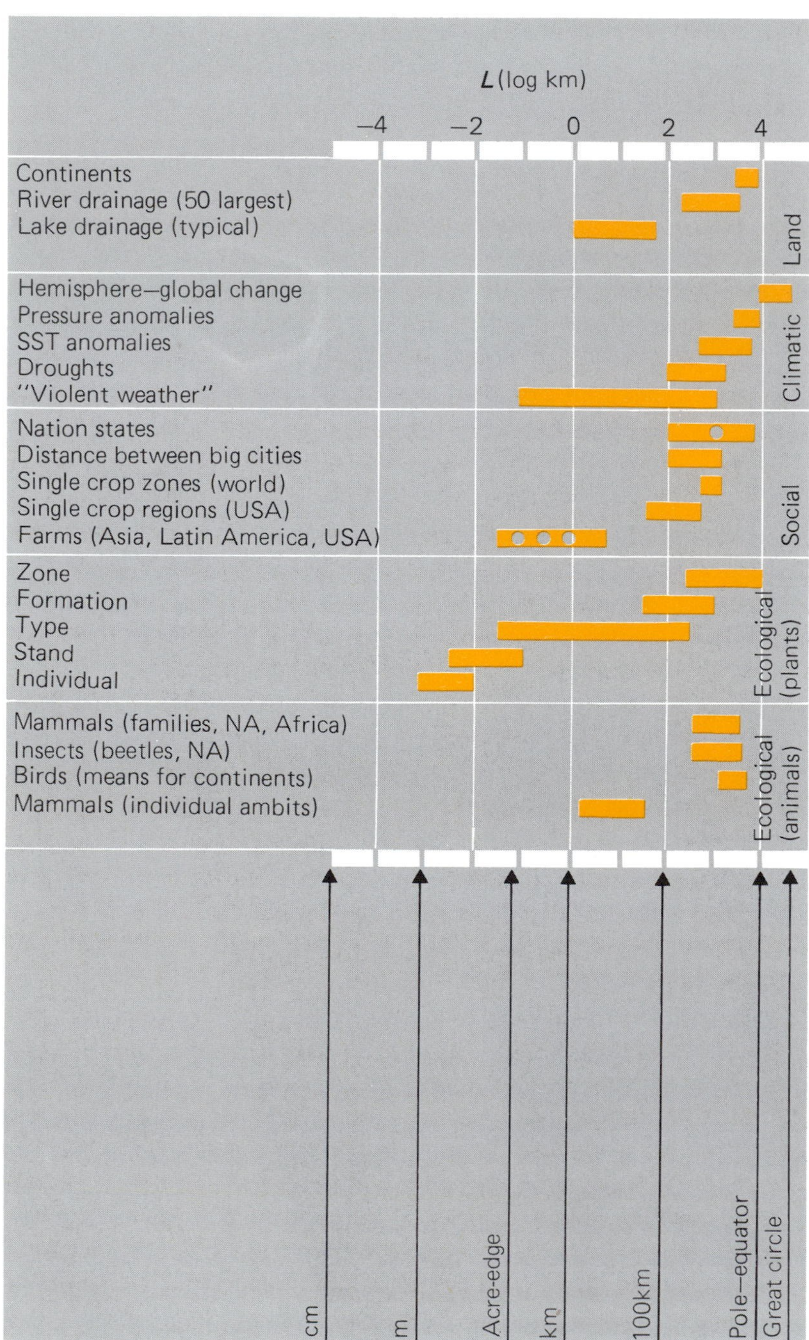

Figure 1.20 Characteristic space scales for climatic, social, and ecological patterns. The space plotted is that of the square root of the area covered by the listed pattern, plotted in logarithmic units of kilometers. Thus, the mean size of the world's nations is about 1000 km across [74].

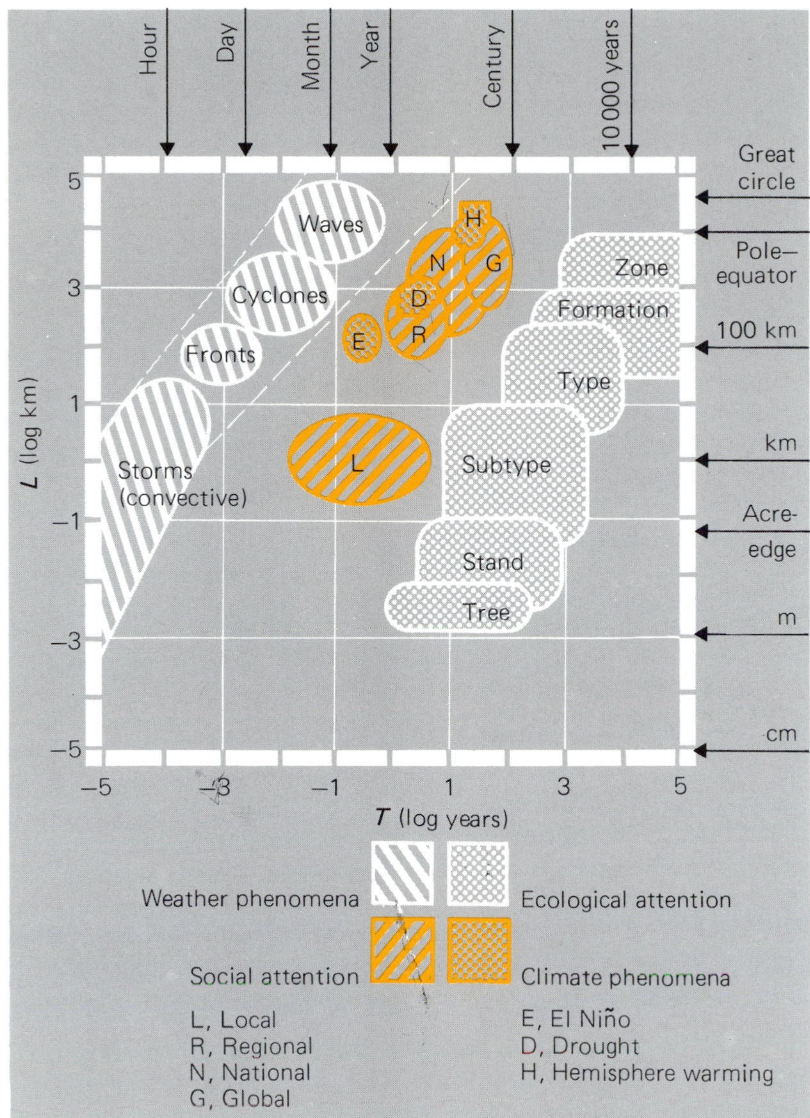

Figure 1.21 Scales of interactions among climates, ecosystems, and societies. The figure combines characteristic time and space scales from *Figures 1.19* and *1.20*. Four bands of phenomena are evident. At a given spatial scale, ecological processes are slowest, meteorological processes fastest, and social and climatic processes overlap at an intermediate time scale [75].

this early stage in the investigation it would be unwise to let our studies of surprise be limited to those approaches that lend themselves most easily to formal mathematical representation. A variety of exploratory approaches, utilizing as wide a range of perspectives as possible, is also needed.

One set of directions that might be pursued has been outlined by Brooks in Chapter 11 on the "typology of surprise". He characterizes the kinds of surprise arising in the interactions of technologies, institutions, and development as follows:

(1) Unexpected discrete events, such as the oil shocks of 1973 and 1979, the Three Mile Island (TMI) reactor accident, political coups or revolutions, major natural catastrophes, accidental wars.

(2) Discontinuities in long-term trends, such as the acceleration of USA oil imports between 1966 and 1973, the onset of the stagflation phenomenon in OECD countries in the 1970s, the decline in the ratio of energy consumption growth to GNP growth in the OECD countries after 1973.

(3) The sudden emergence into political consciousness of new information, such as the relation between fluorocarbon production and stratospheric ozone, the deterioration of central European forests apparently due to air pollution, the discovery of the recombinant DNA technique, the discovery of asbestos-related cancer of industrial workers [76].

Torsten Hagerstrand and Robert Kates have approached the problem somewhat differently, listing a number of techniques (including some discussed above) for constructing surprise-rich scenarios:

(1) Contrariness: How can the surprise-free assumptions be changed?
(2) Perceived expert surprise: What are the tails of the distributions of relevant tasks, events, and outcomes?
(3) Imaging: Given an unlikely future, what sequence of events might be used to reach it?
(4) System dynamics: How could known current trends produce counterintuitive results due to interaction?
(5) Surprise theory: Are there underlying principles that would let us understand unexpected events and developments?
(6) Historical retrodiction: Are the seeds of future surprises always present with hindsight, and how can they be recognized [72]?

One approach to using surprise-rich scenarios in exploring management strategies for future histories of interactions between development and environment is described in the next section.

Future histories of the biosphere: policy exercises for sustainable development

There is a great potential for usable knowledge in the converging research advances sketched in this chapter. But if that potential is to be realized, we will also need the appropriate methods for synthesis (Brewer, Chapter 17). These methods will have to contend with the two central challenges to the efforts to shape the sustainable development strategies that I identified at the beginning of this chapter: the wide range of perceptions concerning what kind of "garden" we want to develop out of the biosphere, and our incomplete scientific understanding of what kind of "garden" we can obtain using alternative management strategies.

Common approaches to synthesis

The two synthesis methods that are commonly used for handling scientific knowledge with policy implications are the expert panel and the formal model. Experience has shown that both have special strengths and special weaknesses.

"Blue Ribbon" panels

The expert or "Blue Ribbon" panel is the longest established and most widely used synthesis method. National academy committees, presidential commissions, international working groups, and their relatives have expressed opinions on most topics that bear on the interactions between development and environment. Surprisingly, little systematic, critical appraisal has been carried out of the performance of such panels. What seems generally clear, however, is that the great strength of the typical "Blue Ribbon" panel is its ability to bring together, at short notice, small, interactive, *ad hoc* groups of the very best people in whatever business is at hand. When a consensus of panel members can be reached, and especially when the consensus is sanctioned by a suitably distinguished convening organization, the *ad hoc* expert group can be an efficient and effective way for producing syntheses of rapidly developing knowledge as it applies to complex problems.

Two major difficulties limit the occasions in which such desirable results emerge. The first and less serious is the expert panel's relatively poor ability to create new knowledge or perspectives rather than just synthesizing what its distinguished members already know. Competing demands on members' time, strong social pressures to defer to colleagues' areas of expertise, and limited panel lifetimes all contribute to this limitation, which can be only partially mitigated by good staff support. The second and more serious difficulty arises when the consensual mode of synthesis is impossible or inappropriate. Unfortunately, this is precisely the case for many of the value laden, incompletely understood issues that arise in the evaluation of strategies for a sustainable development of the biosphere. Reflecting on the history of expert groups that have addressed such issues as desertification or the environmental implications of energy options, one is forced to conclude that the "Blue Ribbon" approach is at risk whenever there is both scientific uncertainty and high political stakes (see Ravetz, Chapter 15).

Formal models

In Chapter 17, Brewer reviews the substantial literature regarding the strengths and weaknesses of formal modeling as a method of synthesizing scientific knowledge for policy purposes. Models clearly have the potential for handling large amounts of technical information in a reproducible and systematic way. Nonetheless, Brewer and many others have concluded that the large computer simulation studies that have become so much a part of the resource management and environmental assessment literature are part of the problem, not of the solution [78].

In large measure, the difficulties arise from the sheer mass of most formal modeling studies that try to tackle significant practical problems. Major models require orders of magnitude more time and resources to build and run through to their conclusions than do even the most expensive expert panels. As a result, even the best intentioned modelers find it difficult in practice to experiment with different perspectives and problem formulations. A single approach to the problem, usually reflecting a single set of underlying assumptions, is often seen to take on a life of its own, demanding continual embellishment and justification. For a program that seeks to illuminate the complex, ill-defined "syndromes" arising from interactions between development and environment, this practical inability of formal models to explore alternative perspectives is a serious drawback. Brewer summarizes the case against large modeling efforts as follows:

(1) Such work is more adversarial than most realize . . .
(2) The work is one-sided; it presents but one perspective on a future rich in potentialities . . .
(3) Quality control and professional standards are nil . . .
(4) Data inputs have obscure, unknown, or unknowable empirical foundations, and the relevance of much data, even if valid, is unknown . . .
(5) [The work ignores] the most interesting aspect of any analysis . . . the set of assumptions used to fashion it . . . [79].

More useful means of practicing how to manage the complex and incompletely understood interactions of the environment and development are badly needed. An extensive review and evaluation of experience with alternative approaches (Brewer, Chapter 17, and Sonntag's *Commentary*) has led us to explore what we have called "policy exercises": organized efforts that bring together decision makers, planners, scientists, and technologists to practice writing "future histories" of how we might manage interactions between societies' development activities and the global environment. What a usable "policy exercise" would actually look like is something we seek to discover through an experimental program of specific case applications and subsequent evaluations.

The history of political gaming

The basic concept of the policy exercise was suggested by approaches developed in support of political–military strategic planning during the late 1950s and early 1960s [80]. At that time, experience had shown that formal models were inadequate to capture the contingencies, the unquantifiable factors, and the contextual richness that seemed central to the main lines of political evolution between the great powers in the period 1955 to 1965. The models also tended to strengthen rather than relax the barrier between analysts and practitioners of political and military strategy. In attempting to design more useful integrative approaches, Herbert Goldhamer of the Rand Corporation realized that his "problem was similar to that confronting historians. He was faced with the task of writing a 'future history' to clarify his ideas about the motives and influences affecting the behavior of great powers, their leaders, and others in the real political world [81]."

The method devised by Goldhamer and his colleagues to write these future histories was dubbed "political gaming". Teams of human (as opposed to computer) participants were confronted with generally realistic problem scenarios and required to work through responses both to the scenario and to the moves made by other teams. The role of "Nature" was played by a control team that determined the impact on conditions of play resulting from the moves of the teams, injected unexpected events, introduced constraints on allowable responses, and so on [82].

Brewer (Chapter 17) describes four "difficult questions that eluded or exceeded the capacity of alternative analytic tools" that were explored by the original political exercises. These questions sufficiently resemble those that confront us in learning to cope with the interactions between development and environment to warrant quoting here:

(1) What political options could be imagined in light of the conflict situations portrayed? What likely consequences would each have?
(2) Could political inventiveness be fostered by having those actually responsible assume their roles in a controlled, gamed environment? Would the quality of political ideas stimulated be as good or better than those obtained conventionally?
(3) Could the game identify particularly important, but poorly understood, topics or questions for further study and resolution? What discoveries flow from this type of analysis that do not from others?
(4) Could the game sensitize responsible officials to make potential decisions more realistic, especially with respect to likely political and policy consequences [83]?

Experience with political gaming indicates that each of the preceding questions can be given an answer of "yes, but only under favorable conditions". The games were very time-consuming to prepare, suffered from acute problems of managing the vast quantities of data involved, and suffered from lack of critical review and evaluation due to the sensitive status of their subject matter. The military moved on to emphasize man–machine and machine–machine games, with less than satisfactory results [84].

From political gaming to policy exercises

Despite the difficulties with early strategic political games, however, some of the basic ideas and experiences involved in their evolution suggest that they might be adapted to complement formal models and expert committees in a new approach to addressing the problems of sustainable development. Brewer (Chapter 17) has called the needed hybrid approach a "policy exercise". Several of the most important potential linkages between political gaming experience and possible exercises for environmental policy are summarized below.

The political games were found to perform better than alternative approaches in studying "... poorly understood dynamic processes ... [and] institutional interactions" and in "... opening participation to many with different perspectives and special competences, on a continuing basis over time" (Brewer, Chapter 17, pp 462–463). The need to accommodate multiple political and environmental perspectives in efforts to learn effective management responses to sustainable development problems has been emphasized throughout this chapter. The "poorly understood" nature of our knowledge of long-term, large-scale environmental processes and institutional responses likewise requires no further elaboration. Policy exercises, whatever they turn out to be, must be designed in ways that will let us address these critical issues.

Reviewing experience with political gaming, Brewer and Shubik concluded

> The selection of competent professionals to participate in the political exercise proved to be critically important. This situation is analogous to that in chess or other games when inferior players tend to consolidate their own bad habits rather than being stimulated to improved or inspired play [85].

To be useful, policy exercises on the interactions of development and environment would almost certainly have to involve several of the very top scientists involved in research on the biosphere, plus their opposite numbers from the world of politics, finance, and industry. The political gaming experience suggests that there is little point in carrying out such exercises using second rate consultants and middle level bureaucrats. Securing the participation of sufficiently senior and innovative people may not be as difficult as it sounds. In carrying out the first phases of the Biosphere Program we have often found the best scientists and best policy people expressing a growing dissatisfaction over their inability to address each other, except through stultifying layers of reports and bureaucracy or in ritualized and guarded public encounters. Carefully designed policy exercises might provide the channel and forum of communication they seek.

Finding a way to order, to evaluate, and ultimately to use the great volume of literature now being generated on the scientific and practical aspects of interactions between development and environment is becoming increasingly urgent. The occasional grand Conference or Commission has a role to play in linking theory to its practical implications, but this role is limited. Goldhamer and Speier commented on their early experience with political exercises that

> [O]ne of the most useful aspects of the political game was its provision of an orderly framework within which a great deal of written analysis and discussion took place . . . [O]ral or written discussion of political problems that arise during the game is one of its most valuable features [86].

The difficulties of managing this flow of data, of analysis, and of discussion turned out to be one of the weak points of realistic political games. The potential for mitigating such difficulties with modern microcomputers has been demonstrated in a number of related exercises (see Sonntag in his *Commentary* to Chapter 17) and provides one more reason for experimenting with hybrid "policy exercise" approaches in the Biosphere Program.

Time sequences of management actions

The "future history" orientation of the exercises' output should make them an excellent vehicle for exploring response options in terms of the time sequences of coordinated action that they imply. Experience suggests that one of the most common and important factors that limits the effectiveness of proposed interventions is likely to be time [87]. The entire Malthusian debate that sets the context for our study can be seen as a question of whether population growth will allow enough time for development activities to mobilize adequate resources to insure acceptable conditions of human well-being. As a specific example, long-term energy studies have consistently concluded that the question is not whether adequate quantities of fuels exist, but rather whether technologies for harnessing them and the infrastructure for supporting them can be developed and implemented on a time scale consistent with the rates of rising demands and falling supplies of alternative fuels [88]. In general, potentially damaging impacts of development on the environment take time to detect in terms unambiguous enough to motivate action. International negotiations on which interventions to adopt can take even longer. Finally, actions, once agreed on, can take a very long time indeed to be implemented over sufficiently large areas to make a difference. As a result, much of the history of efforts to manage large-scale environmental change can be summarized as "too little, too late".

It is therefore of fundamental importance to explore the timing constraints that limit the effectiveness of the technological, institutional, and research management options described earlier. As an example of what can be done, one of the most illuminating approaches to assessing energy options for dealing with the "greenhouse" syndrome has addressed the timing question by showing when various technological strategies would have to be initiated in order to keep future environmental changes to within specified limits [89]. A long tradition of research on the rate and pattern of adoption for technological innovations can be tapped to facilitate explorations of timing constraints in future histories of environment and development. Less well explored, but presenting no obvious barriers to investigation, are comparable studies for institutional innovation (see earlier and Chapter 12).

Very little systematic study has been given to the question of timing constraints on the fruits of research and monitoring programs that deal with interactions between development and environment. The early political exercises did help to refine future research priorities for technical participants and their staffs by exposing them to the kinds of questions that their political masters would need answered, under a range of often unconventional but still plausible future histories. This is an extremely important result since

> [T]he game . . . may under favorable circumstances make more effective use of existing knowledge than other modes of intellectual collaboration, but it would be placing an intolerable burden on it to treat it as a machine that displaces theoretical thought and empirical research [90].

Likewise, policy exercises would in no sense be a substitute for the careful research on basic science and response options that is required to improve the basic tools that can be used in fashioning effective social responses. But the exercises might help the research community to learn which answers – which tools – are likely to be most needed by policy people across a wide range of plausible future histories of the interactions of development and environment [91].

If experience with political gaming is any guide, we should expect that many of the ostensibly useful answers now being sought by scientists and analysts are ones for which no policy people are ever likely to ask the relevant questions. Conversely, we are likely to find a number of urgent questions emerging in the course of the policy exercises that scientists could be studying, but are not. I suggested earlier that such practically important, but understudied, questions might include "row assessments" of the net impact of development-related policy actions across a range of environmental components. More generally, the design of global environmental monitoring and data systems useful for management could benefit from such policy exercises.

Writing future histories

The challenge for the policy exercises will be to design ways of realizing the potential outlined above, thus providing a means of writing useful "future histories" of a sustainable development for the next century. As we envision our first round of experiments, a policy exercise would work through several development scenarios [92], each consisting of a set of trends in population, agriculture, and energy over the next century, set against a background of specific geopolitical and other events (see earlier). The policy people and scientists who participate in the exercise would work through successive time intervals of the development scenario and attempt to manage and maintain what I have called "the kind of garden they want". The particular goals of the Biosphere Program will lead us to emphasize the kinds of gardens that improve human well-being to the greatest extent possible, while assuring that the improvements are ecologically sustainable over the long run and on a global scale.

At each time interval throughout the game (perhaps every 5 or 10 years of game time), participants would evaluate the environmental implications of their "past" and possible "future" development activities, and intervene in the scenario's interactions between development and environment with whatever management activities they judge appropriate. Following the arguments introduced earlier, their responses can draw on a range of possible technological, institutional, or research innovations, each with a characteristic effectiveness and lead-time requirement. The "control" function provided by the organizers of the exercise would update the condition of the scenario's environment in accord with these management actions, drawing on the synoptic framework for environmental assessment described earlier. Explicit allowance would be made for uncertainties in the scientific knowledge and for possible unexpected surprises. The participants would be presented with this updated scenario as a point of departure for their next cycle of play.

The objective of the participants in the exercise would be to arrive at the end of the 100-year gaming period with a sustainable and otherwise desirable garden intact, and an internally consistent "future history" about how it was cultivated. Several cycles of the exercise, based on several different scenarios, would give an opportunity for learning about how to deal with problematical time lags, uncertainties, and surprises and for writing a set of alternative future histories for a sustainable development of the biosphere.

The problem-solving experience and internally consistent sequences of effective management actions resulting from the policy exercises should, we hope, complement the other research initiatives sketched in this chapter and described in more detail throughout the remainder of this volume. With luck, they will also provide an integrative tool for helping us to reach the larger objective of identifying and evaluating strategic interventions through which societies might enhance the long-term, large-scale interactions between development and environment [93].

Notes and references

[1] National Academy of Sciences (1976), Science: a resource for humankind, *Proceedings of the National Academy of Sciences Bicentennial Symposium*, Appendix E, p 4 (National Academy Press, Washington).

[2] For a review of current programs see Malone, T. F. and Roederer, J. G. (Eds) (1984), *Global Change* (Cambridge University Press, Cambridge, UK); Goody, R. M. (1982), *Global Change: Impacts on Habitability*, JPL D-95 (California Institute of Technology, Pasadena, CA); Bolin, B. (1979), Global

ecology and man, in World Meteorological Organization, *Proceedings of the World Climate Conference*, p 24 (WMO, Geneva). A sense of the scope of the science itself can be obtained from Budyko, M. I. (forthcoming), *Evolution of the Biosphere* (Reidel, Dordrecht); Lovelock, J. E. (1979), *Gaia: a New Look at Life on Earth* (Oxford University Press, Oxford); and Bolin, B. and Cook, R. B. (Eds) (1983), *The Major Biogeochemical Cycles and their Interactions*, SCOPE 21 (John Wiley, Chichester).

[3] See Braudel, F. (1984), *The Perspective of the World* (Harper and Row, New York), especially Chapter 1; Wallerstein, I. (1974), *The Modern World-System* (Academic Press, New York); Chisholm, M. (1982), *Modern World Development* (Hutchinson, London).

[4] See pp 31–38.

[5] See pp 38–42.

[6] The goal of my assessment was to outline a research program that would complement, not compete with or duplicate, existing studies of environment–development interactions. To this end, I reviewed plans and products of a number of international efforts, national studies, and individual works of scholarship. The more ambitious of these included work of UNESCO's program on *Man and the Biosphere*; the IFIAS project on *Analyzing Biospheric Change*; the IUCN/UNEP/WWF *World Conservation Strategy* (see the publication of that title published in 1980 in Gland, Switzerland, by the International Union for the Conservation of Nature); the World Meteorological Organization's *World Climate Programme* [see WMO (1979) *Proceedings of the World Climate Conference* (WMO, Geneva)]; the UN Environment Programme's review of *The World Environment 1972–1982* [see the publication of that title, Holdgate, M. W., Kassas, M., and White, G. F. (Eds) (1982) Tycooly International, Dun Laoghaire, Ireland]; The Organization for Economic Cooperation and Development's *Economic and Ecological Independence* [see OECD's publication of that title (1982) (OECD, Paris)]; NASA's *Global Habitability Program* (see Goody, note [2]); SCOPE's study on *The Major Biogeochemical Cycles* (see Bolin and Cook, note [2]); The World Resource Institute's *The Global Possible* [see WRI's forthcoming book of that title (WRI, Washington, DC)]; ICSU's initiative on *Global Change* (see Malone and Roederer, note [2]); and the United Nations's new *World Commission on Environment and Development.*

[7] See especially White, G. F. (1980), Environment, *Science*, **209**, 183–190; Kates, R. W. and Burton, I. (1986), The great climacteric 1798–2048: transition to a just and sustainable human environment, in R. W. Kates and I. Burton, *Geography, Resources and Environment, Vol. 2, Essays on Themes from the Work of G. F. White* (University of Chicago Press, Chicago, Ill); Caldwell, L. K. (1984), *International Environmental Policy* (Duke University Press, Durham, NC).

[8] Quoted in Dubos, R. (1973), *A God Within*, pp 3–4 (Scribners, New York).

[9] For the context of Malthus' work, see James, P. (1979), *Population Malthus: His Life and Times* (Routledge and Kegan, London).

[10] From Kates and Burton, note [7].

[11] Burton, I. and Kates, R. W. (1964), Slaying the Malthusian dragon, *Economic Geography*, **40**, 82–89.

[12] Marsh, G. P. (1874), *The Earth as Modified by Human Action* (Scribner, Armstrong, and Co., New York); Osborne, F. (1984), *Our Plundered Planet* (Little, Brown, Boston, MA); Brown, H. (1954), *The Challenge of Man's Future* (Viking, New York); Carson, R. (1962), *Silent Spring* (Houghton Miffling, Boston, MA); Meadows, D. H., Meadows, D., Zahn, E., and Milling, P. (1972), *Limits to Growth* (Potomac Associates, Washington, DC); US Council of Environmental Quality (1980), *Global 2000 Report to the President* (US Government Printing Office, Washington, DC); Reclus, E. (1876), *La Terre et les Hommes* (Hachette, Paris); Simon, J. and Kahn, H. (Eds) (1984), *The Resourceful Earth* (Blackwells, New York); WRI, note [6].

[13] Weinberg, A. M. (1984), The resourceful earth: a review, *Environment*, **26**(7), 25–27.

[14] An excellent review of the changing perceptions of man–environment interactions is given in White (1980), note [7].

[15] An earlier version of this work, published in 1864 under the title *Man and Nature, or Physical Geography as Modified by Human Action*, is discussed by Williams in his *Commentary* to Chapter 2.

[16] Lotka, A. J. (1924), *Elements of Physical Biology* (Williams and Wilkins, New York). This quotation is taken from p 222 of the reprint edition entitled *Elements of Mathematical Biology*, published in 1956 (Dover, New York).

[17] Vernadsky, W. [V.] I. (1945), The biosphere and the noosphere, *American Scientist*, **33**, 1–12.

[18] Stoppani, writing at the same time as Marsh, argued that man constituted a new geological force, and designated ours as the "anthropozoic era". He wrote

> that "the creation of man . . . was the introduction of a new element into nature, of a force wholly unknown to earlier periods . . . It is a new telluric force which in power and universality may be compared to the greater forces of earth." The translation is from p 607 of Marsh, note [12]. The original source is Stoppani (1873), *Corso di Geologia*, Vol. ii, cap, xxxi, section 1327 (Milan).

[19] Vernadsky, note [17].

[20] Bolin and Cook, note [2], pp 33–35.

[21] This phrase, and much of the thrust of the next several paragraphs, is drawn from a letter by Harvey Brooks to William Clark, 17 September 1984. For a related argument see Wagner, P. L. (1960), *The Human Use of Earth* (The Free Press, New York).

[22] See, for example, the survey of policymakers' needs in Clark, W. C. (1985), *On the Practical Implications of the Carbon Dioxide Question* (World Meteorological Organization, Geneva).

[23] See Golubev, G. N. and Biswas, A. K. (Eds) (1979), *Interregional Water Transfers* (Pergamon Press, Oxford); and Micklin, P. P. (1985), The vast diversion of Soviet rivers, *Environment*, **27**(2), 10–20, 40–45.

[24] See Caldwell, note [7].

[25] Perry, J. (forthcoming) *The Role of International Institutions in the Sustainable Development of the Biosphere* (International Institute for Applied Systems Analysis, Laxenburg, Austria).

[26] The closest model I know of what is needed is the typology of institutional means of calculation and control described by Dahl, R. A. and Lindblom, C. E. (1953), *Politics, Economics, and Welfare* (Harper and Brothers, New York).

[27] Bretherton, F. P. (1985), Earth system science and remote sensing, *IEEE Proceedings*, **73** (6), 118–1127.

[28] See note [2] above, and National Research Council (1984), *Global Tropospheric Chemistry* (National Academy Press, Washington, DC). A few earlier writers did emphasize the important influence of life on the environment, among them W. I. Vernadsky, G. E. Hutchinson, and R. Revelle.

[29] Lovelock, note [2]. See also Lovelock, J. E. and Whitfield, M. (1982), The lifespan of the biosphere, *Nature*, **296**, 561–563.

[30] See note [2], especially Malone and Roederer.

[31] Roederer, J. G. (1984), The proposed International Geosphere–Biosphere Program: some special requirements for disciplinary coverage and program design, in Malone and Roederer, note [2], pp 1–19.

[32] The beginnings of one such framework for water systems is sketched in Douglas, I. (1976), Urban hydrology, *Geographical Journal*, **142**, 65–72. A framework for treating soil systems in the biosphere study is outlined in Harnos, Z. (forthcoming), *The Role of Soils in Sustainable Development of the Biosphere* (International Institute for Applied Systems Analysis, Laxenburg, Austria).

[33] I deal with this issue in more depth in Clark, W. C. (1985), *The Cumulative Impact of Human Activities on the Atmosphere*, paper presented at the Workshop on Cumulative Impact Assessment sponsored by The Canadian Environmental Assessment Research Council and the US National Research Council, Toronto, February 4–7, 1985. The concept of "valued environmental components" is developed in Beanlands, G. E. and Duinker, P. N. (1983), *An Ecological Framework for Environmental Impact Assessment in Canada* (Institute for Environmental Studies of Dalhousie University, Halifax, and Federal Environment Assessment Review Office, Hull, Quebec). The groundwork for both these papers is laid in Holling, C. S. (Ed) (1978), *Adaptive Environmental Assessment and Management* (John Wiley, Chichester, UK).

[34] NRC, note [28]; see also National Research Council (1981), *Atmosphere–Biosphere Interactions: Towards a Better Understanding of Ecological Consequences of Fossil Fuel Combustion* (National Academy Press, Washington); Bolin and Cook, note [2].

[35] Thus changes in ozone concentrations are shown to affect the valued atmospheric component of "ultraviolet energy absorption" because the ozone molecules themselves have the ultimate impact. Halocarbons (e.g., freons) and nitrous oxide, though assuredly relevant to ultraviolet energy absorption, are *not* shown to affect this valued atmospheric component because their action occurs indirectly, by changing the concentration of ozone. Note from *Figure 1.3* that a significant number of chemicals are involved in multiple impacts. Sources of disturbance or intentional policies that affect these chemicals must therefore be assessed in terms of multiple kinds of impacts on the atmosphere.

[36] I have already alluded to an example of such indirect effects in the case of the ozone problem. Industrial processes add halocarbons to the atmosphere. These affect the ultraviolet energy absorbtion only via intermediate impacts of halocarbons on ozone.

[37] An excellent overview of the field is provided in the recent US National Academy of Sciences' report on *Global Tropospheric Chemistry* (NRC, note [28]).

[38] This approach was initially described for general environmental impact assessment in Munn R. E. (Ed) (1975), *Environmental Impact Assessment*, p 72, SCOPE 5 (John Wiley, Chichester, UK).

[39] Examples of "column" assessments are National Research Council (1983), *Changing Climate* (National Academy Press, Washington, DC); Bolin, B., Döös, B. R., Jäger, J., and Warrick, R. A. (forthcoming), *The Greenhouse Effect, Climate Change, and Ecosystems*, SCOPE monograph.

[40] Examples of "row" assessments are National Research Council (1979), *Nuclear and Alternative Energy Systems* (National Academy Press, Washington, DC) and NRC (1981), note [34].

[41] Clark, W. C., note [33].

[42] Crutzen, P. J. (1983), Atmospheric interactions – homogeneous gas reactions of C, N, and S containing compounds, in Bolin and Cook, note [2].

[43] See note [32].

[44] For an overview of key issues, see the special issue of *Atmospheric Environment* [(1984), **18**(3)] on *Air Pollution – Health and Management.*

[45] See also Bach, W. (1984), *Our Threatened Climate* (Reidel, Dordrecht).

[46] See note [3].

[47] Deevey, E. S., Jr. (1960), The human population, *Scientific American*, **203**, 195–204.

[48] United Nations (1979), *Prospects of Population: Methodology and Assumptions*, Population Studies No.

67 (United Nations, New York); Keyfitz, N., Allen, A., Edmonds, J., Dougher, R., and Wiget, B. (1983), *Global Population 1975–2075 and Labor Force 1975–2050,* ORAU/IEA-83-6(M) (Oak Ridge Associated Universities, Oak Ridge, TN); the total population data reported in Keyfitz *et al.* have been republished with some typographical errors corrected in Clark W. C. (Ed) (1982), *Carbon Dioxide Review*: *1982,* pp 461–464 (Oxford University Press, New York). These updated numbers are used throughout the remainder of this chapter. See also Bogue, D. J. and Tsui, A. O. (1979), Zero world population growth, *Public Interest,* **Spring**, 99–113; World Bank (1984), *World Development Report 1984* (Oxford University Press, New York).

[49] Kates and Burton, note [7].

[50] Sources of basic data for *Figures 1.11* to *1.17* are given below. In all cases, individual country data were grouped into contiguous areas of about 5×10^6 km^2 each (see the maps in *Figures 1.12, 1.15,* and *1.16*), and these groupings were used in subsequent calculations. Summary data are presented in the table below this note. The analysis was carried out for this chapter by Anna Clark.

- *Area and Density*: Figures reflect total land area, from World Bank (1984), *World Development Report 1984,* Table 1, pp 218–219 (Oxford University Press, New York).
- *Population*: Data for 1750–1900 from Durand, J. D. (1967), The modern expansion of world population, *Proceedings of the American*

Table Densities of human development (1980).

		Densities[b]		
	Area[a]	*Population*	*Agriculture*	*Energy*
World	133285	33	26	39
China (Turkestan and Tibet)	3600	4	2	0
China (proper)	4000	213	123	77
Manchuria and Mongolias	3565	31	20	2
Japan, Koreas, Taiwan, etc.	628	316	225	756
Indian subcontinent	4490	198	91	29
Southeast Asia	4489	80	51	18
Near East	6927	22	22	19
Europe (excluding USSR)	4874	100	136	350
USSR (Europe and Caucasia)	5642	35	39	130
USSR (Kazakhstan, Central Asia)	3994	10	10	5
USSR (Siberia and Far East)	12766	2	2	1
North Africa	8260	13	8	6
Africa, Sahel States	4822	4	1	0
West Africa	2605	50	26	7
Middle Africa	5329	9	4	1
East Africa	6333	21	10	2
South Africa	2693	12	9	26
Canada (Eastern Provinces)	3148	5	5	37
Canada (Western Provinces)	2912	2	7	6
Alaska (Canadian Territories)	5447	0	0	0
USA (Eastern and Central)	3073	52	83	264
USA (Western and Prairie)	4759	14	36	34
Middle America	2730	46	31	47
Tropical South America	5504	14	8	14
Brazil	8512	14	15	11
Temperate South America	3700	11	19	13
Oceania	8482	3	4	10

[a] *Area* is all land in units of 10^3 km^2.
[b] *Densities* are for all land in 1980; *population* densities in units of people/km^2; *Agriculture* densities in units of 100 US$ value added/km^2; *energy* densities in units of 100 kg coal equivalent/km^2.

Philosophical Society, **111**(3), 137, quoted in Rostow, W. W. (1978), *The World Economy*, Table 1.1, p 4 (University of Texas Press, Austin, TX). Data for 1950–1980 from *United Nations Yearbook 1980*, Table A-2, pp 16–21. Data for 2000–2075 from Keyfitz *et al.*, note [48].

- *Energy*: Data are energy consumption for 1981, from World Bank (1984), this note, Table 8.
- *Agriculture*: Data are total agricultural production for 1979 in dollar values, taken from data tapes of the Food and Agriculture Program at the International Institute for Applied Systems Analysis, Laxenburg, Austria.

Modification of basic data: the following sources were used to divide large national areas into subregions of about 5×10^6 km^2.

- *China*: For land area, McEvedy, C. and Jones, R. (1978), *Atlas of World Population History*, p 166 (Penguin Books, Middlesex, UK). For population and agriculture, Walker, K. R. (1984), *Food Grain Procurement and Consumption in China* (Cambridge University Press, Cambridge, UK) in which Appendix 13 gives the provincial total population, 1980 (p 314), and Appendix 14 gives the provincial total grain output, 1980 (p 319). (Grain, as defined by the Chinese, includes pulses, potatoes, and soya beans, as well as cereals.) No figures for Tibet are included in Walker's data. Since Tibet's size of population and agricultural production is insignificant in relation to China's total figures, data for Tibet have been omitted. Tibet's population was 0.2% of China's total population in 1973 [Lateef, S. (1976), *Economic Growth in China and India*, p 2, EIU Special Report No. 30] and its output of grain was 0.3% of China's total grain output in 1975 [US Central Intelligence Agency (1976), *China: Agricultural Performance in 1975*, CIA Report ER 76-10149]. Energy consumption was distributed according to provincial population because no satisfactory recent data were available.
- *Canada*: For population: Canada (1984), Population of Canada by provinces and territories, *Canadian Almanac and Directory 1984*, p 1008. For land area, agriculture, and energy: Statistics Canada (1981), *Canada Year Book 1980–81* (Statistics Canada, Ottawa), Table 1.8, Total area classified by tenure 1978; Table 11.1, Cash receipts from farming operations by province 1978; Table 13.16, Electricity made available for use in Canada by provinces and territories 1978.
- *United States of America*: US Department of Commerce, Bureau of the Census (1983), *Statistical Abstract of the United States 1982–1983*, Table 338, Area of states, p 199; Table 33, Resident population by states 1980, p 28; Table 1202, Crops – farm value 1981, p 680; Table 977, Total energy consumption 1980, p 574.
- *USSR*: For area and population: Dewdney, J. C. (1982), *USSR in Maps*, Table I, p 28 (Hodder and Stoughton, Sevenoaks, UK). For agriculture and energy, Lewytzky, B. (n.d.), *The Soviet Union Figures–Facts–Data* (K. G. Saur, Munich) Table 5.2, Gross agricultural output by Union Republics 1975; Table 5.1.2.2, Total consumption of electric power 1975.

For a few countries no agriculture or energy data were available. It was assumed that agricultural production and energy consumption densities of these countries approximate the density of their surrounding countries. Neither the area nor consumption or production figures of these "missing data" countries were included in the density calculations. Exceptions: the area of Saudi Arabia (2.150×10^6 km^2) was included for calculation of agricultural densities with zero production; for calculation of energy densities Botswana's per capita energy consumption was assumed to be the same as Angola's, and Namibia's was assumed to be the same as Tanzania's.

[51] For sources, see note [50].

[52] For sources, see note [50].

[53] For perspective, the Serengeti–Mara plains of Africa support about 10 000 kg of ungulates per km^2, which is equivalent in biomass terms to about 200 people per km^2. The most prevalent single species in the Serengeti is present at a density of about 30 individuals per km^2 with a biomass equal to about 30 people per km^2 [Sinclair, A. R. E. and Norton-Griffiths, M. (1979), *Serengeti: Dynamics of an Ecosystem* (University of Chicago Press, Chicago, IL)]. Deer in North America commonly reach densities of 10 individuals per km^2 with a biomass equal to about 20 people per km^2 [Odum, E. P. (1971), *Fundamentals of Ecology*, p 154 (Saunders, Philadelphia, PA).

[54] Rostow, note [50], see p 48 and Appendix A.

[55] World Bank, note [48].

[56] Data from Rotty, R. published in Clark, W. C. (Ed) (1982), *Carbon Dioxide Review: 1982*, pp 457–460 (Oxford University Press, New York).

[57] See, however, Jaeger, J. (1983) *Climate and Energy Systems* (John Wiley, Chichester, UK).

[58] For sources, see note [50].

[59] For sources, see note [50]. An earlier attempt to map energy densities is presented by Llewellyn, R. A. and Washington, W. M. (1977), Regional and global aspects, in National Research Council, *Energy and Climate* (National Academy Press, Washington, DC).

[60] For sources, see note [50].

[61] There is a large literature on the Earth's "carrying capacity" for human population. Among the most recent and careful studies is FAO (1982), *Potential Population Supporting Capacities of Lands in the Developing World*, FPA/INT/513 (FAO, Rome). The difficulty with this approach is that the number of

people that a given region can carry depends not only on the kind of environment involved, but also on how the people conduct their lives. In particular, the carrying capacity depends on the mix of economic activites practiced, the measures that are taken to mitigate the impacts of these activities on the environment, the level of human well-being that is sought, and the degree to which other regions are used as sources of food, energy, and other inputs or as sinks of waste products. Even more important, carrying capacity depends on the inventiveness of the human population in resolving the problems and capturing the opportunities that arise in the continuing interaction of environment and development. Because of these difficulties, we have not as yet felt compelled to adopt the carrying capacity approach in the Biosphere Program, though the option is being considered.

[62] See, for example, Keyfitz, N. (1981), The limits of population forecasting, *Population Development Review*, **7**(4), 579–593; Stoto, M. (1983), The accuracy of population forecasts, *Journal of the American Statistical Association*, **781**, 13–20; Fox, G. and Ruttan, V. W. (1983), A guide to LDC food balance projections, *European Review of Agricultural Economics*, **10**, 325–356; Ausubel, J. H. and Nordhaus, W. D. (1983), A review of estimates of future carbon dioxide emissions, in NRC, note [39], pp 153–185; Ascher, W. (1978), *Forecasting: An Appraisal for Policy-makers and Planners* (Johns Hopkins University Press, Baltimore, MD).

[63] Brooks, Chapter 11, p 326. He cites examples of econometric energy, and environmental models from Greenberger, M. L., Brewer, G. D., Hogan, W. W., and Russell, M. (1983), *Caught Unawares: The Energy Decade in Retrospect* (Ballinger, Cambridge, MA); Office of Technology Assessment (1982), *Global Models, World Futures, and Public Policy: A Critique*, OTA-R-165 (OTA, Washington, DC); and US Council of Environmental Quality, note [12].

[64] Holling, C. S. (1985), *Canada – IIASA Biosphere Project: A Proposal* (unpublished manuscript, Vancouver).

[65] Brooks, Chapter 11, p 326.

[66] McNeill, W. H. (1978), Coping with an uncertain future – historical perspective, in C. Hitch (Ed), *Resources for an Uncertain Future*, pp 59–67 (Johns Hopkins University Press, Baltimore, MD).

[67] Catastrophy theory reached an audience outside of the mathematics community with Thom, R. (1975), *Structural Stability and Morphogenesis* (Benjamin, Reading, MA); related notions of order and chaos are given a nontechnical description in Prigogine, I. and Strengers, I. (1984), *Order Out of Chaos* (Bantam, New York). On the role of catastrophy theory and related ideas as paradigm, see Timmerman, Chapter 16. For a sample of applications to specific problems, see Day, R. (1981), Emergence of chaos from neoclassical growth, *Geographical Analysis*, **13**, 315–327; Lorenz, E. N. (1984), Irregularity: a fundamental property of the atmosphere, *Tellus*, **36A**, 98–110; and Holling, Chapter 10.

[68] Crutzen, P. J. (1986), The role of the tropics in atmospheric chemistry, in R. Dickinson (Ed), *Geophysiology of Amazonia* (Wiley, New York).

[69] Holling, Chapter 10. The original statement of these ideas can be found in Holling, C. S. (1973), Resilience and stability of ecological systems, *Annual Review of Ecology Systematics*, **4**, 1–23.

[70] From Clark, W. C. (1979), Spatial structure relationships in a forest insect system: simulation models and analysis, *Mitteilungen Schweizerischen Entomologischen Gesellschaft*, **52**, 235–257.

[71] Parry, Chapter 14, p 382.

[72] di Castri, *Commentary* to Chapter 10.

[73] Clark, W. C. (1985), Scales of climate impacts, *Climatic Change*, **7**, 5–27.

[74] See note [73].

[75] See note [73].

[76] Brooks, Chapter 11, p 326.

[77] Kates, R. and Hagerstrand, T. unpublished memorandum (International Institute for Applied Systems Analysis, Laxenburg, Austria).

[78] See, for example, Ascher, note [62]; and Holling, note [33].

[79] Brewer, Chapter 17, p 461.

[80] Goldhamer, H. and Speier, H. (1959), Some observations on political gaming, *World Politics*, **12**, 72–83.

[81] Brewer, G. and Shubik, M. (1979), *The War Game: A Critique of Military Problem Solving*, p 101 (Harvard University Press, Cambridge, MA).

[82] Exercises of this sort have been described under the term "free-form, manual games" in the American literature. See Goldhamer and Speier, note [80]; Brown, T. A. and Paxton, E. W. (1975), *A Retrospective Look at Some Strategy and Force Evaluation Games*, R-1619-PR (The Rand Corporation, Santa Monica, CA); Brewer and Shubik, note [81].

[83] Brewer, Chapter 17, p 462.

[84] Brown and Paxton, note [82]; Brewer and Shubik, note [81].

[85] Brewer and Shubik, note [81], p 101.

[86] Goldhamer and Speier, note [80], pp 77–78.

[87] Schelling, T. (1984), personal communication.

[88] Häfele, W., Anderer, J., McDonald, A, and Nakicenovic, N. (1981), *Energy in a Finite World* (Ballinger, Cambridge, MA).

[89] Marchetti, C. and Nakicenovic, N. (1979), *The Dynamics of Energy Systems and the Logistic Substitution Model*, Research Report RR-79-13 (International Institute for Applied Systems Analysis, Laxenburg, Austria); Perry, A. M., Draj, K. J., Sulkerson, W., Rose, D. J., Miller, M. M., and Rotty, R. M. (1982), Energy supply and demand implications of carbon dioxide, *Energy*, **7**(12), 991–1004; Laurmann, J. A. (1985), Market penetration of primary energy and

its role in the greenhouse warming problem, *Energy*, **10**(6), 761–775.

[90] Goldhamer, H. (1973), personal communication reported in Brewer and Shubik, note [81], p 103.

[91] A necessary condition for success of the policy exercises will be our ability to incorporate in them the large amounts of "theoretical thought and empirical research" referred to above. A related difficulty in past exercises has been handling the massive quantities of information necessary to describe and check the consistency of successive states of the game's "play". It seems almost certain that in addressing these problems we will want to make extensive use of formal models that deal with particular aspects of the exercise. In other words, formal models have an important though limited role to play in support of any policy exercise that I can envision.

The crucial distinction between popular computer simulation studies of global environmental impacts and the policy exercise approach envisioned here is thus not one of modeling *per se*. Rather, the policy exercises assign a variety of supporting roles to a variety of specialized models, whereas many previous studies have granted "super-star status" to a single, ostensibly comprehensive formal model. The "bit parts" played by models in these supporting capacities could range from the "scratch pad" exploratory tools, described by Sonntag in his *Commentary* to Chapter 17, to existing state-of-the-art models of key biogeochemical, ecological, and climatic processes, to the artificial intelligence-based simulations of "minor player" responses that have been introduced into some of the most recent political gaming exercises [see Davis, P. K. and Winnerfeld, J. A. (1983), *The Rand Strategy Assessment Center: An Overview and Interim Conclusions about Utility and Development Options*, R-2945-DNA (The Rand Corporation, Santa Monica, CA)]. It can be assumed that some of these supporting models will sometimes offer contradictory results. This is a strength rather than a weakness, however, since it provides a vehicle for the methodological pluralism that is so necessary if we are to create usable knowledge for a sustainable development of the biosphere (see contributions to this volume by Ravetz, Timmerman, Rayner, and Thompson).

[92] For a discussion of the use of scenarios in this context, see Brewer, Chapter 17, and Ascher, W. and Overholt, W. H. (1983), *Strategic Planning and Forecasting* (John Wiley, New York).

[93] My thanks to Anna Clark for performing much of the analysis in the section on Human Development and to R. E. Munn and R. W. Kates for critical readings of earlier drafts of this chapter.

PART TWO

HUMAN DEVELOPMENT

Human development

Contributors

About the Contributors

Jeanne Anderer is a science writer at the International Institute for Applied Systems Analysis, concentrating mainly on energy issues. A graduate of St. John's University, New York, she has authored or coauthored several books, including *Energy in a Finite World: Paths to a Sustainable Future*, the summary report of the IIASA global energy analysis.

Heiko Barnert is Leader of the Department of Nuclear Process Heat at the Nuclear Research Center Jülich (Kernforschungsanlage Jülich, KFA) in the Federal Republic of Germany, as well as Leader (for the FRG) in a joint project with Brazil on the research and development of nuclear process heat and process steam. Educated at Aachem Technical University and MIT, Barnert's interests are in the use of nuclear energy in the form of process heat, in the construction of high-temperature reactors, and in various aspects relating to nuclear safety.

Gordon Baskerville is currently Dean of the Faculty of Forestry at the University of New Brunswick. Educated at this University and at Yale University, he moved to the Canadian Forestry Service as research scientist and project manager, before becoming Assistant Deputy Minister, Forest Resources, with the New Brunswick Department of Natural Resources, Canada.

Pierre Crosson is Senior Fellow at Resources for the Future, Washington, where he has been working since 1965. Prior to this he worked as an economist for the US Departments of Commerce and of the Interior, for the Tennessee Valley Authority, for the Bank of America, and for the National Planning Association. Crosson's current research includes investigation of the economic and environmental consequences of agricultural land use, with major attention on how public policies affect patterns of resource use in agriculture.

Joel Darmstadter is Senior Fellow and Director of the Energy and Materials Division, Resources for the Future, Washington, DC. Educated at George Washington University (B.A.) and New School for Social Research (M.A.), he was an economist for the National Planning Association before moving to Resources for the Future. Darmstadter is also a Consulting Editor for *Environment* and Professional Lecturer at the School of Advanced International Studies, Johns Hopkins University.

Wolf Häfele is Director General of the Nuclear Research Center Jülich (Kernforschungsanlage Jülich, KFA) in the Federal Republic of Germany. His academic training was in physics and engineering. From 1960 to 1972 he directed the joint FRG–Belgium–Netherlands–Luxembourg Fast Breeder Reactor Project. In 1973 he became Head of the Energy Systems Project at IIASA, where he directed the global energy analysis *Energy in a Finite World.* He has received honors for his scientific work from Austria, France, the FRG, Sweden, and USA.

Thomas B. Johansson is Professor of Energy Systems Analysis at the University of Lund in Sweden. Educated at the Lund Institute of Technology, in 1968 he co-founded the Environmental Studies Program at the University of Lund and Lund Institute of Technology. Johansson has published widely on solar and nuclear energy, is a Consultant for the Center of Energy and Environmental Studies at Princeton University, and a Commissioner of the Swedish Nuclear Power Inspectorate.

Leonardas Kairiukstis is Academician-Secretary to the Department of Chemical–Biological Sciences of the Lithuania Academy, Director of the Research Institute of Forestry in Girionys, Kaunas, Lithuanian SSR, and Chairman of the Lithuanian National Committee of the UNESCO International Program "Man and Biosphere". Educated at the Agricultural Academy of the Lithuanian SSR and at the Forestry Institute of the USSR Academy of Sciences in Krasnojarsk, he has produced many publications on the theoretics of forest ecosystem formation. Currently, he is interested in the application of systems analysis in silvicultural investigations, in the modeling and optimizing of specialized wood growth, and in the multiple use of forest resources, as well as in applications of dendrochronology to analyze past and future climatic changes and early indications of forest dieback.

Yoichi Kaya is Professor of Electrical Engineering at the University of Tokyo, a Member of the Club of Rome, and Leader of the Energy System Study of the Government of Japan. He became a Doctor of Engineering in 1962 and his main interest lies in modeling.

Sabine Messner is a Research Scholar with the Energy Development, Economy, and Investments Program at the International Institute for Applied Systems Analysis, Austria. She is a doctoral candidate at the Technical University of Vienna and is working on the energy optimization model MESSAGE II, which deals with the entire energy chain, from primary energy to energy services.

Henry Regier is Professor of Zoology and Environmental Studies at the University of Toronto and is a Canadian Commissioner of the Great Lakes Fisheries Commission. Educated at Queen's University, the University of Toronto, and Cornell University, he has served as research scientist and/or policy advisor on fisheries for a number of government and intergovernmental agencies at all levels, from local to global. Regier has also served on committees, panels, and task forces under the auspices of FAO, UNESCO, and INTECOL.

John Richards is Professor of History at Duke University and is currently heading a major project to delineate land-use changes in South Asia and Southeast Asia over the past two centuries. Educated at the Universities of New Hampshire and California at Berkeley, he has taught a wide range of courses on the history and civilization of India, Pakistan, and Nepal.

Carlisle Ford Runge is a member of the Department of Agricultural and Applied Economics at the University of Minnesota, where he also holds appointments in Forest Resources and in the Hubert H. Humphrey Institute of Public Affairs. Educated at the Universities of Wisconsin and North Carolina, and as a Rhodes Scholar at Oxford University, he was recently selected a Science and Diplomacy Fellow of the AAAS. He currently directs the Future of the North American Granary Project, an interdisciplinary research program on agricultural productivity and environment.

Manfred Strubegger is a Research Scholar with the Energy Development, Economy, and Investments Program at IIASA and a doctoral candidate at the Technical University of Vienna. His recent work focuses on the development of an integrated model of energy–economy interactions to apply the interactive decision-analytical approach elaborated at IIASA.

Michael Williams is a Fellow of Oriel College and University Lecturer in Geography at the University of Oxford. Previously, he was Professor of Geography at the University of Adelaide. Williams is the author of numerous books and articles on the rural settlement and forest and land-use history of Britain, Australia, and the United States.

Chapter 2 World environmental history and economic development

J. F. Richards

Editors' Introduction: Human activities have always modified the natural environment. But the emergence of an integrated world economy during the last few centuries has increased both the intensity and scale of these modifications. Today's environment is not just modified by human action: It is fundamentally transformed.

In this chapter recent efforts of environmental historians to document the human transformation of the biosphere are described, focusing on global changes in patterns of land use over the last 300 years, with special attention given to the processes of deforestation, wetland drainage, irrigation, and conversion of grazing lands. The long-term, large-scale economic linkages that underlie these processes are described.

Introduction

Historians today, writing in the space-age latter decades of the twentieth century, have turned more and more toward a global perspective. It has become increasingly evident that national or even regional histories cannot be written in isolation, since exogenous forces generated by international trends and processes shape local and national histories. Modern history is only comprehensible when we consider both the unique qualities and experiences of national or local units *and* the processes of world change undergone through time. This perception has been heightened by a growing sense that every human history shares an unprecedented interdependence with every other society in the modern world. Terms such as "modern world economy" or "modern world system" are now found in the historical lexicon [1]. At this juncture the view of historical scholarship converges with the everyday perception of common sense that the world has become interdependent in ways unknown to our grandparents. Moreover, the velocity and intensity of interaction has accelerated dramatically in the past few decades. The case for a truly unified world history of mankind is being made by events.

Even the most cursory look at world history of the past three centuries, since 1700, reveals that a number of major processes and changes are underway. These have become truly global in scope and impact, and are seemingly irreversible. They have caused and continue to effect profound changes in human society: none is readily susceptible to control or alteration. Some of the forces that have created our late twentieth-century world economy and world order are:

(1) Expansion of the European frontier of settlement into the New World, the great Eurasian steppe, and Australasia.

(2) A steady growth in human population, from 641 million estimated for 1700 to 4435 million estimated for 1980 [2].
(3) A dramatic growth in the population and spatial extent of the world's cities.
(4) Increased use of fossil fuels and hydroelectric power, which helped to create a revolution in transport, communications, and industrial production.
(5) Development of scientific methods, institutions, and technical means for research and discovery in the biological and physical sciences.
(6) Development of new weaponry with global reach and the capacity for near-global destruction.
(7) Dramatic advances in our ability to cure ill or injured individuals and to control the spread of disease at the collective level.
(8) A steady growth in the scale, efficiency, and stability of large-scale complex organizations (i.e., bureaucracies), in both private and public modes.
(9) The emergence of self-regulating, price-fixing global markets for goods and services.
(10) The emergence of a world division of labor between the developed countries of the North (or core) and the developing countries of the South (or periphery).
(11) The expansion of more intensive sedentary agriculture and the simultaneous compression of tribal peoples engaged in shifting cultivation or in pastoral nomadism.

Historians examining these and other developments through the wide-angle vision of global history have improved our understanding of the changes in human institutions and relationships during the recent past. But these historians have not considered, nor adequately studied, modern man's changing relationship with the natural environment. Most tend to treat the environment as a constant that can safely be ignored, instead of as a significant variable Even economic historians have ignored environmental issues in their concern to tally accumulation of wealth and enhanced human productivity arising from industrialization. It is therefore the business of the environmental historian to study and analyze world environmental change, a task especially important for the post-1700 period. All of the processes of world development listed above have caused sweeping changes in the natural world: the global habitat of the 1980s is vastly different from that of 1700.

A corollary to the construction of a world economy is the commensurate ability of man to intervene in the natural order. When these interventions have been studied and analyzed by historical geographers, forest historians, and others, the emphasis has been on local or regional syntheses, not on the global picture.

Historians working in the nascent field of environmental history may see other tasks before them. They may wish to study changes in the physical environment that are not the result of human intervention, such as changes in climate, which may have had profound consequences for local or regional history. In the exciting area of climate history a consensus seems to be forming: practitioners agree that meaningful climatic changes have occurred in the historical past and that these changes have made appreciable differences in the human condition. They may be traced not only through the long span of human and geological history, but also through the early-modern and modern worlds. Climate, in other words, must be treated as a meaningful variable by environmental historians. Newly identified climatic sequences for any world region must be examined for possible effects on the natural environment and on the human society in that region [3].

Nevertheless, improving our understanding of environmental change in the early-modern and modern worlds has a singular urgency. The cumulative effect of human activity upon water, soils, and vegetation throughout the world has drastically accelerated – we may have already set in motion physical processes of change that are irreversible. For example, the release of CO_2 and other gases into the atmosphere seems to have risen sharply with the intensified burning of fossil fuels. Similarly, trace gases released by human action may also be changing the composition of the atmosphere. To measure these developments we need to know more precisely the causes and the rates of environmental change in each world region.

Another important consideration may be the extent to which development of the modern world economy has relied upon consumption of natural resources. Wild game, topsoil, wood, water, and minerals have all been put to human use in quantities and at rates unimagined in earlier centuries [4]. Detailed measurement of the use of these resources in the post-1700 epoch constitutes a significant challenge for historians of the global environment. Analysis of the implications of this exploitation may suggest issues and questions for future human development. For example, the environmental historian may discover deep-seated, cross-cultural attitudes and values that encourage exploitation and heavy use of "wild" (unowned) resources as opposed to careful husbanding of "domestic" (owned) resources. The whole issue of cultural attitudes toward the environment demands much further exploitation and careful scrutiny.

Expansion of world arable land

For the historian of the world environment a critical long-term trend, active since 1700, is the expansion of arable land which, along with the spread of human rural settlement, has been a global phenomenon. Historians have not hitherto fully recognized the scale or the importance of this change. For every region of the world – whether New World areas of settlement or Old World areas of long-standing peasant cultivation – the growth of arable land has been startling. My estimates for the past 120 years are given in *Table 2.1* [5], which shows that conversion of land to regular cropping was pursued in Eurasia as vigorously as in the New World. Many parts of Asia and even Europe experienced pioneer frontier settlement conditions analagous to those found in North and Latin America. Picturing a Taiwanese, Bengali, or Russian peasant as a sturdy pioneer with axe, gun, and homestead lands is somewhat difficult for a historian of the Americas to comprehend. Nevertheless, the similarities between the nineteenth-century Ohio and Burmese settlement frontiers are much greater than the differences.

Table 2.1 also discloses the unreversed nature of this trend. Only in North America, Europe, and East Asia do we see any substantial reversion of land from cropping. Until quite recently the major form of investment in rural productivity was extensive rather than intensive, with world agricultural investment taking the form of land clearance or reclamation. Only in the post-1920 period has the shift to intensive investment with fertilizers, high-yield

Table 2.1 Land areas converted into regular cropping (10^6 ha).

World region	*First period (1860–1919)*		*Second period (1920–1978)*	
	To crops	*From crops*	*To crops*	*From crops*
Africa	15.9	—	90.5	—
North America	163.7	2.5	27.9	29.4
Central America and Caribbean	4.5	—	18.8	0.4
South America	35.4	—	65.0	—
Middle East	8.0	—	31.1	—
South Asia	49.9	—	66.7	—
Southeast Asia	18.2	—	39.0	—
East Asia	15.6	0.2	14.5	8.4
Europe (excluding Soviet Europe)	26.6	6.0	13.8	12.7
USSR	88.0	—	62.9	—
Australia/New Zealand	15.1	—	40.0	—
TOTAL	440.9	8.7	470.2	50.9
Net area	432.2		419.3	

seeds, and machinery caused land to be removed from cultivation; and only in North America has there been a net loss of cropland. Recent substantial reversion of cropland to pasture, scrubland, or second-growth forest in the highly industrialized USA and Western Europe has led many observers to assume that a similar process will occur elsewhere in the world as development proceeds – highly capital-intensive agriculture requires less cropland. But the ever-rising worldwide demand for food and commodities suggests that this countertrend may not continue in the future. Pierre Crosson and Sterling Brubaker have recently pointed out that, in response to export-led demand, American farmers increased cropland by 24×10^6 ha between 1972 and 1980. They suggest that continued export demand may result in conversion of an additional 24 to 28×10^6 ha of pasture, rangeland, and forest to crops in the next 30 years [6].

The driving impetus behind this spread of agricultural land is not due directly to pressures of population growth. Obviously, swelling numbers of people demand more food and other agricultrual products, but agricultural expansion has not, on the whole, been the work of subsistence farmers. Most land clearance and plowing has been undertaken by peasant proprietors, sharecroppers, or estate and plantation workers – all responding to market forces. The development of an integrated world economic system has generated a vigorous, lucrative demand for foodstuffs, fibers, narcotics, and industrial crops. Cash sale and profits were the motor propelling this trend (after all, it requires some capital to create new arable land). By the nineteenth century provision of global road, rail, and steamship links made the rapid, large-scale physical movement of agricultural products possible; provision of telegraphic and postal services made market information readily available; and provision of banking, credit, and reliable monetary systems accounted for the necessary fiscal element. Admittedly, in more recent decades the correlation between swelling populations and agricultural expansion seems direct. The consuming market – especially for foodstuffs – is much closer to the small-commodity producers.

When aggregated, the estimates of the original study show a total world expansion of arable lands of 852×10^6 ha for the entire 120 year period. The 1979 FAO world estimate for cultivated lands is 1.5×10^9 ha [7], a figure that suggests the 1860 base for world arable land was about 600×10^6 ha, or perhaps more (to allow for some reversion). The 1860 base for total land used in world agriculture is, in itself, the result of an earlier expansive drive. A similar exercise for the 1700 to 1860 period would show huge gains in agricultural land in North America alone. In short, over the last 300 years there has occurred a remarkable surge in the conversion of lands to sedentary agriculture. The most dramatic increases have occurred in the period since 1860.

We may presume, therefore, that unprecedentedly severe and all-encompassing environmental changes in world habitat have occurred since the latter decades of the nineteenth century. For example, the effect of agricultural growth upon the world's soils has been profound – and potentially very grave. The changeover from land covered with natural vegetation to fields causes a pronounced increase in soil erosion. Golubev suggests that "soil erosion on a field is on average two orders of magnitude more than that under a forest and one order of magnitude more than from non-forested landscapes [8]." In a set of calculations for each major climate zone of the world – tropical, subtropical, subboreal, and boreal – he estimates that water-induced erosion of soils has increased to 91.1×10^9 tons of soil worldwide each year. This represents a more than fivefold increase (74.7×10^9 tons) over the rate of erosion in the preneolithic past when no cropland agriculture existed. If we extend this reasoning to the estimated expansion of arable lands from an 1860 base of 600×10^6 ha to the current total of 1.5×10^9 ha, the rate of increase in soil erosion must have risen in proportion. By one very crude estimation the rate of global soil erosion would have progressed as in *Table 2.2*. In brief, the last 120 years has probably seen a near-doubling of the pace of soil erosion in the world, as a direct result of worldwide growth in cropland [9].

Table 2.2 Crude estimates for rates of global soil erosion

	Preagriculture	*1860*	*1978*
Soil erosion ($\times 10^9$ tons/yr)	16.4	46.3	91.1

The recent global increase in cultivated area has meant, quite simply, that more and more areas of land have come under direct and continuous human *control* for the first time. Initially, this may have been in the form of some private tenure. Later the state imposed constraints on private control (or ownership) and in extreme cases, as with forest reserves, the state directly claimed vast areas of wilderness not previously controlled by anyone: used on occasion, yes – controlled, no. Today virtually every hectare of land on the continents can be said to be managed, so we can consider all land as being in human use at all times. In short, the potential for further systematic intervention has increased as settlement and, subsequently, the reach of the state have expanded.

The above assertions regarding global trends can be better supported and developed with a discussion of the most significant forms of agricultural expansion. The mechanics of intervention and land conversion vary according to the ecological conditions encountered, such as forest and woodland; wetland; arid and semiarid land; and grassland. Despite some overlap in these categories, man has evolved characteristic approaches toward intervention in each of these regimes. It is to the nature, scale, and effects of these forms of expansion of world arable land that we now turn.

Deforestation

Throughout the world retreat of wood and forest is concurrent with the advance of arable land. Not all, but much of the land newly brought under the plow has emerged from forest clearance, and urban settlement and transportation requirements for land increase the pressures for clearance. The result has been a massive conversion of woodland into other uses, which thus puts a definitive end to the prospect of regrowth. Abandonment and regrowth is always possible, but thus far (as we have seen) the global trend has not inclined toward reversion of agricultural or settlement lands.

Exploitation and degrading of forests are other aspects of deforestation that cannot be ignored [10]. Demands for fuel, for construction timber, and for forest grazing have mounted steadily, with a resultant sustained and growing pressure on virtually all save the most inaccessible of the world's forests and woodlands. Frequently, forest cutting and consequent degrading has been a prelude to clearance and conversion of land to sedentary agriculture. Easier access and smaller growing stocks facilitate the work of the pioneer settler.

Encroachment upon their habitat by sedentary agriculture has also caused shifting cultivators to change their ways. Reduced or compacted ranges have forced tribal peoples universally to reduce the period of fallow and thereby put heavier pressure on the forest, with the inevitable effects of a less substantial regrowth and a degraded cover. In some cases the period of fallow has been sufficiently reduced to place the shifting cultivators on the verge of sedentary cultivation.

Intensified exploitation and conversion of land to agriculture are the twin engines of modern world deforestation with its attendant consequences. In other words, world economic development has been the direct cause of global deforestation. Just as world development has followed an uneven trajectory, so also has forest clearance and conversion. Just as we can trace the spiraling, expanding pattern of European exploitation, trade, conquest, and settlement, so also can we trace a spiraling, expanding pattern of world deforestation. Western European forests felt the earliest impacts of exploitation for industrial purposes and clearance for agriculture in the sixteenth century. By the eighteenth to nineteenth centuries, unremitting pressures on these woodlands had created the pastoral, managed landscapes of France, England, and Germany, so pleasing to our own sensibilities today. As early as the seventeenth century the forested plains of eastern Germany and Poland underwent steady clearance and plowing as the market for eastern European wheat expanded. By the eighteenth century New World lands in eastern North America and in coastal Brazil felt pressures for marketable wood and arable land. By the early nineteenth century forested covers in India and the midwestern USA were being felled for development purposes, and in the midnineteenth century rapid deforestation had begun in Himalayan India, Australasia, Southeast Asia, South Africa, Manchuria, Taiwan, and elsewhere. The end of the century brought east and west Sub-Saharan Africa, the American far west, and Siberia into this process. The twentieth century has seen a nearly global onslaught on woodlands with, after World War II, swelling pressures for economic development in the era of decolonization.

Clearance of woodland seemingly evokes a common human aesthetic of destruction and removal. Felling trees, or burning, and thus opening up the forbidding woodland for civilized activities has a broad, cross-cultural appeal. The forest is an ominous, forbidding realm, destroyed by the pioneer in order to build and grow again; to open up a larger, more spacious visual horizon is a satisfying achievement. This impulse is reinforced by the clear hierarchy of economic activity maintained by European and all other major world civilizations. The gathering of forest produce and hunting forest game are savage and demeaning activities. Progress involves their replacement by plowing, sowing, and reaping, by stock rearing and other more productive pursuits of the husbandman. Economic stimuli for development are reinforced by weighty cultural norms.

Since 1700 the expanding spiral of global deforestation has not reversed itself. Some lands, fallen out of cultivation, have reverted to forest in western Europe and North America, but the overall trend is toward faster degrading and land conversion over wider, more remote areas. The pressures of population and economic development have mounted rather than diminished. Technological advances and refinements, such as new generations of bulldozers or other advanced land clearance equipment, or even the humble portable chainsaw, ease the cost and difficulty of deforestation.

How do we obtain an accurate measurement of world deforestation during the past three centuries? Conventional history's neglect of environmental and land-use studies has made this a difficult question to answer. Some assistance may be gained from historical geographers who have completed local studies, but an overall synthesis of reliable data is still lacking. A first step might be simply to list those forests that have been severely depleted, or have completely disappeared, as we do not yet possess a reliable, accessible inventory of the missing forests. To set out the causes, the timing, and the ultimate fate of large tracts of land formerly covered by the same mix of trees would, indeed, be instructive. Depending, of course, on scale and definition there could be more than 100 major world forests that no longer can truly be said to exist in recognizable form.

To illustrate this approach, and to support the arguments made earlier, a sample list follows. The first two examples consist of forests subjected to continuing and intensifying pressure over the past four or five centuries. In each case the final spurt of clearance and conversion occurred in the nineteenth and twentieth centuries. The last three case studies are much more dramatic in their timing: each forest was intensively cleared in the last half of the nineteenth century and the first decades of the twentieth century.

The mixed deciduous–coniferous forest of European USSR

The triangle formed by the cities of Petrograd, Moscow, and Kiev dominates the forested heartland of European USSR. This is the land of pine and fir conifers and oak, aspen, and birch deciduous species that have played such a prominent role in the cultural perceptions of generation after generation of Russians. To the north the land is covered by the northern coniferous *taiga* forests; to the south by the increasingly sparse broad-leaved species of the black earth regions.

From as early as the eleventh century a sustained process of Slavic settlement, colonization, and clearance advanced into the mixed forest zone. By the sixteenth to seventeenth centuries Russian peasants had abandoned a form of slash and burn cultivation with long fallows in favor of more intensive three-field farming practices. Steady encroachment of the woodland accompanied the movement of pioneer colonists to the richer soils of the south, onward toward the forest steppe. Industrial demands for timber, charcoal, potash, tar, pitch, turpentine, and other forest products encouraged heavy cutting of the woodlands [11]. After a pause, punctuated by some land abandonment in the seventeenth century, vigorous, sustained clearance and conversion continued from *ca.* 1700 to World War I.

During this period the mixed deciduous–coniferous woodlands of European USSR diminished from six tenths of the total land area to just over one third of the total, while arable land increased by a comparable amount. Tsvetkov has made a thorough analysis of the amount of forest clearance for each province in European USSR from records of official surveys at various dates. He used data back to 1696 for 12 of the 16 provinces in the mixed woods zone (*Table 2.3*). Two centuries of colonization and clearance reduced the woodland by 14.4×10^6 ha in these provinces. For the 16 provinces land clearance between 1796 and 1914 shrank the forests by 13.3×10^6 ha, in which period the forest diminished from one half to one third of the total land area (49.6 to 34.5%) [12]. In the early twentieth century "the forest lands of Russia's southern central strip are a mere dwindling reminiscence of the past, and preserved as a luxury [13]."

Table 2.3 Forest clearance in European USSR [12].

	1696	*1796*	*1914*
Forest area ($\times 10^3$ ha)	37 573	32 836	23 136
Percentage of total land area	60.6	53.0	37.4

Overall, Tsvetkov calculated that the 48 provinces of European USSR lost 67×10^6 ha, about 28%, of its forest and woodlands between the end of the seventeenth and beginning of the twentieth centuries. Deforestation in World War I, the Revolution, and thereafter proceeded without appreciable let up.

The subtropical Atlantic coastal forest of Brazil

In AD 1500 the subtropical forest of the southeastern Brazilian coastal provinces (Espírito Santo, Rio de Janeiro, Minas Gerais, and São Paulo) covered about 500 000 km^2, or about 6%, of the total land area of Brazil [14]. Well watered by the seasonal winds bringing moisture from the tropical South Atlantic, this was a canopied forest closely related to the Amazonian rain forest. The larger trees, brazilwood and ironwood, reached 30 m in height with a closed canopy entwined with lianas and other vines. By the early twentieth century most of this area had disappeared save for "a few patches of the primary forest [14]." Lands once forest have become areas of human settlement, arable land (including sugar and coffee plantations), and rangeland for cattle.

The process of deforestation began with European settlement in the sixteenth century. New Portuguese and French demands for brazilwood (*Caesalpinia echinata*) encouraged intensified cutting by the American Indians of the region. Subsequently, in the seventeenth century, white and mestizo settlers adapted and intensified the native slash and burn regime to a cycle that, because of repeated burning, inhibited forest regrowth. By 1690 the discovery of gold in the region brought in new migrants. Food and fuel requirements for the new economy as well as land requirements for mining encouraged forest burning and clearance on a wider scale. Dean estimates that perhaps 95 000 km^2 of forest was cleared in the 100-year life of the gold fields [15]. By the mideighteenth century a new export crop, plantation sugar, had put further pressure on the forest, and in the 1830s coffee began to overshadow sugar as a money earner. The widely accepted belief that productive coffee bushes must be planted in the soil of recently cleared primary forest lands brought another spurt in forest felling and burning. The post-1860 spread of the railways into the forested zone was the last episode in this sequence, the access to rail lines encouraging clearance and more extensive cattle ranching and agriculture. The railways themselves consumed vast amounts of wood for fuel and sleepers. Only after 1900 were conservationist issues raised and small forest reserves set aside "as biological research stations [16]."

The forests of the Mediterranean: the Sila Forest of Southern Italy

A stubborn impediment to an accurate understanding of the accelerating pace of world deforestation has been the tendency of many authors to assume that the scope of deforestation in earlier epochs has been comparable to that of the modern period – and that often full recovery from the effects of earlier periods of exploitation has never really occurred. Frequently mentioned is the presumed stripping of the Mediterranean forests in classical antiquity. In his recent study of forests in the ancient Mediterranean world, Russell Meiggs points out that it was really only in the last century that deforestation of the Mediterranean lands took place, thus correcting this assumption. Before the midnineteenth century the forests of Italy, Greece, and Turkey, as well as those of the North African countries, were relatively abundant. But the "spectacular growth of population during the nineteenth century" generated an "intensive increase in public and private building" with an accompanying demand for timber. The construction of railways added to this demand, but more significantly made it economically feasible to cut and haul timber from large areas of hitherto inaccessible Mediterranean forest lands [17].

The Sila forest of old Calabria in southern Italy offers one well-documented example of this type of deforestation. The Sila forest consisted of a dense growth of mixed deciduous–coniferous species (pine, fir, oak, beech, and chestnut). Largely untouched because of poor roads and access, it offered a hospitable refuge to bandits and other outlaws, but no incentives for timber merchants. However, in the last decade of the nineteenth century, a railroad was built through Cosenza to the south and a number of roads penetrated the uplands, so the timber of Sila became attractive to contractors for the first time. In the next forty years a "reckless destruction" of this forest ensued [18].

The traveler Norman Douglas spent several years on extended walking tours through the uplands of old Calabria. He wrote in 1915 that "in a few year's time nearly all these forests will have ceased to exist; another generation will hardly recognize the site of them [17]." In his later tours Douglas saw tract after tract of cutover areas that had been luxuriant groves of chestnut or large stands of pine and fir on his earlier visits. Various timber companies – French, German, Italian – had obtained concessions, employed large bodies of workmen, built access roads or even funicular railways to haul the logs, and were busily cutting. After World War I the Sila forest was no longer recognizable to traveler and native alike.

The deltaic rain forests of lower Burma

Prior to British conquest and annexation in 1852, the primary forest of lower Burma consisted of great evergreen rain forests dominated by the *kanazo* tree (*Heritiera fomes*). Along the coasts and near the Irrawaddy river banks the evergreen *kanazo* forest merged with wet mangrove (*Rhizopora*) forests. But by far the largest land cover in deltaic Burma was formed by the *kanazo* forests with, under favorable conditions, trees of up to 45 m in height. This Burmese jungle was the habitat for elephants, tigers, and wild buffalo. Occasional interruptions came in the form of *kaing* or elephant grass (*Saccharum spontaneum*), growing up to 3 m in height.

In 1852 human occupation and penetration of the rain forest was minimal since the preconquest Burmese center for agriculture, settlement, and political life lay in upper Burma. Adas estimates that between 3.5 to 4×10^6 ha of land in lower Burma were under rain forest at the midcentury with only 320 000 ha of land under cultivation [19]. When the British arrived the delta was a jungle destined to become a settlement frontier.

British colonial officials saw the *kanazo* jungle as wasteland that should be converted into agricultural land as quickly as feasible to tap the productive resources of Burma and the Burmese people. The new regime installed new transport, communications, tax policies, and land tenure designed to encourage pioneering land clearance and settlement by Burmese peasants in the south. They encouraged links with the new, booming world market and rising prices for export rice. Free to move because of the new policies, Burmese settlers vigorously cleared and burned the rain forest and planted paddy rice. Within 60 years, by World War I, perhaps several hundred thousand hectares remained, only a remnant of the *kanazo* rain forests of lower Burma. Thus Burma became one of the great rice-exporting countries of the world.

The mallee scrub of South Australia

From the 1840s onward the coastal regions of South Australia were subjected to survey and settlement. By the 1930s much of this fertile land was devoted to grain cultivation, a transformation that came at the expense of woodland. Of the approximately 155 000 km^2 settled, much had originally been under a dense wooded canopy. Over 70% of the area (110 000 km^2) was mallee scrub:

> . . . the characteristic feature of the mallee is the way in which the dominant eucalyptus (*E. dumosa, E. ucinata, E. incrassata*) grow in stunted fashion, the branches splaying out from a thick shallow root in an umbrella-like cluster of stems and canopy. The trees vary in height from 5 to 35 feet, and in density from impenetrable thicket to a more open scrub with an understory of brushes and grass [20].

The mallee woodland comprised a difficult, hated barrier to the expansion of settlement. The earliest settlers discovered that the hard labor and expense required to cut and uproot malleee was often more expensive than the rewards of cultivating the cleared land.

A turning point came in the 1860s with the development of a technique for clearing by dragging a heavy log or metal roller over the scrub. The roller knocked down and uprooted the relatively slender mallee trunks, thereafter the remains could be burnt to permit plowing and sowing. Later developments included much larger rollers and spring-driven stump plows that together could clear even the heaviest scrub growth. The *Scrub Lands Act* of 1866 and later revisions enlarged the areas for individual leases and reduced the costs and effort required to obtain freehold title for settlers.

These incentives, as well as linkage into the world grain markets, set off a sustained process of clearance and land conversion. By 1897, 844 000 ha of land had been taken up under scrub leases, nearly all of which was cleared. But most of the nineteenth century clearance went unrecorded since it was "regarded as a part of the normal farm operations that all new settlers faced when colonizing the land [21]." In 1906, however, the state government began an effort to compile data on the rate of land clearance: between 1906 and 1941 mallee rolling operations cleared 2.9×10^6 ha of scrubland. In all, about half the settled area in South Australia (77 000 km^2) "has had its appearance altered completely by the clearing of woodland, remnants of this once enveloping cover existing only on steep slopes, on the tops of sand ridges, or alongside the country roads . . . [22]."

The five cases of depleted forests listed above represent a much larger number of forested regions that have felt the bite of economic development. The twin processes of degradation and conversion are still at work, but many forest reservations have retained their integrity and some afforestation efforts have been relatively successful. But the overall effect is meagre in the face of global trends toward deforestation.

Completion of a global inventory of this type would be an invaluable first step in better defining the recent forest history of the world. More precise, better documented data on the extent and rate of depletion for each individual forest would do much to correct our perspectives of the global process. Such an historically oriented inventory could then be matched with the contemporary inventory and survey for Latin America, Africa, and Asia completed recently by the FAO [23]. Finally, this inventory would also throw into sharper relief various worldwide efforts of afforestation and land reclamation.

Land reclamation: The drainage of wetlands

In the latter decades of the nineteenth century, the pace and scale of world land reclamation carried out by draining wetlands underwent a dramatic increase. After 1870, new world commodity demands and rising agricultural prices combined with new technologies to encourage swamp and marsh drainage. Large-scale production of cheap clay-tile pipes, development of steam-powered ditching machines, and other devices such as dipper dredges and dragline excavators lowered costs and enhanced returns [24]. Evolution of mechanically powered pumps and more sophisticated hydrological engineering made feasible drainage systems on a larger scale. State activity in the form of various local drainage authorities and boards played an increasingly critical role wherever the agriculturalist confronted inundated swamp, marshlands, or even wet soils that could be drained.

One of the largest efforts at drainage and reclamation occurred in the American midwest. By the midnineteenth century, settlement of the Ohio and Mississippi river valleys was just beginning. Throughout the upper midwest the new settlers found malarial, swampy lands that were potentially rich but too wet to farm. A participant on the 1823 expedition to search out the source of the Minnesota river reported after leaving Chicago that, "the country presents a low, flat, and swampy prairie, very thickly covered with high grass, aquatic plants, and among others the wild rice . . . The whole of this tract is overflowed in the spring, and canoes pass in every direction across the prairie [25]." In recognition of the problem in this region and in the south central states as well, the US Congress passed the *Swamp Lands Acts* (1849, 1850, 1860) that together conveyed 25.9×10^6 ha of wetlands to the states on condition that funds from their sale be used to reclaim the lands for agriculture. By the 1950s private and organized drainage efforts in the north central states had added 10 to 12×10^6 ha of wetlands to arable land. Improved productivity could be found on another 15×10^6 ha of formerly water-logged soils [26]. The 1978 drainage irrigation report from the Agricultural Census showed 21×10^6 ha of drained land in the same region, or 50% of the USA total drained lands [27].

Increasingly effective flood control projects made it possible to extend drainage and land reclamation in the south central region – especially in the Mississippi Delta. Engineering difficulties and costs were greater than in the corn belt states, but drainage continued to grow in the south during the twentieth century well after completion of much of the northern efforts. By 1978 18.5×10^6 ha of wetlands were dried in the south, or 43.4% of the national total [27]. In addition to the Delta and the Mississipi river valley, Florida had 2.6×10^6 ha of land under drainage [27]. The 1978 census reported that 42.6×10^6 ha, or 4.2% of the total land area, were under drainage in the entire USA [27]. It is likely that a somewhat larger area has undergone reclamation efforts in the past century or so, since some private efforts have simply not been recorded or tabulated; and some systems have been abandoned. Wooten and Jones estimated in the mid-1950s that $50–65 \times 10^6$ ha of USA land had been affected by drainage of which $20–25 \times 10^6$ ha was actual swampland [28]. More precise estimates await further research, but land reclamation on this scale has unquestionably made a substantial reduction in North American and, indeed, world wetlands.

An almost identical pattern of land reclamation to that of the midwest USA occured in South Australia. Australian farmer settlers in that region found a large coastal tract of potentially fertile lands with "sheets of standing water, swamps and water-logged grounds [29]." This wetlands area extended over about 1.7×10^6 ha. Between 1864 and 1972 unremitting government and private efforts succeeded in draining the entire coastal swamp, so today virtually all is productive agricultural land or settled. The final episode came in a burst of State Government activity at the end of World War II, when several hundred kilometers of new primary channels and subsidiaries completed the work of artificial drainage [30]. The "watery waste" of the southeast Australian coast is now a carefully controlled agricultural landscape.

Other examples of the large-scale conversion of wetlands can readily be found. Systematic research on reclamation of bogs began in Czarist Russia in the late nineteenth century. Large drainage works begun in 1872 resulted in the emptying of 2.5×10^6 ha of wetlands by 1898 [31], and drainage efforts have continued steadily thereafter. The most recent estimate shows 9.4×10^6 ha of drained organic soils (histosols) in the USSR [32]. In past decades Finland has mounted an impressive drive to drain peat lands, to both improve forest lands and extract peat; by 1980 5.4×10^6 ha of organic soils had been drained [31]. Other European reclaimed organic soils are estimated to total 6.1×10^6 ha [33]. Thus, nearly 21×10^6 ha of European and Russian organic soils in wetlands have been affected by drainage measures. By far the greater part of this activity was carried out during the past century.

Much reclamation of wetlands has occurred in tropical regions as well as in the temperate zones. Incentives identical to those discussed earlier have led agriculturalists and the state to drain swamps and marshlands throughout tropical Asia and Africa. But in many parts of monsoon Asia under wet-rice cultivation reclamation of swamplands has not been as heavily dependent upon lowering of the water table. Instead, cultivators have cleared tree, shrub, and grass cover, constructed dikes and ditches, and regulated the flow of water in order to grow paddy rice. In low-lying, brackish coastal wetlands, such as mangrove swamps, this procedure may involve preventing inflows of sea water and retaining fresh water sources with a relatively high water table. Thus, swamps and marshes may have lost their vegetation and animal populations, but still remain for much of the year a variety of controlled wetlands under rice culture.

The coastal swamplands and mangrove forests of present-day Bangladesh, known as the Sundarbans, fall into this category. This "vast tract of forest and swamp" comprises 1.7×10^6 ha of land stretching along the coast of the Bay of Bengal [34]. The lower part of the Ganges delta was "a tangled network of streams, rivers and watercourses enclosing a large number of islands of various shapes and sizes [35]." In the early nineteenth century the colonial Government of Bengal began to issue land grants for development of the "waste lands" of the Sundarbans, and by 1904 the total settled and cultivated area had reached 522 000 ha [36]. Continued clearance and development throughout the twentieth century reduced the tidal forested area by 1.3×10^6 ha to 400 000 ha in the 1970s [37]. Virtually the entire cultivated area in the Sundarbans is devoted to rice or jute – both wet crops.

Twentieth-century wetland development in southeast Asia followed a similar pattern. Independent Indonesia reclaimed approximately 2.7×10^6 ha of tidal swampland during the past 30 years, two thirds of which is under wet-rice cultivation [38]. Colonial and postindependence reclamation projects in west Malaysia have converted 600 000 ha to sedentary agriculture – mostly paddy [31].

These examples testify to the brisk rate at which the world's swamps, marshes, and bogs have been reclaimed during the past 100 years. Current pressures for productive agricultural lands drive continuing public and private inroads into wetlands – especially in tropical Asia. From the global perspective the massive increase in wet-rice lands under cultivation during the past century may be seen as a counterpoint to the depletion of wetlands. Especially where rice fields are artificially irrigated, as in many parts of east Asia, rice culture is a controlled form of world wetlands.

Land reclamation by drainage has had significant environmental and social effects. The drying up of marshes and swamps releases, among other elements, stored carbon to the atmosphere. Changes in the water table affect local and regional watersheds. When this type of habitat disappears many plant and animal species retreat or disappear. Similarly, tribal communities, forest gatherers, shifting cultivators, and others find their source of subsistence and place of refuge compressed. The last group are frequently forced to seek their livelihood as laborers or sharecroppers in the new agricultural regime.

A preliminary inventory of worldwide land reclamation through draining wetlands is an essential first step for environmental history. Such an undertaking would provide a much clearer picture than has hitherto existed of the scope and pace of wetlands conversion. An inventory could serve as a guide and stimulus for further detailed research. Comprehensive worldwide data is especially important in view of the vast areas of undisturbed wetlands that remain. For example, the undisturbed muskeg (peatland) lands of northern Canada are estimated to stretch 1.3×10^6 km^2 [39]. The USSR has an equally large area of peat and marshlands that could be reclaimed [40]. Economic development may indeed make it desirable to encroach further on both temperate and tropical wetlands. If so, the global context in the past and the present should be better understood.

Land reclamation: Irrigation of arid and near-arid lands

Irrigation is a much more visible and appealing form of land reclamation than any other. Making the deserts bloom responds to some of our deepest aesthetic and cultural instincts. The drama of towering dams, huge turbines, and massive canal systems has made large irrigation systems one index of modernity. Man's potency in halting and diverting great rivers for irrigation and for hydroelectric power renders acceptable the equally massive social investment required to construct these projects. The pace of irrigation projects has surged forward since the late nineteenth century, converting millions of hectares of land in the world's drier regions into productive agricultural land. This is a major environmental change that demands detailed analysis from a global perspective.

Perhaps the best known and certainly one of the largest social investments in irrigation has been in western USA. Projects such as the Imperial Valley Project (begun in 1896) to bring water from the Colorado River; the Twin Falls Project (1903) to divert waters from the Snake River in Idaho; the Boulder Canyon Project (1928) for multiple use of the Colorado River; the Columbia River Project (1935), with the Grand Coulee dam as the centerpiece, were all spectacular engineering feats [41]. By the 1930s the US Bureau of Reclamation and the US Corps of Engineers were planning water diversion and control for entire river drainage basins. The Missouri River Basin Plan (1944) and subsequent legislation called for the eventual irrigation of 2.5×10^6 ha of land by 1950 [42]. These large, planned systems were clear manifestations of the organizational skills, power, and resources of the American state: irrigation on this scale set a new standard of development. It is far from accidental that newly independent postcolonial states placed similarly large irrigation and water control programs at the core of their own developmental planning.

By 1890 western American irrigation was well under way. In that year just over 54 000 farms in the 17 western states (or their territorial equivalent) irrigated nearly 1.5×10^6 ha of land [43]. Over the next 40 years a burst of public and private dam and canal building pushed the total canal-irrigated area to 6.0×10^6 ha for the same region. The total USA system boasted just under 25 000 diversion and storage dams, over 5000 reservoirs, and over 120 000 km of main canals. Another 1.7×10^6 ha of land were irrigated from groundwater by means of wells and pumping systems in the western states [44]. Four decades of further expansion added another 10×10^6 ha to the irrigated arid lands of the 17 western states. In 1978 farmers and ranchers in the western half of the country irrigated 17.7×10^6 ha [45], one of the largest and most complex irrigation systems in the world.

Comparable to, and indeed even surpassing the American efforts, were the massive irrigation systems of colonial India under British rule. As early as the 1850s British engineers began planning and constructing canal systems designed to bring perennial supplies of river water to near-arid agricultural regions. The state's aim was to improve agricultural productivity and to reduce the risk of famine when the monsoon rains faltered. As the early projects proved themselves, and as the technical capacity and confidence of the Public Works Department engineers increased, the scope and size of individual projects grew commensurately. Diversion of the waters of the Ganges, the Godavari, the Krishna, and others of India's major rivers had begun. Curiously, the aim was fixed on irrigation – not on multiple uses, such as hydroelectric power, flood control, or navigation, as in the USA.

In the 1880s British planners conceived and initiated an ambitious grand plan to reclaim the desert lands of northwestern India along the Indus River and its five tributaries. If water could be brought to the dry lands of Punjab and Sind (present-day Pakistan) new lands could be reclaimed by agricultural colonists. Troublesome nomadic pastoralists haunted these tracts, whose only land rights were those of tradition and belligerence. After an initial exploratory project, the lower Chenab canal harnessed the waters of that river and diverted them into channels capable of watering a million hectares of scrub-covered arid lands [46]. By 1895 the first colonists, settled in newly planned villages, had harvested their first crops of wheat.

The centerpiece of the new Punjab canal colonies was the great Triple Canal Project (1905–1917). In this boldly conceived enterprise, three separate linked canals brought surplus water sequentially from the Jhelum, the Chenab, and the Ravi together to irrigate a hitherto inaccessible area in Montgomery District. When completed, three colonies of settlers were able to reclaim 569 500 ha of former wasteland for grain cultivation [47]. The Punjab canal colonies eventually covered 2.2×10^6 ha of reclaimed desert lands that had come primarily under wheat cultivation [48].

At the end of their rule in India, the British could rightly claim their canal irrigation systems as one index of modernization and development. They saw the Punjab canal colonies with their prosperous settler–colonists as one of the finest achievements of the colonial regime.

The total irrigation complex in both north and south India was indeed impressive. At the outset of World War II, official figures showed that 116 112 km of canals watered 11.6×10^6 ha of land in undivided India [49]. The momentum of development intensified after 1947 in both India and Pakistan. Five-year plans, the drive for food self-sufficiency, and foreign aid helped nurture a wide range of massive water-management projects in both countries. By 1978 Pakistan counted 11×10^6 ha under water from canals and India 15.1×10^6 ha. During the same period irrigation by means of ground water tapped through tube wells developed rapidly. Thus, the total noncanal irrigated area for Pakistan was 3.2×10^6 ha; for India 16.6×10^6 ha. In short, some 45.9×10^6 ha of dry lands came under artificial irrigation in 1978 for both countries [50].

Perennial irrigation schemes also commenced early in Egypt. Muhammad Ali (1769–1849) and his successors invested heavily in irrigation schemes to grow the new Jubel variety of long-staple cotton for export. After the British conquest in 1882, the new Cromer regime vigorously planned and developed further irrigation works with the aid of experts from the Government of India [51]. Post-World War II irrigation in Egypt has continued in this tradition with the construction of the Aswan High Dam and other impressive projects. In 1979 the FAO estimated that land under irrigation in Egypt had grown to 2.85×10^6 ha (virtually 100% of the total cultivated land) [52].

Other countries, such as Mexico, Australia, South Africa, and the USSR, developed new river diversion irrigation projects from the late nineteenth and early twentieth centuries. By the post-World War II period, as colonies became independent nations, they turned to state-directed development efforts, mostly aimed at reclaiming dry lands. State sponsored, planned, and executed efforts to water and cultivate arid lands have been a critical part of the development process. In 1979, 14.3% of the world's arable lands (207×10^6 ha) were irrigated [53].

This trend toward controlled watering and cultivation of dry lands is continuing – one projection suggests that the world total will reach 300×10^6 ha by the year 2000 [54]. Few rivers flowing through arid or semiarid regions remain untapped by irrigation schemes. Indeed, the steadily growing quantities of water demanded for irrigation have begun to have an effect on the world hydrological cycle. The runoff of many rivers is substantially depleted and in some cases almost entirely dissipated. The more than 13 000 large-capacity storage reservoirs in the world now have a total capacity of not less than 5500 km^3, or 12% of the entire estimated annual runoff of the world's rivers [55]. The supply of river water may be dwindling, but the supply of groundwater accessible for recovery by wells is very large. Investment in irrigation will continue for the foreseeable future.

The physical changes wrought by irrigation on formerly dry lands are consequential. Waterlogging and increased salinity and/or alkalinity are problems long associated with artificial irrigation which, if not anticipated and corrected, can defeat the very purpose of the irrigation scheme [56].

Placement of irrigation systems on dry-land hydrological systems has had other important ecological effects. Soil fauna in arid lands that have adapted to minimal soil moisture (beetles, sand roaches, carpenter ants, centipedes) have lost their habitat and given way to the earthworms and nematodes of irrigated sandy soils [57]. Rodent populations have experienced phenomenal growths in numbers when restraints imposed by lack of water have been removed [57]. In almost every irrigation zone, significant species shifts in plants can be found. Aquatic weeds in particular have flourished in the reservoirs and channels of the new system. If they are not checked they become a major obstruction to water movement and, in some cases, a human health hazard. Enlarged insect populations, especially malaria-bearing mosquitoes, have been a recurring problem due to increases in their breeding habitat. Numerous outbreaks of malaria have been traced to design and maintenance defects in irrigation systems [58]. Another important human health concern is the spread of schistosomiasis associated with irrigation, because the snails, host to the worm parasite, flourish in sluggish waters. Infection of human populations moving in such waters can easily reach epidemic or pandemic proportions [59].

Finally, for human populations the extension of irrigation brings drastic socioeconomic adjustments. Nomadic pastoralists have had few alternatives but to contract their normal itinerary of seasonal and yearly movements and withdraw to a constricted range. Under this process of compaction (similar to that undergone by shifting cultivators) they press heavily on brush, scrub, and grass cover. Alternatively, they may attempt to settle down and adapt to the new technologies of sedentary agriculture under an irrigation regime. Sometimes the latter has been in response to force imposed by the centralizing modern state. If colonists are brought into the new lands, or if the reclaimed tracts are an extension of already irrigated areas, the transition may be easier. In either case cultivation by means of irrigation does demand a new, more coordinated level of social organization, particularly true for new lands watered by canal systems. The farmer is integrated, involuntarily, into a large-scale complex structure.

The magnitude of world irrigation development over the past two centuries provides strong justification for a systematic, intensive historical review. The physical and ecological changes wrought by continuous water supplies brought to formerly dry tracts are manifold and complex; so also are the human and social consequences of this change. In the aggregate these changes constitute a significant shift in the world environment and in human society.

Grazing lands, large grazing animals, and man

Grazing lands throughout the world are of critical importance in the global economy. Savannas, natural grasslands, shrub lands, forest meadows, and man-made pastures supply most of the nourishment for domestic livestock and for large, wild herbivores. Milk and meat from domesticated animals are primary sources of high-quality protein for man and meat from wild herbivores remains an important supplemental source of protein in many world regions. Since the late nineteenth century (and the advent of refrigeration) meat and milk products have been important components of world economic exchange. In many semiarid parts of the world harvesting either domestic or wild large herbivores may be the most efficient way to use the land to produce food [60].

The relationship between the world's grazing lands, grazing animals, both domestic and wild, and man is tangled and complex. As a result of human economic development it has been a rapidly changing relationship, a changing picture that is of interest to the environmental historian. Certain large-scale worldwide trends are perceptible. Since 1700 a steady reduction has occurred in the extent of the world's grazing lands, for over the past few centuries millions of hectares of the world's great natural grasslands have come under the plow. We can only guess at losses in the American Great Plains, the South African veldt, the Russian steppe, the *campos* of Brazil, or the *pampas* of Argentina. Conversion of grasslands into plowed and cultivated land is a direct outcome of the global spread of arable land discussed earlier.

What might be considered a countervailing trend may also be found. There has been an extension of controlled grazing lands in some parts of the world by means of sown artificial pastures, which constitute a large proportion of rural investment in Australia and New Zealand, for example [61]. Human intervention has inadvertently created grazing and browsing areas in transportation corridors left to grass, but cleared of higher bushes and trees. Thus, highway, railway, and power line rights-of-way occupy hundreds of thousands of hectares of land in North America that provide browsing and grazing lands for many herbivores [62], but the overall balance seems to have been heavily weighted toward diminution.

Another significant worldwide trend has been that of intensifying human intervention in and control of world grazing lands. At the present time no truly uncontrolled grazing lands or wild populations of large grazing animals exist, all being, in fact, managed by self-conscious human agency. Increasingly it is the national state that has acquired that responsibility. Only the definition of goals and the efficiency of management remain an issue.

In part human control of grazing lands has been attained by the forcible displacement of wild herbivores to create pasturage for domestic livestock. Thus, the near demise of bison, elk, and antelope on the American plains and their replacement by cattle and sheep herds is a staple theme of North American history. Similarly, the enormous herds of gazelle, zebras, antelope, and other grazing and browsing animals reported in the journals of early European explorers of Sub-Saharan Africa have all but disappeared [63]. Cattle, sheep, and goats have deposed these vast wild populations on the African plains, a few domestic species thus replacing the wide variety of wild grazing animals. In Africa there has been a perceptible trend toward overgrazing with its attendant consequences: "sharp decreases in standing crop, increases in unpalatable and in poisonous species, denudation and erosion and bush encroachment [64]." The nineteenth century diaspora of European settlers in the New World and Australasia made enormous inroads in world grasslands in a relatively short time.

European invasion of the New World, African, and Australasian grasslands was the product of the sudden growth in ranching or extensive commercial grazing in the second half of the nineteenth century. Steeply rising market demand for beef and wool in the urbanized regions of western Europe and eastern North America spurred the growth of extensive commercial sheep and cattle grazing in areas of new settlement [65]. In some instances pioneer ranchers in South America or, most notably, in New Zealand's South Island, entered grazing lands unused by either larger herbivores or directly by man. Thus, the European settlers of New Zealand brought exotic sheep, cattle, rabbits, and horse species with them to a land that did not contain any native large herbivores. The tussock grassland of the South Island "was a sea of waving bunch grasses dotted sparsely with shrubs or the New Zealand cabbage trees [66]." These 2 to 3 ft tussocks of grass stretched over 6.5×10^6 ha of land when the first European grazing animals were introduced [67]. In the latter case 140 years of heavy, sustained grazing (as well as plowing and conversion) has reduced the area and the luxuriance of the original cover. Cattle and sheep, and New Zealand's exports derived therefrom, have flourished, but the grasslands have been seriously depleted [68].

Despite ranching's explosive success in the nineteenth century the area devoted to extensive grazing of this type has been shrinking. More intensive, more profitable mixed farming has pushed ranchers into the drier regions of poorer vegetation and soils – into the margin [69]. Ranching is an efficient means of exploiting the natural pasture of the semiarid zones, since both carrying capacities and human populations are very light. Nonetheless, ranching at present must be accounted for as one of the sources of pressure upon vegetation in arid lands.

Pastoralists and nomads throughout the world have come to share a common fate with ranchers. Especially during the last century or so sedentary farming and stock raising have preempted the favored habitat of the pastoralist. Whether sheep and goat, horse, camel, or cattle herders; whether fixed transhumance or wide ranging in their migrations, the world's pastoralists have been steadily constricted by the agriculturalist and his ally the state. The pastoralist's best grazing lands have been lost, and the remaining flocks compressed upon smaller and less desirable ranges. Much of the pressure upon semiarid vegetation and the subsequent progress of desertification can be traced to this process. The population and habitat of the nomad and his animals have come under state control, sharply curtailing his economic and political freedoms. Control, and eventually settlement as sedentary farmers or stock ranchers, is the goal of most regimes today.

North Africa under colonial rule offers one important example of this worldwide trend toward compression of pastoralists. The French established their protectorate over Morocco in 1912. Within a relatively brief span, the new colonial regime had expropriated large tracts of the most fertile agricultural and grazing lands from the tribes and religious beneficiaries that occupied and used them. These lands were redistributed to European settlers, while other Europeans purchased and occupied lands as private colonists. By the midcentury, European settlers held 10^6 ha of land in Morocco, or 6.5% of the total farm and grazing land in the colony. The European settlers developed profitable, intensive, diversified farming in the well-watered northwestern area of Morocco (The Chaouia):

> One of the most important results of the development of diversified farming was, of course, to reduce the quality of the land available for the two major occupations of traditional Moroccan agriculture. Grain growing was pushed into the marginal area . . . The herder, in turn, found himself moved even further south onto lands offering highly uncertain pasturage at best [70].

One of the unanticipated consequences of these shifts was the rising prominence of goats, as opposed to cattle or sheep, in the herds of Moroccan pastoralists. Between 1931 and 1952 the percentage of goats in the total animal population (goats, sheep, and cattle) rose from 27.3 to 37.8. Most of the enhancement occurred at the expense of cattle, whose numbers shrank from 16.3% in 1931 to 8.3% in 1952 [71]. In absolute terms the number of goats nearly trebled from 3.2×10^6 in 1931 to 9.8×10^6 in 1952. As Stewart comments:

> This extension of goat raising had its drawbacks, since forest and ground cover were seriously damaged as a result. The herder had little choice, however, relegated as he was by the extension of cultivated areas to lands unfit for raising anything other than the "cow of the poor" [72].

The current impoverished state of most pastoralists in the Third World can be attributed in large measure to the working of the same process of displacement to a less favorable habitat and to grazing ranges severely depleted in area. In the end economic pressures force many nomads to seek employment from settled agriculturalists on a seasonal or permanent basis; thus former pastoralists swell the numbers of the rural proletariat.

Conclusion

Over the past 300 years the Earth's biota has undergone massive changes: world vegetation cover has been profoundly altered, the world's wild land and sea animal populations have been sharply reduced in number and range, and even the physical composition of the oceans and the atmosphere may have experienced subtle changes. At the same time man's control over the natural environment has risen to the point that the natural order is virtually an anthropogenic or man-determined system. By far the most important reason for the changes that have occurred in the biota, the atmosphere, and the oceans is the growing efficiency and global scale of man's economic activity. The extension of world agriculture, producing commodities for a global market, has been a major, but not the sole, contributor to drastic environmental change. Industrial development, rapid urban expansion, mining, large-scale commercial fishing have all contributed to the transformation of the world's environment. These enormous changes in man's habitat over this relatively brief period of time (since 1700) have forced new adaptations in culture and institutions upon human society. Moreover, at this point in time, in the late twentieth century, it does not appear that the powerful forces driving these trends will be diverted or reversed. Man's economic actions will continue to alter drastically the environment in new, and unexpected, ways in the foreseeable future.

The great task for environmental historians is to record and analyze the effects of man's recently achieved control over the natural world. What is needed is a long-term global comparative historical perspective that treats the environment as a meaningful variable. A long-term, three to five centuries perspective is necessary to discover any large, significant trends as opposed to short-term fluctuations or discontinuities. A global vision is essential because the integrating world economic order displays shared patterns of change. After innovation and acceptance new technologies and new institutional forms have been diffused and used throughout the world. Especially after the midnineteenth century the time required for diffusion and deployment of new technologies has shrunk. A comparative vision is necessary to make effective use of detailed local and regional studies in environmental history. Striking localized phenomena may be the product of peculiar local circumstances and run contrary to larger worldwide trends. Comparative examination of analogous circumstances in other localities, regions, or nations may confirm this or, alternatively, a comparative check may reveal the same phenomenon elsewhere.

To be truly effective, a wide-angle global vision and long-term perspective must be supported by more precise and accurate data. Historians must be equipped to measure environmental change over time more accurately than has been previously possible. Along one axis this involves exacting attention to location and to relative scale. Another axis demands meticulous attention to chronology and process. In the end more precision requires basic quantification: How many species of antipastoral plants were introduced into the overgrazed lands of *X* country by which date? How many hectares of mangrove fell to rice cultivation in Burma between 1890 and 1900? Naturally, quantification depends upon the proper and rigorous use of original sources, especially since direct statistical evidence may be incomplete or lacking. Nonetheless, methods long familiar to economic historians permit the construction of reliable estimates and well-grounded time series data.

Obviously, more and better intensive, individual, local, and regional studies are needed. These can give us great insights into the detailed analysis of larger processes of change for it is true that local cultural responses to larger forces vary widely. The value of monographic case studies is immense, but we can also make a case for wide-ranging comparative studies – on a regional or even global basis. Trying to discover, estimate, and quantify changing land-use patterns in a single country or group of countries over a century or more is one example. Trying to study and measure the environmental effects of worldwide technological–institutional innovations, such as the railways or chemical fertilizers, is another. These are large projects beyond the reach of an individual scholar. Some form of sponsored team research might be the best approach.

A long-term global vision is essential for any attempt to discern larger cycles or patterns as well as trends in man-induced environmental change. Thus, the larger fluctuations in world economic cycles have certainly influenced the rhythms of agricultural expansion in many parts of the world. The most striking instance of this is, of course, the impact of the world depression of the early 1930s upon world agriculture. Whether these fall into long-term 50 year Kondratieff cycles, or some variant thereof, is an interesting question yet to be resolved.

More precise, worldwide, quantified historical data on land use, animal populations, human settlement, and plant species distributions are the sort of data that could be meshed with the evolving capacity for remote-sensing monitoring and computerized mapping. In the end, more and better data, presented effectively, will enable us to discover new relationships and to better understand man's interaction with his world in the past and in the future.

Notes and references

[1] The most influential statement of this view is that of Wallerstein, I. (1974), *The Modern World System* (Academic Press, New York) and (1980), *The Modern World System II* (Academic Press, New York). Wallerstein has constructed a model of the dynamic, expansive, and capitalist European economy, an economy that has grown steadily since the sixteenth century to encompass the entire world. This world system is characterized by an unequal division of labor between the countries and regions of the world core (the North) and those of the periphery (the South). W. W. Rostow offers a less radical statement in (1978) *The World Economy, History and Prospect* (Texas University Press, Austin, TX). For a geographer's review of the Wallerstein model see Dodgshon, R. A. (1977), A spatial perspective, *Peasant Studies*, **VI**, 8–18, who argues that a true world system did not emerge until the late eighteenth and nineteenth centuries.

[2] Grigg, D. (1980), *Population Growth and Agrarian Change*, p 1 (Cambridge University Press, Cambridge, UK) for the 1700 world population figure; the 1980 estimate is taken from Food and Agriculture Organization of the United Nations (1981), *FAO Production Yearbook, 1980*, Vol. 34, p 61 (FAO, Rome).

[3] See Chisholm, M. (1980), The wealth of nations, *Transactions, Institute of British Geographers, New Series*, **5**, 273, "Enough has been said to show beyond any reasonable doubt that we can no longer assume the general constancy of the environment during historical times, nor can we rule out the possibility that long-period changes in the relative wealth of regions may be at least partially attributable to the environment." (See also Chapter 14, this volume.)

[4] This is not to deny the essential truth of the definition of natural resources advanced by E. W. Zimmerman and others. It is true, as Zimmerman first observed some 50 years ago, that natural resources are essentially man-made resources. The sum total of natural resources changes as man's technological capacity to use minerals, plants, or other materials expands; See Hunker, H. L. (1964), *Erich W. Zimmerman's Introduction to World Resources* (Harper and Row, New York and London). Nevertheless it is clear, at least to me, that stocks of wild animals, fertile topsoil, important minerals, and densely forested lands have sharply diminished in the past 300 years. Demand for natural materials to be consumed by humans continues to soar.

[5] Richards, J. F., Olson, J., and Rotty, R. M. (1983), *Development of a Data Base for Carbon Dioxide Releases Resulting from Conversion of Land to Agricultural Uses*, ORAU/IEA-82-10 (M), ORNL/TM-8801, p 64 (Oak Ridge Associated Universities, TN).

[6] Crosson, P. R. and Brubaker, S. (1982), *Resource and Environmental Effects of U.S. Agriculture* (Resources for the Future, Washington, DC). Demands for land for human settlement and transportation purposes will, presumably, also increase in industrialized countries.

[7] Food and Agricultural Organization of the United Nations (1981), *FAO Production Yearbook, 1980*, Vol. 34, Table 1, p 45 (FAO, Rome).

[8] Golubev, G. N. (1982), Soil erosion and agriculture in the world: an assessment and hydrological implications, in *Recent Developments in the Explanation and Prediction of Erosion and Sediment Yield, Proceedings of the Exeter Symposiums, July, 1982*, IAHS Publication No. 137, p 282 (International Association of Hydrological Sciences).

[9] This calculation simply takes the total increase, 74.7×10^9 tons, attributable to agriculture, multiplies this by the arable land under cultivation in 1860 (40% of 1978) and adds the preagricultural estimate for annual losses. For a more rigorous analysis,

increases for each of the world regions should be calculated and weighted, as Golubev did. This would change the totals somewhat, but probably would not alter appreciably the magnitude of the increase in erosion. See Golubev, note [8], p 265, for the detailed table.

[10] "Degrading" used in this context implies human interventions that reduce the density and luxuriance of vegetation cover, or the removal of selected species of trees so impoverishing the forest or woodland. For that minority of forest lands protected by rational and strictly enforced sustained-yield forestry practices, one would not necessarily use the term degraded (but see Chapter 3, this volume).

[11] French, R. A. (1983), Russians and the forest, in J. H. Bater and R. A. French (Eds), *Studies in Russian Historical Geography*, Vol. 1, pp 27–30 (Academic Press, London); and French, R. A. (1963), The making of the Russian landscape, *Advancement of Science*, **20**, 44–56.

[12] Based on data provided in Bater and French, p 38, note [11]. The sixteen provinces of the mixed forest zone are listed in Pavlovsky, G. (1930), *Agricultural Russia on the Eve of Revolution*, p 14 (Routledge, London). They include Novgorod, Yaroslavl', Kostroma, Ryazan', Pskov, Tver', Moscow, Kaluga, Vladimir, Smolensk, Vitebsk, Minsk, Mogilev, Chernigov, Kiev, and Volyn. For a map of Russian forest zones see Darby, H. D. (1956), The clearing of the woodland in Europe, in W. L. Thomas, Jr (Ed), *Man's Role in Changing the Face of the Earth*, p 206 (University of Chicago Press, Chicago, Ill.).

[13] Klyuchevski, V. O. (1931), *A History of Russia*, Vol. 5, p 245 (London).

[14] Dean, W. (1983), Deforestation in southeastern Brazil, in R. P. Tucker and J. F. Richards (Eds), *Global Deforestation and the Nineteenth Century World Economy*, p 50 (Duke University Press, Durham, NC).

[15] Note [14], p 59.

[16] Note [14], p 66.

[17] Meiggs, R. (1982), *Trees and Timber in the Ancient Mediterranean World*, pp 385–387 (The Clarendon Press, Oxford, UK).

[18] Note [17], p 387.

[19] Adas, M. (1983), Colonization, commercial agriculture, and the destruction of the deltaic rain forests of British Burma in the late nineteenth century, in R. P. Tucker and J. F. Richards (Eds), *Global Deforestation and the Nineteenth Century World Economy*, p 100 (Duke University Press, Durham, NC).

[20] Williams, M. (1974), *The Making of the South Australian Landscape*, p 11 (Academic Press, London).

[21] Note [20], p 122.

[22] Note [20], p 481.

[23] FAO (1981), *Los Recursos Forestales de la America Tropical*, Informe Technico 1, UN 32/6.1301-78-04 (FAO, Rome); FAO (1981), *Forest Resources of Tropical Asia*, Technical Report 2, UN 32/6.1301-78-04 (FAO, Rome); FAO (1981), *Forest Resources of Tropical Africa, Parts 1 and 2*, Technical Report 3, UN 32/6.1301-78-04 (FAO, Rome).

[24] Wooten, H. H. and Jones, L. A. (1955), The history of our drainage enterprises, in US Department of Agriculture, *Water, The Yearbook of Agriculture*, p 485 (US Department of Agriculture, Washington, DC).

[25] Note [24], p 479.

[26] Note [24], p 478.

[27] US Bureau of the Census, Drainage of agricultural lands, *1978 Census of Agriculture*, Vol. 5, Part 5, Table 2, p 4. Previous drainage censuses, 1920–1950, 1969, and 1974 used two overlapping categories to compile statistical data: "drainage on farms" and "drainage enterprises". Since many of the recorded private efforts on farms were, in fact, included in public drainage enterprises, considerable difficulties exist in creating a consistent time series. The 1978 census avoids this problem by reporting all drained lands regardless of management.

[28] Note [24], p 488.

[29] Note [20], p 178.

[30] Note [20], pp 219–222.

[31] Armentano, T. V., Menges, E. S., Molofsky, J., and Lawlre, D. J. (1984), *Carbon Exchange of Organic Soils Ecosystems of the World*, p 10, Holcomb Research Institute, Paper No. 27 (HRI, Indianapolis). The authors have assembled the most comprehensive data on the world conversion of wetlands known to me, with emphasis upon drainage of organic soils (histosols) and the problems of carbon release. For a discussion of Russian drainage efforts see note [11], French (1963), pp 52–55 and French, R. A. (1959), Drainage and economic development of Poles'ye, USSR, *Economic Geography*, **35**, 172–180.

[32] Note [31], p A-13.

[33] Note [31], p 9.

[34] Government of India (1909), Sundarbans, *The Imperial Gazetteer of India*, XIII, p 140 (Calcutta).

[35] Note [34], p 141.

[36] Note [34], p 145.

[37] Food and Agriculture Organization of the United Nations (1981), *Forest Resources of Tropical Asia*, p 119, UN 32/6.1301-78-04, Technical Report No. 2 (FAO, Rome).

[38] Note [31], calculated from Table B1, pp B2–B3. Table B1, pp B2–B3.

[39] Radworth, N. W. and Brawner, C. O. (1977), *Muskeg and the Northern Environment in Canada*, p vii, (Toronto, Canada).

[40] Note [38], p A-10, "Marsh-ridden area" in the USSR is calculated as 2.1×10^6 km^2.

[41] See Golze, A. R. (1952), *Reclamation in the United States*, pp 144–197 (McGraw-Hill, New York) for descriptions of the projects.

[42] Note [41], p 213.

[43] US Bureau of the Census (1975), *Historical Statistics of the United States, Colonial Times to 1970*, Part 1,

pp 427, 433, Series J-81-91 (Washington, DC). The 17 western states are Arizona, California, Colorado, Idaho, Kansas, Montana, Nebraska, Nevada, New Mexico, North Dakota, Oklahoma, Oregon, South Dakota, Texas, Utah, Washington, and Wyoming.
[44] US Bureau of the Census (1932), Irrigation of agricultural lands, *Fifteenth Census of the United States: 1930*, (Washington, DC). All calculations omit Arkansas and Louisiana, see Table 2, p 15, Table 10, p 20, and Table 21, p 59.
[45] US Bureau of the Census, Irrigation, *1978 Census of Agriculture*, Table 1, p 14 (Washington, DC). Total omits Louisiana where irrigation is primarily for wet-rice cultivation.
[46] Farmer, B. H. (1974), *Agricultural Colonization in India Since Independence*, p 21 (Oxford University Press, Oxford). For an overview of the policy issues, technology, and construction of the Punjab canal system see Michel, A. A. (1967), *The Indus Rivers*, pp 1–98 (Yale University Press, New Haven and London). For a larger view see Whitcombe, E. (1974), Irrigation, in Dharma Kumara (Ed), with Megnad Desai, *The Cambridge Economic History of India*, Vol. 2 (Cambridge University Press, Cambridge, UK).
[47] Note [46], Farmer (1974), p 21.
[48] Note [46], Farmer (1974), p 19.
[49] Department of Commercial Intelligence and Statistics, India (1948), *Statistical Abstract for British India, 1936–37, 1940–41*, pp 382–383 (Calcutta).
[50] Figures for Pakistan are taken from Ministry of Food Agriculture and Cooperatives, Pakistan (1980), *Agricultural Statistics of Pakistan, 1979*, pp 66–67 (Islamabad); and Ministry of Agriculture (1980), *Indian Agricultural Statistics, 1974–75 to 1978–79*, p 57.
[51] Kinawy, I. Z. (1978), The efficiency of water use in Egypt, in E. B. Worthington (Ed), *Arid Land Irrigation in Developing Countries*, pp 371–372 (Pergamon, Oxford, UK).
[52] Food and Agriculture Organization of the United Nations (1981), *FAO Production Yearbook, 1980*, Vol. 34, p 57 (FAO, Rome).
[53] Note [52], p 57.
[54] Kovda, V. A. (1978), Arid land irrigation and soil fertility: problems of salinity, alkalinity, compaction, in E. B. Worthington (Ed), *Arid Land Irrigation in Developing Countries*, p 215 (Pergamon, Oxford, UK).
[55] Golubev, G. N. (1983), Economic activity, water resources and the environment: a challenge for hydrology, *Hydrological Sciences*, **28**, 57–75, see p 60.
[56] Note [54], pp 218–222.
[57] Ghabbour, S. I. (1978), Effect of soil irrigation on soil fauna, in E. B. Worthington (Ed), *Arid Land Irrigation in Developing Countries*, pp 329–330 (Pergamon, Oxford, UK).
[58] Farid, M. A. (1978), Irrigation and malaria in arid lands, in E. B. Worthington (Ed), *Arid Land Irrigation in Developing Countries*, pp 413–419 (Pergamon, Oxford, UK).
[59] Obeng, L. (1978), Schistosomiasis – the environmental approach, in E. B. Worthington (Ed), *Arid Land Irrigation in Developing Countries*, pp 403–405 (Pergamon, Oxford, UK).
[60] For a discussion of the place of large herbivores in grasslands see Van Dyne, G. M., Brockington, N. R., Szocs, Z., Duek, J., and Ribik, C. A. (1980), Large herbivore subsystem, in A. I. Breymeyer and G. M. Van Dyne (Eds), *Grasslands, Systems Analysis, and Man*, pp 270–272 (International Biological Programme, 19, Cambridge University Press, Cambridge, UK).
[61] Sown pasture in Australia has risen from 25 000 ha in 1860 to 26.5×10^6 ha in 1978. The latter figure is more than double the hectarage devoted to crops in Australia in that year, see note [5].
[62] Wilson, C. M. (1961), *Grass and People*, p 31 (University of Florida Press, FL). In 1961 Wilson calculated that power line rights-of-way alone totalled 8250 sq. miles in the USA.
[63] Werger, M. J. A. (1983), Tropical grasslands, savannas, woodlands: natural and manmade, in W. Holzner, J. J. A. Werger, and I. Ikusima (Eds), *Man's Impact on Vegetation*, pp 123–124 (Junk, The Hague).
[64] Note [63], p 125.
[65] Grigg, D. B. (1974), *The Agricultural Systems of the World*, p 241 (Cambridge University Press, Cambridge, UK).
[66] Clark, A. H. (1949), *The Invasion of New Zealand by People, Plants, and Animals*, p 25 (Rutgers University Press, New Brunswick, NJ).
[67] Note [66], p 27.
[68] Note [66], p 210.
[69] Note [65], p 241.
[70] Stewart, C. F. (1964), *The Economy of Morocco, 1912–1962*, p 102 (Harvard University Press, Cambridge, MA).
[71] Note [70], p 94.
[72] Note [70], p 94.

Commentary

M. Williams

It is a cliché that we live in a changing world; nevertheless to understand our changing *milieu* is difficult because we live in the middle of it. It is only when we look back to some historical reference point that our present position on the line of change becomes clear, as do some of our likely futures (and there could be many alternative futures). In his chapter Richards attempts to establish some of the data points for change in the environment and to quantify the magnitude of a number of key processes, such as the expanding extent of arable land, deforestation, the draining of wetlands, the irrigating of dry lands, and the elimination of natural grasslands. The chapter is replete with measures of hectares affected and processes involved.

It is regrettable that many investigations of future global environments have ignored the historical record. This neglect has stemmed in part from a prevailing mood of pessimism, which has predicted doom and chaos as humankind increases and competes for finite resources, and in part from a focus on some future point in time, with a consequent dismissal of the past as irrelevant – unless it illustrates error or how not to do something. The failure to accord the past its proper place in scientific and socioscientific enquiry into the future has led to a disregard of the one reasonably sure record of change that we possess. Consequently, the credibility and scholarship of some studies has been undermined. Thankfully, this is not the case in the current investigation into the sustainable development of the biosphere.

What Richards is saying in this chapter is not controversial; a moment's thought shows that his theme is fairly self-evident. Yet it has not always been so, and it is instructive and illuminating to look at the sociointellectual background to the idea that man is an agent of change in the biosphere.

The idea that man alters his environment rather than the idea that the environment alters man is fairly recent in European (western?) intellectual thought. Until the late seventeenth century the concept prevailed that man was steward who tended and cared for a divinely created world, but altered little in it. However, as scientific knowledge expanded during the eighteenth century, observation, record, and enquiry showed that man was not part of some natural, preordained creation; he had free will and he exerted a powerful influence on the world around him. It is significant that Richards takes 1700 as his starting point, for from about that time onward, not only does the historical record sharpen and the rate of change accelerate, but also the intellectual framework of enquiry begins to alter.

One person who straddled the old and the new modes of thinking was Benjamin Franklin. He observed that woodland clearance on the east coast of North America was proceeding at such a rate that essential supplies were scarce and prices were rising. As he mused about this change he suggested fancifully that if there were inhabitants on Mars or Venus they would see an Earth so bare of vegetation that it would reflect a brighter light (the first global perspective of environmental change?). More important, however, was his warning that not only did mankind change the environment, but also that the change might not be for the best:

> Whenever we attempt to amend the scheme of Providence, and to interfere with the government of the world, we had need to be very circumspect less we do more harm than good [1].

The three revolutions of the late eighteenth century – the industrial in Britain, the social in France, and the political in the USA – introduced new forces and accelerated old ones already at work in the emerging network of trade, colonial expansion, and industrial development, enmeshing the world in a global economy. Perhaps the two outstanding environmentally-oriented counterparts to these political, social, and technical revolutions are Thomas Malthus' *Essay on Population* [2] and Charles Darwin's *Origin of Species* [3]. In the *Essay* it was postulated for the first time that there was some relationship between population numbers and resources, and *Origins* altered the whole concept and dimension of human and biological time by showing that man and his environment were interrelated parts of a long-evolving relationship. Both of these concepts had important implications for man–environment studies.

Lewis Mumford has suggested that an alternative and more useful typology of the last few centuries might be one based on stages of technology [4]. Before the late eighteenth century the world had lived in the dawn age of technology – the eotechnic – when materials used were wood, stone, and fiber, and the motive power was supplied by wind, water, and animals. During the era of ancient technology – the paleotechnic – from about 1800 onward, the characteristic materials used were coal and iron, and the motive power was steam. The railway was the supreme manifestation of this new technology: it broke down regional barriers, facilitated expansion, and moved people and goods cheaply and quickly as never before. Change accelerated and the world shrank.

By the latter half of the nineteenth century the impacts of man, aided by new forms of technology (such as the steam locomotive, the railroad, the steam ship, steam pumps, steam shovels, dams, and farm machinery), as well as the gradual, unrecorded, but cumulatively immense impacts of individual peasants and farmers who cleared the forests and tilled the land (all outlined by Richards) were becoming so evident and so great that George Perkins Marsh was able to write his celebrated *Man and Nature, or Physical Geography as Modified by Human Action* (1864). This is not the place to summarize this *tour de force,* but for the present purposes of achieving perspective it is relevant to note the following points:

(1) It was, said Marsh, the natural order of things that man was a "... disturbing agent. Wherever he plants his foot the harmonies of nature are turned to discords [5]."
(2) Consequently, not everything that happened to the environment was beneficial: man could create, but equally man could destroy.
(3) The scale of change was no longer local, but regional and even global. For example, Marsh speculated on what would happen with the irrigation of the Piedmont plain in northern Italy, the draining of the Maremma and other marshes, the continued deforestation of watersheds, and the probable effects of great engineering projects, such as Suez, Panama, Mediterranean–Dead Sea, and the Caspian Sea of Azov. Above all, he speculated about the climatic and pedological–hydrological effects of deforestation.

Taking a broad view, one must say that Marsh's amazing collection of data and his synthesis of human action and environmental consequences, complemented a little later by the writing of Alexander Ivanovich Woekof in Russia, fell on deaf ears. The triumph of technology, the idea of progress, imperial expansion and hegemony, and the tangible results of the development of resources, left little room for self-doubt about what was happening. It was perhaps only those writers of the early twentieth century concerned with the undesirable effects of urban expansion and urban living that questioned progress.

There can be little doubt that World War II radically changed most previous modes of thinking. First, not one part of the globe was too small or too remote to be attacked or defended – strategy and action was not local, but global. Second, the conflict showed that resources were limited and that supplies needed protection. Third, the end days of the war showed the ability of man not only to alter his world for the worse, but to totally destroy himself and his environment. It was the ultimate action. Fourth, after the war, the unbelievable resurgence of war-torn and war-stressed economies in the developed world, new orders of political and social organization in the less developed world, and better health throughout the world heralded an unprecedented expansion of industrial and agricultural activity, which intensified the global economic interdependence between nations.

In this chapter Richards reemphasizes the role of man in altering his environment through time, but also adds new dimensions by attempting to provide a firm, factual, qualitative basis, in order to calibrate the nature, magnitude, and rate of change in seemingly everyday human activities. Richards has shown that by patient historical scholarship a dossier or inventory of present land-use changes and processes can be built from the monographic works of historians and historical geographers; a dossier that will ultimately give us a surer basis for future speculation than anything we have possessed to date. We may quibble with a figure here or there, but the trend and magnitude of change seems incontrovertible. There are, of course, omissions, but it could not be otherwise in such a wide ranging chapter. "The great task for environmental historians is," says Richards, "to record and analyze the effects of man's recently achieved control over the natural world [p 68]." Thus, seemingly mundane questions such as "How many species of antipastoral plants were introduced into the overgrazed lands of *X* country by which date? [p 68]" or "How many hectares of mangrove fell to rice cultivation in Burma between 1890 and 1900? [p 68]" assume a new importance that is crucial to an assessment of the feasibility of achieving a sustainable development of the biosphere.

Finally, Mumford's typology of technical ages ended with the neotechnic, the new technology, born at about the end of the nineteenth and beginning of the twentieth century. Alloys and chemicals were the characteristic materials of the age; electricity and petroleum its motive powers. However, for all its new constituents, the effects of this age on the biosphere were essentially the same as those of the paleotechnic, although with vastly intensified effects and accelerated trends. Since then the world has entered a fourth age of technology, not adumbrated by Mumford. It is the "biotechnic" or even the "nuclear-technic", where the manipulation by humankind of unseen and microscopic constituents and elements of the atmosphere, soils, and biota, and of static, animal, and even human matter brings man's ability to alter his environment into confrontation with the very essence and meaning of life on the planet. This age also needs documentation and analysis. Some of its characteristics, such as genetic change (plant and animal breeding and adaptation) and chemical fertilizers, have been mentioned, if only briefly, in the chapter, but there are many more new fields of human alteration. We have come to learn, often at great cost (e.g. DDT, acid rain) the truth of Marsh's warning that

> ... we are never justified in assuming a force to be insignificant because its measure is unknown or because no physical effect can now be traced from it [6].

The documentation, quantification, and analysis of these new "forces" is essential to our understanding of environmental change and to the feasibility of achieving a sustainable development of the biosphere. In this difficult task we have an excellent model in the prospectus set out by Richards in this chapter.

Notes and references (Comm.)

[1] See *The Writings of Benjamin Franklin*, edited by A. H. Smyth (New York, 1905–1907), Vol. III, Benjamin Franklin, pp 72–73; and B. Franklin to Richard Jackson, p 133.

[2] Malthus, T. R. (1798) and final version (1830), *An Essay on the Principle of Population*, edited by A. Flew (reprinted, 1970) (Penguin Books, Harmondsworth, UK).

[3] Darwin, C. (1859), *The Origin of Species* (Reprinted, Modern Library, New York).

[4] Mumford, L. (1962), *Technics and Civilization*, pp 109–110 (Harcourt Brace and World Inc., New York).

[5] Marsh, G. P. (1864), *Man and Nature; or Physical Geography as Modified by Human Action*, edited by D. Lowenthal, p 36 (Harvard University Press, Cambridge, MA).

[6] Note [5], pp 548–549.

Chapter 3 Sustainable redevelopment of regional ecosystems degraded by exploitive development

H. A. Regier
G. L. Baskerville

Editors' Introduction: The transformation of the biosphere through human activities has often degraded the productive potential of individual ecosystems and regions. A central challenge for strategies of sustainable development is to discover how such degradation can be reversed through ecosystem redevelopment and rehabilitation.

This chapter is focused on the redevelopment of degraded forestry and fishery resource systems. It is argued that a useful definition of sustainability must specify the quantity and quality of the products and services that society seeks to obtain from an ecosystem, as well as the time period over which performance is to be evaluated. The author's arguments draw on their experiences as Commissioner of the Great Lakes Fisheries Commission and Assistant Deputy Minister of Forest Resources for New Brunswick, respectively.

Introduction

Think globally, act locally

Conventional economic development, which exploits ecological and other features of a *locale*, is usually undertaken with the aim, *inter alia*, that benefits will cumulate and contribute to *regional* economic growth and to "progress", with some consideration that the people of the locale will also receive some benefit from the exploitation involved. *Local* disbenefits, say in the form of bad ecological consequences, also tend to cumulate and to become apparent, eventually, as *regional* degradation. Whether intended or not, conventional local actions and their consequences tend to cumulate powerfully at regional levels, and thus local actions should be viewed in a regional context, with respect to good and bad aspects alike.

We focus our attention on economic development practiced during recent centuries in North America, but especially with respect to two case studies: the New Brunswick forests and the Great Lakes fish (see *Figure 3.1*). Part of the economic development during recent decades in the Third World may resemble, in some ways, the process of North American development outlined here. The resemblance may be weak with some European countries, where renewable resources have not been subjected to so rapid a development pressure and have been managed differently, at least recently.

In his statement on the issues surrounding the "think globally, act locally" philosophy, Holling [1] uses global in the sense of biospheric rather than, say, of some rather arbitrary level of aggregation. As have others, we have difficulty in demonstrating local–biospheric links, though we "know" that such

Figure 3.1 Map of eastern North America showing the Canadian Province of New Brunswick and the binational Great Lakes.

links are almost infinite in number. We deal in this chapter with local–regional links and submit that our contribution may be seen as a small-scale analog of the issue of links across gaps of larger scale.

From the perspective of an interested layman, the conceptual link from the local to the regional may involve a connection between relatively concrete local reality and relatively general and somewhat imprecise regional inferences that are comprehensible only as abstract arithmetic summations or mathematical expressions. A link to a biospheric level might involve a connection to phenomena that may appear so abstract (e.g., computer simulations) as to be almost unreal. How do we demonstrate interactive and cumulative linkages across different levels and modes of cognition in such a way that the new comprehension contributes to the sustainable redevelopment of all these levels within levels? We cannot give clear advice, but we both are involved with regional projects in which attempts are underway to forge useful links between the local and the regional, to be operative in both directions.

Development and redevelopment

Within the context of economic development, *to develop* generally means to use available resources in such a way as to achieve local or regional progress through *increased economic growth*, usually measured by a suitable regional indicator, such as total jobs created, net value added, etc. Underdevelopment may mean that some resources are not used to their full economic potential, with the result that local or regional economic progress is slower than it might otherwise be. Overdevelopment may mean that some important resources are overtaxed, again to the disadvantage of regional economic progress. Misguided or improper development may mean that mistakes have been made concerning the use of some resources, e.g., through the once-only destructive use or sacrifice of a renewable resource important to the sustenance of a future economy. These and other variations on the concept of development have rather close conceptual parallels in the study of the harvest of renewable resources and of the use of the natural environment.

We do not deal with the extraction of nonrenewable resources as related to economic development, except to mention that some general parallels exist in North America between improper and excessive harvesting of renewable resources and improper and wasteful extraction of petroleum and some mineral ores. Both of these major processes, as conventionally practiced, severely disrupted the natural functioning of the affected ecosystem.

We personally experience no euphoria in contemplating recent versions of what is implied, in North America, by economic development, economic growth, and progress. Nevertheless, these terms are widely used the world over and have been accepted as central to our overall study of the sustainable development of the biosphere; we can put them to objective use without necessarily endorsing various connotations subjectively. Perhaps we can help to develop a set of connotations for the term *sustainable redevelopment* that may be acceptable for both objective and subjective purposes.

In many parts of the world the natural environment and its renewable resources have been misused, overused, and abused to the point of severe degradation. *Sustainability*, with respect to some desirable mix of valued uses, has been vitiated in such areas through destructive abuse, overintensive use, or ill-informed practices in general. The only reasonable option in such areas is a redevelopment toward sustainability. Various terms or slogans have been used that relate in some way to a reversal of the degradation, whether ecological, anthropological, social, economic, industrial, or urban, e.g., reform, rejuvenation, remediation, restoration, rehabilitation, reforestation, resettlement, reindustrialization, rezoning, renewal, recovery, revitalization. Taken together, all of these that are relevant in a region might constitute *redevelopment*; ecological rehabilitation of such areas is obviously crucial, else sustainability is an empty or misleading slogan.

Throughout North America in particular, major aquatic, forest, grassland, and cultured ecosystems have been degraded. Major, though perhaps only partial, American initiatives toward ecosystem or regional redevelopment include the Tennessee Valley Authority initiatives in Southern Appalachia, the Soil Conservation Service initiatives on the Great Plains, and regional reforestation in southeastern USA. In subsequent sections we present two case studies of more recent initiatives toward redevelopment: the forests of New Brunswick in eastern Canada and the Great Lakes of central North America. Our cases are similar yet different. The New Brunswick case represents a deliberate provincial effort to redevelop a rural community economy based on forestry before the natural productive base becomes too severely degraded: opinion leaders in New Brunswick have suddenly recognized the associated problems and opportunities. The Great Lakes case represents deliberate international efforts, to date costing perhaps over $10 billion (1985 US dollars), to rehabilitate massively degraded ecosystems, especially those of the Lower Lakes. For various reasons the Great Lakes industrial and agricultural heartland lost its vitality and vigor, and people recognized that ecosystem degradation was part of the problem. It is now widely expected that successful rehabilitation of these lake ecosystems will have both a direct and a catalytic role in the revitalization of this binational heartland, even though the apparent additional monetary benefits of a rehabilitated chain of Great Lakes may be only a small fraction of the wealth created in the Basin [2].

In both cases described here we have greatly simplified the record in the hope of enhancing understanding. We show that integrated regional redevelopment of degraded systems must be served by appropriate information systems, which must explicitly link the local and regional levels of concern. With respect to spatial and temporal scope and scale, and to detail of resolution, a regional information system obviously falls somewhere between biospheric and ecosystemic information systems. Information systems already exist in some, perhaps primitive, form with respect to the redevelopment initiatives mentioned here. Progress in redevelopment will likely depend on the further creative development and thoughtful use of such information systems.

The forests of New Brunswick

The setting and an historical sketch

The Province of New Brunswick is largely forested (85% of its 78 000 km^2) and has been so in the ten millenia since the last continental glacier melted. During the past two centuries the economic development of sparsely populated New Brunswick has depended primarily on the exploitive use of forests and secondarily on the exploitive use of fisheries and agricultural soils. The Province has never served as an industrial heartland to any other region, nor is it likely to do so in the foreseeable future. It is relatively free of the pervasive and massive pollution that attends large industrial and urban concentrations, though the effects of atmospherically transported acid rain and toxic fallout, and of climate warming, all due predominantly to the improper and excessive combustion of fossil fuels, will soon become as apparent there as elsewhere.

In this sketch of the history of conventional exploitive development and of opportunities for sustainable redevelopment in New Brunswick we focus mainly on the forest industry. Consideration of fisheries and agriculture, of human settlements and pollution, and of the outbreaks of the spruce budworm would complicate the account, but would not greatly affect the inferences that can be drawn from a simplified account of the forest industry.

The development of the New Brunswick forests has been a long process [3, 4]. In the early 1800s the forests provided large white pine trees for ship masts, which involved selecting high-value specimens and bringing these out with considerable care. While this generated a step in the development of the local economy, it was at a very low level, and the quality of life associated with it was, in the prosaic sense, hardy. At this time the die for future development was already cast: "for profit is the first motive of all men" as Nicolas Denys, a naturalist-cum-early developer of the early 1700s, wrote in his diary of 1708 [5].

Continued development (read "progress through economic growth") was temporarily arrested by the unavailability of very large, and very old, white pine trees. However, development proceeded with the harvest of white pine to less stringent quality standards in order to produce squared timbers for building purposes: these were mostly exported. This involved considerably more local employment than did the search for masts and resulted in a modest improvement in the quality of life of the pioneer communities.

One course then taken in the development sequence was to broaden the harvest to include younger, smaller trees. If 300-year white pine are harvested faster than they are recruited to that age class, then development may be extended by a shift to younger, smaller trees. This change is illustrated in *Figure 3.2,* where successive steps in the development of the white pine industry began with the high level of minimum standard acceptable for raw material at point *A*, development continuing until the white pine available were of a size characterized by point *B*, which then became the acceptable minimum standard.

By the middle of the 1800s, the next major step in the process of development began with the emergence of a major sawmill industry that produced finished timber, again mostly for export. The work involved in contributing "value added" to the product enabled the local society to gain some additional

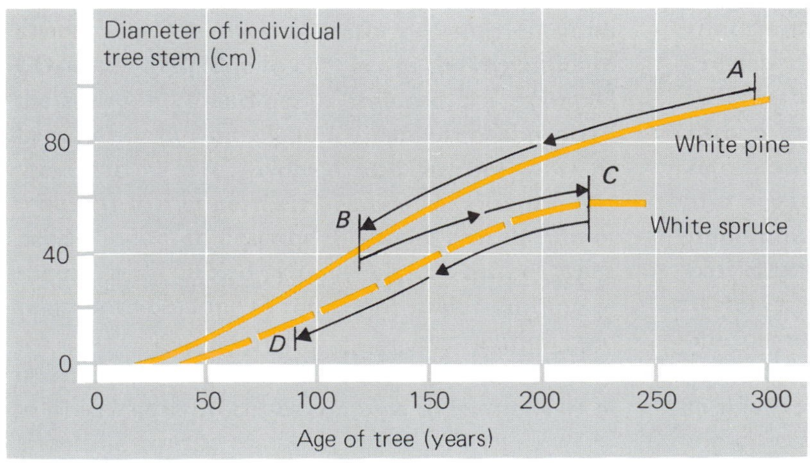

Figure 3.2 The diameter of white pine and white spruce stems as related to age of the stem. Stem size is a crucial factor in sawmill efficiency. Early exploitation began with white pine at point *A*, continued using smaller stems until point *B*, where it was equally efficient to switch exploitation to include large white spruce at *C*. Continued exploitation mined the largest stems until current limits of acceptability were reached at *D*.

economic benefits from the resource, but also the regionally based lumbering industry gained. To develop this lumbering industry further, the source of raw material was again broadened, with respect to minimum standards of acceptability, to include not only the remaining large and small white pine trees available, but also the larger white spruce. This change is characterized in *Figure 3.2* as a movement in the minimum standard acceptable for raw material from *B* to *C*. The sawmill industry continued to grow (develop) in a manner that required quality raw material faster than the forest was producing that quality, with the result that the minimum acceptable raw material gradually moved from *C* to *D* on the lower curve in *Figure 3.2.*

The sawmill industry reached its peak in the late nineteenth/early twentieth centuries and has been declining ever since, with the number of sawmills in the Province today only a fraction of that in the heyday of the industry in the late 1800s. The surviving sawmill industry uses not only the available spruce logs (and the small amount of white pine available), but also balsam fir trees. Currently, the minimum acceptable size for all these species is a fraction of the original.

While the sawmill industry was developed largely on the basis of individual trees, a pulp and paper industry emerged in the province shortly after the beginning of the twentieth century, and this industry was based on whole stands of trees. The arrival of the pulp industry happened to coincide with poor lumber markets and the relative non-availability of large individual stems suitable for efficient use in the kind of sawlog operation that had evolved in the nineteenth century. Pulp mills can use smaller trees, although harvesting of larger trees makes for a cheaper raw material. Just as the sawmill development was based on harvesting only the best individual stems and leaving the poor quality material (hygrading), so pulp and paper development was based on harvesting the stands of best species and lowest logging cost while by-passing lower quality stands (hygrading). Because of the large wood volumes required by pulp mills there was a movement to harvest nearly pure stands of the usable species (spruce and fir), leaving behind stands of species not usable for pulp. Pulp mills marked a major development for the local economy, capturing not only a much larger portion of the value of the finished resource by spin-off employment, but also by generating a better quality of employment.

Initially, the pulp and paper industry used stands of smaller spruce trees and, as these became less available, raw material standards were relaxed to include stands of larger fir trees as well. Since the mid-1950s the industry has adapted technologically so that it can use stands of very small trees of softwood species and some hardwoods. To characterize the impact of the pulp and paper industry on the resource we examine how these utilization standards relate to stand development (i.e., natural production in a stand). *Figure 3.3* shows how a particular softwood stand (i.e., a population/community of trees) might develop in terms of volume per hectare and of average tree size, from its regenerating stages, through maturity, and over-maturity, to break up in old age. Combinations of minimum volume per hectare and maximum trees per cubic meter determine the operability or availability of a stand for economic harvest.

At the beginning of the development of the pulp and paper industry only those stands represented by the range *A* and *A′* in *Figure 3.4* were considered economically usable. These operable stands had reached both the specified economic minimum total volume per hectare and the specified economic minimum average tree size harvestable, but they had not yet broken up to the point where there was less than the minimum volume below which it was not economic to harvest. Clearly, when stands of type *A–A′* were harvested faster than they were recruited to this range, there occurred, over time, a shortage of material for the mills, which could have forced a reduction in their output. Analogous to the case for sawlogs from individual stems, the answer to such a constraint was to change the utilization limits for whole stands to the range *B–B′* in *Figure 3.4*; the advent of a biomass approach suggests extension to the range *C–C′*.

New capital investment and technology – of both hard and soft types – were required to effect a shift from *A–A′* to *B–B′*. This shift then permitted further development of the economy in that the broader utilization standards (at *B–B′*) gave an almost instant increase in the sustainable production. For example, a harvest level of 400 000 m^3/year from a forest was not sustainable when the stand utilization limits were set at *A–A′*, but was sustainable, and could even be increased to

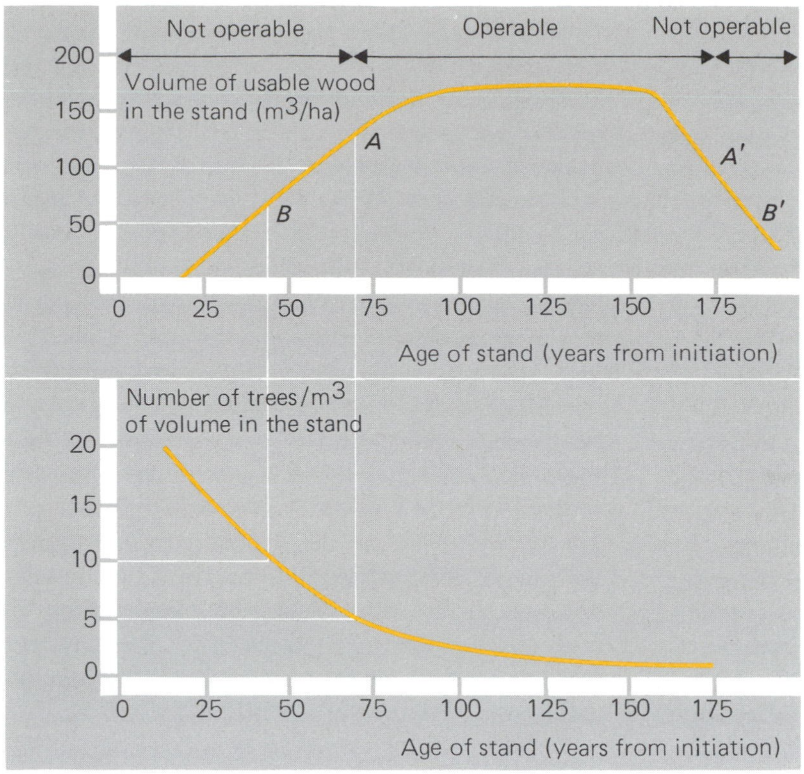

Figure 3.3 Softwood stand development as seen by the developer. Above: The accumulation of volume over time, showing how much wood is available at various points in time. Below: Average tree size as it changes over time. Operability constraints work in two ways: (*a*) the stand must have enough volume per hectare to be worth developing, e.g. 100 m^3/ha, and (*b*) the stand must be made of individual stems large enough to be commercially harvestable, e.g., 5 trees/m^3. The constraints of a minimum of 100 m^3/ha and a maximum of 5 trees/m^3 give the points *A* and *A'*, while the constraints of a minimum 50 m^3/ha and a maximum of 10 trees/m^3 give the points *B* and *B'*. For the first set of constraints the stand is harvestable (operable) between *A* and *A'* and *B* and *B'*, respectively.

600 000 m^3/year when the utilization standards were set at *B–B'* (*Figure 3.5*). Further industrial development was possible by harvestng stands comprising several species only some of which were suitable for pulping (*Figure 3.6*). Harvesting in these mixed stands faced similar utilization standards with respect to total volume and average tree size and had the additional logistical problem of working around the trees of unusable species, but it also gave a larger growing stock and a more sustainable harvest.

As with the pulp and paper industry, the lumbering industry also innovated technologically to use less-valued species, smaller trees, and stems of lower quality in the manufacture of laminated beams, plywood, chipboard, etc. Such innovations are

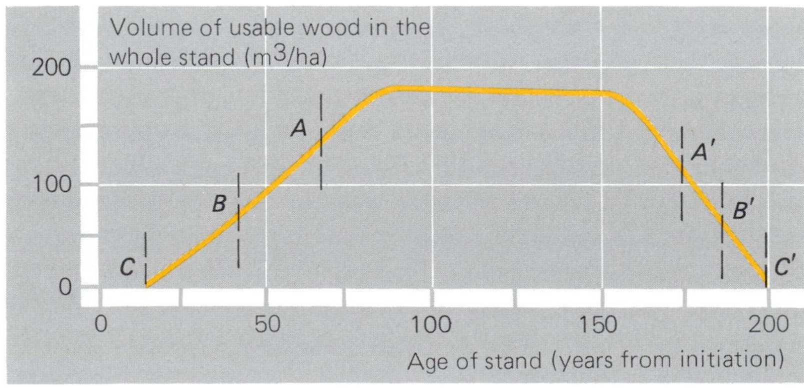

Figure 3.4 Volume development in a spruce–fir stand. The curve shows the merchantable volume per hectare at any point in time. The ranges *A–A'* and *B–B'* are periods when the stand is economically operable. The point *A* is determined by both a minimum volume per hectare and a minimum average tree size. The point *A'* is determined by the minimum volume per hectare. The stand is operable (i.e., eligible for harvest) in the range *A–A'*. If the operability constraints of volume per hectare and average tree size are relaxed then *A* moves to *B* and *A'* moves to *B'*. In this case the stand is operable for the period from *B* to *B'*. The process may continue to some ultimate state identified here by the extremes *C* and *C'*.

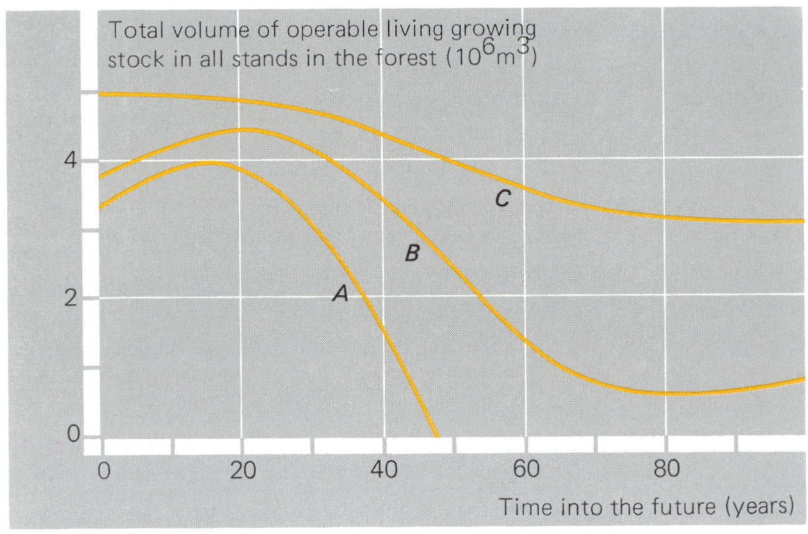

Figure 3.5 Three reasonably possible futures for the same initial forest, but harvested subject to different operability constraints. Curve *A*: With operability limits set at *A–A'* (from *Figure 3.4*) the total operable growing stock increases at first, but cannot sustain a harvest of 400 000 m^3/year. The growing stock goes to extinction in the fifth decade. Curve *B*: With operability limits set at *B–B'* (*Figure 3.4*) the total operable growing stock is larger than in *A* and the larger harvest level of 600 000 m^3/year is sustainable, although the growing stock is driven to a dangerously low level near the end of the forecast period. Curve *C*: With operability limits at *C–C'* (*Figure 3.4*) the total operable growing stock is again larger and the harvest level of 800 000 m^3/year is sustainable. Note that the initial increases in operable growing stock are artifacts in that the wood was always there in case *A* but was not counted because of the operability constraints. Also note that the apparent gains in sustainable harvest are each achieved by harvesting younger (poorer quality) raw material.

capital intensive and involve advanced technology. In effect, capital and technology are used to mitigate reduction in the quality of the resource.

Examining the conventional saw- and pulp-mill examples of development it is clear that both constitute forms of hygrading. The lumber industry was built on hygrading individual stems and the industry declined due to inefficiency when recruitment to these quality standards could not match the level of harvest imposed on the trees of that quality by the hygrading. The pulp and paper industry was built on the principle of hygrading at the stand level, where whole stands of particular species and size characteristics were removed and the utilization standards lowered from time to time, as necessary, to sustain the development [6]. In both cases, it is clear that the forest resource was used to develop the economy. In fact it was a very effective use, in

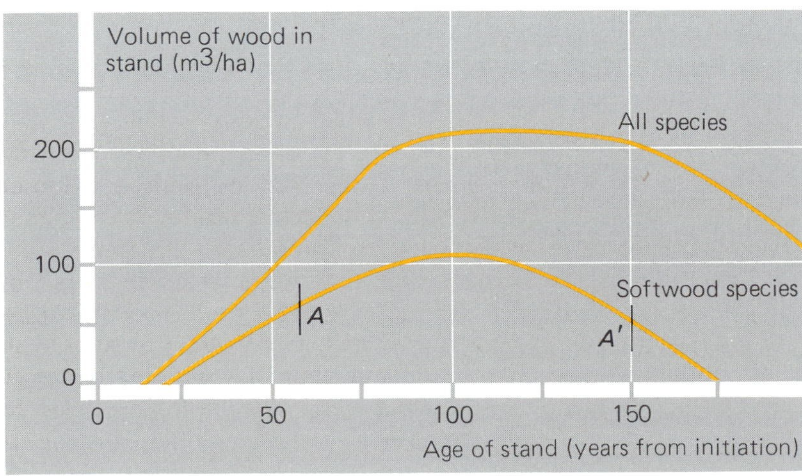

Figure 3.6. Volume development in a mixed stand where only the softwood volume (lower curve) is usable. The softwood volume is subject to operability constraints shown in *Figure 3.3*. When the softwood species are harvested all or nearly all of the other species are left standing.

that each step in the process:

(1) Allocated a larger proportion of the economic benefits to the local society.
(2) Increased the number of jobs locally.
(3) Improved the quality of those jobs.
(4) Improved the quality of life in the local society.

At present, however, the Province of New Brunswick faces a situation where it is no longer possible to develop further in this sense. In fact, it will barely be possible to sustain the forest-based industry even with major forest management efforts of a redevelopment type [7, 8].

Exploitive development

Some further consideration of the concept of exploitive development is appropriate at this point. Clearly, without specifications of quantity, quality, timing, and location the notion of sustainability is meaningless. A sawmill industry has existed for almost 200 years, but no one would claim that product quality was sustained for any significant period of time within those 200 years. What did survive was an industry that used poorer and poorer quality raw material and was strongly dependent on capital-intensive technological innovation for the creation of a quality product out of raw materials of low quality. A forest industry, broadly defined, was sustained, but the productive structure of the *resource* was not.

There are two points to consider with respect to this resource development. First, development was not a single step taken to achieve a specified goal, but rather it involved a continuous expansion of local and regional economic benefits, narrowly defined, that became a goal in itself. In this context there certainly exists for any resource a point in development beyond which sustainability is no longer possible because of biological limits of the resource system. The second point really attacks the nub of the issue. Historically, development in resource-rich areas is driven by the clarion call of government, industry, and enterprising people "to develop our resources." Clearly, it has been primarily the resource-based *industries* that have been developed, secondarily the *regional economy*, and tertiarily the *local economies* – but the *resources* have not been developed at all. The resources have been characteristically run-down or ruined by resource development – surely one of the saddest paradoxes of man. Hygrading has been degrading.

In New Brunswick the forest-based industry may now be characterized as *overdeveloped* and the forest resource is in urgent need of *redevelopment.* The existing economic structure is derived from a sequental lowering of utilization standards and broadening of the target species mix. It now appears that the existing economy could barely be sustained by a further lowering of utilization standards along with a very aggressive management program, but that the limits of development in the traditional economic sense have been reached. The key here is that resource development has been measured throughout this process in terms of the immediate social and economic results and *not at all in terms of the productivity of the resource that was used.* The resource was used to achieve the development, but using the resource altered its dynamic structure. The resource structure changed with respect to:

(1) The size and stature of particular species in the stands.
(2) The species composition of the various stand mixes that were available in the forest.
(3) The age structure of the stands in the forest as a whole.

Instead of progressive overexploitation of the resource to manage (read "expand") the economy, it is now necessary to manage (read "redevelop") the resource in order to maintain the economy. Put bluntly, it is no longer possible to gain further development or even sustain the existing level of development by mining the forest. Maintenance of the existing level is possible with proper management, but any expansionist development must be delayed several decades until the results of new management measures take effect [8].

There are some characteristics of these limits to development that are of interest. Perhaps the most striking is that each of the individual development steps took place at the tree or stand level. Local (tree and stand) actions, which were taken out of context of the regional forest picture, have accumulated to produce an unacceptable regional result. The problem literally was created piece by piece at the local level, and then recognized and felt at the regional level. Further, the design of the solution to problems created by development must proceed from the regional level. The stands that have been hygraded successively for sawlogs now support a mixture of poor quality stems and unusable species. Harvest of only the usable species for pulping has resulted in an increased occupancy of land by non-economic species. Indeed, the only way left to develop in the conventional way is by moving to a lower grade of product which will result in a weaker economic position, based on the massive mechanization required to use the very low-quality natural product.

The result of such development has been a dramatic change in the productive structure of the forest. A plausible age structure for the various stands of the forest prior to development is shown in *Figure 3.7*, along with the age structure for the developed forest. Conventional development has transformed the productive structure into that of a young forest. Given the present age structure either there must be a relaxation of these pressures to allow some time for recovery *or* there must be massive assistance to the natural regrowth of the forest. To achieve some recovery time requires a relaxation of harvesting pressure, which implies a loss of some of the development gained over the last half century. That is not a socially acceptable alternative. Intervention in stand and forest dynamics, to increase the rate of availability of materials, requires major expenditure in terms of forest management. These costs are clearly chargeable to maintenance of the development. Since these costs were not there previously, they substantially reduce the economic value of the accumulated development in conventional economic terms. To hold its own, the Province must give up some of its resource-based *economic* development in order to achieve *resource* redevelopment.

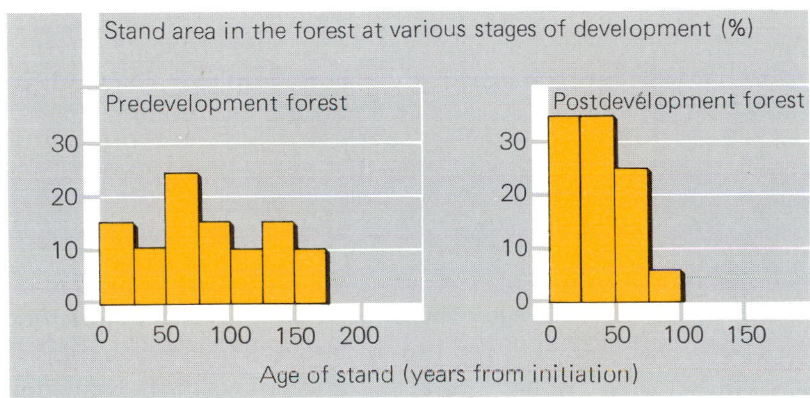

Figure 3.7. Age class structures for pre- and post-development forests. The predevelopment forest shows stands at all stages of development when related to the volume development curve in *Figure 3.4.* The postdevelopment forest has been harvested intensely enough to prevent stands from growing beyond 100 years from initiation. This means that stands in the range 100 to 160 years, which would contain the largest individual trees, no longer occur.

Sustainable redevelopment: design and implementation

Discovering overdevelopment is not easy and designing redevelopment can be downright traumatic. Recognizing incipient overdevelopment is difficult, in part because the necessary information is rarely available until after the event. During the process of nonsustainable development of an economy, data gathering with respect to the *resource* is centered on such features as the total area available for use at the time of data collection and the total volume available for harvest *at the time of data collection.* These inventories were designed to provide a picture of what was available on the ground at that time. They were carried out and reported entirely in the context of the development at issue, i.e., they had an economic (or an accounting) base. Thus, the first unavoidable sign of overdevelopment was when one of these inventories undertaken to justify the newest step in development showed less resource, even after the usual adjustments for lower utilization standards [3].

To discover the reason behind why a conventional inventory showed less material than could be expected on the basis of the historic sequence, it was necessary to examine resource dynamics. Not surprisingly, there was not much information on dynamics, since the information gathering had concentrated on a static inventory of material that possessed immediate economic value. The first approximations of forest dynamics were therefore based on simple assumptions [9, 10]. The emphasis of studies of forest dynamics was on forecasting the availability of particular quantities and qualities of raw material in the future, rather than on what was actually there. A major difficulty was that the problem was perceived at the forest, or regional, level where data on dynamics were particularly poor. It was thus necessary to build a bridge from the relatively data-rich tree and stand levels to the regional forest level. This exercise in problem definition is not simple and is resisted by a considerable amount of rationalization. One hears that "there is always someone forecasting a timber famine and it never comes about, and this one won't either;" that "forecasting is simply not possible;" that "forecasting with models is silly, since everyone knows that with computers it's garbage in garbage out." The realists of the world will state incessantly that "what counts, is what *is*." Then there are those who are certain that technology will permit even lower utilization standards or, indeed, the use of materials other than wood for the production of paper. The point here is that there is significant resistance to the recognition of overdevelopment of the industry and of the need for *resource* redevelopment, because such recognition forces change in the established ways of using the resource for economic gain.

The need for redevelopment of the forest and the forest-based industries in New Brunswick eventually reached the social and political agendas, despite all these forms of resistance. In the end it became clear, even to the most myopic, that the local industry was competing in world trade and that it was inefficient with respect to that market. The inefficiency in trade was traceable in large part to a high cost of raw material, which in turn was the result of the changed forest structure, which of course was the result of development. The gains in development in terms of quality of jobs and the local economy were, in the end, imperiled because of the inefficiency of the resource with respect to the processing part of the system. There were closures of sawmills that could not adapt technologically to the smaller tree size. There were major technological changes in pulp mills and a reduction of interest in maintaining certain pieces of physical plant associated with past development.

When awareness of nonsustainability finally dawned, it was simultaneously accepted that it was no longer possible to manage the economy by mining the resource, so the resource must henceforth be husbanded. This recognition (or cognition) by the decisive players in industry and government was a relatively sudden event, occurring in an interval of about five years, i.e., 1975 to 1980. The turning point came when the decision makers ceased to argue about whether or not there was a problem of maintaining the flow of *quality* raw material. Once that occurred, action toward redevelopment followed rather quickly [11, 12].

Acceptance of a need for change came first to the decisive players in industry and government and not to the general public. The public did, indeed, voice an earlier need for change; however, there was a definite air of unreality about their desires, in that they demanded a continuation of all (or even more) of the economic benefits of development in full measure, while corrective action was to be taken at no cost. Tension developed because industry and government had inadvertently obscured the evidence of the problems of development by the choice of data collected on the resource. This incredible lack of public understanding of the historic linkages between the benefits and the problems persists, indeed has intensified, five years into the redevelopment program.

In the case of the New Brunswick industries based on the forest resource, the program of redevelopment has involved several closely related activities. Once it became clear that the industrial capacity exceeded the ability of the forest to sustain raw material of the desired quality, there was a need to reallocate access to the resource to ensure even (or steady) resource use, and to prevent hoarding or wasteful use. The reallocation of access to the resource was not just in terms of areas of domain, as had been prevalent in the economic development phases, but rather related to access to a particular piece of the productive structure of the forest. Allocation was to a share of the sustainable productivity of the resource, rather than to the standing inventory on some particular piece of ground at the moment of allocation [13].

Recognition of the need to allocate access to productivity required an entirely new approach to the acquisition of information on the productive state of the forest. New surveys of the forest were needed that sought to characterize the resource in terms of its dynamic structure, rather than in terms of a static storehouse of raw material. This required major expenditure in terms of a new form of aerial photography for the Province, with special tests of methods of photointerpretation to discover the best ways to capture information on the stage of development for each stand. Methods for handling this geographic information had to be developed [14]. Not surprisingly, this has proved to be a complex task, and it is already clear that the first approximation of data acquisition on forest dynamics is just that, a first step.

A major characteristic of the recognition of the need for redevelopment is the appearance of concern with forecasting future development on the regional forest scale. During the period of development, interest in the future rarely extended beyond the current-year harvest, and at best to a five-year harvest plan. Those facing the reality of redevelopment have suddenly acquired time horizons of the order of 40 to 50 years. The need to forecast wood availability by amount, quality, location, and timing has resulted in the development of a large array of models that attempt to mimic dynamics at the stand and forest levels. These models were very simple at first, but are quickly developing as more information on dynamics becomes available, and as the necessity for forecasting becomes more clear. To manage a forest it is necessary to be able to make explicit forecasts of the availability of raw material in terms of quantity, quality, location, and time for a variety of patterns of intervention (harvesting, planting, thinning, etc.). The development and acceptance of forecasting tools occurred at a surprising rate. Approaches that were rejected as unnecessary, or unrealistic, five years ago are now embraced and ardently developed to improve their reliability.

The most traumatic element of redevelopment has been the need to actually design *and* implement management interventions. During the earlier period of development there was much experimental work with silviculture and, indeed, a certain amount of what one might term the anecdotal practice of silviculture. All of this effort was characterized by an approach at the local or stand level. The managers would examine a stand and, based only on what they saw at the local level, determine the "best" silviculture action to be taken. Through this approach some considerable experience had evolved with such tools as planting, precommercial thinning, and fertilization. However, with acceptance of the need for redevelopment it also became clear that these local silviculture actions had to be determined in the context of regional forest dynamics [8]. Given that the harvest which can be taken is limited and that money available for silviculture is limited, it is crucial that forestry actions be taken in the right places, in the right amount, and at the right time so that the development of the whole forest is regulated in the manner necessary to reach a goal of true *resource development* (or of economic redevelopment). Thus, the focus of intervention design has moved from the stand level to the forest level [15]. The question is no longer what can be done at this particular place, but rather "what set of local actions taken in what places in the whole forest, and at what times, will cause the regional forest to transform and grow towards a particular chosen goal?" For a generation of managers, accustomed to making local stand decisions out of context of any regional forest picture, the necessity to place their actions in this regional context is proving difficult. A major problem here is that virtually all of the decision aids that have been developed in past decades are aimed at the stand or local level, rather than at the regional forest level. That is, economic decision-analysis methods assumed that, whenever management began, it would be carried out by a series of stand-by-stand local decisions. Major efforts are now underway to develop decision aids that place local actions in the regional context of forest dynamics.

Choosing a plan of management interventions for redevelopment is a complex matter. To be realistic, the plan must recognize that implementation of redevelopment will be local in nature (just as all of the actions of development were local in nature), but that these local interventions must be consistent with a regional pattern of resource development [15]. Management planning becomes an orchestration of local events to achieve a regional goal – what will be done, to what extent it will be done, when, and where – in order that a regional pattern of development, over time, is achieved. Moreover, management must address the new real issues of who pays for management effort and who carries it out.

Making all this happen on the ground is crucial. It is necessary to implement the plan as it is drawn up and to provide a regular audit of performance of the forest. This problem is unique to the stage of redevelopment, since it was *not necessary* during development to worry about matching management platitudes with what was actually happening on the ground. However, during redevelopment it is absolutely essential, for instance, that the actual harvest be taken in the same manner as used in the forecasts for management design. This requires a high degree of geographic control [14]. If, in the actual on-the-ground implementation of the harvest schedule, the oldest first rule is not followed and any younger stands are harvested because of local economic advantage, then some older stands escape harvest and will decay and become nonavailable for harvest. The net effect of this deviation of the real harvest schedule from the planned schedule is a reduction in the actual sustainable harvest. That is, in management, and particularly in management for redevelopment, it is necessary that the way things are done on the ground is the same as the way things were done in the calculations determining sustainability. If it is not possible to implement on the ground the rules as determined in the plan, then the plan must be changed to show an appropriately lower sustainable level of harvest. This requirement for making reality match the plans is perhaps *the* major source of tension in redevelopment [13].

The fisheries of the Great Lakes

The setting and an historical sketch

The Great Lakes lie at the southern edge of the ancient Laurentian Shield where they straddle the Canada–USA border (*Figure 3.1*). Lake Superior is the largest and deepest lake, farthest upstream, and was apparently caused by tectonic events. The other four lakes were formed, or at least deepened and enlarged, by the continental glaciation of geologically recent times. The current ecological reality of the Great Lakes Basin is due to some ten millenia of the natural processes of postglacial succession and some two centuries of the effects of humans of western cultures.

Exploitive human uses, direct plus indirect, have intensified at something like an exponential rate until about a decade ago; today the overall effect may no longer be intensifying, but it remains on balance at a high level. The Great Lakes are downwind of the industrialized Ohio Valley from which airborne pollutants are often carried long distances. In contrast to the size of the area from which pollutants are transported through the atmosphere into the Great Lakes, the watershed of the Basin is small compared to the surface of the water, with no large rivers flowing into these lakes from the surrounding landscapes; hence the lakes are not rapidly flushed. Many of the small rivers and nearshore areas of the lakes are severely polluted, though perhaps not as severely as a decade ago. (The Cuyahoga River flowing into Lake Erie is no longer a fire hazard!)

Over 35 million humans now dwell in the Basin. Most live in large urban concentrations with imperfect sewerage and sewage treatment systems, but most of these are less imperfect than they were a decade ago. Many of the residents work in industries that pollute because of rather obsolescent capital stock erected at a time of great interest in the nonrenewable resources of the Basin (iron, coal, and limestone) and of little concern for those resources that were potentially renewable at sustained levels. Major uses (including abuses) of the Great Lakes have been separated into about 20 classes, of which fishing is but one [16].

By about 1930 the people of the dozen or so largest cities in the Great Lakes Basin had turned their backs to the lake shores and coastal waters that had become degraded; apparently the degradation was taken to be an unfortunate but necessary sacrifice to industrial–economic progress. Cities that thrived had dirty air, foul waters, and some urban slums. Today, however, the situation has changed, and people and institutions in the Great Lakes region are seriously searching for a new basis for the regional economy. In fact, the waters of the Great Lakes, and in particular the coastal waters, are now seen as a great natural resource, at least if their quality can be improved and then maintained.

Somehow an ecologically rehabilitated aquatic ecosystem is expected to contribute directly and also catalytically to the post industrial *redevelopment* of the region [2]. Rehabilitation of the fish together with restoration of high water quality are beginning to be linked by the public to the goal of sustainable redevelopment of the whole Basin's economy. It is coming to be accepted that the state of the fish community is a valid integrative indicator of ecosystem quality [17] and – somewhat more distantly – of the regional quality of life for humans. Some even suggest that the Basin provides the single best indicator of the state of biospheric husbandry as practiced in industrialized North America. The lake trout in the Great Lakes indicate what the white pine indicate in New Brunswick.

The historical sequence related to hygrading in forestry has a close counterpart in the fisheries of the Great Lakes. In fisheries the process is sometimes termed "fishing-up", whether among size or quality classes within a stock or species [18], among locales within an ecosystem or region [19], or among species within an entire fish association [20, 21]. By about 1940 the most preferred species in the Great Lakes were exploited by the fisheries to the point of a risk of overexploitation, if not, in fact, beyond the point of overexploitation (see *Figures 3.8* and *3.9*).

Throughout the process of fishing-up (or progressive hygrading) the enterprising fishermen also progressively added value to their products through salting, refrigeration, freezing, filleting, smoking, precooking, etc. An approximate parallel to the plywood and chipboard products may be canned fish and (as yet experimental) fish sausages. Something comparable to the pulp and paper stage of the lumbering industry has not yet emerged in the Great Lakes, but it has appeared elsewhere with new Japanese technology to produce *surimi* products (such as artificial scallops and crab legs) from the macerated flesh of low-valued finfish species. Serious plans have been drafted to bring this *surimi* technology into the Great Lakes fisheries.

Prior to the period 1940–1955 the fish association of the Great Lakes was strongly and adversely affected ecologically by four stresses due to humans: fishing, blocking of streams by dams, loading of putrescible organic waters into streams and estuaries, and physical destruction of wetlands. During the period 1940–1955 these four stresses were joined, and often superseded in local intensity, by four additional stresses: cultural eutrophication,

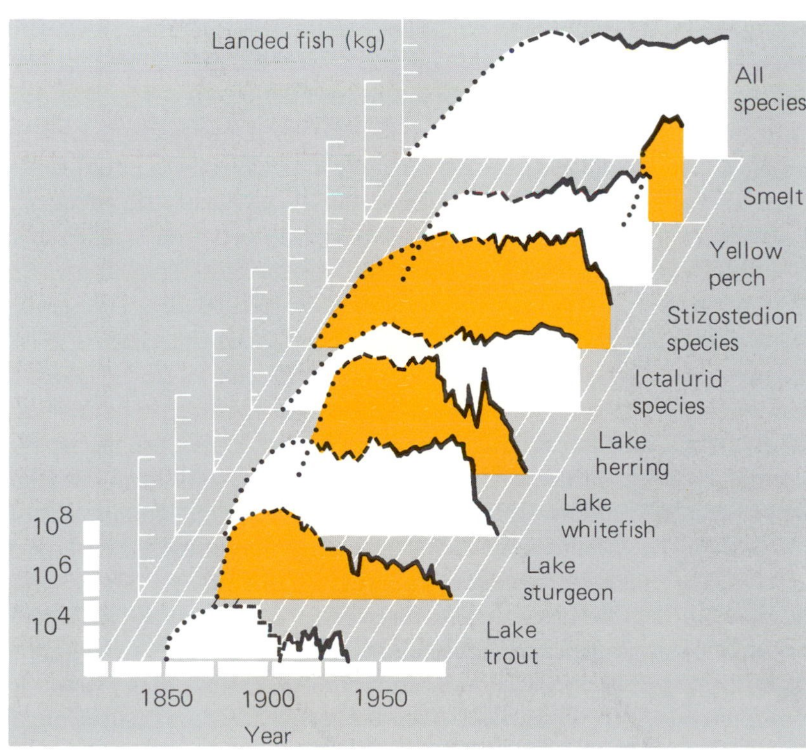

Figure 3.8 Lake Erie fish landings, by selected taxa, in kilograms.

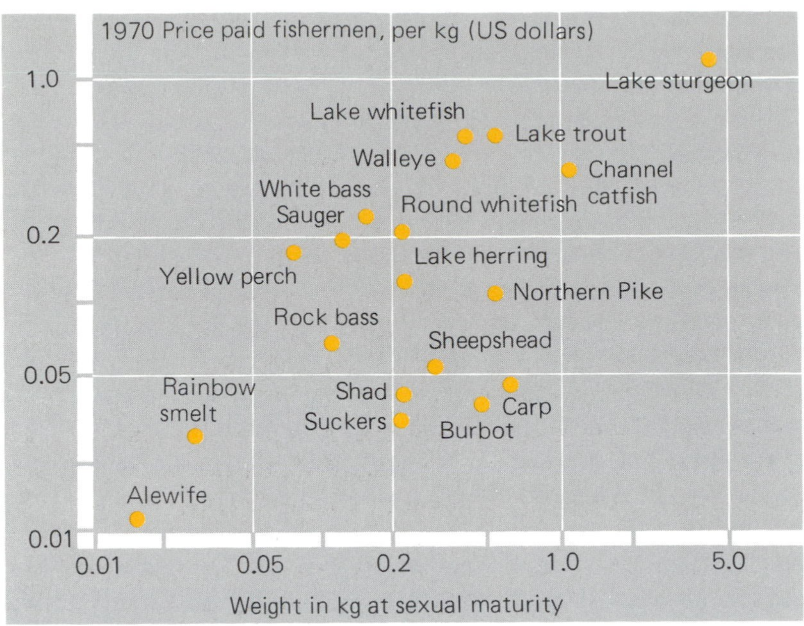

Figure 3.9 Value to fishermen (1970 USA prices) of a kilogram of fish whole weight as a function of the average size of a species at sexual maturation (log scale) [20].

industrial pollution, pesticides and other hazardous man-made chemicals, and expansion of low-valued or harmful exotic fish species, such as the alewife and the sea lamprey. These latter four stresses had appeared locally long before the period 1940–1955, but it was only then that their effects came to be regionally pervasive in many parts of the lakes. Other stresses of various kinds were also acting [16], and intensifying in some cases, but they must be set aside as miscellaneous in this short account.

The year 1955 may be taken as the beginning of growing public cognition that the Great Lakes, as parts of a system, were severely debased. In general, the lower the lake in the overall drainage system the more thoroughly it was debased. The total binational cost of all the relevant studies and corrective efforts since 1955 is uncounted, but might well exceed $10 billion (1985 US dollars). Most of the scientific efforts to understand this overall problem and almost all practical efforts to do something about it have been focused on locales and have attempted to deal with the problem one factor (even one chemical) or one species at a time.

At about this point in our historical sketch, convention would dictate that an account be given of the structures and processes of the governance of these binational Great Lakes, with a strong emphasis on the role of the International Joint Commission [22]. It is true that for a study of sustainable redevelopment of the biosphere we need eventually to deal in depth with the governance institutions. Suffice it here to say that no government agency at any level – binational, federal, provincial, or state – has asserted or been given a mandate for leadership with respect to an integrated program of sustainable redevelopment that is now so necessary. There is no government agency with sufficient responsibility or authority to act locally, think regionally in an integrated, ecosystemic way. Clearly the "think regionally" part of the aphorism is made more difficult to achieve in this multicity, multijurisdiction, binational situation; however, intergovernmental and broader networks are evolving that are beginning to serve this function.

Within the (as yet) piecemeal efforts to correct the causes of the degradation of the Great Lakes, there are many detailed differences in the politics and practices of different jurisdictional governmental and intergovernmental entities. But these differences do not obscure the basic unity within the overall sequence of exploitive development of the two centuries of domination of the Great Lakes by Europeans and their descendants. There may be time lags between jurisdictions, different priorities, internal inconsistencies, etc., that can be very annoying, but *overall* the historical degradative trends and current corrective efforts are sufficiently similar that they can all be subsumed under the generalized concepts of exploitive development and sustainable redevelopment.

Exploitive development

We now examine some of the main *ecological* phenomena related to exploitive development of the Great Lakes, and thus also related to sustainable redevelopment.

Francis *et al.* [16] have sketched some of the main effects on the fish association (and on other highly valued, sensitive ecological subsystems) of the many stresses operating in the Great Lakes. This approach has been fine tuned to the very degraded Green Bay [23] and to the relatively well-conserved Long Point with its bays [24]. The disheartening realization has emerged that most, if not all, of these stresses, acting singly or jointly and to excess, entrain an ecological syndrome of degradation or debasement [25]. Some features of this syndrome, as it relates especially to fish, have been sketched as follows [26, 27]:

(1) The major ecological stresses associated with human uses as conventionally practiced often act synergistically within the ecosystem so as to exacerbate each other's adverse ecological effects; they seldom act antagonistically so as to cancel out adverse effects.

(2) The stresses separately and jointly act to alter the fish association from one that is dominated by large fish, usually associated with the lake bottom and lake edge, to one characterized by small, short-lived, mid-water species. A similar change happens with respect to vegetation; firm-rooted aquatic plants originally nearshore are supplanted by dense suspensions of open-water (pelagic) plankton algae. Further, the association of relatively large benthic invertebrates on the bottom (such as mussels and crayfish) is supplanted by small burrowing insects and worms (such as midge larvae and sludge worms). Broadly similar changes occur in the flora and fauna of the wetlands and nearshore areas bordering these waters.

(3) With the above changes an increased variability in abundance of particular species occurs from year to year, but especially in landings of different fish species by anglers and commercial fishermen. Fluctuations are also more pronounced in the species associations of wetland, benthic, and pelagic areas.

(4) The shift from large organisms to small organisms is not accompanied by a major increase in the total standing biomass of living material, but is accompanied by a reduction in the production of the most preferred species.

(5) In the offshore pelagic region a new fish association, mostly of exotic species, may be created in which the small fish, such as alewife and smelt, may serve as a food resource to large predatory fish, such as salmon, which, however, must be maintained through the capital-intensive technology of fish hatcheries.

(6) Market and sport value per unit of biomass are generally much lower with small mid-water fish species than with large bottom species, and processing costs are higher. Similarly, the aesthetic value to recreationists of the rooted plants nearshore is higher than a turbid mixture of suspended algae and pollutants.

(7) The overall effect on fisheries, in the absence of the put, grow, and take stocking of hatchery-reared predators, is that nearshore, labor-intensive specialized fisheries (sport and commercial) tend to disappear, though highly mechanized, capital-intensive offshore enterprises may persist, if the combined stresses do not become excessive and if the fish are not so contaminated as to become a health threat for those who eat them. Yachtsmen may quickly sail from polluted marinas through the foul coastal water to the less-offensive offshore waters. Beaches are posted as hazardous to health.

(8) The combined effect is one of debasement and destabilization of the system of the natural environment and its indigenous renewable resources with respect to the features of greatest value to humans.

Rapport *et al.* [28] have termed the above a *general stress syndrome* and have inferred that such a syndrome may be observed in terrestrial as well as aquatic ecosystems that are subjected to the stresses typical of conventional exploitive development [25]. Indeed, we see analogs for each of the above symptoms in the New Brunswick forest example. Discovery of this syndrome has rendered obsolete any general policy of managing human uses of ecosystems as though they were ecologically independent. Recall the point made in the preceding section concerning forests: conventional development acted destructively on the underlying natural processes that had generated high-value resources. Some stress and consequent disturbance of the natural generative processes are inevitable when humans intervene, but the extreme, combined destructiveness of the various abuses related to exploitive development was a result of deeply misguided policies and practices, which may have made contemporary sense, but which often involved self-imposed ignorance. Sustainable redevelopment must seek to cooperate more closely with the natural processes that yield resources of high value.

Sustainable redevelopment: design and implementation

Attempts to arrest and reverse the degradation of parts of the Great Lakes ecosystem, including its fisheries, were begun over a century ago. Gross pollution with organic matter such as sawdust and offal was contained, in part. Destructive fishery practices were outlawed, such as dynamiting or setting nets across streams used for spawning. Fishways were made mandatory for dams, but the fishways were seldom effective. With the advent of steam and electrical power many of the small dams across tributaries of the Great Lakes fell into disuse and were washed away. In larger rivers, in this region and elsewhere, bigger and better dams were constructed, often without functional fishways.

Several factors were responsible for the escalation of the degradative practices – an escalation that remained uncontained until very recently. These factors, which must be reformed in a design for redevelopment, were:

(1) An unstated policy, shared binationally, that degradative abuses be addressed only after some particularly abused groups of the public raised a great clamor. Thus, there were always time lags, seldom less than a decade, between the experts' and abused people's awareness of degradation and some beginnings of corrective action, almost always initiated slowly through government action. This problem was not as severe with the New Brunswick forests, perhaps because of the lower population density and the smaller number of nonconsumptive users. Also in New Brunswick there was a more direct feedback that imperiled the economic structure and this may have attracted the attention of business and bureaucracy sooner than in areas where the links are less direct, as with dirty water in the Great Lakes.

(2) Such corrective actions as were taken were designed so as *not to impede* exploitive development to a serious degree, as was also the case in New Brunswick. Corrective actions were seldom fully effective, sometimes hardly effective at all, and sometimes they redirected the problem to other parts of the ecosystem.

(3) There was an unbroken tempo in the advent of new user groups with direct and indirect demands on these ecosystems. The feasibility (or likelihood) that new users were likely to exacerbate the environmental impacts of existing users was generally ignored.

(4) Overall the spatial and temporal scales of the degradative impact of various user groups tended to increase with technological advances related to those uses, but may have become progressively less apparent to the laymen, even the most observant, as with acid rain, atmospherically transported toxic contaminants, leaching from landfill sites, consumptive use of water, etc. The related effects are not readily seen until researchers present generalizations and abstractions of regional impacts.

(5) Corrective action usually consisted of an attempt to reduce the extent and intensity of the abuse with little effort at mitigation or at rehabilitative intervention (ecosystem therapy) to foster recovery processes consistent with the natural healing processes. A conventional engineering approach tends to move a problem to a different place or time period. It does not appreciate that biological systems have a memory or imprint of past actions – that just stopping an abuse does not necessarily lead to self-correction. Corrective intervention is often required.

The time lag between public awareness of a serious problem and public perception of an improvement following corrective action is now about a quarter of a century. Overall, the aggregated level of ecosystemic degradation in the Great

Lakes may have peaked (or ecosystem quality may have "bottomed out") in the early 1980s – or it may not yet have done so. Problems with contaminants leaching from landfill sites, from acids and toxic materials transported atmospherically, and from consumptive use of water appear to be waxing, while those due to nutrient loading, exploitive fishing, and exotic species are waning. Whether or not the aggregated level has peaked, the evidence does indicate that the rate of degradation has been slowed.

The five policy factors sketched above appear to apply to both our cases of regional fishery and regional forestry development. As a policy, sustainable redevelopment must establish a framework that focuses the design of corrective actions on these factors. At the regional level, focused discussion on how to correct them, as a systemic set, has been initiated only recently.

A mandate for a binational policy of sustainable redevelopment for the Great Lakes, with some management responsibility at the binational level, may be inferred from a study of several binational agreements in the context of what we now know about the ecological effects of uses and their interrelations within the Great Lakes ecosystem. These agreements include:

(1) The Boundary Waters Treaty of 1909, served by the International Joint Commission (IJC).
(2) The Migratory Birds Convention (MBC) of 1917 served by an intergovernmental committee.
(3) The Great Lakes Fishery Convention (GLFC) of 1954 served by the Great Lakes Fishery Commission.
(4) The Great Lakes Water Quality Agreements of 1972 and 1978 overseen by the IJC.

Crucial gaps remain:

(1) An agreement on long-range transport of atmospheric pollutants, with its acid rain and toxic fallout.
(2) An agreement on consumptive use and/or extra-Basin diversion of water.
(3) An agreement with authority to take specified, local control actions in a regional control context, i.e., on the water control.

The following ecological features are being managed – in a weak sense of the word – under the four rather general agreements sketched above: water levels, water flows, water quality, and local air quality (IJC), fish quality and quantity, and the predaceous exotic lamprey (GLFC), waterfowl, shorebirds, and, by implication, the wetlands (MBC). Within an ecosystem context, it is obvious that most of these features cannot be managed effectively in the absence of complementary management of some other features. For example, the interrelationships between water quality, fish quality and quantity, the sea lamprey, fish-eating shorebirds, and wetlands as nursery areas for fish and birds is well documented in the scientific literature. Those who have a responsibility for management must have the full perspective.

An ecosystem approach has been endorsed by the two national parties to these agreements, as well as by various lower levels of government. With respect to water quality, the ecosystem approach has been endorsed formally at the regional or Basin level in the 1978 Great Lakes Water Quality Agreement and explicitly, though less formally, at both the Basin and lake levels, as the policy of the Great Lakes Fishery Commission. This ecosystem approach is beginning to be interpreted in the sense of what we have termed Basin-wide sustainable redevelopment [29]. As yet no specific codification of the meaning of the ecosystem approach has been accepted widely. Ecosystem understanding of sub-lake ecosystems is quite advanced [23, 24] compared with that of lake ecosystems or of the entire Basin ecosystem. In the context of think regionally, act locally, few people have yet learned to think regionally, probably because the latter requires that personal (economic) issues be subsumed in a larger community context. That is, we live and see the local context, but the regional context must be some sort of abstraction that cannot be seen or felt, but can only be comprehended.

Public and political commitment, such as it is, to reverse the degradation of the Great Lakes has come at a time of growing awareness that major parts of the old industrial base of the Great Lakes region will likely wane in absolute terms. Examples are the steel and automobile industries and related water-borne transportation services. Urban growth has slackened and some local jurisdictions have experienced net reductions in human population. There is great concern that the region not be relegated to a hinterland of the American Sun Belt (disparagingly called the Parched Belt) to which some of the seeming abundance of fresh water in the Great Lakes might be diverted. How a thriving regional modern economy might receive major benefits from the sustainable redevelopment of the Great Lakes Basin ecosystem is not clear, though many opinion leaders believe that the Great Lakes themselves are the key to such redevelopment. Obviously the choice of indicators of redevelopment is crucial. Think regionally, act locally does not come easily with respect to a Basin ecosystem that is fractured into many jurisdictions and subject to many incompatible uses. Somehow the people of the Basin must together *choose* a future and make a long-term commitment to its realization, rather than just passively wait for something better to come along.

Attempts to achieve sustainable redevelopment of the fishery are dependent on the progress of reform with the fishery and also on rehabilitation of the habitat of the fish, the environment. In a properly managed fishery in a properly managed freshwater ecosystem (of the Great Lakes type) the fish association is dominated by native, self-reproducing, highly valued, bottom-oriented (benthic) species that achieve large size and old age in the natural state. Recall that by 1955 the fish association was well on its way to an inversion, toward dominance by exotic, low-valued, mid-water (pelagic) species that remain small and die at a relatively young age; by 1960 the inversion or debasement was almost complete. The main stresses responsible were excessive fishing of the preferred species (hygrading), invasion by the sea lamprey which had preferences for fish quite similar to those of humans, invasion or introduction of small pelagic fish that thrive in enriched waters, eutrophication through enrichment, pollution of spawning streams, and a miscellany of additional causes.

Corrective local measures began to show promise in the early 1960s. These local measures have generated local responses, but there is as yet only limited, mostly informal coordination at the regional Basin level. These measures include:

(1) Direct control of the sea lamprey through barriers and selective chemical lampricides, with the sterile male technique due for field testing in about 1986.
(2) Reform of fishing practices to permit a recovery of preferred species; in some jurisdictions this was done through a limitation of commercial fishing in favor of sport fishing; the latter is less capable of exerting intense fishing pressure on the valued species.
(3) Hatchery rearing and stocking of native species, such as the lake trout, in an attempt to reestablish self-reproducing stocks where they had been extinguished in recent decades.
(4) Hatchery rearing and stocking of nonnative species on a put, grow, and, take basis, especially salmonids of various species, to suppress the vast stocks of small pelagic species and to supply a highly valued sport fishery.
(5) Environmental programs for reduction of the loadings of phosphates to reverse eutrophication and to foster oligotrophication, which favors the valued native species.
(6) Ecological rehabilitation of some streams that flow into the Great Lakes to provide productive spawning and nursery areas, especially for large, native and some nonnative salmonids.
(7) Banning of the use of some persistent pesticides, such as DDT, which were washed into the lakes and transported and deposited onto the lakes by atmospheric processes, and then interfered with normal development of young fish.

By 1984 all the lakes showed signs that the fish associations were beginning to revert to a state of dominance by large benthic species originally native to the lakes. How quickly or how far this process will go is largely a function of what humans will do next with the Great Lakes Basin ecosystem, and is particularly dependent on how we orchestrate our more local endeavors within the Basin context.

The fact that these preferred native species are so dependent on a healthy aquatic ecosystem has stimulated interest in proposals to use some of them as "indicators of ecosystem quality [17]." A deep oligotrophic ecosystem should support thriving stocks of lake trout, a shallower mesotrophic part of the system should support walleye and yellow perch, and a nearshore part of the system should support black bass and pike. Aquatic systems dominated by such species are likely to provide water of a quality suitable for domestic use after minimal treatment, to be productive of fish species that are highly valued by sport and commercial fishermen, to be healthy for the contact recreation so important in heavily urbanized areas, to be attractive for recreational boating and nature study, and to be naturally self-regulatory to an important degree if all the uses are practiced in a manner that is well-informed in the context of sustainable redevelopment. Sustainability must be defined according to type, intensity, and frequency of use – subject to ecosystemic maxims defined locally and regionally.

Sustainable redevelopment of regional natural resource systems

Degradation and recovery in general

Innis [30], Rea [30], and others have exposed some far-reaching general parallels in the way various renewable resources were exploited within Canada and the USA during the past two centuries for the purpose of what is now termed economic development. Parallels can be traced in the histories of the exploitation of the New Brunswick forests (see above) and of the Great Lakes fisheries, though these parallels may not be as close as between New Brunswick fisheries in the Gulf of St Lawrence and New Brunswick forests, or as between Great Lakes fisheries and Great Lakes forests [31].

The breakdown of ecosystems, as ecosystems, under human influences and their recovery after relaxation or mitigation of those influences is attracting increasing attention [16, 32]. With respect to ecosystem organization in a quite general sense, the closest biotic counterparts in ecosystems like the Great Lakes to the dominant species of trees in a natural forest like that of eastern Canada may be the dominant, large, relatively sedentary species of fish [33]. In the pristine forests of eastern Canada two centuries ago, as in the pristine Great Lakes, much of the total living biomass was contained within such dominant species. Elton's energy pyramid with respect to trophic relationships is sometimes incorrectly taken as indicative of biomass proportions in natural biotic associations of lakes, such as the Great Lakes. In natural lakes the plant species, both the macrophytes and the phytoplankton, have a quick turnover, but some individuals of the large species of terminal consumers and predators survive for decades, such as lake trout and lake whitefish, and some up to perhaps a century, such as sturgeon and snapping turtles. These relative rates (and corresponding biomass proportions) are reversed in forests between, say, white pine and insectivorous birds. Note that the turnover rates of the economically most-valued components are quite slow in each case and hence recovery following cessation of abuse will also be slow in each case.

Incidentally, in natural grasslands and wetlands the relative biomass of producers *versus* consumers plus predators may not be as strongly skewed as in forests or lakes. Grasslands and wetlands appear to be more at the mercy of climate and natural fire with "wipe out" events more frequent and less localized than in either forests or lakes. Neither the plants nor animals tend to grow very large or old in grasslands and wetlands. Large animal species of the forest tend to exploit grasslands and wetlands for forage, and large animal species of the lakes tend to use wetlands for spawning and nursery areas.

Rapport *et al.* [25] have shown that the terrestrial and aquatic degradation syndromes caused by exploitive developments of many kinds are, in some ways, similar ecologically. The degradation syndrome may be seen as a pathological reversal of the usual natural developmental, successional, and morphogenetic processes that have been characterized by von Bertalanffy [34] as follows:

> Bertalanffy's model of hierarchical order was furnished by him with four related concepts: As life ascends the ladder of complexity, there is *progressive integration* in which the parts become more dependent on the whole, and *progressive differentiation*, in which the parts become more specialized. In consequence the [system] exhibits a wider repertoire of [functional] behaviour. But this is paid for by *progressive mechanization*, which is the limiting of the parts to a single function, and *progressive centralization* in which there emerge leading parts that dominate the behaviour of the system.

The science of surprise and the practice of adaptive environmental management of C. S. Holling and his colleagues [35] may relate in the first instance to both the normal natural morphogenetic sequence and the man-caused exploitive degradative sequences, in a general systems context. For example, forest resource degradation is a surprise because we were not collecting data on forest development and degradation – but if proper data had been gathered such a surprise would have not occurred or would have been less serious. One of the implications of Holling's ideas is that ecological sequences have discontinuities that may involve emergent behavior in the natural morphogenetic sequence and, presumably, may involve complementary submergent behavior in the man-caused degradation sequence. Such concepts can be helpful in comprehending the general consequences of opportunistic exploitation and for designing rehabilitative husbandry in a policy of sustainable redevelopment.

Mobilization of public and political support for redevelopment is hindered by a variety of incorrect myths about natural processes and characteristics of ecosystems, such as the following:

(1) *Large species tend to be inefficient producers of resources and should be sacrificed in favor of smaller species with quicker turnover rates.* This ignores the self-regulatory roles of large species that favor other valued features of these ecosystems. It also plays down the difference in per unit value, with economic and aesthetic values of some of the dominant species high in comparison to most of the small species. In any case the relative productivity of harvestable smaller species in aggregate is not likely to achieve twice that of harvestable larger species, and the difference in per unit net values are likely to be greater than twofold, in inverse proportion to the overall rate of production.
(2) *Natural succession of lakes involves eutrophication and thus the preferred species of fish that thrive in less fertile waters are passing phenomena anyway.* This is basically incorrect in that internal ecosystemic processes within lakes generally involve oligotrophication, but eutrophication is generally caused by external processes, such as atmospheric loading with volcanic ash or by humans loading nutrients into the lakes, which override the natural processes that regulate nutrient levels.
(3) *As forests age their susceptibility to natural forest fires, blow downs, and insect pest outbreaks increases greatly.* A natural forest tends to be a rather intricate mosaic of somewhat different associations, all at somewhat different stages of succession, and consequently not all at great risk from natural causes.

General features of sustainable redevelopment

Let us look again at the concept of sustainable redevelopment. One can argue that the pulp and paper industry has been sustained and expanded in New Brunswick, but such sustainability was achieved only by virtue of drastically reducing the utilization standards for raw material. This has driven up the cost of raw material since this smaller material is more costly to harvest and handle (*Figure 3.10*). The word sustainability must be accompanied by definitions of quantity, quality, time, and location to be meaningful. Does the conventional development of a white pine industry for 50 years before it runs out of suitable resources constitute sustainability? Does the existence of pulp mills for 50 years

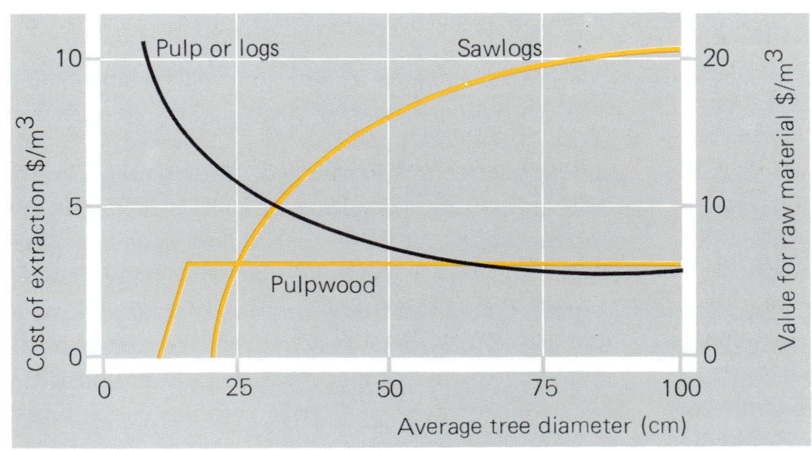

Figure 3.10 The relationship of raw material value (yellow) and cost of logging (black) to the average diameter of trees in a stand. Normally, market pressures limit pulpwood to smaller tree sizes while larger trees are processed for sawlogs. The cost of logging is strongly dependent on tree size, but is little influenced by the intended raw material.

constitute sustainability? Is development still considered sustainable when it results in depreciation of the resource? How is quality of the resource to be specified, in that the quality of raw material determines the quality of the associated industry?

Where the specifications for sustainability include quantity, quality, location, and time, one need expect relatively few resource problems with development. In the absence of such specifications, development inevitably wanders into trouble. Development in our society has meant quick and short-term economic benefit, with the higher the interest rate, the shorter the time horizon. Development *does not* refer to the resource, all brave words at pulp mill or fish plant dedications to the contrary! Frequently, plans for sustainable development include all the highest platitudes with respect to resource management that one could hope to see, but it is rare to see any mechanism that *forces implementation of these fine ideals.*

We could raise the same issues with respect to fisheries or water quality in the Great Lakes, as should now be obvious.

Two characteristics of redevelopment need emphasis. The first is that the transition from development to redevelopment (or from exploitation to husbandry) necessarily involves substantial tension. There is tension between industry (whether forestry or fishery), government, and the public, and this tension is heightened by a mutual lack of trust among these players. Industry feels that neither the government nor the public understand the intensity of world-level competition in which it is engaged. Government does not trust industry to maintain the necessary long-term horizon for a redevelopment program, and fears that the public may not understand that redevelopment will necessarily cost a temporary reduction in the flow of benefits from the resource. The public does not trust either industry or government, because they are seen to be creators of the problem, and hence unlikely candidates to design and execute a solution. None of these players has a very good understanding of the relationship between local actions and regional goals, and consequently they each occasionally produce rather unrealistic action proposals. For example, the public has been so long educated to the effect that a tree should be planted for every one cut, or a fish stocked for each harvested, that the focus on local action is almost exclusive. Public outcry during the design of redevelopment has centered on local events, with the most naive notions about what the forest or lake system in the regional sense will do if their local actions are followed, with no sense of the orchestration of these local actions. A most counterproductive feature of this tension is the desire of some members of all parties to demonstrate that it is some other "they" who are to blame. In fact *all three* – industry, government, public – applauded, and shared in, the development that has caused the problem.

The second characteristic of redevelopment worth noting has to do with the geographic reality of planning the context of acting locally while thinking regionally (or globally). Many may *talk* about management of a resource and, indeed, about specific management actions, without ever specifying the actual location of these actions or even recognizing the need for such specification. Some may actually *plan* resource management without dealing with the regional context of local actions, although it is not possible to escape the implications of context. That is, a plan that states that 5000 ha will be planted every year, 2000 ha precommercially thinned, and 3000 ha harvested, implies that someone knows where, in geographic location, each of these activity sums will be accumulated. (A similar statement applies to lake trout stocking in the Lakes.) When it comes to *implementing* a management plan on the ground or in the water, however, it is not possible without explicit geographic reference in a regional context and consequently we have few real redevelopers. To implement any plan it is necessary to implement a series of local actions, which together cumulate to the desired regional effect. If the plan does not specify the geographic pattern of these local events, then the plan cannot be implemented. This is perhaps the greatest learning experience of the redevelopment phase.

So important is geographic control to the implementation of management that, in the historical absence of a technical capability for geographic control, exploitation was all that was possible. Certainly, where it is not possible to relate the parts of a geographically dispersed system in terms of *both* attributes and location, only the broadest form of management control is possible. Fortunately, such limitations are being rapidly erased by computerized mapping systems embodying relational data bases [14]. For the first time these systems allow the planner not only to design a regional resource redevelopment strategy, *but also* to identify readily the geographic locations where particular tactical actions must be taken to accomplish the regional strategy. In systems where broad geographic extent adds to both the cost and complexity of resource exploitation, and to the cost and complexity of resource management, the ability to analyze the same resource data at appropriate scales for strategic and tactical designs is a major technological breakthrough. Characteristically, the exploitation interests have been as aggressive as management interests in implementing these expensive systems – this is true of the New Brunswick forestry case, but not of the Great Lakes fisheries case. In resources with many users and many managers with disparate responsibility and/or authority it is crucial that a common geographic data base be employed to avoid antagonistic strategies and to permit coherent management design.

There has been much *talk* about management in the past and, indeed, a certain amount of *planning*, but little of this has come to grips with the fact that *to make a plan happen on the ground and in the water, in the forest and lakes where it really counts, it is necessary to specify where the local events will take place in order to achieve a specified regional effect.* It is not an overstatement to say that what separates real management (or husbandry) from those highest platitudes of development is a specification of the geographic control of actions in the sense of: do this, at this time, to cause this to happen at this place in the future, so that the whole forest or lake ecosystem will develop along this desired pattern. Management is then acting locally while thinking regionally.

Local and regional decision making

In North America the nature of rights to the direct use of a renewable resource and to the indirect use of the habitat that generates the renewable resource is very complex [27, 36]. The heuristic schema in *Figure 3.11* may aid understanding of rights from the perspective of how these rights are exercised by the putative owners of them, and of how society administers the allocation and supervises the exercise of them.

A right to determine the use of a designated part of a renewable resource and/or its habitat may be held exclusively by designated individuals or by duly constituted groups of individuals, or may be shared nonexclusively with others. From a complementary but orthogonal perspective, such rights may be transferred or exchanged between individuals or groups largely at their own initiative, or they may not be transferable [37].

Figure 3.11 shows eight different combinations of these two criteria. The four outer corners are rather sharply defined, hard or formal manifestations of the four possible combinations, while the four inner corners are less clearly defined, soft or informal counterparts of the outer set. It may be noted that exercise of an illegitimate right has been included, in part to complete the schema, but also to take note of the fact that such illegal actions are not uncommon and do affect the exercise, by others, of their legitimate rights.

Regier and Grima [36] have suggested that the set of four hard elements,where they are condoned in practice by society, may become organized into a kind of competitively interdependent complex. In contrast, the soft elements may develop into a more mutual interdependent complex. In situations where a particlar complex is dominant the interdependent process may link with the relevant social mores to effect a kind of positive feedback reinforcement of the dominance, e.g., free enterprise and exploitation.

In some very general way, the interdependent hard set may be employed by the forces that dominate conventional exploitive development, especially as imposed from a metropolis onto a hinterland. This may happen with private capitalism in which the free market is a dominant element within the set, or with state capitalism in that the administrative element (linked to the international market) is dominant.

At the very local level, in an established, healthy human community, much of the exercise of rights occurs within the general framework of the soft, informal set. This is seen most clearly with respect to sharing within a family, whether nuclear or extended. This more informal approach to the identification and practice of rights may be severely distorted by cultural invasion of the more formal approaches associated with conventional development. Also, some of the smaller scale, less informal

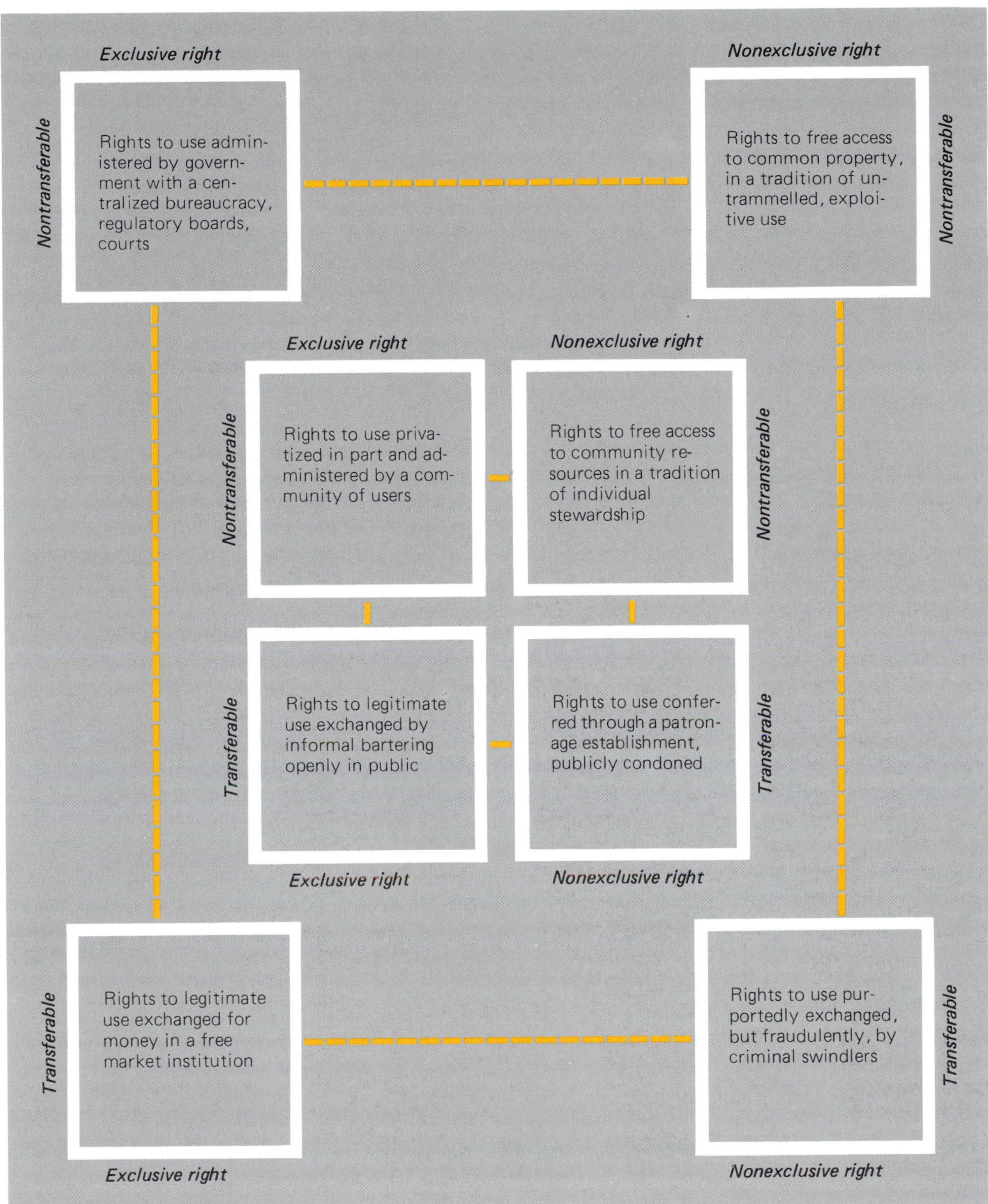

Figure 3.11 A perspective, expanded from a schema by Dales [37], on the variety of ways in which rights to the use of fish and similar resources are managed in North American society. In the four inside characterizations the exclusivity and transferability of user rights are satisfied in a practical manner. User rights in these four inside types are less sharply defined than those in the four corners. The schema may be viewed to have a soft core with a more sharply defined, hard shell or edge.

Table 3.1 Institutional or policy mechanisms for managing aquatic ecosystems and for allocating the use of fish and their habitats. Items at the top are largely administrative while those at the bottom have a prominent role for the market system. Examples relate to the Great Lakes of North America.

Mechanism	*Instrument of control*	*Purpose or observed consequence*
Prohibition	Exclusion of sport fish from commercial harvests	Improve recreational opportunities for anglers
	Specification of zero discharge of some toxics or contaminants	Reduce exposure of biota and humans to poisons
Regulation	Specification of low phosphorus concentrations in sewage effluents	Control eutrophication which, if intense, degrades the aquatic ecosystem
	Specification of gear and area, in fishing	Reduce fishing intensity to prevent overfishing
Direct government intervention	Control nonnative sea lamprey by lampricide, dams, etc.	Foster recovery of lake trout and other preferred species to benefit fishermen
	Development of islands and headlands with fill and dredge spoils	Provide recreational facilities and spawning areas to benefit anglers, boaters, etc.
Grants and tax incentives	Subsidy to industry for antipollution equipment	Lower pollution levels and distribute costs more widely
	Subsidy to commercial fishermen to harvest relatively undesirable species	Reduce competitive undesirable species to benefit preferred species and their users
Buy-back programs	Government purchase and retirement of excess harvesting capacity	Reduce excess fishing capacity and compensate owners of the excess capacity
Civil law	Losers enabled to sue despoilers in civil court	Preserve ecosystem amenities for broader public, recompense losers
Insurance	Compulsory third party insurance for claims of damage	Reduce pollution loadings because insurance premiums are scaled to loading levels
Effluent charges	Charge for waste disposal, either direct cost of treatment or indirect cost of impacts on ecosystem	Reduce pollution and/or allocate resources to high-value and/or profitable uses
License fees	Tax or charge on harvesters, scaled to level of use	Foster efficient use of resource by discouraging overcapitalization, recovering fair return for the owners (populace) of the resource
Demand management	Rates involve marginal cost pricing and/or peak responsibility pricing	Improve overall efficiency of use and foster conservation
Transferable development rights in land use planning	Limited rights to develop one area exchanged for broader rights to develop another	Direct the development to areas preferred by government
Specific property rights, as with transferable individual quotas	Purchase of pollution loading rights to predetermined loading levels	Limit pollution and foster efficient use of resources
	Harvest rights to explicit quantities to be purchased	Limit effective fishing effort and allocate resources to high-value and/or profitable uses

rights enjoyed by a community may be vitiated by the exercise of the exploitive rights of the developers.

North American societies have had great difficulty in dealing with the injustices and inequities visited on local communities by external developers. The wrongs have often been rationalized with respect to a rather simplistic utilitarian ethic of the greatest good for the greatest number, with some abstract logic about the desirability of Pareto-optimality, if only it could be implemented. We think of what has been done at federal and provincial levels with respect to the objective of maximizing the net value added, with less consideration of how the resulting benefits are distributed among people. Is value added a good measure of progress?

Currently, a swing to neoconservatism has brought with it a preference for the market as an allocator of rights to use and has brought antipathy for the administrative function which had come to be dominant. With respect to *Figure 3.11*, such a change in preferences may be interpreted as a shift in the center of gravity of the whole interactive complex, now involving both the hard and soft sets. The shift is toward the lower left of the figure and away from the upper left and also away from the central soft set. Green parties, as yet miniscule in North America, would presumably favor some dominance by the soft set over the hard set.

Compromise intergrades of various kinds have been developed within the spectrum of strictly administrative and strictly market methods on the left side of *Figure 3.11*, see *Table 3.1*. Some of these could also be adapted to serve as intergrades between the hard and soft sets. To be most effective in redevelopment it is crucially important that the selection of mechanisms be chosen and designed in a manner that utilizes a constructive feedback loop to make developers *want* to manage because it is in their interests. All else may be futile.

With respect to thinking regionally and acting locally, in the context of rehabilitative husbandry for sustainable redevelopment of regional ecosystems, the complex regime of rights should be organized so that a well-functioning interdependent soft set of local allocative methods should not be overridden destructively by an interdependent hard set of allocative methods serving primarily those interests defined at a regional level.

Acknowledgements

We thank A. P. Grima, J. P. Hrabovsky, and R. E. Munn for helpful criticisms. J. Retel typed numerous versions of the manuscript.

Further reading

Cluter, J. and Fortson, J. C. (1983), *Timber Management: A Quantitative Approach* (Wiley, New York).

Loftus, K. H. and Regier, H. A. (1972), Proceedings of the symposium on salmonid communities in oligotrophic lakes, *Journal of the Fisheries Research Board of Canada*, **29**, 611–986, sponsored by the Great Lakes Fishery Commission in July 1971.

Colby, P. J. and Wigmore, R. H. (1977), Proceedings of the Percid international symposium, *Journal of the Fisheries Research Board of Canada*, **34**, 1445–1999, sponsored by the Great Lakes Fishery Commission in September 1976.

Berst, A. H., Simon, R. C., and Billingsley, L. W. (1981), Proceedings of the stock concept international symposium, *Canadian Journal of Fisheries and Aquatic Sciences*, **38**, 1457–1923, sponsored by the Great Lakes Fishery Commission in October 1980.

Annual Reports of the Great Lakes Fishery Commission (Ann Arbor, Michigan).

Annual Reports of the Great Lakes Water Quality Board and *Annual Reports of the Science Advisory Board of the International Joint Commission* (Great Lakes Regional Office, Windsor, Ontario).

Notes and references

[1] Holling, C. S. (1984), Ecological interdependence, *Options*, **1984/1**, 16 (International Institute for Applied Systems Analysis, Laxenburg, Austria).

[2] Thurow, C., Daniel, G., and Brown, T. H. (1984), *Impact of the Great Lakes in the Region's Economy* (The Center for the Great Lakes, Chicago, Ill.

[3] Tweeddale, R. E. (1974), *Report of the Forest Resources Study 1974* (Province of New Brunswick, Department of Natural Resources, Fredericton, Canada).

[4] Wynn, G. (1980), *Timber Colony: A Historical Geography of Early Nineteenth Century New Brunswick* (University of Toronto Press, Toronto, Canada).

[5] Denys, N. (1908), *The Description of Natural History of the Coast of North America (Acadia)*, translated, with memoir to author, by W. F. Ganong (Toronto Champlain Society, Toronto, Canada).

[6] Swift, J. (1983), *Cut and Run: The Assault on Canada's Forests* (Between the Lines, Canada).

[7] Reed, F. L. C. (1978), *Forest Management in Canada, Vol. 1*, Information Report FMR-S-102 (Forest Management Institute, Canadian Forestry Service, Canada Dept. of Agriculture, Ottawa, Canada).

[8] Baskerville, G. L. (1983), *Good Forest Management, A Commitment to Action* (Province of New Brunswick, Department of Natural Resources, Fredericton, Canada).

[9] Baskerville, G. L. (1976), *Report of the Task Force for Evaluation of Budworm Control Alternatives* (New Brunswick Cabinet Secretariat in Economic Development, Fredericton, Canada).

[10] Hall, T. H. (1978), *Toward a Framework for Forest Management Decision Making in New Brunswick*, TRI-78 (Province of New Brunswick, Department of Natural Resources, Fredericton, Canada).

[11] Fellows, E. S. (1980), *Final Report of the Task Force on Crown Timber Allocation to the Minister of Natural Resources* (Province of New Brunswick, Department of Natural Resources, Fredericton, Canada).

[12] Ker, J. W. (1981), *Final Report of the Task Force on Forest Management to the Minister of Natural Resources* (Province of New Brunswick, Department of Natural Resources, Fredericton, Canada).

[13] Hanusiak, R. E. (1985), Forest policy and forest management in New Brunswick, *Report of the 1984 CLA Meeting* (Canadian Lumbermen's Association, Ottawa, Canada).

[14] Erdle, T. and Jordan, G. A. (1984), Computer-based mapping in forestry: a view from New Brunswick, in *Proceedings of the 1984 CPPA Woodlands Section Annual Meeting* (Canadian Pulp and Paper Association, Woodlands Section, Montreal, Canada).

[15] Hall, T. H. (1981), Forest management decision making, art or science, *Forestry Chronicle*, **57**, 233–238.

[16] Francis, G. R., Magnuson, J. J., Talhelm, D. R., and Regier, H. A. (1979), *Rehabilitating Great Lakes Ecosystems*, Technical Report 37, Great Lakes Fishery Commission (Ann Arbor, MI).

[17] Ryder, R. A. and Edwards, C. J. (Eds) (1984), *A Proposed Approach for the Application of Biological Indicators for the Determination of Ecosystem Quality in the Great Lakes Basin*, Report to the Science Advisory Board of the International Joint Commission (Great Lakes Regional Office, Windsor, Ontario, Canada).

[30] Innis, H. A. (1938), The lumber trade in Canada, in A. R. M. Lower, W. A. Carrothers, and S. A. Saunders (Eds), *The North American Assault on the Canadian Forest*, pp vii–xvii (Ryerson Press, Toronto, Canada); reprinted in M.Q. Innos (Ed) (1956), *Essays*

oligotrophic lakes, *Journal of the Fisheries Research Board of Canada*, **29**, 959–968.

[20] Regier, H. A. (1973), The sequence of exploitation of stocks in multi-species fisheries in the Laurentian Great Lakes, *Journal of the Fisheries Research Board of Canada*, **30**, 1992–1999.

[21] Regier, H. A. (1979), Changes in species composition of Great Lakes fish communities caused by man, in *Transactions of the 44th North American Wildlife and Natural Resources Conference*, pp 550–556 (Washington Wildlife Management Institute, Washington, DC).

[22] Willoughby, W. R. (Ed) (1979), *The Joint Organizations of Canada and the United States* (University of Toronto Press, Toronto, Canada).

[23] Harris, H. J., Talhelm, D. R., Magnuson, J. J., and Forbes, A. (1982), *Green Bay in the Future – A Rehabilitative Prospectus*, Technical Report 38, Great Lakes Fishery Commission (Ann Arbor, MI).

[24] Francis, G. R., Grima, A. P., Regier, H. A., and Whillans, T. H. (1985), *A Prospectus for the Management of the Long Point Ecosystem*, Technical Report 43, Great Lakes Fishery Commission (Ann Arbor, MI).

[25] Rapport, D. J., Regier, H. A., and Hutchinson, T. C. (1985), Ecosystem behaviour under stress, *American Naturalist*, 617–640.

[26] Paloheimo, J. E. and Regier, H. A. (1982), Ecological approaches to stressed multispecies fisheries resources, in M. C. Mercer (Ed), *Multispecies Approaches to Fisheries Management Advice*, Canada Department of Fisheries and Oceans, *Special Publication – Fisheries and Aquatic Science*, **59**, 127–132.

[27] Regier, H. A. and Grima, A. P. (1984), The nature of Great Lakes ecosystems as related to transboundary pollution, *International Business Lawyer*, **June**, 261–269.

[28] Rapport, D. J., Regier, H. A., and Thorpe, C. (1981), Diagonosis, prognosis and treatment of ecosystems under stress, in G. W. Barrett and R. Rosenberg (Eds), *Stress Effects on Natural Ecosystems*, pp 269–280 (Wiley, New York).

[29] Research Advisory Board and International Joint Commission (1978), *The Ecosystem Approach*, 47 pp (Great Lakes Research Advisory Board and the International Joint Commission, Windsor, Ontaro, Canada). International Joint Commission (1982), *First Biennial Report Under the Great Lakes Water Quality Agreement of 1978* (International Joint Commision, Windsor, Ontario, Canada). Lee, B. J., Regier, H. A., and Rapport, D. J. (1982), Ten ecosystem approaches to the planning and management of the Great Lakes, *Journal for Great Lakes Research*, **8**, 505–579.

[30] Innis, H. A. (1938), The lumber trade in Canada, in A. R. M. Lower, W. A. Carrothers, and S. A. Saunders (Eds), *The North American Assault on the Canadian Forest*, pp vii–xvii (Ryerson Press, Toronto, Canada); reprinted in M. Q. Innos (Ed)

(1956), *Essays in Canadian Economic History*, pp 242–251 (University of Toronto Press, Toronto, Canada). Rea, K. J. (1976), *The Political Economy of Northern Development*, Background Study No. 36, SS21-1/36 (Science Council of Canada, Ottawa, Canada).
[31] Flader, S. L. (Ed) (1983), *The Great Lakes Forest, an Environmental and Social History* (University of Minnesota Press, Minneapolis, MN).
[32] Holdgate, M. W. and Woodman, M. J. (Eds) (1978), *The Breakdown and Restoration of Ecosystems* (Plenum Press, New York). Cairns, J., Jr. (Ed) (1980), *The Recovery Process in Damaged Ecosystems* (Ann Arbor, MI). Barrett, G. W. and Rosenberg, R. (Eds) (1981), *Stress Effects on Natural Ecosystems* (Wiley, New York).
[33] Regier, H. A. (1972), Community transformation – some lessons from large lakes, *University of Washington, Seattle, Publication in Fisheries, New Series*, **5**, 35–40.
[34] Davidson, M. (1983), *Uncommon Sense: The Life and Thought of Ludwig von Bertalanffy* (J. P. Tharcher, Houghton Mifflin, Los Angeles, CA).
[35] Holling, C. S. (Ed) (1978), *Adaptive Environmental Assessment and Management* (Wiley, New York). Holling, note [1] and Chapter 10, this volume. Regier, H. A. (in press), On the concepts and methods of Holling's science of surprise, in V. MacLaren and J. B. R. Whitney (Eds), *New Approaches to Environmental Assessment* (Methuen Press, Toronto, Canada).
[36] Regier, H. A. and Grima, A. P. (1985), Fishery resource allocation: an exploratory essay, *Canadian Journal of Fisheries and Aquatic Science*, **42**, 845–859.
[37] Dales, J. H. (1975), Beyond the market place, *Canadian Journal of Economics*, **8**, 483–503.

Commentary

L. Kairiukstis

The process of depletion of renewable resources and the degradation of natural ecosystems has grown steadily in recent decades due to intensive landscape transformation and increasing industrial–agricultural impacts on the environment. Two such examples are the Great Lakes and the forests of New Brunswick, as discussed by Regier and Baskerville. The authors, using the general approach suggested by Holling to *Think Global, Act Local* [1] have described the historical exploitation of forest resources and fisheries that led to serious degradation of both these ecological systems. In determining the reasons for such negative effects, the authors have concluded that there has been much talk about management and, indeed, a certain amount of planning, but little of this has come to grips with the fact that to make such plans happen in the lakes and in the forest, where it really counts, it is necessary to specify where the local events will take place in order to achieve a specified regional effect. The authors also critically appraise and specify the measures neccessary to redevelop those areas.

In response, I would like to support the idea of cross-national comparisons of experiences in linking global perspectives and policies on the interactions of environment and development to the multitude of local actions required to implement real change [2]. Bearing in mind the experience gained in the Lithuanian SSR, I also suggest some ideas and possible ways of maintaining the sustainability of a forest system, securing the long-term use of resources for many purposes, without the need for redevelopment of regional ecosystems. These suggestions concern the following two problems:

(1) Sustainability of forest resources and wood use.
(2) Regional development optimization.

The suggestions are made in accordance with the main concept of adaptive environmental management, developed by Holling and his colleagues [1, 3], and can be helpful in designing rehabilitative husbandry in a policy of sustainable redevelopment such as that suggested by Regier and Baskerville.

Sustainability of forest resources and wood use

The European history of forest depletion in recent centuries has shown that the sustainability of forest resources is simultaneously a precondition and a consequence of the sustainability and continuity of the *use* of forest resources. Government laws and the use of strong regulating forces and programs for afforestation have been adopted in many countries. In the Lithuanian SSR, a decrease of forest area, which had been continuous during the last eight centuries [4], was halted in the middle of this century. At that time, the wooded areas in Lithuania reached a critical level and comprised less than 20%

of the territory. Now, the forested area is steadily increasing and already (1984) comprises 27.6% of the territory. To achieve such results, strong governmental laws limiting the amount and methods of clearance, as well as supporting forest regeneration, protection, and afforestation, were adopted. Monitoring of forest resources dynamics was also improved. Modern techniques for forest inventory were established by the Lithuanian Research Institute of Forestry [5].

To ensure an increasing continuity and sustainability of wood use, an area method based on rotation and age-class distribution of the management classes (*Betriebsklasse*) was developed [6]. These factors are combined at the level of a forest enterprise after the stand inventory has been carried out and they are characterized by the dominant tree species, its rotation, and areas of age classes in ten-year intervals [7].

Regional development optimization

Owing to the increasing level of landscape cultivation of large territories, natural boundaries of forest growth are being moved and the partition between forest, agriculture, urban settlements, and areas of other land use is being changed. As described by Regier and Baskerville, this has also occurred in North America. On the other hand, in recent decades the socioeconomic value of individual territories, including forest areas and water basins, has been reappraised. Society now needs to use lands in new ways (e.g., different kinds of business, recreation, reservation, etc.). The year 1955, as pointed out by Regier and Baskerville, may be taken as the beginning of growing public cognition that the Great Lakes Region, as part of a system, was severely debased. The total binational cost of all the relevant studies and corrective efforts since 1955 is over $10 billion (1985 US dollars).

In the light of this, the optimization of regional development, based on physical planning of the landscape's main functional allocation, including specialization of forest growth, agriculture, etc., is of paramount importance. In countries where land is state owned, e.g., the USSR, these problems have been more or less solved with the help of the general planning of land use. Nevertheless, when optimizing the land use of separate regions, rather intricate problems arise in achieving an adequate improvement of living conditions of the local inhabitants, better development of the social infrastructure of the region, and development of separate industrial areas.

Owing to the difficulties of objective coordination of the interests of sectoral development versus those of regional development, a number of socioeconomic development problems remain insufficiently solved: for instance, the choice of construction sites, the determination of the maximum capacity of large plants in the chemical and power industries, and the use of water transport can negatively affect the ecological equilibrium of the environment of a region. Sometimes, this leads to a disproportionate overexploitation of natural resources, which also severely debases a region.

The difficulty in elaborating coordinated decisions, integrating the needs of separate branches and the region as a whole, is caused by at least two factors:

(1) Shortcomings in management.
(2) Lack of a reliable set of tools for simulation analyses and the selection of variants for optimum decisions.

The shortcomings in management can, to some degree, be mitigated by accumulating finances from the profit funds of enterprises, according to the degree of negative impact on the region, and directing them to resource regeneration and environmental restoration. To fill the other gap, a system of models to evaluate the quality and reproduction of natural resources that are being exploited and a system of balanced interbranch models of optimization on a regional level are being constructed by the Lithuanian Committee of the UNESCO Program *Man and Biosphere.* Taking the socioeconomic and environmental development of a region as a complex economic–ecological system, we are dealing with two kinds of optimization: regional optimization and sectoral optimization.

For instance, a preliminary analysis of the territorial transformation trends in Lithuania conditioned by production output and the ecological and social consequences of this production enabled an optimal distribution of the territory according to its function to be established approximately [8]. Accordingly, agricultural territories must occupy

55–58% (including 17% of meadows and pastures), forest areas 30–33%, water and boggy territories 5–6%, and industrial–urban areas 5–7%.

The forest and agricultural territories, apart from their own forest and agricultural areas, include areas designated for other functions (recreation, conservation, etc.). Owing to the contiguity limits and mutual scale of overlap of these territories, buffer zones are established with respect to the chief adjacent economic sectors. The forestry sector, for example, under the pressure of agricultural, recreational, industrial–urban, etc., sectors in a specific territory, becomes greatly specialized. Under the pressure of contiguous economic activities other sectors are planned rationally: agriculture, water, etc., which means that the adverse impact of competing functions (production, recreation, conservation) can be avoided and the territory distribution more precisely determined.

The forestry sector of a national economy, guided by a given purposeful function of territorial units and forestry science and by economic sector possibilities, specializes in forest growth [9]. At the same time, the forest sector is primarily responsible for total wood supply and other forest products to the national economy. Thus, by optimizing forest growth, a forester foresees measures for productivity increments of forest areas. He is directed toward this by the inherent laws of economic development. Moreover, this is in conformity with the main energy accumulative function of the forest in the biosphere. According to the pattern of the territorial ecosystem "Lithuania", forest growth is optimized as follows:

(1) Industrial–exploitative forests, including plantation forests – 52.7%.
(2) Agricultural–protective forests – 18.5%.
(3) Recreational forests – 11.2%.
(4) Conservation forests – 10.6%.
(5) Sport (recreational) hunting forests – 7.0%.

Optimal standards of forest growth are established for each specialized sector [10].

Notes and references (Comm.)

[1] Holling, C. S. (1984), Ecological interdependence, *Options*, **1984/1**, 16 (International Institute for Applied Systems Analysis, Laxenburg, Austria).

[2] Clark, W. C. and Holling, C. S. (1984), *Sustainable Development of the Biosphere: Human Activities and Global Change.* Paper presented at the First ICSU Multidisciplinary Symposium on *Global Change*, Ottawa, Canada (25–27 September 1984) (ICSU Press).

[3] Holling, C. S. (1984) (Ed), *Adaptive Environmental Assessment and Management* (Wiley, New York).

[4] Historical data was collected by Professor P. Matulionis at the beginning of this century and processed by Professor M. Jankauskas at the Lithuanian Research Institute of Forestry, Kaunas.

[5] Kenstavicius, J. and Brukas, A. (1984), *The Results of Forest Inventorization Based on Soil Sites in Lithuanian SSR and Recommendations for Improvement* (Vilnius, Kaunas, USSR; in Lithuanian).

[6] Deltuvas, R. (1982), *Timber Production Planning in Lithuanian SSR*, Research Notes N16 (Helsinki Department of Forest Mensuration and Management, Helsinki).

[7] Mizaras, S. B. (1979), *Recommendations for Forest Cadastres* (Lithuanian Research Institute of Forestry, Kaunas; in Russian); Rutkauskas, A-V. M. (1980), A subsystem of production models of systems analysis and prediction of reproductive processes of gross national product of the Republic, in *Regional Modeling of Reproduction Processes* (Vilnius, Kaunas, USSR; in Russian).

[8] Kairiukstis, L. A. (1982), *Outline of Environmental Optimization in Lithuanian SSR by Modelling Regional Development* (Vilnius, Kaunas, USSR; in Russian).

[9] Kairiukstis, L. A. (1981), *Optimization of Forest Growth* (Academy of Sciences of the Lithuanian SSR, State Committee of Forestry of the USSR, Kaunas, USSR).

[10] Kairiukstis, L. A., Judovalkis, A., Jonikas, J., and Barkauskas, A. (1980), *The Formation of Maximally Productive Standard Stands* (Lithuanian Research Institute of Forestry, Vilnius, Kaunas, USSR; in Lithuanian).

Chapter 4 Agricultural development – looking to the future

P. Crosson

Editors' Introduction: As shown by Richards in Chapter 2 of this volume, agricultural activities have been the prime transformers of the Earth's environment throughout most of human history. In this chapter the future of agricultural development is portrayed as an interacting system of resources, technologies, institutions, and environments. It is suggested that for the next 20 years or so, the environmental constraints to agricultural development will be strongly shaped by the present resource endowments of different world regions, especially their potential for productivity improvements and for expansion of arable land. In the long term, however, the determining factor is likely to be the ability of institutions and technologies to respond flexibly to changing conditions. The use of "induced innovation" theory to illuminate such sociotechnical responses to environmental constraints is explored in detail.

Factors external to agriculture

The direction, pace, and character of world agricultural development will be strongly conditioned by several factors that bear generally on all economic sectors, not just agriculture. Some discussion of these external factors provides a necessary perspective for the subsequent analysis of the internal dynamics of agricultural development.

Among the most fundamental of these are the growth and regional distribution of world income and population, and the world political climate, especially as it affects international trade. These factors are interrelated. If the political climate moves the nations of the world toward increasing protectionism, then world income, if not population, grows more slowly than if movement is toward a more open trading system. If, for reasons apart from trade, income growth slows significantly, pressures for more trade protection will likely rise. Should the political climate induce increasing friction and military confrontation among nations, the resulting diversion of resources into armaments would likely slow income growth as well as promote protectionism. The impact of nuclear war on world population, income, and trade needs no elaboration.

In this chapter I assume there will be no nuclear war and that the world political climate will not induce either significantly more or less protectionism than at present. My guess is that if there is a change over the next 100 to 200 years it will be toward more open trade, for which I can provide no convincing argument. I base it on the belief that over the long term a more open system promotes everyone's interests and on the perhaps too sanguine assumption that people and governments will, in time, recognize this and act accordingly.

Energy and other resources

If nuclear war and increased protectionism are avoided, then world population and income can continue to grow unless impassable resource barriers are encountered or the processes of growth impose disastrous environmental stress. The critical resources are energy and human ingenuity. With enough of these, other resource constraints are likely to be overcome and environmental stress contained. Darmstadter (see Chapter 5, this volume) indicates no insuperable limits to energy supply, although he expects real energy prices to increase more or less steadily well into the twenty-first century. Experience since the early 1970s provides strong evidence that societies can adjust to much sharper energy price increases than Darmstadter foresees, without seriously jeopardizing income growth.

Goeller and Zucker [1], on the assumptions of "reasonably stable political conditions, a continuing supply of energy, continuing availability of capital, and, most importantly, vigorous and successful research in the field of materials [1, p 456]," argue that the supply of nonenergy material resources will be adequate to support an ultimate world population of 8.5×10^9 people, reached in 2100, at a standard free of squalor and hunger, and with acceptable environmental consequences.

Goeller and Zucker's stress on capital and research highlights the importance of human ingenuity as a resource. I think it fair to say that this resource is the key to attaining sustainable development of the biosphere. To raise the question of sustainability is to ask whether human beings can devise, in timely fashion, the technical and institutional instruments needed to remove resource and environmental limits to mankind's aspirations for a better material standard. By assuming continuous growth of world population and income, I implicitly assume that the limits will be continuously extended. More particularly, I implicitly assume that barriers to world agricultural development will be pushed back. So important is agriculture, especially in the developing countries, that a sustained world economic advance requires successful agricultural development. There is, therefore, a certain circularity in the argument presented here, but it is more apparent than real. It disappears if the question is presented thus: What resource and environmental conditions must be met if mankind is to achieve aspirations for higher and indefinitely sustainable levels of food and fiber production? In this chapter we explore those conditions.

Precise projections of population and income growth are not essential to the exploration. Higher projections might, but would not necessarily, imply more pressure on resources and the environment than lower projections. In particular, high income does not necessarily generate greater environmental damage, since some of the most severe environmental stress occurs in the poorest countries. But within any consensus range of projections the resource and environmental conditions necessary for successful agricultural development will not vary greatly.

Population growth

Table 4.1 and *Figure 4.1* give population estimates for the current developed and developing countries from 1850, with projections to 2075. Note the steadily increasing share of the developing countries since 1940 and into the future. The World Bank report [2] gives no figures for developing country populations beyond 2050, but contains graphics [2, p 75] showing that for those countries given projections between 1980 and 2100 most growth will occur by 2050. Thus the Keyfitz *et al.* [3] and World Bank projections exhibit the same pattern of growth for the developing countries. The World Bank graphics indicate very little population growth in the developing countries beyond 2100, with estimates for that year ranging from about 7×10^9 to about 10×10^9. The lower estimate assumes a rapid fall in fertility and a "standard" decline in mortality, whereas the higher estimate assumes a "standard" decline in fertility and a rapid fall in mortality.

The World Bank's projections for the developing countries bracket those of Keyfitz *et al.*; it does not provide projections for developed countries, but it is safe to assume they would differ from those of Keyfitz *et al.* by much less than the differences for developing countries. Both sets of projections coincide in indicating that world population growth will virtually cease by 2100, with relatively little growth after 2050, and with a final total population of between 8.5×10^9 and 11.5×10^9, 82–87% of whom will live in the current developing countries.

Table 4.1 World population (millions).

Year		Developed countries[a]	Developing countries[b]	Percent in developing countries
1850	Barrie [4]	302	869	74
1900		510	1098	68
1920		605	1255	67
1930		677	1391	67
1940		729	1565	68
1950		751	1764	70
1960		854	2144	72
1975	Keyfitz *et al.* [3]	1092	2844	73
2000		1274	4619	78
2025		1380	5984	81
2050		1415	6782	83
2075		1434	7012	83
1982	World Bank [2]	1106	3413	76
2000			4835	
2050			8313	

[a] Developed countries are the USA, Canada, western Europe, USSR, eastern Europe, Japan, Australia, and New Zealand.
[b] Developing countries are all those not listed in *a*.

Income growth

At the present levels of per capital income in developed countries additional income stimulates little additional demand for basic food commodities. In technical terms, the income elasticity of demand for basic foods, such as grains, oilseeds, and root crops, in these countries is very low. Any increase that does occur is primarily in animal products, which indirectly spurs demand for feed grains and oil meals. As income in the developed countries grows further even this source of demand will diminish as diets reach saturation with animal products. Consequently, whatever the growth of per capita income of developed countries it will add little to world demand for basic food commodities. I venture the judgment that beyond the middle of the twenty-first century it will add virtually nothing.

The matter is quite different in the developing countries. Because per capita income in these countries is low, additions to income, if widely shared among the poor, stimulate relatively large increases in demand for basic foods; that is, income elasticities of demand for food are high. Thus, in the poorest countries higher income directly stimulates demand for food grains and other basic foods. In middle-income developing countries demand is shifting toward more animal products, thus also indirectly increasing demand for feed grains and oil meals.

Since income elasticities of demand for food are high in developing countries, rates of growth in food demand in these countries are sensitive to rates of per capita income growth. Between 1970 and 1982 the real per capita gross national product (GNP) in the developing countries grew at an average annual rate of 3.2% [2]. It was 3.0% in the poorest countries (average per capita GNP 1982 of $280), 3.6% in middle-income countries (average per capita GNP in 1982 of $1521), and 5.6% in four high-income oil-exporting countries with an average per capita GNP of $14 820 (Libya, Saudi Arabia, Kuwait, and the United Arab Emirates) [2]. Despite the run up of oil prices and associated balance of payments problems in the 1970s, the developing countries, apart from the oil exporters, achieved only a marginally less per capita GNP growth in that decade than in the 1960s.

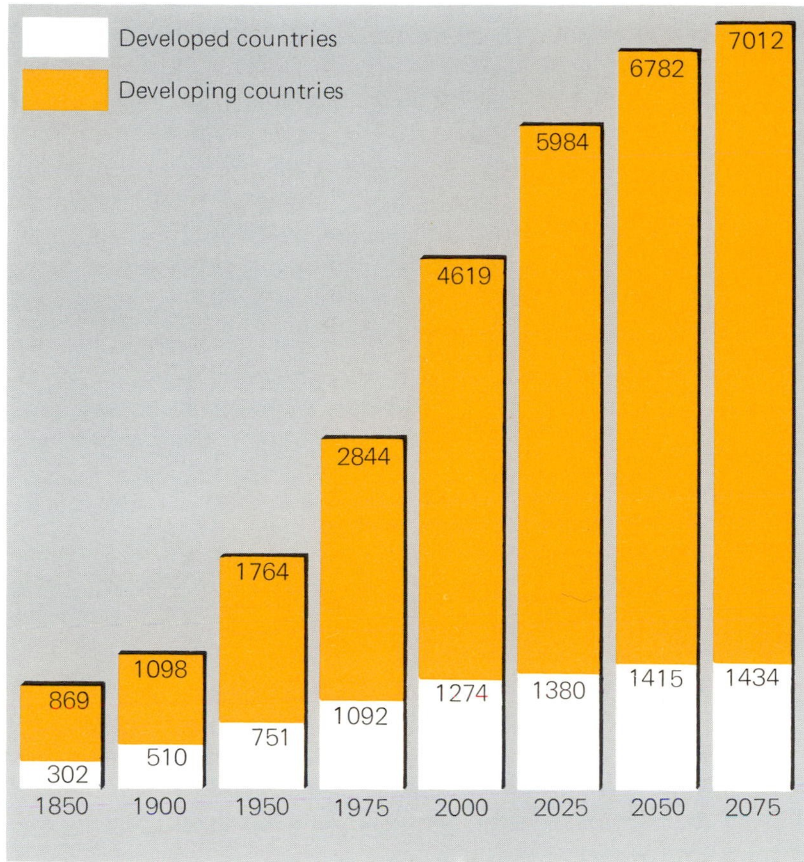

Figure 4.1 Cumulative population in developed and developing countries ($\times 10^6$) [3, 4].

I assume a per capita GNP growth in the developing countries of not less than 2% annually for the indefinite future, and that the poorest countries will share fully in this (*Figure 4.2*). At this rate, per capita GNP in the poorest countries will be $1765 in 2075 (1982 prices) and $2900 in 2100. The latter is about the level of South Africa and Yugoslavia in 1982 and 15% above that for Argentina in the same year. Middle-income countries would have per capita GNPs of $9600 in 2075 and $15 700 in 2100, a little less than that of Switzerland in 1982 [5].

At first glance, since per capita GNP in even the poorest countries increased at an annual rate of 3.0% from 1960 to 1982, the projection of a sustained 2% growth does not seem unreasonable. The 3.0% figure, however, conceals widely diverging growth among the poorest countries. According to the World Bank report [2] China's per capita GNP grew by 5% annually from 1960 to 1982 [6], whereas the average annual growth for all other countries in the poorest group was 1.4%, with 1.3% in India. In eight of the poorest countries, seven of them in Africa, per capita income declined over the period.

These data indicate that, apart from China, the poorest countries (with a population of 1258×10^6 in 1982) would have to do significantly better in the future than they did in the past to achieve a sustained 2% per capita GNP growth. What reasons are there to believe that they can do it?

One is that other countries have done it, starting from similarly low levels of per capita GNP, as *Table 4.2* indicates. Another reason admittedly involves a leap of faith. If the poorest countries fail to achieve a sustained per capita GNP growth of at least 2% annually they will almost surely fall short of meeting the aspirations of their people for a better life. The figure 2% carries no special magic and the world will not end if per capita GNP in the poorest countries grows by less than that. But over 118 years (1982 to 2100) the difference between 2% and the 1.4% recently experienced (excluding China) is substantial. If the latter rate prevails, per capita GNP in the poorest countries (excluding China) would

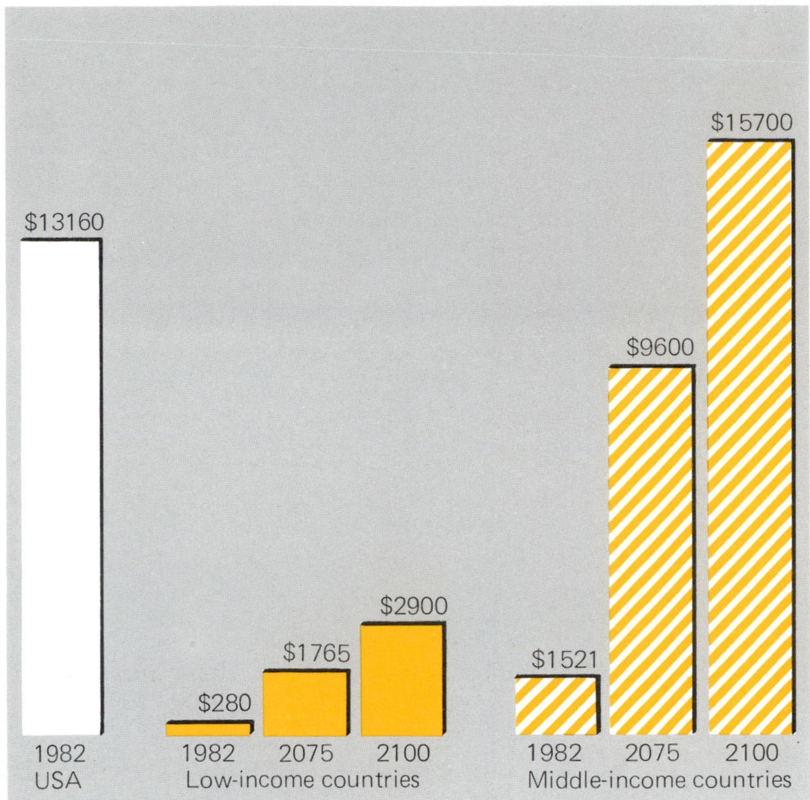

Figure 4.2 Per capita income in the USA and developing countries (1982 dollars); the projections assume a 2% average annual growth [2].

be $1320 (1982 prices) in 2100 instead of $2900 that 2% will give. There is no guarantee that the higher GNP would satisfy the aspirations of the people of these countries, but it surely would be more likely to than if GNP is less than half the amount. I believe the consequences for social stability of a difference that large will have a profound impact on the effort that political leaders of the poorest countries will devote to economic development. They may fall short, hence the leap of faith required to believe that they will not. However, the consequences of failure will be severe, and this, perhaps, is the strongest reason for believing the leap is justified.

Climatic change

An apparent consensus has emerged that increasing atmospheric concentrations of carbon dioxide (CO_2), and perhaps other gases, will result in significant warming of the Earth's atmosphere, with the largest increases in the polar regions [7, 8, 9, 10]. Most estimates agree that atmospheric CO_2 will be double the present levels sometime between 2050 and 2075, but there is less agreement about the resultant increase in temperature. Hare [7] cites a source which indicates a rise in mean global temperature of between 1.5 and 6.0 °C, with the most likely increase being 3–4 °C. Cooper [11], drawing on work of others reported in [8], assumes an increase of 2–3 °C, and the National Research Council [9] asserts that the more sophisticated models of climatic change indicate an increase of 1.5–4.5 °C. Dickinson (Chapter 9, this volume) suggests that the warming effects of increased CO_2 concentrations and of certain other trace gases will more than offset a slight cooling effect from increased surface albedo and aerosol concentrations, resulting by 2100 in global temperatures 6 °C higher than those in preindustrial times (Dickinson, p 270). However, emphasizing the high uncertainty attached to such estimates, Dickinson places the 6 °C increase in the range +1

Table 4.2. Sustained increases in GNP of the poorer countries [2].[a,b]

Country	*Per capita GNP in 1960 (1982 $)*	*Average annual growth 1960–1982 (%)*
China	106	5.0
Sri Lanka	182	2.6
Pakistan	207	2.8
Thailand	300	4.5
Egypt	317	3.6
Philippines	447	2.8
Republic of South Korea	468	6.6

[a] The list is not exhaustive of countries in the poorest group that achieved an annual per capita GNP growth of at least 2% from 1960 to 1982.

[b] National income accounting systems, from which the above and all other per capita GNP estimates are derived, do not include the social and environmental costs of income growth. If they did, estimates of growth might be lower. A main objective of this chapter is to emphasize the importance of environmental costs in considering the indefinite sustainability of measured income growth. Social costs are addressed only in connection with equity aspects of income distribution.

to +10 °C. In his judgment, the chances are 95 out of 100 that the actual increase by 2100 will fall within these limits.

All analysts who have studied the matter agree that increasing atmospheric concentrations of CO_2 and the associated warming have important implications for world agriculture, but they agree less about the nature of the implications. Wittwer [12, 13] and Waggoner [14] argue that larger atmospheric concentrations of CO_2 would increase the rate of photosynthesis, particularly in wheat and other C_3 plants, thus stimulating production. Cooper [11] and Rosenberg [10, 15] are skeptical, arguing that the supply of nutrients other than CO_2 may generally be the factor that limits plant growth [16].

Cooper [11] states that various climate models predict that rising CO_2 concentrations will increase rainfall as well as temperatures, but that the models are weak in predicting which regions will be most affected (except that polar temperatures will increase more than the average and equatorial temperatures less). Dickinson (Chapter 9, this volume) also emphasizes the difficulty of estimating the regional impacts of global warming. Waggoner [14] assumes that CO_2-induced warming would be accompanied by less, not more, precipitation, at least in western USA. And a model developed by Manabe and Wetherald, discussed by Cooper [11], shows that doubling the CO_2 concentration would increase the difference between precipitation and evaporation between latitudes 15° and 35°, leave the difference about the same between latitudes 35° and 40°, decrease it between latitudes 40° and 52°, and increase it north of 52°.

Cooper [11] argues that present knowledge of the effects of doubling atmospheric CO_2 is too limited to determine whether world agricultural production would be increased or decreased. Rosenberg [15] agrees, as does Schelling [17] in an appraisal of the main findings reported in [9]. All three authors, and others as well, emphasize, however, that while agriculture in some regions will likely benefit, it may suffer in others. Some regions now marginal for crop production because of climate may become even more so. More warming in northern latitudes should extend crop growing seasons, but this would probably benefit Canada less than might be expected, and less than the USSR, because in the more northern latitudes of the country the thin Canadian soils are more limiting than the growing season. However, forest productivity in these latitudes should increase because of the warmer temperatures [11].

Parry (Chapter 14, this volume) argues that changes in mean temperatures and other climate characteristics may be less significant for agriculture (and perhaps for other activities) than changes in the magnitude and frequency of extreme events. An event is *extreme* only by comparison with some standard, such as the mean of the distribution of events. Parry suggests that in organizing their operations, farmers give little weight to extreme events, e.g., those with a probability of occurrence of 0.05 or less. If, after a lengthy period of stability, mean values (e.g., of temperature) rise, the frequency of extremely hot periods is likely to increase, judged by the mean temperature to which farmers have become adjusted. In time they will find a new pattern of adjustment to the changed (warmer) climatic regime; but the interim may include an unusual number of difficult growing seasons, i.e., those in which, by historical experience, production is unusually suppressed or stimulated by weather events.

None of the writers cited foresees an overall disasterous agricultural consequence of CO_2-induced warming. They note the time available to make adjustments and the demonstrated ability of mankind to develop agricultural systems that are productive under climatic conditions even more variable than those portended by a doubling of atmospheric CO_2. Schelling [17] emphasizes that through history people have been willing and able to move from less to more favorable climates. Rosenberg [15] points out that through adaptive plant breeding and farm-management improvements the growing area for hard, red winter wheat in the USA has spread since 1920, both north and south, across climatic changes wider than those forecast for CO_2-induced warming. All writers stress the importance of increased research, both to improve predictions of the likelihood and consequences of warming and to develop plant varieties adapted to changing climatic conditions. Wittwer [13] makes the point that because of high uncertainty as to the agricultural consequences of warming, societies should concentrate on building the capacity of researchers to respond quickly to unexpected change, avoiding commitments to lines of research which close off this option.

Clearly, CO_2-induced climatic change on the scale expected now may have important consequences for world agricultural development. However, on present evidence it appears that the major consequences will be shifts in the regional distribution of production rather than an increase or decrease in total production. Given present uncertainties about the regional effects, there is no point in further speculation along this line. It is clear, however, that the prospective climatic change compounds the many uncertainties that are already an integral part of future world agricultural development.

Increasing urbanization

One of the clearest lessons of human history is that economic development involves a relative shift of people and resources out of agriculture into a widening array of nonagricultural activities. Because these activities typically require much less land per unit of output than agriculture, development inevitably results in increasing concentrations of people, that is to say, in urbanization.

Since urbanization is a response to the increasing importance of nonagricultural output, it is not surprising that the more strongly industrialized economies are more heavily urbanized than those less industrialized. Thus in 1975 about two thirds of the people in developed countries (those of North America, Europe, USSR, Japan, and Oceania) lived in urban areas, compared to 28% of the people in the developing countries of Latin America, Africa, and Asia [18].

While data do not enable an accurate dating of the beginnings of urbanization, it is clear that the process has been under way for centuries. The industrialization of Europe in the nineteenth century resulted in rapid urbanization of that region, the population of major cities in Belgium, Denmark, France, Germany, Italy, the Netherlands, Spain, and the UK rising from 5.8% of the population in 1850 to 10.5% in 1900. The urbanization process continued well into the twentieth century, but at a slower rate: by 1960, the same set of cities had 13.5% of the total population of the named countries [19, 20].

Data for Latin America and the USA demonstrate the relationship between industrialization and the process of urbanization. In 1850 8.8% of the USA population lived in towns of 20 000 or more, while in Latin America the comparable percentage was only a little less, 6.3. By 1940, 47% of the USA population lived in such towns, compared to 19.5% in Latin America. Since 1940 industrialization and related activities have increased relatively faster in Latin America than in the USA, and Latin American urbanization has also occurred at a faster rate (*Table 4.3*). The developing countries of Asia and Africa have also experienced a relatively rapid urbanization since 1950 (*Table 4.4*). The projections of continued growth in population and income in developing countries imply increasing urbanization as well, although probably at a slower percentage rate than in the last several decades. Some additional urbanization is also likely in the developed countries. However, since in all of these countries most people already live in urban areas and prospective population growth is small, urbanization will proceed much more slowly than in the developing countries.

Table 4.3 Percentage of the USA and Latin American populations living in towns of 20 000 or more, various years [21].

	1850	*1900*	*1920*	*1940*	*1960*	*1970*
Latin America	6.3	10.7	14.7	19.5	29.1	35.2
USA	8.8	25.9	42.0	47.0	58.5	64.0

Table 4.4 Percentage of the population in urban areas, developed and developing countries, various years [18].

	1950	*1960*	*1970*	*1975*
World	29.0	33.9	37.5	39.3
Developed countries	52.5	58.7	64.7	67.5
Developing countries	16.7	21.9	25.8	28.0
Latin America	41.2	49.5	57.4	61.2
Africa	14.5	18.2	22.9	25.7
South Asia	15.7	17.8	20.5	22.0
China	11.0	18.6	21.6	23.3
Other East Asian countries (except Japan)	28.6	36.3	47.5	53.4

Urbanization and associated increases in urban income are essential to successful agricultural development [22]. Growing urban markets for food and fiber, and for labor, provide the incentives farmers need to produce more, and they make feasible the adoption of technologies which raise the productivity of farm labor. As this suggests, urban income growth is likely to provide more stimulus to agricultural development if it is widely shared among all urban income classes than if it is narrowly concentrated. Urbanization also expands the supply of goods and services – fertilizer, machinery, credit, etc. – essential to agricultural development. Experience suggests that this process is more effective if urban growth is regionally dispersed rather than concentrated in a few very large metropolitan centers.

The point here is that agricultural development is integrally connected to nonagricultural development, and awareness of the connection is essential to clear thinking about the pace and direction of agricultural change. In particular, government policies that emphasize the broad sharing of urban income gains among all income classes and promote regionally dispersed urban growth are more likely to encourage strong agricultural development than policies of opposite tendencies.

Rising economic value of labor

Sustained increases in per capita income and urbanization imply a rising economic value of labor, which does not necessarily mean a less unequal distribution of income. Indeed, the relationship is consistent with widening relative differences between income classes. However, it is unlikely that the per capita income in any country could rise for long periods at a rate as low as even the 2% I have assumed for the developing countries without most of the working population sharing in the increase. Real wages rose substantially in all of the presently developed countries as they made the transition from agricultural–rural to industrial–urban economies. A study by the Brookings Institution [23] indicates that in the developing countries the poor shared, although often not equally, in the income gains achieved since the 1950s. Indeed, the report asserts that "...for many of the poor, income growth has been rapid. General economic growth and deliberate policy efforts have interacted to produce some development for the poor [23, p 33]."

The rising economic value of labor is significant for agriculture because it conditions the technological choices that farmers make. Over time they will shift toward more labor-saving technologies, the rate of shift depending to a large measure on the scarcity or abundance of farm labor relative to other resources and on the rate of increase in urban employment and income [24]. Where the farm labor force is large relative to land and capital, as in India, China, and other Asian countries, the income of farm workers will likely grow only slowly for several decades, even with continued agricultural development, retarding the shift toward more labor-saving technologies. Where farm labor is less abundant the shift should be more rapid, but in a perspective on agricultural development extending over many decades the rising economic value of labor should be taken as a basic factor moving farmers more or less steadily toward more labor-saving technologies.

Equity as a social objective

Agricultural development is a social process, profoundly affecting and affected by other social processes. The notion of equity in income distribution and, more broadly, in access to emerging opportunities opened up by the process of development seems increasingly to focus the attention of large segments of societies everywhere. It finds increasing expression even in the utterances of public officials. To be sure, the notion is usually vaguely defined or defined in different ways, and the utterances are as often rhetoric as expressions of serious purpose. Yet I think if we look back over a period of many decades, or a century, we observe that many societies have, in fact, moved toward a more equitable treatment of their more disadvantaged members, and that in most societies that movement continues today, halting though it may be. Indeed, in some the movement is indiscernable and in many, if not most, it often appears a matter of two steps forward and one step back. Yet for a long perspective on the future I expect the notion of equity to acquire increasing force, if for no other reason than that in the long run the welfare of the few cannot be disentangled from that of the many.

If I am correct in this, the notion of equity will increasingly affect patterns of agricultural development, especially patterns of technological change in agriculture. Technologies perceived to have inequitable consequences for income distribution or to impose inequitably distributed environmental costs on present or future generations will meet more resistance than technologies not so perceived.

Such resistance is observable already. The Green Revolution technology used in developing countries has come under heavy criticism from many who perceive it to favor the relatively few more wealthy farmers and to result in rising rural unemployment [e.g., 25, 26]. Much evidence disputes this view [e.g., 27, 28], but its accuracy is not the point here. One presumes that in the long run truth will prevail, but meanwhile perceptions of the truth count.

Other examples abound of the role of perceptions of equity in technological change. Concern about the unintended consequences of pesticide use stems from equity considerations, and erosive farm practices are disapproved because they threaten to impose inequitable cost increases on future generations. A striking, recent example of the power of equity concerns is the legal action taken by a farm union and others against the University of California for its role in developing a mechanical tomato harvester [29]. The charge is illegal use of public funds to benefit private interests, but the driving force behind the suit is concern about equity: the harvester displaced thousands of workers formerly employed in picking tomatoes.

The strengthening commitment to equity in economic development is not peculiar to agriculture – it cuts across the whole development process. But that it will condition patterns of agricultural development all around the world seems increasingly likely.

Induced technical and institutional change

Agricultural development can be viewed as a cumulative process of change in technology and institutions, understanding of which requires some concept of what drives the technological and institutional change. The induced innovation hypothesis of Hayami and Ruttan [30], building on work of Schultz [31], is such a concept. The hypothesis states that emerging resource scarcities are signaled by rising prices, and these provide incentives for agriculturalists to find substitutes for the increasingly expensive resources. In the short term, substitution occurs within existing technology by simply using more of the less-expensive resources relative to the more expensive for a given amount of output. Unit production costs increase but less than they would without the substitution. Eventually – the time period is variable – new technologies are developed that involve the use of new resources and permit replacement, on an even larger scale, of those resources that have become more expensive. The total quantity of resources required to produce a given output declines and typically unit production costs fall.

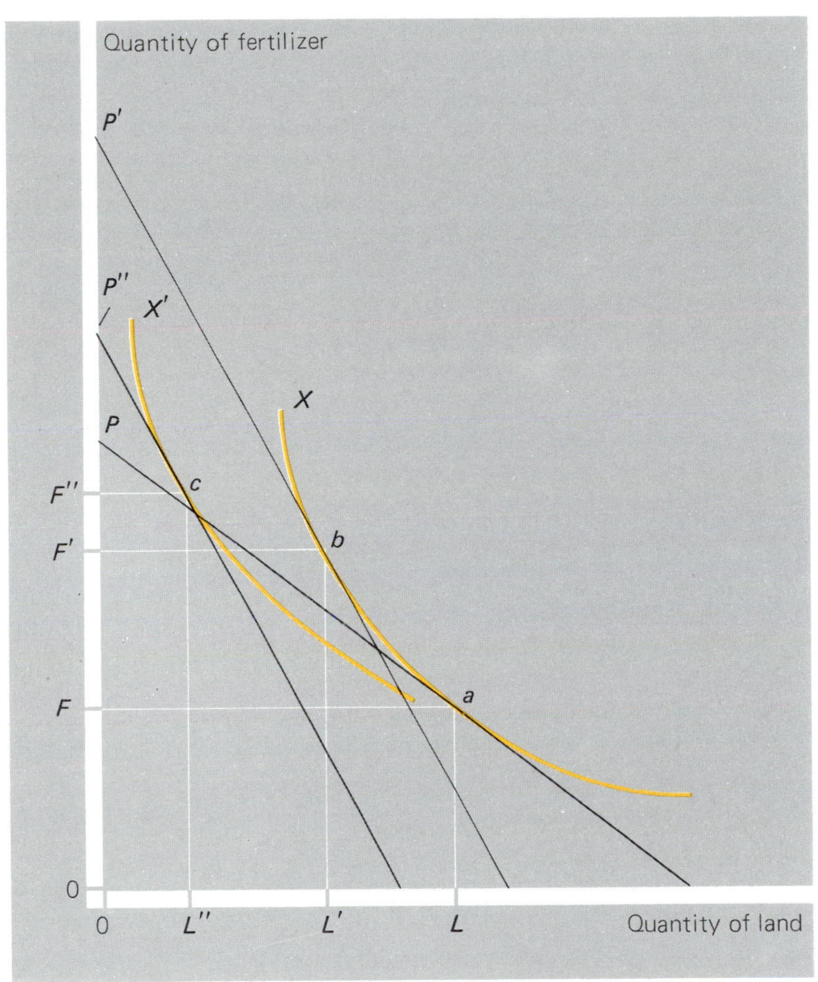

Figure 4.3 Substitution of less expensive for more expensive resources in agricultural development.

Figure 4.3 is a simplified illustration of these processes. The figure deals with only two resources, fertilizer and land, but inclusion of labor and other resources would not change the principles involved. The price of fertilizer relative to the price of land in the initial period is given by the slope of the price line *P*. Curve *X* shows the various combinations of fertilizer and land that will produce the maximum amount of some product, say rice, with existing technology. It is called a production possibility curve. Given the relative prices of fertilizer and land, the most economical combination of the two resources is given by the point of tangency (indicated by *a*) of price line *P* with production possibility curve X. So efficient farmers will combine OL ha of land with OF kg of fertilizer to produce the given amount of rice.

With rising population and income the demand for rice increases, putting pressure on the supply of land. Its price rises relative to that of fertilizer, the new price relationship being described by price line P'. Farmers respond by substituting some of the relatively cheaper fertilizer for some of the relatively more expensive land. To produce the given amount of rice the most efficient resource combination now is OL' ha of land and OF' kg of fertilizer (point *b*).

The substitution of fertilizer for land makes farmers and consumers of rice better-off than if the substitution had not occurred. However, consumers are less well-off than they were before the price of land began to rise because the price of rice now is higher [32]. Moreover, continuing increases in demand for rice portend even higher costs and prices unless new technology is developed to reduce pressure on the land. This is the situation which, according to the induced innovation hypothesis, stimulates demand for technological change. Farmers wishing to acquire land may convey their distress through extension agents or their political representatives, and the latter may also fear rising consumer unrest because of higher food prices. Scientists and those responsible for management of agricultural research respond to these pressures by allocating more resources to development of land-saving technologies, but they may respond also to their own direct perception of the increasing scarcity of land. If the response of the research establishment is judged inadequate by those agitating for relief from increasing land scarcity, pressure rises to force the institutional changes needed to obtain the wanted response. If land tenure or rural credit systems impede adoption of the new technology after it has been developed, further pressure is built-up to remove these obstacles. Thus the induced innovation hypothesis seeks to explain institutional change as well as technological advance.

Referring again to *Figure 4.3*, X' is the production possibility curve representing a new technology developed in response to rising land scarcity. The shift from X to X' illustrates the shift from traditional rice varieties to the short stem, high yield, fertilizer responsive varieties developed at the International Rice Research Institute (IRRI) in the Philippines. In the figure it is assumed that relative prices of fertilizer and land have not changed since the initial run up in land prices (price lines P'' and P' have the same slope). With the higher yield rice varieties the efficient combination of resources is OL'' ha of land and OF'' kg of fertilizer. Production of the given amount of rice now requires far less land than originally or than after the initial substitution of fertilizer for land. Total fertilizer consumption has increased substantially.

Since its initial elaboration by Hayami and Ruttan [30] the induced innovation hypothesis has been much discussed in agricultural development literature [33]. Hayami and Ruttan will soon publish a revision of their book which will incorporate both extensions and modifications of the theoretical argument, as well as substantially more empirical material than in the original book.

In my judgment, the hypothesis provides keen insights into the processes of technological and institutional change in agricultural development, but it is more successful in explaining the demand for change than in explaining its supply. The hypothesis was first used to examine agricultural development in the USA and Japan, where increasing scarcity of labor in the first instance and of land in the second seemed to generate appropriate technological and institutional responses. But there are other areas, e.g., Sub-Saharan Africa, where the increasing scarcity of agricultural resources is notorious and has excited much concern in the international community, as well as in the African countries themselves. However, appropriate technological and institutional innovations have been slow in forthcoming or not forthcoming at all. Similarly, the increasing scarcity of firewood for home cooking and heating is well documented in the developing countries, but so far development of substitute fuels lags badly.

The induced innovation hypothesis depends on prices to signal emerging resource scarcity. Consequently, it has difficulty in explaining technological and institutional responses to increasing scarcity of unpriced resources, such as land, water, and air where these are used as cost-free receptors of agricultural, industrial, and municipal wastes. For the same reason the hypothesis, at least in its present form, is of limited use in accounting for the role of equity concerns in shaping technological and institutional change in agriculture. The social costs that arouse these concerns are real, but they are not priced. Finally, the hypothesis seems to have little relevance to technological change based on scientific inquiry spurred by nothing more than the questing human mind. Few would deny that attacks on scientific ignorance simply "because its there" have played a role in the history of technological advance. The induced innovation hypothesis assumes utilitarian underpinnings of the demand for knowledge, not a disinterested or even playful search for the fun of it.

Despite these limitations the induced innovation hypothesis provides a useful approach to understanding technological and institutional change in agricultural development. I adopt a broad formulation of the hypothesis, asserting that changes in agricultural technology and institutions are induced by perceptions of emerging scarcities of land, water, labor, and other agricultural resources, of mounting environmental stress, and by concerns about the equity impacts of development.

Time horizon

The concern here is with interactions between agriculture and the biosphere over the long term. But how long is that? The answer I adopt is: as long as agricultural pressure on the resource base and environment is increasing significantly. I believe most of the increase that will ever occur will occur in the next 75 to 100 years, for two reasons. One is that world population growth will have virtually ceased by then, removing one of the main driving forces behind rising demand for food. The other is that by then per capita income in all but the very poorest countries will have reached a level at which additional income will generate very little additional demand for basic foods. Recall the earlier projections of $1765 per capita income (1982 prices) in the poorest developing countries by 2075 and of $2900 by 2100. At present, the income elasticity of demand for basic foods at these levels of per capita income is 0.1 to 0.2 for food grains and 0.3 to 0.4 for feed grains, by way of demand for animal products. If these elasticities also hold 100 years from now, then a 2% annual growth in per capita income in the poorest countries in the last quarter of the twenty-first century would generate 0.2 to 0.4% and 0.6 to 0.8% annual growths in demand for food and feed grains, respectively. Increased population would add to the demand, but not by much. In the present developed countries (17% of the world population in 2075, according to Keyfitz *et al.* [3]), neither population nor income growth will add anything to the demand for these foods. In the present middle-income developing countries (26% of the world population in 1982, but probably less in 2075) population growth will probably add even less to the additional food demand than it does in the poorest countries, and per capita income will be so high ($9600 in 2075 – see the discussion of income projections above) that income elasticities of demand for basic foods will be close to zero.

By this line of argument virtually all the increase in world demand for basic foods in the last quarter of the twenty-first century will be in the present poorest countries. Growth of annual world demand for feed grains might be 0.6% (0.8% to account for income in the poorest countries, plus 0.2% for population growth there times, say, 0.6 for their fraction of world population). Annual world demand for food grains would be about half that for feed grains.

In the decade up to 1982, annual increases in world demand for rice, wheat, and feed grains were 2.5, 2.3, and 1.9%, respectively. Projections of population and income growth in the developing countries indicate that these rates of demand growth will probably not decline much before about 2050. The drastically lower rates likely by the last quarter of the twenty-first century thus point to the earlier stated conclusion that almost all the increase in demand for basic foods the world will ever see will be witnessed within the next 75 to 100 years.

Note the emphasis throughout the discussion on demand for "basic foods", meaning direct consumption of rice, wheat, and other food grains and both direct and indirect (by way of animal products) consumption of maize, grain sorghum, and other feed grains. These commodities, and oilseeds used as animal feed, occupy most of the world's cropland, employ most of the world's labor engaged in agriculture, take virtually all the fertilizers and pesticides applied by agriculturalists, and account for almost all the world's use of water for irrigation.

Moreover, much of the deforestation and resultant environmental damage that occurs in the world is due to land clearance for production of these crops. For these reasons, increasing demand for these crops is responsible for most of the rising pressure on the resource base and for the environmental stress that stems from world agricultural development. I suspect, although present data are inadequate to prove, that in comparison with these crops the resource and environmental pressures originating from the production of forest products are at least an order of magnitude less.

If most of the additional demand growth for food- and feed-grains (and oilseeds) occurs in the next 75 to 100 years it is likely that most of any additional pressure on the world resource base and environment will also occur in this period. Accordingly, the relevant time horizon for this analysis, insofar as world food production is concerned, is the next 75 to 100 years. For forestry the horizon may be longer because the income elasticity of demand for lumber and furniture does not decline with rising income as does the demand elasticity for basic foods. However, population growth is also an important component of rising demand for forest products and the virtual cessation of world population growth after 2100 will tend to significantly slow the growth of demand for these products after that date.

This line of argument implies that beyond the next 100 years agriculture's draw on resources for the production of basic food commodities will be primarily for maintenance and replacement of existing production systems, not for expansion. The level of resource demand will be much higher than now, but it will not be increasing. The share of land, water, and ordinary labor in the resource mix will be less than now and the share of science-based manufactured resources and human capital will be greater. Problems of resource management in agriculture will not disappear and they may be quite different than at present. My guess is that the cessation of growth in demand will make the problems easier to handle, whatever they may be. In any event, limiting this analysis to the period in which demand for basic foods and fiber ceases to grow still leaves plenty of work to do. If we can manage the problems generated by growth we should be well positioned to deal with those that arise in the subsequent period of no growth.

When growth in demand for basic foods and fiber ceases in the developing countries, the quantity of resources used to process, package, and deliver food and fiber to final consumers will increase relative to the quantity devoted to production on the farm and in the forest. This tendency is already far advanced in the developed countries. (In the USA the farmer receives less than 40 cents of each dollar consumers spend on food.) High income promotes demand for restaurant food and the high value of labor and leisure spurs demand for time-saving forms of food preparation in the home. The high value of labor also stimulates demand for more processed wood materials used in housing and building construction. There is every reason to believe that these tendencies will continue in the developed countries and will acquire increasing momentum in the developing countries as their incomes increase.

Thus the quantity of resources devoted to the delivery of agricultural commodities beyond the farm gate and the forest will likely continue to increase indefinitely beyond the next 100 years, which may pose mounting resource and environmental problems. But these are problems of economic development in general. They are not peculiar to agricultural development and so are not considered here.

The argument that world demand for agricultural resources will show little growth beyond the next 75 to 100 years assumes no massive substitution of biomass for fossil and other sources of energy over the next 75 to 100 years and beyond. Should this happen, the declining resource and environmental pressure associated with declining growth in food demand could be offset by increasing pressure from rising demand for biomass energy. How likely is this?

No one knows. The answer depends on the long-term behavior of the economic and environmental costs of fossil and other energy sources relative to those of biomass sources. Under present conditions biomass, unless subsidized, is not competitive with fossil energy in most uses, as demonstrated by the currently small share that biomass contributes to the total world energy supply. However, there is a consensus that the costs of fossil energy will rise indefinitely, although there are differences as to the rate of rise (see Darmstadter, Chapter 5, this volume). Such an indefinite rise would strengthen the competitive position of biomass energy unless its costs rise proportionately or more, or unless other nonfossil sources (nuclear, photovoltaics, and other direct solar) become even more competitive. This is a guessing game, but guess we must until improved data and analysis permit us to do better. In his review of the best qualified energy guessers Darmstadter does not give major attention to the prospects for biomass. However, he says of a study at IIASA [34] that in its view "... small scale solar and other renewable energy sources, while assumed to play a growing role, are likely to satisfy only a modest fraction of the total global energy demand beyond the first quarter of the twenty-first century" (Darmstadter, Chapter 5, p 158).

This view supports the conclusion that most of any increased resource and environmental pressure from world agricultural development will occur over the next 75 to 100 years. The future of biomass fuels, however, may contain one or more of those "surprises" which shock conventional linear thinking, and set us into unexpected and poorly understood patterns of resource use. If this happens, the future course of world agricultural development will be different from that assumed in this inquiry: so be it.

Components of agricultural systems

Nature of the systems

It is useful to think of agricultural systems as consisting of four interdependent components: resources, technology, environment, and institutions (R–T–E–I systems). The function of the systems is to produce food, fiber, and forest products in response to the effective demand for them. The systems perform the function by combining land, water, labor, fertilizer, and so on (resources) in specific ways (technology). The production process always generates effluents which may have detrimental impacts on the receiving media (environment). These impacts often affect income distribution because the people who bear them cannot exact compensation from the farmers and foresters who impose them. The production process also affects the distribution of income between workers, landowners, and suppliers of purchased resources. Income distribution, and particularly changes in distribution, inherently raise issues of equity which every society must contend with, like it or not. Institutions, the fourth component of R–T–E–I agricultural systems, are social creations designed to deal with these issues. Apart from equity, however, institutions govern the quantity and kinds of resources that flow into agriculture and the disposal of the products that flow out. Institutional performance thus profoundly affects agriculture at any time, as well as its development over time.

Relation of system components

The quantity, quality, and terms of availability of resources affect the kinds of technologies agriculturalists choose to employ, but these choices also affect the terms of availability of resources. For example, the wide adoption of land-using technologies (say because of rising fertilizer prices) will likely increase land prices, while the sustained growth of a land-saving technology, such as irrigation, will, in time, increase the relative scarcity of water. As noted, these technological choices also affect the environment, setting up feedback effects on both resource and technology components of the system. For example, if the chosen technologies are highly erosive, the productivity of the land will eventually be impaired, increasing the relative scarcity of land. In addition, accelerated sedimentation of downstream reservoirs diminishes capacity to provide irrigation, electric power, and flood control. These impacts in turn will likely trigger public policies to induce or require agriculturalists to employ less erosive technologies and management practices. If the existing institutional structure is inadequate to accomplish this, pressure for a more effective structure will rise.

Actors in R–T–E–I systems

The R–T–E–I systems [35] are driven by demand for food, fiber, and forest products, and by the decisions of a hierarchy of actors. The prime mover is the "agriculturalist", defined as whoever decides how to respond to demand. The agriculturalist chooses what and how much to produce, on which specific piece of land, the resources to be employed, and what particular combination. The agriculturalist may be a single individual or family living on and working the land or a committee in a local, regional, or national office. What distinguishes the agriculturalist from other actors in the systems is that "his" decisions directly set the production process in motion.

The empirical evidence is overwhelming that agriculturalists are rational in the sense that they will seek to maximize the returns through time of the resources they control. In market economies the return is best represented by the net return from agricultural production. In centrally planned economies the return may be in higher salaries or bonuses, opportunity for advancement, enhanced prestige for a job well done, or the accumulation of savings for reinvestment in the agricultural enterprise.

In either type of economy rational decision making requires looking beyond the immediate future. This is obviously true in forestry or tree crops, but it is true also for animal production and even for production of annual crops. In all these cases the agriculturalist is aware that the decisions taken now have consequences many months or even years ahead. To the extent that those consequences affect the return to the agriculturalist he will try to take them into account, discounting them implicitly or explicitly so that alternative decisions can be compared. If the consequences do not affect the discounted net return to the agriculturalist he will ignore them, even though they may adversely (or beneficially) affect other parts of the R–T–E–I system or the larger society of which it is a part. In particular, environmental consequences of production are likely to be ignored because institutional structures bring them within the agriculturalist's decision making purview only imperfectly or not at all.

Rational decision makers also take account of risk. Agriculturalists are particularly exposed to risk because of the relatively long time (at least a single growing season for annual crops, and years for managed forestry) between the beginning of production and collection of the output. In that time the weather is unpredictable, output prices may fall or input prices rise, credit may become more expensive or not available at all, supplies of crucial inputs may be interrupted, strikes may occur at harvest time, and so on. Agriculturalists have differing attitudes toward risk, but all will take it into account, seeking to reduce their exposure to its adverse consequences.

For poor agriculturalists this may mean continued reliance on traditional seed varieties and cropping systems even though exclusive cropping of higher yield varieties would seem to offer a higher return. But the new technology is likely to be riskier than the traditional one. It would involve unfamiliar practices and the monocultural system may be more vulnerable to attack from insects and disease than the traditional system, so there is as great a likelihood of loss as of gain from the new system. But the poor agriculturalist operates so close to the margin of subsistence that the consequence of loss would be disaster. So he weights the probability of loss more heavily than the probability of gain and opts for the traditional system.

Of course poor agriculturalists will respond to new technology, as demonstrated by the spread of the Green Revolution technology in Asia. Risk may slow the response, but this does not reflect irrational behavior. Close analysis of agriculturalists' responses to new technology usually shows that when they fail to adopt it is not in their interest to adopt, with high risk often a major reason.

The nonagriculturalists in the hierarchy of actors affecting R–T–E–I systems do so indirectly by influencing the conditions that determine the return to agriculturalists from the production process. These actors include extension agents and vendors of inputs, managers of irrigation systems, bank lending officers, ministers of agriculture and finance, researchers in national and international agricultural research institutions, officials in environmental protection agencies, and the collectivity of anonymous individuals whose behavior in world and national markets affects the prices of agricultural commodities and inputs.

Nonagriculturalists transmit signals, deliberately or not, which guide agriculturalists in deciding what and how much to produce and which technologies to use. Many of the signals are incidental results of actions taken with no thought of their agricultural impact. Oil pricing decisions by OPEC, for example, evidently are made with little if any consideration of their important long-term implications for resource use in world agriculture. Credit policies designed to restrain inflation will likely increase interest rates or result in credit rationing, reducing the attractiveness of investment in new agricultural technology. Policies to protect domestic industry against imports increase the prices of agricultural inputs.

These unintended signals may drown out others designed by public officials to induce action by agriculturalists to improve performance of the R–T–E–I systems. In fact, this probably explains many instances of the failure of agriculturalists to respond to public policies; they do not get the message because contrary signals come through more loudly and clearly. It is unlikely that these unintended signals can be completely turned off. Where they exist they will likely constrain, sometimes severely, the ability of policymakers to deliberately change the behavior of R–T–E–I systems.

Future of R–T–E–I systems

The several external factors, discussed on pp 104–115, will shape and constrain the behavior of R–T–E–I systems everywhere. The prospective increase in energy prices is especially important because agricultural development inevitably requires increasing amounts of effectively used energy per hectare and per unit of labor. Much of the additional energy will be in the form of nitrogen fertilizer. It is not certain that more expensive energy will increase prices of nitrogen fertilizer – advances in fertilizer production technology may compensate – but the likelihood surely is greater than if the outlook were for constant or declining energy prices.

Much irrigation now is from pumped groundwater. In fact most of the additional irrigation installed during the last 20 years in the USA was from groundwater and groundwater was of major importance in the spread of the Green Revolution in the Punjab region of India and Pakistan. In many parts of the world groundwater is an important potential source of future irrigation, but rising energy prices render tapping the potential less economically attractive than it otherwise would be.

Phosphorus is an essential nutrient for all agricultural production. Concern has been expressed from time to time about its long-term supply, but in a recent assessment Goeller and Zucker [1] concluded that, with world economic growth similar to that projected here, current and extended reserves of phosphorus would be only 28% depleted by 2100. They do not estimate phosphorus production costs, but the large reserves estimated to be available in 2100 suggest that costs would not rise sharply. Production of phosphate fertilizer is not nearly as energy intensive as that of nitrogen fertilizer.

The short-to-medium term

Although all systems will be conditioned by energy and other external factors, their responses to these factors will differ according to their specific circumstances. Over the short-to-medium term, say two decades, the key circumstances are the systems' present endowment of resources and their potential for productivity gains with currently known agricultural technologies. Over the longer term the current circumstances become less relevant and the issue increasingly concerns the capacity of countries to devise institutions needed to channel resources into development of appropriate technologies and to deal with emerging environmental stress and equity concerns. This distinction between short- and long-term circumstances does not mean that present institutional capacities are irrelevant in the short run or resource endowments in the long run. The distinction is analytically useful, but like all analytical constructs it does not pretend to faithfully reflect reality in all its complexity.

In thinking about the short-to-medium term behavior of R–T–E–I systems I find it useful to put countries or groups of countries into one of the following four categories (*Figure 4.4*):

(1) Low productivity potential–high land potential.
(2) Low productivity potential–low land potential.
(3) High productivity potential–high land potential.
(4) High productivity potential–low land potential.

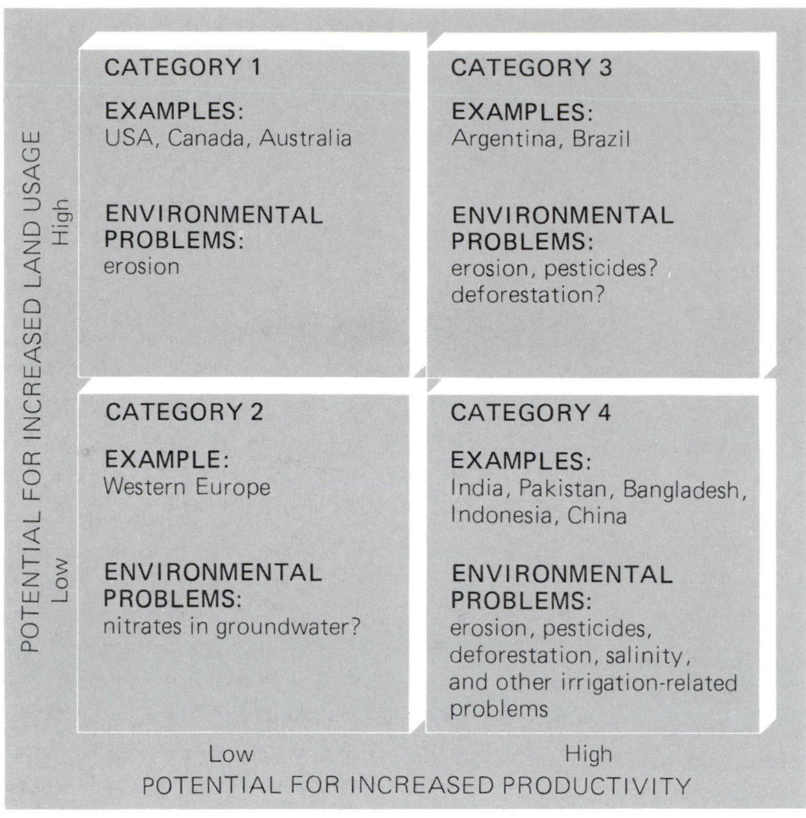

Figure 4.4 Potential for increasing land and technological productivity, with representative countries and environmental problems.

Recall that productivity potential is measured with respect to currently known technologies. This means technologies currently in wide use as well as those the use of which is spreading as agriculturalists become more aware of their potential. Land potential refers to the supply of land relative to the supply of other agricultural resources, particularly labor.

Category 1 countries

The USA is the largest agricultural producer in this category, to which Canada and Australia probably also belong. I believe the situation in the USA is similar enough to that in these other countries that it can be taken as representative.

Agricultural land is abundant in the USA with respect to category 2 and 4 countries. USA agricultural productivity is high, but there is reason to believe that with present technology its potential for future gain is low relative to the potential 10 or 20 years ago and relative to the potential currently available in category 3 and 4 countries. The rate of increase in USA crop output per acre (yields) and in total productivity (ratio of total output to total input) slowed significantly following the run up in energy prices in 1973 [36, 37]. Other evidence indicates that the marginal productivity of fertilizer (output increase per unit addition of fertilizer) is substantially less than it was a decade or two ago [38].

This combination of relatively much land and relatively limited productivity potential suggests that, confronted with rising energy prices, USA agriculturalists will respond to increasing crop demand by adopting technologies that incorporate increasing amounts of land. Use of fertilizers and pesticides will increase also, but not by nearly as much, relative to land, as in the 1950s and 1960s.

The additional cropland will be converted from land now in pasture, range, and forest, with pasture providing the greater part. The amount converted is not likely to seriously impinge on land needed for livestock and forestry production, given modest improvements in productivity of land in these uses. However, much of the land likely to be converted is inherently more erosive than land already in crops and, moreover, land in crops inevitably erodes more than land in grass or forest. For these reasons, the prospective pattern of agricultural expansion in the USA over the short-to-medium term is likely to greatly increase erosion [36]. To date, erosion in the USA appears to have had a small effect on productivity of the land. (Erosion reduced the growth of maize and soybean yields by 4% between 1950 and 1980 [39].) And analysis indicates that continuation of current rates of erosion would reduce yields by only 5–10% in 100 years if there were no offsetting advances in technology [40]. Should erosion increase substantially, the yield impact would no doubt be greater than this.

Off-site damages of erosion originating in USA agriculture now cost several billion dollars annually, significantly more than the cost of soil productivity loss [41]. These damages – siltation of lakes and reservoirs, costs of dredging of rivers and harbors, increased flooding, impaired recreational values, and so on – of course would increase with more erosion.

Erosion damage, especially off site, now appears to be the most severe environmental impact of USA agricultural production. By comparison, damages imposed by fertilizer, pesticides, and salinity with irrigation are relatively small, and likely to remain so [36]. Historically, insecticides have received most of the attention of those concerned with environmental impacts of agriculture. However, insecticide use in the USA is decreasing rapidly, primarily because of the adoption of integrated pest management practices in cotton production and a shift in production from the southeast and Mississippi Delta, where insect problems are severe, to Texas, where they are much less so. Moreover, the insecticides of greatest environmental concern, the persistent chlorinated hydrocarbons, e.g., DDT, are now of minor importance.

Herbicide use is increasing rapidly in USA agriculture. Present evidence indicates that the most commonly used herbicides are not particularly threatening to the environment, but not all the ways in which herbicides might impact the environment have been well studied. Should subsequent research reveal damage not now apparent, the judgment that erosion will present the major environmental threat of agriculture may have to be revised.

There are grounds for some confidence that USA institutions involved with agriculture have enough flexibility to cope reasonably well with the short-to-medium term situation depicted here. Farmers generally adapted well to the run up in energy prices in the 1970s, rapidly adopting conservation tillage, a less energy-intensive practice than tillage with the moldboard plow. And in the arid and semiarid west the increasing energy costs of pumping induced various practices to increase the efficiency of irrigation water use, e.g., private and public institutions provided farmers with information about soil moisture and other conditions, enabling them to irrigate at optimum times and in optimum amounts [42]. The federal Environmental Protection Agency, although frequently subject to heavy criticism, nonetheless did a creditable job in restricting use of the most environmentally threatening insecticides. And the development of integrated pest management techniques was a direct response of public institutions to perceived threats, both economic and environmental, posed by the continued heavy reliance on insecticides in cotton production.

The main grounds for concern about the institutional response to emerging environmental stress concerns off-site erosion damage, arguably the most important source of stress. Historically, erosion impacts on soil productivity have been the main concern of policymakers in the US Department of Agriculture. Off-site damage now is commanding more attention, but the habitual view of the erosion problem still dominates.

Institutional response in dealing with equity concerns also seems to have been reasonably good in the USA. Real per capita disposable income of farm people increased twice as much as that of the nonfarm population between 1950–1951 and 1980–1981, almost achieving parity by the latter date, *Figure 4.5* [43]. Within agriculture, the largest gains were among the smallest farmers, measured by annual sales of farm output [44]. Both of these developments reflected the expansion of job opportunities in nonagricultural activities after World War II, which permitted a massive migration of labor out of agriculture, increasing the earnings of those who remained. Just as important, if not more so, expansion of nonfarm activity in villages and towns throughout much of rural America provided both full and part-time job opportunities for people remaining in agriculture.

The principal emerging issue of equity is an apparent tendency toward a bimodal distribution of farm income, with the largest and smallest farmers doing reasonably well (the latter by virtue

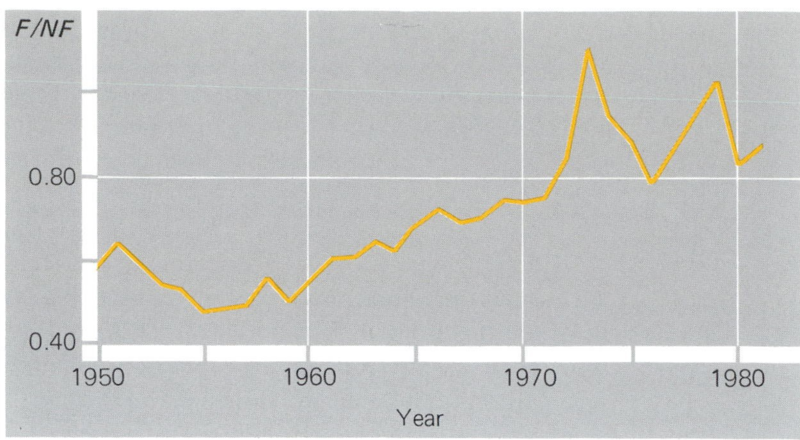

Figure 4.5 Ratio (F/NF) of per capita income of the farm population (F) to that of the nonfarm population (NF) in the USA, 1950–1981.

of off-farm employment), but those of medium size lagging behind. How deep-seated this tendency is and what it portends are not clear. Nor is it clear whether market and other institutions will yield a satisfactory outcome or whether some form of public intervention will come into play.

Category 2 countries

Western Europe is the major region classified in category 2: low productivity potential with existing technology–low land potential [45]. Current technologies in western Europe are highly land-saving, befitting the region's resource endowment. Fertilizer use per acre is higher than in the USA and mechanization is well advanced, labor is more abundant relative to land than in the USA, and insecticide applications, per acre and in total, are much less. Cotton and maize are the principal recipients of insecticides in the USA and elsewhere, but cotton is not grown in western Europe and maize occupies much less land than in the USA.

The high per acre use of fertilizer in western Europe suggests that its marginal productivity is relatively low, probably little if any above its price [46]. Consequently, if rising energy prices force up the price of fertilizer, farmers will seek ways to economize its use. Unlike American farmers, however, they are not likely to move toward more land-using technologies because of the already relatively high value of land. Mixed crop–livestock farms are more common in western Europe than in the USA and this may increase the feasibility in Europe of organic farming as a response, should nitrogen fertilizer prices rise.

Thinking about likely short- and medium-term technological responses of western European farmers to rising prices of energy, and possibly of fertilizer, is complicated by uncertainty about the future of the European Community's (EC) Common Agricultural Policy (CAP). There is growing concern in the Community about the high cost of CAP and declining real support prices in the early 1980s suggest that the Community is responding to that concern. One consequence is that less efficient farmers have been leaving agriculture, permitting an increase in average farm size and continuing mechanization. This, and genetic improvements in crop varieties, have helped to offset the impact of declining commodity prices and rising energy prices. If these two tendencies continue, western European agriculture would eventually be tightly squeezed. What the institutional response might be is uncertain. Commodity prices might be raised despite the high cost of this policy or more research may be devoted to relieving the land and energy constraints. The outcome will determine how long the present expansion of western European agriculture can continue [47].

Environmental impacts of erosion and agricultural chemicals do not now seem to be major problems in western Europe, although there is growing concern in some areas about nitrates in groundwater. By comparison with environmental problems in category 1, 3, and 4 countries, however, those in western Europe are and likely will remain small.

Category 3 countries

These are countries with high productivity potential and high land potential. Argentina, Brazil, and most other Latin American countries appear to fit this category, as do those in Sub-Saharan Africa. These countries are much more amply endowed with unused, or only lightly used, arable land than western Europe or category 4 countries (the Indian subcontinent, China, and Indonesia). In Latin America only about 15% of the arable land is cultivated and in Africa less than 25%. By contrast, 77% of the arable land in densely populated Asia is now under cultivation [48].

The abundance of arable land in category 3 countries may be deceiving. In Latin America most of the 535×10^6 ha of potentially arable land is in the remote, lightly populated Amazon basin. It is a gigantic frontier area, and opening it up would require massive investments in land clearance and in transportation to move production supplies into the area and the resultant output to the principal Brazilian markets along the southern coast. In Africa a significant percentage of the 565×10^6 ha of potentially arable land is unusable because of tsetse fly infestation.

In addition to these problems, the dominant soils in tropical Latin America and Africa are the so-called acid-infertile oxisols and ultisols. These soils generally have good physical properties (good structure and water holding capacity), but poor chemical properties, being both acidic (low pH) and low in nutrients. Although they support luxuriant growth, much of the nutrient supply is tied up in vegetation and ground litter, and most of that stored in the soil is near the surface. When the land is cleared for crop production the nutrient supply is diminished and that stored in the soil is readily leached or carried by runoff of the heavy rainfall typical of the humid tropics.

It was long believed, and perhaps still is commonly accepted, that these acid-infertile tropical soils could not economically be kept in continuous crop production. Recent work gives some promise that the soil limitations can be overcome with production systems that combine skilled management with the heavy use of fertilizers [49]. But management with the requisite skills is generally in short supply in these regions and, with prospective increases in energy prices, fertilizer may become too expensive.

Thus the abundance of land in these countries is conditional upon the availability of substantial amounts of capital, human and otherwise, without which the "high land potential" is more apparent than real. Nevertheless, even if the real potentially available arable land in Latin America and Africa is substantially less than the numbers above indicate, it almost surely is significantly more in relation to population than in densely populated Asia. Accordingly, in responding to rising demand and increasing energy prices, agriculturalists in Latin America and Africa are likely to adopt more land-using technologies than their counterparts in Asia.

Category 3 countries are also said to have high productivity potential on the basis that average yields for crops, animals, and forest products are low relative to those in developed countries. More to the point, yields are low relative to those achieved by a few farmers and foresters in the countries themselves, using improved technology and management practices.

Tapping the high productivity potential of these technologies and practices requires a commitment of resources to provide the mass of agriculturalists with the information they need to adopt the improvements and to assure that the requisite components of the technologies – improved seeds, fertilizers, etc. – are available, when, where, and in the quantities needed at prices the agriculturalists can afford. This is a tall order. Indeed, meeting the commitment is one of the principal challenges of agricultural development. Explaining how the commitment is met, and why it is not in some instances, is also a major challenge to the induced innovation hypothesis that I have adopted as a guide to thinking about the future of R–T–E–I agricultural systems. It indicates that the existence in category 3 countries of technologies with high productive potential will, in time, induce the institutional changes necessary to exploit the potential. Assuming this happens, what will be the pattern of technological change and its environmental consequences?

Per hectare use of fertilizer is much less in category 3 countries than in the USA and western Europe, suggesting that the marginal productivity of fertilizer is relatively high. And, indeed, per hectare use of fertilizer has been increasing much more rapidly in developing countries (including those in category 3) than in the USA and western Europe. As long as the marginal productivity of fertilizer remains relatively high in category 3 countries, agriculturalists there will have less incentive to respond to high fertilizer prices by shifting toward land-using technologies than those in the USA.

It seems that category 3 agriculturalists have more options than those in category 1, 2, or 4 countries. The presence of high-potential technologies makes viable a land-saving response to rising demand even if fertilizer prices rise. And the relative abundance of land also opens the possibility of a land-using response. My guess is that over the short-to-medium term category 3 countries will follow a middle course, adopting technologies that involve more rapid increases in per hectare fertilizer use than in the USA and other category 1 countries and incorporating more land per unit of output growth than in category 2 or 4 countries.

This pattern of response will mean clearance of forests for crop and animal production, much of it in tropical areas, a process that has already aroused much concern about its possible environmental consequences. Additional clearance will likely add to the concern.

Clearance of the tropical forests raises two questions: How much is occurring and what are, or will be, the consequences? After careful consideration of the available, inadequate information, Sedjo and Clawson [50] conclude that the widely held view that the world's tropical forests are rapidly disappearing is questionable. They cite recent studies by the UN Food and Agricultural Organization and others that coincide in showing that the annual rate of tropical forest clearance is some 11 to 12×10^6 ha, about 0.6% of the total. Sedjo and Clawson note that the rate of clearance in some areas of Africa and elsewhere is much higher, but conclude that as a global phenomenon the problem has been exaggerated.

The rate of tropical deforestation and the severity of its environmental consequences, of course, are not independent. Attention has been focused particularly on four sets of consequences:

(1) Change in local climate.
(2) Contribution to build-up of atmospheric CO_2.
(3) Species loss.
(4) Increased soil erosion.

Hamilton [51] reviewed the literature concerning local climatic effects of forest clearance. He found some work [52] suggesting that the tropical Amazon forest regenerates some of its own rain, giving reason to believe that large-scale, permanent clearance in the region could reduce or alter rainfall in parts of it. On balance, however, Hamilton concludes that although "there are very many compelling, scientifically sound, and philosophically rewarding reasons for trying to preserve a large amount of our remaining primary tropical rainforest . . . fear of reduced rainfall is not one of them [51, pp 3–4]."

The climatic consequences of tropical deforestation depend in part on what kind of ground cover replaces the forest. Dickinson's discussion of the issue (Chapter 9, this volume) suggests that if the forest is replaced with well managed silvicultural or agricultural systems, the climatic consequences will not be marked. However, if the soil is inhospitable to developed agriculture or silviculture, e.g., such as some lateritic soil in the humid tropics, then the shift from forest to low vegetative cover may significantly alter the regional climate.

To date forest clearance has contributed significantly to the accumulation of atmospheric CO_2, but according to the *Carbon Dioxide Review: 1982* [8] this effect will be swamped in the future by the contribution of fossil fuel combustion.

Loss of species following tropical deforestation has excited much concern, but as with most issues of tropical deforestation very little is known about the rate of species loss or its significance [53]. Sedjo and Clawson [50] argue that while extreme statements about the rate of species loss have little basis in historical experience, there nonetheless is cause for serious concern. They report estimates that 25 to 50% of the world's species are in tropical forests and view these genetic resources as "an unopened treasure chest in that their long-term value to humanity is quite uncertain [50, p 34]." Precisely because of the great uncertainty, investigation of the species-loss consequences of tropical deforestation should command high priority in the world scientific community. And the consequences should weigh in assessing the costs of the land-using mode of response to rising demand for agricultural output in category 3 countries.

Deforestation has aroused concern also because of its consequences for increased erosion. Erosion is an issue in both category 3 and 4 countries, and can result from the cropping of any land, not just that recently deforested. For these reasons I discuss erosion as a pervasive problem in developing countries, not as a special consequence of the land-using mode of response in category 3 countries.

There is much anecdotal evidence that erosion rates are high in developing countries and its consequences severe [54], but reliable, comprehensive, quantitative information about the amount of erosion and the costs it imposes in soil productivity loss and off-site damage are not available [55, 56]. After a review of the available evidence, I concluded that nothing could be confidently said about the importance of erosion-induced productivity losses in developing countries, but that the evidence for high off-site damage is strong [57].

Unless remedial action is taken, the land-using mode of response to rising demand is sure to exacerbate the erosion problem in category 3 countries and the problem will remain severe if not worsen in category 4 countries as well. Because this and other environmental impacts of agriculture are common to both categories I discuss institutional responses to the impacts after considering modes of response to rising demand in category 4 countries.

Category 4 countries

As noted above, India, Pakistan, Bangladesh, China, and Indonesia are the principal countries in this category. The ratio of population to arable land is much higher than in category 1 and 3 countries, and about three quarters of all arable land is under cultivation [48]. The extreme land scarcity makes it certain that these countries will adopt land-saving technologies in response to rising demand for agricultural production. Fortunately, the land-saving mode offers promise even in the face of rising energy, and possibly fertilizer, prices. In contrast to the situation in category 1 and 2 countries, levels of fertilizer and energy use in category 4 countries are low relative to land and labor, which suggests relatively high marginal productivities of fertilizer and energy if adequate supplies of high-yield seed varieties and irrigation water are available. There are good prospects that these conditions can be met. The International Rice Research Institute (IRRI) pioneered the development of high-yield rice varieties suitable for Asian conditions and continues to do so. In the last decade India and Indonesia have made major strides toward building their own capacity for this kind of research [58].

Irrigation has long been an integral part of agriculture in this part of the world. India and China now have more land under irrigation than any of the other countries, with plans for additional expansion. There is much evidence that irrigation water in these (and other) countries is inefficiently used and that its distribution is often inequitable [59]. Nonetheless, the presence in category 4 countries of large supplies, both existing and potential, of irrigation water, combined with a continuing stream of new high-yield varieties of wheat and rice and the present low per hectare use of fertilizer and energy suggests that over the short-to-medium term these countries have the potential for a strong land-saving response to rising demand for agricultural output.

Concern has been expressed that the development of these technologies may expose farmers to increased risk of yield failure. The reason is that, at least so far, the technologies have involved the substitution in a single field or across large regions of a small number of high-yield crop varieties for a much larger number of traditional varieties. If an insect or disease attacks the new variety it may sweep through the entire area planted to it, which could not happen in areas planted with the more genetically diverse traditional varieties. The loss in 1970 of some 15% of the USA maize crop to a kind of leaf blight is frequently cited as an example of the greater vulnerability of high-yield varieties.

In a careful study of grain yield variability over time in India and the USA, Hazell [60] found that variability had increased in both countries, and that narrowing the crop genetic base contributed something to this. Other possible contributing factors were weather, increasing variability in fertilizer prices, which induced variability in the amounts farmers applied, and variability in the supply of electricity (in India).

Taking a quite different approach from Hazell's statistical analysis, Duvick [61] surveyed 101 plant breeders in the USA and collected information about changes in crop genetic diversity over time. The responses indicated that elite adapted lines of cotton, soybeans, wheat, sorghum, and maize were among the most important sources used for breeding pest resistance. Duvick writes that this

> directly contradicts commonly heard statements to the effect that gene pools of elite materials have been so narrowed by successive generations of selection for yield that they no longer contain the diversity needed to counter new diseases and insect problems.

The issue of whether the movement toward higher yield crop varieties has increased yield variability is important because risk is important to farmers. If they are forced to trade-off increased risk of annual yield loss against higher average yields over time, their rate of adoption of the new varieties will likely be less than if there were no additional risk. The apparently conflicting results of Hazell and Duvick suggest that this is an issue needing more study.

Whatever the outcome, adoption by category 4 farmers of land-saving technology will increase the potential for significantly more environmental damage. As already noted, erosion is now and promises to remain a problem in these countries, but the land-saving mode of response suggests that problems associated with heavier use of fertilizers, pesticides, and irrigation are likely to become more severe.

As for erosion damages, evidence for those due to fertilizers, pesticides, and salinity in developing countries are anecdotal. That the damage sometimes is severe is attested by the death of at least 2000 from gas leaking from a pesticide manufacturing plant in India in late 1984. Less dramatic, but more common, are reports of fish kills from pesticide poisonings; also salinization and public health problems associated with the spread of irrigation are frequently noted. Algal blooms in and accelerated eutrophication of reservoirs and lakes because of fertilizers in runoffs and sediment are also widely observed. Reports of nitrate poisoning of animals and humans are much less common.

However severe the present environmental impacts of fertilizers, pesticides, and irrigation may be in category 4 countries, the land-saving mode of response that these countries will follow is likely to exacerbate them over the short-to-medium term. Much will depend on the ability of these countries to develop institutions for effective control of the use of the damaging materials and mitigation of their impacts. Whatever is done in this respect will raise equity issues, as will the income distribution consequences of the land-saving mode of response. Because these issues will arise in essentially the same form in category 3 and 4 countries I discuss them together.

Institutional responses

As noted earlier, the Green Revolution technology – the prototype of the land-saving response that category 3 countries will adopt – has been severely criticized as inequitable on the grounds that only larger, more wealthy farmers can afford it and that it increases landlessness and unemployment among the rural poor. The charge seems overdone. Studies by IRRI [27] of the spread of the Green Revolution technology among rice farmers in Asia demonstrate that both large and small farmers who adopted it have enjoyed an increased income. And the World Bank [62] reports that in the Indian Punjab the Green Revolution in wheat production was first adopted by large farmers, but small farmers and tenants soon took it up. Within only 6 years (1966 to 1972) farm incomes doubled. The growth of production and income stimulated nonagricultural activity in the area and many landless farm workers moved into nonfarm jobs. Per capita income in the region has been growing by 3 to 3.5% annually for the last two decades.

The Green Revolution technology is not a panacea, but income distribution has been moving most strongly against the rural poor in Asia in places where the technology was *not* adopted. Hayami and Kikuchi [28] found this in their study of population growth and technological change in Asia. Their work strongly suggests that under conditions of high population density and growth, the development and spread of land-saving technology is the only way to prevent increasing inequality of income distribution among landowners, tenants, and landless workers. In the absence of such technologies the relentless growth of population and food demand forces food prices higher and the resultant economic rents are captured by landowners.

Because it can be used by small as well as large farmers and favorably affects farm employment, the Green Revolution promotes "unimodal" patterns of agricultural development, i.e., those in which all farm income classes can share. Mellor and Johnston [63] have argued persuasively that the more equal income distribution associated with the unimodal pattern not only directly benefits the rural poor, but also ensures rising demand for a wide range of nonagricultural consumer goods, the production of which tends to be labor intensive. Thus, the unimodal pattern of agricultural development promotes rising employment and more equal income distribution in nonagricultural activities as well as in agriculture.

Hence land-saving technologies of the unimodal type are appropriate not only for the land-scarce conditions of category 4 countries, but also for meeting equity criteria for development. The challenge is to ensure that these technologies are widely available on terms all agriculturalists can afford. Meeting the challenge almost surely will require institutional innovations relating to the development and adoption of new technology. There are no clear guidelines of how to do this, but studies of the successful development and spread of the Green Revolution, e.g., in Taiwan and the Punjab region, offer clues [64].

The income distribution issue, of course, arises in category 3 countries also. Per capita national income in Latin America is higher than in Asia [2], and this must mean that per capita agricultural income is also higher. However, land is distributed much more unequally in Latin America than in Asia or Africa, so income of the rural poor in Latin America may be no greater than that in the other two regions.

A question arises: Can the relatively land-using mode of expansion likely to be followed by category 3 countries be shaped to incorporate more equal distribution of agricultural income? I cannot pursue the question here in detail, but a line of thought suggests itself. I stated earlier that the large areas of presently uncultivated arable land in Latin America and Africa are frontier regions. On grounds of economics, climate, and soil type much of these frontiers appears unpromising for development. Yet it is worth recalling that well into the nineteenth century the frontier region in the USA between the Mississippi River and California was called "the Great American Desert". Today, of course, much of that "desert" blooms, and the development of it provided rising employment and income opportunities for the poor and disadvantaged in the eastern states (*Figure 4.6* [65]).

One must be cautious about analogies. Yet I suggest it is worth thinking seriously about the possibility that the frontier areas of Latin America and Africa today could play the same role in promoting development and improvement of the lot of the poor in those regions as did the American frontier in the nineteenth century. This is not a new idea. Schemes to colonize the Latin American tropics have been tried, usually with indifferent success or outright failure. What I am suggesting is that if the land-using mode of expansion is appropriate for these countries and policymakers wish to find a form that also moves toward more equal income distribution, then incorporating the large-scale development of their frontiers as an integral part of the whole development process deserves serious consideration. The high capital cost and the potential environmental damages have already been touched on. In addition, research on tsetse fly control in Africa and on how to permit continuous intensive use of the acid-infertile soils of both Africa and Latin America will have to be pushed. The work described by Sanchez *et al.* [49] appears to offer promising leads for the acid-infertile soil problem. Mobilizing resources for such an undertaking and protecting against irreparable environmental damage would surely require important institutional innovations, which may or may not happen. But if study shows that, in fact, opening up the frontiers has high promise, this in itself should spur the needed institutional change.

Dealing with the environmental impacts of agricultural development in category 3 and 4 countries also calls for institutional innovation. In an important sense these impacts cause problems because of the absence of institutions requiring or inducing agriculturalists to take account of both short- and long-term consequences outside their immediate purview of the ways in which they manage the land and other resources. The basic problem is a lack of well-defined and enforceable property rights in the resources. Sedjo and Clawson [50] cite this difficulty as a main cause of tropical deforestation in areas where deforestation excites concern. Where the land and forests are owned in common by a village or tribe, but rules of access do not limit cutting, or where the resources do not clearly belong to anyone, or owners for whatever

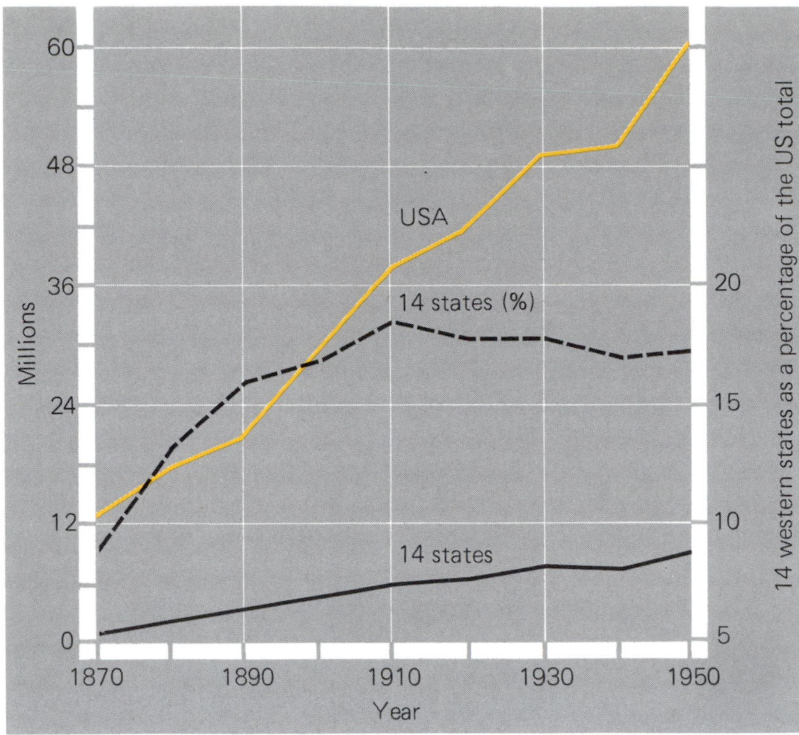

Figure 4.6 Gainful workers in the USA (yellow) and 14 western States (black), 1870–1950 (excluding California, Oregon, and Washington).

reasons cannot limit access, then no individual has the incentive to limit cutting to preserve future values, because if he holds back today he has no assurance that the trees will be there tomorrow. This is the "commons" problem discussed in the famous article by Garrett Hardin [66]. The problem arises also where many people have unlimited access to groundwater.

Excessive off-site erosion damages also arise from the absence of well-defined or enforceable property rights. Where stretches of river are unowned, or owned but poorly monitored, agriculturalists have no incentive to avoid dumping soil, animal wastes, and agricultural chemicals in the water, despite downstream damages.

If the absence of clearly defined, enforceable property rights is a major contributor to environmental problems, then in principle the solution is to correct the property rights deficiency. In some circumstances this is relatively easy. For example, the state of Sonora, Mexico, has developed a highly productive agriculture overwhelmingly dependent on groundwater irrigation. In the late 1960s and early 1970s it became apparent that uncontrolled pumping was threatening the aquifer by permitting intrusion of seawater. The Mexican government, which by law owns all groundwater resources in the nation, used its until then unused authority to limit the drilling of additional wells and to regulate rates of pumping. It is too early to say whether the threat to the aquifer has been contained, but the case illustrates the importance of unambiguous property rights to resources as the basis for more socially efficient management [67].

Usually, however, the property rights deficiency is not so easily remedied, such as for the problem of off-site erosion damage. The damaging sediment usually originates over wide areas, perhaps tens of thousands of hectares, and results from the land-use practices of perhaps hundreds of thousands of farmers and foresters. Property rights in the sediment-damaged reservoirs, lakes, and harbors may be as well defined as the Mexican government's right to groundwater, but enforcement of the right as a practical matter is impossible.

Even perfectly clear and enforceable property rights assigned to members of the present generation will not necessarily protect the interest of future generations in how resources are used. This is an exceptionally difficult issue because who is to define the interest of future generations? Nonetheless, the issue will not go away, being at the heart of concern about erosion-induced losses of soil productivity and about species loss resulting from tropical deforestation. Present owners of the land and of the forest can be assumed to manage those resources in a way optimal for them. But what assurance is there that the result is also optimal across generations? There is no easy answer. The issue can be clarified by research to better establish the facts: just how much are we losing in soil productivity and through species extinction? But even if these facts are perfectly known, owners of the resources still may differ with those who would speak for future generations in judging whether the losses are excessive. There is no analytical standard in economics or ecology for deciding this issue, which is ultimately a political decision to be made using political criteria [68].

The long term

In this section I deal only with category 3 and 4 countries. If agricultural development in these countries proceeds in a way consistent with the long-term projection of not less than a 2% annual per capita income growth, then import demand for food in these countries will begin to decline sometime in the twenty-first century, probably sooner rather than later. This will reduce whatever pressure agricultural production may have added to the resource base of the USA and other exporting countries, while growth of domestic demand for basic foods in these countries will be negligible.

Looking beyond the next couple of decades, technological responses in category 3 countries are likely to move increasingly toward the land-saving mode as increasing population reduces the now relatively abundant supply of land. This will occur later rather than sooner if these countries decide to develop their frontier areas, but that it will occur is likely. In the present land-scarce category 4 countries, continuing population increase will strengthen the already pronounced move toward the land-saving mode of response.

This will not necessarily increase the environmental impact of agricultural chemicals in category 3 and 4 countries. By sometime in the next century, probably in the first quarter, we will likely be able to make better use of the sun's energy in photosynthesis and also to have developed plants and related microorganisms to extract substantially more nitrogen from the atmosphere by biological processes. Escape of nitrogen fertilizer to the environmental media would be reduced. Similarly, integrated pest management techniques for dealing with insects should by then have greatly reduced the use of chemical insecticides, as they already have done in the USA.

Environmental impacts of erosion are more likely to diminish than increase as we move through the twenty-first century, but I do not count on institutional reforms to accomplish this. The difficulty of regulating the land-use practices of vast numbers of people over vast areas seems too indefinitely intractable to expect this. If improvement comes it will more likely be because of the development of such high-yield technologies that crop demand can be satisfied by a diminishing commitment of land. Moreover, the development process by then (say the second or third decade of the twenty-first century) combined with slower population growth will finally begin to reduce the number of people in agriculture, even in the most densely populated countries. With fewer people trying to scrabble together a living on the land, and less land devoted to crop production, the most erodible land can be converted to pasture, or forest, or some other less erodible use.

This suggests that the rate of deforestation will diminish in category 3 and 4 countries, but much depends on success in finding alternatives to present ways of providing fuel for home heating and cooking. Rising prices for fossil energy arouse skepticism that these fuels will substitute widely for wood, although improvement in internal transportation systems would help. Sedjo and Clawson [50] see great potential for plantation forestry in tropical areas. It is not beyond imagination that this development could go far in substituting for the current environmentally destructive systems that provide fuel wood in those areas. Again, however, improvements in internal transportation would be necessary to make this work, but since economic development of the sort envisaged implies improvement in these transportation systems anyway, the plantation forestry alternative for fuel wood supply may be worth serious consideration.

The development process also implies a long-run movement toward the equity objective, if this is defined as widely distributed gains in per capita income. These gains should increase more rapidly for rural people in category 3 countries than for those in category 4, particularly if the former pursue successful development of their frontiers. The gains will be slower in the densely populated category 4 countries until that time in the twenty-first century when slower rural population growth and expanding urban employment opportunities induce an absolute reduction in the number of people engaged in agriculture.

But if equity is defined as a more equal income distribution, movement toward it may be delayed well into the twenty-first century, particularly in category 4 countries. Recall that only in the 1970s did per capita income of farm people in the USA approach rough equality with that of the nonfarm population. And among groups of countries the prospect is for increasing inequality of income distribution, the poorest developing countries falling increasingly behind those in the middle-income group as well as those in developed countries.

Equity is in the eye of the beholder. If the relevant comparison for the poor is their position "now" with what it was 5 or 10 years ago, then steady increases in income over time should go far to satisfy their sense of equity. But if the relevant comparison is with the better-off at a point in time, then the poor's expectations for equity will be disappointed. This raises issues of social psychology and political science for which I have no professional competence. I suggest, however, that these issues must be addressed when thinking about the sustainability of the development process. Violation of ecological, or biospheric, imperatives is not the only threat to the process. Social imperatives also must be observed, although defining them is probably even more difficult than defining those imposed by the biosphere.

As we move further and further into the long term, present resource endowments and technologies become less and less relevant, whereas issues of institutional flexibility in responding to emerging resource scarcities, environmental stress, and equity concerns become increasingly relevant. Institutions must have the capacity to generate both the technical responses needed to relieve rising pressures on the natural system and the social responses that adequately satisfy equity objectives. Given the great inherent uncertainty about the distant future, the specific problems most likely to challenge the capacity for institutional response cannot confidently be foreseen. Precisely for this reason resiliency is a quality of high value in institutional structures. By resiliency I mean the capacity to absorb and react creatively to unexpected shocks to the natural and social systems. I cannot specify all the characteristics that distinguish resilient from nonresilient institutional structures, but surely resiliency requires high levels of technical expertise across the whole range of natural and social sciences, as well as political structures that promote openness in the exchange of ideas and in access to the levers of political power. Beyond this I am not prepared to go, but I am convinced that just as we seek agricultural systems that combine resiliency with high productivity, we must learn to impart these characteristics to systems of institutions as well [69].

The veil which shrouds the long-term future of agricultural R–T–E–I systems is opaque, but not completely so. In this chapter I have suggested some lines of inquiry worth pursuing as we consider how best to prepare ourselves for the shocks and surprises that the future surely holds. I summarize them in no particular order of priority.

Regional impacts of climatic change

Perhaps emphasis should be given to study of the question as to whether, in fact, long-term climatic change is likely, but I am impressed with the strength of the consensus that has formed on the question. Given that, the more important issue when considering the long-term future of agricultural R–T–E–I systems is the change in regional climates that are likely to accompany global warming. Until this is better understood we have little basis for judging where the impacts are likely to most severely stress the systems or for considering alternatives to ease the stress.

Economic, environmental, and social conditions for the successful opening of the Latin American and African frontiers

Let it be stressed immediately that any inquiry along this line would have to be conducted in an atmosphere free of any suggestion of intellectual or any other kind of imperialism. A fundamental condition for this is that political leaders as well as natural and social scientists in the countries involved be convinced that the inquiry is in their own personal, professional, and national interests; and that they participate fully in it. To this outsider looking in, it appears that in the long-term perspective opening these frontiers has great potential for a pattern of agricultural development that combines rising production with increasing equity for the rural poor. However, unless this perception is shared and welcomed by the people most directly concerned, the suggested inquiry cannot be successfully pursued.

Several of the issues requiring study can be discerned. The work cited earlier and reported by Sanchez *et al.* [49] suggests that the continuous, long-term sustainable cropping of the structurally strong but nutrient-poor acid-infertile soils (oxisols and ultisols) of Africa and Latin America is possible with greatly increased use of fertilizers and a high level of management. Consideration should be given to further research along these lines to examine both the scientific basis of the argument and the conditions necessary for large-scale mobilization of the physical and human capital indicated to be necessary.

Evidently, control of the tsetse fly is a condition for opening the African frontier, indicating the desirability of increased research to achieve this. But the performance of agricultural R–T–E–I systems in Africa over the previous several decades indicates that much more is required than control of the tsetse fly. Among the developing regions of the world, only in Africa has per capita food production not increased since the end of World War II. Indeed, it has declined. In an analysis of this performance Eicher [70] takes note of the tsetse fly problem, but leaves no doubt that institutional failure in the broadest sense is crucial to understanding why African agriculture has lagged. Distribution of land does not seem to be part of this. With the exception of Zambia and Zimbabwe, land distribution in Sub-Saharan Africa is "remarkably egalitarian" according to Eicher, most farms being between 2 and 6 ha [70, p 153]. Nor is lack of willingness of African farmers to respond to economic incentive part of the problem. On the contrary, research has conclusively demonstrated that African farmers respond as willingly to economic incentive as farmers in the most agriculturally developed nations [70, p 159]. It is precisely the lack of incentive that explains a major part of the poor performance. Governments in Sub-Saharan Africa have placed major emphasis on industrial development, neglecting investment in agricultural research, farm to market roads, and education of farm people. And they have adopted pricing policies for farm commodities that systematically discriminate against domestic producers in favor of imports. Clearly, close investigation of African institutional performance should receive major attention in any inquiry into the prospects for opening up the African frontier as part of a strategy for agricultural development.

Opening the African and Latin American frontiers would be a massive undertaking in which governments would have to take the major initiating and sustaining role. Finding the right degree and kind of government intervention would be crucial to success. The problem would vary from country to country, but experience suggests that as a general rule the government's role is best limited to supporting the research needed to develop technologies appropriate to the highly diverse soil and climate conditions in category 3 and 4 countries and to providing physical and institutional infrastructures which give private agriculturalists reasonable security of expectations about supplies of inputs and marketing services, credit availability, prices, and land tenure [71, 72].

Capitalizing on local knowledge of R–T–E–I systems

There is a common perception that peasant agriculturalists in category 3 and 4 countries are unskilled managers of their resources. The evidence accumulated since T. W. Schultz sharply questioned this perception in *Transforming Traditional Agriculture* [31] leaves no doubt that the perception is inaccurate. It is now clear that as a group peasant farmers are, in fact, highly skilled within the constraints imposed on them by natural, technical, and institutional conditions. The World Bank [72] provides numerous illustrations of this from developing countries all around the world.

In thinking about the long-term future of R–T–E–I systems in category 3 and 4 countries, and about how to make this future brighter, attention should be given to ways of making more use of peasant farmers' knowledge of local conditions. This knowledge may be – I think it is – an enormously valuable form of capital which so far has been little recognized and greatly underestimated. The world agricultural research establishment has much to teach peasant farmers, but it also has much to learn from them.

Learning about species extinction

The literature cited by Sedjo and Clawson [50] indicates that we presently know too little about species extinction around the world to accurately estimate either its rate or its consequences. Yet Sedjo and Clawson agree that since extinction is irreversible there is cause for concern even if we cannot currently demonstrate its importance. Some may argue that maintaining natural genetic diversity will become less important over the very long term as advances in genetic engineering make possible substitution of genes among species. However, the time when we might acquire this capacity for plant species, if we ever do, now seems to be quite far ahead [73]. To bank on the timely development of this capacity may prove to be a costly gamble with the future of world R–T–E–I agricultural systems.

Yield variability and new crop varieties

Concern was noted earlier about the possibility that the development of higher yield crop varieties is narrowing the genetic base on which crop breeders depend, resulting in greater interannual yield variability. There are apparently conflicting views over this (e.g., [60] and [61]). The matter seems of sufficient importance to justify further investigation to determine the implications of current plant-breeding practices for the long-run diversity of the global plant gene pool.

Erosion-induced loss of soil productivity

Erosion reduces soil productivity by carrying away soil nutrients and organic matter and by reducing the water-holding capacity of soil. Farmers can replace nutrients and organic matter, although at a cost, but there seems to be a consensus among soil scientists that losses in the water-holding capacity of soil are permanent. Whether this means permanent in the sense that species extinction is permanent or "permanent" in the sense of replaceable only at unreasonably high cost is not clear. In either case, the threat as a practical matter is to the long-term sustainability of R–T–E–I agricultural systems.

The present lack of quantitative knowledge of the threat in both developed and developing countries was noted above. Promising work to overcome this lack has recently been reported in the USA and other countries [39, 56], but the surface has only been scratched. Much more needs to be done, particularly in category 3 and 4 countries, to permit judgments as to the severity of the long-run threat.

Research capacity to develop new technology

The capacity to develop a continuous stream of appropriate new technology is a necessary condition for maintaining sustainable R–T–E–I agricultural systems. Without this capacity the systems will eventually be overcome by the mounting pressure imposed by population and income growth. The Malthusian specter at last will take harsh command.

That the world community can develop an impressive capacity to produce new agricultural technology is not in doubt. It has been done and institutionalized in the Consultative Group on International Agricultural Research (CGIAR), and in national research institutions of both developed and developing countries [74]. The formation of CGIAR and related national research agencies must be regarded as one of the major institutional innovations of modern times. Its positive impact on the welfare of people involved, both as producers and consumers of agricultural commodities, has already been enormous.

This success offers high promise that the research and development capacity required for long-term sustainability can be created and maintained, but more needs to be done to develop the national capacities in Latin America and Africa [58]. The lack of this capacity as a factor in the poor performance of African agriculture has already been noted; the CGIAR system cannot fill this gap. Developing technologies appropriate to the highly diverse soil and climate conditions of R–T–E–I systems in developing countries requires a decentralized research capability. CGIAR can develop technologies with potential for adaptation across diverse systems, but the work of adaptation must be performed locally.

The study of institutional change

This, perhaps, is the most challenging line of inquiry of all. Institutional flexibility and resilience are as important to the long-term sustainability of R–T–E–I systems as the capacity to develop new technology. But we have little understanding of why some institutional structures have these qualities and others do not. Nor do we understand well the processes of institutional change, so we are ill-equipped to develop institutional flexibility and resilience where it is lacking.

We greatly need a systematic study of these processes if we are to anticipate and act to counter major emerging pressure points on global R–T–E–I systems. The induced innovation hypothesis asserts that increasing pressure on these points will bring forth appropriate technical and institutional responses, but no one argues that the hypothesis is more than a fruitful beginning toward the better understanding of these processes of change. Under what circumstances can we expect the appropriate responses to be forthcoming without special social intervention and when is intervention essential to avoid system breakdown?

Perhaps the world needs an equivalent of CGIAR to study processes of institutional change in R–T–E–I systems. Or, less ambitiously, perhaps this line of inquiry could be added to the research CGIAR institutions now do on the development of new technology. Whatever the vehicle, I suggest that investigation of institutional processes should be an integral part of any study of the long-term sustainability of world R–T–E–I agricultural systems.

Further reading

Clark, W. C. (Ed) (1982), *Carbon Dioxide Review: 1982* (Clarendon Press, Oxford, UK, and Oxford University Press, New York).

Mellor, J. W. and Johnston, B. F. (1984), The world food equation interrelationships among development, employment and food consumption, *Journal of Economic Literature,* **XXII**, 531–574.

International Rice Research Institute (1978), *Economic Consequences of the New Rice Technology* (IRRI, Laguna, Philippines).

Crosson, P. and Stout, A. T. (1983), *Productivity Effects of U.S. Crop Erosion* (Resources for the Future, Washington, DC).

Sedjo, R. A. and Clawson, M. (1984), Global forests, in J. Simon and H. Kahn (Eds), *The Resourceful Earth* (Basil Blackwell, Oxford, UK).

Crosson, P. (1984), *Soil Erosion in Developing Countries: Amounts, Consequences and Policies* (Center for Resource Policy Studies, University of Wisconsin, Madison, WI).

World Bank (1982), *World Development Report 1982* (Oxford University Press, New York and London, for the World Bank).

Plucknett, P. L. and Smith, N. J. H. (1982), Agricultural research and third world food production, *Science,* **217**, 215–220.

Notes and references

[1] Goeller, H. E. and Zucker, A. (1984), Infinite resources: the ultimate strategy, *Science,* **223**, 456–462.

[2] World Bank (1984), *World Development Report 1984* (Oxford University Press for the World Bank, New York and London).

[3] Keyfitz, N., Allen, E. L., Douglas, R., and Wiget, B. (1981), Estimates of population, 1975–2075, world and major countries and of labor force, 1975–2050, *CO_2 Assessment Program Contribution No. 82-6* (Institute for Energy Analysis, Oak Ridge Associated Universities, Oak Ridge, TN).

[4] Barrie, W. D. (1970), *The Growth and Control of World Population* (Weidenfeld and Nicolson, London).

[5] The numbers in this paragraph are derived from [2].

[6] China uses a different methodology for calculating national income to that of the USA and other market economies. Analysts at the International Food Policy Research Institute have suggested that this may impart an upward bias to Chinese income growth compared to growth in market economies. However, the World Bank adjusts the Chinese figures (and those of other centrally planned economies) in an effort to achieve comparability among all the GNP estimates that the Bank publishes.

[7] Hare, K. (1980), Climate and agriculture: the uncertain future, *Journal of Soil and Water Conservation*, **35**(3), 112–115.

[8] Clark, W. C. (Ed) (1982), *Carbon Dioxide Review: 1982* (Clarendon Press of Oxford University Press, New York and Oxford, UK).

[9] National Research Council (1983), *Changing Climate*, Report of the Carbon Dioxide Assessment Committee (National Academy Press, Washington, DC).

[10] Rosenberg, N. J. (1984), *Climate, Technology, Climate Change and Policy: The Long Range*, paper given at the National Conference on the Future of the North American Granary, University of Minnesota, St Paul, June 18–19.

[11] Cooper, C. F. (1982), Food and fiber in a world of increasing carbon dioxide, in Clark, note [8].

[12] Wittwer, S. (1982), *Commentary on C. F. Cooper*, in Clark, note [8].

[13] Wittwer, S. (1980), Carbon doxide and climate change: an agricultural perspective, *Journal of Soil and Water Conservation*, **35**(3), 116–120.

[14] Waggoner, P. E. (1983), Agriculture and a climate changed by more carbon dioxide, in National Research Council, note [9].

[15] Rosenberg, N. J. (1982), *Commentary on C. F. Cooper* in Clark, note [8].

[16] For a comprehensive presentation of current thinking about the effect on plant growth of concentrations of atmospheric CO_2 up to 600 p.p.m., see Lehman, S. R. (Ed) (1983), *CO_2 and Plants*, AAAS Selected Symposium 84 (Westover Press, Boulder, CO).

[17] Schelling, T. (1983), Climate change: implications for welfare and policy, in National Research Council, note [9].

[18] Ross, J. A. (Ed) (1982), *International Encyclopedia of Population* (The Free Press, New York).

[19] The data for both cities and countries are from note [20]. The data for most countries and cities are within a year or two of the stated dates of 1850, 1900, and 1960. The 1960 percentage of population in the named cities understates the rate of urbanization in western Europe after 1900 because the numbers for Paris are for the city only. The Paris metropolitan area grew much faster than the city itself. If population for the entire area is included the 1960 percentage for the named cities is 15.5.

[20] Mitchell, B. R. (1981), *European Historical Statistics 1750–1975,* 2nd revised edn (Facts on Files Publishers, New York).

[21] Wilkie, J. W. and Haber, S. (Eds) (1981), *Statistical Abstract of Latin America, Vol. 21* (UCLA Latin American Center Publications, University of California, Los Angeles, CA).

[22] For excellent discussions of the interrelationships between agricultural and nonagricultural development see note [63]; Johnston, B. F. and Mellor, J. W. (1961), The role of agriculture in economic development, *American Economic Review,* **51**(4), 566–593; and Johnston, B. F. and Kilby, P. (1975), *Agriculture and Structural Transformation: Economic Strategies in Late Developing Countries* (Oxford University Press, New York).

[23] Frank, C. R. and Webb, R. C. (1977), *Income Distribution and Growth in the Less Developed Countries* (The Brookings Institution, Washington, DC).

[24] Labor-saving technologies do not necessarily lead to less employment of labor. Technologies are labor saving if they involve an increase in the quantity of other resources relative to labor. If demand for the output of the technology is increasing, demand for labor may rise; the Green Revolution technology is labor-saving in this sense. In many areas where it was adopted the resultant increase in output stimulated increased demand for labor and higher real incomes of farm workers [27, 28].

[25] Griffin, K. (1979), *The Political Economy of Agrarian Change: An Essay on the Green Revolution* (Macmillan Press, London).

[26] de Janvry, A. (1981), *The Agrarian Question and Reformism in Latin America* (The Johns Hopkins University Press, Baltimore, MD, and London).

[27] International Rice Research Institute (1974), *Changes in Rice Farming in Selected Areas of Asia* (IRRI, Laguna, Philippines).

[28] Hayami, Y. and Kikuchi, M. (1981), *Asian Village Economy at the Crossroads* (University of Tokyo Press, Tokyo, and the Johns Hopkins University Press, Baltimore, MD).

[29] Sien, M. (1984), Weighing the social costs of innovation, *Science,* **223**, 1368–1369.

[30] Hayami, Y. and Ruttan, V. W. (1971), *Agricultural Development: An International Perspective* (The Johns Hopkins University Press, Baltimore, MD).

[31] Schultz, T. W. (1964), *Transforming Traditional Agriculture* (Yale University Press, New Haven, CT).

[32] Farmers as a group may be better-off because the price elasticity of demand for rice is less than 1. This means that as the price of rice increases, demand falls less than proportionately, increasing total rice revenues. However, landless farm workers do not share in the gains and any who wish to enter farming by acquiring land find it more difficult to do so because of higher land prices.

[33] For references to this literature and a fruitful application of the induced innovation hypothesis see note [28].

[34] Häfele, W., Anderer, J., McDonald, A., and Nakicenovic, M. (1981), *Energy in a Finite World, Vol. 1. Paths to a Sustainable Future* (Ballinger, Cambridge, MA).

[35] This section draws heavily on Crosson, note [36].

[36] Crosson, P. R. (1979), *Resources, Technology and Environment in Agricultural Development,* Working Paper WP-79-103 (International Institute for Applied Systems Analysis, Laxenburg, Austria).

[37] Crosson, P. and Brubaker, S. (1982), *Resource and Environmental Effects of U.S. Agriculture* (Resources for the Future, Washington, DC).

[38] Heady, E. (1982), The adequacy of agricultural land: a demand–supply perspective, in P. Crosson (Ed), *The Cropland Crisis: Myth or Reality?* (Johns Hopkins University Press, Baltimore, MD, for Resources for the Future).

[39] Crosson, P. and Stout, A. T. (1983), *Productivity Effects of U.S. Cropland Erosion* (Resources for the Future, Washington, DC).

[40] Larson, W. E., Pierce, F. J., and Dowdy, R. H. (1983), The threat of erosion to long-term crop production, *Science*, **219**, 458–465.

[41] Clark, E., II, Havercamp, J., and Chapman, W. (1985), *Eroding Soils: The Off-Farm Impacts* (The Conservation Foundation, Washington, DC).

[42] Frederick, K. and Hanson, J. (1982), *Water for Western Agriculture* (Resources for the Future, Washington, DC).

[43] US Department of Agriculture (1972, 1983), *Agricultural Statistics* (Government Printing Office, Washington, DC).

[44] US Department of Agriculture (1981), *Economic Indicators of the Farm Sector Income and Balance Sheet Statistics 1980*, Statistical Bulletin 674 (Economic Research Service, Washington, DC).

[45] Japan is also a category 2 country, but on the world scale its agricultural production is negligible and will remain so.

[46] Over time the price of fertilizer and the value of the marginal product of fertilizer will approach equality. If the price is higher, then farmers earn less from an increment of fertilizer than they pay for it. They will cut back its use and the value of the marginal product will rise. If the price is less, then farmers earn more from additional fertilizer than they pay for it and so they will use more until the value of the marginal product falls to equality with the price.

[47] The EEC has sought to dispose of its mounting agricultural surpluses by pursuing a policy of vigorous export expansion. Over the long term a policy of subsidized export expansion would bring the Community into increasing conflict with the USA and other major exporters. This conflict is already troublesome.

[48] National Research Council (1977), *Supporting Papers: World Food and Nutrition Study, Vol. II*, pp 46–49 (National Academy of Sciences, Washington, DC).

[49] Sanchez, P. A., Bandy, D. E., Villachica, J. H., and Nicholaides, J. J. (1982), Amazon Basin soils: management for continuous crop production, *Science*, **26**, 821–827.

[50] Sedjo, R. A. and Clawson, M. (1984), Global forests, in J. Simon and H. Kahn (Eds), *The Resourceful Earth*, pp 128–170 (Basil Blackwell, Oxford, UK).

[51] Hamilton, L. S. (1983), *Removing Some of the Myth and Mis-es* (Mis*interpretation*, Mis*information* and Mis*understanding*) *About the Soil and Water Impacts of Tropical Forestland Uses* (unpublished paper, Environment and Policy Institute, East–West Center, Honolulu).

[52] Salati, E. (1981), *Precipitation and Water Recycling in Tropical Rainforests with Special Reference to the Amazon Basin* (unpublished paper presented to a workshop at the Centro de Energia Nuclear na Agricultura, Peracicaba, Sao Paulo, Nov. 10–13).

[53] Harrington, W. and Fisher, A. (1982), Endangered species, in P. R. Portney (Ed), *Current Issues in Natural Resource Policy* (Johns Hopkins University Press, Baltimore, MD, for Resources for the Future).

[54] Eckholm, E. (1976), *Losing Ground* (W. W. Norton, New York).

[55] El-Swaify, S. A., Dangler, E. W., and Armstrong, C. L. (1982), *Soil Erosion by Water in the Tropics* (University of Hawaii, Honolulu).

[56] Rijsberman, F. R. and Wolman, M. G. (Eds) (1984), *Quantification of the Effect of Erosion on Soil Productivity in an International Context* (Delft Hydraulics Laboratory, Delft, The Netherlands).

[57] Crosson, P. (1984), *Soil Erosion in Developing Countries: Amounts, Consequences and Policies* (Center for Resource Policy Studies, University of Wisconsin, Madison; WI).

[58] Plucknett, D. (1984), International agricultural research, in D. S. Bendahmane (Ed), *Science, Technology and Foreign Affairs* (Center for the Study of Foreign Affairs, US Department of State, Washington, DC).

[59] Crosson, P. (1975), Institutional obstacles to expansion of world food production, *Science*, **188**, 519–524.

[60] Hazell, P. B. R. (1984), Sources of increased instability in Indian and US cereal production, *American Journal of Agricultural Economics*, **66** (3), 302–311.

[61] Duvick, D. (1984), Genetic diversity in major farm crops on the farm and in reserve, *Economic Botany*, **38**(2), 161–178.

[62] The World Bank (1982), *World Development Report 1982*, p 70 (Oxford University Press, New York and London, for the World Bank).

[63] Mellor, J. W. and Johnston, B. F. (1984), The world food equation: interrelationships among development, employment and food consumption, *Journal of Economic Literature*, **XXII**, 531–574.

[64] The article by Mellor and Johnston [63] has useful references to such studies.

[65] US Department of Commerce (1975), *Historical Statistics of the United States: Colonial Times to 1970, Part 1* (US DOC, Washington, DC).

[66] Hardin, G. (1968), The tragedy of the commons, *Science*, **162**, 1243–1248.

[67] Burke, K. J., Cummings, R. G., and Muys, J. C. (1983), *Interstate Allocation and Management of Nontributary Groundwater*, pp 151–152 (Western Governor's Association, Denver, CO).

[68] *Editors' note*: See also Regier and Baskerville, Chapter 3, this volume, pp 96–99.
[69] *Editors' note*: See also Holling, Chapter 10, this volume.
[70] Eicher, C. (1982), Facing up to Africa's food crisis, *Foreign Affairs*, **61**(1), 151–174.
[71] Nelson, M. (1973), *The Development of Tropical Lands* (Johns Hopkins University Press, Baltimore, MD, for Resources for the Future).
[72] The World Bank (1982), *World Development Report 1982* (Oxford University Press, New York and London, for the World Bank).
[73] Duvick, D. (1986), North American grain production – Biotechnology research and the private sector, in C. F. Runge (Ed), *The Future of the North American Granary: Politics, Economics, and Resource Constraints in North American Agriculture* (Iowa State University Press, Ames, IW), argues that genetic engineering techniques for plants confront sizable barriers of scientific ignorance. Most traits of economic importance in the world's main food crops are governed by very large numbers of genes, or so it appears. At this time genetic engineering cannot handle more than one unlinked gene at a time. Major gaps in scientific knowledge must be filled before this difficulty can be overcome.
[74] For an account of the work of CGIAR and national agricultural research institutions see Plucknett, D. L. and Smith, N. J. H. (1982), Agricultural research and Third World food production, *Science*, **217**, 215–220.

Commentary

C. F. Runge

The theoretical core of Pierre Crosson's analysis is the induced innovation model originating from Sir John Hick's *Theory of Wages* [1] and applied to technological and institutional change in agriculture by Hayami and Ruttan [2]. Because this somewhat abstract model drives Crosson's analysis, resulting in conclusions oriented around the technical and institutional basis of agricultural progress, it is appropriate to focus this brief commentary on the model and its relevance. The model's importance extends to several other chapters in this volume, notably those of Majone (Chapter 12) and Darmstadter (Chapter 5). Three issues seem germane: the induced innovation model itself, Crosson's interpretation of it, and elaborations of the model in some recent work.

The model

The "induced innovation hypothesis" maintains that both technological and institutional innovations are driven by resource constraints [3]. These constraints establish a demand for changes in both techniques of production and the rules that coordinate production activity. The supply of new techniques and new institutions responds to this demand. The theory can be used to explain different historical paths of agricultural development arising from different resource and factor endowments. In the case of technology, it predicts that where factors are relatively scarce, technological innovations are likely to conserve on the scarce factors and be biased toward use of abundant ones. The constraints imposed by a relative shortage of labor in countries such as the USA, Canada, and Australia, for example, have been offset by technical advances leading to the substitution of animal and mechanical power for labor, as well as by techniques that utilize relatively abundant resources such as fertilizer or mineral fuels. In Japan and Taiwan, in contrast, relative shortages of land have led to the substitution of fertilizer and labor-intensive cultivation practices for scarce land resources.

On the side of institutions, research and education produce relevant examples of induced innovations. Where research and education in agriculture are relatively scarce, implying an institutional constraint, a demand exists for knowledge that can inform technical and social decisions so that they are based on more than trial and error [4]. This information can only be acquired in a systematic fashion by the development of new institutions, manifest in innovations such as the Land Grant University system.

Induced innovation theory also suggests a dialectical relationship between technical and institutional change. Institutions create constraints to technical progress, while technological bottlenecks may delay institutional reforms. Both existing technology and the institutional setup create constraints for one another, and define the joint possibilities for change. While this brief description oversimplifies the richness of the approach, its thrust is quite straightforward: scarcity creates constraints and innovations in technology and institutions are designed to overcome these constraints.

Interpretations of the model

Owing in part to the emphasis given by economists to prices as the principal signals of resource scarcity, many have perceived the induced innovation hypothesis to be restricted to the role of perfectly competitive market prices in economic development. It is, in fact, far more general, and does not rely solely on market prices as signals for emerging scarcity. This is significant, because market prices, especially in less developed economies, are seldom accurate signals. Developed economies, too, face price distortions, making markets only one, often unreliable, signaling device. Even so, both in theory and practice, constraints on factors and other resources send signals, theoretically defined as "shadow prices", which reflect underlying resource scarcity [5]. In a perfectly competitive economy without distortions, these signals happen to correspond to market prices. But they do not need to do so for the induced innovation model to be relevant. The model can still be used to guide second best choices in which market prices do not reflect marginal values, as long as relative resource scarcity can be determined. In the USA, it is not necessary to have land market price data to recognize the relative abundance of land as a factor of production and its role in the technological and institutional development of agriculture.

In contrast to Crosson's observation, this makes the theory even more powerful as an explanation for technological and institutional responses to unpriced resources with a public good quality, such as air, water, and land, where it is obvious that market prices do not necessarily signal a need to conserve. The theory suggests that even though imperfectly reflected by markets, this scarcity can be discovered. Incentives will then exist to develop technologies and institutions that conserve scarce natural resources and utilize abundant factors. IIASA itself represents the use of abundant information-processing technology in order to conserve scarce public goods, such as land and water.

In short, with or without the price mechanism, increasing scarcity, if known, creates incentives for conservation. If a private market for all resources existed, part of a perfectly functioning set of markets, then in theory the entire signaling process would be accomplished through market prices. But the fact that a perfectly competitive market does not exist for most things does not prevent their scarcity from being known and creating incentives for change. Of course, new technology or institutions will not always respond to these signals, especially in light of the problem of discovery mentioned above. Indeed, the inability to respond costlessly and with perfect foresight to resource scarcity by inventing new techniques and new institutions is part of what makes these resources scarce to begin with. The problem is to discover which resources are most scarce and to design techniques and institutions that respond to this scarcity.

Recent elaborations of the theory

The general problem of discovering which resources are most scarce raises the interesting and significant role played in the induced innovation process by expectations. Many recent debates over resource scarcity, especially of highly aggregated global assessments, depend critically on some form of resource optimism, resource pessimism, or linear extrapolations of current trends. These are in essence expectations of the future, rather than facts. At a more microeconomic level of decision, the relative capacity of individual agents to anticipate resource scarcity and to profit from the resource once its scarcity becomes generally known is based largely on expectations [6]. This capacity, sometimes referred to as entrepreneurship, occurs in an environment of considerable price uncertainty, relying on other forms of knowledge, insight, and intuition. As John R. Commons argued, "all value is expectancy [7]." The supply of both new technology and new institutions is powerfully affected by this entrepreneurial capacity [8].

Recent declines in world oil prices are instructive and relate to some issues raised in Chapter 5 by Darmstadter. Since these prices are the result of oligopolistic distortions in the world market, they are not competitive market prices that reflect the true scarcity value of oil. Declining world prices result in large part from declining discipline within OPEC. However, geological discoveries also inform both private and public judgments about oil availability, and knowledge of production constraints creates incentives for technologies and institutions promoting both more efficient oil extraction and use as well as overall policies of conservation and substitution. These technical and institutional innovations continue, despite temporarily falling prices, because they are based on expectations of future scarcity. By recognizing that falling market prices are misleading signals of relative scarcity, primarily reflecting changes in market structure, rational agents respond according to expectations that correctly discount short-term price movements.

More generally, evidence in support of the capacity of individuals to recognize longer run interdependence and to innovate institutions and technologies to overcome common problems, suggests that the "logic of the commons" to which Crosson and Darmstadter refer may be overstated [9]. This is not to suggest that such problems are always overcome. Often the knowledge, foresight, and capacity to overcome them may not be at hand. Nonetheless, the idea of induced innovation helps us to recognize that the constraints posed by living in an increasingly interdependent world raise the rewards to international cooperation discussed by Majone. All of the chapters in this volume may be understood as a response to these incentives.

The themes of innovation and expectations also merge in the issue of institutional innovation and its relationship to fairness. Perhaps the most fundamental issue raised by Crosson's discussion of induced innovation concerns the role of equity, because both technological and institutional changes have distributive as well as efficiency impacts. I have argued elsewhere that the demand for change (notably on the institutional side) is a direct function of distributive concerns [10]. As long as we are willing to admit that economic efficiency is only one (often rather minor) concern of rational individuals [11], the theory of induced innovation is, as before, even stronger in explaining behavior.

Consider, for example, the fundamental issue of property institutions, including those that are developed to overcome commons problems. Property rights are the particular characteristics of property institutions that channel streams of net benefits to particular agents in a given situation. They extend to individual contracts and deeds, agreements between corporate entities, and even international agreements between countries. In addition to the increases in efficiency that they provide, property rights grant assurance with respect to the distribution of benefits over time.

My property right to exclude others (or to be rightfully included) depends on an established expectation. Yet, if my claim on land, air, or water is not perceived to be fair, my entitlement may not be respected, and its security may be overturned through institutional innovations that redistribute entitlements. Changing resource endowments or technical changes, for example, may create disequilibrium in the structure of expectations established according to an older set of institutions. This disequilibrium will be manifest in judgments that the current structure of rights is inefficient or unfair, as when traditional land tenure patterns are overturned in favor of agrarian reform. Such disequilibria create incentives for technological and institutional entrepreneurs to propose alternative arrangements more consistent with a conjectured structure of both future efficiency and equity. It is crucial to recognize that because constraints differ from time to time and place to place, the particular arrangement of technology and institutions most responsive to these constraints also will change and differ across environments. There is no uniquely optimal arrangment. This finding is broadly consistent with much ecological theory.

In sum, the induced innovation model provides a logical yet surprisingly flexible framework within which to interpret both microlevel and macrolevel innovations in technology and institutions. If these innovations are demonstrably responsive to the particular constraints faced by individuals seeking greater control over the long-term future of the environment, then the elaborated theory provides a clear message. Knowledge concerning environmental constraints will be aided by market signals, but cannot depend on them exclusively: nonprice information is also key. Moreover, since the constraints to which individuals respond differ radically in various parts of the world, existing patterns of innovation should be studied carefully. The very definition of both efficiency and equity is relative to the constraints that must be faced – locally and globally. Continued research into similarities and differences in these constraints is thus fundamental to the design of new technological and institutional responses.

Notes and references (Comm.)

[1] Hicks, Sir John (1964), *The Theory of Wages*, 2nd edn (St. Martin's Press, New York).

[2] Hayami, Y. and Ruttan, V. W. (1971), *Agricultural Development: An International Perspective* (Johns Hopkins University Press, Baltimore, MD).

[3] Binswanger, H. and Ruttan, V. W. (Eds) (1978), *Induced Innovation: Technology, Institutions and Development* (Johns Hopkins University Press, Baltimore, MD).

[4] Ruttan, V. W. (1984), Social science knowledge and institutional change, *American Journal of Agricultural Economics*, **66**(5), 549–559.

[5] Dixit, A. K. (1976), *Optimization in Economic Theory* (Oxford University Press, London).

[6] Rosenberg, N. (1976), On technological expectations, *Economic Journal*, **86**, 523–535.

[7] Commons, J. R. (1924), *Legal Foundations of Capitalism* (Macmillan, New York).

[8] Farrell, K. R. and Runge, C. F. (1983), Institutional innovation and technical change in American agriculture: the role of the New Deal, *American Journal of Agricultural Economics*, **65**, 1168–1173.

[9] Runge, C. F. (1981), Common property externalities: isolation, assurance, and resource depletion in a traditional grazing context, *American Journal of Agricultural Economics*, **63**, 595–606; Runge, C. F. (1984), Institutions and the free rider: the assurance problem in collective action, *Journal of Politics*, **46**, 154–181.

[10] Runge, C. F. (1984), Strategic interdependence in models of property rights, *American Journal of Agricultural Economics*, **66**, 807–813.

[11] Sen, A. K. (1977), Rational fools: a critique of the behavioral foundations of economic theory, *Philosophy and Public Affairs*, **6**, 317–344.

Chapter 5 Energy patterns – in retrospect and prospect

J. Darmstadter

Editors' Introduction: Energy production and consumption in support of human development have increased over the past century to the point where they now exceed even agriculture as the major transformers of many aspects of the Earth's environment.

In this chapter long-term, large-scale patterns of energy use are reviewed, highlighting the interplay of changing technology and resource endowments. The successive rises and declines of energy systems based on animal power, wind and water, wood, and coal that have led to the present mix of fuels embedded in the global energy economy are described. The possible relationship of these substitution patterns to "long wave" Kondratieff cycles is discussed. These historical patterns and the response of the world energy system to various shocks of the 1970s are drawn on in the discussion of long-term forecasts of energy use and its environmental impacts.

Introduction

The focus of this chapter is on the broad sweep of global energy development, with subsidiary attention paid to its implications for the environment. The issue that gives rise to this theme is almost self-evident: how to provide energy in the amounts and at a cost compatible with sustained economic progress without inflicting major harm on environmental integrity.

The discussion begins with a review of energy trends during the last half century, including the significant demand–supply adaptations that occurred in the wake of the two major energy "shocks" during the 1970s. The review of long-term energy developments provides an opportunity, as well, to comment on efforts to deduce recurrent and predictable energy-system cycles. Following this evaluation, I next turn to energy projections, starting with a section on the difficulty and uncertainty inherent in such undertakings, but indicating – in the ensuing discussion of prospective trends to the year 2000 – that the very adaptation evident in recent years lends a degree of robustness to such a mid-term projection. I then review a number of studies that project alternative energy paths well into the twenty-first century, singling out those which focus, in particular, on the CO_2 implications of those trajectories.

While CO_2 has been accorded much recent attention in long-term global energy modeling, it is but one of a multiplicity of energy-related environmental impacts. In this chapter I suggest that a clue to countries participating in *global* environmental management may emerge from their *national* commitment. The conceptual thread linking both dimensions is the "commons" problem, which dilutes incentives to make environmental concern an inherent element in decisions on energy projection, delivery, and use. Yet the environmental record – even in some of its transnational manifestations – is perhaps not so flawed as to rule out the prospects for long-term accommodation between energy needs and environmental imperatives. A recent contribution to that issue by Thomas Schelling, stressing adaptations as much as avoidance, is briefly touched upon.

At various places in the chapter – being largely an exploratory one – I raise questions for further research. These are brought together at the end.

There is considerable emphasis throughout the chapter on the presentation of, and comments on, different viewpoints about energy development paths. As part of a volume designed to serve as a springboard for, rather than the final word on, research and its implications, the chapter stops short of hard-and-fast conclusions.

Past energy trends

Whether one's fascination with the story of civilization reverts back to prehistory or extends only over the past several centuries, the role of energy is crucial. At the beginning, human and animal energy were the mainstays for accomplishing mechanical tasks – lifting, pulling, pushing – that helped build pyramids, dig furrows, and construct the earliest irrigation works. Wind, water, and firewood – renewables were in the picture long before they came back into fashion in the 1970s! – followed, accommodating more refined needs, both stationary and mobile. The emergence and intensification of the industrial revolution saw fuel wood and water power give way to coal and coal-based technology in iron smelting and steam engines. Textiles, steel, and railroads were among the significant manifestations of increased and increasingly versatile energy resources. Electrification and the petroleum age sustained the process into the twentieth century with developments – for example, in transport, chemicals, and metallurgy – too familiar to need much recounting here.

As in earlier times, patterns of energy demand and supply during the present century have evolved within the complex interplay of demographic, economic, and technological phenomena. Given the difficulty of explaining in any precise fashion the way in which these forces have driven the pace and character of past development, it goes without saying that attempts to plot future energy pathways are an infinitely more formidable challenge. The wreckage of dozens of energy forecasts – mine included – that have foundered with the collapse of their surprise-free assumptions, are gloomy testimony to that state of affairs. Yet, if we are to gain even an approximate sense of what future energy trends may look like – and what such trends, in turn, imply for a variety of environmental conditions – some historical perspective is indispensable.

The 55-year period 1925–1980 would appear to represent a sufficiently lengthy interval to permit some broad generalizations about long-term energy developments. Consider, however, some of the extraodinary subperiod features of those five and a half decades: global depression throughout the 1930s, World War II, a prolonged period of postwar reconstruction, and major readjustments in energy markets following the 1973–1974 oil shock. In all, scarcely more than 20 out of those 55 merit the label "normalcy".

Those observations are worth bearing in mind in reviewing the statistical highlights presented in *Figures 5.1–5.8* and *Table 5.1.* What are some of the more outstanding features of the quantitative record? Energy consumption growth – total and per capita – has been a pervasive worldwide phenomenon during this half century as a whole. With the caveat that we are referring to aggregate measures for broad geographic areas, it is clear that energy growth among developing nations and centrally planned economies (CPEs) has persistently exceeded that for the highly industrialized market economies of North America, Western Europe, and the Far East. That finding is no surprise. The pace of energy-intensive industrialization was beginning to abate in the West while, in the centrally planned group the build-up of heavy industry – reinforced by political strategy – was still very much under way. Among developing economies – many of them barely emergent from a primitive, preindustrial society – the rudiments of a modern infrastructure spelled a disproportionately fast growth in demand for commercially traded fuels and power, which were beginning to supplant traditional forms of

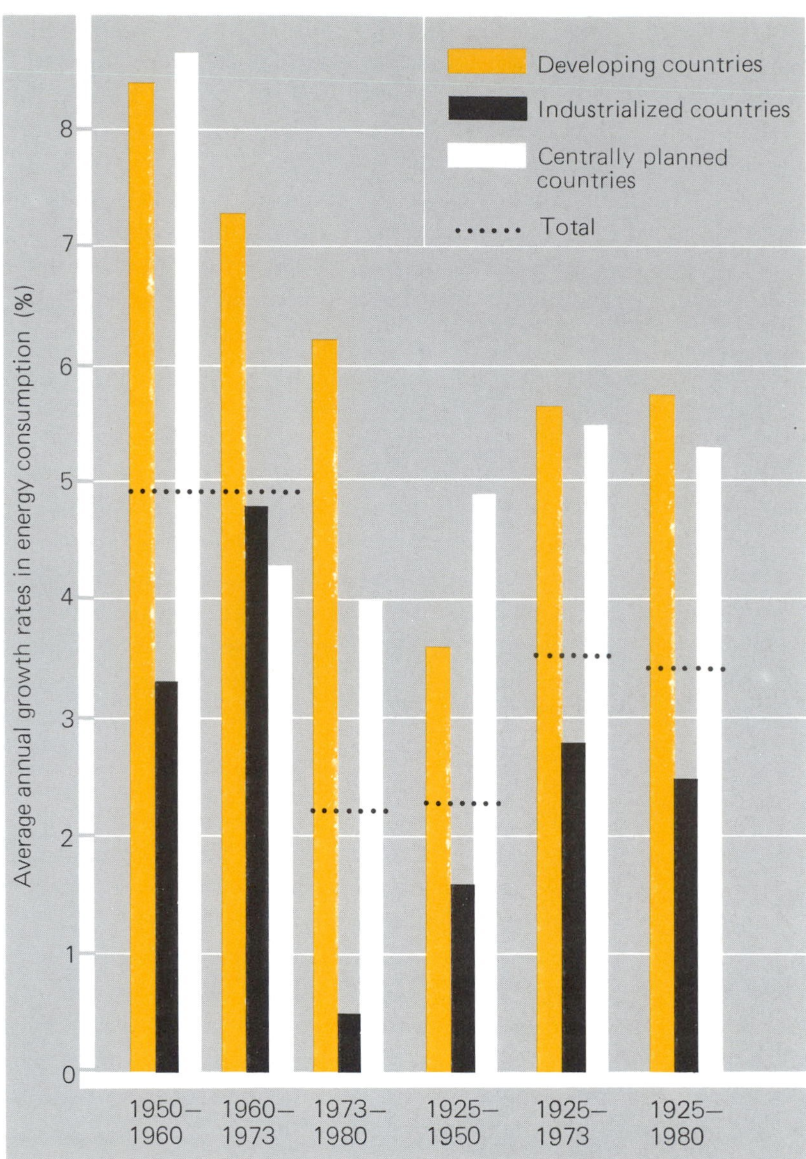

Figure 5.1 Average annual percentage growth rates in energy consumption, selected periods, 1925–1980 [1]. Among the developing countries in the period 1973–1980, oil exporters' consumption grew at a 7.7% annual rate while nonoil exporters' consumption grew at a 6.0% annual rate.

Table 5.1 World oil production, 1960–1982 [3, 4, 5].

Country group	*1960* (10^3 barrels/day)	(%)	*1973* (10^3 barrels/day)	(%)	*1982*[a] (10^3 barrels/day)	(%)
OPEC	5269[4]	20	31351	56	19150	36
USSR	3300[5]	13	8465	15	12000	23
Western Europe	288[5]	1	450	1	3063	6
North America	7553[5]	29	11008	20	9890	19
Mexico	278[5]	1	465	1	2749	5
All other	9549[5]	36	3935	7	6310	12
Total	26237	100	55674	100	53162	100

[a] Preliminary.

energy, such as wood and animal wastes. [The fact that the data in *Figures 5.1–5.8* and *Table 5.1* include only commercial energy sources contributes to the high, measured growth rates for the less-developed countries (LDCs); *Box 5.1* (p 156) contains a discussion of noncommercial energy.] Differential growth rates aside, it is worth noting that, in 1980, per capita energy consumptions for not only the LDCs, but also the CPEs, were substantially below levels achieved by industrial countries in 1925. Even without the Asian CPEs, which weight the CPE-wide total substantially downward, the 1925 USA per capita figure was about 25% greater than recent levels in the USSR and Eastern Europe.

Shifts among energy sources during these 55 years mirrored economic, technological, and geological changes that occurred throughout the world. Some parallel developments were serendipitous. Large-scale petroleum discoveries, perfection of the internal combustion engine, rising incomes in numerous (though, of course, by no means all) countries all insured that the automobile age would be one of the revolutionary fixtures of the century. Dieselization plus the adequacy of oil led to the gradual replacement of coal-burning steam locomotives, with a concurrent enhanced transport efficiency. Agricultural mechanization was similarly aided by the joint contribution of technology and

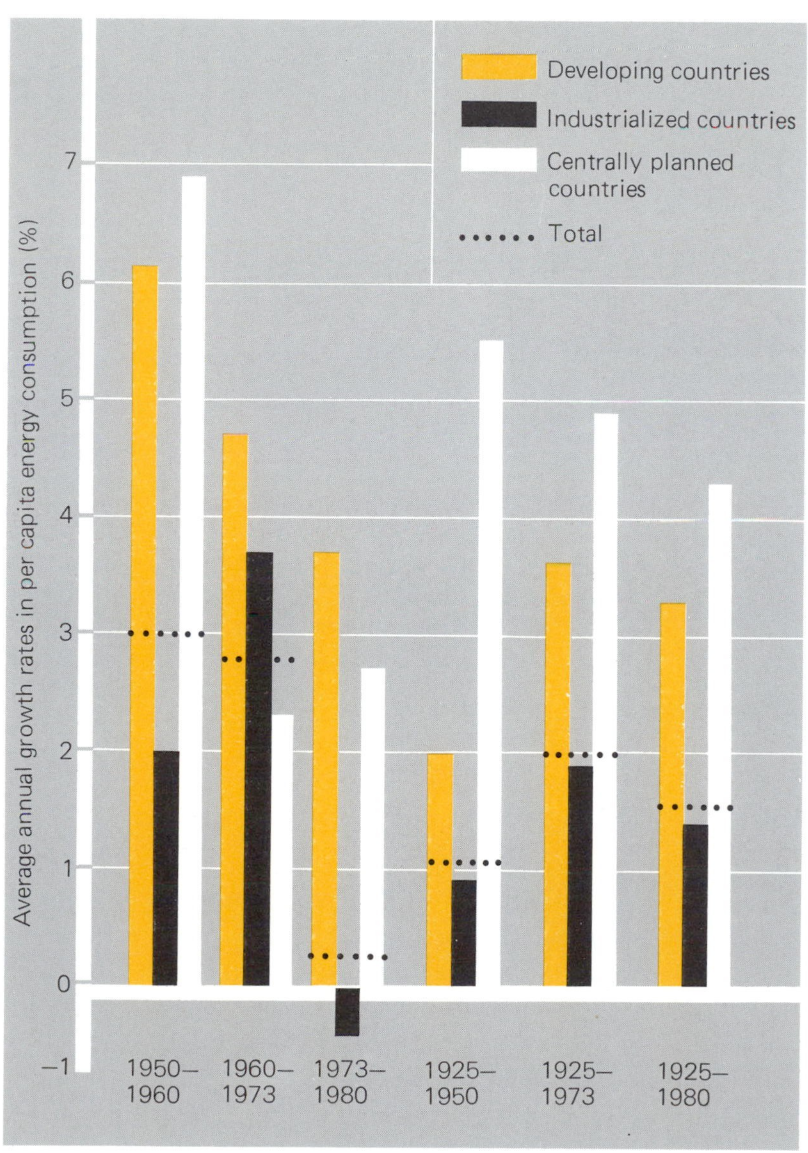

Figure 5.2 Average annual percentage growth rates in per capita energy consumption, selected periods, 1925–1980 [1].

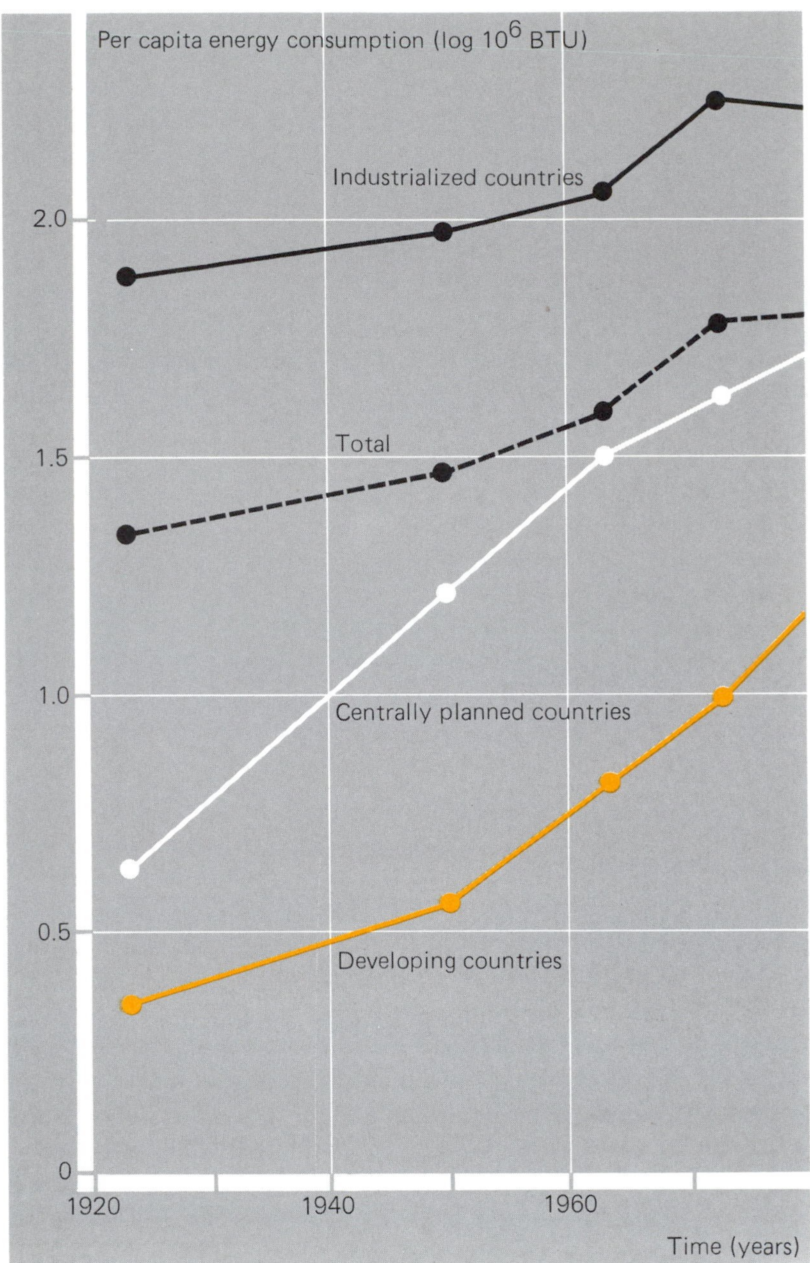

Figure 5.3 Levels of per capita energy consumption (log scale) 1925–1980 [1, 2]. (Since only selected benchmark years are plotted, intervening years should not be construed as lying along the lines connecting those benchmarks.)

liquid fuel supplies. Where incomes and fuel availability permitted, gaseous and liquid energy became the fuels of choice in residential energy use. As we see in *Figure 5.6,* the long-term reduction in coal's relative importance is one of the striking energy trends of the half century we are surveying. An equally important development has been the persistently faster growth of electricity than of primary energy in the aggregate. This long-term trend is reflected in *Figure 5.8,* showing the steadily rising proportion of primary energy supplies being used at generating stations for electric power production. Along with gaseous and liquid fuels, increased availability of the electric form of energy has been of enormous significance in the shaping of twentieth century economic life. It goes without saying that what has been highlighted here applies to a different degree – in some cases markedly so – to different

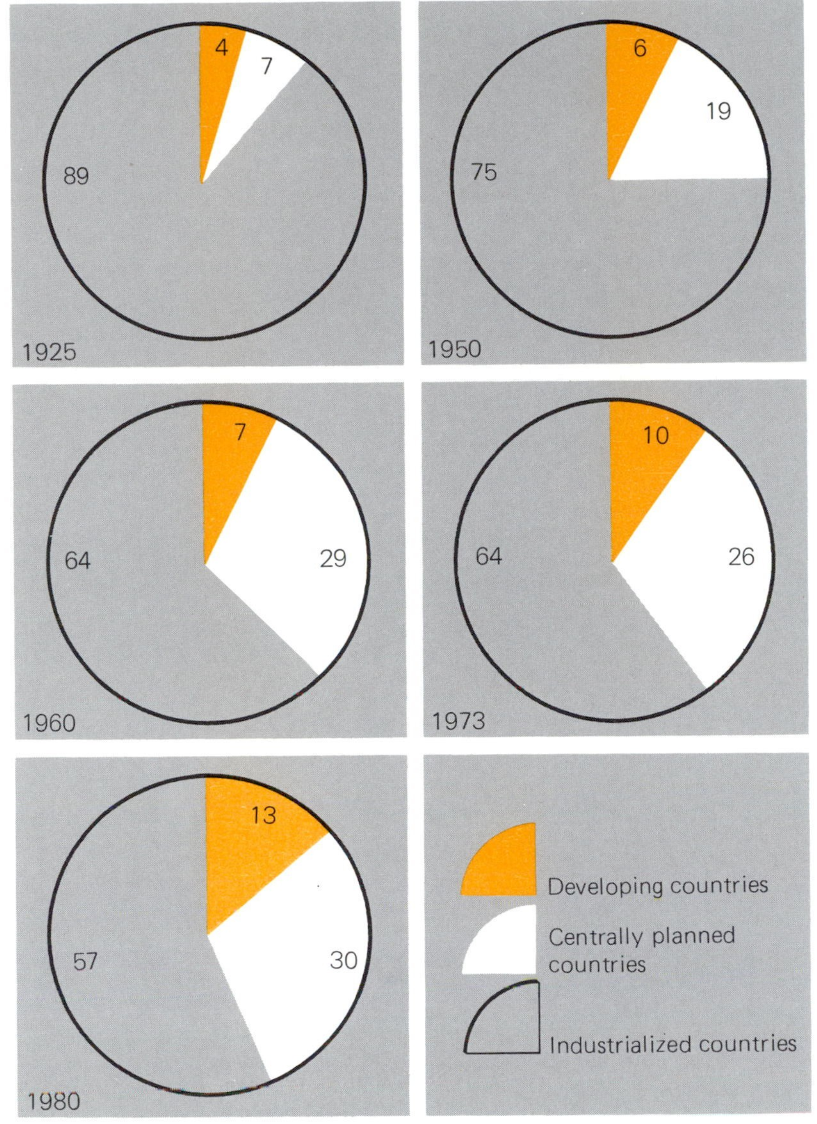

Figure 5.4 Percentage share by regions of worldwide energy consumption, selected years, 1925–1980 [1].

parts of the world. The earlier reference to disparate levels of per capita energy consumption underscores this point, to which we will need to revert in pondering future global energy requirements.

Over most of the 55-year period we are looking at the developments outlined above were accompanied by, and benefited from, declining or at least stable real energy prices. Indeed, in reviewing past global energy developments, it has become customary to denote 1973 as a watershed year: the turning point that separates a period of energy abundance, price stability, and relative political tranquility – at least as regards energy – from a period of perceived scarcity, rising real prices, and political upheaval in major oil-producing regions. Related to this, 1973 serves as a benchmark that symbolizes the emergence of Middle Eastern oil as an indispensable and major contributor to world energy supplies. Without questioning the usefulness of considering 1973 as a milestone for comparing prior and ensuing energy trends and developments – particularly for analyzing the economic dynamics of energy demand and supply – I suggest that one ought not to push the landmark significance of that year to an extreme degree.

For one thing, notwithstanding the virtually instantaneous threefold increase in the OPEC oil reference price (in real terms), abundance and

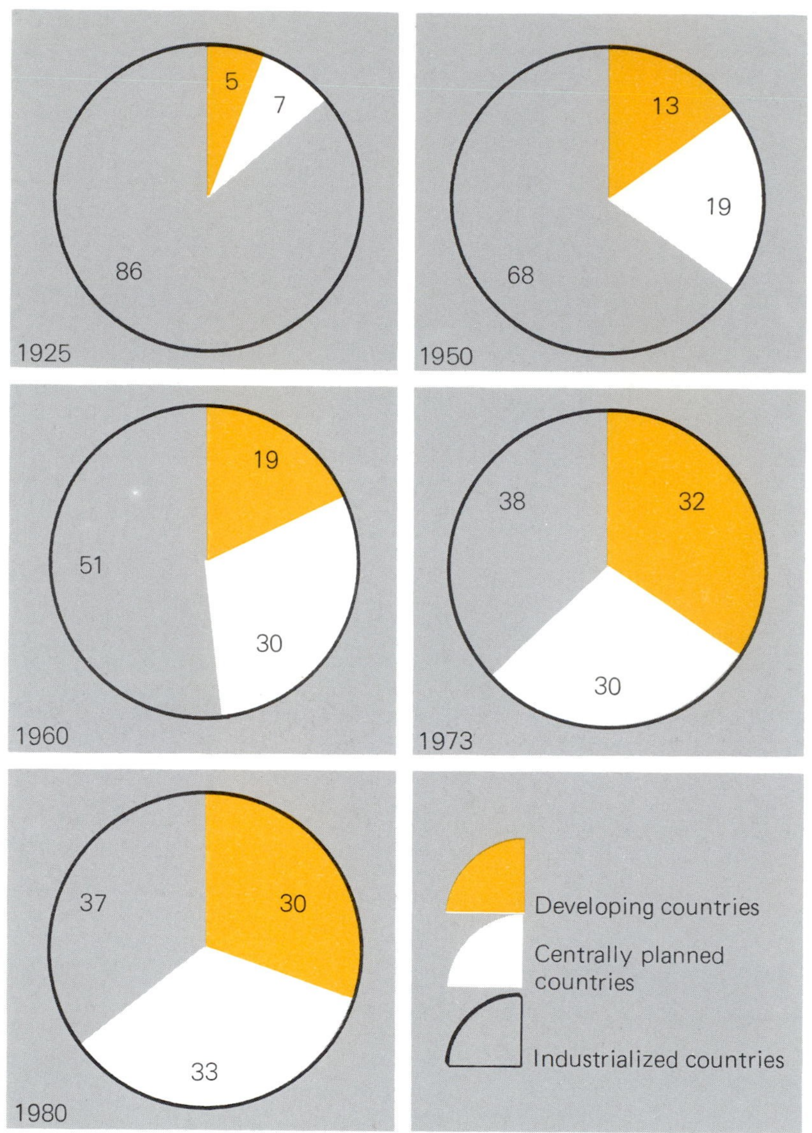

Figure 5.5 Percentage share by region of worldwide energy production, selected years, 1925–1980 [1].

scarcity could not possibly be demarcated by a single, overnight event. Indeed, as an economic purist, one accepts the trivial fact that oil, an economic good bearing a price, has always been "scarce" along with other economic goods traded in the market. Correspondingly, the production cost of oil – least of all the lifting cost – could obviously not have skyrocketed overnight. Which is not to deny that the circumstances of 1973 may, for the first time, have provided overseas producers with the market environment for exacting a part of the opportunity cost (reflecting the value of a depleting asset) that, in prior years, had been preempted to a significant degree by the concessionary companies.

What was that 1973 market environment? Oil demand in the major industrial countries had, for a number of years, been rising rapidly. (In the USA, import controls had channeled most of this demand onto domestic producers who, as a result, gradually eroded their spare producing capacity.) The coal industry was in a state of deterioration. Boom economic conditions coincided around the globe. OPEC confronted an inviting state of affairs for flexing its muscle, but the temptation to see 1973 and what followed exclusively in terms of monopolization of oil supply must be tempered by keeping in mind the particular conditions prevailing at the time.

Keeping in mind, therefore, the arbitrariness of interval selection, one may still glean some useful insights from reviewing major energy trends pre- and post-1973. Note, from *Figure 5.1*, the virtual halt to energy consumption growth among the

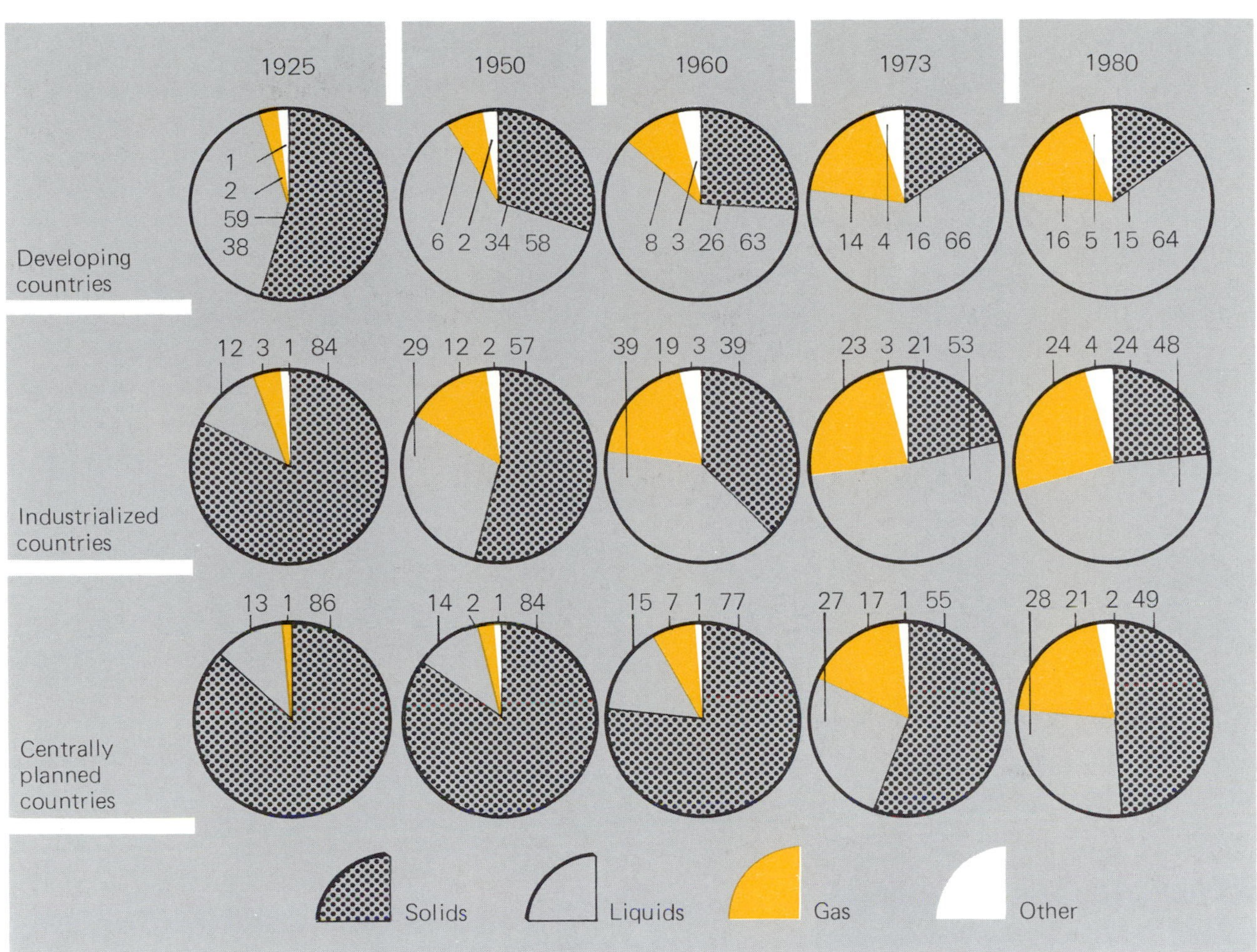

Figure 5.6 Regional energy consumption, by source, selected years, 1925–1980 [1]. The source "other" was dominated by hydroelectricity prior to 1950 and by hydroelectricity and nuclear power thereafter. (Figures rounded to nearest percent.)

industrial market economies. Elsewhere, the slowdown in demand growth was fairly small. In the case of the LDCs, it was, in fact, the sustained flow of recycled petroleum dollars that enabled these countries to reasonably maintain their pace of economic growth and, concomitantly, their use of energy throughout most of the post-1973 period. Of course, the very success of that process contributed significantly to the ensuing LDC debt management problem which, in the early 1980s, provoked a rude economic retrenchment in a number of countries.

The slowdown in energy consumption growth among industrial market economies (whose weight is large enough to shape the aggregate global trend) can be ascribed to two major forces:

(1) An unprecedented rise in world oil prices – averaging 28% annually in real terms between 1973 and 1980 – followed by "sympathetic" increases in other energy prices.
(2) The slowdown in economic growth – itself brought on to a large measure by the energy shocks of 1973–1974 and 1979–1980. The International Energy Agency has estimated oil-induced income loss for the OECD member countries from just the second oil shock at over $1 trillion in current prices [6, p 64].

The World Bank has highlighted the income and price effects on energy consumption with two simple

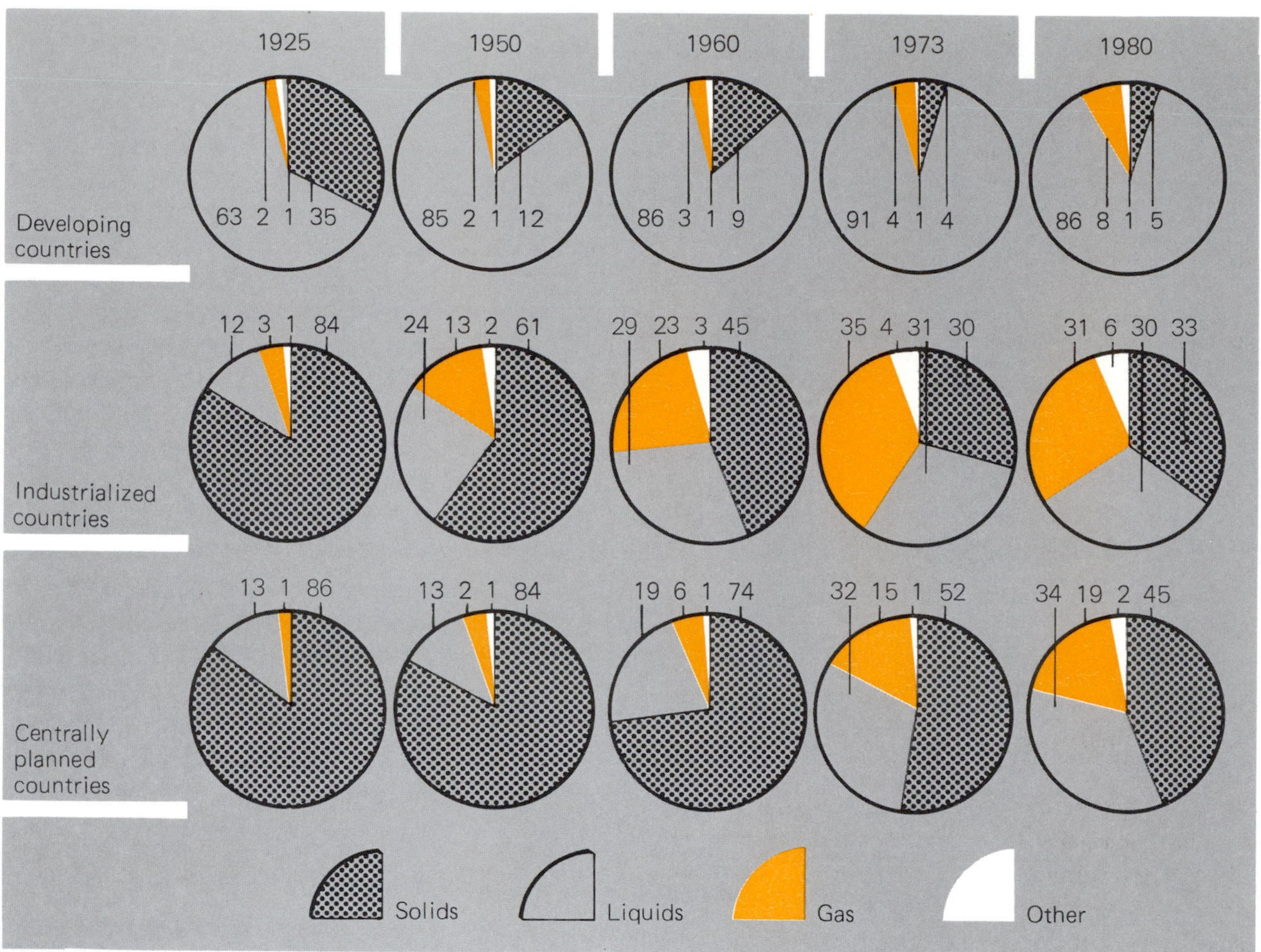

Figure 5.7 Regional energy production, by source, selected years, 1925–1980 [1]. The source "other" was dominated by hydroelectricity prior to 1950 and by hydroelectricity and nuclear power thereafter. (Figures rounded to nearest percent.)

schematics, reproduced here as *Figure 5.9.* The diagrams (for industrial and developing countries) rest on the proposition that annual increases in energy use can be represented as a function of (and statistically largely explained by) changes in income and prices. The equation expressing this relationship is:

energy growth =

$$a \times (\text{percentage income growth}) - b \times (\text{percentage price increase or decrease}).$$

a is defined as the income elasticity, the rate at which energy consumption increases relative to increases in real GDP; while b is the price elasticity, the rate at which energy consumption decreases (increases) as it becomes more (less) costly.

Income elasticities are thought to be near 1.0 in industrialized countries and around 1.3 in LDCs experiencing the early stages of industrialization. Price elasticities – much more flimsy estimates – are believed to be around 0.3 or 0.4 for both groups of countries. An important distinction must be drawn between short- and long-term price elasticities: the longer the adjustment period for responding to energy price increases – that is, the longer one allows for replacement of such energy-associated fixed assets as machinery, buildings, and transport equipment – the higher the price elasticities that can be assumed to govern. (For example, if the vehicle stock requires a minimum of 10 years to turn over, the long-run price elasticity of demand for motor fuel would not be fully reached much before that interval.)

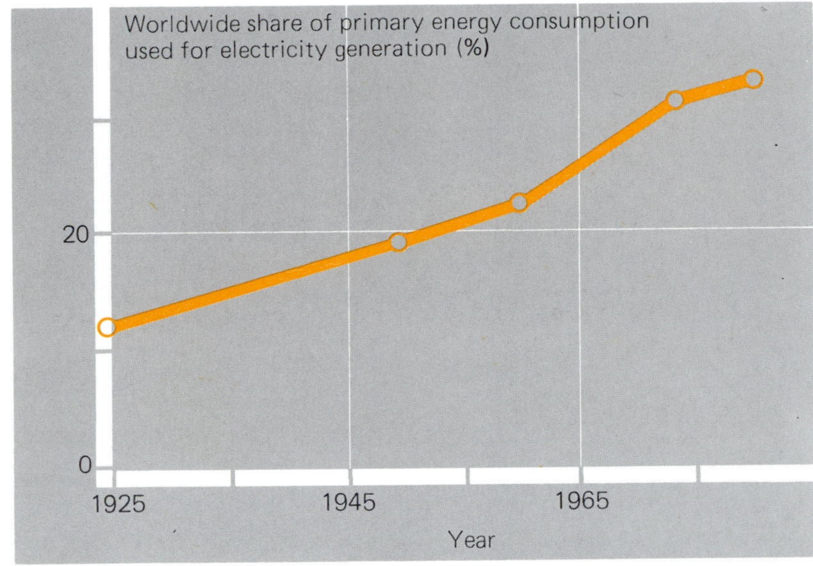

Figure 5.8 Worldwide share of primary energy consumption used to generate electricity, 1925–1980, approximations derived from [1]. Obtaining the figures involved ascribing an estimated worldwide generating efficiency factor to total electricity generation, and dividing the resultant estimate of energy inputs at power stations by primary energy consumption (since only selected benchmark years are plotted, intervening years or periods should not be construed as lying on the lines connecting those benchmarks).

The World Bank has estimated that, with no price changes, industrialized countries in 1980 would have consumed 10 million barrels/day oil equivalent (MBDOE) more than they did. These price-induced "savings" are represented by the yellow areas of *Figure 5.9.* For the oil-importing LDCs, the income effect tends to dominate the price effect so that their savings (an estimated 1.0 MBDOE) were less pronounced. For the world as a whole, energy savings calculated within this construct were estimated as 13 MBDOE. If this type of analysis is even approximately correct, it suggests that energy use and income growth show no evidence of having been fundamentally "uncoupled", *once price change is factored into the equation.* On the other hand, the limited data base from which this finding is derived underscores the importance of gaining greater insight into developments that could mean significant long-term structural change with respect to the role of energy (telecommunications is often cited as one such energy-saving prospect). We touch on this point again in our review of projections.

Figures 5.1–5.8 and *Table 5.1* show a number of other interesting pre- and post-1973 comparative energy trends. Being largely self-evident, they require little discussion here, other than to emphasize just a few highlights. In the industrial West, the precipitous pre-1973 decline in the share of coal in total energy used appears to have been arrested; the sharp rise in oil reversed. In contrast to fairly marked shifts in fuel shares prior to 1973, these shares have been broadly stable in the years since for each of the major regions, on both the consumption and production sides. In the worldwide distribution of oil output, however, several striking trends (shown in *Table 5.1*) emerge: OPEC's share went from 20 to 56% between 1960 to 1973; that proportion subsequently receded sharply to 36%. And both Western Europe's and Mexico's shares are now 5–6 times their 1960 global proportion of 1%. One may argue about how enduring an impact the events of the 1970s will eventually have; for now, it is hard to avoid the conclusion that these upheavals induced significant changes in world energy patterns.

Order or disorder in long-term energy trends? Some critical comments

As noted, the 55-year period on which we are focusing is one characterized by major episodes of global turmoil, including a period of precipitously higher energy prices. Arguably, such events dominate any independent and underlying forces shaping the energy characteristics of prolonged phases of economic history that may exist. Yet some noted analysts demur, insisting that whatever the nature of upheavals, such as those described, it is possible to discern pronounced and systematic long-term

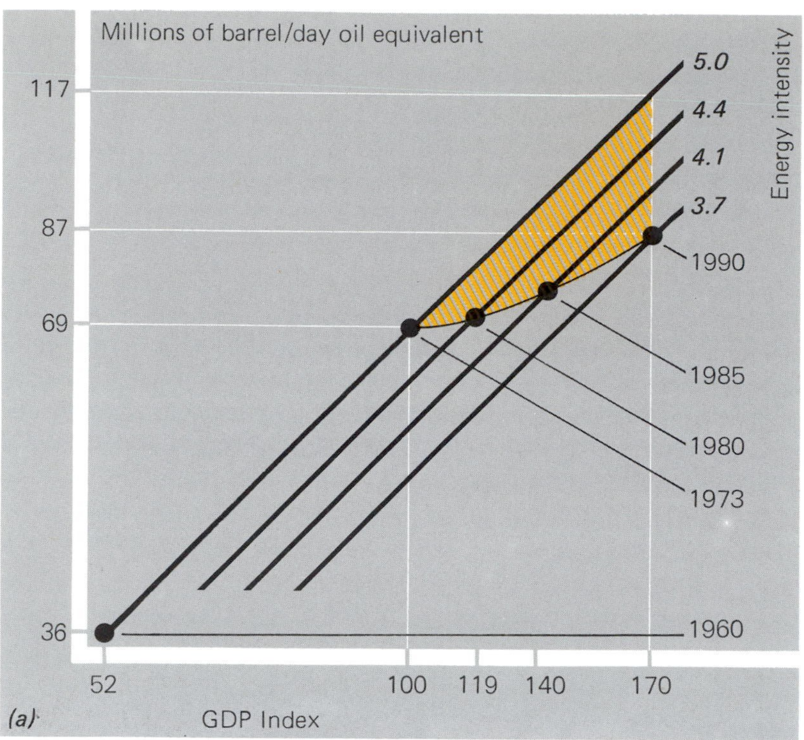

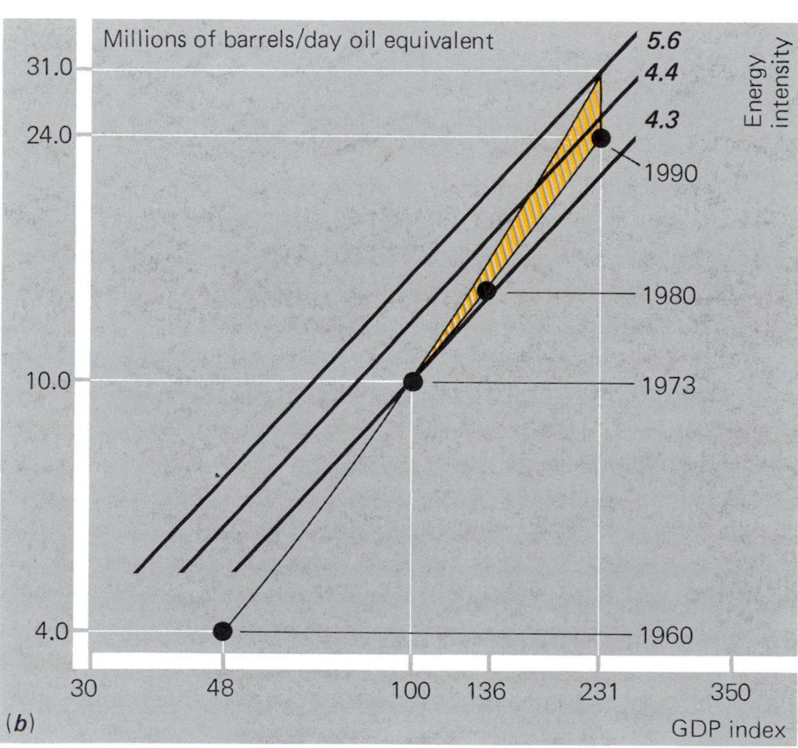

Figure 5.9 Effects of income and price on energy consumption in (*a*) industrial countries and (*b*) oil-importing developing countries, 1960–1990 [7]. Numbers in italic are in units of barrels/$1000 GDP; yellow area shows the reduction in consumption due to price increases. **Note:** Since the GDPs of developing countries are believed to be significantly underestimated, their *level* of energy intensity is probably greatly overstated.

periods of economic change, each bearing the stamp of distinct energy systems and technologies.

A notable expression of that viewpoint appears in the IIASA study *Energy in a Finite World* [8], in which it is argued that there prevails a predictable, regular pattern of one primary energy system substituting for another over time. "(E)vents such as wars, economic depression, or the recent oil embargoes . . . produced only small deviations from the long-term substitution paths [8, p 101]." In spite of acknowledged data gaps of serious proportions, the model is characterized as robust and capable of generating "stable estimates of so-called 'takeover times' that are relatively constant for any given system and that extend over several decades." A stylized representation of the process is given in *Figure 5.10* (similar to one in the IIASA book, except that it includes more recent information and has a more distant time horizon). The figure shows global primary energy substitution between the middle of the nineteenth century and recent years, and plots a projected path to the year 2100. (We discuss energy projections in the following section.) What the figure seeks to convey is the extent to which one can impose a simple function onto the actual path of market penetration. (The transformation on the left is the ratio of a given energy source's market share to the share of all other sources. Thus, for example, coal's market share between 1880 and 1920 increased from about 50 to 60% and the transformation factor increased from 1.0 to 1.5; the corresponding logarithmic representations are 0 and 0.18, respectively. Hence the 1880 and 1920 plots on the left-hand scale are 10^0 and $10^{0.18}$, respectively.)

If validated, the thrust of the IIASA analysis would be highly significant. It would mean, for example, that the peaking of coal in 1920 could have been predicted many decades earlier; similarly, the peaking of oil around 1970 (if that turns out to have been the case) could have been anticipated. And most important, future waves of energy innovation and penetration could be pinpointed with a fair degree of assurance.

The prime exponent of the applicability of the substitution model is Cesare Marchetti, whose specialized work in the area the IIASA authors have incorporated into their study. Marchetti is not only confident of the explanatory power of the substitution model with respect to energy; he generalizes its relevance to all kinds of technological change and product development. On energy, he deems concepts of prices and resources entirely unnecessary to describe the system, describing the logistic representation of the data "snug" for very long periods of time. "To a physicist's eye, present-day econometric models still look much like toddling and stuttering. What I think is most dangerous and misleading is their blind devotion to monetary concepts [10]."

In turning his attention from energy to an analysis of technological change in general, Marchetti preserves his repudiation of prices, capital, and other economic variables. Indeed, all his time series involve physical quantities, such as number of objects or tons of coal. Focusing primarily on the putative bunching of innovations, their penetration rate, and the ultimate market absorptive capacity for the various resultant goods and services – and these can be as disparate as "margarine, vacuum cleaners, or the theory of relativity" – Marchetti states that, in this way, he has succeeded in cutting "the Gordian knot of congested econometric hypotheses, where the result usually depends on the investigators' opinions, or more often on those of the sponsor [9]." The materials introduced in support of his conclusions are spotty and hard to unravel. But Marchetti is persuaded that there exists an overarching time constant that describes the coincident appearance and ultimate saturation of products in the marketplace: ". . . the time for a basic innovation to go from 10% to 90% of the available market is usually 50 years [9]." Curiously, primary energy – which, one would have thought, would be one basic economic flow conforming to the time dimensions of the generalized model – is, by Marchetti's own admission, empirically somewhat anomolous. The time constant is described as more like a century; moreover, as *Figure 5.10* shows, market shares for different energy sources turn down far short of 90%.

The principal agent or agents that cause industries and products to flourish and recede in the characteristic pattern adduced in the model are never clearly articulated, though references to human population waves as an analog give the analysis a quasi-anthropological cast. These ambiguities notwithstanding, Marchetti is able to state that the "present innovation rush will formally start in 1984 with 10% of the basic innovations

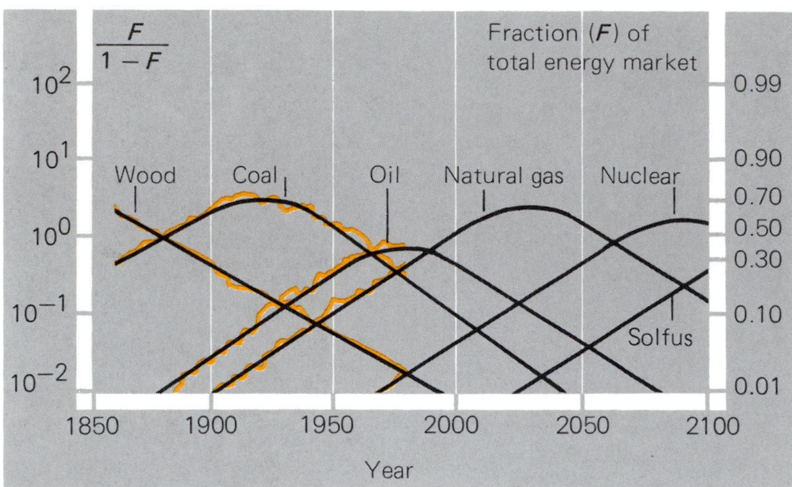

Figure 5.10 World primary energy substitution, Marchetti representation, 1850–2100 [9, p 335]. Fraction (F) refers to market shares of given energy sources; $F/(1-F)$ refers to the corresponding (logarithmic) transformation function. "Solfus" means solar and fusion energy. The yellow curves are from the data and the black curves are extrapolations based on the logistic model. For a discussion, see text.

introduced and end in 2002 with 90% of the basic innovations introduced;" and that, "in the field of energy, electricity consumption will pick up again in the nineties [9]."

To help put Marchetti's assertions into context, a brief excursion into long-wave theory may be instructive. Marchetti is certainly not alone in reviving interest in long-wave hypotheses of technological and economic cycles – an interest which, in any case, was never stilled, either before or after Kondratieff's work of 50 years ago: witness the work of Joseph Schumpeter, Simon Kuznets, and – more recently – W. W. Rostow. Where Marchetti appears to stand alone is in the utter certainty of his judgment about the unambiguous existence of long waves and their precise regularity; his emphatic repudiation of economic factors as an intrinsic (rather than merely resultant) factor in such cycles; and his almost unqualified confidence in the fact and timing of long waves for long-term forecasting purposes.

A useful reconnaissance article by Rostow depicting the intellectual development of long-wave hypotheses appeared a decade ago [11]. In his review, Rostow traces, in particular, those contributions of Schumpeter and Kuznets involving the importance, in cycles of economic growth, of leading sectors and technological innovations. Rostow's synthesis, focusing on concepts "cumulatively generated from Kondratieff on forward," revolves around the critical role of:

(1) The sequence of leading, disaggregated growth sectors – e.g., factory-manufactured textiles, transport, and major branches of the modern chemical industry.
(2) The food sector.
(3) Raw material inputs to the industrial system.
(4) Housing and social infrastructure.

However, in contrast to Marchetti (who, to be sure, cites many of the same technological milestones), Rostow, and those whose work he reviews, regards movements in prices, investment profitability, and production – in other words, monetary phenomena – as critical elements in whatever dynamic process gives rise to long-term cyclical patterns.

Another, more critical, contemporary perspective of long-wave hypotheses is provided by Rosenberg and Frischtak [12]. Asking "What conditions would need to be fulfilled in order for technological innovation to generate long cycles of the periodicity postulated by [Kondratieff] and his disciples?" Rosenberg and Frischtak focus on four key issues:

(1) *Clustering.* Proponents of technologically driven long-wave cycle theories would need to show why spurts of innovations, giving rise, in turn, to intensified stages of investment and disinvestment, occur in the clustered fashion described by their models.
(2) *Timing.* "The process of technological innovation involves extremely complex relations

among a set of key variables – inventions, innovations, diffusion paths, and investment activity. A technological theory of long cycles needs to demonstrate that these variables interact in a manner that is compatible with the peculiar timing requirements of such cycles [12]." If macroeconomic factors could be shown to play a dominant role in explaining the timing of new innovations, that could provide one important and plausible clue to support the theory.

(3) *Economy-wide repercussions.* To produce the aggregate quantitative features of long-wave upswings and downswings, the effect of pervasive interindustry flows of materials, equipment, and technology would need to be demonstrated.

(4) *Recurrence.* It is not sufficient to demonstrate causal linkage from innovation through investment and growth to retardation. Careful to exclude events explained by historical accident, one needs to demonstrate regular recurrent patterns in this schema, including the turning points inherent in the process. (It would, for example, be necessary to explain the introduction and diffusion of nuclear power as a major energy source as an inevitable and intrinsic element in the shaping of a 50-year long wave, and not – as some might argue – as the fortuitous purusit of nuclear fission in the course of a world war.)

Rosenberg and Frischtak are not disdainful of the fascination with long waves; indeed, they underscore the importance and value of more research. But they are skeptical that the four conditions which would affirm long waves as a distinct phenomenon have been met.

This brief discussion suggests that we are probably a long way from having at hand the empirical and logical underpinnings to substantiate long-wave hypotheses. Whether, confronted with "uncanny" hints of cyclical regularity either for energy systems or clusters of technological innovation generally, one is nonetheless justified in incorporating such a presumptive schema into long-range forecasting seems very questionable. A contrary view, however, is easy to appreciate. In Isaiah Berlin's words: "One of the deepest of human desires is to find a unitary pattern in which the whole of experience, past, present, and future, actual, possible, and unfulfilled, is symmetrically ordered." And elsewhere in the same essay: "The notion that one can discover large patterns or regularities in the procession of historical events is naturally attractive to those who are impressed by the success of the natural sciences in classifying, correlating, and above all predicting [13]." Yet, notwithstanding the problematic nature of theories underlying long-range projections, the fact remains that such projections are indispensable for a variety of purposes, a topic taken up in the following section.

Projection issues

If the preceding section has shown us to be wary of imposing a generalized model of energy development, this does not mean that there is *any* energy projection schema that does not pose major hazards and uncertainties. Yet, projections, once committed to print, often acquire a sanctity that no caveats can totally dispel, notwithstanding the fragility of such exercises – a fragility heightened by the long future time span and the degree of detail one frequently seeks to embody in the effort. Projections are hypothetical forecasts that depict the estimated consequences of specified assumptions. Accept or reject the assumptions, and you accept or reject what flows from these assumptions. Assumptions can involve judgments regarding future policy decisions – for example, measures governing plutonium reprocessing or decisions to impose tightened controls on fossil-fuel combustion; they can refer to behavioral factors, such as future trade-offs between work and leisure or between goods and services consumed; they can be technological in character, both as regards manufacturing processes and the appearance of new goods. Last, but not least, there must be explicit or implicit assumptions concerning economic variables, such as income, prices, and other factors.

To specify assumptions is not to accept the likelihood of the assumed trend or event occuring; rather, it is to explore what would follow given that it does happen. (Nor does it mean that a phenomenon *excluded* from consideration because its consideration would be intractable may not, on the grounds of probability, be seriously deserving of attention; one ignores the chances of worldwide conflict, in spite of the ominous position of the *Bulletin of Atomic Scientists* doomsday clock, an American magazine's symbol of approaching or – less frequently – receding nuclear danger.) Thus, in long-term energy projections, it is routinely assumed that the per capita income of today's poor countries will minimally grow at around 2% yearly over the next several decades. Although a more dismal course of events can hardly be ruled out, its likelihood is ignored. Perhaps there is an unwillingness on the part of analysts to contemplate the consequences, or perhaps it is assumed that the international community will forestall a lower level of achievement. In the ambiguous comments of the International Energy Agency, these countries:

> ... may continue to suffer from the effects of higher oil prices and agricultural shortfalls. The projected [economic] growth rates assume, however, that improvements will be achieved in domestic policies with emphasis on increasing exports to OECD countries as well as to other developing countries [6, p 160].

The last several decades have witnessed some notable instances of faulty projections. In the USA, for example, the 20-year growth in nuclear power capacity that had been predicted in the 1960s turned out to be wildly exaggerated; foreign oil, rather than being the force expected to *moderate* upward energy price pressures, became the *source* of this pressure. If one can rationalize those particular misjudgments – after all, OPEC can be shrugged off as an "exogeneous" unanticipatable phenomenon and one can excuse the nuclear optimism for lack of an historic trend line – it is less easy to explain an inability to spot even those developments one would have associated with a strong degree of momentum and continuity. Yet, within a year or so after the US Bureau of the Census had projected USA population growth at the postwar rate of 1.6% yearly, a slide in the birth rate brought the figure down to 1% or less [14].

Yet, however prone to error and uncertainty, long-term energy projections are necessary for investment, research, and policy decisions in both private and public sectors. The commitment of scarce resources to, say, a multigigawatt hydro project in a developing economy can hardly ignore the demand for – and alternative ways of producing – electricity at the time such a facility begins to go on stream and over its expected life span. The pursuit of atmospheric and climatological research must, at least in part, be guided by prospective demand for different energy sources. But the need for projections to assist with these and other questions should not blind us to the tenuousness of such analyses, especially the longer the time horizon being considered. When one thinks about some of the (nonmilitary) societal and technological developments in the past half century – in telecommunications, computerization, urbanization, transportation – one cannot help but wonder in what degree of detail it makes sense to speculate for 50–100 years from now. (As a technical aside, it is not even conceptually possible to allow for a fundamentally altered GNP mix. This arises from the "index number problem" as it relates to weighting schemes.) Perhaps there *is* some level of aggregation – say, total household energy demand or energy use by industry – at which, because of fortuitous offsets within the aggregate or because of other reasons, there is sufficient observable regularity to lend credence to the bounds which one can reasonably project. A better handle on this question might help channel energy modeling efforts in more rewarding directions than their somewhat obsessive numerological bent.

A final point about projections takes on particular relevance to the present project. Most multicountry or global analyses surveyed in connection with this chapter appear to make no or only limited allowance for any "feedback" effects or adaptations occasioned by the potential environmental perturbations or other spillover effects posed by the projected level of energy use and output. Thus, if a given projection of fossil fuel use implies an unacceptable degree of environmental damage, without that fact being fed back (iterated) to yield a revised (or alternative) projection incorporating a response to that damage, the outcome is a projection whose fuzziness as regards important policy issues must be recognized.

Projections of national, regional, and global energy-demand abound. Here it is possible to touch on just a few such efforts, and these only cursorily. Energy projections vary greatly in their time horizons, purpose, level of detail, and degree of methodological sophistication.

Projections of five years or less tend to be predictions; that is, underlying assumptions are expected to hold. Those of 10–20 years veer toward the type of hypothetical forecasts discussed earlier. Projections many decades into the future may embody features which also characterize the short- and medium-term efforts, but that – in the form of scenarios – frequently have an exploratory or probing objective.

Projections may have the purpose of delineating a future that is likely to unfold or – normatively – a future the analyst would like to see unfold or, at least, test as to its tenability. (Scenarios examining, say, a world without nuclear power, or constrained by limits on CO_2 release, or subsisting largely on renewable forms of energy fall into this category.)

The shorter the time horizon, the greater the degree of detail the researcher is willing to specify. Few researchers would be emboldened to project categories of end use (such as household or commercial energy consumption) forward to the year 2050 for major world regions.

Lastly, one should note the variety of projection techniques employed. Formal econometric models have the virtue of positing behavioral and structural relationships in a disciplined manner and of indicating the robustness of the estimating equations by testing them (backcasting) against actual historical trends. Key assumptions, such as those concerning price, income, and technological change, must be explicitly stated. For projections extending many years into the future, these virtues diminish and may, indeed, be an impairment. There is no reason to expect past patterns to endure indefinitely; yet formal models cannot easily allow for the effects of speculative events and departures. Not surprisingly, a recent survey showed that, for the more formal energy projection models, technological and certain other qualitative insights tended to take second place to methodological and computational manipulation [15]. No wonder: "optimization modelers" are most comfortable when they can specify all critical inputs and relationships in unambiguous quantitative terms. (If, instead of his blanket condemnation of econometrics and econometricians noted earlier Marchetti had focused on such specific shortcomings, his points would have greater force.) Thus, penetration of synfuels and, more broadly, of backstop technologies is frequently treated in a disembodied fashion, in which their appearance is tightly governed by relative supply prices and other variables and parameters. Yet, observing, for example, the chronic roller coaster behavior of synfuels projects in the USA, might not such technological components of a future energy system be treated more effectively outside, or as an adjunct to, a formal model?

Other projection efforts embrace a more eclectic approach, intermingling analytical constructs with numerous judgmental elements. The IEA's *World Energy Outlook* [6] and the World Bank's *World Development Report 1981* [7] fall into this group. An undertaking by the World Energy Conference chose to be rather more casual, stating "[p]rice-elasticity was not introduced, but the general climate resulting from the . . . scenarios is implicitly characterized by a gradual upward trend of prices in the long run . . . [16, p xxiv]."

Energy projections to the year 2000

For projections to the end of the century one has the choice of numerous studies to consult. Here we cite some findings of the IEA's *World Energy Outlook* [6] as fairly typical of other such efforts of the early 1980s. In this work, most of the analysis is directed to the energy demand–supply prospects for the 24-member OECD group of countries. In the more formal stages of its procedure, the IEA presents two alternative projections:

(1) *High energy demand.* This model is characterized by economic growth of about 3% per year between 1980–2000, falling real oil prices between 1980–1985, and a stable level thereafter. For the entire 20 years, energy demand growth rises at 2.1% yearly.
(2) *Low energy demand.* This model assumes economic growth of around 2.5% and the 1980–1985 oil price decline to be succeeded by an annual rise of 3% per year. For the 20 years, energy demand growth averages 1.5% yearly.

Finding that even the low energy demand alternative produces an undesirable level of "excess" OECD oil demand in the year 2000 of nearly 10 million barrells per day (MBD) – the high demand variant results in an excess of over 20 MBD – the IEA analysts constructed a "Reference Case" projection which, though producing total energy demand levels higher than the low energy demand case, features marked retrenchment in oil demand. This preferred variant, in which energy demand during the 1980–2000 interval rises at 1.9% yearly, judgmentally incorporates factors that are more policy- than market-driven to achieve greater oil demand–supply equilibrium by 2000. Fulfillment of the "Reference Case" depends on an enormously increased contribution of coal and nuclear power to energy demand growth: an incremental 1980–2000 share of 90% compared to their combined base-year share of 26%.

For non-OECD regions of the world, the IEA's projections approach is less refined (necessarily so for nonmarket economies). The IEA's chapter on the energy outlook for developing countries reviews the factors that are likely to produce sustained pressure on oil supplies in particular, and energy in general:

(1) Economic growth targets (of 4–5% yearly) designed to provide some margin of increased per capita well-being above continued high rates of population growth (projected at no less than 2% yearly for some years to come).
(2) The shift from locally traded noncommercial fuels (such as firewood, crop residues, animal wastes, and animal draught power) to commercial fuels, notwithstanding the potential for more rational exploitation of the former fuel category (*Box 5.1*).
(3) A relatively high income elasticity of energy demand characterizing regions that pursue enhanced agricultural efficiency and mechanization, motorization, expanded manufacturing

Box 5.1 A note on noncommercial energy

This fuel category is dominated by firewood, but comprises also other forms of biomass, animal wastes, and human and animal power. ("Traditional" would be a more appropriate adjective than "noncommercial", since at least a portion of such fuel is sold in the marketplace.) Worldwide, noncommercial energy has been estimated as representing around 5% of total energy consumption in 1980, though for less developed countries the share may be over 25% and vastly higher still for a number of individual countries. But the poor efficiency with which such fuels are used lowers their effective relationship to total energy. Data on such energy forms and their use are fragmentary; hence, quantitative analyses of energy trends tend to treat them at best in an adjunct fashion. Although careful management in their exploitation and more efficient utilization can enable these energy forms to perform an important role, historically their use has also contributed significantly to deforestation, erosion, depletion of soil fertility, siltation, and desertification of arid and semiarid regions, brought about by the inhabitants' increasingly arduous search for fuel wood. Pachauri singles out for particular notice 45 developing countries with a fuel-wood problem and a heavy dependence on oil imports. Their combined population numbers 329 million people [17]. China and India are excluded because they do not meet the oil-import dependence criterion; but they have major fuel-wood problems as well. India's rural population has been estimated by Revelle to be dependent on noncommercial sources for 90% – and on fuel wood alone for 40% – of its total energy use [18].

Future trends in noncommercial energy use and their ecological consequences are hard to project. The World Bank sees their consumption in LDCs growing by less than 1% annually between 1980 and 1990; this compares to a projected growth of 6% for commercial energy and represents a decline in the former's share of the total from 26 to 17% [7, pp 36–38]. Like the Bank, Dunkerley, of Resources for the Future, expects the proportionate importance of noncommercial fuels to fall. But her projection shows the absolute level of noncommercial fuel demand growing between 2.5 and 3.5% yearly during the period 1980–1995 [19]. The high end of the range is based on the possibility of substituting a limited amount of biomass fuels for imported oil in electricity generation, transport, and selected industrial activities.

Undoubtedly, the trend that finally evolves will depend, among other things, on price- or policy-induced constraints on fuel wood exploitation, enhanced income and education, relative price changes for alternative energy sources, rural out-migration, and programmatic initiatives of various kinds. (The Brazilian government-sponsored ethanol program is an example.)

While forest management in general may arguably have a role to play in dealing with an emerging CO_2 problem, it seems unlikely that noncommercial fuel use can, by itself, have much of an effect in that respect.

activity, and the establishment of basic infrastructural facilities.

For LDCs in the aggregate, the IEA projects an energy consumption growth during 1980–2000 that ranges between 4.6 and 5.6% annually. This comprises both OPEC and non-OPEC countries; and, for that combined total energy demand–supply balance appears ensured. The wherewithall to finance the energy requirements of the non-OPEC LDCs is the more formidable problem.

Energy consumption growth for the CPEs as a group is projected by the IEA as ranging between 1.8 and 2.3% for the 1980–2000 period. The principal energy challenge facing the CPEs revolves around the imperative for interfuel substitution as the USSR's capacity for expanded production and net export of oil (to both Eastern and Western Europe) diminishes. The best supply prospects involve USSR natural gas. China's oil potential is still problematic. Eastern Europe appears to enjoy only limited opportunity for energy supply expansion.

A summary of the IEA's regional energy consumption projections is compared with the long-term historical record in *Table 5.2.* Note that even the high end of the 1980–2000 ranges implies a deceleration in growth rates. The slowdown is especially marked for the planned economies.

Aggregating these regional perspectives, the IEA sees the implications of its exercise as pointing to a fairly tenuous world energy picture over the next several decades. OPEC maximum oil production in 2000, for example, is estimated at 28 MBD and no increase in maximum sustainable capacity beyond something over 30 MBD is foreseen. Then, too, the robust expectations for coal and nuclear energy, on which the Agency had pinned such great hopes, have dimmed. The prospects for quantitatively significant penetration by synthetic fuels and solar energy (slighted in the IEA analysis) are poor. Reemergence of a tight oil market by the end of the century could be one consequence. In their worldwide implications, the IEA projections point to primary energy demand–supply balances in 2000 ranging from an approximate equilibrium, achieved by modestly higher real prices compared to levels in the early 1980s, to a supply shortfall as high as 50 MBDOE (equivalent to about 100×10^{15} BTU and roughly 20% of worldwide energy consumption that year), implying presumably strong upward price pressures.

Others may see things less disturbingly. Market adjustments induced by energy price on both the demand and supply sides have surprised skeptics who saw active policy intervention in energy markets as necessary. In the USA, for example, conservation has been a pervasive, economy wide phenomenon. Then, too, the prospects for USA gas supply expansion look better than the IEA posits in even its reference case. Non-OPEC developing countries (other than Mexico) may develop greater oil-producing potential than the IEA – conservatively – assumes.

Table 5.2 Summary of IEA's regional energy consumption projections [6].

Region	*Average annual rates of growth (%)*	
	1925–1980	*1980–2000*
Developing countries	5.7	4.6–5.6
Industrialized countries	2.5	1.5–2.1
Centrally planned countries	5.3	1.8–2.3
Total	3.4	2.1–2.8

Projections beyond the year 2000

In pondering energy pathways well beyond the turn of the century, one is well advised to dispense with fine-tuned estimating refinements and to concentrate on some overarching questions: What sort of a world – economically and socially – is subsumed by different rates and patterns of energy demand growth? What do such trends imply for claims on the energy resource base? What is implied for worldwide income inequality? What issues are posed by potential environmental conflicts or constraints? What is the prospect for adapting to, or sidestepping, such barriers?

If the IEA type of exercise provides cautious reassurance that global energy demand and supply can be in balance at tolerable prices in 2000, analyses looking beyond that milestone tend to be far less sanguine. For example, the World Energy Conference (WEC), in connection with its triennial 1983 meeting, concluded that the cumulative claims on global energy resources implied by its demand projection can be satisfied only if one assumes a marked shift in demand patterns away from oil (or conversely, assumes the feasibility of large-scale conversion into liquids of coal and other synthetic feedstocks), recourse to renewables, and/or the exploitation of uranium resources in advanced reactor technologies [16]. Worldwide, the WEC projects energy demand growth at a rate which slows to 2.1% yearly during the period 2000–2020. (It calls this projection a normative one, in distinction to a lower-case projection termed undesirable in its implications for LDC economic prospects.)

A study of a somewhat different character, already referred to at some length, was the IIASA modeling effort released in 1981 [8]. In its own words, "The study explores the possible features of global energy systems that are not just post-oil systems, but *post-fossil* systems." And it explores how, realistically, the world might successfully negotiate the transition to such systems. The IIASA study implies that the resources and technology are available to sustain the anticipated increase of the world's population to a level of 8 billion in the year 2030; but it offers no assurance that the requisite political, economic, and institutional hurdles toward that end can be overcome. One reason for the study's somewhat pessimistic tone is the Marchetti notion outlined above: that penetration of new energy systems or resource technologies materializes in characteristically ponderous, long-term swings, which show little historic evidence of being manipulatable, in part because they occur within the more general context of recurrent long waves of global economic expansion and contraction.

Our earlier remarks raised a number of critical questions about this line of argument. Since we are now concerned with judgments about the unfolding of future events, it is particularly important to reflect on the extent to which we are locked into a rigid, inevitable process; or face, at worst, considerable uncertainty; or, at best, some measure of flexibility and choice. Even in terms of the IIASA study's own substantive explorations and discussion, the last case is the more persuasive one. For example, worldwide coal reserves are so huge that, if their use is not environmentally constrained, coal's eroding market share shown in *Figure 5.10* might arguably be reversed. Unconventional gas resources are very likely also large, though not in the coal league. Breakthroughs in extraction technology, coupled with improvements in seaborne transport, could make sustained gas use at least as likely as fusion, pictured as taking off early in the twenty-first century. Even as a stylized representation, *Figure 5.10,* by focusing exclusively on energy sources rather than on the technological facets of their exploitation, rules out the possibility of time profiles that could be quite different from those plotted. Then, too, there are sufficient references, in both the IIASA study and Marchetti's writing (referred to earlier), of the extent to which policy intervention can deflect energy-system patterns from preordained paths and cycles. Research and development (R & D) initiatives, active nuclear energy promotion or moratoriums, and major infrastructural projects fall into this category. That France seems to have made a clear-cut choice for nuclear generation while Denmark has reverted to coal-based power is evidence of how decisively policy departures can attenuate, rather than reinforce, the technological forces that might have been expected to dominate electricity production trends.

But it is only fair to point out that the IIASA authors' judgments are not governed dogmatically by the schematics of the model, however rigid the model may itself appear to be. Thus, in its more pragmatic slant, the study finds that full use of all available energy resources will be required as we look ahead to the first several decades of the next century. Dirtier and more expensive fossil resources and vast quantities of synfuels will have to be developed transitionally, along with both nuclear breeder reactors and large-scale solar plants. In the view of the IIASA authors, small-scale solar and other renewable energy sources, while assumed to play a growing role, are likely to satisfy only a modest fraction of the total global energy demand beyond the first quarter of the twenty-first century.

Underlying the concern expressed in reports such as those of WEC and IIASA are statistical comparisons between the estimated size of the resource stock, on the one hand, and the cumulative consumptive claim on that stock, on the other. For example, the cumulative energy consumption total between 1980 and 2030 at, say, a 2% rate of growth worldwide is 24×10^{18} BTU. Yet, as *Table 5.3* shows, the entirety of the existing oil and gas reserves plus the estimated resources of these fuels that might become reserves would be absorbed by this consumption stream; while, to be useful, resources in abundant supply would need to be convertible into forms suitable to the evolving pattern of end-use demand.

Table 5.3 World recoverable reserves and resources of conventional mineral fuels, as estimated in the late 1970s [20].

Fuel	*Reserves (10^{18} BTU)*	*Resources (10^{18} BTU)*
Coal	18	141
Oil	4	8–12
Natural gas	3	9–10
Uranium:		
without breeders	7	17
with breeders	443	1003

The historical dynamics of resource development, the conceptual dubiousness of physical scarcity, and the capacity for substitution in resource use and supply all argue against the running out idea as a central analytical construct in energy planning. Technological change induced by rising resource costs – for example, development of coal liquefaction processes triggered by rising oil prices – need not be any less likely to occur in the energy field than, say, in agriculture. (See the discussion of induced technical change by Crosson, Chapter 4 of this volume.) Nonetheless, the figures just cited do serve a useful illustrative purpose insofar as they prod examination of policy initiatives and processes – e.g., in the areas of synfuels and solar R & D – that could help smooth future energy transitions in a timely fashion.

Three other projection studies were instigated by, in particular, concern over long-term environmental implications of energy use. These were prepared by the Institute for Energy Analysis (IFEA), Massachusetts Institute of Technology (MIT), and the National Academy of Sciences (NAS). Brief comments on each follow.

The long-term regionalized energy model developed at the IFEA, Oak Ridge, Tennessee, tests the sensitivity of CO_2 releases to assumed rates of growth in total energy consumption and to interfuel use patterns [21]. In a base-case representation, the authors project the period growth rates in global energy consumption and associated CO_2/energy elasticities, see *Table 5.4.* For most of the post-World War II period, growth in CO_2 release relative to energy growth was about 0.92, with successive penetration into the fuel mix of oil and gas at the expense of coal signifying some progressive diminution in carbon content per unit of aggregate energy. While the CO_2 implications of energy growth are quite muted for the balance of this century (influenced, to a considerable extent, by nuclear energy's gain in market share), the situation will change in the twenty-first century as heavy reliance on coal and shale intensify CO_2 loadings.

Table 5.4 IFEA projections for CO_2 release and energy consumption [21].

Period	*Energy consumption growth rate (%)*	(ΔCO_2 release) / (*Δ energy consumption*)
1975–2000	2.5	0.61
2000–2025	2.6	0.88
2025–2050	2.4	1.30

The consequences for atmospheric CO_2 concentrations depend on what is absorbed by various sinks (oceans, biomass) and what remains as the airborne fraction. But the authors suggest that (CO_2 release)/(energy consumption) coefficients and the airborne fractions would have to settle at values well outside what they regard as plausible ranges to prevent a near doubling of CO_2 concentration occurring by 2050 or soon thereafter. The 600 p.p.m. implied (compared to an estimated 339 p.p.m. in 1975) has become a kind of reference point for the likelihood of measurable, global average surface temperature increase [22].

The IFEA authors conclude their analysis by recognizing the commons nature of the CO_2 problem – i.e., where the absence of mutual restraint mechanisms removes any *single* country's willingness to avoid fouling the global nest (see pp 164–165). The study probes several alternative control devices (ignoring, however, the effects of a major shift to solar energy), ranging from a purely USA CO_2 tax (a naive case, in the IFEA authors' own characterization) to a global CO_2 tax accompanied by USA coal export curtailment. However, such a severe restraint measure (which reduces global CO_2 emissions by 40% compared to the base case in 2050) does not forestall a doubling of CO_2 concentration beyond around 2080. And since the perceived economic sacrifice of such environmental discipline may be more than many countries, such as China with its large coal resources, would tolerate, the prospects for successful collective action are dim. However, the authors sought primarily to lead the reader through some quantitative trends and relationships rather than explore the political dynamics of achieving environmentally desirable outcomes.

The MIT study [23] is not wholly independent of the IFEA effort (indeed, few pairs of energy analyses are totally free of incest), insofar as its series of energy scenarios are cast within the IFEA group's model structure and depend on the latter's exogenous specification of demographic and economic variables. But the MIT report explores a wider array of energy demand–supply paths to the year 2050 in order to test the feasibility of less threatening CO_2 outcomes. Indisputably, the lower the evolving ratio of energy use to GNP (where that lowering is assumed to occur due to inherent, non-price driven efficiency improvements and other changes), the lower the degree of environmental insult of all kinds. The question is whether such long-term energy/GNP improvement is likely. The MIT authors argue that it is, citing the historic decline in the USA energy/GNP ratio during prolonged intervals when relative energy prices were *falling*. The prospect of that phenomenon occurring on a worldwide level over the next 65 years cannot be dismissed, but the presumption would benefit from some concrete research. Can a declining energy/GNP ratio be projected into perpetuity, implying steadily less attractiveness in energy as an economic good? Perhaps, if an energy–food analog holds; beyond some ultimate level of caloric intake, further consumption, even if not subject to outright saturation, could rise proportionately less than income.

At the very least, discrete study of different country groupings, according to the stage of economic development, is vital in gaining a better understandng of the energy–economic growth relationship and its global implications. If saturation and the growing role of service industries are, arguably, energy-dampening characteristics of the most advanced countries, much remains to be understood about the extent to which energy use by LDCs is amenable to the stages-of-growth paradigm, however useful a construct this may be in other connections. (To assume that Chad will display energy–economy characteristics reminiscent of industrial countries at a similar stage of development is to disregard the fact that numerous technologies available to Chad – e.g., electrification – were, at best, in their infancy in the earlier Western case.)

Aside from the contribution that improved energy efficiency can make, the MIT study examines, and finds substantial promise in, several other low-CO_2 scenarios. One of these involves the successful development of solar photovoltaics and a resurgence of economic nuclear power. The two sources combined imply an intense degree of electrification and, therefore, the use of electricity in a much wider range of applications than at present. Technological progress in economic electric storage systems is likewise implied by this scenario. In such a future, the constrained use of fossil fuels signifies a conspicuous divergence between the energy trend line and the much more muted rate of change in CO_2 releases.

The NAS–National Research Council report on climate change includes an analysis by Nordhaus and Yohe of alternative energy futures and their CO_2 implications [24]. Their technique – probabilistic scenario analysis – involves the specification of a range of possible values for the major variables or parameters in a global economic–energy–carbon cycle model. The outcome of the effort is not only a best guess trajectory of the critical indicators, but also a set of alternatives characterized by each one's likelihood of occurrence. Notwithstanding the subjective elements in the estimating process, the NAS authors underscore the importance and utility of the probabilistic features of the model for various policymaking options: guidance for deciding on more research to narrow the uncertainty range; avoidance of policy-driven paths whose outcome is judged too risky; or simply a basis for proceeding with policies governed by an expected value.

The NAS authors' time horizon extends to the year 2100. They describe their results as generally pointing to a lesser degree of carbon emissions and build up than predicted in earlier studies they surveyed, citing two underlying reasons: deceleration of global economic growth and price-driven substitution away from the fossil fuels as these become scarcer. At the same time, Nordhaus and Yohe emphasize how strikingly these two factors contribute to the overall uncertainty about CO_2. Consequently, they express the need for more research on worldwide economic growth and on interfuel substitution. They nevertheless pose a 1-in-4 possibility of a doubling in atomospheric CO_2 concentration occurring before 2050 and even odds for it happening during 2050–2100. Like the authors of the IFEA study, Nordhaus and Yohe perform simulations to test the sensitivity of long-run (end of the twenty-first century) CO_2 concentration to worldwide fossil-fuel taxes. They conclude that these would have to be quite forceful to have a significant impact. A unilateral national tax would clearly have to be steeper yet.

One hesitates to reproduce the most likely set of outcomes from the Nordhaus–Yohe effort, for their contribution – however insightful – is more a blending of conceptualization and statistical techniques than of empirical investigation or technology assessment. Indeed, there are a number of abstract qualities and simplifying assumptions in the Nordhaus–Yohe schema which, without further development, tend to limit its applicability to the real world, including issues of environmental perturbations that affect the biosphere. For example, the model is necessarily very aggregated since the use of probability distributions for even a limited number of input variables quickly translates into a major computational task. That is why, in contrast to the IFEA model, the world is treated as one region (some attention focused on different regions' fuels and fuel-using technologies would surely be required); only two energy sources – fossil and nonfossil – are consumed; all nonenergy inputs (capital, labor, other resources) are largely dealt with in a consolidated fashion; and unitary income elasticity is assumed indefinitely. (Large-scale synfuel programs – signifying, for example, rising primary BTUs of coal per BTU of final energy – would alone render the last assumption suspect.) And, just as Marchetti spurns economically determined outcomes, so the Nordhaus–Yohe model is intimately governed by production theory and the economics of resource exhaustion.

Table 5.5 Most likely annual growth rates of critical variables, Nordhaus–Yohe projections, 1975–2100 (%/year) [24, p 138].

Variable[a]	*1975–2000*	*2000–2025*	*2025–2050*	*2050–2075*	*2075–2100*
GNP	3.7	2.9	1.5	1.5	1.5
Energy consumption	1.4	2.7	1.2	1.1	1.2
Fossil fuel consumption	0.6	2.5	0.9	0.5	0.4
Nonfossil fuel consumption	5.6	3.1	1.8	2.0	2.0
Price of fossil fuel	2.8	0.3	1.2	2.9	1.1
Price of nonfossil fuel	0.5	0.1	0.1	0.1	0.1
CO_2 emissions	0.6	2.6	1.2	0.9	0.4
CO_2 concentrations	0.3	0.6	0.8	0.8	0.8

[a]Absolute values for 1975 were: GNP, $\$6.4\times10^{12}$; energy consumption, 244.3×10^{15} BTU; fossil fuel, 224.9×10^{15} BTU; nonfossil fuel, 19.4×10^{15} BTU; fossil price, $\$1.26/10^6$ BTU (\$2.74); nonfossil price, $\$4.25/10^6$ BTU (\$4.25); CO_2 emission, 4.59×10^9 metric tons/year; CO_2 concentration, 340 p.p.m. The parenthetical figures are prices in 1981 expressed in the general 1975 price level. Conversions into BTU were made at the rate of 27.76×10^6 BTU/metric ton coal equivalent.

Yet one of the Nordhaus–Yohe tables may be worth including. Having previously narrowed the trajectories to be considered to a random sample of 1000 out of a possible 3^{10} different trajectories, and assigned probabilistic factors to each of these 1000 runs, the NAS authors presented the annual growth rates of the most likely path for the major projected indicators, as shown in *Table 5.5.*

Keeping in mind the substantial – and for some critical indicators, ballooning – uncertainty surrounding these most likely values, what sort of story emerges from *Table 5.5*?

(1) That both global economic growth and energy consumption are expected to decelerate over the long term – the former more sharply than the latter.
(2) That doubling of atmospheric CO_2 concentration may occur around 2065.
(3) That the triggering of backstop technology – making feasible a significant shift toward non-carbon fuels – is surprisingly long delayed, gathering momentum, at the earliest, after the middle of the twenty-first century; a gradually rising trend in the fossil/nonfossil price ratio – delayed, for a time, by the considerable size of the fossil resource base – does occur, but this is deemed insufficient to counteract the assumed presence of technological barriers that impede smooth nonfossil for fossil substitution.

Environmental dilemmas

The foregoing review of medium- and long-term energy projections suggests that:

(1) Global energy demand and supply, composed predominantly of today's conventional energy sources and forms, can probably be balanced in the year 2000 at modestly higher prices compared to prevailing levels.
(2) After the turn of the century, scarcity of conventional oil and gas will necessitate a progressive shift toward exploitation of more amply available resources, among which coal and nuclear power seem major candidates, and oil shale and solar energy promising, but somewhat less certain, contenders.
(3) Energy consumption growth, while decelerating, will persist well into the twenty-first century and will be accommodated predominantly by fossil fuels before the feasibility of backstop technologies, coupled with rising fossil fuel prices, induces major reliance on nonfossil sources.
(4) There is a high probability of a significant rise in global CO_2 concentration by the middle of the twenty-first century, though some projections use that finding as a feedback trigger to explore the feasibility, though not necessarily the likelihood, of environmentally more benign energy-growth paths and patterns.

We have focused on the consequences for CO_2 emissions and concentrations of long-term energy growth not because they are in any sense a proxy for the multiplicity of environmental constraints and conflicts which such growth may encounter, but simply because the CO_2 problem is the one which has received the most analytical attention from a global perspective. Obviously, energy–environmental conflicts and problems are much more pervasive. They arise from the facts that:

(1) High levels of economic activity signify high levels of energy (and material) flows whose concentrated state in nature gives way to a dilute and degraded state in the environment.
(2) Sophisticated measurement tools and scientific command over cause-and-effect relationships continue to give us ever greater insight into potential hazards to health and safety.

Thus, the list of major and minor environmental perturbations feared from, or actually provoked by, energy production, transport, and use is wide ranging:

(1) Combustion products from power plants, transport, and industrial boilers and their damaging impact on human health, property values, and (by way of acid rain) water bodies and vegetation.
(2) Radioactive substances and fission products from various stages of the nuclear fuel cycle.
(3) Spills and dumping in maritime oil transport.

(4) Land disturbance and water contamination from coal mining and synfuel projects.
(5) Pollution from wood combustion and, as noted earlier, the deforestation, erosion, and loss of soil fertility associated with the use of renewable energy forms throughout numerous parts of the world.

And that is an incomplete accounting.

It must quickly be noted, however, that these and other energy-caused environmental problems involve widely varying degrees of severity. For one thing, given the absorptive or regenerative capacity of the atmosphere, water, or soil, numerous abuses to the environment may be transitory rather than enduring. Then, too, control and corrective actions instigated by regulatory or other policies, and reflected in tens of billions of dollars spent on pollution abatement, have succeeded in fostering improved environmental management, probably with very modest costs in GNP and other macroeconomic measures [25]. To be sure, such success is spread unevenly around the globe because of the motivational differences that arise from disparities in income, from political milieu, and from cultural tradition, as well as from the extent to which different countries or regions are dependent on energy systems posing more or less intractable environmental dilemmas.

In this light, an analysis of the willingness or unwillingness of different nations to embrace environmentally driven energy constraints could prove illuminating. It could provide some clue, however suggestive, as to how far one can expect the world to move as a *collective* of nations, and in what time scale, toward strategies that face up to potential environmental barriers or dangers. These strategies could countenance modification in energy associated activities in view of tolerances that risk being breached. Or they could revolve around initiatives that seek to make these tolerances less constricting. (Whether or not liming lakes, sometimes proposed as an alternative to SO_2 emission reductions, turns out to be a practical way to avoid sulfur-causing acidification, the option illustrates that acceptable outcomes may be achieved by quite different approaches.)

An effort to appraise individual countries' records in energy-related environmental improvements and to draw lessons from those records about the potential for transnational cooperation would not be an easy task. One would want to probe a number of possible explanatory (and interrelated) factors:

(1) *Income.* Presumably the higher the level of per capita income, the greater the financial commitment to environmental improvement. But how tidy is that relationship? There are countries that have evolved from a historic ethic of subduing, rather than living with, nature. Does this fact fault the income hypothesis?
(2) *Political system.* While one of the flaws in the conventional market transaction is a failure to account for use of common-property resources, such as air, there is no compelling evidence – ideological claims to the contrary notwithstanding – that CPEs, with their historic emphasis on production quotas, are any more sensitive to environmental objectives.
(3) *Social discount rates.* A factor no doubt closely related to standards of living. The more a country is attentive to conditions 50, 100, or 200 years hence, the more it is likely to discount the future minimally – that is to place a high value on environmental preservation for multiple generations into the future.
(4) *Ethics, values, biases.* Again, factors likely to be closely related to (1) and (3). But perhaps there are instances where dedications to environmental objectives can be shown to have a unique ethical basis.
(5) *Amelioration costs.* Heavy coal-using nations, such as the GDR (and, perhaps even more, low-income ones like India or China), could face proportionally more burdensome improvement costs than those on whom natural endowment (or policy) has conferred a less polluting energy profile. The fact that, *ex post,* pollution-control expenditures might turn out not to have a severe impact on measured GNP (let alone GNP adjusted for environmental quality benefits) may be insufficient inducement to bear the costs involved.
(6) *Risk and benefit perception.* Attitudes vary widely, irrespective of objective evidence. Some countries may, for whatever reason, be far more willing than others to subject themselves to some environmental hazards, while acting decidedly risk-averse in other cases.

(7) *Intracountry regional conflict.* The acid rain debate in the USA, pitting the alleged polluter (the Ohio Valley and Great Lakes region) against the apparent victim (New England) illustrates the problem which, in the absence of negotiation and compromise on behalf of broader national goals, can be a formula for paralysis.

Whether factor analysis, or some other organizing framework, could enable the estimation of a country's overall propensity toward sound environmental management deserves study. But even short of such a systematic undertaking, it seems clear from points (1)–(6) listed above that if some countries lack incentives to deal with localized or national, energy-induced environmental impacts, they will likely be still more deterred to contribute to transnational efforts. That brings us to the commons problem.

The commons problem

The conceptual thread that links the disincentive characteristics just discussed is the commons phenomenon which surrounds many dimensions of economic activity – both nationally and internationally – but is particularly important in the case of energy. Energy yields valuable services whose allocation in the economic system is – like goods and services in general – effectively handled through the institutions of the marketplace and private property rights. (At least, this is what occurs in a market-oriented society.) At the same time, the residual mass of these energy resources flows back into the common-property resources, within or across countries, that make up much of the environment – e.g., air sheds [26] and water bodies. When access to such waste refuges is unimpeded and free, overuse and degradation is the inevitable result. Valuable as they are, free commons deter ameliorative action by individual users since that would merely confer benefits to nonpaying free riders. Thus, what the last several decades of environmental concern has alerted us to is a great asymmetry in the working of a system of economic incentives: it works well in promoting the production and distribution of resources but, to date, has in large measure failed conspicuously in controlling the disposal of residuals to common-property resources. The insidiousness of this asymmetry may, however, be registering with growing numbers of the public, legislative bodies, and governments. Market-like tools for internalizing the external costs of environmental pollution have begun to be employed, alongside more strictly command-and-control proscriptions of such activity.

If overuse and degradation of environmental resources are frequently the inevitable consequence of the weaknesses of traditional, free-market regimes, centrally planned economic systems do not seem to have produced a formula for rational environmental management either. A picture of the factory manager, wedded to meeting ambitious production targets while subordinating environmental objectives, may be part of the clue to the puzzle, but there may also be elements of caricature in that explanation. After all, as noted above, to date many environmental protection measures, even in capitalist societies, have not – to the chagrin of some, but not all, economists – relied on the emerging system of market-like incentives and disincentives to achieve desirable results. And the standards, regulations, and penalties commonly employed as the alternative are obviously not absent from the nonenvironmental walks of socialist life, where compulsion has long been part of the daily scene. So the question arises: what impels authorities in CPEs to prescribe, or avoid prescribing, environmental regulations? There is clearly much to be learned about the comparative management of environmental and commons issues on different sides of the world's ideological frontiers.

It may be a long leap from the willingness of a community or nation – linked by the kinship of common laws and values – to fashion arrangements for managing the commons, to assume readiness on the part of *the world as a whole* to construct an incentive or cooperative framework within which various countries can participate. In spite of limited successes, as with dumping-at-sea conventions, Antarctica, communications treaties, fishing accords, or partial arms control agreements (and their converse: military pacts), prospects for significant global environmental cooperation have still to prove themselves. On the other hand, merely being able to list numerous areas showing halting movement by nations that recognize a basis for acting in concert may be a positive sign.

There may be instances, though they are probably uncommon, where a single nation is willing to bear the cost of a collective benefit when just its share of that benefit justifies the price it is paying. As an example, USA restrictions on – and, thereby, the loss of sales revenue from – the export of civilian nuclear materials and technology fall into that category. Of course, if one country's restraint is simply offset – in this case, through expanded nuclear sales by others – the former's initiative, rare in any event, becomes quite pointless.

Adaptation

As a thought exercise, rather than building a firmly anchored scenario, one can envisage a multiplicity of ways to deal or live with energy-induced environmental constraints. Suppose the search for scientific, technical, and economic breakthrough pays off. Success with low-risk nuclear reactor systems, in perception and in fact, solar photovoltaics, hydrogen systems, and nuclear fusion could, singly or in combination, ensure that energy and its environmental impacts need not, in themselves, be a drag on wealth, leisure, and mobility. Alternatively, unwillingness to pursue, or disappointing results from pursuing, such options could mean that environmental constraints, if adhered to, translate into the loss of some amount of total real income as well as the relinquishing of some control over disposition of that income. If, say, automotive nitrogen oxide emissions stand incriminated as a major contributor to acidification, the cost of meeting control strategies could exact, as a penalty, some forfeiture of economic well-being – at least as conventionally measured – as well as some encroachment on traditional transport patterns. (Not surprisingly, the 1973–1983 USA record suggests a far greater willingness, initially, to give up horsepower, style, and size than attachment to the individual car. But adaptation clearly occurred even in this short time span.) Incidentally, the phrase "as conventionally measured" is important; the greater the amount of gross product surrendered to achieve enhanced environmental quality (which is often *not* reckoned in social accounting schemes), the more we will need to worry about the adequacy of our economic and environmental measurement practices.

The problem of forestalling, or adapting to, more distant perturbations – CO_2 being the obvious example – is especially vexing. Effects may creep up on us; points of irreversibility may be breached. Thomas Schelling, in his contribution to the NAS study, reflects on some basic aspects of such longer term adaptation [27]. Specifically with respect to CO_2 and the threat of global climate change, he points out that effective defensive strategies need not be ruled out. (Nor should defensive strategies be viewed in an exclusively *governmental policy* context; *individual actions* may also come into play.) The Dutch, more than half of whom reside below sea level, have protected themselves by dikes for centuries. A 5 m rise in sea level at Boston could arguably be defended against by dikes costing less than the value preserved. But, as Schelling recognizes, such a calm assessment is scarcely applicable to, say, Bangladesh, with a wretched level of income and extensive episodes of coastal flooding. Thus, the problem of the distributional burden of a defensive strategy arises immediately.

But, for much of the world, it is also important to appreciate that, if steady increases in real income and independent factors that enhance agricultural productivity, as discussed by Crosson (Chapter 4, this volume), are reasonable assumptions for the long-term future, the depressive effect of climate change on food production might signify no more than a few percentage points in world gross product below the levels that otherwise prevail. This perspective does not belittle the importance of technological research programs on long lead-time issues, such as water conservation and agricultural development. Nor does the inevitability of adaptation to some degree of warming render efforts at nonfossil fuel development immaterial, for such efforts are likely to prove valuable under any circumstances. But economic growth can surely make the burden of adaptation easier.

Research suggestions

At various points in this chapter we have identified subjects worthy of further research. Of course, a discussion revolving around the future is inherently full of speculation and uncertainties that research has limited capacity to overcome. Nonetheless, there are some topics that a focused research effort may help to illuminate. Here is a brief, six-part listing:

(1) The lower the level of energy demand (consistent with broad social and economic goals), the lower the energy-caused environmental problems. The last decade has alerted us to substantial adaptability in energy use in response to price changes. What we know little about is the long-term demand for energy as a function of economic growth in different areas of the world. The estimates of income elasticity of energy demand cited earlier in the text are derived from limited geographic and historical experience. If, over many decades, energy were to follow the path which many foresee for food, its trend might be one tending toward saturation. Alternatively, energy demand tracking income growth would mean much higher consumption levels. A reasoned analysis of the various possibilities is needed.

(2) Under the normal course of events, fossil fuel development is likely to precede a major turn to the nonfossil option. If a significant rise in CO_2 concentrations is probable, and its consequences unwelcome – e.g., per unit of deliverable energy, coal-based synthetics emit roughly twice the CO_2 of conventional oil or gas – could that prospect heighten support for the accelerated and acceptable development of noncarbon-based energy sources, such as nuclear breeders, solar energy, and nuclear fusion?

(3) Even *within* nations it has been difficult to come to grips with the fact that access to a free commons deters ameliorative action by polluting enterprises, since that would merely confer benefits to nonpaying free riders. *Transfrontier* and *global* environmental management would be more formidable still. Research should focus on the propensity of different countries (facing different population pressures, possessing different resource endowments, characterized by different development imperatives, and having different political systems) to submit to international environmental management.

(4) In market economies, market-like incentives or disincentives are gaining considerable acceptance as the instruments of choice in optimal environmental management of industrial activity, though such mechanisms coexist with more tightly framed proscriptive measures of various kinds. What approaches can yield desirable outcomes in centrally planned systems?

(5) International compacts, which impose greater or lesser constraints on national behavior, exist in a number of fields: oil spills, law of the sea, radio frequencies, civil aviation, nuclear safeguards. The proclivity for cooperation in these areas no doubt owes much to factors that may not be applicable to a long-term global problem, such as CO_2. (Notably, CO_2 is associated with gradual, unobservable damage; the other items can involve sudden, visible effects.) Nevertheless, to establish what it is that differentiates impulses in the listed areas from what is needed in the latter case might make for an instructive inquiry. A bonus would be a lesson that might actually be helpful for the long-term problems discussed in this chapter.

(6) In this chapter I have reviewed sharply divergent approaches to mapping possible future energy pathways – notably, reliance on the historic long-wave paradigm, at one extreme, and on the application of econometric techniques, at the other. Inherently an ungratifyingly risky undertaking, probing long-term energy trends remains nonetheless an integral element in studying the dynamics of the biosphere. Therefore, there should be efforts to more resolutely assess the virtues, deficiencies, and mutually complementary features of different projection techniques.

Further readings

Except for the Kneese–d'Arge reference, the suggested readings listed below are discussed critically and at some length in the text. Kneese and d'Arge provide an insightful analytical framework which discloses the complexity in achieving solutions to intercountry environmental disputes and problems.

Edmonds, J. and Reilly, J. (1983), Global energy and CO_2 to the year 2050, *The Energy Journal*, **4** (July), 21–47.

Häfele, W., Anderer, J., McDonald, A., and Nakicenovic, N. (1981), *Energy in a Finite World, Vol. 1, Paths to a Sustainable Future* (Ballinger, Cambridge, MA, for the International Institute for Applied Systems Analysis, Laxenburg, Austria).

Internation Energy Agency (1982), *World Energy Outlook* (OECD, Paris).

Kneese, A. V. and d'Arge, R. C. (1984), Legal, ethical, economic and political aspects of transfrontier pollution, in T. D. Crocker (Ed), *Economic Perspectives on Acid Deposition Control*, pp 123–133 (Butterworth, Boston, MA).

Nordhaus, W. D. and Yohe, G. W. (1983) Future paths of energy and carbon dioxide emissions, in National Research Council, *Changing Climate, Report of the Carbon Dioxide Assessment Committee* (National Academy Press, Washington, DC).

Rose, D. J., Miller, M. M., and Agnew, C. (1983), *Global Energy Futures and CO_2-Induced Climate Change*, report prepared for the National Science Foundation; a summarization appears in Rose, D. J., Miller, M. M., and Agnew, C. (1984), Reducing the problems of global warming, *Technology Review*, **87**(4), 49–58.

Schelling, T. C. (1983), Climatic change: implications for welfare and policy, in National Research Council, *Changing Climate, Report of the Carbon Dioxide Assessment Committee* (National Academy Press, Washington, DC).

Notes and references

[1] Data for 1925–1960 from Darmstadter, J., Teitelbaum, P. D., and Polach, J. G. (1971), *Energy in the World Economy.* (Johns Hopkins, Baltimore, MD, for Resources for the Future); later data from United Nations (1982), *1980 Yearbook of World Energy Statistics* (United Nations, New York). A break in the series slightly impairs pre- and post-1960 comparisons. The country groupings for 1925 are based, as far as possible, on 1980 classifications. This means that for statistical purposes, a given "developing" country is locked into that status for the 55-year period. Pre-World War II Germany could not be allocated between FRG and GDR, and so is included in the "industrialized" category.

[2] Data in cited sources [1] are expressed in coal-equivalent terms. In *Figure 5.3* they are presented in terms of 10^6 BTU, using a conversion of 27.3×10^6 BTU per metric ton coal equivalent.

[3] Source for *Table 5.1:* Department of Energy, Energy Information Administration (1983), *International Energy Annual 1982*, Table 8, p 14 (DOE/EIA, Washington, DC), but also see [4] and [5].

[4] *Table 5.1:* these figures from Congressional Research Service (1980), *The Energy Factbook* (CRS, Washington, DC).

[5] *Table 5.1:* these figures from American Petroleum Institute (1983), *Basic Petroleum Data Book*, Vol. III, No. 3, Sect. IV, Table 2 (API, Washington, DC).

[6] International Energy Agency (1982), *World Energy Outlook* (OECD, Paris).

[7] World Bank (1981), *World Development Report 1981* (Oxford University Press, New York).

[8] Häfele, W., Anderer, J., McDonald, A., and Nakicenovic, N. (1981), *Energy in a Finite World, Vol. 1. Paths to a Sustainable Future* (Ballinger, Cambridge, MA, for the International Institute for Applied Systems Analysis, Laxenburg, Austria).

[9] Marchetti, C. (1983), Recession 1983: ten more years to go?, *Technological Forecasting and Social Change*, **24**, 331–342. Other quoted passages come from this article.

[10] Marchetti, C. (1980), Society as a learning system: discovery, invention, and innovation cycles revisited, *Technological Forecasting and Social Change*, **18**, 280–281.

[11] Rostow, W. W. (1975), Kondratieff, Schumpeter, and Kuznets: trend periods revisited, *The Journal of Economic History*, **XXXV** (Dec), 719–753.

[12] Rosenberg, N. and Frischtak, C. R. (1983), Long waves and economic growth: a critical appraisal, *American Economic Review*, **73**(2), 147–151.

[13] The two statements appear in Berlin, I. (1969), Historical inevitability, in his *Four Essays on Liberty*, pp. 107, 43 (Oxford University Press, Oxford, UK).

[14] For a postmortem on USA economic and energy projections prepared in the early 1960s, see Landsberg, H. H. (1985), Energy in transition: the view from 1960, *The Energy Journal*, **April,** 1–18.

[15] Darmstadter, J., McDonald, D., and Coda, M. (1983), *Review of Post-2000 Energy Projections,* Report to the Gas Research Institute by Resources for the Future (Resources for the Future, Washington, DC).

[16] World Energy Conference, Conservation Commission (1983), *Energy 2000–2020: World Prospects and Regional Stresses* (Graham and Trotman, London).

[17] Pachauri, R. K. (1982), Financing the energy needs of developing countries, *Annual Review of Energy*, **7**, 109–138.

[18] Cited in Dunkerley, J., Ramsay, W., Gordon, L., and Cecelski, E. (1981), *Energy Strategies for Developing Nations*, p. 52 (Johns Hopkins, Baltimore, MD, for Resources for the Future).
[19] Research in progress for the US Department of Agriculture and the Agency for International Development.
[20] Schurr, S. H., Darmstadter, J., Perry, H., Ramsay, W., and Russell, M. (1979), *Energy in America's Future: The Choices Before Us* (Johns Hopkins University Press, Baltimore, MD, for Resources for the Future). Reserves are also included in resources. See their table on pp 242–243 and the accompanying text for detailed documentation and explanation of the concepts.
[21] A concise account appears in Edmonds, J. and Reilly, J. (1983), Global energy and CO_2 to the year 2050, *The Energy Journal*, **4** (July), 21–47.
[22] See Dickinson, Chapter 9, this volume.
[23] Rose, D. J., Miller, M. M., and Agnew, C. (1983), *Global Energy Futures and CO_2-Induced Climate Change*, report prepared for the National Science Foundation; a summarization appears in Rose, D. J., Miller, M. M., and Agnew, C. (1984), Reducing the problems of global warming, *Technology Review*, **87**(4), 49–58.
[24] Nordhaus, W. D. and Yohe, G. W. (1983), Future paths of energy and carbon dioxide emissions, in National Research Council, *Changing Climate, Report of the Carbon Dioxide Assessment Committee* (National Academy Press, Washington, DC).
[25] For support of this conclusion, applicable to the USA, see Peskin, H. M., Portney, P. R., and Kneese, A. V. (Eds) (1981), *Environmental Regulation and the U.S. Economy* (Johns Hopkins University Press, Baltimore, MD, for Resources for the Future).
[26] An airshed is that part of the atmosphere within which airborne emission of pollutants from nearby industrial or other sources may occur in potentially hazardous concentrations.
[27] Schelling, T. C. (1983), Climatic change: implications for welfare and policy, in *Changing Climate, Report of the Carbon Dioxide Assessment Committee*, National Research Council (National Academy Press, Washington, DC).

Commentary

T. B. Johansson

This volume clearly shows the multitude of threats to the biosphere, of which the environmental impact of energy systems is but one. In fact, energy is related to many of the other problems that form the present predicament of man and biosphere. The energy system should not only evolve in a way that is sensitive to the environmental issues, but also in response to other problems [1]. These other global issues include, in addition to the environmental issues, development of developing countries, North–South conflicts, global security and the risks of nuclear war, and nuclear weapons proliferation. Each of these global issues must be resolved and each has strong links with energy. Accordingly, they have to be included in the search for and analysis of future energy systems for a sustainable world; it is not enough just to consider a sustainable energy system.

In the traditional view of the energy problem, articulated in studies such as those by IIASA, World Energy Conference, International Energy Agency, and others quoted by Darmstadter, global energy demand will continue to grow, and perhaps double over the next few decades. But, on analysis, one finds that such energy system developments would aggravate the environmental and many other global problems, because of a continued or increased dependence on Persian Gulf oil, and a greatly expanded use of coal and nuclear power.

The energy system is only a means for providing the desired energy services – space heat, transportation, light, mechanical drive, etc. Some work to establish how and for what energy is used has been done over the last decade: the results provide a new starting point for the energy debate. In fact, there are good reasons to be hopeful, because cost-effective options exist for avoiding global risks and for providing energy for the needs of developing countries. The results from our on-going global energy project [1, 2, 3] and other work [4, 5, 6] suggest that the energy demand/economic growth link is weaker than was formerly thought and that in the future the opportunities for decoupling are even larger, especially for the highly industrialized countries. There are at least three major components behind this conclusion.

First, the composition of the economy in industrialized countries is shifting toward more service industries, which are far less energy intensive than goods production. For example, the output of the goods-producing sector – measured by gross product originating (GPO), or value added – grew at only 0.83 and 0.60 times the rate of the gross national product (GNP) in the period 1970 to 1980 for the USA and Sweden, respectively [1].

Second, the composition of industrial output is shifting from production of basic materials to activities that involve more fabrication and finishing [2, 7]. Relative to total industrial output, basic materials declined by 10% and fabrication and finishing increased by 35% in the USA during the period 1947–1983 [1]. This implies important shifts in overall intensity of industrial energy. In USA manufacturing, the basic materials-processing industries accounted for 87% of final energy use in 1978, but only 32% of value added; for Sweden the figures were 92% and 51%, respectively [1]. Thus, shifts to fabrication and finishing, which typically require an order of magnitude less energy per unit of output, can have a profound effect on industrial energy use. For the USA, such shifts accounted for an annual rate of decline in industrial energy use of 1.0% per year during 1973–1982, relative to the situation with no such shifts [8].

Third, and most important, energy price increases and technological developments have rendered widely available cost-effective opportunities for the much more efficient use of energy. Some examples include space heating requirements for single family dwellings, where a new construction is offered by Swedish firms that uses less than 15% as much energy per square meter per degree day than the average for the housing stock; new automobiles are available with fuel economies twice as high as the average European fleet rate of fuel consumption; new technology with energy requirements reduced by tens of percent are available for the processing of basic materials. Steel manufacturing in the USA, for example, could cut its energy use per ton of steel by more than half if new steelmaking processes, such as Plasmasmelt or Elred, were widely employed. Finally, new products and new materials are being developed that could reduce energy demand further. Incidentally, the impact of technology on the historical development of societies has been impressive, but it appears that Darmstadter has not taken into account such technological developments in his analysis (*Figure 5.9*) [9, 10].

The potential future impacts on total energy demand from structural changes and the use of more energy-efficient technology, along the lines discussed here, have been investigated for Sweden [2] and the USA [3]. The results indicate that even with increases in per capita consumption of goods and services in the range of 50 to 100%, final energy use per capita could be reduced by approximately one half due to the net effect of on-going structural shifts and the exploitation of opportunities for cost-effective investments in energy efficiency.

Much of what has been learned from the analyses of the Swedish and USA situations is probably applicable to most other industrialized countries as well [4, 5, 6], so that these studies serve as an "existence proof" of what might be achieved in most industrialized countries.

For developing countries, the role of technology is even more important. Most commercial energy is used in the modern sector, and much of what is relevant for industrialized countries is relevant here. As Darmstadter points out, they need not follow the old path of industrialization. Investment in energy efficiency makes even more sense here because net capital savings can be achieved [11]. The traditional sector depends on noncommercial energy – one half of all energy use in developing countries – which is used at very low efficiency, primarily for cooking. The potential for energy efficiency improvement is very large. It has been shown, for example, that if one assumes that the best presently available technology or technology in the advanced stages of development were used throughout the developing world, along with a shift to modern energy carriers, and a per capita standard of living similar to that of Western Europe in the 1970s, per capita final energy use would be 1 kWyr/yr (kW), only slightly higher than the present level [1].

Energy availability *per se* is no constraint on development, although availability of capital and other resources needed to bring this technology into wide use may be limited. These capital requirements would be less than if the same energy services were provided by conventional methods, as has been illustrated for the Brazilian electricity sector [11].

These observations suggest that it would be feasible for the world to develop an energy system for continued economic growth using about 11 TW of primary energy in 2020 [1], compared to 10.3 TW in 1980, and IIASA's 22–25 TW for 2030, WEC's 19–25 TW for 2020, MIT's 12–18 TW for 2025, and IEA/ORAU's 19–27 TW for 2025. A low level of energy demand renders the energy supply problem far more manageable than it would otherwise be. Contrary to the assertion quoted by

Darmstadter, it would not be necessary to make "*full* use of *all* available energy resources" (Darmstadter, p 158, emphases added). It becomes possible to avoid or minimize dependence on the more costly and troublesome options. This means that there exist routes of development for the global energy system that are compatible with, and perhaps even supportive of, the development of a sustainable world.

While the technical and economic prospects are good for bringing about major reductions in energy use via investments in more efficient end-use technology, the question remains as to the rate at which these opportunities will or can be seized. In a time frame of 20 to 40 years there is ample opportunity to introduce energy-efficient technology at the rate of introducing new capital. The choice of technology in an investment situation is determined by the economics based on present prices, consumer expectations about future energy prices, the opportunities for financing, the information available to the consumer about investment opportunities and their cost-effectiveness, and other factors.

Some of the opportunities will be realized under present conditions, but the directed policies of governments will speed up the process and also increase the probability that these opportunities will be seized. One way is to institute policies that cause investment decisions to focus on the needed energy services instead of the traditional focus on energy supply. The available policy instruments include energy price adjustments, taxes and subsidies for particular technologies, regulations, and the provision of information. The most appropriate policy instruments will probably differ from one cultural context to another, but must be based on a detailed understanding of the technical options and local conditions. This is an area where more work is needed.

Because cost-effective options exist, we may not be pushed into taking the chance that nature will tolerate all abuse or that adaptation, e.g., building dikes around Boston, will prove feasible and more attractive than adjusting energy use habits. The belief that energy use must grow to provide for peoples' needs leads to defeatist attitudes. These are dangerous, because the problems are taken less seriously and the search for ways to cope with the future becomes muted.

A major challenge for energy planners in the decades ahead is to bring about a global energy system that is compatible with a sustainable world society. A future of low energy demand is a key feature of a global energy strategy for the realization of that goal. Fortunately, this is an achievable goal, because the future of the energy system in industrialized countries is a matter of choice to a much larger degree than has hitherto been thought possible.

Notes and references (Comm.)

[1] Goldemberg, J., Johansson, T. B., Reddy, A. K. N., and Williams, R. H. (1985), An end-use oriented global energy strategy, *Annual Review of Energy*, **10**, 613–688; and Goldemberg, J., Johansson, T. B., Reddy, A. K. N., and Williams, R. H. (1985), Basic needs and much more with one kilowatt per capita, *Ambio*, **14**, 190–200.

[2] Johansson, T. B., Steen, P., Borgren, E., and Fredriksson, R. (1983), Sweden beyond oil, the efficient use of energy, *Science*, **219**, 355–361.

[3] Williams, R. H. (1985), *A Low Energy Future for the U.S.*, PU/CEES Report No. 196 (Princeton University, Princeton, NJ).

[4] Leach, G., Lewis, C., Romig, F., Foley, G., and Buren, A. V. (1979), *A Low Energy Strategy for the United Kingdom* (Science Reviews, London).

[5] Krause, F., Bossel, H., and Muller-Reissman, K.-F. (1980) *Energie-Wende, Wachstum und Wohlstand ohne Erdol und Uran* (S. Fischer, Frankfurt).

[6] Brooks, D. B., Robinson, J. B., and Torrie, R. D. (1985), *2025: Soft Energy Futures for Canada* (Friends of the Earth, Canada).

[7] Larson, E. D., Williams, R. H., and Bienkowski, A. (1984) *Material Consumption Patterns and Industrial Energy Demand in Industrialized Countries*, PU/CEES Report No. 174 (Princeton University, Princeton, NJ).

[8] Marlay, R. C. (1984), Trends in industrial use of energy, *Science*, **226**, 1277–1283.

[9] Solow, R. M. (1957), Technical change and the aggregate production function, *Review of Economics and Statistics*, **34**, 312–320.

[10] Berg, C. A. (1979), *Energy Conservation in Industry: the Present Approach, the Future Opportunities*, Report for the President's Council on Environmental Quality (Washington, DC).

[11] Goldemberg, J. and Williams, R. H. (1985), *The Economics of Energy Conservation in Developing Countries: The Consumer Versus the Societal Perspective – A Case Study for the Electrical Sector in Brazil*, PU/CEES Report (Princeton University, Princeton, NJ).

Chapter 6

Novel integrated energy systems: the case of zero emissions

W. Häfele,
H. Barnert,
S. Messner,
M. Strubegger,
with J. Anderer

Editors' Introduction: The combustion of fossil fuels for energy production has already resulted in significant modifications of the Earth's environment, primarily through the emission of carbon dioxide, sulfur dioxide, nitrogen oxides, and certain heavy metals (see Chapter 8). Many possible future scenarios of energy development involve even greater use of fossil fuels, and many depict an environment "dirtier" than today's (see Chapter 5).

In this chapter one long-term technological strategy that might limit pollutant emissions sufficiently to permit an ecologically sustainable development of the world's fossil fuel resources is explored. Many questions of technical and economic feasibility must still be addressed. The chapter is included here as an illustration of creative strategic thinking about long-term technological responses to large-scale environmental constraints.

Introduction

The terms of the energy debate have changed dramatically over the past decade. Whereas the size of the fossil fuel resource base was the overriding concern of the 1970s, we are now faced with the mounting challenge of exploiting these energy sources in an environmentally benign manner.

The world is not short of energy resources. In terms of energy content there are at least four potential Persian Gulfs: the Athabasca region in Canada, the state of Colorado in the USA, the Orinoco area in Venezuela, and the Oleneik Siberian deposits of the USSR. Recent advances in drilling technologies have led to upwardly revised estimates of both conventional and unconventional gas resources [1]. Innovation and technological progress along the entire chain of energy activities, from exploration to end-use conversion, are also easing the problems of energy supply.

It is helpful to consider in such terms the likely pattern of energy supply and demand over the next 50 years, as revealed by the global scenarios of the International Institute for Applied Systems Analysis (IIASA) [2]. With the assumptions of an almost doubling of the world population and a modest rate of per capita energy consumption, the IIASA study concludes that by 2030 the global demand for primary energy would have increased some threefold over the current figure of some 6×10^9 tons of oil equivalent [(t.o.e.); 1 ton of oil equivalent is equal to 10.7 gigacalories (GC) or 44.76 gigajoules (GJ)]. As the data in *Table 6.1* suggest, over this period the global energy system will shift to an even greater reliance on fossil fuels, albeit lower grade fuels of a predominately solid nature. By 2030, it is unlikely that the global energy system will have advanced beyond the initial stage of its (ultimate) transition to a more sustainable nature, in which nonfossil energy sources will be the primary inputs and electricity and (liquid) hydrogen will serve as clean and complementary secondary energy forms [4].

Table 6.1 Global primary energy demand, 2030 [3].

Source	Relative (%)	Absolute (10^9 t.o.e./yr)
Oil	20	4
Gas	25	5
Coal[a]	20	4
Nuclear	25	5
Other[b]	10	2
Total	100	20

[a] Includes other solid primary fuels, such as tar sands and oil shales. [b] Includes solar and other renewable energy sources.

The enhanced use of fossil fuels in the current energy system is therefore likely to intensify the already serious environmental problems associated with the emissions of energy conversion. Essentially, there are two levels of emissions from fossil fuel use. The combustion of carbon and hydrogen atoms results in so-called mainstream emissions of carbon dioxide (CO_2) and (environmentally harmless) water (H_2O). Accompanying this combustion process are side-stream emissions of sulfur dioxide (SO_2), nitrogen oxides (NO_x), and certain heavy metals.

The salient point for emission control is the time scale. As the IIASA study indicates, there are some 10 to 20 years remaining in which humankind can deal effectively with the environmental hazards of side-stream emissions, and possibly 50 years in which we can effectively respond to the potential threat posed to the climate system by high atmospheric levels of CO_2. The problems of acid deposition suggest a more pressing time scale within which to deal with side-stream emissions. Viewed through the lens of known ecological damage, solutions should have been found yesterday. Clearly, the handling of these emissions is *the* energy problem.

Conventional approaches to emission control generally involve abatement measures (e.g., stack gas cleaning) for the postcombustion process. While such techniques can limit the level of emissions, they cannot eliminate emissions completely. Moreover, there are enormous costs associated with such measures, particularly when a high level of reduced emissions is desired. The case of lignite use for electricity generation in the FRG, following the establishment of stringent SO_2 emission standards, underscores the difficulties one can expect with such practices [5].

A new energy system is therefore needed, and in this chapter we analyze conceptually a novel system that, in principle, can eliminate side-stream emissions and keep the problem of CO_2 emissions in focus. Moreover, the infrastructure of this new system would support a more flexible use of fossil fuels, as well as ensure a smooth transition to a sustainable energy system. We begin by briefly explaining the advantages of this new system *vis-à-vis* the existing system, and review some of the technologies required for its introduction. Thereafter we consider this novel system under static conditions and then under dynamic conditions, as its introduction proceeds competitively with the existing system, in response to emission reduction schemes.

A novel energy system

In order to understand the features desirable in a new energy system we must first describe the

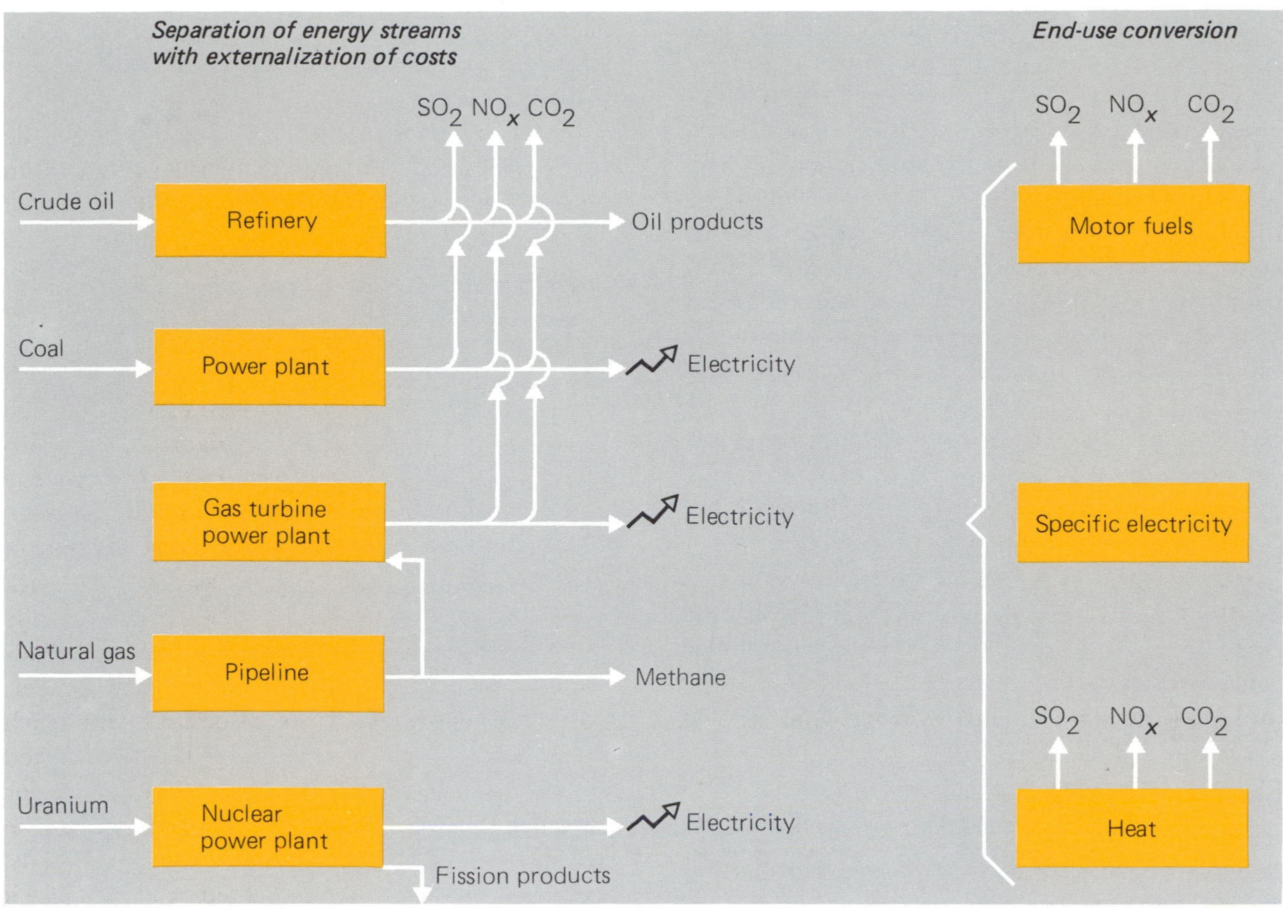

Figure 6.1 The current energy system, illustrating inputs and outputs.

existing one. In the current system, illustrated in *Figure 6.1*, crude oil, coal, natural gas, and uranium form parallel but separate streams of energy. Except for natural gas, these primary energy sources must be converted into secondary energy forms before they can be transported and distributed as final energy for consumer end uses [6]. Natural gas, which is transported mainly by pipelines, may be used directly by the consumer or converted into electricity in gas turbine power plants.

The chain of activities that connects these primary energy inputs with end uses is, to some extent, integrated managerially and optimized, thus far with remarkable success. The notions of upstream and downstream applicable to the oil industry indicate this vertical integration. Horizontally, these energy streams compete with one another in meeting energy requirements – for example, coal and nuclear energy for electricity. However, vertical integration and horizontal separation limit the possibilities of interfuel substitution. It is only in the case of electricity supply that substitution is possible prior to the consumer end. In all other instances, substitution can take place only at the point of consumption, say, by switching from an oil burner to a gas or coal burner. The recent energy crises have demonstrated this major shortcoming of the current system.

A striking and increasingly alarming feature of the current system is the emission of pollutants to the atmosphere and hydrosphere that occurs at energy conversion or end use, or both. In the case of oil, there are SO_2, NO_x, and CO_2 emissions, first at the refinery and then as the oil products are used, say, as motor fuels. The generation of environmentally clean electricity by coal or gas restricts the responsibility for pollutants to the power station. In the case of nuclear-generated electricity, fuel conditioning (i.e., chemical conversion of uranium ores and isotopic enrichment) takes place before energy conversion, so that the fission products generated at the power station are largely contained. On-site generation of heat results in SO_2, NO_x, and CO_2 emissions. Side-stream emissions, as well as the mainstream emissions, externalize costs of the current energy system.

Let us now consider a novel approach to an energy system that, in principle, can achieve the goal of zero emissions and enhance system flexibility through horizontal integration. The basic idea behind such a novel integrated energy system (NIES) is to decompose and purify the primary fossil energy inputs *before* combustion, to integrate these decomposed (clean) products *horizontally*, and to allocate them *stoichiometrically* in line with the requirements for final energy (*Figure 6.2*).

We begin with the system inputs: air, water, coal (or other solids), and natural gas [7]. The desired products of decomposition are carbon monoxide (CO), hydrogen (H_2), and oxygen (O_2). For heuristic purposes, we identify the stoichiometry of final energy as 80% methanol (CH_3OH) and 20% electricity produced, for example, from fossil carbon primary energy.

In this illustrative case the fictitious molecule characterizing the final demand for energy (the so-called demandite [8]) is:

$$0.8\,\text{kcal}\,\frac{CH_3OH\ \text{mol}}{180\,\text{kcal}} +$$

$$0.2\,\text{kcal}\,\frac{CO_2\ \text{mol}}{90\,\text{kcal}} \times \frac{1}{e}\,, \tag{6.1}$$

where e is the efficiency of electricity conversion.

Normalizing this to one carbon atom and assuming an efficiency of one third for electricity generation, we obtain

$$C \times H_{1.6} \times O_{1.6}\,. \tag{6.2}$$

Allocating CO, H_2, and O_2 to fit the stoichiometry

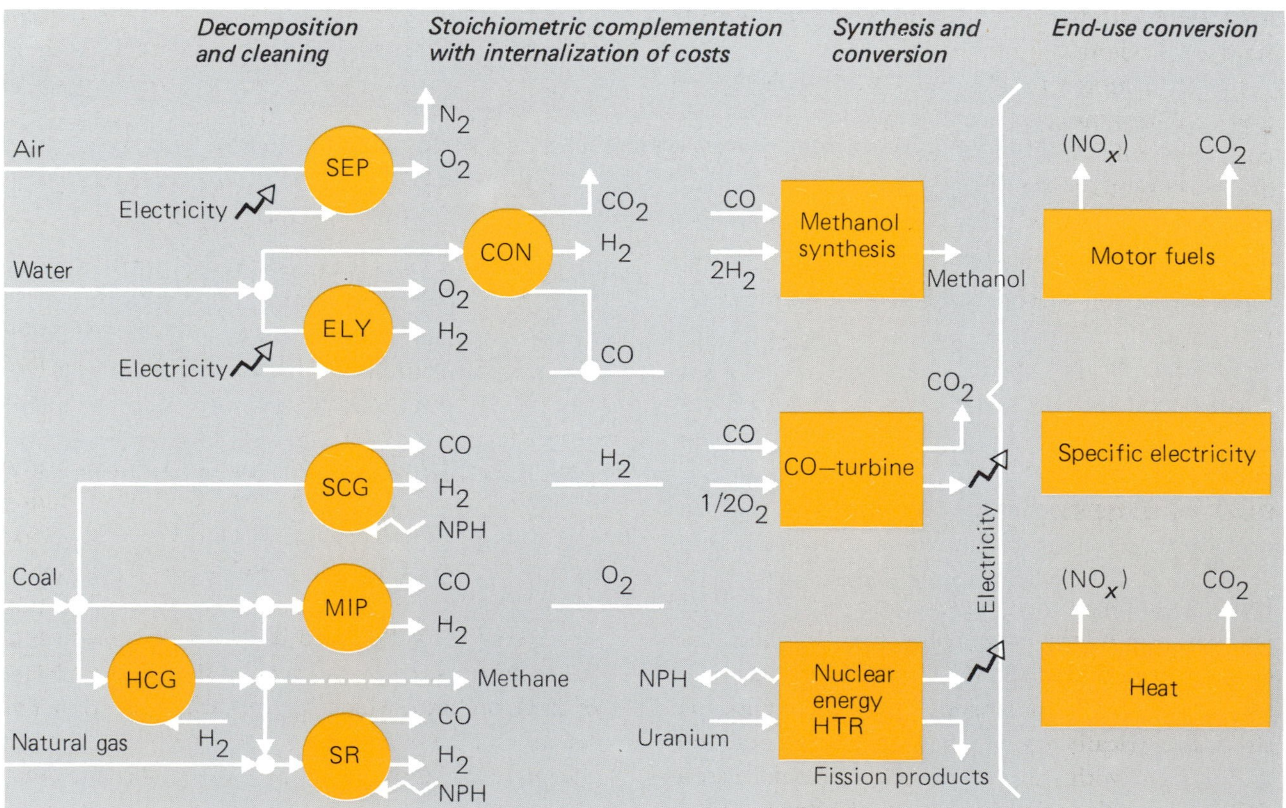

Figure 6.2 A novel integrated energy system: SEP, separation; CON, conversion; ELY, electrolysis; SCG, steam coal gasification; MIP, molten iron process; HCG, hydrogen coal gasification; SR, steam reforming; NPH, nuclear process heat; HTR, high temperature reactor.

of demandite, we obtain

$$CO + 0.8H_2 + 0.3O_2 \,. \tag{6.3}$$

In this example we avoid the generation of unnecessary CO_2 (hereafter referred to as non-demandite CO_2) and use the carbon atom prudently at the stage of final energy [9]. Those CO_2 emissions that occur as the final energy is consumed can be attributed, in the case of methanol, to the combustion of its carbon content (some 40%), or, in the case of electricity, to heat production by means other than nuclear energy.

We are not suggesting the use of methanol as the only liquid energy form in NIES. For example, the Mobil Oil process and the Fischer–Tropsch process of methanol and gasoline synthesis could provide other suitable forms of secondary energy for NIES. Methanol is used here as representative of such fuels [10]. Similarly, we do not exclude direct uses of natural gas. However, since gas combustion in air generates NO_x emissions, we pursue herein more novel ways of burning gas with oxygen that are in line with the system goal of eliminating side-stream emissions.

In the sections that follow we seek to elaborate and analyze a novel integrated energy system as a concept only. We restrict ourselves, where appropriate, to considering the use of methanol in the heat market and of CO_2 for electricity generation. To support this, we first review the candidate technologies associated with this novel energy system.

Technologies of decomposition and synthesis

In reviewing the technologies associated with NIES, we bear in mind the system objectives of generating clean intermediary fuels (CO, H_2, O_2) that, through synthesis and conversion, can be allocated stoichiometrically to meet the final energy demand.

We begin with the decomposition of air. Essentially this involves the separation of nitrogen from oxygen and hence the elimination of some two thirds of the NO_x emissions associated with today's open-cycle combustion. Air separation as such is an established technology. But in view of the relevance of this technology for fossil fuel purification, it would appear worthwhile to explore the feasibility of higher efficiencies and/or lower capital costs.

Generally speaking, hydrogen is in short supply in NIES and a second system input – H_2O – can serve as a rich source of this carrier. The conversion of H_2O into hydrogen and oxygen by means of CO (the so-called shift reaction) is also a well-established technology. It is considered here as a means for overcoming imbalances at the stage of stoichiometric allocation, but at the expense of generating nondemandite CO_2. To obtain hydrogen without unnecessary emissions we include in the scheme the technology of electrolysis, whereby electricity is used to split water. Electrolysis is an established technique that likewise suffers from relatively low efficiencies and high capital costs. Research at the Nuclear Research Center Jülich (KFA–Jülich) and elsewhere seeks to surmount these problems, possibly through the application of high temperature process heat [11, 12].

There are a number of suitable technologies for the decomposition of coal and other solid fossil resources, such as tar sands and oil shales. A good example is steam coal gasification, which requires high temperature process heat [13]. Bergbau Forschungs GmbH, FRG, has developed this process as part of the project Prototype Plant Nuclear Process Heat [14]. There have been good results from the process tests conducted since 1976 at a semitechnical pilot plant with a carbon throughput of 230 kg/h (equivalent to some 2 MW). The recent use of catalysts has improved the process, but more work is needed to bring this technology to the stage of commercial feasibility.

Alternatively, the molten iron process can be used to decompose solid fuels. Essentially, the application of high temperature heat (of about 1400 °C) breaks down the coal lattice and produces a gaseous mixture of H_2 and CO. The undesired combustion products (i.e., sulfur and other pollutants, such as the heavy metal content) form a slag on the surface of the iron bath, which can be easily removed. The salient feature of this process is the environmentally acceptable level of the sulfur content of the product gas, which is no more than 20 p.p.m. and on average 5 p.p.m. Moreover, the molten iron process does not require an exogenous supply of high temperature process heat, since the production of CO through the partial oxidation of the solid fuel generates heat. Some of this heat can be recovered by using the high temperature of the product gas for other applications. Currently there are two versions of the molten iron process at the demonstration stage: the first, developed by Klöckner, FRG, operates at atmospheric pressures [15, 16]; Klöckner, Humboldt-Deutz, FRG, developed the second, which works at pressures of 30 bar [17].

Hydrogen, as well as oxygen, can be used for decomposing solid fuels. Rheinische Braunkohlenwerke AG, FRG, has developed the process of hydrogen coal gasification as part of the project Prototype Plant Nuclear Process Heat [18]. A semi-technical pilot plant with a throughput of 320 kg/h of dry lignite (equivalent to 160 kg/h of anthracite and some 1.5 MW) was used to test the process from 1976 to 1982. Good experimental results supported the development of a scaled-up pilot plant with a capacity of 10 tons/h of dry lignite (equivalent to some 50 MW), which began operation in 1983. Inherently, this process advances the goal of zero emissions, but currently there are a few side-stream emissions that need recycling.

In addition to the above coal refinement processes there are other candidate technologies for decomposing solids in line with the reasoning presented here. One example is the Texaco process of converting hard coal into fuel gas and methanol [19]. Other feasible techniques exist for purifying fossil fuels, particularly crude oils of all American Petroleum Institute (API) grades. At a later stage of analysis, we plan to incorporate the results of research on nonresidual fuel refineries, which is being done at Shell International in the Hague, Netherlands [20]. *De facto*, such a development would add a dimension of its own to NIES.

In considering the process of steam reforming for decomposing natural gas, we recall the tendency of NIES to be hydrogen deficient and hence the importance of stoichiometrically engaging the relatively high hydrogen content of natural gas (with a hydrogen to carbon ratio of 4 to 1). Indeed, in the illustrative equation (6.2), demandite requires only 1.6 hydrogen atoms per carbon atom. Moreover, steam reforming operates with high temperature process heat. For these reasons, we consider allothermal steam reforming technologies rather than conventional autothermal processes. Compared with conventional methods, the use of allothermal processes would decrease the amount of gas (or coal) needed to produce a given amount of fuel by the order of one third to one fourth. As a consequence, CO_2 emissions would also be reduced to between one third and one fourth the level associated with autothermal methods.

KFA–Jülich, together with industry, has developed the EVA–ADAM scheme in which such heat can be applied exogenously. This is an adaptation of an existing technology to the new system requirements. A 10-MW demonstration plant (the EVA II) has been operational at KFA–Jülich since 1981 [21] and has demonstrated the feasibility of allothermal steam reforming and its applicability to the process of raw gas decomposition.

Recall that both steam coal gasification and steam reforming of natural gas require an exogenous supply of high temperature heat. Such temperatures (900 °C and above) are generally not feasible for conventional nuclear reactors, but in principle can be achieved with a high temperature reactor (HTR), operating with the coolant helium. For the past five years the HTR at KFA–Jülich (the AVR) has operated successfully in a ceramic environment at temperatures of 950 °C or more. The fuel elements of the AVR are graphite pebbles, 6 cm in diameter.

Difficulties in achieving such helium outlet temperatures in a metallic environment have spurred efforts to develop higher temperature alloys, as well as structural design principles. The alloy Inconel 716 is a prime candidate. Test periods of some 25 000 h yielded sufficiently good results to permit extrapolation of up to 75 000 h and the identification of data for the design of more efficient heat exchangers and steam reformers [22]. This indicates the feasibility of designing heat exchangers that would allow helium (at temperatures of 950 °C) to be extracted from the ceramic environment of a HTR core and then transferred to a metallic environment.

To demonstrate this, efforts are underway at KFA–Jülich to remodel the existing AVR (50MW_{th}) [23], the results of which are expected during the second half of the 1980s. Once this has been realized, reactors of this type could become an integral part of NIES, allowing nuclear power to contribute to the goal of zero emission energy systems.

Admittedly, the decomposed product CO could be recycled as a way of generating high temperature process heat. However, in line with the conceptual design of NIES, this would not represent a prudent use of the carbon atom. Indeed, the consumption of the carbon atom would increase by a factor of between two and four, exacerbating the problem of CO_2 emissions.

As for the technologies of synthesis and conversion, the synthesis of CO and H_2 into methanol is an established industrial process. Nevertheless, as currently designed these processes generate emissions, particularly from the combustion of the purge gases that result from incomplete chemical processes. Work at KFA–Jülich seeks to identify possible schemes for recycling purge gases and other emissions.

The availability of CO and O_2 as intermediary fuels in NIES would support the use of CO-turbines for electricity production. Essentially, these turbines operate with an isothermal expansion of CO and O_2, which supports both thermal efficiencies of the order of 60% and significant reductions in CO_2 emissions. The combustion product CO_2 could be used for industrial purposes, such as tertiary recovery in oil fields. Moreover, as the introduction of NIES advances, the need for these secondary applications of CO_2 would diminish. The Energy Laboratory of the Massachusetts Institute of Technology (MIT) is engaged in developing such turbines and is cooperating with KFA–Jülich in exploring their use in NIES [24, 25].

We have refrained from discussing technologies of energy end use, since this requires a different set of expertise. However, they are considered implicitly in the dynamic analysis of NIES (pp 182–189). For example, we assume the use of methanol in automobiles, which could result in a 50% reduction in NO_x emissions in the transport sector alone.

The static case

There are five degrees of freedom in NIES: the moles of the four system inputs (oxygen from air, water, coal, and natural gas) and the system output (nondemandite CO_2). Together they constitute one mole, say, of methanol. If nondemandite CO_2 is to be avoided, there are only four degrees of freedom. There are also three balances (carbon, hydrogen, and oxygen), so one can optimize. *Table 6.2* lists the stoichiometric relations (first-order approximations) of the related system technologies. As the relations are linear, a linear programming (LP) approach was used to handle the vast number of activities and constraints to be accounted for in the analysis. The system costs (i.e., investment, operation and maintenance, and fuel) were chosen as the objective functions to be optimized. In so doing, we obtained information, such as shadow prices, that enhanced our understanding of how the system operates. Rough cost estimates of NIES-related technologies are given in *Table 6.3*.

As *Figure 6.3* illustrates, there are major differences in the ratios of hydrogen-to-carbon and oxygen-to-carbon for the fossil primary energy inputs, which define the quality of the fuel. For

Table 6.2 Stoichiometric relations of NIES-related technologies.

Technology	*Reactant*[a]	*Product*
Air separation	Air	O_2
Electrolysis	H_2O	$H_2+0.5O_2$
Molten iron process	$CH_x+0.5O_2$	$CO+(0.5x)H_2$
Hydrogen coal gasification[b]	$CH_x+(1-0.5x)H_2$	$0.5CH_4+0.5C$
Steam coal gasification[c]	CH_x+H_2O	$CO+(1+0.5x)H_2$
Steam reforming[d]	CH_4+H_2O	$3H_2+CO$
Conversion	$CO+H_2O$	H_2+CO_2
Methanol synthesis	$CO+2H_2$	CH_3OH
CO-turbine	$CO+0.5O_2$	CO_2

[a] CH_x represents fossil fuels, particularly coal. [b] The degree of gasification is 50%. [c] Product also contains CH_4 and CO_2. [d] Product also contains CO_2, CH_4, and H_2O.

Table 6.3 Costs and efficiencies of components in NIES.

Technology	*Efficiency* (%)	*Investment* (DM/kW)	*Operation and maintenance* (% of investment cost)
Air separation		2600[a]	7
Electrolysis	80	350	6
Molten iron process	59[b]	650	8
Hydrogen coal gasification	90[c]	700	5
Steam coal gasification	77[d]	530	12
Steam reforming	95[e]	330	7
Conversion	100	200	8
Methanol synthesis	87	250	10
CO-turbine	65	1250	1.5
High temperature reactor (thermal)[f] (HTR)	100	1200	15
Steam cycle for HTR	42	250	7
HTR power plant	42	3100	14

[a] Investment in DM/kW(e) (input). [b] Excluding O_2 production and H_2 throughput. [c] Degree of gasification (in C) is 50% (65% in energy units). [d] Use of process heat: 0.48 per unit output. [e] Use of process heat: 0.28 per unit output. [f] Includes 0.007 DM/kW fuel cost.

example, natural gas has the highest H/C ratio (4:1), while that of solids is low, ranging from 2:1 to 0.1:1. Thus, in analyzing coal as a system input it is necessary to identify three types of coal according to the H/C ratio: hard coal (0.3:1), imported hard coal (0.8:1), and lignite (1.2:1). A distinction was also made for coal prices [26].

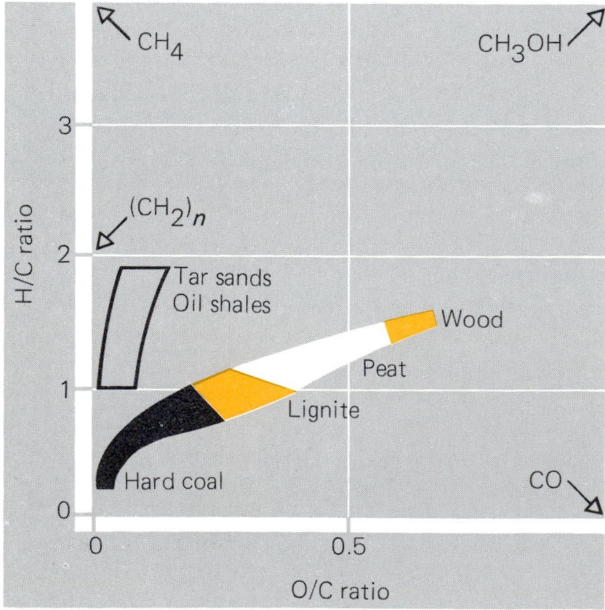

Figure 6.3 Ratios of hydrogen to carbon and of oxygen to carbon for fossil resources. [Hydrogen balance includes water content; $(CH_2)_n$ represents liquid hydrocarbons only.]

Based on these assumptions calculations were made for a stylized final energy demand of 100 GWyr/yr, of which 80% is supplied by methanol and 20% by electricity. In the static case, end uses are represented as final energy. We assume the use of HTRs for generating both electricity and high temperature heat. Obviously, for a more realistic analysis other types of reactors would have to be considered for electricity generation. Since the interplay of nuclear reactors is a well-researched subject, we sought to avoid duplicating these efforts; such duplication is not the point of this analysis.

Table 6.4 gives the results of four LP runs that allow both the use of natural gas and the generation of nondemandite CO_2. As the data indicate, the system requirement for natural gas exists only in Case *A* (hard coal with H/C ratio of 0.3:1). In Case *B*, the system operates entirely on imported hard coal (H/C ratio of 0.8:1) and nuclear energy, as the hydrogen content of this coal enhances the hydrogen balance and constrains the shadow price of gas to the low level of 0.037 DM/kWh. Indeed, it is only at this price that the system would be ready to buy

Table 6.4 The static case: reference cases *A–D*[a].

Value	*Dimension*	*Cases*			
		A	*B*	*C*	*D*
Coal H/C ratio	1	0.3	0.8	0.8	1.2
Coal price	DM/t.c.e.	250	150	250	100
Natural gas price	DM/kWh	0.05	0.05	0.05	0.05
Gas; coal as percentage of primary energy (rest: nuclear)	%	27; 19	0; 54	0; 47	0; 75
CO_2 production	10^4 mol/kWh	0	1.4	0.8	3.2
Thermal efficiency, e^b	%	76	73	72	68
Fuel efficiency, c^c	%	176	135	155	91
Objective function (system costs)[d]	DM/kWyr	634	552	627	501
Investment costs	DM/kW	2032	2205	2203	2224
Share of investment costs in system costs	%	28	35	31	39
Shadow price of methanol	DM/ton	386	313	379	269
Shadow price of gas	DM/kWh	0.05	0.037	0.05	0.03

[a] Final energy demand is 100 GWyr/yr (80 GWyr/yr methanol; 20 GWyr/yr electricity). CO_2 production and natural gas use are permitted. [b] Includes nuclear energy in the ratio of demandite to required primary energy inputs. [c] Excludes nuclear energy in the ratio of demandite to required energy inputs. [d] System costs comprise investments, operation and maintenance, and fuel costs.

a marginal kilowatt hour of gas. Case *C* represents an upper limit on the use of hard coal as the sole fossil input. Since both the shadow price of gas and its actual price are the same (0.05 DM/kWh), a slight increase in the price of hard coal would enable gas to enter the system immediately. The case of lignite (Case *D*) is remarkable. Because of the hydrogen richness of this coal and its low cost *vis-à-vis* other coal types, we found that lignite could generate some 75% of the energy required. Accordingly, in this case both the shadow price of gas (0.03 DM/kWh) and the annual system costs (501 DM/kWh) are the lowest. This implies a shadow price of methanol of 269 DM/ton (equivalent to US $108/ton).

The investment costs for all four cases were found to be in the neighborhood of 2200 DM/ton (equivalent to US $880/kW), which we consider a reasonable figure. Likewise, the relative shares of these capital costs in the total system costs are in the acceptable range of 28–30%. From the perspectives of thermal efficiency and fossil fuel efficiency, Cases *A* and *C* represent the most prudent uses of the carbon atom.

The stoichiometries of Cases *A* to *D* are given in *Tables 6.5–6.8* and demonstrate the high interfuel substitutability at the producer side and the adaptability of this novel energy system to demand requirements. For Case *A* (*Table 6.5*), only steam coal gasification and steam reforming of natural gas are called into play, in addition to methanol synthesis. The contributions of coal and natural gas to the CO balance are about equal, with natural gas supplying the lion's share of hydrogen. The demandite in Case *B* (*Table 6.6*) differs from that of equation (6.1), since the CO-turbine generates only a part of the electricity required. In this case only one technology – steam coal gasification – is needed to supply all of the CO and H_2. In view of the composition of the demandite and its CO_2 component, the hydrogen content of hard coal (0.8 : 1) is sufficient to satisfy the hydrogen requirements of methanol. The use of the more expensive domestic hard coal in Case *C* (*Table 6.7*) results in a relatively larger role for nuclear energy in electricity generation; accordingly, the demandite also differs from that of equation (6.1) and a conversion technology is needed. In Case *D* (*Table 6.8*), the relatively large CO_2 component of the demandite and the use of the molten iron process result in a comparatively lesser role for nuclear-generated electricity.

Table 6.5 Allocation of final energy demand to primary energy demand: Case *A*.

Chemical reactions	*Technology*	*Balance* H_2	CO
$0.54(CH_{0.3} + H_2O = CO + 1.15H_2)$	Steam coal gasification	0.62	0.54
$0.46(CH_4 + H_2O = CO + 3H_2)$	Steam reforming	1.38	0.46
		2.00	1.00
$1.00(CO + 2H_2 = CH_3OH)$	Methanol synthesis	−2.00	−1.00
$0.54CH_{0.3} + 0.46CH_4 + H_2O = CH_3OH$		0	0

Table 6.6 Allocation of final energy demand to primary energy demand: Case *B*.

Chemical reactions	*Technology*	*Balance* H_2	CO
$1.43(CH_{0.8} + H_2O = CO + 1.4H_2)$	Steam coal gasification	2.00	1.43
		2.00	1.43
$1.00(CO + 2H_2 = CH_3OH)$	Methanol synthesis	−2.00	−1.00
$0.43(CO + 0.5O_2 = CO_2)$	CO-turbine	0	−0.43
		−2.00	−1.43
$1.43CH_{0.8} + 1.43H_2O + 0.215O_2 =$ $CH_3OH + 0.43CO_2$		0	0

Table 6.7 Allocation of final energy demand to primary energy demand: Case *C*.

Chemical reactions	*Technology*	*Balance* H_2	CO
$1.25(CH_{0.8} + H_2O = CO + 1.4H_2)$	Steam coal gasification	1.75	1.25
$0.25(CO + H_2O = CO_2 + H_2)$	Conversion	0.25	−0.25
		2.00	1.00
$1.00(CO + 2H_2 = CH_3OH)$	Methanol synthesis	−2.00	−1.00
$1.25CH_{0.8} + 1.5H_2O = CH_3OH + 0.25CO_2$		0	0

Table 6.8 Allocation of final energy demand to primary energy demand: Case *D*.

Chemical reactions	*Technology*	*Balance* H₂	CO
$1.11(CH_{1.2}+0.5O_2=CO+0.6H_2)$	Molten iron process	0.66	1.11
$0.84(CH_{1.2}+H_2O=CO+1.6H_2)$	Steam coal gasification	1.34	0.84
		2.00	1.95
$1.00(CO+2H_2=CH_3OH)$	Methanol synthesis	−2.00	−1.00
$0.95(CO+0.5O_2=CO_2)$	CO-turbine	0	−0.95
		−2.00	−1.95
$1.95CH_{1.2}+0.84H_2O+1.32O_2=CH_3OH+0.95CO_2$		0	0

The interesting results of Case *D* prompted an alternative run, Case *D*′, in which the generation of nondemandite CO_2 was not permitted, the technology of steam coal gasification was assumed to be nonoperative, and gas use was constrained to a maximum level of 25% for primary energy. As the data in *Table 6.9* suggest, the system would operate differently for the two cases. For example, in Case *D*′ the relatively high shadow price of gas suggests that the system would like to buy gas, but is constrained. As a consequence, the contribution of nuclear energy is higher here than in Case *D*. The system costs for Case *D*′, as well as the shadow price of methanol, are also higher than those of Case *D* (some 28 and 63%, respectively).

Table 6.10 is helpful for understanding the differences between Cases *D* and *D*′. In the latter, the use of hydrogen coal gasification yields coke,

Table 6.9 The static case: reference Case *D* and alternative Case *D*′.

Value	*Dimension*	*Reference case D*[a]	*Alternative case D*′[b]
Lignite H/C ratio	1	1.2	1.2
Coal price	DM/t.c.e.	100	100
Natural gas price	DM/kWh	0.05	0.05
Gas; coal as percentage of primary energy (rest: nuclear)	%	0; 75	25; 22
CO_2 production	10^4 mol/kWh	3.2	0
Thermal efficiency, e[c]	%	68	72
Fuel efficiency, c[d]	%	91	152
Objective function[e]	DM/kWyr	501	643
Investment costs	DM/kW	2224	2419
Share of investment costs in system costs	%	39	33
Shadow price of methanol	DM/ton	269	438
Shadow price of gas	DM/kWh	0.03	0.085

[a] CO_2 production and natural gas are permitted; steam coal gasification is operative. [b] CO_2 production is not permitted; natural gas use is limited to 25% of primary energy needs; steam coal gasification is not operative. [c] Includes nuclear energy in the ratio of demandite to required primary energy inputs. [d] Excludes nuclear energy in the ratio of demandite to required primary energy inputs. [e] System costs comprise investments, operation and maintenance, and fuel costs.

Table 6.10 Allocation of final energy demand to primary energy demand: Case *D′*.

Chemical reactions	*Technology*	*Balance* H_2	CO
$0.55(CH_{1.2}+0.4H_2=0.5CH_4+0.5C)$	Hydrogen coal gasification	−0.22	0
$0.27(C+0.5O_2=CO)$	Molten iron process	0	0.27
$0.73(CH_4+H_2O=3H_2+CO)$	Steam reforming	2.18	0.73
$0.04(H_2O=H_2+0.5O_2)$	Electrolysis	0.04	0
		2.00	1.00
$1.00(CO+2H_2=CH_3OH)$	Methanol synthesis	−2.00	−1.00
$0.55CH_{1.2}+0.45CH_4+0.77H_2O+0.115O_2=CH_3OH$		0	0

which can be used in the molten iron process. Although natural gas is decomposed, the hydrogen balance is not met, necessitating the use of electrolysis. All electricity is nuclear generated in Case *D′*.

We repeat: these numerical results are meant to be suggestive only, and should therefore not be construed as final or definitive of the system requirements and associated costs. These results have been used here only to illustrate the concept of a novel integrated energy system operating under static conditions. In the next section we consider NIES under more dynamic conditions, in which technologies, such as steam coal gasification and molten iron process, compete and, in fact, NIES – as a system – competes with the existing energy system over the 50-year period, 1980–2030.

The dynamic case

Realistically, the novel integrated energy system described in this chapter must compete with the existing energy system. As concern for combustion emissions mounts, so the current system responds by adopting abatement measures, such as stack gas cleaning. While such measures can limit the quantity of atmospheric pollutants, they cannot achieve the NIES system goal of zero emissions. Moreover, such measures imply additional costs, which increase significantly as the level of desired emission reduction escalates. Rough estimates of the costs of such measures are given in *Table 6.11.* As the data suggest, costs could rise some thirtyfold as the degree of emission reduction increases from 50 to 95%.

Table 6.11 Estimated costs of side-stream emission reduction in central conversion facilities of the existing energy system (in DM/kW; base year 1980).

Emission	*Relative costs of reduction* 50%	80%	95%
SO_2	60	500	2000
NO_x	60	500	2000

It is a simple fact that without any required reduction of emissions the costs of the existing energy system would be considerably lower than those associated with NIES. But as observed earlier, side-stream emissions externalize costs of the current system.

We therefore analyzed the dynamics of introducing NIES competitively with the existing energy system, considering different levels of emission reduction. The characteristics of the dynamic case are summarized in *Box 6.1. Figure 6.4* provides an overview of how these systems would compete and of the associated emissions.

The four cases of emission reduction are described in *Table 6.12* and *Figure 6.5.* In sum, we identified a weak level of emission reduction (Case

Box 6.1. Characteristics of the dynamic case

Time horizon:	1980–2030
Objective function:	Sum of discounted costs
Discount rate:	12%/yr (unless otherwise indicated)
Final energy demand:	200 GWyr/yr (see note [27] for data typical for the FRG in 1981)
motor fuels (final energy)	42 GWyr/yr (constant over the period)
specific electricity (final energy)	22 GWyr/yr increasing to 31 GWyr/yr
heat (useful energy)	60 GWyr/yr (decreasing to 48 GWyr/yr because of efficiency improvements)

System constraints:
Primary energy demand:
hard coal $\geq$ 57 GWyr/yr
lignite $\leq$ 30 GWyr/yr
$\geq$ 26 GWyr/yr

Investments for NIES:
500×10^6 DM/yr beginning in the late 1980s and increasing thereafter to a final value of 5×10^9 DM/yr.
Increases of maximum annual investments, I, were calculated as follows:
$I_t \leq (I_{t-1} \times 1.02) + 5 \times 10^6$ DM/yr.

Costs and efficiencies of components of existing energy system [28]
Costs and efficiencies of components of the NIES [28]
Categories of end use considered:

Methanol cars Gasoline (or diesel fuel) cars	Transport sector
Electricity	Specific electricity (e.g., for light, power)
Oil burners Coal burners Gas burners Methanol burners Electrical boilers	Heat market

1), a medium level (Case 2), and a strong level (Case 3). Case 0 represents no change in the level of emissions compared with current conditions. Strategies for reducing emissions were treated exogenously in the analysis. The design of such strategies is a political task, which ideally could be aided by analyses of this kind.

The quantitative results of this dynamic analysis are vast and in this chapter we have necessarily been selective. We begin with examples of the influence of the four emission reduction cases on the behavior of the energy market. We then look at Case 2 in some detail to gain additional insights.

Emission reductions and energy market dynamics

Figure 6.6 illustrates the levels of coal and nuclear energy used for generating electricity in the four cases over the period 1980–2030. In Case 0, nuclear energy increases its share of the electricity market slowly, accounting for some one third of the electricity generated in 2030. Progressing from Case 1 to Case 3, we observe an accelerated and intensified application of nuclear energy for electricity generation, at the expense of both hard coal and lignite.

Figure 6.7 suggests that as emission requirements

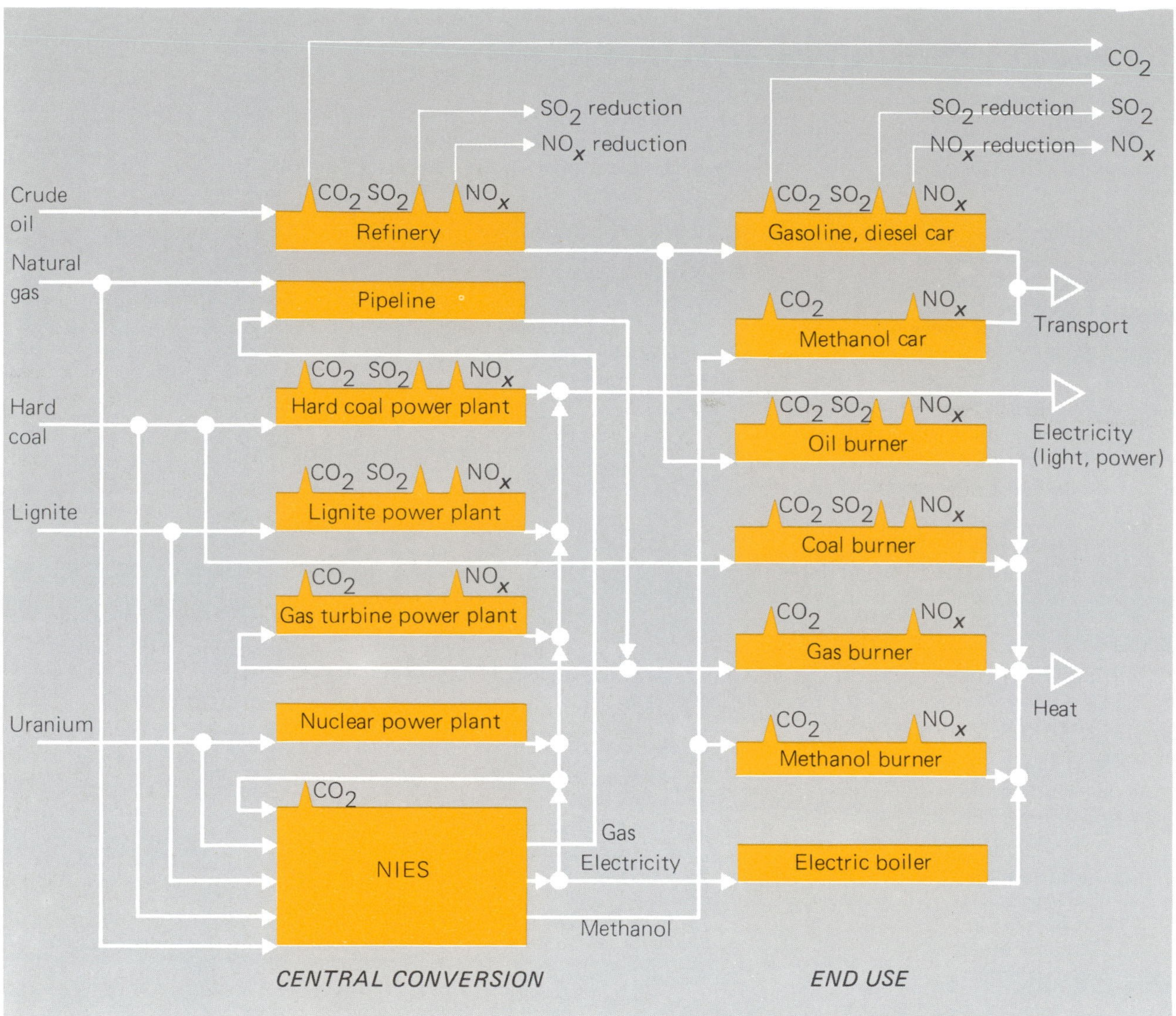

Figure 6.4 Dynamics of the introduction of NIES alongside the current energy system.

become more stringent, solid resources serve increasingly as inputs for the three novel technologies being considered here (i.e., molten iron process, steam coal gasification, and hydrogen coal gasification). This is paralled by increases in the amount of high temperature process heat generated by HTRs. Comparing the results of *Figures 6.6* and *6.7*, we note that compliance with increasing emission reductions results in a decrease in the use of hard coal and lignite for electricity generation and a concomitant increase in their application in NIES-related technologies.

The impact of such developments on methanol production is shown in *Figure 6.8.* No methanol is produced in Case 0. Methanol production begins around the year 2000 for weak reductions, in the early 1990s for medium reduction requirements, and in the early 1980s when strong emission reductions are the criterion. In the latter case, production peaks around 2020 since conditions then necessitate the substitution of electricity for methanol in the heat market.

The results of analyzing the interplay of oil products and methanol in the heat market are shown in *Figure 6.9.* For Case 0, only oil products are used for heating purposes. Progressing from Case 1 to Case 3 we note that the methanol share of the market grows accordingly. However, to meet the

Table 6.12 The dynamic case: four emission reduction cases.[a]

Case	*Side-stream emissions* SO_2	NO_x	*Mainstream emission* CO_2
Zero (0)	No reduction	No reduction	No reduction
Weak (1)	Medium reduction for central conversion facilities	No reduction	No reduction
Medium (2)	Medium reduction for central conversion and end-use facilities	Medium reduction for central conversion facilities	Slow reduction for central conversion plants
Strong (3)	Fast reduction for central conversion and end-use facilities	Medium reduction for central conversion facilities	Slow reduction for central conversion facilities

[a] Relative emissions as a function of time are shown in *Figure 6.5*.

emission reduction requirements of Cases 2 and 3, methanol use in the heat market peaks at about 2020 and 2000, respectively, as electricity is used increasingly for heating purposes.

We also considered the interplay of oil products and methanol in the transport sector (see *Figure 6.10*). The features of this situation are qualitatively similar to those for these fuels in the heat market. Basically, the difference lies in the timing, with the introduction of methanol occurring relatively later in the transport sector and only in Cases 2 and 3. We attribute this to the assumption that the substitution of methanol for oil products in the heat market is more efficient in terms of reduced side-stream emissions than in the transport sector. Moreover, we assumed a greater efficiency for gasoline (60%) than for methanol (50%), measured in terms of the motive power produced from the final energy consumed in the transport sector. In so doing, we implicitly accounted for the structural changes

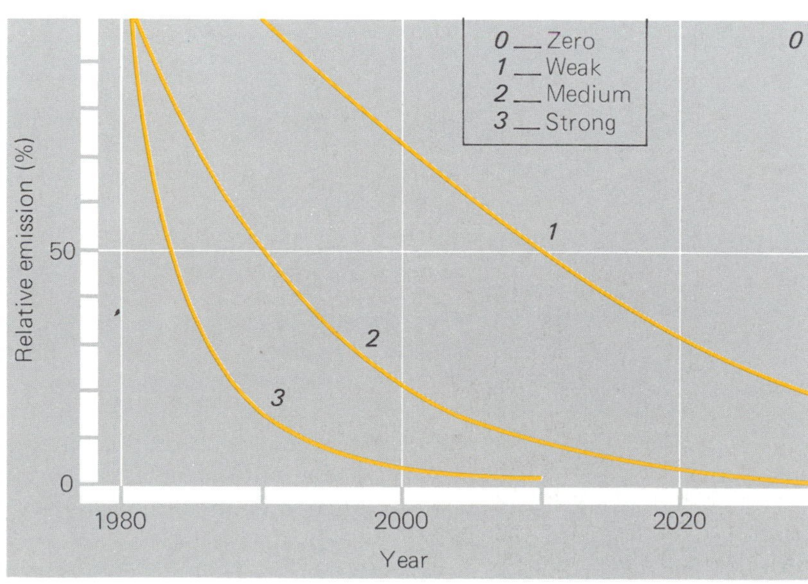

Figure 6.5 Relative emission reductions (strong, medium, weak, and zero) as a function of time, 1980–2030.

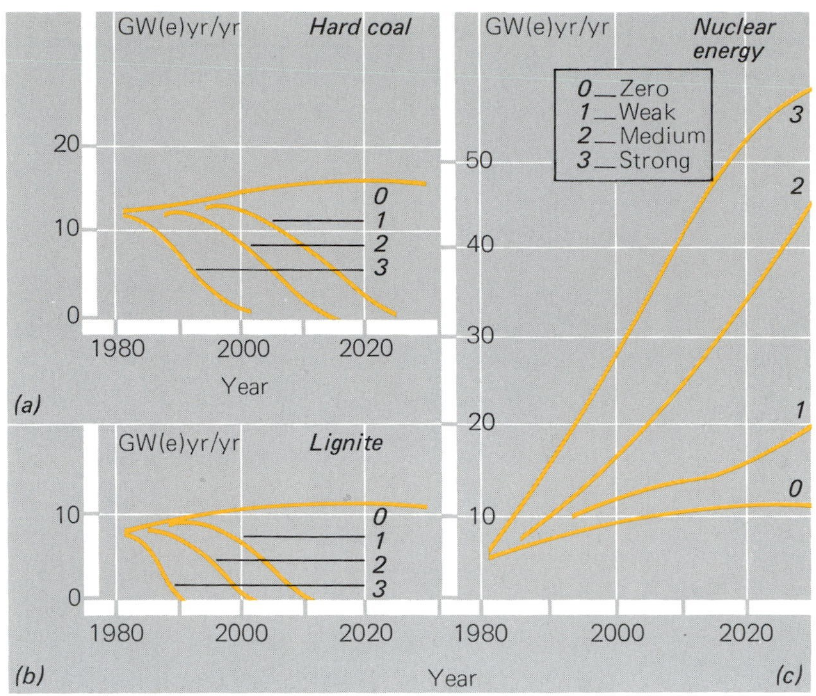

Figure 6.6 Emission reductions and use of coal and nuclear energy for electricity generation, 1980–2030. (a) Hard coal; (b) lignite; (c) nuclear energy.

essential to the introduction of methanol in this sector. Understandably, the greater thermal efficiency of methanol *vis-à-vis* gasoline would have to be considered in a more detailed analysis of the use of methanol as a transport fuel. Work along these lines is underway at KFA–Jülich.

The case of medium-level emission reductions

To gain further insights on the response of the energy market to reduced emission requirements, we considered the case of medium-level emission

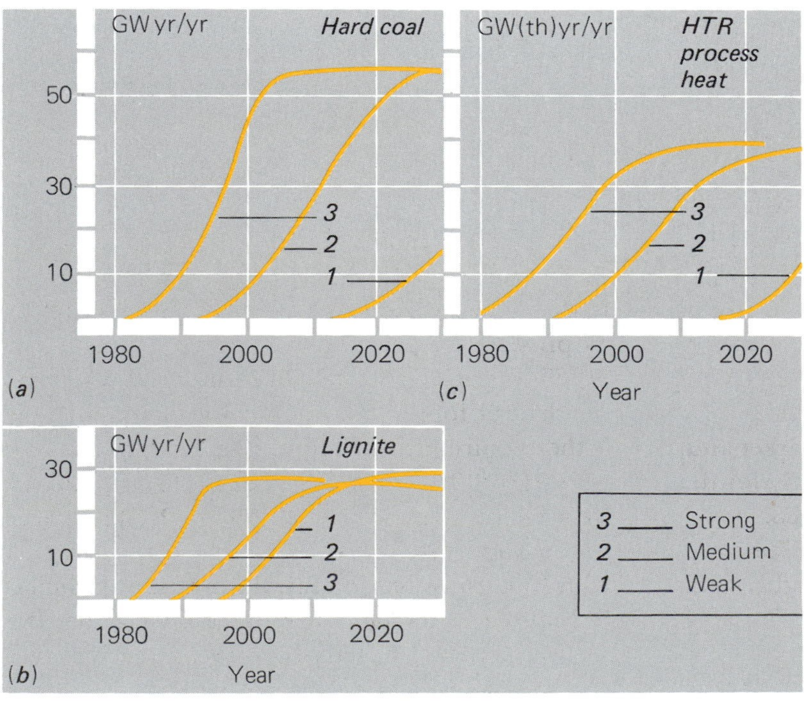

Figure 6.7 Emission reductions and uses of coal and nuclear energy for novel energy processes, 1980–2030. (a) Hard coal; (b) lignite; (c) HTR process heat.

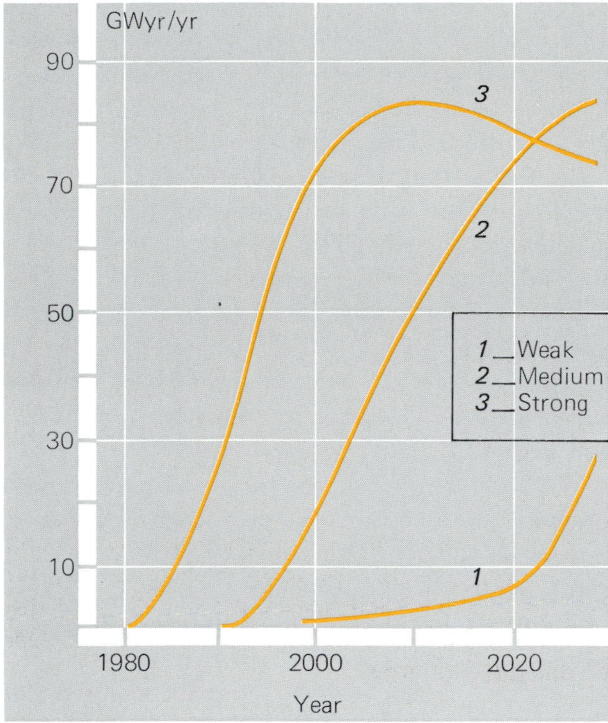

Figure 6.8 Emission reductions and methanol production, 1980–2030.

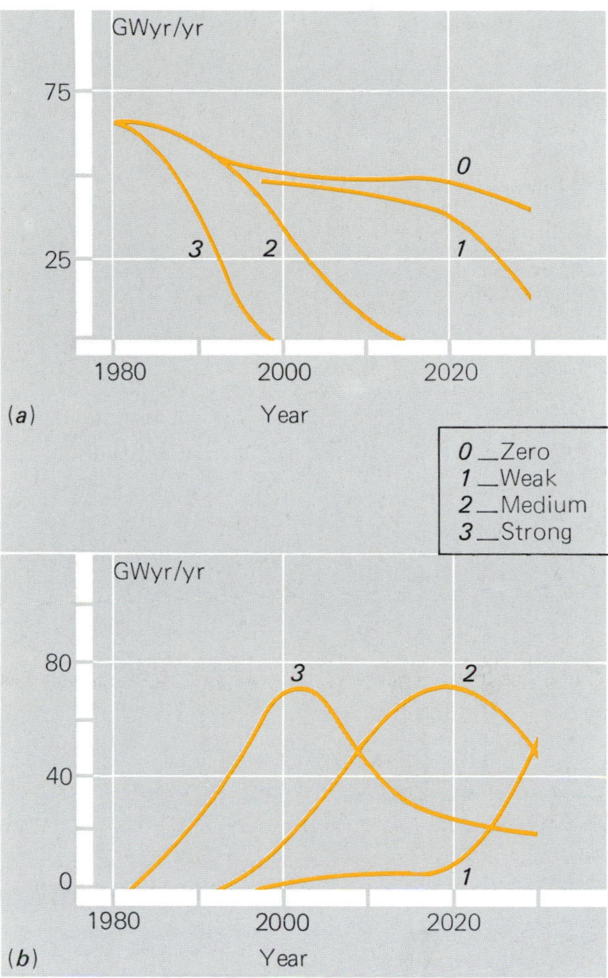

Figure 6.9 Emission reductions and (a) oil products and (b) methanol in the heat market, 1980–2030.

reductions (Case 2) in some detail. When no reduction is the criterion there is no significant change in the mix of primary energy sources over the period. As *Figure 6.11* illustrates, however, in compliance with the requirements of Case 2 the primary market undergoes significant changes: as the share of oil declines significantly so nuclear energy gradually becomes *the* primary souce of energy supply. As a condition, the shares of hard coal and lignite were kept relatively constant and natural gas was assigned the residual supply role. The rise of nuclear energy in the primary energy market reflects compliance with the requirements of Case 2 in two ways: first, nuclear energy substitutes for hard coal and lignite in the electricity market and, second, nuclear energy is used increasingly for the refinement processes associated with NIES.

Changes in the final energy market are illustrated in *Figure 6.12.* Again, Case 0 implies roughly static conditions in the market over the period. For Case 2, we observe a changed pattern, with methanol's share in the final energy market increasing substantially, beginning around the 1990s. By 2030, methanol holds some 50% of the final energy market, along with electricity and gaseous fuels. Methanol's substitution of solids (coal) takes place relatively early and is complete; its replacement of oil products begins somewhat later and, while it is significant, it is not total. The salient point is that the driving force for change in the energy market is the requirement of reduced emissions. Thus, in Case 2, by 2020 side-stream emissions are reduced by a factor of 20 compared with 1980 values for SO_2 in central conversion and end-use facilities, and with NO_x emissions in central conversion facilities. It is also a remarkable feature of Case 2 that the use of nuclear-generated process heat for fuel refinement reduces CO_2 emissions.

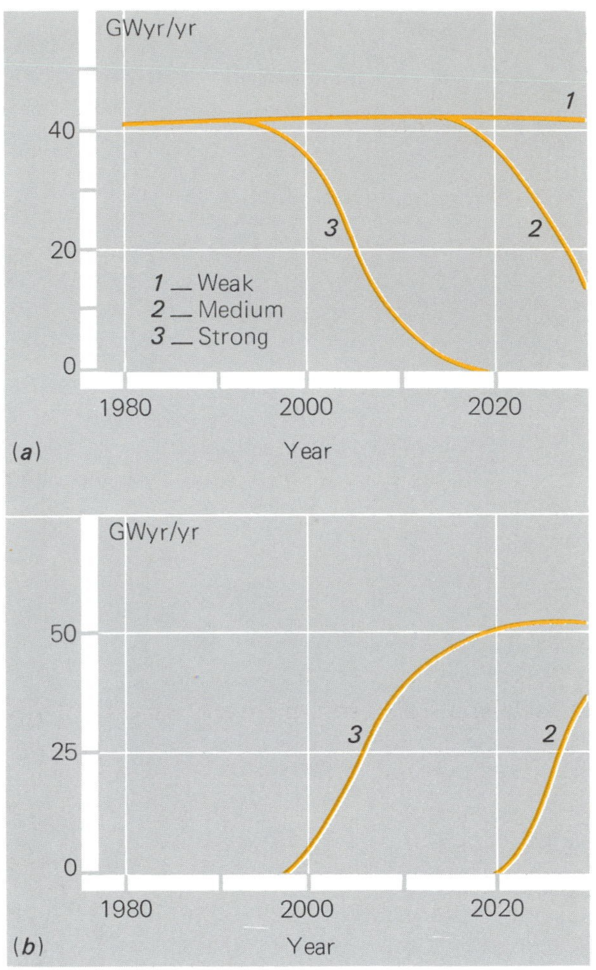

Figure 6.10 Emission reductions and (*a*) oil products and (*b*) methanol in the transport sector, 1980–2030.

Figure 6.13 shows that the shift from hard coal and lignite to nuclear energy in the electricity market is complete by around 2010. Gas increases its share slowly, as both natural gas and syngas are used for electricity generation.

For the heat market (see *Figure 6.14*), the replacement of oil products and coal by methanol occurs around 2015 and 2025, respectively. By 2030, methanol holds the major share in the market, along with gas and electricity. The increase in electricity's share of the heat market is explained by the need to comply with Case 2 emission reduction requirements and the resultant substitution of electricity for methanol.

With the assumption that such changes as those outlined above occur, we then considered their impact on the uses of hard coal and lignite. As a condition, both primary energy sources were kept relatively constant. *Figure 6.15* shows that these primary solid fuels are used decreasingly for heating and electricity purposes and increasingly for methanol production. The transition to the novel use of (relatively cheaper) lignite occurs somewhat earlier than for hard coal.

Finally, we considered the features of the transition to the use of coal in three novel technologies (i.e., molten iron process, steam coal gasification, and hydrogen coal gasification). Recall that the latter two processes require HTR process heat. The results are shown in *Figure 6.16*. By 2030, some

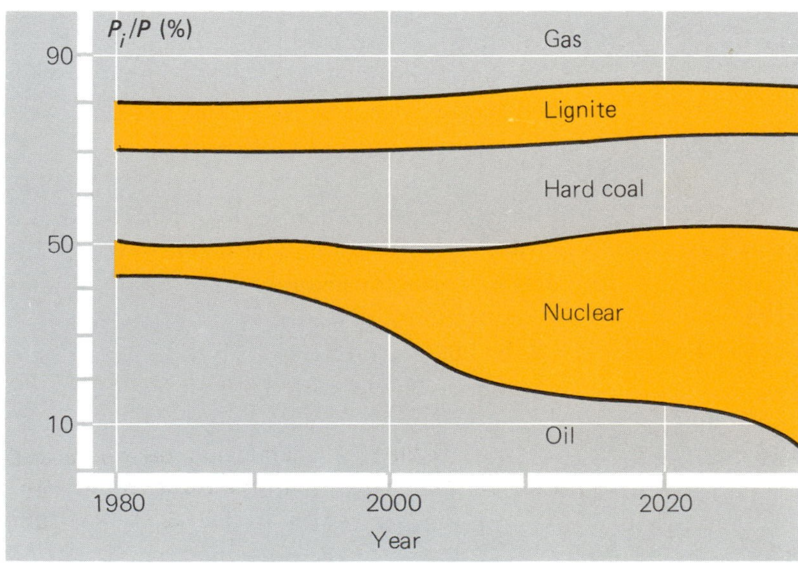

Figure 6.11 Emission reduction and the primary energy market, 1980–2030: medium (Case 2) reductions.

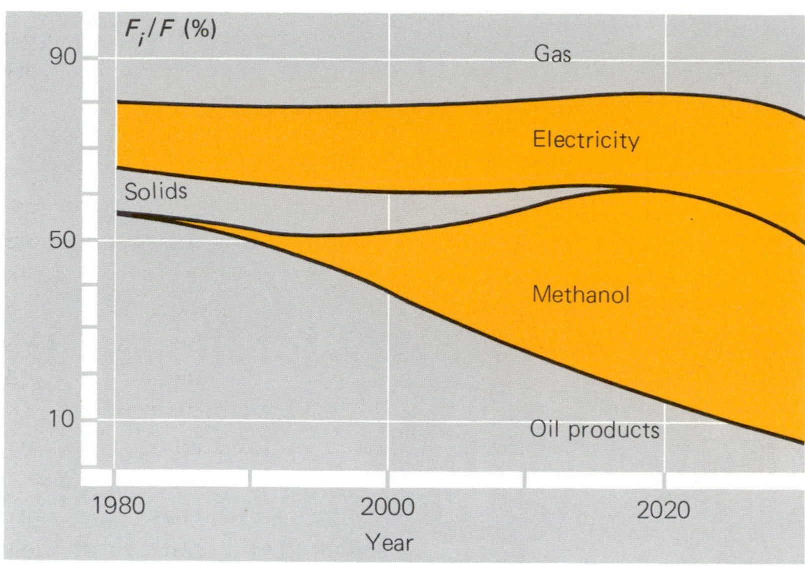

Figure 6.12 Emission reductions and the final energy market, 1980–2030: medium (Case 2) reductions.

57 GWyr/yr of hard coal, 26 GWyr/yr of lignite, and 40 GWyr/yr of HTR process heat are used in the production of some 80 GWyr/yr of methanol. As the dynamic case is typical for the FRG it is worthwhile noting that the absolute size of the HTR application in Case 2 represents some one third of all nuclear-generated electricity measured in thermal power.

Final remarks

In this chapter we have used the case of zero emissions to analyze conceptually a novel integrated energy system. Robust conclusions about the dynamics of introducing this new system must await the results of detailed systems analyses to be carried out collaboratively with members of the energy community and other research partners. We consider such analyses mandatory.

Still, the present conceptual analysis has revealed certain criteria and likely changes in the energy market with respect to the introduction of NIES. In compliance with the pressing need to reduce side-stream emissions, it would appear that the introduction of NIES would result in practically zero emissions when converting primary energy

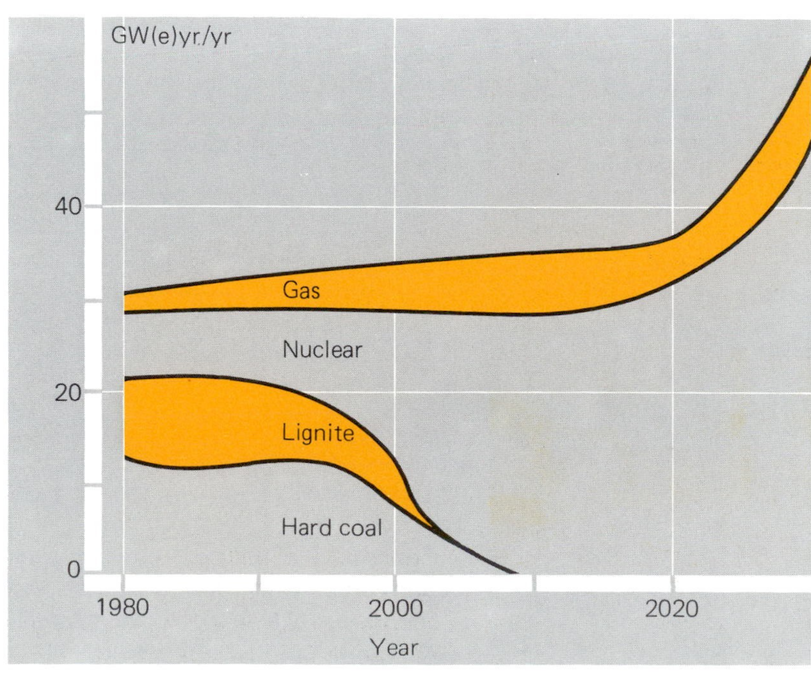

Figure 6.13 Emission reductions and cumulative contributions to the electricity market, 1980–2030: medium (Case 2) reductions.

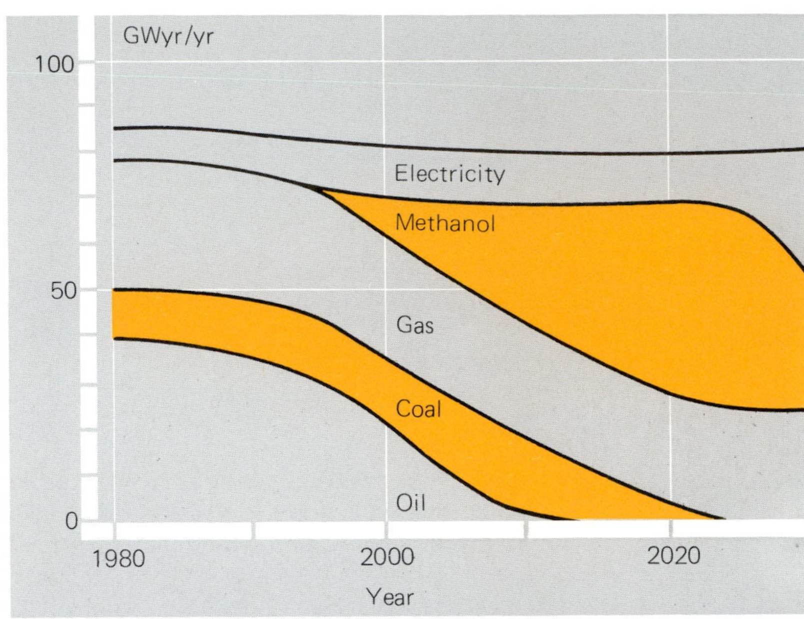

Figure 6.14 Emission reductions and cumulative contributions to the heat market, 1980–2030: medium (Case 2) reductions.

into secondary energy, a total elimination of SO_2 emissions at end use, and a substantial reduction in end-use emissions of NO_x. Compared to the present situation, this would represent a drastic step toward an environmentally benign energy system. More specifically, we observe that:

(1) More stringent standards for emission levels would support the introduction of NIES as a more economic measure, compared to the continuous increase in abatement measures for the existing energy system over the study period.

(2) The introduction of NIES would result in a major shift in the use of lignite and, to a lesser extent, of hard coal away from electricity generation and toward methanol production. The relatively low cost and high hydrogen richness of lignite make it particularly attractive for such purposes.

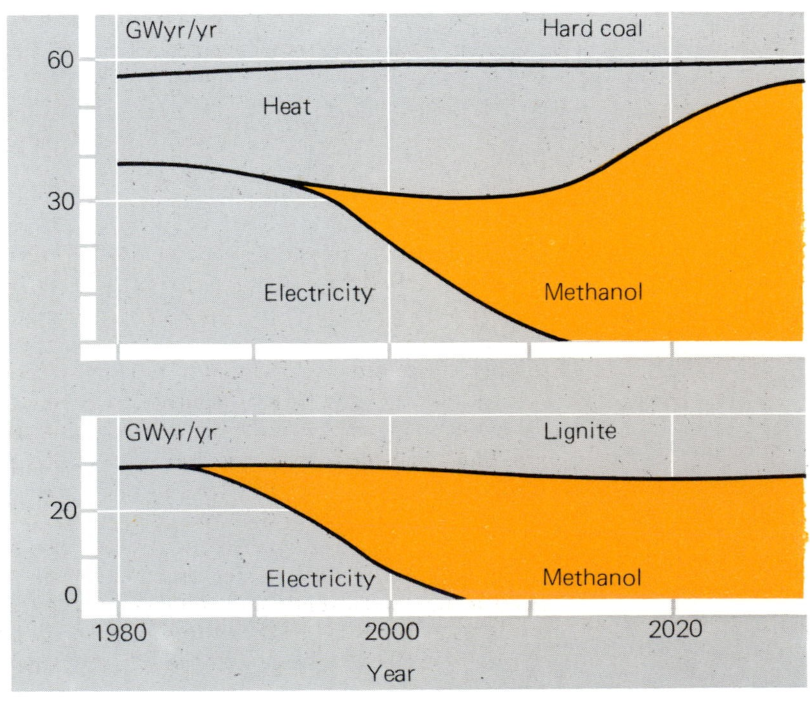

Figure 6.15 Emission reductions and cumulative uses of hard coal and lignite for novel energy processes, 1980–2030: medium (Case 2) reductions.

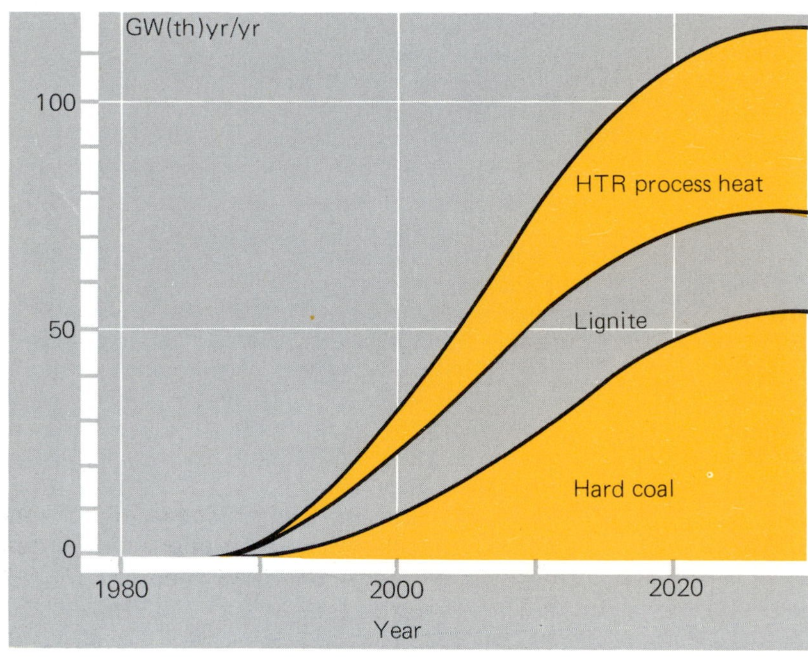

Figure 6.16 Emission reductions and cumulative uses of coal and nuclear energy for novel energy processes, 1980–2030: medium (Case 2) reductions.

(3) A large share of primary energy input could remain with hard coal, even at relatively high costs (in the range of 250 DM/ton).

(4) The share of nuclear energy in the electricity market would increase substantially with the introduction of NIES; there would be an increasing need for high temperature process heat from HTRs, beginning at about the period 1995–2000.

(5) There is a whole set of candidate technologies that can support the introduction of NIES. Their development over the next 10 years is both meaningful and feasible. These are: steam coal gasification; high temperature reactors for generating high temperature process heat; methanol synthesis with zero emissions; hydrogen coal gasification; steam reforming of natural gas with exogenous process heat; the molten iron process; air separation facilities; CO-turbines; modern, high-temperature electrolysis.

Notes and references

[1] Halbouty, M. T. (1984), Reserves of natural gas outside Communist Block countries, in *Proceedings of Eleventh World Petroleum Congress, August 28–September 2, 1983, London* (John Wiley & Sons, Chichester, UK).

[2] Häfele, W. (1981), *Energy in a Finite World. Volume 2: A Global Systems Analysis*, Report by the Energy Systems Program Group of the International Institute for Applied Systems Analysis (Ballinger, Cambridge, MA).

[3] Estimates based on [2] and expert judgment.

[4] Häfele, W., with N. Nakicenovic (1984), The contribution of oil and gas for the transition to long-range novel energy systems, in *Proceedings of Eleventh World Petroleum Congress, August 28–September 2, 1983, London* (John Wiley & Sons, Chichester, UK).

[5] Der Bundesminister des Inneren (1983), 13. Verordnung zur Durchführung des Bundes-Imissionsschutzgesetzes (Verordnung über Grossfeuerungsanlagen – 13. BlmSchV), *Bundesgesetzblatt, Teil I.*, 25th June 1983 (Bonn, FRG).

[6] This report distinguishes between primary, secondary, final, and useful energy. Primary energy refers to the energy inputs in their natural form; secondary energy forms are those to which primary energy is usually converted (e.g., electicity, gasoline). Final energy refers to the form in which energy is consumed at the consumer end (e.g., electricity at the wall plug, gasoline at the gas station). Useful energy is that share of final energy used for a service, such as moving a car or heating a room.

[7] For this analysis we assumed dwindling supplies of conventional crude oil resources and the concurrent need to tap unconventional resources, such as tar sands and oil shales. These unconventional oil resources are considered under the rubric "solids".

[8] Goeller, H. E. and Weinberg, A. M. (1975), *The Age of Substitutability*, Eleventh Annual Foundation Lecture, United Kingdom Science Policy Foundation's Fifth International Symposium, *A Strategy for Resources*, 18 September 1975 (Eindhoven, The Netherlands).

[9] Häfele, note [2], pp 804–805.

[10] Asinger, F. (1983), Methanol auf Basis von Kohlen: Ubersicht über einen alten Chemie-und künftigen Energierohstoff, in 3 Teilen: Teil I: Herstellung; Teil II: Methanol als Vergaser- und Dieselkraftstoff; und Teil III: Uberführung von Methanol in Gase, Erdöl und Kohle, *Erdöl und Kohle – Erdgas – Petrochemie*, **36**, Teil I: Heft 1, 28; Teil II; Heft 3, 130, und Teil III: Heft 4, 178.

[11] Dönitz, W., Schmidberger, R., Steinheil, E., and Streicher, R. (1980), Hydrogen production by high temperature electrolysis of water vapour, *International Journal of Hydrogen Energy*, **5**(1), 55–63.

[12] Nürnberg, H. D., Divisek, J., and Struck, B. D. (1983), Modern electrolytic procedures for the production of hydrogen by splitting water, in G. S. Bauer and A. McDonald (Eds), *Nuclear Technologies in a Sustainable Energy System, Selected Papers from an IIASA Workshop* (Springer Verlag, Berlin, FRG).

[13] Kirchhoff, R., van Heek, K. H., Jüntgen, H., and Peters, E. (1984), Operation of a semi-technical pilot plant for nuclear aided steam gasification of coal, *Nuclear Engineering and Design*, **78**(2), special issue: *The High Temperature Reactor and Nuclear Process Heat Applications*, 233–239.

[14] Barnert, H., von der Decken, C. B., and Kugeler, K. (1984), The HTR and nuclear process heat applications, *Nuclear Engineering and Design*, **78**(2), special issue: *The High Temperature Reactor and Nuclear Process Heat Applications*, 91–98.

[15] von Bogdandy, L. (1983), *Praxisbezug der universitären Forschung, dargestellt an einem Beispiel aus der deutschen Stahlindustrie: Kohlevergasung im Eisenbadreaktor*, China Symposium der TU Berlin, 10 October 1983.

[16] Klöckner-Werke, A. G. (1983), Umweltfreundliche Energietechnik, das Klöckner-Verfahren, eine führende Technologie zur Kohleveredelung, *Pütt und Hütte*, **32**, 1.

[17] KHD Humboldt Wedag AG (1984), *Kohlervergasung im Eisenbad nach dem MIP–Verfahren* (KHD Humboldt Wedag, AG, Cologne, FRG) (English translation forthcoming).

[18] Scharf, H.-J., Schrader, L., and Teggers, H. (1984), Results from the operation of a semi-technical test plant for brown coal hydrogasification, *Nuclear Engineering and Design*, **78**(2), special issue: *The High Temperature Reactor and Nuclear Process Heat Applications*, 223–231.

[19] Guy, C., Ruprecht, R., Langhoff, J., and Dürrfeld, R. (1980), Stand der Texaco-Kohlevergasung in der Ruhr-Chemie/Ruhr-Kohle-Variante, *Stahl und Eisen*, **100**(7), 388–392.

[20] Arbogast, G., Peutz, M. G. F., and van der Zijden, A. J. A. (1984), Refinery system, alternative energy sources and environmental matters in a nonresidual fuel refinery, in *Proceedings of the Eleventh World Petroleum Congress, August 28–September 2, London* (John Wiley & Sons, Chichester, UK).

[21] Singh, J., Niessen, H. F., Harth, R., Fedders, H., Reutler, H., Panknin, W., Müller, W. D., and Harms, H. G. (1984), The nuclear-heated steam reformer-design and semi-technical operating experiences, *Nuclear Engineering and Design*, **78**(2), special issue: *The High Temperature Reactor and Nuclear Process Heat Applications*, 251–265.

[22] Nickel, H., Schubert, F., and Schuster, H. (1984), Evaluation of alloys for advanced high-temperature reactor systems, *Nuclear Engineering and Design*, **78**(2), special issue: *The High Temperature Reactor and Nuclear Process Heat Applications*, 251–265.

[23] Kernforschungsanlage Jülich GmbH (1983), *Memorandum zum Umbau des AVR zu einer Prozesswärmeanlage*, October 1983 (Jülich, FRG).

[24] Smith, J. L. and El-Masri, M. A. (1981), *The High-Temperature Carbon Monoxide-Combustion Turbine Cycle* (MIT, Cambridge, MA).

[25] Magnusson, J. H. (1982), *A Gas Turbine Combined Cycle Approaching Minimal Irreversibility* (MIT, Cambridge, MA).

[26] Domestic hard coal (both types) was assumed to cost 250 DM/t.c.e. [1 ton of coal equivalent (t.c.e.) is equal to 7 gigacalories (GC) or 29.31 gigajoules (GJ); 250 DM/t.c.e. is equivalent to US$100/t.c.e.]; imported hard coal, 150 DM/t.c.e. (equivalent to US$60/t.c.e.); and lignite, 100 DM/t.c.e. (equivalent to US$40/t.c.e.). We estimated the price of gas as 0.05 DM/kWh (equivalent to 20 US mills/kWh).

[27] The dynamic case: primary and final energy demand in FRG, 1981 (in GWyr/yr).

Energy form	*Primary energy*	*Final energy*
Oil/refinery products	118.2	108.7
Hard coal	60.5	21.4
Lignite	28.0	0
Natural gas	52.0	39.0
Nuclear	14.1	0
Electricity	0	31.0
Total	275.0	200.0

[28] Costs and efficiencies of components of the existing energy system.

Technology	*Efficience* (%)	*Investment* (DM/kW)
Hard coal power plant[a]	33–40	1200
Lignite power plant[a]	32–38	1660
Gas turbine power plant[a]	31–42	800–950
Gas pipeline[a]	95	642
Refinery[a]	92	245
Gas heating[b]	70	100
Electric heating[b]	95	50
Oil heating[b]	60[c]	130
Coal heating[b]	50	200
Methanol heating[b]	50[c]	130

	Operation and maintenance	
Technology	*Fixed* (DM yr/kW)	*Variable* (DM/kWyr)
Hard coal power plant[a]	80	80
Lignite power plant[a]	83	100
Gas turbine power plant[a]	40	180
Gas pipeline[a]	13	0
Refinery[a]	12.5	9
Gas heating[b]	0	3
Electric heating[b]	0	2.4
Oil heating[b]	0	4
Coal heating[b]	0	6
Methanol heating[b]	0	2

[a] Efficiency = secondary/primary energy.
[b] Efficiency = useful/final energy.
[c] Differences in efficiency of oil and methanol use in the heat market (as well as the transport sector) represent losses in methanol production that have not been accounted for elsewhere.

Commentary

Y. Kaya

Two types of antipollution measures have been employed to date. One is technological compensation, i.e., elimination of emitted pollutants by use of antipollution facilities. The other is prevention, i.e., regulation of the use of pollution sources themselves. Desulfurization at electricity power plants is a typical example of the former, and prohibition of the use of DDT one of the latter.

From the above viewpoint the idea raised in this chapter apparently seems close to the former type of measure, but the concept is much more drastic than those of conventional antipollution measures. The essence of the concept lies in building a new system in which emission of pollutants is restricted mainly to the stage of final use, so that the total amount of pollutants may be greatly reduced. In current energy systems coal, natural gas, and sometimes even crude petroleum are burnt with air and hence emit SO_x and NO_x, while in the system NIES proposed here all primary energy resources except nuclear are first put into conversion processes which produce CO and H_2 as the main energy carriers. Some conversion processes may also emit pollutants, but careful selection of these will give low emission levels and, furthermore, if a certain amount of pure O_2 is made available by air separation, zero emission will be achieved.

Since emission of SO_x and NO_x has induced a number of environmental problems, such as photochemical smog and acid rain, the concept proposed here is highly attractive in the sense that the entire energy system is designed to be environmentally robust in structure. I am convinced that NIES is one of the most interesting ideas to have appeared in the field of energy systems in the last 10–15 years.

The concept is, however, at present rather crude and should be elaborated in more detail. The following four aspects at least must be investigated at further stages of the study.

Comparison with the present energy system

It is implicitly assumed that NIES will become, sooner or later, economically more attractive than the present energy system, given the required level of environmental protection. It is, however, very uncertain whether this will be so.

In examining the Japanese energy system, for example, we see that moderate economic growth has been attained in harmony with proper environmental protection. The concentrations of SO_x and CO in the air have been steadily decreasing since 1973, and the concentration of NO_x in the air began to decline in 1979. The success is mainly due to the installation of antipollution measures in electricity power plants, industrial boilers, and automobiles. Although the cost of these measures is high – in the case of coal-fired power plants the cost of desulfurization and of denitrogenization are 13 and 3% of the total system costs, respectively – all utility companies and industries introduced these measures and produced both a low level of pollutant emission and the moderate economic growth of the country. This fact indicates that even the present energy systems may be harmonized with the environment, with the necessary antipollution investments, and with the economic development of society. In other words, the significance of NIES can be validated only when its long-term advantages, both in economic and environmental terms, relative to the present system become recognized. The discussion in this chapter is only an introduction to this end. (The costs of novel conversion processes given in *Tables 6.4* and *6.9* and those on antipollution facilities given in *Table 6.11* are still controversial.)

System robustness in the changing international situation

The present energy systems depend heavily on supply conditions, namely resource availability and their prices. As most developed countries import much of their energy resources, they are always keen on reinforcing their energy systems to be flexible enough to adapt to changes in the energy supply in the world. NIES seems to have potential flexibility, since the conversion processes in the system can be a good buffer between a varying primary energy supply and the secondary energy carriers. Results of the optimization scheme for selection of the cheapest conversion processes that satisfy demand constraints are shown in *Tables 6.5–6.8* and *6.10*, but selected processes are very different from case to case. This suggests that other types of optimum solutions should be sought that are relatively insensitive to changes in supply situations and yet still sustain reasonable economic feasibility. I expect that investigation of this will reveal another interesting (probably attractive) feature of NIES.

Assessment of CO, H_2, and methanol as energy carriers

We have experienced in the past CO-rich city gas made from coal, but neither H_2 nor methanol as energy carriers. Even CO, which may be used as a single energy carrier in NIES, and methanol are so toxic that handling of these fluids may induce new environmental and social problems, and require additional investment and operation costs. Investigation of this is essential before implementation of NIES.

Evaluation of dynamic feasibility and modification of NIES so as to increase its feasibility

Since the structures of NIES and of the present energy system differ very much, the key question a decision maker may raise, even if he or she has a sympathy with the NIES concept, is how easily can we move from the present system to NIES? A preliminary study on the dynamic feasibility has been reported in this chapter, but we know that practice is far different from the results of calculations of this sort. We should notice that revolutionary changes have happened once in a while in many technological areas, but that the changes we are seeking here lie in the huge system on which our daily activities rely. Since the costs of new technologies in this area are still very uncertain, we will necessarily be prudent in introducing new facilities and systems. A number of in-depth studies are required to fill the gap between the present system and NIES.

I hope the above comments will be seriously taken into consideration in further studies of this novel concept.

PART THREE

THE WORLD ENVIRONMENT

The world environment

Contributors

About the Contributors

Paul Crutzen is Director at the Max-Planck-Institute for Chemistry in Mainz, Federal Republic of Germany. He studied meteorology at the University of Stockholm, Sweden, obtaining a Ph.D. in 1968, and worked there until 1974. He then became Research Scientist, and later Division Director, at the National Center for Atmospheric Research in Boulder, Colorado. Crutzen's main research interest is in atmospheric chemistry and its role in biogeochemical cycles.

Francesco di Castri is Director of the Centre of Ecology at Montpellier of the National Research Council (CNRS) of France. He is also the President of the French interministerial group for evaluation of and programming for the environment. Previously he was Secretary of the Council of the Man and Biosphere Programme and Director of the Division of Ecological Sciences of UNESCO, Vice-President of the Scientific Committee on Problems of the Environment (SCOPE), and Director of the Institutes of Animal Production and of Ecology, Universities of Santiago and Valdivia, Chile.

Robert Dickinson is Deputy Director of the Atmospheric Analysis and Prediction Division of the National Center for Atmospheric Research (NCAR, which is sponsored by the National Science Foundation) in Boulder, Colorado. Educated at Harvard and the Massachusetts Institute of Technology, he has worked for NCAR since 1968, as well as holding academic and teaching affiliations at the Universities of Washington and Colorado, and at Colorado State University.

Thomas Graedel is a Distinguished Member of the Technical Staff at ATT-Bell Laboratories, New Jersey. Educated at the Universities of Washington State, Kent State, and Michigan, his research interests are in atmospheric chemistry and physics, the measurement and chemical modeling of air pollutants, and the corrosion of materials by atmospheric contaminants.

Crawford Holling is Professor in the Department of Zoology at the University of British Columbia, Canada. Educated at the Universities of Toronto and British Columbia, he has been Director of the Institute of Animal Resource Ecology and Director of the International Institute for Applied Systems Analysis. He is a Fellow of the Royal Society of Canada.

James Lovelock is an independent scientist, working from his home in the UK, and paying for his research from the proceeds of his inventions. He has cooperated with NASA in their space program, is a Fellow of the Royal Society, and visiting Professor in the Department of Cybernetics at Reading University. He invented the electron capture detector, which has revolutionized environmental analysis.

Michael McElroy is Professor of Atmosphere Sciences in the Division of Applied Sciences at Harvard University. Educated at Queen's University, Belfast, he was a physicist at the Kitt Peak National Observatory before taking his present post. McElroy has served in advisory capacities to a number of government groups, including NASA and the US Congress, and is currently a member of the Center for Earth and Planetary Sciences at Harvard University.

Ernõ Mészáros is Director of the Institute for Atmospheric Physics of the Hungarian Meteorological Institute and President of the Meteorological Commission of the Department of Earth Sciences and Mining of the Hungarian Academy of Sciences. Educated at the L. Eötvös University, Budapest, he has produced many papers on the physics and chemistry of the atmosphere and takes an active part in the environmental program of the World Meteorological Organization. He directs the WMO Training Center on background air pollution monitoring.

Tom Wigley is Director of the Climatic Research Unit at the University of East Anglia, UK. He trained as a meteorologist in Australia, after obtaining an undergraduate degree in theoretical physics, and then completed a Ph.D. in plasma kinetic theory at the University of Adelaide. While teaching mathematics and meteorology in the Department of Mechanical Engineering at the University of Waterloo, Canada, he developed an interest in carbonate geochemistry and isotope chemistry.

Chapter 7 Change in the natural environment of the Earth: the historical record

M. B. McElroy

Editors' Introduction: An understanding of how the biosphere works, and of how it responds to human activities, must be based on two facts. The first is the biosphere's natural variability – the continuously changing patterns of climate, geochemistry, and ecology that characterize the historical record. The second is the "boundedness" of that variability – the fact that the Earth's changing environment has remained within the narrow limits congenial to life for more than 4 billion years.

In this chapter is presented an overview of the nature, origins, and limits of the biosphere's natural variability. How astronomical factors and ocean dynamics account for some of this variability is described. Special emphasis, however, is given to elements of the "Gaia" hypothesis – the fundamental roles played by the Earth's biota in regulating its environment.

Introduction

We are accustomed to think of the environment for life on Earth as relatively stable, at least on a global scale. Change, especially that induced by human activity, is thus regarded as threatening, and perhaps it is. It is important, though, to maintain a sense of perspective. There was a time, not so long ago, when vast areas of the northern hemisphere were covered in ice, when the sea level was more than 100 m lower than it is today; ice ages have come and gone; the Earth has been both warmer and colder; ecosystems have evolved and disappeared; even the composition of the atmosphere has varied: change is the norm, not the exception, for life on Earth. No less than the social scientists, we physical scientists need a sense of history if we are to chart a wise course to the future. The challenge is to develop an understanding of the powerful forces that shape our environment. The record of the past, preserved in terrestrial deposits, in lakes, in peat bogs, in polar ice, and in deep sea sediments is an invaluable, perhaps even an essential, stimulus for a research program designed to meet this objective.

We focus here on changes to the Earth which may arise on time scales extending from decades to hundreds of thousands of years. The emphasis is on subtleties, on interactions between the physical, chemical, and biological environments that involve positive feedbacks, on instances where relatively small disturbances are suspected to lead to potentially significant large-scale changes in the overall environment of the planet. The importance of positive feedback is readily illustrated by considering the enormous change that has taken place on the Earth over the past 125 000 years, despite a minimal change in the incident flux of solar radiation. It was warm 125 000 years ago, warmer than today: then, it began to cool. Ice sheets advanced and, with occasional reversals, continued to do so for about 110 000 years. The trend reversed abruptly 15 000 years ago: the planet began then to warm and in less than 10 000 years evolved from a configuration in which major portions of the land areas were covered in ice to the present, essentially ice-free condition. The presence of ice appears to have promoted further cooling, but there was a limit to the spread of the ice sheets, and their ultimate demise was quite dramatic. The sequence has been repeated approximately eight times over the past 750 000 years [1]. It is obviously important that we identify the mechanisms responsible for the vast shifts in climate which characterize the recent past of our planet. It may be particularly urgent to clarify the processes responsible for the relatively abrupt nature of glacial to interglacial transitions, and for the comparatively brief duration of interglacial epochs such as the one in which we now live.

There is also evidence of variability in the transport mechanisms that shuffle life-essential elements back and forth between the biosphere, soil, atmosphere, ocean, and lithosphere, and of changes in the biogeochemical cycles that mobilize H_2O and major nutrients, such as carbon, nitrogen, and phosphorus. The concentration of atmospheric CO_2 was relatively low, about 200 p.p.m., during the last ice age [2, 3], jumping abruptly, to about 280 p.p.m., at the end of the ice age. There is evidence that the change in CO_2 concentration may have led the change in climate 1500 years ago, and evidence that this may not be coincidental, that it may be characteristic of many of the major climatic shifts of the past [4]. This result is quite unexpected. We are accustomed to think of the biogeochemical cycles as steady, being described conventionally in terms of simple diagrams. Reservoirs (the biosphere, soil, atmosphere, upper ocean, lower ocean, and sediments, for example) are indicated by boxes and numbers define the content of individual reservoirs. Fluxes between reservoirs are described by numbers attached to arrows connecting the boxes. The simplicty of the diagrams lulls us readily into a false sense of security. It is easy to forget that the diagrams are constructed usually on the basis of observations of the Earth as it is now. Fluxes are derived from incomplete data or from imaginative extrapolations constrained to force closure, to ensure a steady state. Observations of the variability in CO_2 levels call this approach into serious doubt, at least for time scales longer than a few thousand years. We return to this matter later (pp 204–207) and argue that the steady-state assumption may be questioned not only for carbon, but also for nitrogen and phosphorus. In fact, nutrient cycles are inextricably linked, with changes in the distribution of one element possibly leading to important fluctuations in others. The box and arrow approach is at best a pedagogical device, useful for purposes of illustration, but dangerous if used to excess.

We begin with a brief review of the observational evidence for changes in ancient climate, emphasizing primarily changes that are relatively persistent, having endured for periods of centuries or longer. For the recent past we rely on the historical record, on pollen that reveals shifts in vegetation patterns, and on a variety of geomorphological studies that record changes in alpine and continental glaciers. Studies of deep sea cores extend the record to the more distant past but with a somewhat coarser resolution in time. Patterns of change appear more regular on longer time scales and, indeed, there is convincing evidence that changes in the orbital elements of the Earth play a major role in modulating climate at frequencies of 19 000, 23 000, and 41 000 years. An astronomical control was suggested first by Adhemar almost 150 years ago [5]. The theoretical basis was extended significantly by Croll [6] and reached a level of maturity some 50 years ago in a remarkable series of papers by the Yugoslav mathematician Milankovitch [7].

The elements of Milankovitch's theory, and the data which support it, are discussed on pp 203–204, and problems posed by the dominance of the 100 000 year signal in the climate record are treated on pp 204–207. We argue that changes in CO_2 concentration, driven by changes in marine productivity, provide the most plausible explanation for this phenomenon. The oceans can also have an important influence during shorter time scales, as discussed on pp 207–208.

On pp 208–209 we briefly indicate important contemporary changes in the concentrations of atmospheric CH_4, N_2O, CO, and O_3, discussed in more detail in Chapter 8, this volume. This completes our survey, which demonstrates that changes of the past persist to the present. Indeed, rates of change may be larger today than at any time in recent history. Our task, to provide the breadth and depth of knowledge needed for a comprehensive assessment of the complex impacts of modern technology, is urgent, but feasible. It requires a combination of skills from different disciplines and nations. It also requires the committment of governments and the enthusiastic support of our scientific colleagues. There is a will, indeed an impatience, to proceed. We can ill afford to delay.

The record of change

As ice sheets accumulate on land they draw water from the ocean. Light water evaporates more readily than heavy, so sea water is enriched in heavier isotopes, such as ^{18}O, during climatic cooling. Conversely, the relative abundance of ^{18}O in sea water is expected to decrease during a period of climatic warming, as light water is added to the ocean from the melting of glaciers. Shells growing in the ocean are composed in part of the mineral calcite, $CaCO_3$, and the isotopic composition of these shells reflects the composition of the water from which they form. A fraction of the shells fall to the ocean bottom and are preserved in sediments, providing a record of the climate. Such sediments have been sampled from large areas of the world's oceans during the past decade.

Figure 7.1 shows changes in the relative abundance of ^{18}O obtained from shells of planktonic organisms isolated from cores of sediment drawn from five distinct regions of the ocean [1]. The data reflect an average of results from individual cores and are expected to provide a reasonable estimate for the mean abundance of oceanic ^{18}O. This is defined by the quantity $\delta^{18}O$, proportional to the ratio of concentrations ^{18}O to ^{16}O. Values of $\delta^{18}O$ are related approximately to changes in sea level as follows: a shift of 1 in δ (expressed in units of per thousand and written as ‰) denotes a change in sea level of about 90 m. Larger values of δ reflect an isotopically heavier ocean, indicating the presence of important quantities of ice on land and, presumably, a relatively cold climate. It is immediately clear from *Figure 7.1* that major changes in climate occur about every 100 000 years. Associated fluctuations in sea level exceed 100 m.

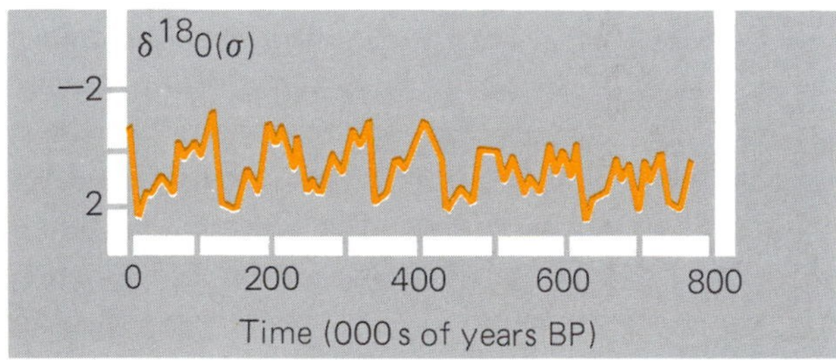

Figure 7.1 Variations of oceanic $\delta^{18}O$ (‰) over the past 782 000 years inferred from measurements of oxygen in the shells of organisms preserved in deep sea cores. The data reflect an average of measurements from five cores. As noted in the text, a change in $\delta^{18}O$ of 1‰ is thought to correspond to a change in sea level of about 90 m. Adapted from Imbrie [1].

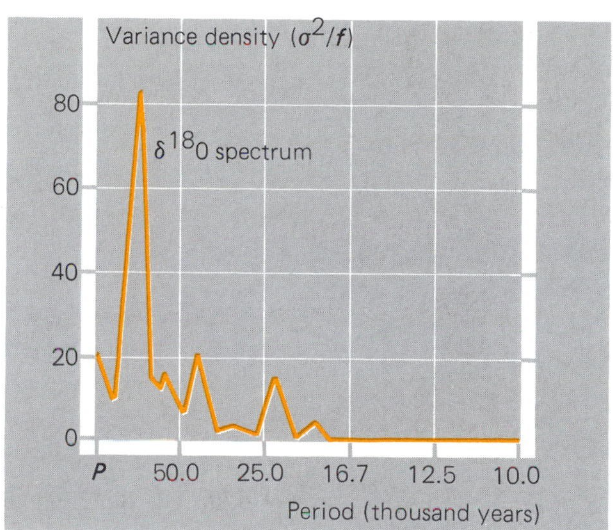

Figure 7.2 A power spectral analysis of the data in *Figure 7.1*, indicating major peaks in the record corresponding to periods of 100 000, 59 000, 41 000, 23 000, and 19 000 years. Adapted from Imbrie [1].

Figure 7.2 presents a power spectral analysis of the data in *Figure 7.1*. In addition to the peak at 100 000 years we see significant peaks at 41 000 and 23 000 years and subsidiary peaks at 59 000 and 19 000 years. The peaks at 41 000, 23 000, and 19 000 years almost certainly reflect an astronomical forcing of the climatic system by changes in the seasonal pattern of insolation. These are driven by changes in the orbital parameters of the Earth, specifically due to variations in the tilt of the rotation axis and due to secular precession of the rotation axis. The tilt of the axis varies by about 1.5° on either side of its average value of 23.5°, describing a complete cycle every 41 000 years. The axis describes a complete precessional cycle in about 26 000 years. The effect of precession, combined with slow rotation of the orbit, leads to a change in the position of the Earth relative to the sun for any particular season. At present the Earth is relatively close to the sun on December 21 during the northern winter. The opposite situation prevailed 11 000 years – northern winter occurred when the Earth was relatively far from the Sun. The position of the Earth at any given seasonal epoch, say, at the spring equinox, precesses along the orbit, describing a complete cycle in about 22 000 years. A spectral analysis of the resultant changes in insolation, at 65 °N in August, for example, indicates significant power every 23 000 and 19 000 years. The power in the climatic record at frequencies corresponding to these periods almost certainly reflects the effect of equinoctial precession. In the same fashion, power in the climate record at 41 000 and, perhaps, at 59 000 years can be attributed to changes in seasonal insolation driven by variations in the tilt of the rotation axis. We return to these matters later on.

Changes in insolation are expected also over longer periods, at 100 000 and 413 000 years, due to changes in the ellipticity of the Earth's orbit. These changes are small, however, compared with those induced by the variations in axial tilt and equinoctial precession. The dominance of the 100 000 year signal in climate requires a separate explanation, addressed on pp 204–207.

A summary of climatic data for more recent times is presented in *Figure 7.3* [8]. The ice sheets reached their maximum extent about 18 000 years ago, with the planet starting to warm up about 15 000 years ago. The ice sheets had essentially disappeared 10 000 years later, but the course of climate change was quite irregular. There is evidence for about five episodes of significant cooling over the past 10 000 years. The first, known as the Younger Dryas period, is particularly notable and is discussed in some detail below. The last, referred to as the Little Ice Age, began in the early part of the fourteenth century and continued until the late 1800s, as illustrated in *Figure 7.4* [9].

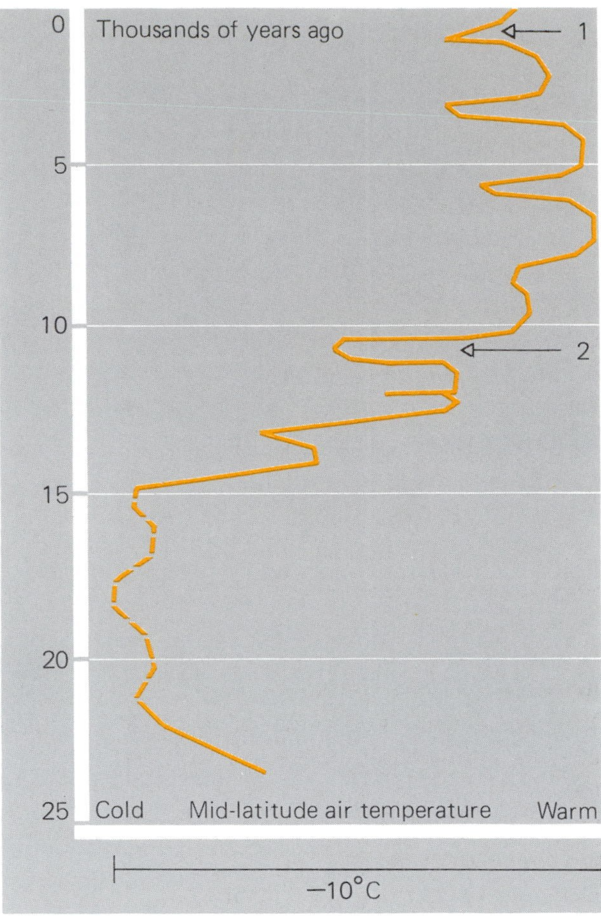

Figure 7.3 Estimates for the variations of atmospheric temperature at mid latitudes over the past 20 000 years. The data are based on a variety of primary sources as summarized in note [8]. The Little Ice Age episode and the Younger Dryas cold period are indicated by arrows 1 and 2, respectively.

Deep sea cores record a dramatic temperature fluctuation in the North Atlantic, beginning about 13 000 years ago. The boundary marking the transition from temperate to polar waters shifted north, from its glacial position to close to that of today. It stayed there for about 1000 years. It then reversed, returning to the glacial configuration, where it remained for several hundred years before finally moving north to take up the position it has occupied for most of the Holocene. The fluctuation is evident also in pollen records from Europe and from the Maritime Provinces of Canada. As the polar front shifted north, during the so-called Allerod period,

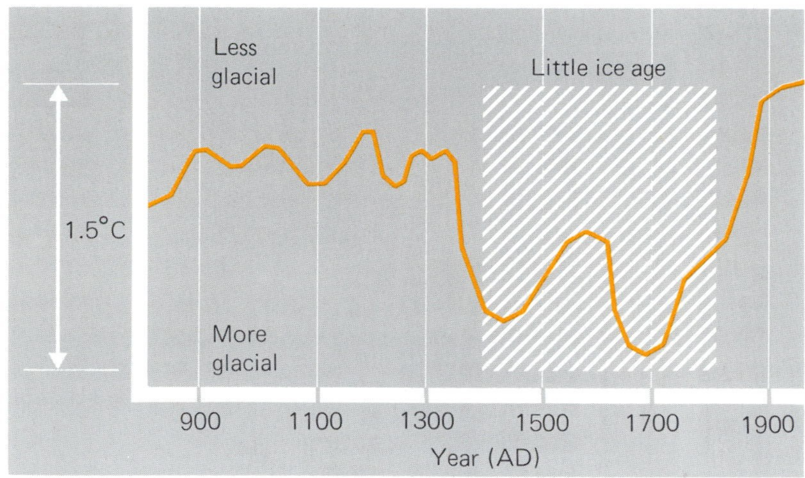

Figure 7.4 Estimates for the variation of winter temperatures in Eastern Europe since the ninth century. Data based mainly on manuscript records discussed by Lamb [27]. Adapted from [9].

the European climate became milder. The grasses and sparse vegetation of the glacial period were replaced by forests, which disappeared rapidly, however, as cold conditions resumed during the Younger Dryas. The remarkable nature of these changes is that they appear to have been relatively localized. They are evident most clearly in pollen taken from regions bordering the North Atlantic, but there is no evidence for the Allerod–Younger Dryas epoch in the pollen record of the USA [10].

The significance of the Allerod–Younger Dryas transition is considered further on pp 207–208. A strong case can be made that it must involve changes in the circulation of the North Atlantic, but the central issue concerns the factors responsible for this change and the extent to which they might be involved in the other excursions of climate depicted in *Figure 7.3.*

The Milankovitch hypothesis

The thesis advanced by Milankovitch was basically very simple. He had the idea that changes in insolation in summer would play a critical role in determining whether ice sheets would advance or recede. He had the view that snows would come no matter what in winter, with the central issue being the extent to which the winter snows melted in summer.

In a period of decreasing summer illumination, the ice line would gradually advance, he suggested, facilitated by the higher reflectivity of a larger area of snow. Change would feed on itself, an example of positive feedback. There is little doubt that Milankovitch's essential hypothesis, that changes in the seasonal pattern of insolation should lead to significant variations in climate, is basically correct. It is equally clear, however, that the mechanism whereby changes in illumination combine to effect a change in climate is assuredly more complex than the simple ice albedo feedback he invoked.

The reality of the astronomical influence on climate is strongly supported by the presence of peaks in the climatic spectrum at frequencies directly related to the astronomical forcing, as shown in *Figure 7.2.* The data, at least at the higher frequencies, can be described by a relatively simple linear model. We assume that the temperature, or ice volume, response to a forcing function $p(t)$ is given by

$$\frac{\mathrm{d}T}{\mathrm{d}t} = p(t) - \alpha T \tag{7.1}$$

If the forcing function is sinusoidal with frequency, the temperature response, T, should also be periodic, and the associated frequency should be identical to that of $p(t)$. We expect that the response function, T, should be phase shifted with respect to p by an amount proportional to the intrinsic time constant of the climatic system, represented here by α^{-1}.

An analysis of the climatic data, *Figure 7.1,* has been carried out by Imbrie [1] making use of these simple concepts. He assumed that the isotopic signature in any given frequency band could be described by a function $gp(t-\phi)$ where $p(t)$ is a measure of the astronomical forcing, g is a constant amplitude, and ϕ is a constant phase. He sought to determine the quantities g and ϕ directly from analysis of the data given in *Figure 7.1*. The record was divided into two intervals, from the present to 400 000 years before present (BP), and from 400 000 years BP to 780 000 years BP. He considered frequency bands corresponding to changes in orbital eccentricity, with periods of 123 000, 95 000, and 59 000 years, changes in axial tilt, with a period of 41 000 years, and changes associated with equinoctial precession, with periods of 23 000 and 19 000 years. The analysis revealed a remarkable pattern. The response to changes in axial tilt and precession is regular throughout the entire time domain. Data in each of the relevent frequency bands can be given an essentially constant amplitude and phase. The response of climate to change in axial tilt is shifted in phase with respect to orbital forcing by about 8500 years. The response with respect to axial precession is shifted in phase by about 5500 years at 23 000 years, and by about 3500 years at 19 000 years. The fit is excellent throughout the time domain. There is no evidence for drift in either amplitude or phase for the high frequency response, at periods of 19 000, 23 000, and 41 000 years, over almost 800 000 years. In contrast the response to forcing associated with changes in ellipticity was less regular. Imbrie's statistical approach indicated a phase shift of 5000 years for elliptical forcing at 123 000 years over the most recent 400 000 years. The analysis of the earlier record implied a phase shift of 14 000 years, of opposite sign, however.

The statistical analysis strongly supports, therefore, the reality of the astronomical influence on climate, at least for changes associated with variations in axial tilt and precession. The nature of the variability on longer time scales is more problematical. We argue here that changes on the 100 000 year time scale are more likely to reflect changes in the internal function of the climatic system, and that they may be induced, at least in part, by changes in the manner in which nutrients are cycled through the ocean. The potential importance of the biogeochemical connection is suggested by measurements of CO_2 in air trapped in ancient ice in Greenland and Antarctica [2, 3]. The level of CO_2 was relatively low, about 200 p.p.m., during the last ice age. It shifted quite abruptly, to about 280 p.p.m., near the end of the ice age, as illustrated in *Figure 7.5*. A variation of this magnitude on such a brief time scale must involve an important change in the physical, chemical, and biological properties of the ocean [11].

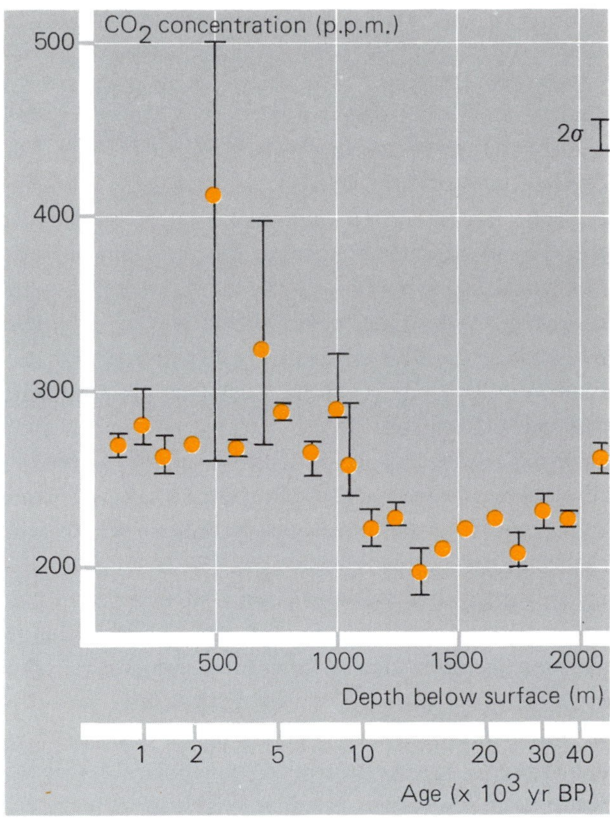

Figure 7.5 Variations in atmospheric CO_2 inferred from measurements of air trapped in bubbles of ice at Byrd Station in Antarctica (semi-log scale). Adapted from Neftel *et al.* [28].

Biogeochemical processes in the ocean

Most of the ocean's volume is occupied by cold, dense water formed at the surface in localized regions at high latitude, in the Weddel and Norwegian Seas, for example. Cold water reaches the surface as part of the general circulation of the ocean, driven for the most part by the contrast in temperature and density between the surface and depth. Deep water is warmed by the sun as it reaches the surface and distributed globally through a vast system of wind-driven currents. Residues of the surface currents reach high latitudes where the water cools and its salt content increases, reflecting the combined influence of evaporation and formation of sea ice. The cycle is completed as water sinks back from the surface to depth. The overall trip through the deep sea takes a long time, nearly 1000 years at the present epoch. Regions in which bottom water forms occupy but a small fraction of the ocean's surface area, less than 10% of the total. Most of the ocean is characterized by slow upwelling of waters from depth.

Upwelling plays an important role in the life cycle of the sea. Biological productivity is limited, for the most part, by the availability of nutrients, particularly nitrogen and phosphorus. These elements are present mainly at depth, as nitrate and phosphate, and are supplied to the surface in conjunction with the overturning of the ocean – the thermohaline circulation. Alfred Redfield and his associates recognized, 25 years ago, that organisms in the sea have a more or less constant appetite for nitrogen and phosphorus: approximately 105 atoms of carbon are fixed in biological tissue for every 15 atoms of nitrogen and for every atom of phosphorous [12]. As organisms grow they draw carbon from surface waters. The quantity of carbon thus fixed is controlled for the most part by nitrogen and phosphorus, the limiting nutrients, in waters that reach the surface. As organisms die, they fall and decay at depth releasing their constituent carbon, nitrogen, and phosphorus. Oxygen is consumed in the process. Biological activity acts, therefore, to lower the carbon content of waters at the surface and to raise that at depth. Roger Revelle refers to the process as a biological pump. The higher the efficiency of the pump, the greater the discrepancy between the carbon content of waters at the surface and at depth. The level of atmospheric CO_2 is set primarily by exchange of gas across the air–sea interface. Release of CO_2 by the ocean at low latitudes is balanced, at least in the steady state, by uptake at high latitudes. A change in the efficiency of the biological pump is reflected in a change in the carbon content of surface waters, leading to a relatively immediate adjustment in the rate at which carbon is transferred across the air–sea interface. If the pumping efficiency were to increase, carbon would be drawn from the atmosphere to the surface ocean and transferred from there to the deep. The level of atmospheric CO_2 would fall and the system would evolve to a new steady state, characterized by a higher concentration of carbon in the deep sea, with lower concentrations in surface waters and in the atmosphere. An efficient pump reflects a productive ocean. The flux of organic detritus from surface to depth is large, and so the demand for dissolved oxygen is correspondingly high. We expect, thus, a direct connection between CO_2 in the air and O_2 in the deep sea. When CO_2 is low, so also is O_2; conversely, when CO_2 is high, the marine biota are relatively ineffective, and the concentration of dissolved O_2 should rise, approaching the limiting value set by equilibrium with the air.

In its simplest form we can think of the biological pump as a cycle in which water rich in nutrient moves from depth to the surface. A fraction of its dissolved carbon is converted into organic material and falls back to the deep, leaving nutrient- and carbon-depleted water at the surface. The upward flow of water occurs over relatively large areas of the world's oceans and is balanced by a return flow from localized regions at high latitudes. The nutrient content of water in the return flow represents unutilized biological potential. Redfield referred to it as preformed nutirent, noting, on the basis of analysis of the deep sea oxygen budget, that approximately half of the nutrients at depth are formed *in situ* by decomposition of detritus. The balance is carried down (preformed) with waters in the descending mode of the overall ocean circulation. The presence of preformed nutrient attests to an inefficiency in the biological pump. Its significance in the context of CO_2 was recognized only recently, by groups at Harvard, Princeton, and Berne [13, 14, 15]. A low concentration of preformed nutrient, with a specified total nutrient content, implies an efficient pump and, consequently, a low value for the concentration of atmospheric CO_2. A variation in preformed nutrient over its theoretically possible range could result, according to recent calculations, in a swing in CO_2 as large as 100 p.p.m. with a large associated change in dissolved O_2.

What processes combine to determine the level of preformed nutrient? Observations of nitrate and phosphate in surface waters provide an important clue. Concentrations of nitrogen and phosphorus are small in surface waters at latitudes below about 50°, reflecting, presumably, the ease with which nutrients are taken up by the biota. They climb rapidly at higher latitudes, where it is clear that uptake of nutrients must be limited by factors other than supply. Light is an obvious possibility. The concentration of nutrients in surface water that sinks to depth reflects the origin and history of this water. To the extent that it is supplied by surface currents from latitudes below 40° the nutrient level tends to be low. To the extent that it is derived from upwelling at higher latitudes, where biological activity is less efficient, the nutrient concentration could be quite high. The level of preformed nutrient is set thus, in part, by physical processes and, in part, by biological influences in the high latitude ocean. We expect biological activity at higher latitudes to respond to changes in illumination, as indeed, also might physical mechanisms. A higher than average flux of solar radiation favors enhanced uptake of nutrients and could lead to a decline in the concentration of preformed nitrogen and phosphorus, and an associated reduction in the level of CO_2.

There are several mechanisms, therefore, through which the climate system can respond to changes in the seasonal pattern of illumination at high latitude. The Milankovitch process, operating through variations in snow and ice on land and ice in the sea, allows a direct effect. Changes in high latitude illumination could alter the metabolism of the marine biota, affecting preformed nutrient, thus CO_2, and, consequently, the energy budget of the atmosphere. Changes in preformed nutrients and CO_2 could arise due to changes in the dynamics of the atmosphere and ocean, induced, for example, by variations in snow and ice. Most probably all of these factors are involved and further work is required to define their relative importance. We need a model to describe the ocean and the atmosphere as a coupled system. A great deal of work remains to be done to refine our understanding of the processes that regulate the distribution and utilization of nutrients in the ocean and the manner in which these processes combine to set the level of atmospheric CO_2.

The low value for the concentration of CO_2 in the glacial environment suggests that the pumping mechanism was more effective then than now: a larger fraction of the carbon that reached the surface in glacial times appears to have returned to depth as a component of detrital material. There are three possible explanations (11). We might suppose that the abundance of nutrients in the glacial ocean was much larger than today. We might surmise that the ecology of the glacial ocean was different, that organisms could fix larger amounts of carbon for a given supply of nitrogen and phosphorus. Or we might argue that a larger fraction of upwelling nutrients in the glacial ocean was converted into organic material – that the abundance of preformed nutrient was less than today. The first two of these suggestions have implications for the ocean as a whole, while the third involves only processes occurring at high latitude. We expect shells of organisms that form at depth to reflect the isotopic composition of carbon in the deep sea. Planktonic species, on the other hand, should record the isotopic composition of the surface ocean. Carbon in surface water is relatively heavy, enriched in ^{13}C, reflecting the preferential uptake of ^{12}C by photosynthesis and the subsequent removal of ^{12}C by gravitational settling of detritus. The abundance of ^{13}C relative to ^{12}C is specified by a quantity, $\delta^{13}C$, proportional to the ratio of concentrations $^{13}C/^{12}C$. The difference in $\delta^{13}C$ between surface and depth should reflect the efficiency of the pumping mechanisms. A large difference would imply an efficient pump, consequently a low level for CO_2, and a large demand for dissolved O_2. Measurements of carbon isotopes in deep sea cores can be used, therefore, as a surrogate for CO_2 levels. Shackleton *et al.* [16] showed that the trends in CO_2 content of deep sea cores over the past 60 000 years are generally consistent with the data obtained from ice cores. A more extensive analysis of the deep sea carbon data [4], displayed in *Figure 7.6*, indicates significant power in the spectrum at 100 000 years and suggests that variations in CO_2 can play an important role in regulating the behavior of climate, particularly on longer time scales [17].

How might these changes arise? The lifetime of phosphate, defined as the time required to fill the ocean to its current content at present rates of supply from the land, is about 100 000 years in the contemporary ocean, curiously similar to the time

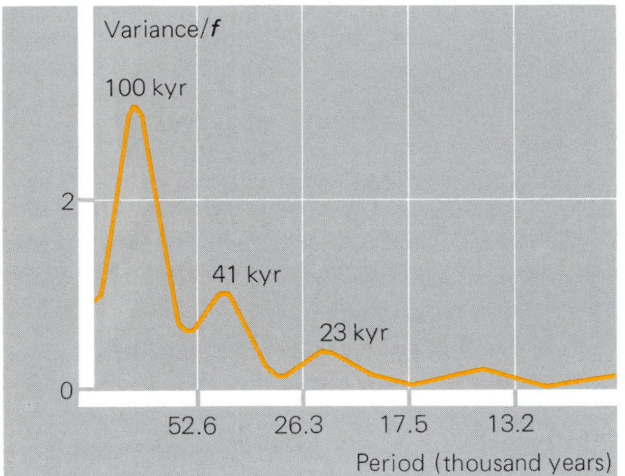

Figure 7.6 A power spectral analysis of the variability of atmospheric CO_2 inferred from measurements of the difference in the isotopic composition of carbon in the shells of benthic and planktonic foraminifera isolated from a deep sea core (V19–30). The concentration of CO_2 appears to vary on time scales of 100 000, 41 000, and 23 000 years. The peak at 100 000 years is particularly significant in light of the large variation in climate observed at this period (see text). Adapted from [4].

constant associated with major changes in climate. Suppose that the budget of phosphate in the ocean is variable, that the ocean gains phosphate from the land during cold conditions when the stand of sea level is low. Phosphate could be stored in coastal sediments when sea level is high and eroded subsequently as sea level falls (11). The level of oceanic phosphorus, and presumably that of nitrogen, would rise then during glacial time, the ocean would become more productive, and the efficiency of the biological pump would be enhanced. The level of CO_2 would fall and with it the level of dissolved oxygen. Carbon dioxide and dissolved oxygen would be modulated by changes in preformed nutrients driven by changes in climate associated with changes in the orbital parameters discussed earlier. Conceivably, changes in preformed nutrients could arise directly as a result of changes in illumination at high latitudes. Eventually the productivity of the ocean would rise to the point where the supply of O_2 would be inadequate to keep pace [18]. Aerobic respiration would be inhibited in this case and denitrification could become prevalent. We might expect a rapid drop in the level of nitrate – the lifetime of dissolved NO_3^- is only about 10 000 years even today. Removal of phosphate could also accelerate, with high productivity and local anoxia favoring the delivery of a high concentration of organic material to sediments and the associated formation of mineral phosphates, such as apatite. This would lead to a rapid drop in productivity. Oxygen would return, CO_2 would be released, the climate would warm, sea level would rise, and the sequence could begin anew. The scenario is admittedly speculative. It has elements, however, that must hold true. There is a limit to the productivity of the ocean, imposed by the supply of O_2. A rapid loss of NO_3^- could account for a drop in productivity and a rise in CO_2, and could account for the rapidity of the final glacial to interglacial transition.

Role of the ocean on shorter time scales

The possible importance of changes in the circulation of the ocean was noted earlier in connection with the Allerod–Younger Dryas phenomenon. Broecker *et al.* [10] suggested that this fluctuation may have arisen because of changes in the rate at which deep water is formed in the North Atlantic. Formation of deep water in this region today involves a flux of approximately 20 Sverdrups (about 2×10^7 m^3/s), supplied by near-surface advection from a source region characterized by a temperature of about 10 °C. As the water cools and sinks it releases an amount of energy equal to approximately 30% of the total solar energy reaching the surface of the Atlantic at latitudes north of 35 °N. It is obviously important to define the mechanisms responsible for deep water formation. Broecker *et al.* suggest that they relate ultimately to the distribution of salinity in the surface ocean [10].

Waters in the North Pacific are relatively fresh, and even if allowed to cool to 0 °C they could sink to only intermediate depths. The high salinity of the North Atlantic is more favorable to deep water formation. Deep water can form also in localized regions of the Southern Ocean, where export of fresh water from semienclosed basins, such as the

Weddel Sea, can lead to production of relatively dense surface water.

The suspension of deep water formation in the North Atlantic during the Younger Dryas, postulated by Broecker *et al.* [10], could reflect an input of fresh water associated with the melting of continental ice. It raises, however, a larger question. At the present time deep water flows south in the North Atlantic, mixing with waters of Antarctic origin, before entering the Indian and Pacific Oceans, where it flows north before eventually returning to the surface. Is this the only stable mode for deep water circulation? There is evidence that deep water did not form in the Atlantic during the last ice age [19]. Is it possible that the direction of deep water flow could reverse under some conditions, with the North Pacific providing the dominant source? The implications for climate, for marine chemistry, and for the biota are staggering and point to the need for a much better understanding of thermohaline circulation in the ocean. The relative inertia of the deep water circulation (approximately 1000 years being taken to describe a complete loop at the present epoch) suggests that the implications of a switch could be relatively long-lived. It would be interesting to explore the extent to which the changes in climate depicted in *Figure 7.3,* other than the Allerod–Younger Dryas, might relate to possible changes in the circulation of the ocean; it would also be interesting to search for evidence of associated fluctuations in rates of biogeochemical cycling. Continuing study of the composition of trapped gases in ice cores could shed important light on this issue.

Other evidence for changes in biogeochemical processes

We have emphasized to this point changes in biogeochemical cycles on time scales longer than a few thousand years. There is evidence, however, for changes on much shorter time scales. The level of atmospheric CH_4 has risen from about 0.7 p.p.m. in AD 1600 to about 1.6 p.p.m. today [20, 21]. Contemporary evidence suggests that the increase continues, at between 1 and 2% per year [22]. The concentration of atmospheric N_2O is increasing, at about 0.2% per year [23], and there are reports of changes in CO [24] and O_3 [25], in addition to the well documented rise in CO_2 [26]. A selected summary of data is given in *Figures 7.7–7.9.* The human activities contributing to these increases and the

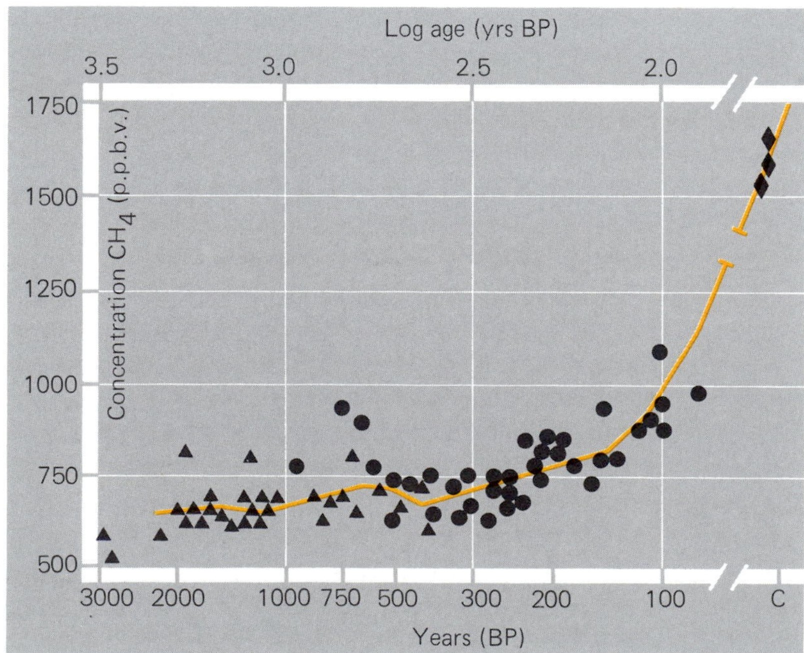

Figure 7.7 Concentrations of atmospheric CH_4 measured in air trapped in bubbles of ice in Greenland (circles) and at the South Pole (triangle). Also indicated is the range of concentrations observed today (diamonds). Adapted from [22].

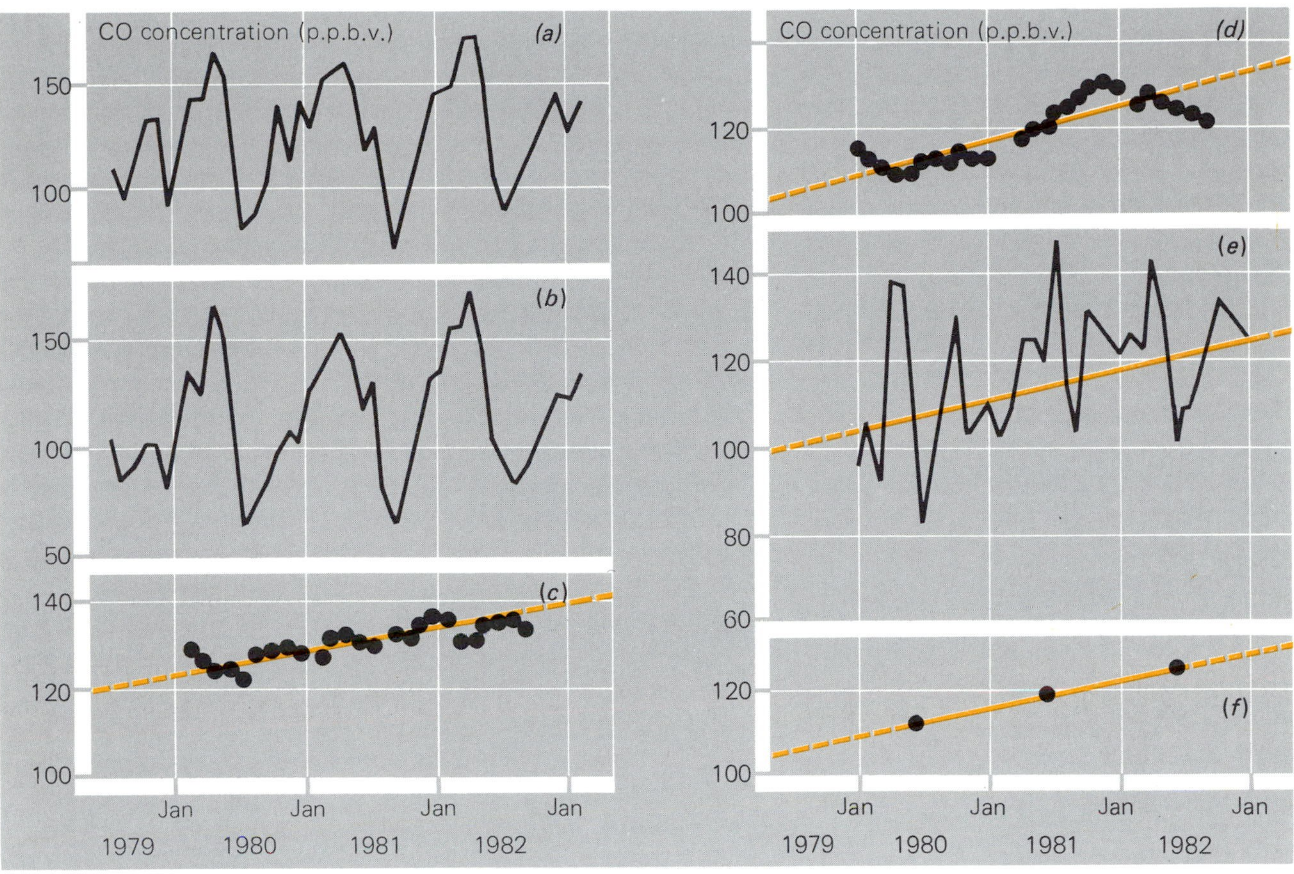

Figure 7.8 Variations in the concentration of CO in air at Cape Meares, Oregon. Adapted from [23].

chemical processes involved are discussed by Crutzen and Graedel (Chapter 8, this volume).

Implications for future work

We need a program of research designed to improve our understanding of the factors responsible for contemporary changes in atmospheric composition. The rise in CO_2 provides a case in point. The concentration of CO_2 was about 280 p.p.m. in the latter half of the nineteenth century. It rose to about 315 p.p.m. in 1958 and has climbed since to a value near 345 p.p.m. (27, 28). Ice-core data [2, 3] suggest that the natural range is between 200 and 300 p.p.m., with low values applying during glacial times and high values appropriate to interglacial conditions. We have moved, in this century, outside the range of what might be considered natural variability. Most of the rise since the industrial revolution can be attributed to release of CO_2 associated with combustion of fossil fuel.

The fossil source, vented initially to the air, is taken up eventually by the ocean. Transfer into the ocean is a slow process, however, limited ultimately by the sluggish circulation of the sea and by the need to dissolve sufficient quantities of carbonate minerals to neutralize acid introduced with CO_2. The ocean is supersaturated with respect to calcite down to depths of about 4 km. Enhanced uptake of CO_2 is possible only when acid-rich water reaches this depth. In the meanwhile, approximately 60% of the carbon released by the burning of fossil fuel is expected to remain in the air.

The contemporary rise in atmospheric CO_2 represents about half of the source from fossil fuel. Is there an unidentified sink for carbon over and above that attributed to the ocean? In particular,

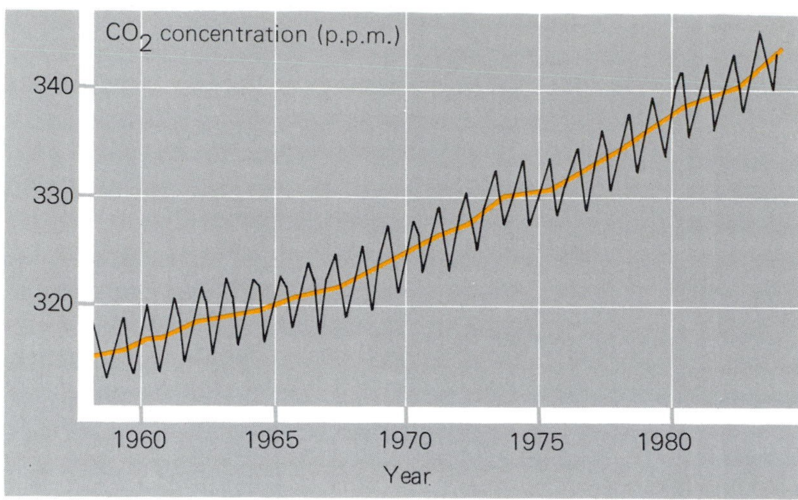

Figure 7.9 Mean monthly concentrations of CO_2 at Mauna Loa, Hawaii. Adapted from [25].

what is the role of the terrestrial biosphere? Are plants growing faster, stimulated by anthropogenic sources of nitrogen and phosphorus? What are the direct and indirect effects of higher levels of CO_2? A higher level of CO_2 could induce a stomatal response, in some plants at least, resulting in the enhanced conservation of H_2O. Are there implications here for regional climate or for the growth of plants? Should we consider the possibility of a significant shift in the distribution and variety of species in specific ecosystems? Careful studies of selected systems could enhance our ability to monitor, understand, and eventually predict effects of environmental change. At a minimum, we need to document the gross effects of human activity, to monitor the areas of land cleared annually for agriculture, and to define the associated changes in the storage of carbon in both biomass and soils. We need a better model and better data to describe the uptake of carbon by the ocean. Models in use now rely heavily on observations that defined the infusion of radioisotopes introduced as a consequence of nuclear testing in the 1960s. How representative are these data? Is the circulation of the ocean constant during the time scales of interest to the assessment of the impact of fossil fuel? We need, eventually, a prognostic model to treat the ocean, atmosphere, and biosphere as an interactive system. This model must incorporate all of the relevant, potentially significant, feedbacks.

The preceeding discussion suggests that interactions are often subtle. Development of models should proceed hand-in-hand with the design and implementation of observational strategies intended to elucidate processes, to place them in an appropriate global context. Studies of the past can play a valuable role in focusing attention on important issues, which might otherwise escape attention.

Our needs are not confined to problems connected with assessment of the effects of anthropogenic CO_2. Similar issues arise for CH_4, CO, N_2O, for oxides of nitrogen, and for O_3. The atmosphere gives a clear, early signal of the widespread effects of human activity. Humankind is now a global factor in the mobilization of life-essential elements, such as carbon, nitrogen, phosphorus, and sulfur. A large part of the landscape is under our influence, either directly or indirectly.

Assessment of the complex effects of human activity requires a broad perspective. There is a general, current recognition of the need for an interdisciplinary approach, and a sense of urgency. The Global Habitability Program launched by NASA, IIASA's Program on Sustainable Development of the Biosphere, and the Global Change Program under consideration by ICSU are essential steps to progress. They must be encouraged and should proceed without delay.

Notes and references

[1] Imbrie, J. (1985), A theoretical framework for the Pleistocene ice ages, *Journal of the Geological Society, London*, **142**, in press.

[2] Berner, W., Stouffer, B., and Oeschger, H. (1978), Past atmospheric composition and climate. Gas parameters measured on ice cores, *Nature*, **276**, 53–55.

[3] Delmas, R. J., Ascencio, J. M., and Legrand, M. (1980), Polar ice evidence that atmospheric CO_2 29 000 BP was 50% of the present, *Nature*, **282**, 155–157.

[4] Pisias, N.G. and Shackleton, N. J. (1984), Modelling the global climate response to orbital forcing and atmospheric carbon dioxide changes, *Nature*, **310**, 757–759.

[5] Adhemar, J. A. (1842), *Revolutions de la mer*, privately published in Paris, quoted in Imbrie and Imbrie, note [9].

[6] Croll, J. (1875), *Climate and Time* (Appleton and Company, New York).

[7] Milankovitch, M. (1941), Kanon der Erdbestrahlung und seine Andwendung auf das Eiszeitenproblem, *Royal Serbian Academy, Special Publication*, **133**, 1–633 [English translation available (1969) from Israel Program for Scientific Translation, US Department of Commerce, Washington, DC].

[8] US Committee for the Global Atmospheric Research Program (1975), *Understanding Climatic Change: A Program for Action* (National Research Council, Washington, DC).

[9] Imbrie, J. and Imbrie, K. P. (1979), *Ice Ages: Solving the Mystery* (Enslow, Short Hills, NJ).

[10] Broecker, W. S., Pateet, D., and Rind, D. (1985), Does the ocean–atmosphere system have more than one stable mode of operation? *Nature*, **315**, 21.

[11] Broecker, W. S. (1982), Glacial to interglacial changes in ocean chemistry, *Progress in Oceanography*, **11**, 151–197.

[12] Redfield, A. C., Ketchum, B. H., and Richards, F. A. (1963), The influence of organisms on the composition of sea water, in *The Sea, Vol. 2*, pp 26–77 (Wiley, New York).

[13] Knox, F. and McElroy, M. B. (1984), Changes in atmospheric CO_2: influence of the marine biota at high latitude, *Journal of Geophysical Research*, **89**, 4629–4637.

[14] Sarmiento, J. L. and Toggweiler, J. R. (1984), A new model for the role of the oceans in determining atmospheric P_{CO_2}, *Nature*, **308**, 621–624.

[15] Siegenthaler, U. and Wenk, T. (1984), Rapid atmospheric CO_2 variations and ocean circulation, *Nature*, **308**, 624–626.

[16] Shackleton, N. J., Hall, M. A., Line, J., and Shuxi, C. (1983), Carbon isotope data in core V19–30 confirm reduced carbon dioxide concentrations in the ice age atmosphere, *Nature*, **306**, 319–322.

[17] Preliminary observations (Fairbanks, 1985, personal communication) of $\delta^{13}C$ in shells of planktonic organisms formed at high latitude suggest that the abundance of ^{12}C relative to ^{13}C was higher in the surface polar ocean in glacial time than today, i.e., $\delta^{13}C$ was less. This behavior is in the opposite sense to that observed at low latitude [16]. It suggests that the abundance of preformed nutrient may have been higher in glacial time than today, in contradictory to assertions of most of the models cited earlier [13–15]. It is obvious that further work is required to accommodate this new information. The general result invoked here, however, that differences between $\delta^{13}C$ in surface and deep waters should reflect CO_2, is unlikely to be affected by this complication. It is similarly clear that changes in preformed nutrient, irrespective of sign, must be considered in any attempt to balance the budget of carbon.

[18] Supply of O_2 is limited by the capacity of the ocean to draw O_2 from the air, i.e., it is limited by the solubility of the gas at the coldest realizable surface temperature. The ocean could supply a much larger quantity of organic carbon to depth than it does today if its nutrient content were higher. Eventually, productivity would be limited by light, but the limitations discussed here would arise earlier. Demand for O_2 is appreciable even for the contemporary ocean, as indicated by the relatively low concentration of O_2 observed at intermediate depths over an extensive region of the sea.

[19] Boyle, E. and Keigwin, L. (1982), Deep circulation of the North Atlantic over the last 200 000 years: geochemical evidence, *Science*, **218**, 784–787.

[20] Craig, H. and Chou, C. C. (1982), Methane: the record in polar ice cores, *Geophysics Research Letters*, **9**, 1221–1224.

[21] Rasmussen, R. A. and Khalil, M. A. K. (1984), Atmospheric methane in the recent and ancient atmospheres: concentrations, trends and interhemispheric gradient, *Journal of Geophysical Research*, **89**, 11 599–11 605.

[22] Khalil, M. A. K. and Rasmussen, R. A. (1983), Sources, sinks and seasonal cycles of atmospheric methane, *Journal of Geophysical Research*, **88**, 5131–5144.

[23] Weiss, R. F. (1982), The temporal and spatial distribution of nitrous oxide, *Journal of Geophysical Research*, **86**, 7185–7195.

[24] Khalil, M. A. K. and Rasmussen, R. A. (1984), Carbon monoxide in the earth's atmosphere, *Science*, **224**, 54–56.

[25] Logan, J. A. (1985), Tropospheric ozone: Seasonal behavior, trends and anthropogenic influence, *Journal of Geophysical Research*, submitted.

[26] Machta, L. (1983), The atmosphere, in National Research Council, *Changing Climate*, pp 242–251 (National Academy Press, Washington, DC).

[27] Lamb, H. H. (1969), Climatic fluctuations, in H. Flohn (Ed), *World Survey of Climatology, 2, General Climatology*, pp 173–249 (Elsevier, New York).

[28] Neftel, A., Oeschger, H., Schwander, J., Stauffer, B., and Zumbrunn, R. (1982), Ice core sample measurements give atmospheric CO_2 content during the past 40 000 years, *Nature*, **295**, 220–223.

Commentary

J. E. Lovelock

In many ways our relationship with the Earth is like that between the eighteenth century physician and his patients. We share with him a vast ignorance illuminated only by an instinct that warns us not to act precipitately, for such action is potentially as disastrous as inaction. Wisdom comes slowly and painfully by trial and error.

Early in the history of medicine, and long before the sciences of bacteriology and biochemistry had made their crucial but specific contributions, this wisdom was enlarged by the science of physiology. The discovery by Harvey of the circulation of blood and by Paracelsus that the poison is the dose were two such enlightenments that gave a scientific basis to the practice of medicine.

The great physiologist Walter Cannon wrote "Organisms, composed of material which is characterized by the utmost inconstancy and unsteadyness, have somehow learned the methods of maintaining constancy and keeping steady in the presence of conditions which might reasonably be expected to prove profoundly disturbing." This wisdom that later he called "homeostasis" is in various ways a prime characteristic of life. It is a property of simple bacteria as well as the largest plants and animals. The ecologists Eugene Odum and C. S. Holling both recognized that some of this homeostasis is also exhibited by ecosystems. These collectives seem to regulate their environment to the benefit of their member species. When the constancy of the Earth's environment during the 3.6 billion years of life is considered, it seems that homeostasis may also be a property of that largest collective, the biosphere.

This view of the Earth as a physiological system is hypothetical and many scientists prefer to believe that the evolution of life and the evolution of the physical and chemical environments are separate and only loosely linked by biogeochemical cycles and biogeophysical processes. Perhaps they are right and perhaps they will show us how to live in harmony with the Earth when their questions have all been answered. But this may take a long time and by then we may be in no position to use the knowledge. The present condition of the Earth is like that of a patient whose illness requires a research program for the development of the cure. We may not have the time for the leisurely and exacting progress of academic science. We need, in addition, the less rigorous and more empirical approach that characterizes the practice of science in times of war.

The need for a pragmatic approach in planetary science is indicated by the choice of titles for many recent international programs for the study of interlinked global systems; for example, Global Habitability, The Geosphere–Biosphere, Global Change, or Earth Science System. These titles invoke, even if their authors did not intend it, a down-to-earth science. A practical science that is global in another sense, namely, that it is neither expensive to practice nor cloistered in an academy. A science that can be practiced and shared with the barefoot scientists of the third world.

It is often forgotten that in medicine vast improvements in the length and quality of life came from practical engineering based on a dubious empiricism. The Romans thought that malaria was a consequence of breathing the smelly air of the swamps. They drained them and the malaria declined: it hardly mattered that their theory was wrong. In a similar way the drinking water of Europe was made clean to drink long before the microbial theory of disease was established. The medieval farmer who rotated his crops and kept a balanced ecosystem within and around his farm did so without the benefit of theoretical ecology. In the complex and nonlinear real world, theory is often the *post hoc* explanation of a practical success and time is needed before it bears fruit in the way of useful predictions.

It takes a war or a great catastrophe to quieten the chattering from the cloisters. The consequences of the changing surface and atmosphere of the Earth seem to loom ominously and in many ways science is again about to be conscripted. At such times there can be an enjoyment in the down-to-earth approach that marks our escape from school and the beginning of an adult life as practicing geophysiologists.

Chapter 8

P. J. Crutzen
T. E. Graedel

The role of atmospheric chemistry in environment–development interactions

Editors' Introduction: The mobility and relatively low mass of the atmosphere make it the component of the Earth's environment most susceptible to large-scale transformations through human activities.

In this chapter is presented a synoptic framework for understanding how chemical processes link human activities, such as fossil-fuel combustion, agriculture, and industrial production, to important properties of the atmosphere. The atmospheric properties included in the framework are absorption of ultraviolet energy, thermal radiation balance (climate), photochemical smog, global oxidation efficiency, precipitation acidity, visibility, and corrosivity. A graphical format is developed that displays the net impacts of multiple sources of disturbance across multiple atmospheric properties. This synoptic approach could serve as a model for similar impact assessments focused on the biosphere's soils, waters, and biota.

Introduction

It is now recognized that the activities of mankind produce detectable and deleterious impacts upon the atmosphere and biosphere. Prudent planning requires that these impacts be well understood so that further developments will not perturb the atmosphere and biosphere beyond their carrying capacities. At present, these human activities rival the global natural inputs into the atmosphere of sulfur, nitrogen, and many other elements. For example, in the local and regional vicinities of industrial centers the atmospheric deposition of many chemical compounds is strongly dominated by such human activities. Indeed, agricultural fires in the tropics are an important regional and global source of air pollution. Furthermore, the synthesis and release of special products may have surprising consequences, as is shown by the effects of chlorofluorocarbons on the stratosphere.

In this chapter we describe the role of the atmospheric chemical system in biogeochemical cycles. We then discuss the most prolific emission sources to the atmosphere, as well as several potential atmospheric impacts related to anthropogenic perturbations. In the final section we assemble the impact information in tabular form and demonstrate that only a relatively few types of human activities are related to most of the potential effects.

Chemical processes in the lower atmosphere

The flow of particles and gases through the atmosphere

The effects of particles and gases emitted into the atmosphere, naturally or as a result of man's activities, can be assessed only by following their chemical pathways. These pathways are illustrated in *Figure 8.1*, which deals with both gases and condensed-phase (liquid or solid) species. We consider first a gas emitted by industrial activities with flux (α). Each gas has five possible fates:

(1) If the gas is unreactive, or nearly so, it spreads through the troposphere and is transported to the stratosphere and upper atmosphere [*Figure 8.1*, (1)]. At these higher altitudes it may function as an infrared radiation absorber or "greenhouse gas" or it may influence the ozone (O_3) layer. The effects of such gases are discussed later (pp 235–236).

(2) If the gas is water soluble, some portions are incorporated into cloud water droplets [*Figure 8.1*, (2A)], deposited onto wet particle surfaces [*Figure 8.1*, (4A)], or deposited onto the ground [*Figure 8.1*, (5A)]. As it turns out, few precursor (original) gases are sufficiently soluble to be directly affected by these processes.

(3) If the gas is reactive in the atmosphere, it is oxidized to a variety of products [*Figure 8.1*, (3)]. Those with high vapor pressures remain in the gas phase, but in most cases oxidation leads to greater solubility and lower vapor pressure.

(4) Gas-phase reaction products that have low vapor pressures and hydrophilic characteristics can act as nucleating centers for new atmospheric particles [*Figure 8.1*, (4B)]. ("Low vapor pressure" may be operationally defined as $< 10^{-10}$ atmospheres.)

(5) Some of the gas is deposited on surfaces [*Figure 8.1*, (5A)], particularly if the supply of effective oxidizing gases is diminished due to low levels of solar radiation, time of year, time of day, etc. This deposition is commonly termed "dry deposition", even when it occurs on a wet surface, such as the ocean, and is a function of the aqueous solubility of the molecule.

We then consider the fates of the gaseous products resulting from chemical reactions of the precursor gases. These fates are similar to those of the

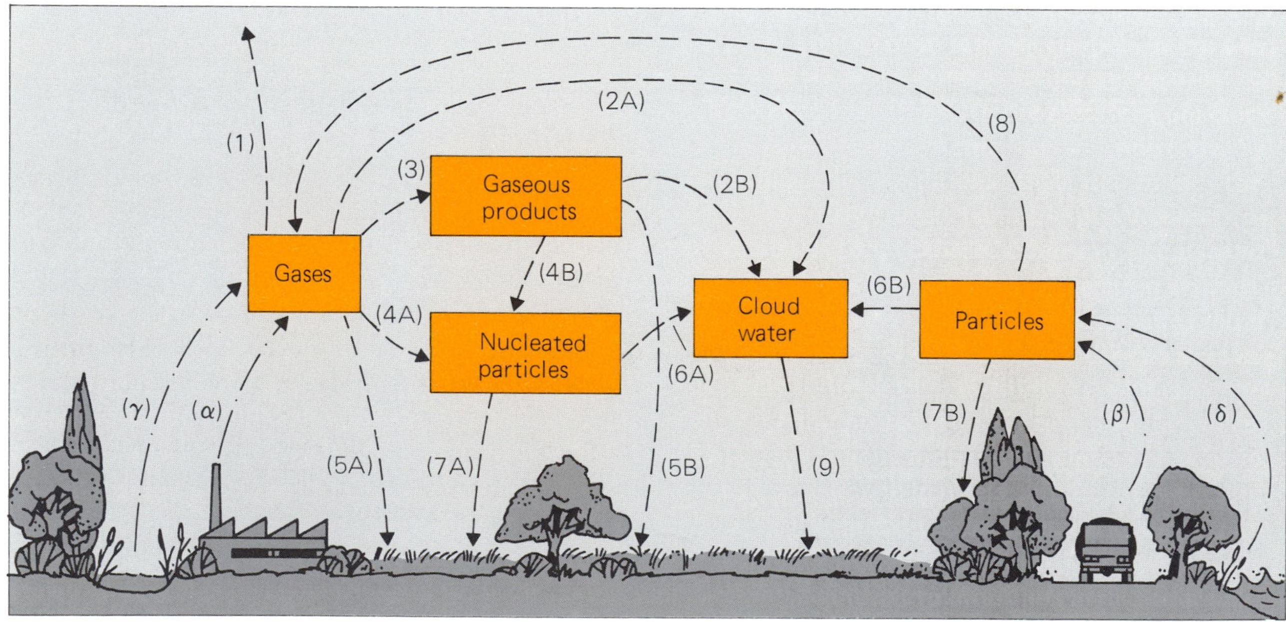

Figure 8.1 Chemical and physical pathways for gases and particles in the atmosphere. The numbers and Greek letters represent fluxes (mass or number of molecules per unit time and unit area), which are discussed in the text.

precursors, but differ in the proportion of gas that undergoes each process, because the first-stage products are invariably more highly oxidized than their precursors. This property frequently renders them more soluble in water, which favors uptake both on atmospheric water droplets [*Figure 8.1*, (2B)] and on water frequently present on surfaces [soil, lakes, particles, materials, etc., *Figure 8.1*, (5B)]. In addition, oxidized products are usually of higher molecular weight and lower vapor pressure than their precursors, which makes them more likely to act as nucleating centers for new particles or to deposit onto existing particles, thus preventing the oxidized products from diffusing to higher altitudes. In most cases, therefore, flux (2B) > flux (2A), flux (4B) > flux (4A), and flux (5B) > flux (5A) (*Figure 8.1*).

Atmospheric particles, whether emitted as such or created by nucleation, occur in a wide range of sizes. As they interact, coagulation produces a multimodal size spectrum, as shown in *Figure 8.2*. Those in the "accumulation mode" (i.e., with particle diameter in the range 0.1 μm $\leq D_P \leq$ 1 μm) have atmospheric lifetimes of at least several days. Since they are similar in size to the wavelengths of visible photons, they are responsible, to a large extent, for reductions in visibility. Particles in the "coarse mode" (i.e., with $D_P \gg$ 1 μm) have large gravitational settling velocities and short lifetimes, and appear to take little part in atmosphereic chemistry.

Accumulation mode particles are eventually removed from the atmosphere either by nucleation or incorporation into cloud or rain water [*Figure 8.1*, (6A)] or by dry deposition [*Figure 8.1*, (7A)]. It seems reasonable to suppose that the efficiency of these processes depends on the degree of hydrophilicity of the particle surfaces, but studies are needed to confirm this supposition.

Atmospheric particles are either injected directly into the atmosphere or created by the transformation of gaseous molecules. The injection process [*Figure 8.1*, flux (β)] produces particles of a quite different chemical nature to those arising from the gas phase; they include as constituents transition

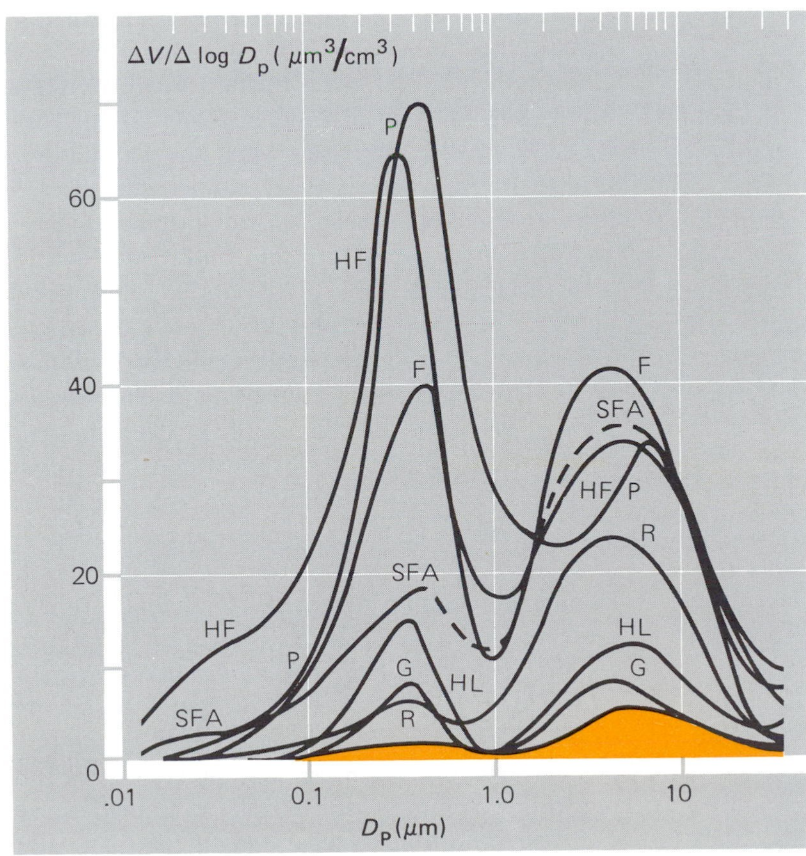

Figure 8.2 In situ atmospheric aerosol particle size distributions established from measurements at various sites in California, USA. The "clean continental background" data (yellow) are for Fort Collins, Colorado. F, Fresno; G, Goldstone; HF, Harbor Freeway; HL, Hunter–Liggett; P, Pomona; R, Richmond; SFA, San Francisco Airport; $\Delta V/\Delta \log D_P$ is the change in aerosol volume per change in aerosol diameter. From Whitby [1]; reproduced by permission of the New York Academy of Sciences.

metal compounds, refractory oxides, soot, and sulfate particles. The atmospheric lifetimes of these particulates are limited by water droplet scavenging [*Figure 8.1*, (6B)] and dry deposition [*Figure 8.1*, (7B)]. In a few cases supersaturated gases may evaporate from cloud droplets or airborne particles into the atmosphere, e.g., hydrogen chloride (HCl), or nitric acid (HNO_3) from sulfuric acid-containing aerosol [*Figure 8.1*, (8)].

Another process is the deposition of impure cloud water as rain or snow [*Figure 8.1*, (9)]. This process is theoretically capable of scavenging particles and gases during the fall from cloud base to ground, but to date the evidence indicates that most impurities are present in the cloud water prior to the formation of raindrops and snowflakes.

It is important to recognize, particularly when considering biogeochemical processes on a global scale, that not all the gases and aerosol particles entering the atmosphere come from man's activities. In several cases the natural emission fluxes of gases (γ) and particles (δ) (*Figure 8.1*) exceed those from anthropogenic sources. A careful assessment of the flux magnitudes is thus important, although in some cases (e.g., lightning as a source of nitric oxide, NO) the assessment can be very difficult.

The difference between rates of production and loss determines the concentration changes with time of an atmospheric species. If its production grows, the concentration of a species in the atmosphere increases with time. This has definitely been the case for several important atmospheric gases, such as carbon dioxide (CO_2) [2], nitrous oxide (N_2O) [3], and methane (CH_4) [4], whose atmospheric abundances are currently increasing by 0.3, 0.2–0.3, and 1–1.2% per year, respectively. Large increases are taking place as well in the abundances of many industrial chlorofluorocarbons [5]. For CO_2, measured volume-mixing ratios increased from 315 p.p.m. in 1958 to 340 p.p.m. in 1982 [2]; analyses of air trapped in ice cores [6] suggest that the atmospheric CO_2 concentration in 1850 may have been as low as 260 p.p.m. For CH_4, Craig and Chou [7] have estimated mixing ratios for several hundred years from the analysis of air trapped in Greenland ice cores. They find a linear increase in CH_4 concentrations from 0.7 p.p.m.v. in 1550 to 1.1 p.p.m.v. at the beginning of the twentieth century, followed by a much faster increase to 1.6 p.p.m.v. by 1980.

A different budget situation occurs if rates of production and loss for a specific atmospheric species are in balance and so give a stable atmospheric concentration. This situation currently applies to oxygen (O_2), for example. Since it is believed that O_2 concentration evolved during the early history of the Earth, we are reminded here that budget assessments apply to specific time periods, and thus that a balanced budget is always potentially subject to factors that may throw it out of balance.

Finally, if species production declines or is terminated the species concentration decreases with time. Examples are artificial radioactivity following the ban of atmospheric nuclear tests in the early 1960s [8] and stratospheric aerosol following volcanic eruption [9].

Tropospheric photochemistry and the budgets of atmospheric trace gases

Although only about 10% of all atmospheric O_3 is located in the troposphere, its presence there is of fundamental importance for the composition of the atmosphere. This is because the very reactive hydroxyl radical ($OH^{\cdot}$) is produced through the absorption of ultraviolet radiation by O_3 and the subsequent reaction of excited oxygen atoms with water vapor (see Appendix 8.1). The radical $OH^{\cdot}$ reacts with many gases which would otherwise be inert in the troposphere, including hydrocarbons and many organic halogen and sulfur compounds. Thus $OH^{\cdot}$ initiates the oxidation of these compounds to carbon monoxide (CO) and CO_2 and to inorganic acids that are removed from the atmosphere by deposition on vegetation, on the Earth's surface, on particulate matter, or by precipitation. These reactions initiated by O_3 strongly determine tropospheric photochemistry, as indicated in *Figure 8.3*. Without O_3 in the troposphere, the composition of the atmosphere and the conditions for life on Earth would be very different. (The other essential ecological role of O_3 occurs mainly in the stratosphere, namely to filter out hazardous ultraviolet solar radiation.) However, as too much O_3 in the troposphere is harmful to plants, it may be possible that tropospheric O_3 is important in regulating the proper functioning of the biosphere. Indeed, there are some indications of substantial increases (several

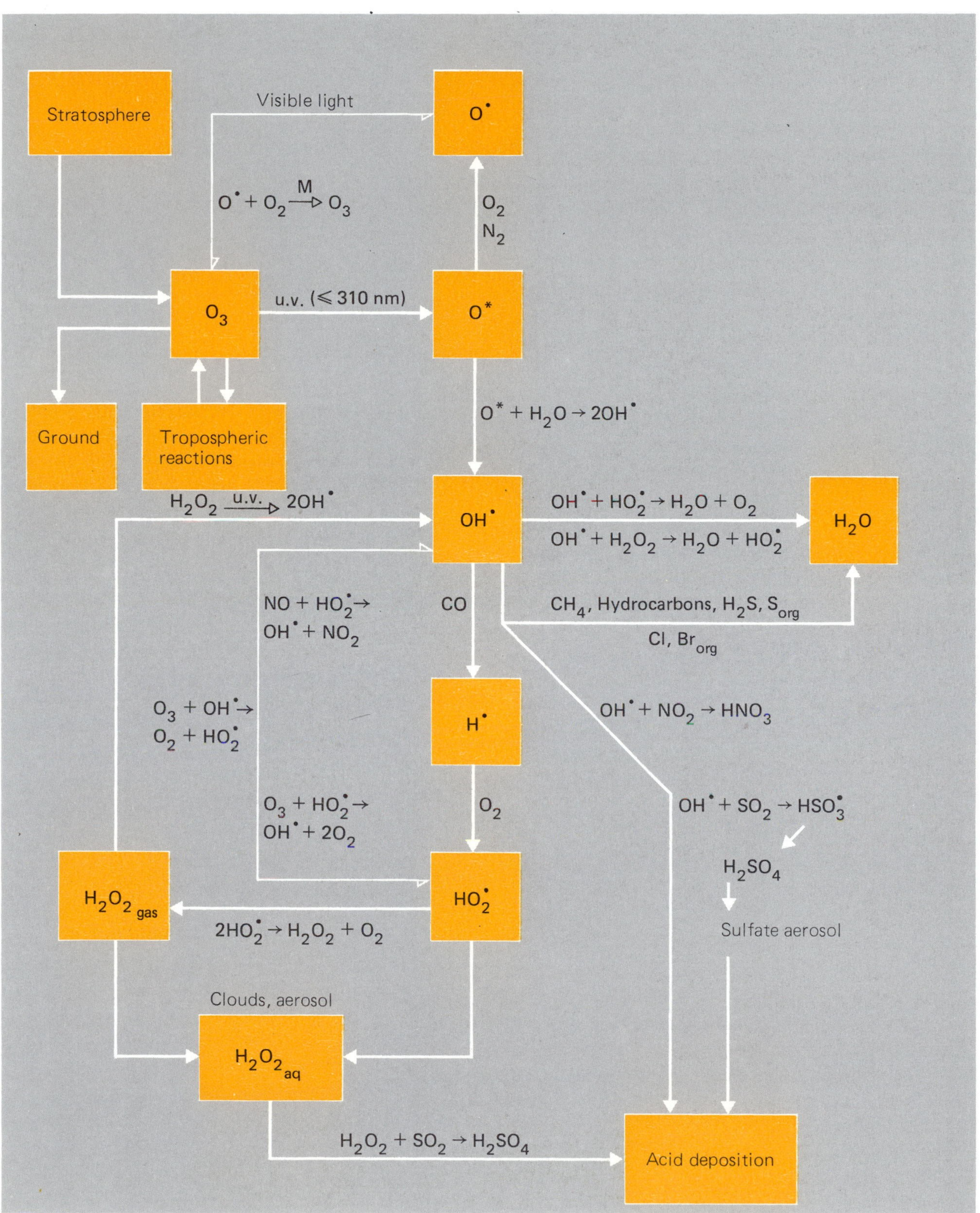

Figure 8.3 Some important homogeneous tropospheric reactions.

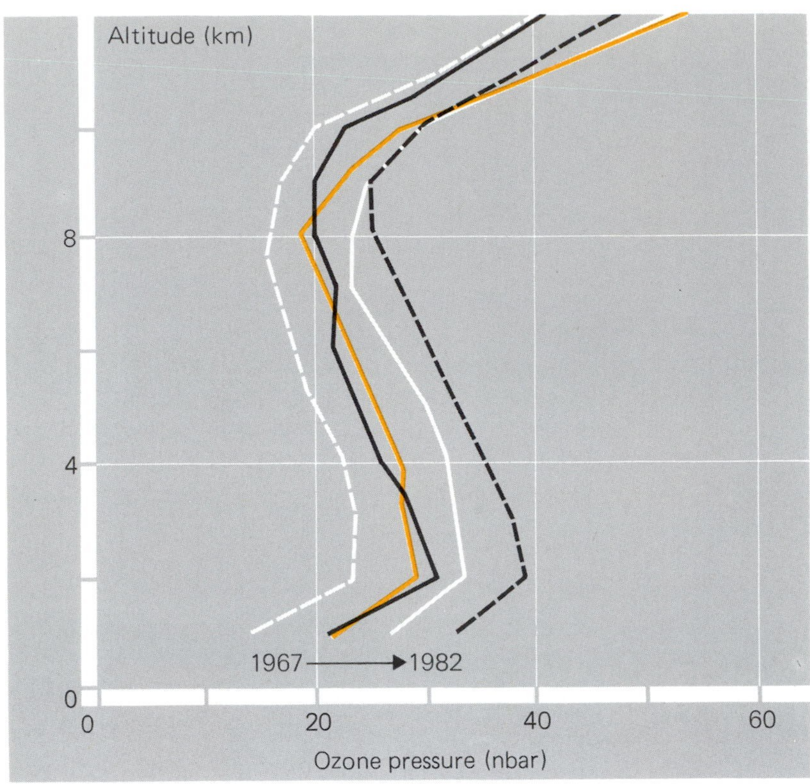

Figure 8.4 Tropospheric O_3 profiles measured at Hohenpeissenberg, FRG, during the past 15 years. 1967, dashed white; 1974, yellow; 1976, full black; 1981, full white; 1982, dashed black. (Attmannspacher [11]; reproduced with permission).

tens of percent) in tropospheric O_3 concentrations above the mid-latitude zone of the Northern Hemisphere during the past 20 years (*Figure 8.4*) [10, 11]. Such increases in tropospheric O_3 are not surprising if we consider the effects of greatly increased NO_x emissions into the atmosphere from anthropogenic activities (see Appendix 8.1). It is unfortunate that few long-term O_3 records are available from other parts of the world to confirm the observed trends in Europe.

On the basis of the photochemistry of O_3, $H_2O_{(g)}$, CH_4, CO, and NO and NO_2 (nitrogen dioxide), it is possible to make some rough estimates of the $OH^{\cdot}$ distribution in the atmosphere. This distribution can be tested against the observed global distribution of methylchloroform (CH_3CCl_3), since CH_3CCl_3 is produced by industrial processes only, ones that have rather well-known historical production rates, and is removed from the atmosphere only by reaction with $OH^{\cdot}$. The calculated daytime-average $OH^{\cdot}$ concentration distributions for the four seasons, shown in *Figure 8.5,* indicate maximum $OH^{\cdot}$ concentrations in the tropics. The globally and diurnally averaged $OH^{\cdot}$ concentration equals about 5×10^5 molecules/cm^3 [12]. Therefore, although the Earth's atmosphere contains almost 21% molecular oxygen, its oxidative power is determined mainly by the ultraminor constituent $OH^{\cdot}$, which is 10^{13} times less abundant. Questions should, therefore, be asked as to whether the $OH^{\cdot}$ concentrations are stable or whether they could be changing because of anthropogenic activities. This is discussed briefly in the next section and in Appendix 8.2. This information, combined with knowledge of the global distribution of other trace gases that react with $OH^{\cdot}$, enables atmospheric budgets of these gases to be roughly established. For other gases, source and sink terms may be estimated by other means; currently estimated atmospheric transformation rates and sources and sinks at the Earth's surface for the most important gases are given in *Figure 8.6.* The indicated global fluxes are still quite uncertain, although our knowledge has clearly improved since 15 years ago, when several of the estimates were incorrect by more than an order of magnitude. The sources of the most important compounds and estimates of their contribution are shown in *Table 8.1.* Despite the uncertainties, *Figure 8.6* and *Table 8.1* consolidate important information and should be consulted frequently

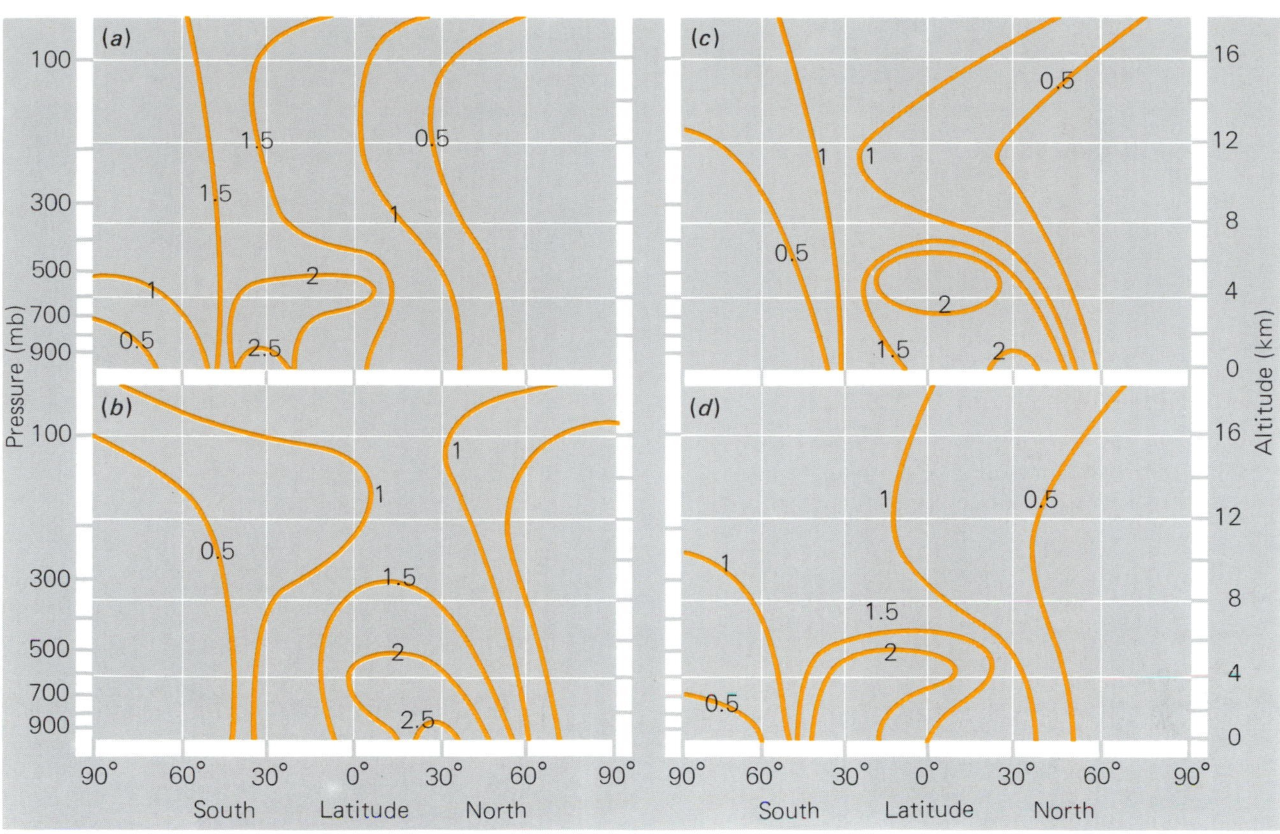

Figure 8.5 Calculated meridional daytime distributions of OH˙ (10^6 molecules/cm^3) for (*a*) January, (*b*) July, (*c*) April, and (*d*) October [12].

while reading the following text. They show in almost all cases the substantial and growing importance of man's activities.

Carbon compounds

The principal components of the atmospheric carbon cycle are shown in *Figure 8.7*, which is the first of several figures designed to help examine the atmospheric processing of compounds that contain a specific element. The carbon compounds in the atmosphere arise largely from combustion of fossil fuels and from natural emissions.

When carried to chemical completion, the combustion of fossils fuels produces a flux with CO_2 as the carbonaceous product [*Figure 8.7*, flux (α_1)]. CO_2 is unreactive in the lower atmosphere, where its chief fates are atmospheric transport and upward diffusion [*Figure 8.7*, (1A)] and uptake by plants on land areas [*Figure 8.7*, (5)]. Incomplete fossil-fuel combustion, which inevitably occurs to some degree, forms CO [*Figure 8.7*, flux (α_2)] and a very large family of hydrocarbon compounds [*Figure 8.7*, flux (α_3)]. A few of these gases, chiefly CO and CH_4, are long lived and are added to the global atmospheric background [*Figure 8.7*, (1B)], although slow indirect conversion to CO_2 also occurs [*Figure 8.7*, (2C)]. The annual budgets of CO and CH_4, determined from their observed global distributions, as shown in *Figures 8.8* and *8.9*, and those calculated for OH˙ indicate maximum sources and sinks for these gases in the tropics [12]. This demonstrates the importance of the tropics in atmospheric photochemistry. The growth in atmospheric CH_4, currently observed at 1–1.2% per yr [4], may lead to lower hydroxyl concentrations and a reduction in the oxidative power of the atmosphere. It is postulated [13] that global average OH˙ concentrations will decrease, while in the upper troposphere and at mid-latitudes, continential boundary layer O_3 concentrations will increase. This is explained in more detail in Appendix 8.2. In the general case,

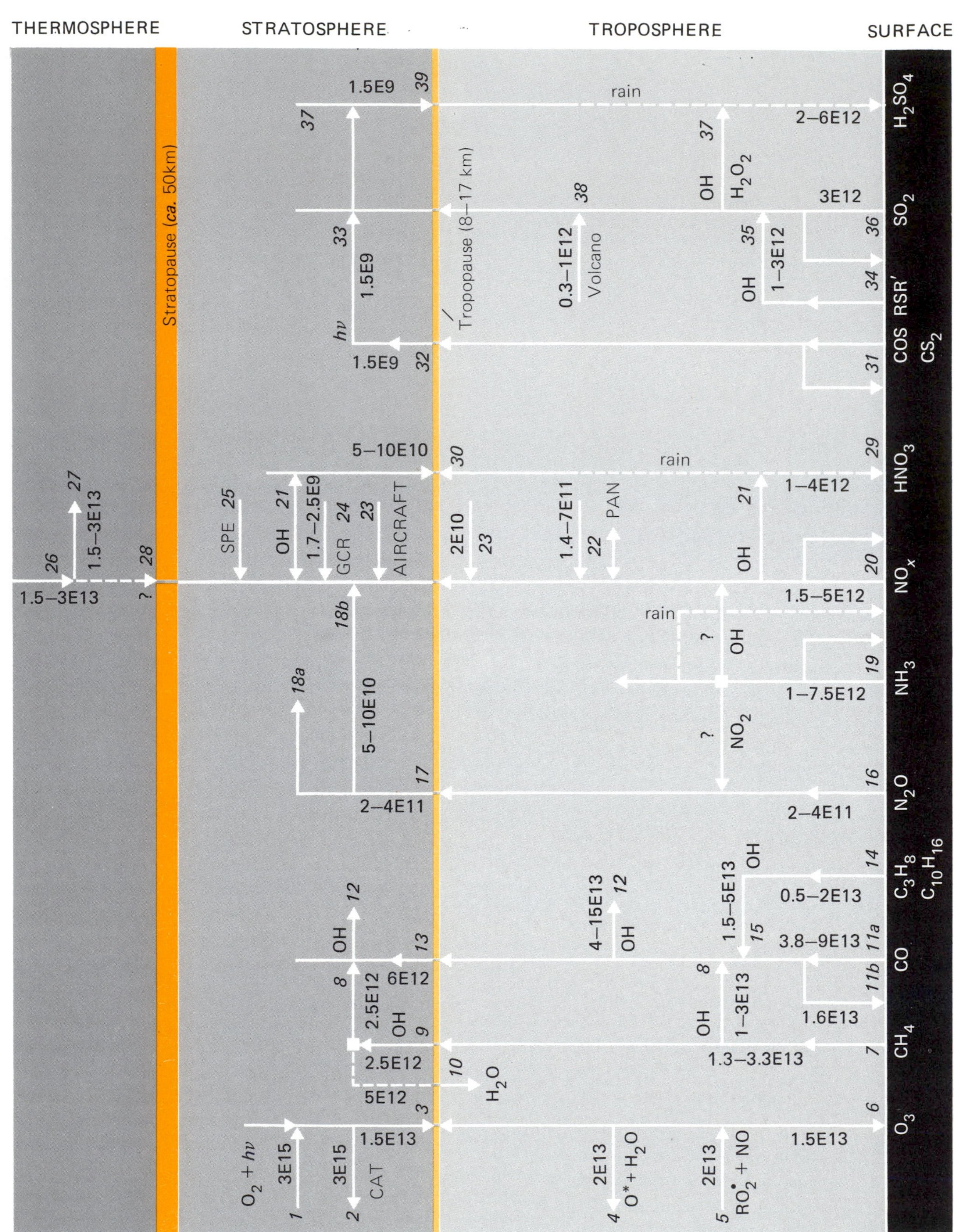

THERMOSPHERE
STRATOSPHERE
TROPOSPHERE
SURFACE
Stratopause (*ca.* 50km)
Tropopause (8–17 km)
O3
CH4
CO
C3H8
C10H16
N2O
NH3
NOx
HNO3
COS
CS2
RSR′
SO2
H2SO4
rain
Volcano
PAN
SPE
GCR
AIRCRAFT
CAT

1. $O_2 \xrightarrow{h\nu} 2O^{\cdot}$

$O^{\cdot} + O_2 \xrightarrow{M} O_3\ (2\times)$

$3O_2 \rightarrow 2O_3$

2. $O_3 \xrightarrow{h\nu} O^{\cdot} + O_2$

$XO + O^{\cdot} \rightarrow X^{\cdot} + O_2$

$X^{\cdot} + O_3 \rightarrow XO + O_2$

$2O_3 \rightarrow 3O_2$

X = NO, Cl, OH (catalysts)

3. Downward flux to troposphere (small difference between *1* and *2*).

4. $O_3 \xrightarrow{h\nu} O^* + O_2$

$O^* + H_2O \rightarrow 2OH^{\cdot}$

$HO_2^{\cdot} + O_3 \rightarrow OH^{\cdot} + 2O_2$

5. $RO_2^{\cdot} + NO \rightarrow RO^{\cdot} + NO_2$

$NO_2 \xrightarrow{h\nu} NO + O^{\cdot}$

$O^{\cdot} + O_2 \xrightarrow{M} O_3$

$RO_2^{\cdot} + O_2 \rightarrow RO^{\cdot} + O_3$

R = H, CH_3, etc., from CO and hydrocarbon oxidation, e.g.,

$CO + OH^{\cdot} \rightarrow H^{\cdot} + CO_2$

$H^{\cdot} + O_2 \xrightarrow{M} HO_2^{\cdot}$

6. Ozone destruction at the ground: difference between *3, 4,* and *5.*

7. Release of CH_4 at the ground by a variety of sources with range 1.3–3.3E13 moles/yr; 2.5E13 calculated with average $OH^{\cdot}$ concentration of 6E5 molecules/cm^3.

8. $CH_4 + OH^{\cdot} \rightarrow CH_3^{\cdot} + H_2O$

$CH_3^{\cdot} + O_2 \xrightarrow{M} CH_3O_2^{\cdot}$

$CH_3O_2^{\cdot} + NO \rightarrow CH_3O^{\cdot} + NO_2$

$CH_3O^{\cdot} + O_2 \rightarrow HCHO + HO_2^{\cdot}$

$HCHO \xrightarrow{h\nu} CO + H_2$

and other oxidation routes.

9. Flux of CH_4 to the stratosphere.

10. Flux of H_2O to the troposphere from CH_4 oxidation.

11a. Release of CO from a variety of sources, mostly man-made.

11b. Uptake of CO by microbiological processes in soils.

12. $CO + OH^{\cdot} \rightarrow H^{\cdot} + CO_2$. Global loss of CO of 4–15E13 moles/yr. 7E13 calculated with $[OH^{\cdot}]$ = 6E5 molecules/cm^3.

15. Isoprene (C_5H_8) and terpene ($C_{10}H_{16}$) oxidation to CO following reaction with $OH^{\cdot}$; oxidation mechanism and CO yield not well known.

16, 17. Release of N_2O to atmosphere by a variety of sources; no significant sinks of N_2O discovered in troposphere; stratospheric loss estimated by model calculations.

18a. $N_2O \xrightarrow{h\nu} N_2 + O^{\cdot}$

$N_2O + O^* \rightarrow N_2 + O_2$

18b. $N_2O + O^* \rightarrow 2NO$

19. Release of NH_3 to atmosphere by a variety of sources; redeposition at the ground; most NH_3 removed by rain, but some NO_x loss and N_2O formation possibly by $NH_2^{\cdot} + NO_x \rightarrow N_2O_{x-1} + H_2O$, while NO_x may be formed via $NH_2^{\cdot} + O_3 \rightarrow NH_2O^{\cdot} + O_2$.

20. Release of NO_x at the ground by a variety of sources; redeposition at the ground.

21. $NO_2 + OH^{\cdot} \rightarrow HNO_3$

$HNO_3 \xrightarrow{h\nu} OH^{\cdot} + NO_2$

$HNO_3 + OH^{\cdot} \rightarrow H_2O + NO_3$

22. NO_x produced by lightning.

23. NO_x produced by subsonic aircraft.

24. NO_x from galactic cosmic rays (GCR).

25. NO_x from sporadic solar proton events (SPE), maximum production recorded in August 1972 event, 1 E10 moles.

26. NO production by high-speed photoelectrons in thermosphere and by auroral activity.

27. $NO + N^{\cdot} \rightarrow N_2 + O^{\cdot}$

28. Downward flux of NO to stratosphere; small difference between *26* and *27,* may be important.

31, 32. COS destruction in stratosphere calculated with model: uptake of COS in oceans and hydrolysis may imply an atmospheric lifetime of only a few years and a source of a few E10 moles/yr. Very little is known about the sources and sinks of COS and CS_2.

33. $COS \xrightarrow{h\nu} S^{\cdot} + CO$

$S^{\cdot} + O_2 \rightarrow SO^{\cdot} + O^{\cdot}$

$SO^{\cdot} + O_2 \rightarrow SO_2 + O^{\cdot}$

34. Release of H_2S, $(CH_3)_2S$, and CH_3SH by biological processes in soils and waters. R = H, CH_3, etc.

35. Oxidation of H_2S, $(CH_3)_2S$, and CH_3SH to SO_2 after initial attack by $OH^{\cdot}$.

36. Industrial release of SO_2.

37. SO_2 oxidation to H_2SO_4 on aerosols, in cloud droplets, and by gas-phase reactions following attack by $OH^{\cdot}$ and H_2O_2.

38. Volcanic injection of SO_2 (average flux over past centuries).

Figure 8.6 Compilation of the most important photochemical processes in the atmosphere, including estimates of flux rates (expressed in moles per year) between the Earth's surface and the atmosphere, and within the atmosphere (*a*E*b* means $a \times 10^b$; CAT is catalysis; PAN is peroxyacetyl nitrate). Numbers in italics refer to the key above.

the hydrocarbons are oxidized by photochemical reactions to form carbon monoxide, peroxides, aldehydes, and ketones [*Figure 8.7*, (3)]; some formation of soot and organic aerosol particles also occurs [*Figure 8.7*, (4A) and (3)+(4B)]. From a carbon budget standpoint, biogenic emissions of hydrocarbons [*Figure 8.7*, flux (γ)] appear to dominate globally. On local and regional scales, however, anthropogenic activities, such as fossil-fuel combustion, industrial activity, forest cutting, and biomass burning, may be predominant.

Although hydrocarbons are rather insoluble in water, their oxidized derivatives have relatively high solubility and are thus more readily incorporated into cloud droplets [*Figure 8.7*, (2B)] or on to aerosol particles [*Figure 8.7*, (4B)]. (For example, formaldehyde, HCHO, is commonly found in atmospheric droplets [14].)

Combustion of hydrocarbons generally produces condensed-phase material [*Figure 8.7*, flux (β)], as well as the combustion gases. The former may contain elemental (or graphitic) carbon (i.e., soot) or

Table 8.1 Budgets of carbon, nitrogen, and sulfur species[a] [29].

Gas	*Direct input* (g/yr) *Source type*	*Secondary input* (g/yr) *Source type*	*Removal by*[b]	*Atmospheric lifetimes*[b]	*Transport distances* Δx, Δy, Δz (km)[c] *Volume mixing ratios in remote troposphere*[d]
CO	$4–16 \times 10^{14}$ CO Biomass burning	$3.7–9.3 \times 10^{14}$ CO CH_4 oxidation	30×10^{14} CO $OH^{\cdot}$	2 month	4000, 2500, 10 50–200 p.p.b.v.
	6.4×10^{14} CO Industry	$4–13 \times 10^{14}$ CO C_5H_8, $C_{10}H_{16}$ oxidation	4.5×10^{14} CO Uptake by soils		
	$0.2–2 \times 10^{14}$ CO Vegetation				
CH_4	$0.3–0.6 \times 10^{14}$ CH_4 Rice paddy fields		4×10^{14} CH_4 $OH^{\cdot}$	7 year	Global 1.5–2.0 p.p.m.v.
	$0.3–2.2 \times 10^{14}$ CH_4 Natural wetlands				
	0.6×10^{14} CH_4 Ruminants				
	$<1.5 \times 10^{14}$ CH_4 Termites				
	$0.3–1.1 \times 10^{14}$ CH_4 Biomass burning				
	0.2×10^{14} CH_4 Gas leakage				
C_5H_8, $C_{10}H_{16}$	8.3×10^{14} C Trees		8.3×10^{14} C $OH^{\cdot}$	10 hour	400, 200, 1 0–10 p.p.b.v.
NO_x ($NO+NO_2$)	$12–20 \times 10^{12}$ N Industry	$1.0–1.5 \times 10^{12}$ N Oxidation of N_2O	$25–85 \times 10^{12}$ N $OH^{\cdot}$ Deposition on soils and oceans	1.5 day	1500, 400, 1.0 1–100 p.p.t.v.
	$10–40 \times 10^{12}$ N Biomass burning				
	$1–10 \times 10^{12}$ N Lightning				
	$1–15 \times 10^{12}$ N Soils				
	0.15×10^{12} N Ocean				
	0.25×10^{12} N Jet aircraft				

Table 8.1 (*Cont.*)

Gas	*Direct input* (g/yr) *Source type*	*Secondary input* (g/yr) *Source type*	*Removal by*[b]	*Atmospheric lifetimes*[b]	*Transport distances* $\Delta x, \Delta y, \Delta z$ (km)[c] *Volume mixing ratios in remote troposphere*[d]
HNO_3		$15\text{–}85 \times 10^{12}$ N $NO_2 + OH^{\cdot}$	Rain	3 day	3000, 600, 1.5 10–300 p.p.t.v.
N_2O	1.8×10^{12} N Fossil fuel burning $1\text{–}2 \times 10^{12}$ N Biomass burning $1\text{–}2 \times 10^{12}$ N Oceans, estuaries $1\text{–}3 \times 10^{12}$ N Cultivation of natural soils $<3 \times 10^{12}$ N Fertilized fields ? Natural soils		$6\text{–}11 \times 10^{12}$ N Stratospheric photolysis	*ca.* 100–200 yrs	Global *ca.* 300 p.p.b.v.
NH_3	$10\text{–}20 \times 10^{12}$ N Domestic animals $2\text{–}6 \times 10^{12}$ N Wild animals $<3 \times 10^{12}$ N Fertilized fields $<30 \times 10^{12}$ N Natural fields $4\text{–}12 \times 10^{12}$ N Coal burning $<60 \times 10^{12}$ N Biomass burning		Rain	<9 day	<9000, 1000, 3 0–3 p.p.b.v.
SO_2	64×10^{12} S Coal burning 26×10^{12} S Petroleum burning 11×10^{12} S Nonferrous ores $10\text{–}30 \times 10^{12}$ S Volcanoes	$40\text{–}100 \times 10^{12}$ S Oxidation of H_2S, $(CH_3)_2S$	$OH^{\cdot}$ Rain	5 day	5000, 700, 2.5 10–200 p.p.t.v.
H_2S, $(CH_3)_2S$, CH_3SH	$<4 \times 10^{12}$ S Agricultural fields $31\text{–}42 \times 10^{12}$ S Open ocean 10×10^{12} S Coastal waters 16×10^{12} S(?) Tropical forests 24×10^{12} S(?) Wetlands		$OH^{\cdot}$	2 day	2000, 500, 1.5 0–100 p.p.t.v.

[a] COS and CS_2 are not listed because too little is known about their sources and sinks. The industrial input of CS_2 is about 2×10^{11} g S/yr. [b] Lifetimes and removal rates are calculated with $[OH] = 6 \times 10^5$ molecules/cm^3. [c] Δx, Δy, and Δz are, respectively, east–west, south–north, and vertical diffusion directions over which concentrations in the free atmosphere are reduced to 30% by chemical reactions. [d] 1 p.p.m.v. $= 10^{-6}$; 1 p.p.b.v. $= 10^{-9}$; 1 p.p.t.v. $= 10^{-12}$.

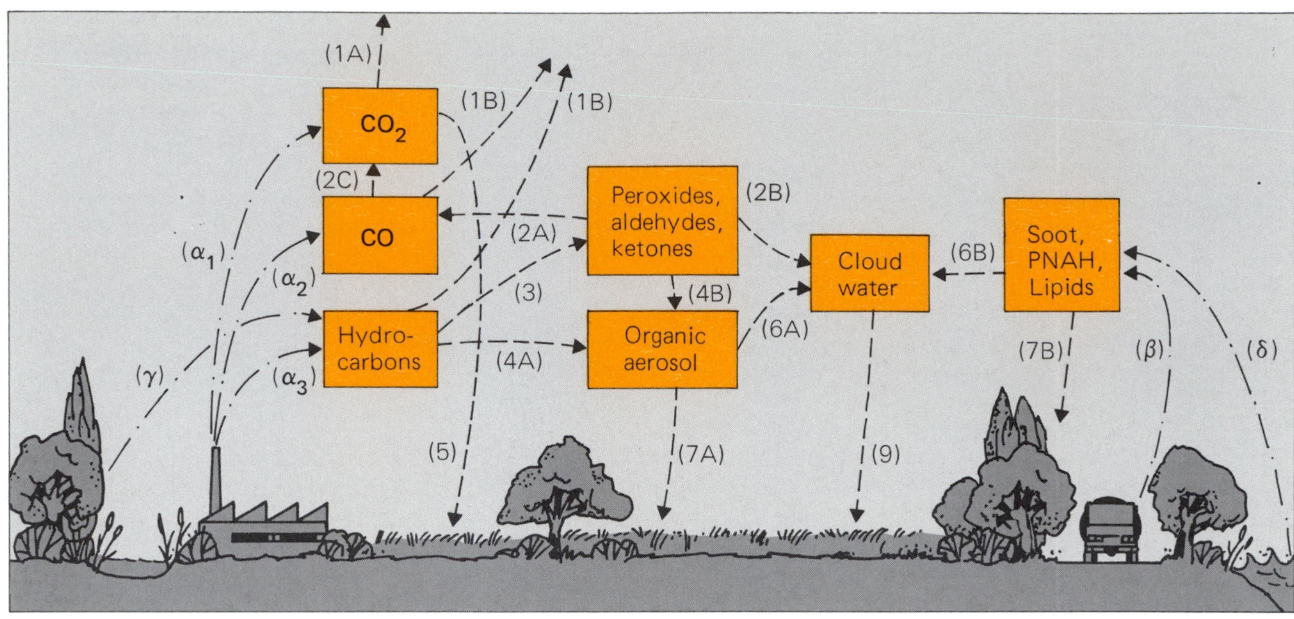

Figure 8.7 Atmospheric carbon compounds: the principal species and chemical and physical pathways. This figure, as well as *Figures 8.11–8.14*, are versions of *Figure 8.1* that are specific to the cycle for a particular element. Compounds and pathways regarded as minor by the authors have been omitted in the interest of clarity.

fuel-derived polynuclear aromatic hydrocarbons (PNAH). The size of most of these particles is such that they have a significant negative impact on visibility, enhanced for soot particles by strong absorption of sunlight. Long-range transport of soot particles is possible because of their long atmospheric residence times, especially during winter and at high latitudes. This contributes to the "Arctic haze" phenomenon, which may itself have climatic consequences. The combustion-produced particles, together with organic products formed from gaseous reactions and oxidized hydrocarbons, constitute the organic components available for cloud water chemistry [*Figure 8.7,* (2B, 6A, 6B)]. The possibility of coalescence of these components to form thin surface films on the droplets has been discussed [15], but definitive measurements have not yet been made.

At this point it is useful to comment on the nature of the aerosol particles. As shown in *Figure 8.10,* they generally possess a water shell (except at very low humidities). The presence of this aqueous medium effectively excludes reactions between gaseous molecules and dry surfaces. At the same time, to understand the chemistry we must acquire a deeper understanding of the atmospheric chemistry of the liquid phase than has yet been achieved. The potential presence of a surface organic layer is indicated, and may be common on aerosol particles; if present, such layers strongly inhibit gas–liquid transfer and the subsequent liquid-phase chemistry.

The removal of solid and liquid material by deposition on to the Earth's surface is a complicated function of particle size, chemical nature, and boundary layer meteorology [*Figure 8.7,* (7A), (7B), (8)]. For the solid aerosol particles (generally with a water shell), typical atmospheric lifetimes are days to weeks. Reaction times are therefore unlikely to limit chemical processes on particles, and particles deposited on surfaces or incorporated into droplets are often at an advanced stage of chemical aging; aerosol particles often also serve as condensation nuclei. Cloud droplets have lifetimes of a few hours at most; they disappear either by reevaporation or by coalescence into raindrops; reevaporation is at least 10 times more likely. Raindrops survive for only a few minutes before they impact on the Earth's surface. The time spans as well as the chemical constituents of particles, cloud droplets, and raindrops

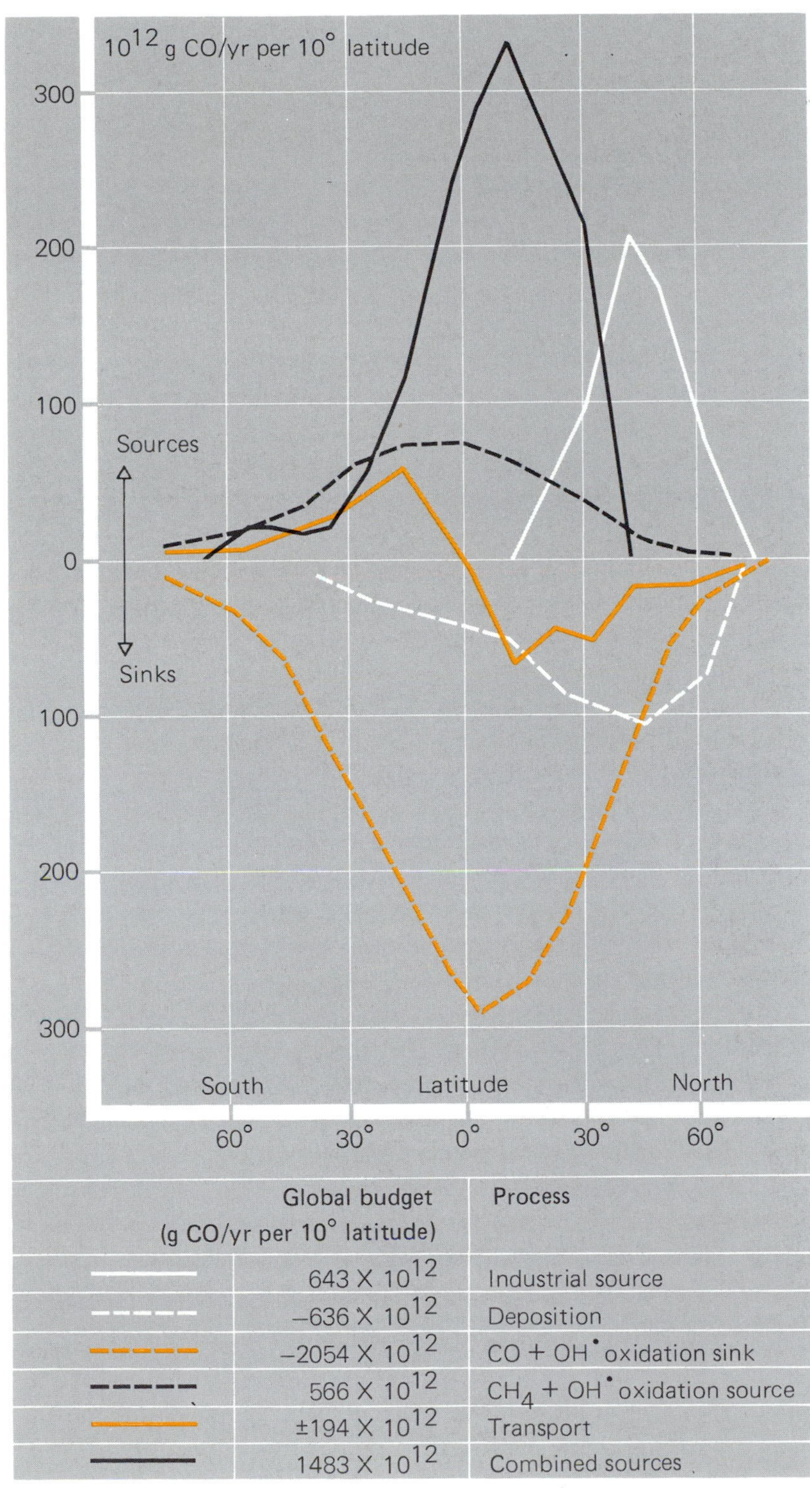

Figure 8.8 The annual CO budget calculated with the OH˙ distributions of *Figure 8.5* and observed CO distributions [12].

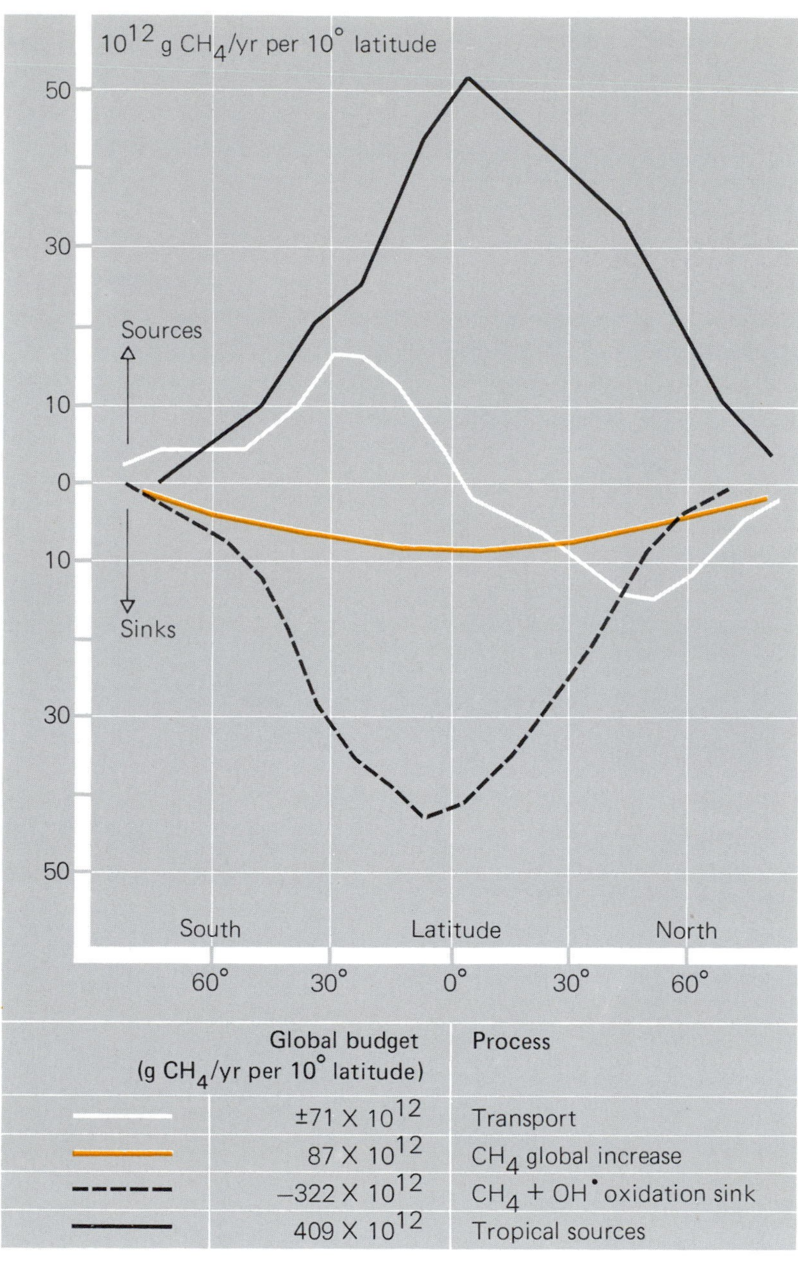

Figure 8.9 The annual CH_4 budget [12].

are thus quite different. It is this complexity that makes a detailed understanding of condensed-phase atmosphere chemistry so difficult.

The oxidation of hydrocarbons in the atmosphere is a complicated matter and very uncertain. It is initiated by reaction with $OH^{\bullet}$ or O_3, but further oxidation steps depend on the prevailing concentrations of NO and on reactions that involve aerosol particles and hydrometeors; these are very difficult to quantify. As the simplest example, we discuss CH_4 oxidation in Appendix 8.2.

Nitrogen compounds

Anthropogenic nitrogen-containing gases are principally the oxides of nitrogen produced by high-temperature combustion [*Figure 8.11*, flux (α)]. NO and NO_2, the so-called NO_x gases, play important roles in atmospheric chemistry, mostly through catalytic reactions (Appendix 8.3). NO is an efficient reducer of peroxyl radicals ($HO_2^{\bullet}$; $RO_2^{\bullet}$). The reduction reactions regenerate the highly reactive alkoxy radicals ($OH^{\bullet}$, $RO^{\bullet}$) which propagate many

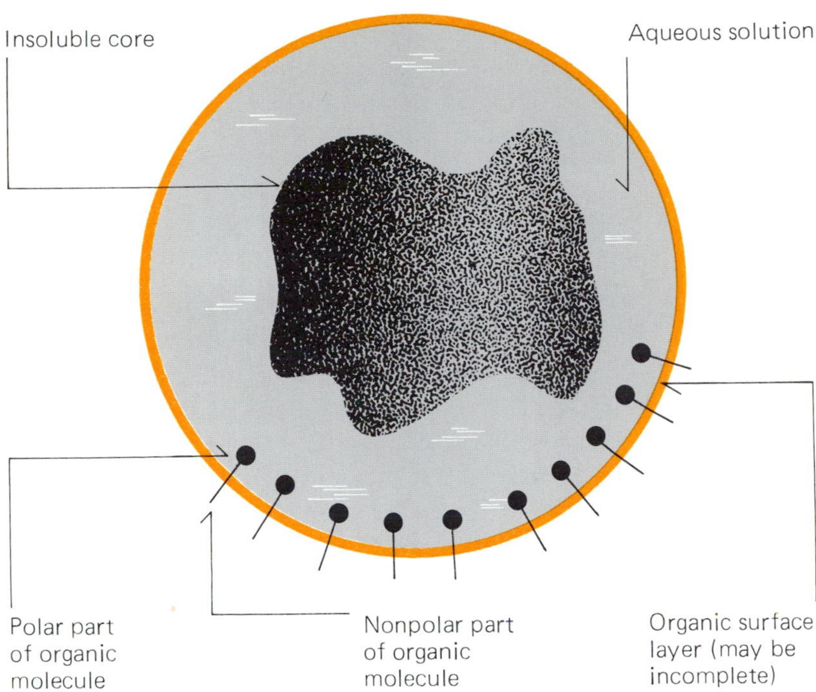

Figure 8.10 Possible structure of an aged atmospheric aerosol particle.

atmospheric reaction sequences. The most important role of NO_2 in atmospheric chemistry is the absorption of solar ultraviolet radiation, which is followed by NO_2 dissociation and subsequent O_3 production. Net O_3 production takes place during hydrocarbon and CO oxidation if sufficient NO is present. However, at low NO concentrations (< 10 p.p.t.v. at ground level) oxidation of CO leads to O_3 loss. Anthropogenic NO_x emissions are probably at least as large as natural NO_x production. Because of the short lifetime of NO_x (only a few days), O_3 production is especially expected at mid-latitudes

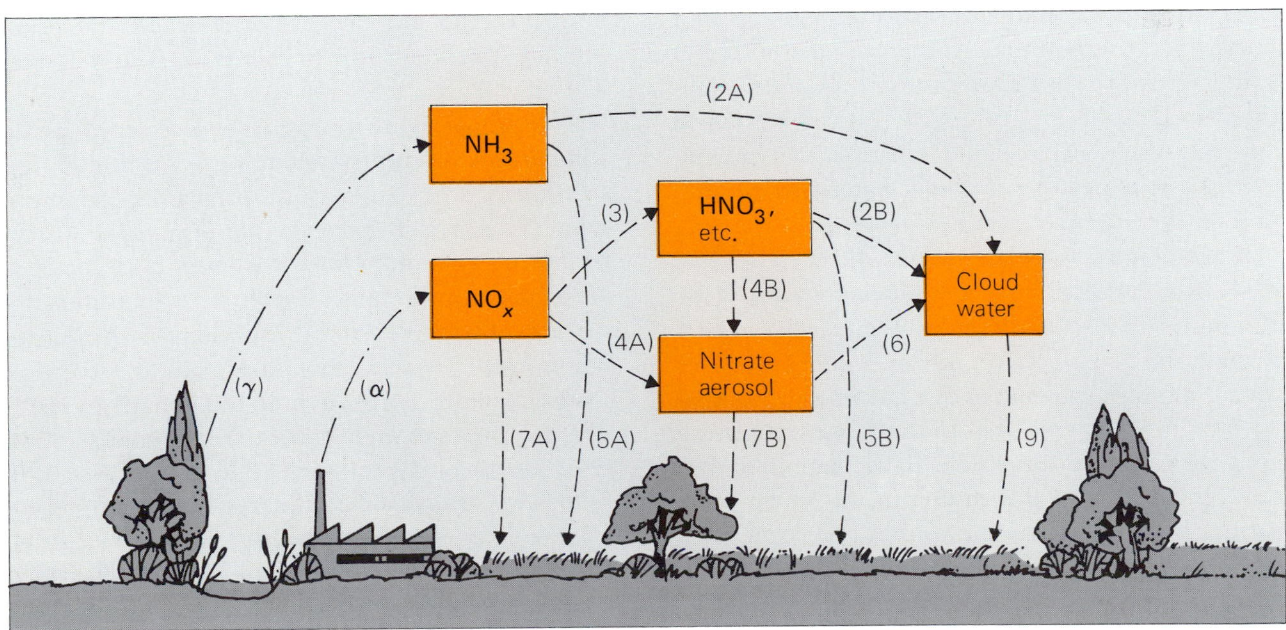

Figure 8.11 Atmospheric nitrogen compounds: principal species and chemical and physical pathways.

in the northern hemisphere. Most likely this explains the large growth in O_3 concentrations over the entire troposphere observed at some European stations [10, 11]. The observed global increase in CH_4 may also have played an important role in this (see Appendix 8.2).

The catalysts NO and NO_2 are removed from the atmosphere by the formation of nitric acid (HNO_3) through gas-phase reactions or reactions that involve the deposition of nitrogen trioxide (NO_3) and nitrogen pentoxide (N_2O_5) on wetted aerosol particles or cloud droplets [*Figure 8.11*, (3, 4A)]. The dry deposition of NO_2 on vegetation [*Figure 8.11*, (7A)] is also of great importance in the NO_x budget [16]. As a result, transport of NO_x from the troposphere does not occur to any significant degree. Lightning is the most important natural source of NO; the amount produced by this source is still very uncertain. Owing to its short atmospheric lifetime, background levels of NO_x in the troposphere are often determined by lightning frequency, which strongly peaks over the tropical continents. Unfortunately, NO_x observations are still very scarce, especially in the tropics.

Nitric acid is soluble in water droplets [*Figure 8.11*, (2B)], in which form it greatly contributes to precipitation acidity and, to some extent at least, to corrosion near industrial areas. It is also an important source of fixed nitrogen for the biosphere. The organic nitrates, commonly found on aerosol particles, include potent lacrimating agents and mutagenic compounds. Anthropogenic nitrogen compounds are thus sources of compounds implicated in biospheric productivity on the one hand and photochemical oxidant formation, acid precipitation, public health risks, and materials degradation on the other.

Biogenic emissions of reactive nitrogen [*Figure 8.11*, flux (γ)] consist largely of ammonia (NH_3), although soils also emit significant quantities of NO. Even in the case of NH_3, anthropogenic influences are substantial as a result of coal burning and animal husbandry. NH_3 emissions to the free troposphere are weak from soils covered with vegetation, so that most emission probably occurs in the warm early spring, before new vegetation has emerged. NH_3 is the only common gas potentially capable of partly neutralizing the acid gases. It dissolves readily in water droplets and is efficiently sorbed by wetted surfaces. A minor fraction (about 10%) reacts with $OH^{\cdot}$ in the gas phase and its oxidation may lead to production or destruction of NO_x, depending on uncertain reaction paths and the prevailing concentrations of NO_x.

Sulfur compounds

The atmospheric sulfur cycle [17] is presented in *Figure 8.12*. A major, though not adequately quantified, portion of atmospheric sulfur compounds is emitted by natural biological processes [flux (γ), *Figure 8.7*, and *Figure 8.12*]. These compounds tend to contain sulfur in a reduced valence state, principally hydrogen sulfide (H_2S) from land and dimethyl sulfide (CH_3SCH_3) from marine sources. In the atmosphere, photochemical oxidation reactions paralleling those of the carbon compounds produce sulfur dioxide (SO_2) and various other products [*Figure 8.12*, (3) and Appendix 8.4]. Alternatively, the reduced sulfur gases may be deposited on surfaces [*Figure 8.12*, (5B)]. These gases are a major cause of corrosion and degradation for many metals and alloys [18]. Although carbonyl sulfide (COS) is emitted at relatively low rates from a variety of sources, it is removed very slowly by atmospheric processes, so on a global basis it is the most abundant atmospheric sulfur gas. Global sources and sinks of COS have been estimated by Khalil and Rasmussen [19]. A substantial anthropogenic influence is possible, although there are large uncertain aspects about the natural sources of COS.

The primary anthropogenic source of SO_2 is the combustion of sulfur-containing fossil fuels [flux (α), *Figure 8.12*]. In large industrial regions, input from this normally exceeds that of natural sources by a large amount. There are three fates for SO_2, all of them important. The first is deposition on surfaces [*Figure 8.12*, (5A)], especially on soils, vegetation, and waters, but also on construction materials, a process thought to be primarily responsible for the deterioration of masonry and statuary [20]. The second is the formation of sulfate aerosol following atmospheric oxidation to sulfuric acid [*Figure 8.12*, (4)]. This aerosol is a major factor in limiting visibility in the atmosphere [21]. The third is the absorption and oxidation of SO_2 in cloud water [*Figure 8.12*, (2)], which, together with direct sulfuric acid aerosol incorporation, constitutes a

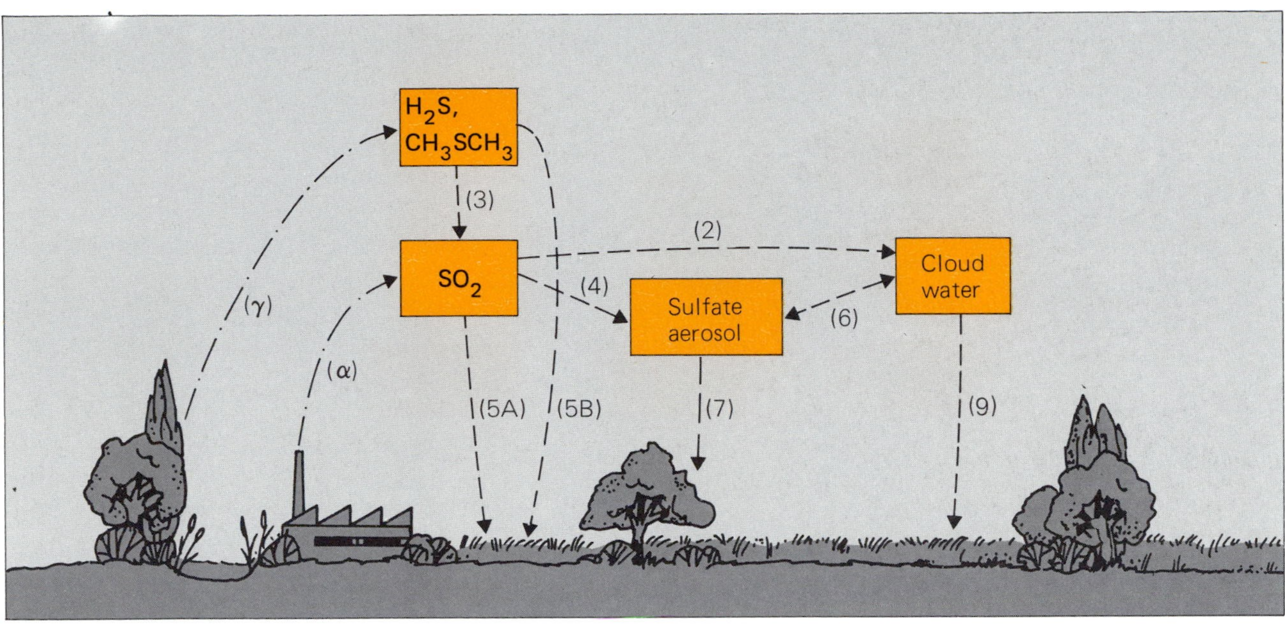

Figure 8.12 Atmospheric sulfur compounds: the principal species and chemical and physical pathways.

primary factor in determining the acidification of precipitation. These fast photochemical reactions and rainfall in the troposphere prohibit the transport of all sulfur gases, except COS, to the stratosphere. The photochemical conversion of COS into SO_2 and H_2SO_4 explains the persistence of the stratospheric sulfate layer, even during periods with no eruptive volcanic activity [22]. In the troposphere, cloud water evaporation is also a source of sulfate aerosol [*Figure 8.12*, (6)].

Halogen compounds

The principal components of the atmospheric chlorine cycle are shown in *Figure 8.13*. (Concentrations of compounds containing halogens other than chlorine are much smaller in the lower atmosphere and are not considered in detail in this chapter.) Methyl chloride (CH_3Cl) is emitted primarily from the ocean, but also from biomass burning [flux (γ) *Figure 8.13*]; it is destroyed in the atmosphere by reaction with $OH^{\cdot}$ and yields HCl through a sequence of atmospheric chemical reactions [*Figure 8.13*, (3)]. CH_3Cl is the main natural source of chlorine in the stratosphere. In addition, and today more importantly, chlorofluorocarbons (CCl_4, $CFCl_3$, CF_2Cl_2, CH_3CCl_3, etc.) are emitted from anthropogenic sources [flux (α_1), *Figure 8.13*] and reach the stratosphere in increasing amounts [*Figure 8.13*, (1A)]. There they are photochemically transformed into active chlorine compounds which attack O_3 catalytically. Final removal of these compounds is through a return flow of HCl to the troposphere [*Figure 8.13*, (1B)], followed by rainfall.

Two other tropospheric sources of HCl are much larger than those already mentioned. One is direct emissions from coal-burning power plants [flux (α_2), *Figure 8.13*]. The second is volatilization from sea-salt aerosol, influenced by the uptake of H_2SO_4 or HNO_3 from the atmosphere [*Figure 8.13*, (8)]. These sources are difficult to study, so the tropospheric sources and sinks of HCl are not well established.

HCl is very soluble in water, a property that is reflected by the high concentration of chloride ion commonly found in rain. This high solubility prevents tropospheric HCl from reaching the stratosphere. The deposition of atmospheric chlorine, whether in water drops [*Figure 8.13*, (9)], aerosol particles [*Figure 8.13*, (7)], or the gas phase [*Figure 8.13*, (5)], is also of interest since many metals and alloys are susceptible to chlorine corrosion [23].

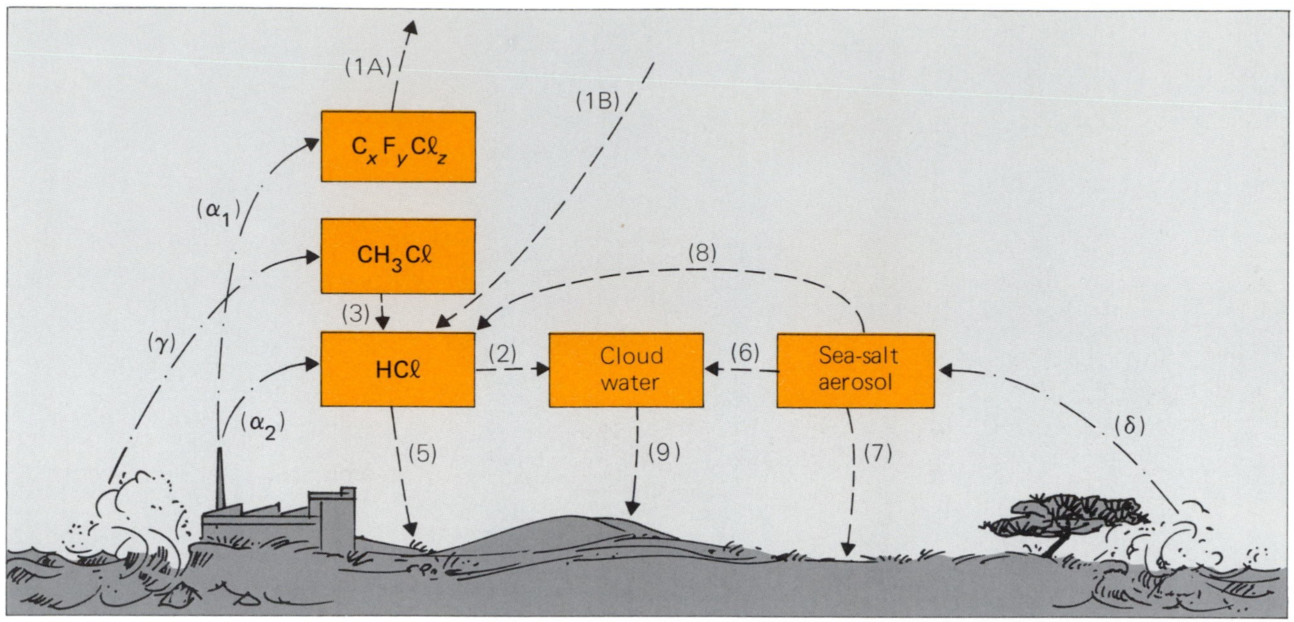

Figure 8.13 Atmospheric halogen compounds: principal species and chemical and physical pathways.

Chemistry in atmospheric water droplets

The study of chemical reactions in cloud and fog droplets and in raindrops is in its infancy. At present it appears that we can identify some of the major products of these reactions, but the reaction paths that link the products to their precursors are very little known [24].

A central concern in aqueous atmospheric chemistry is the fate of H_2SO_4 and HNO_3. In solution, each acid is in equilibrium with its ions (see Appendix 8.5) and the acids are referred to as "strong" because the equilibria strongly favor the ionic forms. The generation of the acids in solution must involve oxidizing species. Hydrogen peroxide (H_2O_2), with an aqueous-phase concentration of a few micromoles/liter, appears to have a central chemical role: it serves both as an oxidant and as a source of solution $OH^\cdot$ when it is photolyzed. O_3 dissolved in water is also a source of $OH^\cdot$ and is probably a significant radical source when H_2O_2 concentrations are low.

The initial acidity of atmospheric water droplets is established by the equilibrium between gaseous CO_2 and solution bicarbonate ions (HCO_3^-). Except for this function, the presence of the bicarbonate ion does not appear to affect solution chemistry in any significant way. Among other common trace components are the chloride salts. They are relatively unreactive and thus chloride's greatest influence is likely to be as an electrolyte, that is, as a contributor to the overall ionic strength of the solution.

NH_3 (and the ammonium ion, NH_4^+) and HNO_3 (and the nitrate ion, NO_3^-) are the most important inorganic nitrogen compounds in atmospheric water droplets. Ammonia is the principal species that neutralizes the strong acid anions, as evidenced by the large concentrations of ammonium salts found in aerosols. This leads to a lowering of the proton concentrations (acidity) of rainfall.

Sulfur compounds are important constituents of atmospheric aqueous solutions. The most abundant are the sulfates, including H_2SO_4, hydrogen ammonium sulfate (NH_4HSO_4), and ammonium sulfate [$(NH_4)_2SO_4$]. However, SO_2 is soluble in water and chemically active. Its principal reactions (in the customary bisulfite form, HSO_3^-) are probably with H_2O_2 and O_3.

Transition metals are not abundant atmospheric species, but their ability to function as catalysts in the oxidation of SO_2 to give H_2SO_4 makes them potentially important [25]. As shown in *Figure 8.14*,

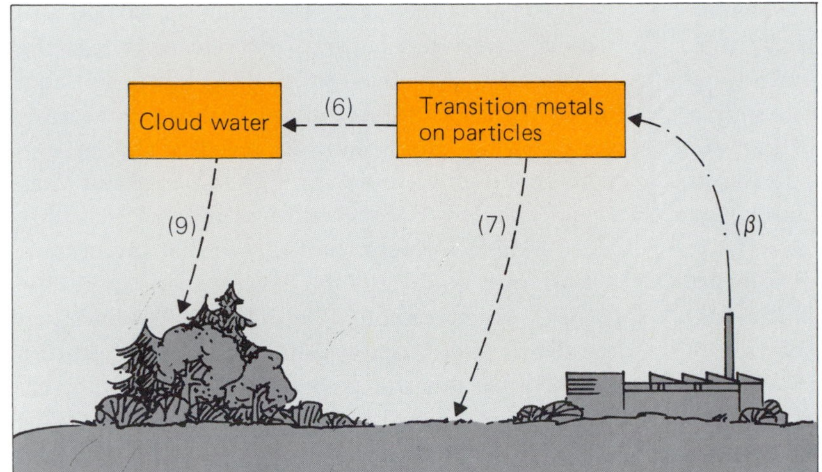

Figure 8.14 Atmospheric transition metal compounds: principal species and chemical and physical pathways.

they enter the atmosphere in particulate form, generally through emissions from smelters or coal-combustion plants [flux (β), *Figure 8.14*]. In the latter case, at least, they tend to concentrate at the particle surfaces [26], so that their effects may be much more significant than their concentrations would indicate. The role of soot particles in SO_2 oxidation reactions is also potentially important [25], but the mechanism and magnitude of the effect are still not well-enough known.

Oxidized organic compounds may be important participants in atmospheric solution chemistry, but have thus far received relatively little study. What research has been done indicates that organic acids are common. Organic hydroperoxides (ROOH) are readily formed in the gas phase, and may be precursors for acids. Precursors likely to be more important are the aldehydes (RCHO), which are hydrolyzed in solution to the glycol form [$RCH(OH)_2$]. Oxidation of these precursors gives the organic acids, but the precise mechanisms need to be defined.

In solution, the organic acids [RC(O)OH] are only partly ionized and are thereby termed "weak". Nonetheless, organic acids can be the major acid constituent of atmospheric water droplets if the precursors of the strong inorganic acids are present only at very low concentrations. Such conditions may exist in the more remote parts of the Earth that are biologically highly productive, such as tropical rain-forest atmospheres [27].

Present knowledge does not enable us to say definitively whether incorporation from the gas phase or solution oxidation is the most important for generation of the acids found in clouds and rain. It is likely that both processes operate in all situations, one dominating the other under different conditions. Current evidence suggests that solution reactions are often important for the formation of H_2SO_4 [25] and less important for HNO_3 [28]. However, during the nighttime hours NO_3 and N_2O_5 are formed in the gas phase, because NO_3 is not photolyzed by solar radiation; their uptake by water droplets and wet aerosol surfaces leads to the formation of nitric acid. Unfortunately, the organic acids have been too little studied for a dominant mechanism to be specified. We repeat that the primary emitted NO_x and hydrocarbon gases have low water solubilities, so that oxidation reactions in the atmosphere must occur before most aqueous-phase chemical reactions become operative.

Principal sources of atmospheric trace species

Oceans and estuaries

The marine waters are significant contributors of chemical species strongly involved in important atmospheric chemical processes. In most cases, species emission arises as a consequence of biological processes, such as for N_2O, which is a biological nitrification and denitrification product. The largest fluxes are from coastal waters and estuaries or from oceanic regimes of high productivity, which are generally found in upwelling regions. Various processes on land also contribute significantly to atmospheric N_2O. Estimates of the global source strength are discussed by Crutzen [29] and are reproduced in *Table 8.1* and *Figure 8.6.* Reduced sulfur gases are also emitted from the marine environment as by-products of biological activity. The oceanic and estuarine flux of H_2S appears small, but that of CH_3SCH_3 is substantial [30, 31].

Large emissions of chlorine from the ocean occur due to the generation of sea-salt particles by breaking waves; Blanchard and Woodcock [32] summarized the various flux estimates. HCl is thought to be volatilized as a consequence of the uptake of H_2SO_4 and HNO_3 by sea-salt aerosols, but no flux estimates appear to have been made.

Biological activity plays an important role in the production of various halogen organic compounds. For example, the biological formation of CH_3Cl provided the main source of chlorine to the stratosphere prior to the synthesis of CCl_4, $CFCl_3$, and CF_2Cl_3 by man. The oceans also play a very important role in the biogeochemical cycling of CO_2 (as reviewed by McElroy, Chapter 7, this volume).

Vegetation

Vegetation is a source of many organic compounds, as is immediately obvious when we recall the diversity of floral scents. As with other atmospheric species, the vegetative emissions are subject to reaction and transformation in the atmosphere. Although the ultimate product of the oxidation reactions is CO_2, a large number of intermediate products play an important role in atmospheric photochemistry. These compounds, including CO, aldehydes, ketones, organic peroxides, and various carboxylic acids, may occur in the gaseous or liquid phase. Formic and acetic acids have been detected in precipitation in unpolluted regions [27]. If anthropogenic NO_x is present, gas-phase photochemical oxidant formation is possible. The potential role of terpenes in aerosol formation and visibility deterioration has recently been reassessed. Contrary to previous belief, Altshuller [33] claims that the terpenes play only a minor role, but the more recent studies of Hooker *et al.* [34] emphasize the opposite.

Forests are a very large reservoir for carbon and one effect of forest harvest is to release CO_2 to the atmosphere. Woodwell *et al.* [35] proposed that forest clearance in the tropics could be a net source of CO_2 to the atmosphere larger than that arising from fossil fuel burning. This viewpoint was strongly contested, especially by Broecker *et al.* [36], because measurements had shown that the oceanic uptake rate could not be larger than about 2×10^{15}g C, which is about 40% of the carbon released by fossil fuel combustion. Additional studies [37, 38] now put the carbon release from tropical forest clearance in the range $1\text{–}3 \times 10^{15}$g C/yr. This release rate to the atmosphere may be at least partially balanced by reforestation in other parts of the world and by production of decay-resistant charcoal from biomass burning in tropical savannas [37]. The most recent study on this issue is that of Goudriaan and Ketner [39], but the issue is far from settled. Remote sensing from satellites will give important information in the future. Historical analyses of land-use changes may also prove extremely useful, as indicated by Richards (Chapter 2, this volume). Estimates of CO_2 release from soils and forest cutting are important for the interpretation of the observed CO_2 increase in the atmosphere over the past century and for extrapolation of the increases expected because of increased fossil fuel burning. For more thorough discussions see Clark [40], McElroy (Chapter 7, this volume), and Liss and Crane [41].

The animals

The animal kingdom [42] is a source of several atmospheric trace gases, but apparently not a large source. CH_4 is produced by enteric fermentation in ruminants and by the digestion of wood by termites [29]; although difficult to quantify, these fluxes from the animal kingdom are probably small compared with other CH_4 sources. Animals also produce NH_3 as a volatile product of excrement, a source that may be relatively important [43]. Estimates of the relative role of animals in the production of CH_4 and NH_3 are given in *Table 8.1*, which shows the substantial role of domestic ruminants in CH_4 production, but a lesser one with regard to NH_3. The tendency for these animals to be gathered together in feedlots and pastures can cause substantial local impacts.

The number and mass of wild animals on Earth has almost certainly decreased because of the steady expansion of human influence. The effects of this decrease on emissions to the atmosphere of CH_4 and NH_3 are, however, overwhelmed by the growth in domestic animal populations and by the concomitant conversion of natural lands, especially forests, into pasture and agricultural lands under highly productive cultivation. The domestic animal population is related to the human population, for which the evolution since 1930 is shown in *Figure 8.15.* For comparison, that for domestic animals is shown in *Figure 8.16.*

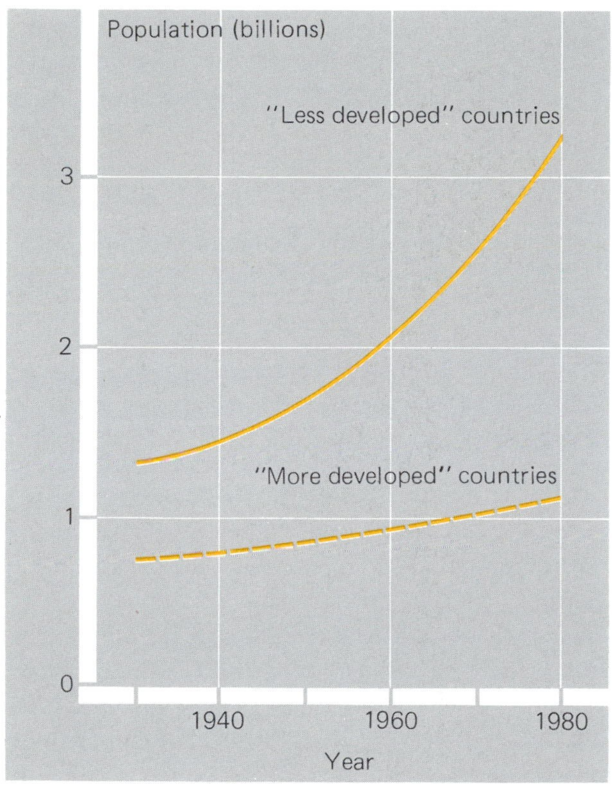

Figure 8.15 Populations of the developed and developing world over the past half century. The population figures and the division of countries into the two categories are from the United Nations Population Division [44].

Biomass burning

The combustion of biomass (trees, brush, vegetation, grasses, etc.) is a rich source of atmospheric material, but has received relatively little study. In the results of one extensive field investigation [45] emission fluxes of CO_2, CO, NO_x, N_2O, CH_4, and soot are presented and related to the global budgets of these species. Much more work is needed to determine how representative these results may be.

Biomass burning may be natural, as in fires started by lightning, or it may be anthropogenic, as in fires started to clear land for agriculture. The latter is much more common, so the emissions from this source are largely under human control. An ancillary factor is the extensive efforts made over the past half century to control natural forest fires. For example, it is estimated [46] that the acreage burned annually in the USA decreased by a factor of 10 between 1930 and 1980. More recently, foresters in mid-latitude regions have recognized the beneficial effects of natural fires on forests and have tended to control fires only near populated areas. The annual acreage burned has thus oscillated significantly over the past few decades, but has almost certainly decreased in the mid-latitude forest regions. Conversely, the increasing population and agricultural activity in the tropics suggest an increase in biomass burning compared with past times. Agricultural burning was very common during the nineteenth century in Europe and North America, during which period, and for the first quarter of this century, it was the main source of air pollution, in these regions at least. No integrated worldwide historical assessments of biomass burning have been performed.

Crop production

Crop production involves changes in land use and in soil chemistry. As such, it has the potential to affect the atmosphere. Perhaps the greatest potential effect is on CO_2, which is released to the atmosphere when soil organic matter is oxidized by tilling. Until 30 years ago this effect was greater than that due to the harvesting of forests [38] and was the main cause of the CO_2 increase in the atmosphere, more important than fossil fuel combustion. It may also have led to substantial emissions of N_2O, as implied by studies in Florida [47] which show that 2.7% of the fixed nitrogen lost in drained, cultivated soils in South Florida appeared as N_2O (see *Table 8.1*).

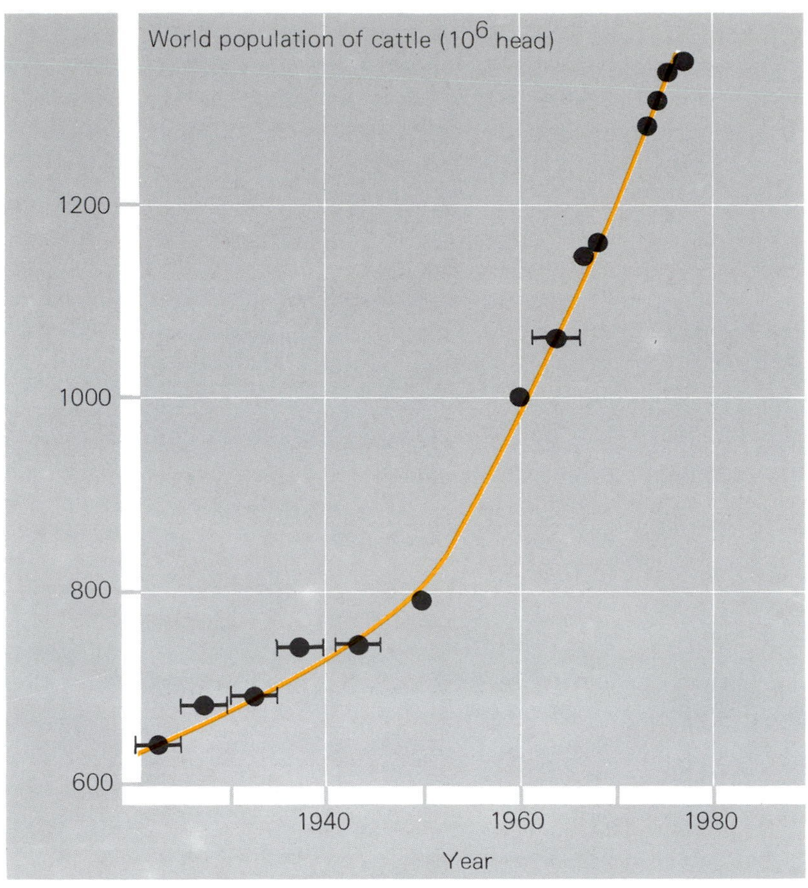

Figure 8.16 Estimated world population of cattle over the past half century.

Fertilization of cultivated land is a common practice and results in the emission of N_2O. Contrary to early belief, this source of atmospheric N_2O is not the only one, but is one which may grow with increasing use of fertilizer nitrogen, especially due to the nitrification of NH_3-containing fertilizer.

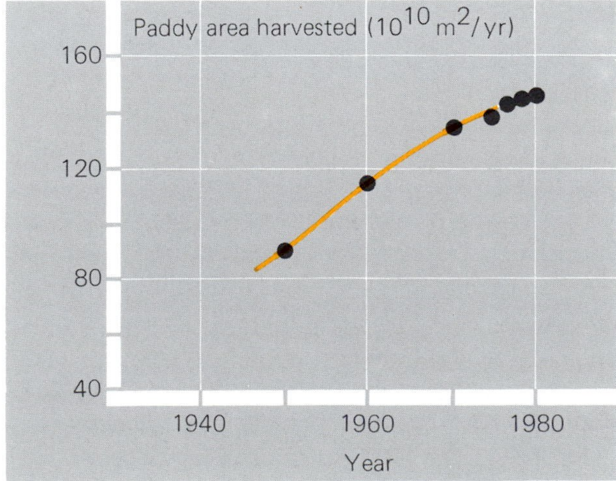

Figure 8.17 Estimated world area of paddy field lands used for rice cultivation.

Certain crops, especially rice, grow under anaerobic conditions, thus forming a variety of reduced gases. CH_4 emission is of the greatest concern because the flux is relatively large [29]. Not only has the total area in rice production increased over the past half century (*Figure 8.17*), but multiple cropping has increased the total yield per unit area. As a consequence, rice production has contributed significantly to the observed atmospheric increase in CH_4, an effect offset only slightly by the drainage of wetlands. More measurements and historical analyses are needed to better estimate these effects (see Richards, Chapter 2, this volume).

Fossil fuel combustion

The combustion of petroleum in its various forms is a very substantial source of atmospheric trace gases, many of which are potentially detrimental. The primary gaseous emission is CO_2, whose source flux has been estimated most recently by Rotty [48]. Incomplete combustion also produces CO, together with a wide variety of hydrocarbon compounds. Flux estimates have been given for CO by Seiler [49] and Logan *et al.* [50] and for the non-CH_4 hydrocarbons by Duce [51]. The relative magnitudes of these emissions to those of biomass burning and natural hydrocarbon emissions are shown in *Table 8.1* and *Figure 8.5.*

High-temperature combustion inevitably produces various oxides of nitrogen. NO and NO_2 are highly active catalysts in most atmospheric chemical chains, since they are involved in vital oxidizing reactions that lead to O_3 formation in the troposphere (see Appendix 8.2). Source strengths of NO_x have been estimated by several authors [50, 52], and the anthropogenic contributions are large (see *Table 8.1*). N_2O is inert in the lower atmosphere, but important in stratospheric chemistry since it releases NO_x after photochemical reactions [3]. This leads to O_3 loss above about 25 km in the stratosphere and O_3 gain below (see Appendix 8.6). The flux of N_2O from petroleum combustion is quite significant.

The sulfur content of petroleum varies widely and, as a result, so do the emissions of SO_2 which, on a global basis, are important [30, 53]. Regionally, they dominate by far the natural emissions and are a major component of acidic precipitation.

Petroleum burned in inefficient combustors is a prolific producer of carbon soot, which is involved in visibility degradation and in condensed-phase atmospheric chemistry. Carbon soot may also influence the Earth's radiation budget by absorption of sunlight, especially during the early spring period in the Arctic [54].

Coal combustion also results in large and diverse emissions into the atmosphere. As with petroleum, coal combustion is a major source of CO_2, CO, hydrocarbons, NO and NO_2, N_2O, SO_2, and soot. In addition, coal combustion produces HCl [55] and a variety of transition metals [56]. The importance of coal combustion will almost certainly increase in the future, because of the large reserves of coal; CO_2 and SO_2 emissions are of especial concern.

Industrial processes

The most diverse group of emissions into the atmosphere is due to the wide variety of industrial processes and subsequent uses of the products where fuel combustion is not the primary source. Three classes of emissions can be identified for special attention. The first is the family of halocarbon propellants ("Freons") involved in stratospheric O_3 depletion [57]. A second emission of concern is SO_2, which is generally emitted in ore smelting and a variety of other industrial processes [53]. The third class is the transition metals, produced by a wide variety of materials-processing industries, for which Nriagu [56] has made flux estimates.

Manifestations of atmospheric degradation

Modification of the stratosphere

The temperature structure and dynamic processes in the stratosphere are, to a major degree, determined by the absorption of solar ultraviolet radiation by O_3. The total amount of O_3 in the stratosphere is remarkably small; it is equivalent to the number of air molecules contained in a 3 mm thick layer in the lower atmosphere. The troposphere contains only 10% of the atmospheric O_3, so most O_3 is located in the stratosphere (i.e., between 10 and 50 km). Atmospheric O_3 also plays an important ecological role, since it filters out most ultraviolet solar radiation between about 210 and 310 nm. However, penetration of ultraviolet radiation to ground level starts at wavelengths of about 300 nm and increases by orders of magnitude up to 320 nm. It is this radiation that may be biologically harmful: its intensity at the ground is strongly enhanced by reduction of the O_3 column. It has been estimated that a 1% reduction in total O_3 would produce a 2–4% increase in biological effects. The effects on the biosphere have been reviewed [58], but great uncertainties exist, especially on the ecosystems level. As we have already noted, the penetration of ultraviolet radiation at wavelengths near 310 nm to the troposphere leads to the production of $OH^{\cdot}$, which initiates most oxidation processes in the atmosphere, producing compounds that are effectively removed from the troposphere. Changes in the stratospheric O_3 column, therefore, also affect tropospheric photochemistry.

Stratospheric O_3 may be particularly affected by compounds that are relatively inert in the troposphere. These compounds (e.g., N_2O and several chlorocarbon gases, such as natural CH_3Cl and industrially produced $CFCl_3$, CF_2Cl_2, CCl_4, and CH_3CCl_3) have low solubilities in water, slow photolysis rates, and slow reaction rates with $OH^{\cdot}$. Stratospheric chemistry is also influenced by the direct injection of material into the upper troposphere and stratosphere by thunderstorms, volcanic eruptions, solar proton events, auroral activity, meteoritic impacts, nuclear weapons testing, and emissions from jet aircraft. The solar proton event of August 1972 led to large depletions of O_3 in the middle stratosphere at high latitudes, which was observed by satellite instruments [59]. There are strong indications that the recent eruption of the El Chichon volcano in Mexico caused global O_3 depletions of several percent [60], perhaps as a result of HCl injection.

The chemistry of the stratosphere is complex (see Appendix 8.6) and the effects on O_3 concentration of different emitted species are difficult to predict and to monitor. It now appears that the major causative species are the chlorofluorocarbons used as propellants and refrigerants. N_2O, produced largely from natural sources, is also important in these processes.

The predicted severity of these effects has varied substantially as computer models of the atmosphere have been refined, new transformation reactions have been discovered, and better rate constants have become available. Current predictions of total O_3 depletions at the contaminant input rates of 1978 are about 4%. If other factors, such as CH_4 and CO_2 increases, are also taken into account, predicted total O_3 depletions become even smaller and a total O_3 increase seems possible. This could have undesired climatic implications, as a large loss of O_3 above 30 km is compensated for by an O_3 increase at lower altitudes, enabling the lower stratosphere to receive more solar heating and to more efficiently trap outgoing heat radiation. Also, O_3 depletions may become especially large if the halocarbon release rates increase so much that chlorine mixing ratios in the stratosphere become of the order of 20 p.p.b.v., which requires only about twice the present manufacturing rates [61]. The chemical and thermal stability of the stratosphere remains, therefore, an issue of continuing and serious concern.

Modification of the atmospheric radiation balance

The interaction of radiation and climate is discussed in detail by Dickinson (Chapter 9, this volume). We mention it here in passing because of increasing evidence that atmospheric trace gases play important roles in establishing the radiation budget. The most important of these gases are H_2O and CO_2. O_3 is an important absorber, especially in the upper troposphere and stratosphere, while significant effects are also caused by $CFCl_3$, CF_2Cl_2, N_2O, and CH_4. The concentrations of these four gases are all increasing in the Earth's atmosphere because of anthropogenic activities, and the same is probably true of tropospheric O_3. Their combined heating effect on the Earth's surface approaches that of CO_2 [5, 40].

Soot produced by inefficient combustion is another potential cause of radiative modification, because it absorbs solar radiation efficiently; the magnitude of its effect has not yet been adequately assessed. Currently, its impact has been especially considered with respect to the observed pollution of the Arctic atmosphere in the spring season [62]. The expanding industrialization of the northern USSR may have a substantial effect in the future. The importance of soot for the atmospheric radiation balance has been considered dramatically in connection with several recent studies on the climatic and ecological effects of a large nuclear war, because in such an event huge quantities of black smoke would be produced from fires in industrial and urban centers [63]. These studies indicate the possibility of substantial alterations of global meteorological conditions, with continental surface temperatures dropping far below 0 °C even during the summer.

Photochemical oxidant formation

Smog (a combination of *smoke* and *fog*) is an undesirable mixture of atmospheric trace species produced in sunlight by high fluxes of anthropogenic emissions. The severity of smog is most often judged by the ground-level concentration of O_3. As already noted, O_3 is produced by reactions involving hydrocarbons and oxides of nitrogen. The chemical processes undergone by these precursors are the result of a complex interplay of the mixture of reactive compounds, time of day, season, and meteorological conditions.

The trace species (the "photochemical oxidants"), of which O_3 is the main representative, also include peroxides, organic nitrates, aldehydes, and a variety of other compounds. They are frequently implicated in damage to crops [64], eye irritation [65], decreased lung function [66], and the degradation of works of art [67]. Amelioration of smog by regulatory measures is complicated by the simultaneous involvement of many chemical compounds. Regulations may not be universally advisable, though measures leading to reductions in automotive emissions are likely to be of substantial benefit.

The appearance of photochemical smog is now a rather common, regional-scale phenomenon in many parts of the world, such as Southern California, Western Europe, Mexico City, and Sydney. The widespread damage to, in 1984, about half of the German forest trees (an increase from one third in 1983) may be due to the impact of photochemical oxidants, in combination with a wide range of physiological factors [68]. If this suspicion is confirmed, the critical role of tropospheric O_3 in the environment will be even more strongly emphasized.

Acidic precipitation

The presence of the strong inorganic acids H_2SO_4 and HNO_3 in precipitation is widely recognized as detrimental, although the effects of acidic precipitation on lake and forest ecosystems, statuary, etc., are still subjects of active debate. Deleterious effects on fish populations in Scandinavian and North American lakes seem clearly related to an increased acidity of rainfall [69], but it has proved difficult to relate forest damage to acid precipitation. In fact, forest damage seems to be especially pronounced in central Europe, a region whose alkaline soils could be expected to provide an effective buffer to acidic precipitation. This suggests that the damage results from changes in concentrations of photochemical oxidants (see above) rather than from changes in the chemistry of precipitation.

The deposition of H_2SO_4 and HNO_3 is due largely to SO_2 and NO_x emissions by industrial processes and automotive transport. Transboundary transport often occurs, leading to international, scientific, and legal disputes. A case study of sulfur transport by a group of Swedish scientists was undertaken for the United Nations conference on the human environment, held in Stockholm in 1972 [70]. Recent regulatory actions in the FRG have concentrated on a reduction of NO_x emissions from motor vehicles with the aim of diminishing photochemical oxidant formation. There is no certainty that the proposed measures will rapidly produce the desired results, since the cause and effect relationships are not convincingly clear. Drastic reductions in all emissions from fossil fuel combustion by all nations are probably needed to prevent major ecological effects.

Visibility degradation

Visibility in the atmosphere decreases when solar radiation in the visible band (about 0.3–0.7 μm) is scattered by particles or gases. The most efficient scattering particles have sizes similar to those of visible light. Particles in this size range have an atmospheric lifetime of tens of days, since the particles are too large for efficient agglomeration and too small for rapid gravitational settling and precipitation scavenging. As shown in *Figure 8.18*, the reduction in visibility may be quite severe. In the past few years it has become clear that sulfate aerosol particles are the predominant scatterers, at least in regions near or downwind of industrialized areas [71]. The graphitic soot generated by incomplete combustion also has a significant influence [72]. In urban areas with large emissions gaseous NO_2 can also contribute to visibility degradation. Naturally emitted hydrocarbons from vegetation, especially terpenes, were regarded for many years as likely visibility inhibitors [73], but from recent research appear to be less important on a continental scale than the sulfate and soot aerosol particles [33].

Combustion of fossil fuels in power plants is a major source of light-scattering particles. Another energy source that has recently become more widely used than previously, especially in North America, is wood for home heating. Wood burning is generally rather inefficient and produces relatively large fluxes of small particles. Visibility degradation due to agricultural biomass burning is common and severe during the dry season in the tropical savanna regions of the world. It has been estimated [45] that $3–5 \times 10^{15}$ g of biomass are now burned annually, mostly in the tropics, releasing about 2×10^{14} g of

Figure 8.18 (*a*) New York City with good visibility. The buildings in the foreground are the towers of the World Trade Center during construction. (*b*) New York City with visibility impaired by high levels of atmospheric particulate matter.

smoke. This practice was previously also very common at temperate latitudes and is still an important contribution to visibility deterioration in Southern Europe.

Atmospheric corrosion of materials

The corrosion of materials exposed to the atmosphere, both indoors and out, is a problem of truly massive proportions. The cost to society in the USA alone was estimated in 1975 at over $70 billion per year [74]. The most common problems include the chloridization and sulfidization of iron and its alloys, the chloridization of aluminum, the sulfidization of copper (*Figure 8.19*) and its alloys, and the sulfidization of marble and masonry. The active agents include the reduced sulfur gases (principally H_2S and CH_3SCH_3, and perhaps also COS), SO_2, HCl, and the sea-salt aerosol.

The processes of materials degradation that involve atmospheric trace gases remain obscure, despite extensive research. It is apparent that water is needed if corrosion is to proceed vigorously [23], which implies that electrochemical cells may be set up on the surface of the corroding material [75].

Figure 8.19 Detail of torch of the Statue of Liberty, New York Harbor (reproduced with permission of Statue of Liberty–Ellis Island Foundation, Inc.).

The corrosive agent may be delivered to the corroding surface in either of two ways [76]: as a gas (e.g., SO_2) or as a component of precipitation (e.g., SO_4^{2-} ion), but examination of the corroded material does not enable us to distinguish between the two.

Unlike most other effects of atmospheric trace species, in many regions corrosion is due in large part to natural processes. Within heavily populated or industrialized areas, containing most capital investments and monuments of art, anthropogenic emissions may be the most important. A synergism which makes quantitative assessment difficult is the recent discovery that O_3 and solar radiation together can enhance atmospheric corrosion processes [77]. Much work remains to be done to determine the details of these interactions. For the moment, it may be most expedient to devise protective techniques for the materials rather than to attempt to control the atmospheric concentrations of species involved in corrosion.

Discussion

Atmospheric chemistry has been a recognized speciality for only a few decades, a very short time by comparison with many other scientific fields. It is also complex and interdisciplinary, combining meteorology and the physics of fluid flow and radiation with a chemical inventory known to contain more than 2000 distinct compounds [78]. As a result, discoveries of compounds not previously known to be present in the atmosphere, or of their unexpected effects, have been frequent. However, it now appears likely that the most important atmospheric species and their impacts have been determined. Source strengths are much less well known and in many cases atmospheric distributions of gases and particles are almost totally unstudied, especially in the tropical and marine environments.

Although our information is admittedly fragmentary in some cases, it is useful to present our best comparative assessments of atmospheric sources and impacts. In *Figure 8.20* the rows represent the principal sources of atmospheric trace species and the columns the major consequences of emissions. The top two sources are totally natural, the middle two have both natural and anthropogenic components, and the bottom three are totally anthropogenic. At each point in the figure where any source contribution exists, we indicate both the relative magnitude of the consequences produced by emissions from that source and the reliability of the magnitude assessment. (The assessments are made for 1985; prospective and retrospective assessments would also be valuable, but are beyond the scope of the present chapter.) *Figure 8.20* shows that the sources of most general concern, as indicated by their impact ratings, are almost wholly anthropogenic in nature: fossil fuel combustion, biomass combustion, and industrial processes. Emissions from crop production, especially CH_4 from rice paddies, and from estuaries near heavily populated areas may have future impacts. Some sources (the animal kingdom and vegetation) have

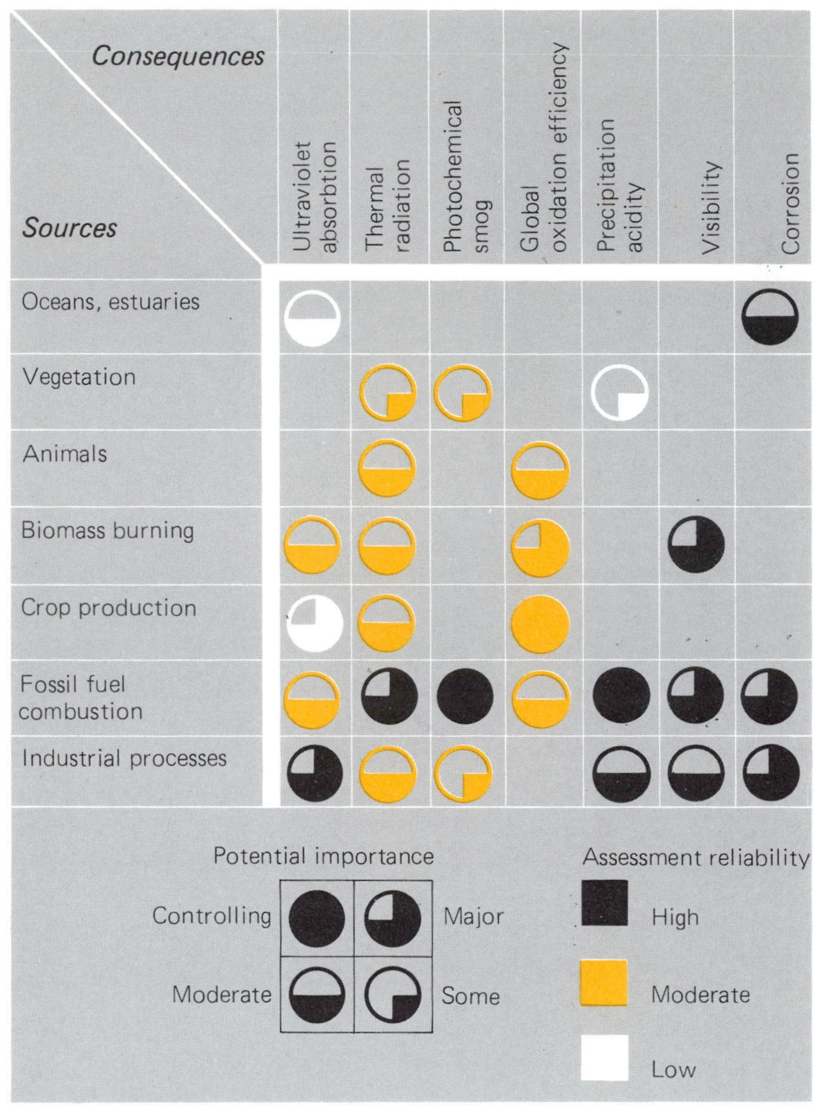

Figure 8.20 Sources of atmospheric trace gases and the potential consequences of emission from those sources.

sufficiently small effects on atmospheric processes that they need not cause much concern, even if their emissions should increase.

Figure 8.20 also suggests areas in which useful interdisciplinary studies could be performed. These are areas that have significant potential importance, but whose assessment reliability is low to moderate. The two best candidates for such studies are biomass burning (thought to have a great significance for tropical photochemistry, the atmospheric radiation balance, and visibility) and vegetative emissions (potentially linked to photochemical oxidant formation, the radiation balance, and precipitation acidity).

A cause of concern among those who are interested in the atmospheric sciences, but are not specialists in them, has been the "phenomenon of the month": the frequent appearance of new effects or compounds of concern. Although it is possible that other major emission sources or new processes will be identified, it seems more likely that increased knowledge of the intricacies of the chemistry or of projected changes in emission rates will change our understanding about the relative importance of the causative species. The way in which this may happen is illustrated in *Figure 8.21*, in which the effect produced by a species is related to its emission flux, to its rate of transformation into another species, and to the degree of certainty with which the effect is predicted. The flux assessments are related to type and degree of anthropogenic activity and to the technology involved. Usually assumptions must

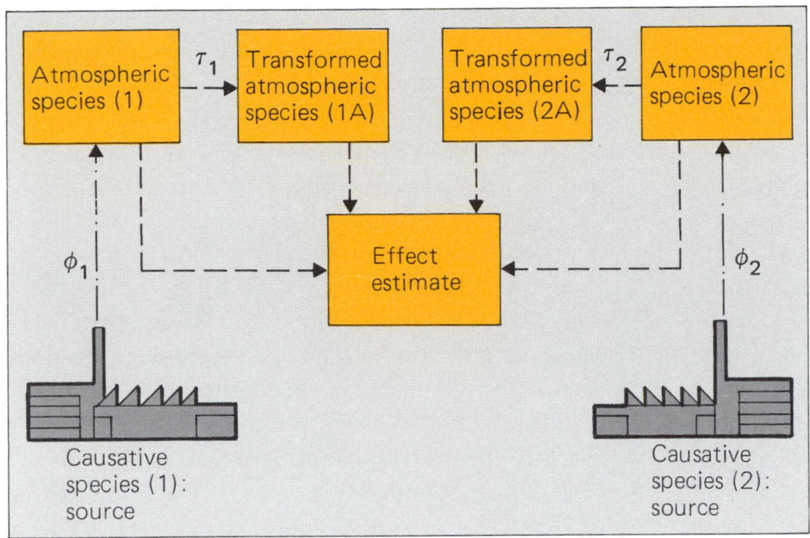

Figure 8.21 Assessment of an atmospheric effect resulting from anthropogenic emissions of two causative species: ϕ is the emission flux and τ the transformation rate.

be made, particularly if one is attempting to visualize a few decades into the future. Estimates of transformation rates depend on the degree to which the atmospheric chemical processes are understood, and evolve in proportion to the intensity of the research effort and the extent to which new chemical regimes are considered. Gas-phase chemistry was not comprehensible until the importance of $OH^{\cdot}$ was recognized in 1971. A similar educative process now appears to be happening for photochemistry in water droplets. The chemistry and physics of atmospheric aerosol particles are not well understood, and may involve crucial species and processes as yet unidentified.

Finally, estimates of effects are difficult to make even when trends are known, since the interlocking global processes are so difficult to comprehend. It seems likely that the present major uncertainties have to do not with the atmosphere as a separate regime, but with the processes that link the atmosphere to the biological and nonbiological systems on the Earth's surface.

As can be seen from *Figure 8.20*, the diversity in the primary anthropogenic sources of causative species is surprisingly small, fossil and biofuel combustion and industrial processes being dominant. There is enough evidence to state that a significant decrease in the rate of fossil fuel combustion would stabilize the atmospheric radiation balance, improve visibility, hinder smog formation, and minimize acidic precipitation and its effects. Stratospheric modification is best constrained by devising alternatives to the use of chlorofluorocarbons as propellants and refrigerants. The involvement of transition metal chemistry in droplets and aerosol particles is not yet understood, but better control of their emission from combustion and smelting operations may be desirable to alleviate acidic deposition problems. The problem of corrosion would have to be handled differently since natural sources are responsible for a large portion of the corrosive agents. It might thus be most effective to reduce corrosion by improved selection and treatment of materials rather than by control of emission sources.

Many of the thoughts expressed in this chapter are not sufficiently quantified and tested by atmospheric observations, especially in the important tropical regions. It seems useful, therefore, to finish with a list of recommendations for future research in atmospheric chemistry:

(1) Many of the analytical techniques in atmospheric chemistry are highly sophisticated and require the measurement of very low concentrations of highly reactive species. To obtain reliable data these techniques need to be verified and compared. International programs fulfill a very important function here by supporting technology transfer and cross-calibration experiments [79].

(2) The sources of atmospheric constituents resulting from biological processes and related activities (e.g., biomass burning) must be better quantified. Biological sources should not be treated as "black boxes"; it is essential that the

processes responsible for these emissions be studied so that we can make the step from a descriptive to a predictive understanding. Studies are especially important in the tropical regions, since current evidence suggests that large biological emissions occur in these regions.

(3) The fate of trace species injected into the atmosphere should continue to be investigated. The loss processes include both chemical transformations in the atmosphere and the physical processes of transport (especially long-range transport) and of scavenging and deposition. Attention must be paid both to homogeneous (gas-phase) processes and to heterogeneous processes that involve the interaction of gases, particles, and cloud droplets.

(4) Models are essential in helping us to understand complex interactive systems like the chemical processes in the Earth's atmosphere. We have to develop models that attempt to represent the entire physical and chemical system that regulates the composition of the atmosphere, comparable to the general circulation models used in meteorology. Submodels, e.g., those that describe photochemical reactions in the troposphere, need to be developed or refined. It is essential that field experiments be devised to provide critical tests of these models and that a continuous exchange of information between model development and critical experiments be maintained.

(5) In order to understand global change, we need to investigate the historical record of atmospheric composition preserved in ice cores, tree rings, and other media, and to establish measurement programs to document current changes in the atmospheric concentrations of key substances. The explanation of the records requires the involvement of environmental historians. Richards (Chapter 2, this volume) has given ample evidence of man's role in changing the world's ecosystems, a role that has altered the sources of several important atmospheric species, such as CO_2, CH_4, and N_2O.

(6) Studies of the impacts of changing atmospheric chemistry on the biosphere are of vital importance. Problems like the complicated interactions that lead to the dramatic forest losses in central Europe can only be solved by such collaboration. Major breakthroughs in understanding cannot be predicted and various research approaches should be attempted.

Atmospheric chemistry is clearly an international and strongly interdisciplinary science with substantial economic and political importance, and a science that deals with an essential common resource for humanity. As a recent branch of the atmospheric sciences, its foundations are still being built. Many questions need to be answered by long-term, dedicated research. A major, international, collaborative program in global tropospheric chemistry has recently been proposed by a panel of the US National Academy of Sciences [79]. The International Council of Scientific Unions (ICSU) has taken the initiative for an even broader interdisciplinary program on biogeochemical cycles [80]. It is our hope that programs such as these will contribute not only to a much better understanding of the interactions between humankind and our world, but also to greater understanding and collaboration among nations.

Appendixes:

Atmospheric reaction sequences

Appendix 8.1. Tropospheric oxidizers

The principal oxidizing species in the lower atmosphere are O_3 and $OH^\cdot$. The former is produced in the troposphere by a reaction sequence initiated by the peroxyl radical oxidation of NO:

$$NO + RO_2^\cdot \rightarrow NO_2 + RO^\cdot$$

continued by the photolysis of NO_2:

$$NO_2 \xrightarrow{h\nu} NO + O^\cdot \qquad (\lambda < 420\ \text{nm})$$

and concluded by

$$O^\cdot + O_2 \xrightarrow{M} O_3$$

O_3 is, however, also transported downward from the stratosphere where it is formed through dissociation of molecular O_2 (see *Figure 8.3* and Appendix 8.6). The contributions from *in-situ* tropospheric O_3 formation and stratospheric transport may be comparable, but much uncertainty exists [29]. Tropospheric destruction of O_3 also occurs and this constitutes the primary source of $OH^{\cdot}$ [81]:

$$O_3 \xrightarrow{h\nu} O_2 + O^* \qquad (\lambda \leq 310\text{ nm})$$

$$O^* + H_2O \rightarrow 2OH^{\cdot}$$

A second $OH^{\cdot}$ source is oxygenated organic compounds. HCHO, the simplest example, photolyzes to produce hydroperoxyl radicals ($HO_2^{\cdot}$):

$$HCHO \xrightarrow{h\nu} H^{\cdot} + CHO^{\cdot} \qquad (\lambda < 335\text{ nm})$$

$$H^{\cdot} + O_2 \xrightarrow{M} HO_2^{\cdot}$$

$$CHO^{\cdot} + O_2 \rightarrow HO_2^{\cdot} + CO$$

The $HO_2^{\cdot}$ radicals give $OH^{\cdot}$ by disproportionation and photodissociation of the resultant H_2O_2:

$$HO_2^{\cdot} + HO_2^{\cdot} \rightarrow H_2O_2 + O_2$$

$$H_2O_2 \xrightarrow{h\nu} OH^{\cdot} + OH^{\cdot} \qquad (\lambda < 350\text{ nm})$$

and by oxidation of NO:

$$NO + HO_2^{\cdot} \rightarrow NO_2 + OH^{\cdot}$$

The alkyl peroxyl radicals ($RO_2^{\cdot}$), produced by similar chemical chains from the higher aldehydes and from ketones, also generate $RO^{\cdot}$ and oxidize NO.

Appendix 8.2. Hydrocarbon oxidation

Oxidation of CH_4 is not only a potentially significant source of CO, but the atmospheric concentrations of $OH^{\cdot}$ may largely be determined by the particular reaction paths that are followed during CH_4 oxidation [29, 82]. NO plays an important role in determining the favored reaction sequence. With enough NO present (>10 p.p.t.v.) the reaction path, leading to HCHO, is as follows:

$$CH_4 + OH^{\cdot} \rightarrow CH_3^{\cdot} + H_2O$$

$$CH_3^{\cdot} + O_2 \xrightarrow{M} CH_3O_2^{\cdot}$$

$$CH_3O_2^{\cdot} + NO \rightarrow CH_3O^{\cdot} + NO_2$$

$$CH_3O^{\cdot} + O_2 \rightarrow HCHO + HO_2^{\cdot}$$

$$HO_2^{\cdot} + NO \rightarrow OH^{\cdot} + NO_2$$

$$2(NO_2 \xrightarrow{h\nu} NO + O^{\cdot}) \qquad (\lambda < 420\text{ nm})$$

$$2(O^{\cdot} + O_2 \xrightarrow{M} O_3)$$

$$\text{net: } CH_4 + 4O_2 \rightarrow HCHO + H_2O + 2O_3$$

With little NO present, CH_4 oxidation may follow the pathway:

$$CH_4 + OH^{\cdot} \rightarrow CH_3^{\cdot} + H_2O$$

$$CH_3^{\cdot} + O_2 \xrightarrow{M} CH_3O_2^{\cdot}$$

$$CH_3O_2^{\cdot} + HO_2^{\cdot} \rightarrow CH_3O_2H + O_2$$

$$CH_3O_2H \xrightarrow{h\nu} CH_3O^{\cdot} + OH^{\cdot} \qquad (\lambda < 350\text{ nm})$$

$$CH_3O^{\cdot} + O_2 \rightarrow HCHO + HO_2^{\cdot}$$

$$\text{net: } CH_4 + O_2 \rightarrow HCHO + H_2O$$

However, the photolysis of methyl hydroperoxide (CH_3O_2H) is slow, resulting in a residence time of one week. Thus CH_3O_2H may be rained out of the atmosphere or react with the Earth's surface or aerosol particles. If this is the case, the oxidation of CH_4 leads to the loss of two odd hydrogen radicals ($OH^{\cdot}$ and $HO_2^{\cdot}$) and no CO is formed in the atmosphere. An even larger loss of odd hydrogen in NO-poor environments occurs because of the fast catalytic subcycle:

$$CH_3O_2^{\cdot} + HO_2^{\cdot} \rightarrow CH_3O_2H + O_2$$

$$CH_3O_2H + OH^{\cdot} \rightarrow CH_3O_2^{\cdot} + H_2O$$

$$\text{net: } OH^{\cdot} + HO_2^{\cdot} \rightarrow 2O_2$$

Depending on the concentrations of NO, the further oxidation of HCHO to CO and CO_2 may follow one of two routes:

$$HCHO \xrightarrow{h\nu} H^{\cdot} + HCO^{\cdot}$$

$$H^{\cdot} + O_2 \xrightarrow{M} HO_2^{\cdot}$$

$$HCO^{\cdot} + O_2 \rightarrow HO_2^{\cdot} + CO$$

$$CO + OH^{\cdot} \rightarrow H^{\cdot} + CO_2$$

$$H^{\cdot} + O_2 \xrightarrow{M} HO_2^{\cdot}$$

$$3(HO_2^{\cdot} + NO \rightarrow OH^{\cdot} + NO_2)$$

$$3(NO_2 \xrightarrow{h\nu} NO + O^{\cdot})$$

$$3(O^{\cdot} + O_2 \xrightarrow{M} O_3)$$

$$\text{net: } HCHO + 6O_2 \rightarrow CO_2 + 2OH^{\cdot} + 3O_3$$

or

$$HCHO \xrightarrow{h\nu} H^{\cdot} + HCO^{\cdot}$$

$$H^{\cdot} + O_2 \rightarrow HO_2^{\cdot}$$

$$HCO^{\cdot} + O_2 \rightarrow CO + HO_2^{\cdot}$$

$$CO + OH^{\cdot} \rightarrow H^{\cdot} + CO_2$$

$$2(H^{\cdot} + O_2 \xrightarrow{M} HO_2^{\cdot})$$

$$3(HO_2^{\cdot} + O_3 \rightarrow OH^{\cdot} + 2O_2)$$

$$\text{net: } HCHO + 3O_3 \rightarrow CO_2 + 3O_2 + 2OH^{\cdot}$$

Additional oxidation routes are also possible. However, the important implications are that in the presence of enough NO there is a net production of $3.7O_3$ molecules and $0.5OH^{\cdot}$ radicals per CH_4 molecule oxidized, while otherwise a net loss of $1.7O_3$ and 3.5 hydroxyl radicals may occur.

More complex hydrocarbons follow analogous but even more extensive reaction sequences. If we use the symbol $R^{\cdot}$ to represent any saturated hydrocarbon chain radical [e.g., the methyl radical ($CH_3^{\cdot}$), the ethyl radical ($CH_3CH_2^{\cdot}$), etc., also called alkyl radicals], we can write the reaction between the hydrocarbon molecule and $OH^{\cdot}$ as

$$RH + OH^{\cdot} \rightarrow R^{\cdot} + H_2O$$

In the Earth's atmosphere, the $R^{\cdot}$ will promptly add molecular oxygen to form an alkylperoxyl radical, $RO_2^{\cdot}$:

$$R^{\cdot} + O_2 \xrightarrow{M} RO_2^{\cdot}$$

$RO_2^{\cdot}$ radicals are not particularly reactive and the vigor of the chemistry depends upon the rate at which they can be recycled to oxidizing species, which can be accomplished by disproportionation

$$RO_2^{\cdot} + R'O_2^{\cdot} \rightarrow ROOR' + O_2$$

followed by photolysis

$$ROOR' \xrightarrow{h\nu} RO^{\cdot} + R'O^{\cdot} \qquad (\lambda < 350\,\text{nm})$$

A potentially much more effective reaction is that with NO:

$$RO_2^{\cdot} + NO \rightarrow RO^{\cdot} + NO_2$$

but this requires the presence of sufficient NO, as in the CH_4 oxidation chain. The $RO^{\cdot}$ is rapidly converted into a carbonyl molecule (a molecule with an oxygen atom doubly bonded to a carbon atom) by reaction with molecular oxygen:

$$RO^{\cdot} + O_2 \rightarrow R_{-1}CHO + HO_2^{\cdot}$$

Thus, if the initial saturated hydrocarbon is ethane (CH_3CH_3), this reaction sequence produces acetaldehyde (CH_3CHO); if propane ($CH_3CH_2CH_3$), the product will be acetone [$CH_3C(O)CH_3$] or propionaldehyde (CH_3CH_2CHO), depending on the point of initial $OH^{\cdot}$ attack.

In addition to alkanes (compounds with saturated hydrocarbon chains), the atmosphere contains olefins (compounds with unsaturated hydrocarbon chains) and aromatics (compounds with aromatic ring structures). These compounds arise from both natural and anthropogenic sources and have relatively short atmospheric lifetimes, especially during summer, and intricate chemical reaction sequences [83]. Large natural emissions of isoprene, terpenes, and other olefinic hydrocarbons emanate from vegetation, and are of the same magnitude as those for CH_4. These fluxes are probably decreasing due to deforestation in the tropics.

Appendix 8.3. Atmospheric nitrogen oxide chemistry

The reactions of NO and NO_2 (together denoted NO_x) in the atmosphere are diverse and very important [84], because NO_x molecules play an important catalytic role in many photochemical reactions. In the troposphere, NO_x enhances the formation of O_3; in the stratosphere, the opposite is the case. HNO_3 is formed during daytime according to the reaction

$$NO_2 + OH^{\cdot} \xrightarrow{M} HNO_3$$

This is an important sink or temporary reservoir for the catalysts NO and NO_2. During nighttime, HNO_3 forms through a mixture of gas-phase and surface reactions:

$$NO_2 + O_3 \rightarrow NO_3 + O_2$$
$$NO_3 + NO_2 \leftrightarrow N_2O_5$$
$$N_2O_5 + H_2O_{(aq)} \rightarrow 2HNO_3$$

An interesting, and potentially important, interaction between carbon and nitrogen gases in the atmosphere is the formation of various organic nitrates during the oxidation of hydrocarbons [29, 84, 85]. The most important of these are probably the peroxyacyl nitrates, especially peroxyacetyl nitrate [PAN, $CH_3C(O)O_2NO_2$]. PAN thermally decomposes in the atmosphere according to the reaction:

$$CH_3C(O)O_2NO_2 \rightarrow CH_3C(O)O_2^{\cdot} + NO_2$$

liberating "captured" NO_x radicals. PAN is rather stable against thermal decomposition above about 3 km, so that long-range transport through the free troposphere may occur. PAN is often found in substantial concentrations in polluted atmospheres and is also formed in the oxidation cycle of C_2H_6 in the upper troposphere, probably through the involvement of NO_x produced by lightning [85].

It has been proposed that NO is also generated by a reaction sequence initiated by $OH^{\cdot}$ attack on NH_3, but this is probably not very significant. NH_3 is more important as a neutralizing compound in rainwater and as a nutrient gas for the biosphere.

Appendix 8.4. Atmospheric sulfur chemistry

The principal naturally emitted atmospheric sulfur compounds, H_2S and CH_3SCH_3, react to produce SO_2. For

H_2S the proposed reaction sequence is [86]:

$$H_2S + OH^{\cdot} \rightarrow HS^{\cdot} + H_2O$$

$$HS^{\cdot} + O_2 \xrightarrow{M} HSO_2^{\cdot}$$

$$HSO_2^{\cdot} + O_2 \rightarrow SO_2 + HO_2^{\cdot}$$

The reaction chain for CH_3SCH_3 is [87]:

$$CH_3SCH_3 + OH^{\cdot} \rightarrow CH_3SCH_2^{\cdot} + H_2O$$

$$CH_3SCH_2^{\cdot} + O_2 \xrightarrow{M} CH_3SCH_2O_2^{\cdot}$$

$$CH_3SCH_2O_2^{\cdot} + NO \rightarrow CH_3SCH_2O^{\cdot} + NO_2$$

$$CH_3SCH_2O^{\cdot} \rightarrow CH_3S^{\cdot} + HCHO$$

$$CH_3S^{\cdot} + O_2 \xrightarrow{M} CH_3SO_2^{\cdot}$$

$$CH_3SO_2^{\cdot} \xrightarrow{\text{intermediate steps}} CH_3SO_3H, SO_2$$

The most common photochemical sequence for gas-phase SO_2 oxidation has been identified [88] as:

$$SO_2 + OH^{\cdot} \xrightarrow{M} HSO_3^{\cdot}$$

$$HSO_3^{\cdot} + O_2 \rightarrow SO_3 + HO_2^{\cdot}$$

$$SO_3 + H_2O \rightarrow H_2SO_4$$

Appendix 8.5. Atmospheric droplet chemistry

All reactions described in Appendix 8.5 occur in aqueous solution. Chemical processes in liquid water droplets are governed in large part by the prevailing acidity. This is initially established by equilibration of the droplet with atmospheric CO_2:

$$CO_{2(g)} \rightleftharpoons CO_{2(aq)}$$

$$CO_{2(aq)} + H_2O \rightleftharpoons H^+ + HCO_3^-$$

The addition of strong mineral acids raises the acidity (i.e., lowers the pH), since the acid equilibria produce protons and the correspondng anions:

$$H_2SO_4 \rightleftharpoons H^+ + HSO_4^-$$

$$HSO_4^- \rightleftharpoons H^+ + SO_4^{2-}$$

$$HNO_3 \rightleftharpoons H^+ + NO_3^-$$

$$HCl \rightleftharpoons H^+ + Cl^-$$

Similar processes occur with the organic acids known to be present in the droplets:

$$RC(O)OH \rightleftharpoons H^+ + RC(O)O^-$$

but the equilibria are such that the acids are only partly ionized.

Once the electroytic properties and the acidity of the solution are fixed, then the equilibria between the liquid and gas phases are established and further reactions are governed by the oxidants present. H_2O_2, which may be the most important, serves both as an oxidant and as a source of solution $OH^{\cdot}$ when it is photolyzed:

$$H_2O_2 \xrightarrow{h\nu} HO^{\cdot} + HO^{\cdot} \quad (\lambda < 380\ \text{nm})$$

O_3 dissolved in water is also a source of $OH^{\cdot}$:

$$O_3 + HO_2^{\cdot} \rightarrow HO^{\cdot} + 2O_2$$

$$O_3 \xrightarrow{h\nu} O_2 + O^* \quad (\lambda \leq 310\ \text{nm})$$

$$O^* + H_2O \rightarrow H_2O_2$$

Appendix 8.6. Stratospheric chemistry

Since the beginning of the 1970s it has become increasingly clear that a number of human activities can lead to global changes in the amount of stratospheric O_3. Initial attention was directed to pollution of the stratosphere by direct injections of NO from high-flying aircraft. Earlier, Crutzen [89] had proposed that NO_x ($NO + NO_2$) would catalyze the destruction of O_3 and limit its stratospheric abundance by a simple set of photochemical reactions:

$$O_3 \xrightarrow{h\nu} O^{\cdot} + O_2 \quad (\lambda < 900\ \text{nm})$$

$$O^{\cdot} + NO_2 \rightarrow NO + O_2$$

$$NO + O_3 \rightarrow NO_2 + O_2$$

$$\text{net:}\ 2O_3 \rightarrow 3O_2$$

In the stratosphere below 40 km this catalytic chain of reactions largely balances the formation of O_3, which occurs through the reaction sequence:

$$O_2 \xrightarrow{h\nu} 2O^{\cdot} \quad (\lambda < 240\ \text{nm})$$

$$2(O^{\cdot} + O_2 \xrightarrow{M} O_3)$$

$$\text{net:}\quad 3O_2 \rightarrow 2O_3$$

This 10–40 km altitude region contains almost all the O_3 of the atmosphere. The main source of NO_x in the stratosphere is probably the oxidation of N_2O via

$$O_3 \xrightarrow{h\nu} O^* + O_2 \qquad (\lambda \leq 310\,\text{nm})$$

$$O^* + N_2O \rightarrow 2NO$$

NO is formed as an intermediate in biological nitrogen transformations (nitrification or denitrification) and is also produced during various combustion processes (see *Table 8.1).* The oxides of nitrogen, NO_x, play a remarkable catalytic role in the O_3 balance of the atmosphere. Above about 25 km the net effect of NO_x additions to the stratosphere is to lower the O_3 concentration by the set of reactions presented already. However, below about 25 km, NO_x protects O_3 from destruction. An important reason for this is the set of reactions:

$$HO_2^{\cdot} + NO \rightarrow OH^{\cdot} + NO_2$$

$$NO_2 \xrightarrow{h\nu} NO + O^{\cdot}$$

$$O^{\cdot} + O_2 \xrightarrow{M} O_3$$

$$\text{net:} \quad HO_2^{\cdot} + O_2 \rightarrow OH^{\cdot} + O_3$$

In the lower stratosphere the chain of reactions tends to counteract the destruction of O_3 by the catalytic reaction pair:

$$OH^{\cdot} + O_3 \rightarrow HO_2^{\cdot} + O_2$$

$$HO_2^{\cdot} + O_3 \rightarrow OH^{\cdot} + 2O_2$$

$$\text{net:} \quad 2O_3 \rightarrow 3O_2$$

by deferring it into the chain:

$$OH^{\cdot} + O_3 \rightarrow HO_2^{\cdot} + O_2$$

$$HO_2^{\cdot} + NO \rightarrow OH^{\cdot} + NO_2$$

$$NO_2 \xrightarrow{h\nu} NO + O^{\cdot}$$

$$O^{\cdot} + O_2 \xrightarrow{M} O_3$$

no net chemical effect.

NO_x in the stratosphere also interacts with $Cl^{\cdot}$ and $ClO^{\cdot}$. As with NO_x, chlorine atoms and chlorine monoxide molecules participate in an effective catalytic chain of reactions that converts ozone back to molecular oxygen:

$$O_3 \xrightarrow{h\nu} O^{\cdot} + O_2$$

$$O^{\cdot} + ClO \rightarrow Cl^{\cdot} + O_2$$

$$Cl^{\cdot} + O_3 \rightarrow ClO^{\cdot} + O_2$$

$$\text{net:} \quad 2O_3 \rightarrow 3O_2$$

whereby one mole of $ClO^{\cdot}$ is about three times more efficient than one mole of NO_2. The abundance of $ClO^{\cdot}$ radicals in the stratosphere is increasing rapidly with the release of chlorofluorocarbons. The presence of NO in the stratosphere makes the $ClO^{\cdot}$ catalytic cycle less effective, as the reactions

$$NO + ClO^{\cdot} \rightarrow Cl^{\cdot} + NO_2$$

$$Cl^{\cdot} + CH_4 \rightarrow HCl + CH_3^{\cdot}$$

transform the catalysts $Cl^{\cdot}$ and $ClO^{\cdot}$ into HCl, which does not react photochemically with O_3. Furthermore, since the titration reaction

$$ClO^{\cdot} + NO_2 \xrightarrow{M} ClNO_3$$

combines some $ClO^{\cdot}$ and some NO_2 as nonreactive $ClNO_3$, it is clear that O_3 removal by additons of chlorine to the stratosphere is mitigated by NO_x interference in the chlorine cycle. If additions of chlorine to the stratosphere do not override those of NO_x, the reductions in total stratospheric O_3 are calculated to be not too large, because O_3 reductions above 30 km are compensated by increases below. The potentially significant climatic consequences of these processes have been discussed. At high chlorine concentrations in the stratosphere, severe O_3 depletions are foreseen [61]. Expansions in the manufacturing of chlorocarbons worldwide should, therefore, be watched with utmost concern.

The photochemistry of the stratosphere is complicated because the odd chlorine, nitrogen, and hydrogen radicals react with each other. These reactions are especially important below about 30 km, and lead to the formation of HNO_3, pernitric acid (HNO_4), hypochlorous acid (HClO), and chlorine nitrate ($ClNO_3$). Except for HNO_3, none of these compounds has been detected in the lower stratosphere, so an important test of stratospheric photochemistry is lacking.

Notes and references

[1] Whitby, K. T. (1980), Aerosol formation in urban plumes, *Annals of the New York Academy of Sciences,* **338**, 258–275.

[2] Keeling, C. D., Bacastow, R. B., and Whorf, T. P. (1982), Measurements of the concentration of carbon dioxide at Mauna Loa Observatory, Hawaii, in W. C. Clark (Ed), *Carbon Dioxide Review: 1982,* pp 377–385, (Clarendon Press, Oxford, UK).

[3] Weiss, R. F. (1981), The temporal and spatial distribution of tropospheric nitrous oxide, *Journal of Geophysical Research,* **86**, 7185–7195.

[4] Rasmussen, R. A. and Khalil, M. A. K. (1981), Atmospheric methane (CH_4): trends and seasonal cycles, *Journal of Geophysical Research,* **86**, 9826–

9832; Blake, D. R., Mayer, E. W., Tyler, S. C., Makide, Y., Montague, D. C., and Rowland, F. S. (1982), Global increases in atmospheric methane concentration between 1978 and 1980, *Geophysical Research Letters*, **9**, 477–480.

[5] World Meteorological Organization (1982), *Report on Potential Climatic Effects of Ozone and Other Minor Trace Gases*, Report No. 17 (WMO, Geneva).

[6] Neftel, A., Oeschger, H., Schwander, J., Stauffer, B., and Zumbrunn, R. (1982), Ice core sample measurements give atmospheric CO_2 content during the past 40 000 years, *Nature*, **295**, 220–222.

[7] Craig, H. and Chou, P. C. (1982), Methane: the record in polar ice cores, *Geophysical Research Letters*, **9**, 1221–1224.

[8] Larsen, R. J. (1984), *Worldwide Deposition of ^{90}Sr through 1982,* US Department of Energy Report EML-430 (DOE, Washington, DC).

[9] Castleman, A. W., Jr., Munkelwitz, H. R., and Manowitz, B. (1973), Contribution of volcanic sulphur compounds to the stratospheric aerosol layer, *Nature*, **244**, 345–346.

[10] Warmbdt, W. (1979), Ergebnisse Langjähriger Messungen des bodennahen Ozons in der DDR, *Zeitschrift für Meteorologie*, **29**, 24–31.

[11] Attmannspacher, W. (1982), The behavior of atmospheric ozone during the last 15 years, based on results of ozone sounding at Hohenpeissenberg, in *Proc. Workshop Biological Effects of UV-B Radiation*, (BPT Berichte/GSF, München, FDR).

[12] Crutzen, P. J. and Gidel, L. T. (1983), A two-dimensional photochemical model of the atmosphere. 2: The tropospheric budgets of the anthropogenic chlorocarbons, CO, CH_4, CH_3Cl, and the effects of various NO_x sources on tropospheric ozone, *Journal of Geophysical Research*, **88**, 6641–6661.

[13] Crutzen, P. J. (1986), The role of the tropics in atmospheric photochemistry, in R. Dickinson (Ed), *The Geophysiology of Amazonia Vegetation and Climate Interactions* (Wiley, New York).

[14] Grosjean, D. and Wright, B. (1983), Carbonyls in urban fog, ice fog, cloudwater, and rainwater, *Atmospheric Environment*, **17**, 2093–2096.

[15] Gill, P. S., Graedel, T. E., and Weschler, C. J. (1983), Organic films on atmospheric aerosol particles, fog droplets, cloud droplets, raindrops, and snowflakes, *Reviews of Geophysics and Space Physics*, **21**, 903–920.

[16] Hill, A. C. (1971), Vegetation: a sink for atmospheric pollutants, *Journal of the Air Pollution Control Association*, **21**, 341–346.

[17] Ivanov, M. V. and Freney, J. R. (Eds) (1983), *The Global Biogeochemical Sulphur Cycle*, SCOPE Report No. 19 (John Wiley, Chichester, UK).

[18] Rice, D. W., Cappell, R. J., Kinsolving, W., and Laskowski, J. J. (1980), Indoor corrosion of metals, *Journal of the Electrochemical Society*, **127**, 891–901; Graedel, T. E., Franey, J. P., and Kammlott, G. W. (1983), The corrosion of copper by atmospheric sulphurous gases, *Corrosion Science*, **23**, 1141–1152.

[19] Khalil, M. A. K. and Rasmussen, R. A. (1984), Global sources, lifetimes, and mass balances of carbonyl sulphide (COS) and carbon disulphide (CS_2) in the earth's atmosphere, *Atmospheric Environment*, **18**, 1805–1813.

[20] Gauri, K. L. and Holdren, G. C., Jr., (1981), Pollutant effects on stone monuments, *Environmental Science and Technology*, **15**, 386–390.

[21] Barnes, R. A. and Lee, D. O. (1978), Visibility in London and the long distance transport of atmospheric sulphur, *Atmospheric Environment*, **12**, 791–794.

[22] Crutzen, P. J. (1976), The possible importance of CSO for the sulfate layer of the stratosphere, *Geophysical Research Letters*, **3**, 73–76.

[23] Barton, K. and Bartonova, Z. (1969), Mechanism and kinetics of corrosion in a humid atmosphere in the presence of hydrogen chloride vapours, in Ya. M. Kolotyrkin (Ed) *Proceedings Third International Congress on Metallic Corrosion*, pp 483–496 (Moscow, distributed by Swets-Zeitlinger, Amsterdam).

[24] Graedel, T. E. and Weschler, C. J. (1981), Chemistry in aqueous atmospheric aerosols and raindrops, *Reviews of Geophysics and Space Physics*, **19**, 505–539; Chameides, W. L. and Davis, D. D. (1982), The free radical chemistry of cloud droplets and its impact upon the composition of rain, *Journal of Geophysical Research*, **87**, 4863–4877.

[25] Martin, L. R. (1983), Kinetic studies of sulfite oxidation in aqueous solution, in J. G. Calvert (Ed), *Acid Precipitation, SO_2, NO, and NO_2 Oxidation Mechanisms: Atmospheric Considerations* (Ann Arbor Science Pubs., Ann Arbor, MI).

[26] Keyser, T. R., Natusch, D. F. S., Evans, C. A., Jr., and Linton, R. W. (1978), Characterizing the surfaces of environmental particles, *Environmental Science and Technology*, **12**, 768–773.

[27] Keene, W. C., Galloway, J. N., and Holden, J. D., Jr. (1983), Measurement of weak organic acidity in precipitation from remote areas of the world, *Journal of Geophysical Research*, **88**, 5122–5130.

[28] Lee, Y.-N. and Schwartz, S. E. (1981), Evaluation of rate of uptake of nitrogen dioxide by atmospheric and surface liquid water, *Journal of Geophysical Research*, **86**, 11971–11983.

[29] Crutzen, P. J. (1983), Atmospheric interactions – homogeneous gas reactions of C, N, and S containing compounds, in B. Bolin and R. B. Cook (Eds), *The Major Biogeochemical Cycles and Their Interactions*, SCOPE Report No. 21 (John Wiley, Chichester, UK).

[30] Granat, L., Rodhe, H., and Hallberg, R. O. (1976), The global sulphur cycle, in B. H. Svensson and R. Söderlund (Eds), *Nitrogen, Phosphorus, and Sulphur – Global Cycles*, SCOPE Report No. 7., *Ecology Bulletin (Stockholm)*, **22**, 89–134.

[31] Bonsang, B. (1980), *Cycle atmosphérique du soufre d'origine marine,* Thése de Doctorat (Université de Picardie, France); Andreae, M. O. and Raemdonck, H. (1983), Dimethylsulfide in the surface ocean and the marine atmosphere: a global view, *Science,* **221**, 744–747; Bingemer, H. (1984). *Dimethylsulfid in Ozean und mariner Atmosphäre–Experimentelle Untersuchung einer natürlichen Schwefelguelle für die Atmosphär* (Berichte des Instituts für Meteorologie, Frankfurt University, FDR).

[32] Blanchard, D. C. and Woodcock, A. H. (1980), The production, concentration, and vertical distribution of the sea-salt aerosol, *Annals of the New York Academy of Sciences,* **338**, 330–347.

[33] Altshuller, A. P. (1983), Review: Natural volatile organic substances and their effect on air quality in the United States, *Atmospheric Environment,* **17**, 2131–2165.

[34] Hooker, C. L., Westberg, H. H., and Sheppard, J. C. (1985), Determination of carbon balances for smog chamber terpene oxidation experiments using a carbon-14 tracer technique, *Journal of Atmospheric Chemistry,* **2**, 307.

[35] Woodwell, G. M., Whittaker, R. H., Reiners, W. A., Likens, G. E., Delwiche, C. C., and Botkin, D. B. (1978), The biota and the world carbon budget, *Science,* **199**, 141–146.

[36] Broecker, W. S., Peng, T.-H., and Engh, R. (1980), Modeling the carbon system, *Radiocarbon,* **22**, 565–598.

[37] Seiler, W. and Crutzen, P. J. (1980), Estimates of gross and net fluxes of carbon between the biosphere and the atmosphere from biomass burning, *Climatic Change,* **2**, 207–247; Olsson, J. S. (1982), Earth's vegetation and atmospheric carbon dioxide, in W. C. Clark (Ed), *Carbon Dioxide Review, 1982,* pp 388–395 (Clarendon Press, Oxford, UK).

[38] Moore, B., Boone, R. D., Hobbie, J. E., Houghton, R. A., Melillo, J. M., Peterson, B. J., Shaver, G. R., Vörösmarty, C. J., and Woodwell, G. M. (1981), A simple model for analysis of the role of terrestrial ecosystems in the global carbon budget, in B. Bolin (Ed) *Carbon Cycle Modelling,* SCOPE Report No 16, (John Wiley, Chichester, UK).

[39] Goudriaan J. and Ketner, P. (1984), A simulation study for the global carbon cycle, including man's impact on the biosphere, *Climatic Change,* **6**, 167–192.

[40] Clark, W. C. (Ed) (1982), *Carbon Dioxide Review: 1982* (Clarendon Press, Oxford, UK).

[41] Liss, P. S. and Crane, A. J. (1983), *Man-made Carbon Dioxide and Climate Change* (Geo Books, Norwich, UK).

[42] We use "animal" here in the sense of a "living organism endowed with voluntary motion". This definition is meant to include microbes and insects and to exclude plants.

[43] Lenhard, U. and Gravenhorst, G. (1980), Evaluation of ammonia fluxes into the free atmosphere over Western Germany, *Tellus,* **32**, 48–55.

[44] United Nations Population Division, personal communication, 1984.

[45] Crutzen, P. J., Heidt, L. E., Krasnec, J. P., Pollack, W. H., and Seiler, W. (1979), Biomass burning as a source of atmospheric gases CO, H_2, N_2O, NO, CH_3Cl, and COS, *Nature,* **282**, 253–256.

[46] Ward, D. E., McMahon, E. K., and Johansen, R. W. (1976), *An Update on Particulate Emissions from Forest Fires,* Paper 76–22, 69th Annual Meeting, Air Pollution Control Association, Portland, OR.

[47] Terry, R. E., Tate, R. L., III, and Duxburg, A. (1980), *Nitrous Oxide Emissions from Drained, Cultivated Organic Soils of South Florida,* 73rd Annual Meeting, Air Pollution Control Association, Montreal, June 22–27.

[48] Rotty, R. M. (1983), Distribution of and changes in industrial carbon dioxide production, *Journal of Geophysical Research,* **88**, 1301–1308.

[49] Seiler, W. (1974), The cycle of atmospheric CO, *Tellus,* **26**, 117–135.

[50] Logan, J. A., Prather, M. J., Wofsy, S. C., and McElroy, M. B. (1981), Tropospheric chemistry: a global perspective, *Journal of Geophysical Research,* **86**, 7210–7254.

[51] Duce, R. A. (1978), Speculation on the budget of particulate and vapor phase non-methane organic carbon in the global troposhere, *Pure and Applied Geophysics,* **116**, 244–273.

[52] Söderlund, R. and Svensson, B. H. (1976), The global nitrogen cycle, in B. H. Svensson and R. Söderlund (Eds), *Nitrogen, Phosphorus, and Sulphur – Global Cycles,* SCOPE Report No 7, *Ecological Bulletin (Stockholm),* **22**, 23–73.

[53] Cullis, C. F. and Hirschler, M. M. (1980), Atmospheric sulphur: natural and man-made sources, *Atmospheric Environment,* **14**, 1263–1278; Möller, D. (1984), Estimation of the global man-made sulphur emission, *Atmospheric Environment,* **18**, 19–27.

[54] Information on the structure and atmospheric cycle of soot is given in Ogren, J. A. and Charlson, R. J. (1983), Elemental carbon in the atmosphere: cycle and lifetime, *Tellus,* **35B**, 241–254.

[55] Block, C. and Dams, R. (1975), Inorganic composition of Belgian coals and coal ashes, *Environmental Science and Technology,* **9**, 146–150.

[56] Nriagu, J. O. (1979), Global inventory of natural and anthropogenic emissions of trace metals to the atmosphere, *Nature,* **279**, 409–411.

[57] Summaries of global emissions of halocarbon propellants are issued annually by the Chemical Manufacturers Association, e.g. CMA (1983), *World Production and Release of Chlorofluorocarbons 11 and 12 through 1982* (CMA, Washington, DC).

[58] National Research Council (1976), *Halocarbons* (National Academy Press, Washington, DC).

[59] Heath, D. F., Krueger, A. J., and Crutzen, P. J. (1977), Solar proton event: influence on stratospheric ozone, *Science*, **197**, 886–889.

[60] Heath, D. F. and Schlesinger, B. M. (1984), Evidence for a decrease in tropical stratospheric ozone following the eruption of El Chichon, in *Proceedings of the Quadrennial Ozone Symposium, 3–7 Sept.* (Kassandra, Greece).

[61] Prather, M. J., McElroy, M. B., and Wofsy, S. C. (1984), Reductions in ozone at high concentrations of stratospheric halogens, *Nature*, **312**, 227–231.

[62] Clarke, A. D., Charlson, R. J., and Radke, L. F. (1984), Airborne observations of arctic aerosol, 4. Optical properties of Arctic Haze, *Geophysical Research Letters*, **11**, 405–408; Rosen, H. and Hansen, A. D. A. (1984), Vertical distribution of particulate carbon, sulfur, and bromine in the Arctic haze and comparison with ground-level measurements at Barrow, Alaska, *Geophysical Research Letters*, **11**, 381–384; Valero, F. P. J., Ackerman, T. P., and Gore, W. J. Y. (1984), The absorption of solar radiation by the Arctic atmosphere during the hazy season and its effect on the radiation balance, *Geophysical Research Letters*, **11**, 465–468.

[63] Crutzen, P. J. and Birks, J. W. (1982), The atmosphere after a nuclear war: twilight at noon, *Ambio*, **11**, 114–125; Turco, R. P., Toon, O. B., Ackerman, T. P., Pollack, J. B., and Sagan, C. (1983), Nuclear winter: global consequences of multiple nuclear explosions, *Science*, **222**, 1283–1292; Ehrlich, P. R., Harte, J., Harwell, M. A., Raven, P. H., Sagan, C., Woodwell, G. M., Berry, J., Ayensu, E. S., Ehrlich, A. H., Eisner, T., Gould, S. J., Grover, H. D., Herrera, R., May, R. M., Mayr, E., McKay, C. P., Mooney, H. A., Myers, N., Pimentel, D., and Teal, J. M. (1983), Long-term biological consequences of nuclear war, *Science*, **222**, 1293–1300; Covey, C., Schneider, S. H., and Thompson, S. L. (1984), Global atmospheric effects of massive smoke injections from a nuclear war: results from general circulation model simulations, *Nature*, **308**, 21–25; Crutzen, P. J., Galbally, I., and Brühl, C. (1984), Atmospheric effects from post-nuclear fires, *Climatic Change*, **6**, 323–364; Pittock, A. B., Ackerman, T. P., Crutzen, P. J., MacCracken, M. C., Shapiro, C. S., Turco, R. P., Harwell, M. A., and Hutchinson, T. C. (1986), *Environmental Consequences of a Nuclear War. Vol. 1: Physical and Atmospheric Effects. Vol. 2: Ecological and Agricultural Effects*, SCOPE Report 28 (Wiley, Chichester, UK).

[64] Taylor, O. C. (1969), Importance of peroxyacetyl nitrate (PAN) as a phytotoxic air pollutant, *Journal of the Air Pollution Control Association*, **19**, 347–351.

[65] National Research Council (1977), *Ozone and Other Photochemical Oxidants* (National Academy Press, Washington, DC)

[66] Campbell, K. I., Clarke, G. L., Emik, L. O., and Plata, R. L. (1967), The atmospheric contaminant peroxyacetyl nitrate. Acute inhalation toxicity in mice, *Archives of Environmental Health*, **15**, 739–744.

[67] Shaver, C. L., Cass, G. R., and Druzik, J. R. (1983), Ozone and the deterioration of works of art, *Environmental Science and Technology*, **17**, 748–752.

[68] Forschungsbeirat Waldschäden (1984), Zwischenbericht, Dezember.

[69] Odén, S. (1968), *The Acidification of Air and Precipitation and its Consequences in the Natural Environment*, Ecology Committee Bulletin No. 1, Stockholm (Translation Consultants Ltd, Arlington, VA); Cowling, E. B. (1982), Acid precipitation in historical perspective, *Environmental Science and Technology*, **14**, 110A–123A.

[70] Bolin, B., Granat, L., Ingelstam, L., Johannesson, M., Mattsson, E., Odén, S., Rodhe, H., and Tamm, C. O. (1971), *Air Pollution Across National Boundaries. The Impact on the Environment of Sulfur in Air and Precipitation* (Norstedt, Stockholm).

[71] Stevens, R. K., Dzubay, T. G., Shaw, R. W., Jr., McClenny, W. A., Lewis, C. W., and Wilson, W. E. (1980), Characterization of the aerosol in the Great Smoky Mountains, *Environmental Science and Technology*, **14**, 1491–1498.

[72] Rosen, H., Hansen, A. D. A., Dod, R. L., and Novakov, T. (1980), Soot in urban atmospheres: determination by an optical absorption technique, *Science*, **208**, 741–744.

[73] Went, F. W. (1960), Blue hazes in the atmosphere, *Nature*, **187**, 641–643.

[74] National Bureau of Standards (1978), *Economic Effects of Metallic Corrosion in the United States*, Special Pub. NBS 511-1 (NBS, Washington, DC).

[75] Mattsson, E. (1982), The atmospheric corrosion properties of some common structural metals – a comparative study, *Materials Performance*, **23**(7), 15–25.

[76] Yocom, J. E. and Baer, N. S. (1984), Effects on materials, Section E7, *The Acidic Deposition Phenomenon and Its Effects*, Environmental Protection Agency 600/8–83–016B (EPA, Washington, DC).

[77] Graedel, T. E., Franey, J. P., and Kammlott, G. W. (1984), Ozone- and photon-enhanced atmospheric sufidation of copper, *Science*, **224**, 599–601.

[78] Graedel, T. E. (1978), *Chemical Compounds in the Atmosphere* (Academic Press, New York).

[79] National Research Council (1984), *Global Tropospheric Chemistry – A Plan for Action* (National Academy Press, Washington, DC).

[80] Malone, T. F. and Roederer, J. G. (Eds) (1984), *Global Change* (Cambridge University Press, Cambridge, UK).

[81] Levy, H., II (1971), Normal atmosphere: large radical and formaldehyde concentrations predicted, *Science*, **173**, 141–143.

[82] Crutzen, P. J. (1973), A discussion of the chemistry of some minor constituents in the stratosphere and

troposphere, *Pure and Applied Geophysics*, **106–108**, 1385–1399.

[83] Detailed discussions of these sequences are given in Atkinson, R., Carter, W. P. L., Darnall, K. R., Winer, A. M., and Pitts, J. N., Jr. (1980), A smog chamber and modeling study of the gas phase NO_x–air photo-oxidation of toluene and the cresols, *International Journal of Chemical Kinetics*, **12**, 779–836; Lloyd, A. C., Atkinson, R., Lurmann, F. W., and Nitta, B. (1983), Modeling potential ozone impacts from natural hydrocarbons – 1. Development and testing of a chemical mechanism for the NO_x–air photooxidation of isoprene and α-pinene under ambient conditions, *Atmospheric Environment*, **17**, 1931–1950; Killus, J. P. and Whitten, G. Z. (1984), Isoprene: a photochemical kinetic mechanism, *Environmental Science and Technology*, **18**, 142–148.

[84] Crutzen, P. J. (1979), The role of NO and NO_2 in the chemistry of the troposphere and stratosphere, *Annual Review of Earth and Planetary Sciences*, **7**, 443–472.

[85] Singh, H. B. and Hanst, P. L. (1981), Peroxyacetyl nitrate (PAN) in the unpolluted atmosphere: an important reservoir for nitrogen oxides, *Geophysical Research Letters*, **8**, 941–944.

[86] Thiemens, M. W. and Schwartz, S. E. (1978), *The Fate of HS Radical Under Atmospheric Conditions*, paper presented at 13th Informal Conf. on Photochem., Clearwater, FL, Jan. 4–7 (proceedings not published).

[87] Hatakeyama, S. and Akimoto, H. (1983), Reactions of OH radicals with methanethiol, dimethyl sulfide, and dimethyl disulfide in air, *Journal of Physical Chemistry*, **87**, 2387–2395; Niki, H., Maker, P. D., Savage, C. M., and Breitenback, L. P. (1983), An FTIR study of the mechanism for the reaction $HO + CH_3SCH_3$, *International Journal of Chemical Kinetics*, **15**, 647–654.

[88] Stockwell, W. R. and Calvert, J. G. (1983), The mechanism of the $HO–SO_2$ reaction, *Atmospheric Environment*, **17**, 2231–2235.

[89] Crutzen, P. J. (1970), The influence of nitrogen oxides on the atmospheric ozone content, *Quarterly Journal of the Royal Meteorological Society*, **96**, 320–325.

Commentary

E. Mészáros

The studies carried out during the last two decades of biogeochemical cycles of different elements (compounds) have produced important scientific achievements. These studies have partly revealed the complexity of environmental interactions and partly demonstrated that nature is an indivisible and integral entity. It has become clear that biogeochemical cycles not only make life on the Earth possible by providing nutrients for ecosystems in various environmental media, but also that this continuous material flow in nature maintains constant the composition of the atmosphere. This constant composition, controlled partly by the biosphere, is indispensable for the present form of life on our planet.

The Earth's climate also depends on the chemical composition of the atmosphere, because the chemical composition controls short-wave and long-wave radiation transfers in the air and consequently the energy balance. In this way atmospheric composition regulates not only the temperature distribution, but also the general circulation of the atmosphere caused by differences in energy balance between different territories.

The stability of atmospheric composition is currently of crucial importance, since human activities are now capable of disturbing this natural equilibrium by releasing different substances into the atmosphere. It follows that to consider the future development of our society, possible atmosphere–development interactions must be accounted for.

Among the examples of atmospheric degradations enumerated by Crutzen and Graedel I stress the importance of the modification of atmospheric radiation balance, the climatic consequences of which are discussed by Dickinson (Chapter 9, this volume). It seems to me that the increases of CO_2 concentration and that of other radiatively active trace gases in the atmosphere (N_2O, CH_4, halocarbons, etc.) pose the most important threat to the stability of climate. Unfortunately, the possible response of the hydrosphere and biosphere to the increase of atmospheric CO_2 is speculative. The synergism of CO_2 effects and other anthropogenic modifications (e.g., the increase of aerosol concentration) is also an open question.

It is evident that anthropogenic pollutants modify not only the state of the atmosphere. They also influence considerably the composition of atmospheric depositions and consequently biochemical processes in aquatic and terrestrial ecosystems. Since the atmosphere is a mobile dynamic system these detrimental effects can be observed far from urban and industrial pollution sources, the distance being a function of the transformation rate and residence time of the pollutants considered. The formation of acid precipitation is a good example of this atmospheric action. I believe that the acid rain problem is the second most important environmental damage due to atmospheric degradations. It should be mentioned, however, that acid precipitation is only a part of the problem. The dry deposition of acid-forming gases (SO_2 and NO_x) and of acids themselves (e.g., HNO_3) can be as important as wet deposition caused by acid precipitation. The dry deposition, due mostly to turbulent transfer in the air, is particularly strong in the vicinity of the sources. Thus, in central Europe dry and wet sulfur and nitrogen depositions are of comparable magnitudes.

Crutzen and Graedel also discuss photochemical oxidant formation. I emphasize in this respect that acid formation is closely related to oxidant formation in air and precipitation. This indicates that damage in lakes and forests is related not only to the sulfur emission rate, but also to the concentration of free radicals formed by photochemical processes. Thus, I think that even in this case synergism is an important factor. It is also possible, of course, that photochemical oxidants influence directly the state of forests, as discussed in this chapter.

At the end of their correct and compact chapter Crutzen and Graedel note that the number of primary anthropogenic sources of causative species is small. Their discussion seems to suggest that for this reason the solution of the problems is not too complicated: "There is enough evidence to state that a significant decrease in the rate of fossil fuel combustion would stabilize the atmospheric radiation balance, improve visibility, hinder smog formation, and minimize acidic precipitation and its effects [pp 240–241]." This statement is true in itself. However, at least at the moment energy production by humankind is mainly due to fossil fuel combustion. Alternative energy sources are needed to be able to decrease significantly the rate of fossil fuel burning and consequently the deleterious effects of atmospheric pollution. The necessity to find other energy sources is one of the greatest challenges for the human race.

The aim of air chemistry is to study the atmospheric part of the biogeochemical cycles of substances in nature. Thus, I agree with Crutzen and Graedel that atmospheric chemistry should play an important role in the estimation of environment–development interactions. For this reason, as discussed at the end of the chapter, further research is needed in this important field to understand more correctly the interrelation between human activity and our atmospheric environment. Such a deeper understanding is indispensable for planning the technological development of our society.

Chapter 9 Impact of human activities on climate – a framework

R. E. Dickinson

Editors' Introduction: Climate is one of the most important environmental properties identified from the general perspectives of Chapters 7 and 8.

In this chapter a detailed framework for assessing the potential overall impacts of human activities on the world's climate is presented. Continuing a theme introduced in Chapter 7, Dickinson begins with a characterization of the spatial and temporal patterns of climatic variation that arise from natural processes on time scales of decades to centuries. The magnitude and extent of human activities that could impose significant additional variation on these natural patterns are examined next. Alternative methods of assessing the climate's response to perturbation are critically discussed. The chapter is concluded with a novel, probabilistic assessment of the net global climatic change that might result from human activities by the end of the twenty-first century.

Introduction

Sometimes, as I look out of my window at a summer shower, I recall the bright sun that shone a few hours previously. Here in Boulder, Colorado, hot weather in April often precedes snowstorms by a few days, and our coldest winter weather is sometimes accompanied by summer-like weather in Seattle.

Climate is the aggregate of such day-to-day weather patterns. Climatology involves studying the local microclimate for grass growing under Ponderosa pine, the length of the growing season for Kansas wheat, or the global average of surface temperature observations. It includes establishing the statistics of extreme and perhaps hazardous events, such as tornadoes, the seasonal variation of pack ice around Antarctica, the frequency of cloud occurrence over a potential site for solar collectors, or the wind characteristics at a potential site for wind-power collection. Warm equatorial oceans and cold polar oceans, a multiyear drought in Sub-Saharan Africa, spring flooding in the Rockies – these are all part of the climatic system.

Those who realize how difficult it is to correctly decide which day to carry an umbrella will certainly wonder how it is possible to identify any underlying patterns and processes in such a complex system as climate. A careful analysis of the examples of climate given above, however, suggests a few basic underlying variables and processes, i.e., temperatures, humidities, winds, and the fluxes of energy and water. Fluctuations in these parameters occur on various scales of time and space. For example, energy converts water from liquid into vapor; this energy of "latent heat" is released when the water condenses. Even the kinetic energy released by water falling under gravity, although much less in total than the latent heat energy, can be of considerable practical importance.

Research on the climate system over the last few decades has helped quantify the relationships between temperatures, humidities, and winds on the one hand, and fluxes of energy and water on the other. With increased understanding has come the realization that many human activities modify the environment on a sufficient scale to compete with natural processes. Here I intend to provide a general framework needed for intelligent assessment of the potential impacts of future human activities on climate. The orientation of the framework is primarily global, with time scales of decades to centuries, and both past and future are considered.

The basic source of energy for most terrestrial processes is the absorption of solar radiation. This is certainly true for the climate system. For climate to be in equilibrium, the absorbed solar radiation must be balanced by outgoing thermal radiation. The partial trapping of this thermal radiation by radiative absorbers – particles or molecules – within the atmosphere elevates surface temperatures by several tens of degrees Centigrade above what they would be in the absence of an atmosphere [1], as illustrated in *Figure 9.1.*

Much of the current concern over the impact of human activities on climate evolves from our capabilities to alter the amount of absorbed solar energy or the leakage of thermal radiation. Another broad class of problems involves the possibility of significant modifications of the surface or atmospheric hydrological cycles. From the societal viewpoint, threats to the climate system are mainly due to the human uses of energy or to land-use changes. Some industrial activities can also change the atmosphere's radiative properties, e.g., by adding radiatively active trace gases to the atmosphere.

The climate system has a large degree of natural variability, which both limits the detectability of human perturbations and greatly influences any evaluation of the practical consequences of such perturbations (see pp 254–263). To understand where human activities have leverage in the climate system, I also consider the basic processes that determine global radiative energy fluxes and the hydrological cycle (pp 263–266).

With such a suitable conceptual framework, we can consider the magnitudes and scales of human activities, past, present, and future, in terms of their disruptions of fluxes of energy and water in the climate system. Future human activities are, of course, basically unpredictable. However, we can simply extrapolate present trends to create

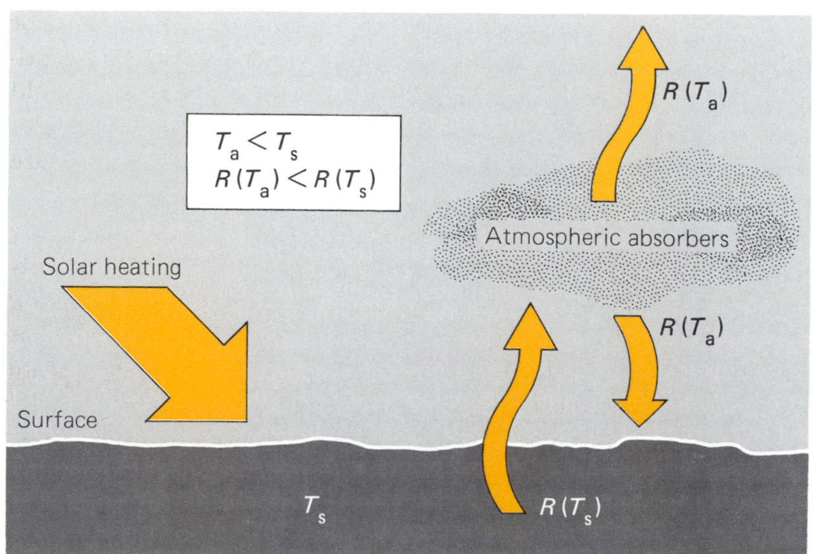

Figure 9.1 Atmospheric radiative trapping (greenhouse effect). The surface at temperature T_s emits upward infrared radiation, $R(T_s)$. The amount of thermal infrared radiation emitted increases in proportion to the fourth power of surface temperature T_s. Some fraction of the surface emission is absorbed by atmospheric absorbers, depending on their absorptivity. These absorbers also emit thermal radiation $R(T_a)$ at a rate proportional to their absorptivity and dependent on their temperature T_a. If their temperature is less than that of the surface, as is usually the case, they emit upward less thermal radiation than they absorb. The surface is, in turn, warmed by downward thermal radiation and coupled to the overlying atmospheric layers by convection. If more absorbers are added, the system warms up, thus providing the net thermal radiation flux to space that is required to balance absorbed solar radiation.

"surprise-free" scenarios that provide some basis for attaching relative importance to different future human activities.

Given the possible future human disruptions of energy fluxes in the climate system, it is necessary to assess the effects of these disruptions on climatic processes. Unfortunately, there are no close laboratory analogs to the climate system, and historical analogs are also of limited use because of the lack of distinction between internal fluctuations and external disruptions of energy fluxes. Thus, the primary tools for studying future climate are mathematical models. On p 268 I introduce the simplest such model, for global energy balance, which is used to compare the relative effects of different possible human disruptions on global fluxes or radiative energy. The global energy-balance viewpoint is ineffective for examining possible climatic effects that involve a disruption of the hydrological cycle or the internal redistribution of energy in the climate system (see pp 272–275). On pp 275–277 I give a brief review of the content, use, and limitations of the most complete computer models of the climate system to date (1986).

The period we are interested in here as regards human effects is of the order of 10 to 100 years. A global spatial scale is primarily considered. Where I need to be more specific, I consider the possible climatic change by the year 2100 over areas of at least $1–10 \times 10^6$ km^2 (10×10^6 km^2 is 2% of the surface area of the Earth). There is nothing special about 2100; our current information does not suggest that any major discontinuities would occur in the climate system between now and then. However, focusing on the year 2100 helps illuminate a wide range of possibilities. Possible changes of climate at more immediate dates can then be considered using interpolation between the present and 2100.

Besides pursuing the question of future climatic change of maximum or at least reasonable likelihood, we should also consider what nasty surprises might be realized. To refrain from too extreme flights of fancy, I consider primarily those possibilities that seem to me to have at least a 1% probability of occurring in the next century or two (pp 277–279). Such considerations not only suggest possible aspects of climate system vulnerability, but also help to identify its especially stable or resilient characteristics.

Scales and patterns of natural atmospheric variability

Time variability

The theme of this chapter is the impact of human activities on climate. However, the implications of human-mediated climatic change can only be properly evaluated and procedures for its detection be devised in the context of natural climatic variations. Thus, I intend to provide a conceptual framework for nonmeteorologists to enable a better understanding of the large natural variability of weather and climate. Those who are totally unfamiliar with concepts of random time series and nonlinear system dynamics may wish to skip or read lightly this section.

Natural climatic and meteorological variations occur over a wide range of time and space scales. Examining the time variations of one variable, e.g., surface temperature, at one location reveals that the time structure of variations consists of periodic and stochastic components. The periodic components are seasonal and diurnal cycles, largely sine wave variations with one-year and one-day periods. There are, in addition, some higher harmonics, e.g., twice-a-year and twice-a-day components. The diurnal and seasonal cycles are sometimes the largest contributors to the total time variance of a meteorological variable [2], although frequently they are only recognized by data analysts as the piece of a time series that needs to be removed in order to look at the more intriguing stochastic components. Often the time mean and seasonal variations are lumped together by using the long-time average of monthly or three-monthly average fields (e.g., January means or winter means). *Figure 9.2* [3] illustrates the June-to-August and December-to-February average surface air temperature, and *Figure 9.3* [4] shows the seasonal range of surface air temperature.

Global temperature

Global climatic change is often discussed in terms of global and annual average surface air temperature. This is a good example of a stochastic climatic time series and is illustrated for the last 100 years or so in *Figure 9.4* [5]. The time series starts

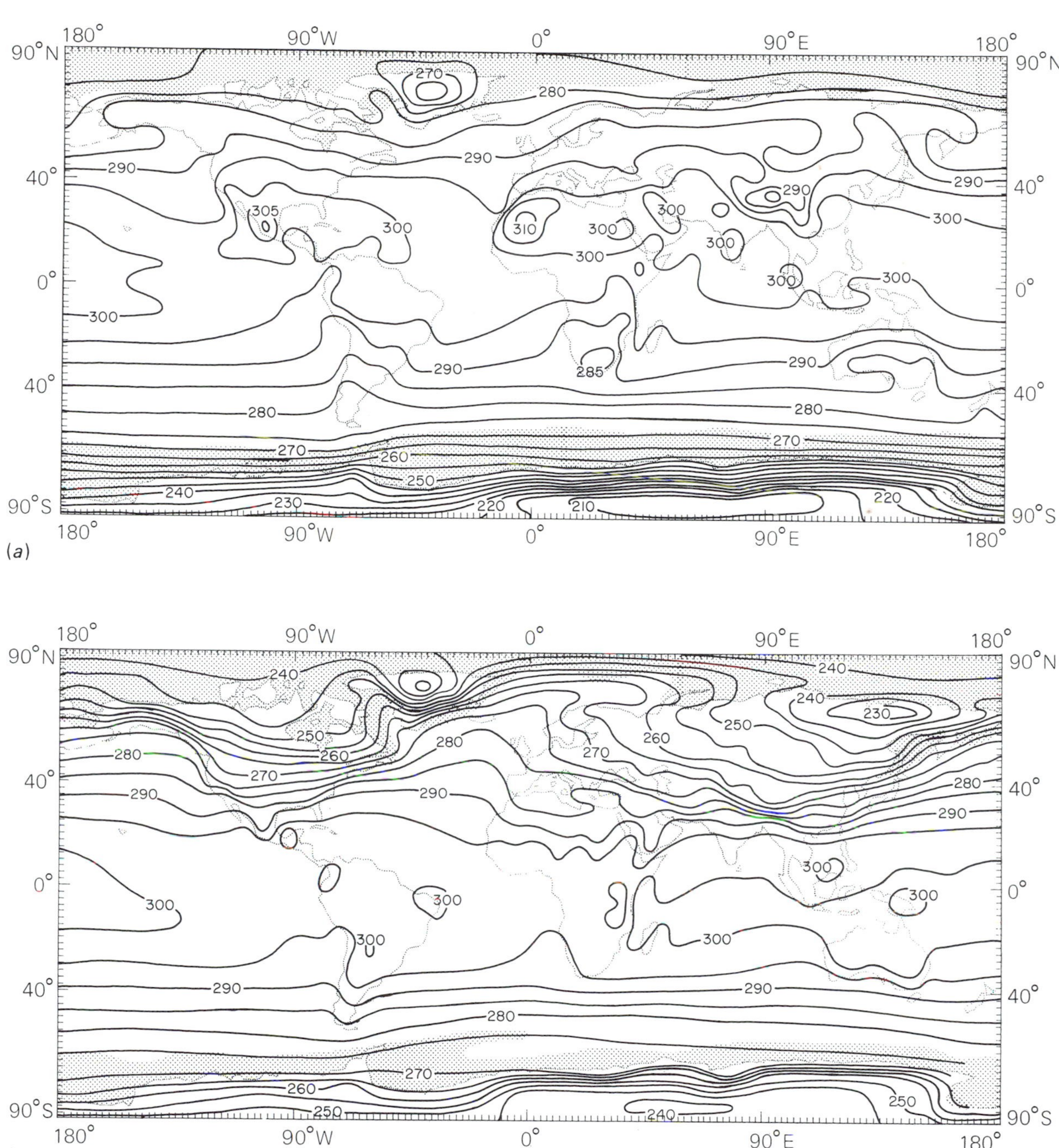

Figure 9.2 Geographical distribution of observed average surface air temperature for (*a*) June–July–August and (*b*) December–January–February (units are degrees Kelvin; subtract 273.15 to convert into degrees Celsius) [3]. Contours are drawn every 5 °C. Note that outside the tropics the mid-continent land temperatures in summer are typically about 10 °C warmer than the ocean temperatures, and in winter 20 °C or more colder than the ocean temperatures. Figure (*a*) shows that average temperatures below 210 K (almost −80 °F) are found in central Antarctica during winter, and above 310 K (about 100 °F) in the Sahara during summer. In dry continental areas, the daytime high can be 20 °F warmer than the day–night average, and daytime ground temperatures can be much warmer again.

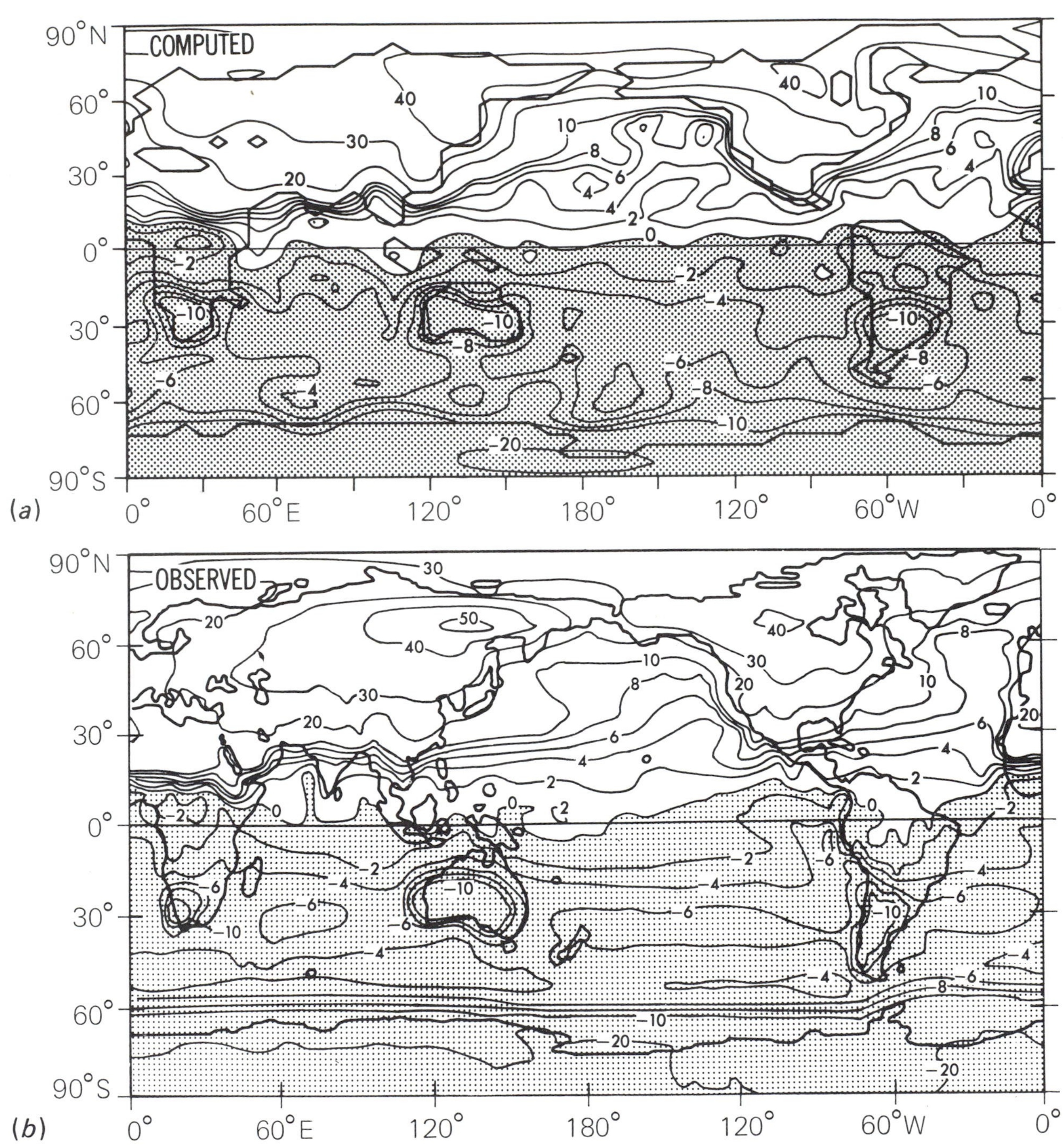

Figure 9.3 The geographical distribution of surface air temperature difference (degrees Kelvin) between August and February: (*b*) Observed values and (*a*) values calculated by a general circulation climate model [4] (see pp 275–277). Note the >40 °C (72 °F) range of temperature for some high-latitude regions in the Northern Hemisphere and the <10 °C variations over most ocean areas. The seasonal cycle of solar heating is perhaps the largest forcing term that the climate system is exposed to. The good agreement between model and observations indicates that the model responds reasonably to this forcing.

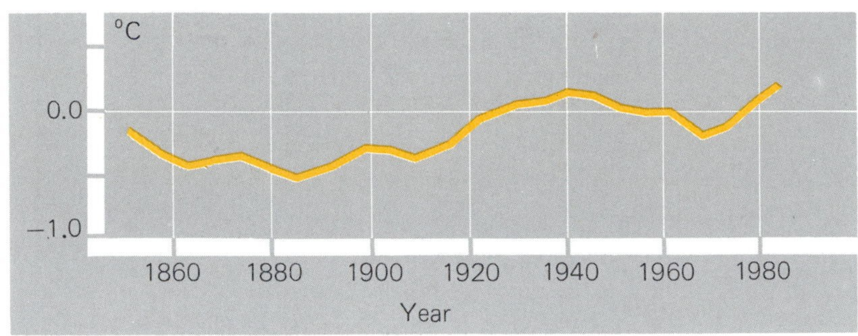

Figure 9.4 Northern Hemisphere average annual average surface air temperatures over the last 120 years. The curve shows a smoothed version of the individual annual averages [5].

from the time when temperature records were maintained at enough stations to allow determination of this average. Various proxy records, such as tree ring thickness, pollen counts, and historical records, are used to estimate earlier climates of the last few thousand years. Other geological data are used to estimate climate variations over the last 0.1×10^6 to 500×10^6 years.

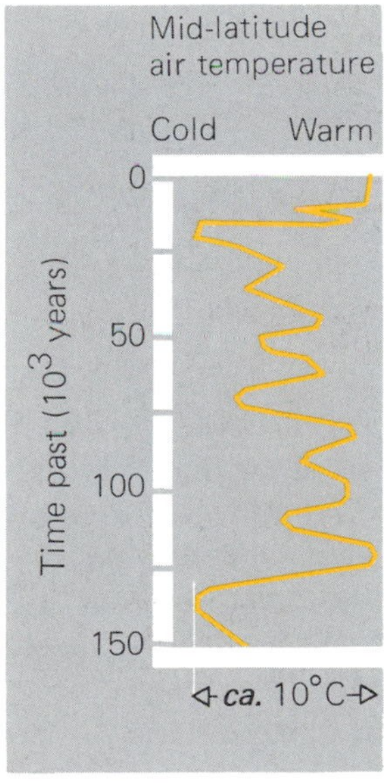

Figure 9.5 Generalized Northern Hemisphere temperature trends over the last 150 000 years [7]. Warmest temperatures occurred about 125 000 years ago and over the last 8000 years. The most severe cold (ice age) occurred about 20 000 years ago. Peaks and dips in temperature occur about every 20 000 years with a range of about 3 °C, and the largest temperature dips occur about every 40 000 years. The total range of temperature variation over the last 100 000 years has been about 10 °C. Similar patterns have occurred every 100 000 years for the last million years.

A large amount of information is available on climates over geological time. These past climates are only of interest here in that they help to provide a frame of reference for present-day climatic variations. For example, the Cretaceous era, about 100×10^6 years ago was much warmer than the present-day climate. Tropical temperatures were perhaps 5 °C warmer than present and in high latitudes temperatures are estimated to have been 10 to 20 °C warmer than present [6]. From that time, temperatures have gradually declined. Over the last million years, the climate system has experienced numerous ice ages, with temperatures as warm as or warmer than the present ones occurring for only a small fraction of the time. *Figure 9.5* [7] illustrates the estimated behavior of global temperature over the last 100 000 years or so. The warmest climate period in the present "interglacial" occurred from 5000 to 9000 years ago when summer temperatures in the Northern Hemisphere may have been about 2 °C warmer than present [8].

Spectral peaks and red noise

A highly prized and much sought-after feature of the stochastic components of climatic variation is large amounts of variability in specific frequency bands. If such spectral peaks were found to represent a large part of the total variability, they would provide an effective means for statistical time-series forecasting. In the geological record, such peaks are pronounced for periods of approximately 20 000, 40 000, and 100 000 years. These variations are related to variations in the Earth's orbital parameters that have similar periods [9]. Both these very long-period and short-period periodicities form but a small fraction of the climatic variability that occurs over a few hundred years or less. The most commonly found meteorological example of a short-period spectral peak is that of quasi-biennial oscillations [10]. The occurrence of 11- and 22-year variations in sunspots is one of the most well-known examples of a large spectral peak in a random geophysical time series. Much effort has been applied to relating climatic variations to these sunspot variations, but with little success [11].

The time structure of climatic variability is largely that of "red noise" [12], with time variations and spectra such as those in *Figure 9.6.* Red-noise variability has a power spectrum that is constant at low frequency (i.e., white noise) and decays monotonically with increasing frequency at higher frequencies. It can be imagined as consisting of a superposition of pulses that are turned on at random points in time with initial amplitudes that are random and decay with various decay scales [13]. This illustrates a conceptualization of the internal dynamics of the atmosphere and its coupling to other elements of the climate system to which I return later.

A counterintuitive feature of a pulse excitation, such as shown in *Figure 9.6,* is the excitation by very short pulses of disturbances whose spectral power *increases* with *decreasing* frequency, that is, consists of red noise. For periods long compared to t_c, the spectral power associated with that time scale becomes independent of frequency. Hence, short time-scale weather events cause much of the year-to-year variations in monthly average climates and even the year-to-year variations in a given location [14]. Only by forming global averages and taking multiyear running means is the variability due to short-term weather events largely eliminated. Thus, in considering long-term regional climatic variations, it is important to establish the contribution to this variability from short-term weather events.

Conversely, although very long time-scale processes may cause large variations over the length of their period, they contribute little to climate trends at a specific time. This is evident because the magnitude of trends is given by the amplitude of the variation divided by its time scale. The long period ice-age variations can be characterized by a 4 °C temperature variation on a 20 000 year time-scale, suggesting a trend of 0.02 °C/century. By contrast, the last 100 years of temperature records indicate a change of about 0.5 °C or a trend of 0.5 °C/century, as inferred from *Figure 9.4.* Thus, the overall temperature variations associated with past ice ages are too small to affect current climate significantly.

The large natural variability must also be recognized in estimating impacts of climatic change on human activities. The most deleterious impacts of weather and climate are frequently a result of threshold events rather than of means of the diurnal and seasonal cycles (Parry, Chapter 14, this volume) [15]. Thus, changes in impacts are likely to result from shifts in the numbers of threshold events, either because of changes in climatic means or changes in variances. The effect of a change in climatic mean on the frequency of threshold events is illustrated in *Figure 9.7.* Changes in climatic variances are more difficult to recognize than changes in means. Parry, Chapter 14, this volume, emphasizes further the possible importance of changes in climate variance.

Spatial structures

Before considering further the time structure of natural variability, it is necessary to consider the spatial structures of weather systems and climate processes. Spatial patterns can be associated separately with the periodic and with the stochastic weather–climate time variations. Again, considerable spatial variation is associated with the simplest time structures, i.e., the climatic mean and annual cycle patterns, as illustrated for surface temperature in *Figures 9.2* and *9.3.* Many other meteorological variables, e.g., rainfall, are also of practical interest, and their seasonal and longitudinal distributions are discussed in geography texts on climate [16].

The climatic mean and annual cycle patterns involve all space scales, with small-scale variability more pronounced near and at land surfaces because of the rapid spatial variations in terrain and microclimatological environments. At higher levels in the atmosphere, spatial variations become largely global [17]. Climatologists study in detail the zonal mean (that is, longitudinally averaged) and time mean east–west wind systems as well as the north–south mean meridional circulation (the Hadley cell in the tropics). These features have horizontal spatial scales of at least 10° latitude. The most pronounced continental-scale systems are the Asiatic winter and summer monsoons [18]. Distinct dynamical structures are also ascribed to the planetary waves. These

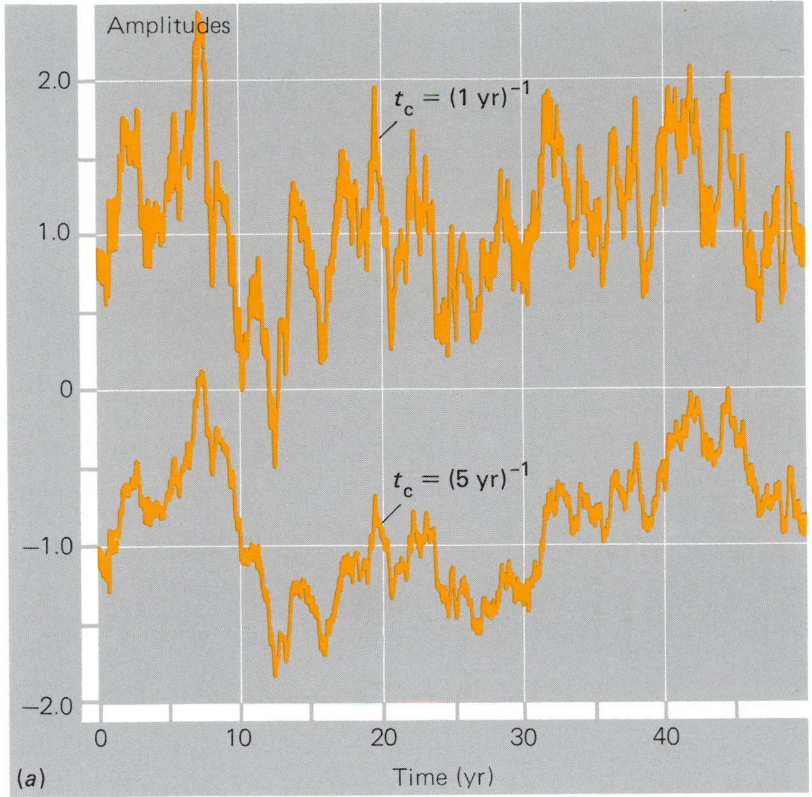

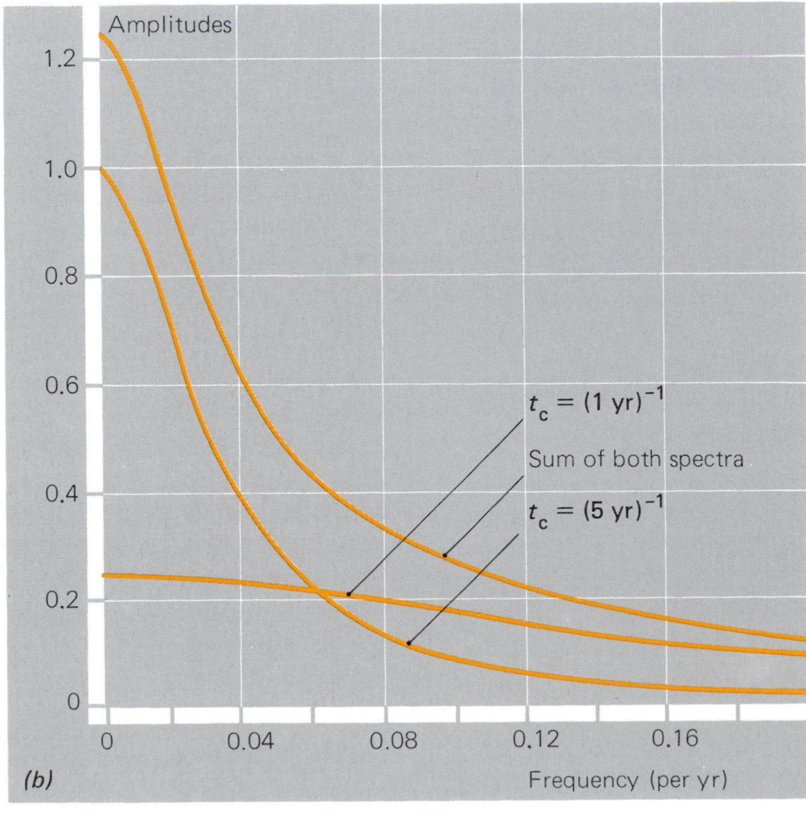

Figure 9.6 Examples of (*a*) "red-noise" stochastic time series and (*b*) the corresponding power spectra. The time scale is arbitrary but can be imagined to represent years. The time series are constructed using the equation

$$X^{n+1} = (X^n + 0.1 t_c^{-1} F^n)/(1 + 0.1 t_c^{-1})$$

where F^n is a random number with equal probability of lying anywhere between 1.0 and −1.0, such as can be obtained from the BASIC of most home computers, X^n is the value of the time series at point n, and X^{n+1} its value at point $n+1$. The points are assumed to be spaced 0.1 units apart. The parameter t_c gives the damping time scale of the time series. Two examples are given: a series for $t_c = 1$, and a series for $t_c = 5$.

waves are obtained observationally as the lowest order components of expansions of monthly mean climatic fields into longitudinal Fourier series [19].

The spatial patterns of atmospheric variables with stochastic time variations take on a number of interesting structures best known to devotees of the science of weather prediction. These structures can be divided into mesoscale systems, synoptic scale systems, and global systems, with typical scales of 10–200 km, 200–2000 km, and >2000 km, respectively.

Mesoscale systems range from frontal systems and squall lines to individual, large cumulus-cloud systems. These systems provide much of the world's violent weather and typically have lifetimes of less than a day. They occur mainly in the moist tropics or over continents during summertime [20].

Synoptic systems include especially the extratropical cyclonic storms and anticyclonic systems that bring much of the winter weather variations outside of the tropics. These systems typically have time scales of about three days [21]. Stochastic systems on a planetary scale include fluctuations in the longitudinally averaged and the planetary wave systems. Some distinct planetary wave spectral peaks have been identified that correspond to free normal modes of the global atmosphere [22]. Although of considerable interest to the dynamic meteorologist, these peaks are probably of little importance to questions of climatic change. Except for these peaks, which have periods of 5 and 16 days, atmospheric variations on time scales of days to weeks or more appear to be of the red-noise variety.

Climate variability as the balance between excitation and decay processes

The red-noise atmospheric variations originate through the growth of instabilities that release the potential energy of the atmosphere, which had been previously created by differential radiative heating. Convective processes, which dominate on the mesoscale, are driven either by the upward motion of air made buoyant by surface heating or by the release of latent heat during precipitation processes.

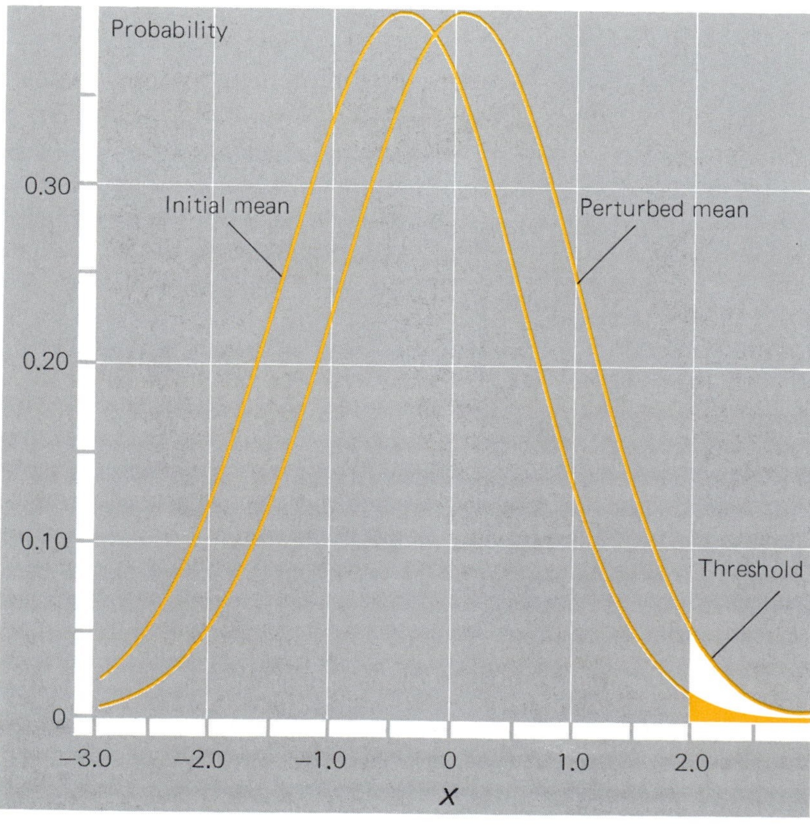

Figure 9.7 Deleterious effects of climate change may be manifested in terms of a change in the frequency of threshold events, such as temperature extremes. For example, the day-to-day variations of temperature may be summarized in terms of a probability distribution of given values. The figure shows two probability distributions identical except for a shift in mean. The yellow area represents the probability of exceeding a hypothetical threshold value for an original climate and the white area the increase in the probability of exceeding that threshold for a hypothetical change in climate. Note that a small change in mean conditions can lead to a large increase in the probability of exceeding the threshold.

The synoptic scale and global systems are, by contrast, driven by the pole-to-equator temperature gradient. Instabilities that develop on the mean atmospheric state grow to finite amplitude, then are dissipated by nonlinear dynamic processes, as well as by frictional and radiative damping and coupling to the surface. As a simple model, we can think of the initial unstable growth as instantaneously injecting disturbances.

The nonlinear damping requires about a day or less to dissipate mesoscale systems, typically 5 days for synoptic scale and at least 5 days for global scale disturbances. Frictional damping requires about 10 days [23] and radiative damping about 40 days [24]. Land surfaces have little heat capacity and can equilibrate to atmospheric conditions without dissipating significantly the atmospheric disturbance [25]. Over the oceans, on the other hand, rapid fluxes of sensible and latent heat keep the surface layers of the ocean in thermal equilibrium with the atmosphere [26]. This air–sea thermal coupling, in transferring the disturbance from the atmosphere to the ocean, largely removes the atmospheric disturbance because of the atmosphere's relatively low heat capacity. The adjustment time for this process is about 2 to 10 days, depending on the differences between air and sea temperatures and water–vapor mixing ratios. All the above time scales are approximate and vary with the situation. The nonlinear damping will normally dominate large-amplitude disturbances. Friction is especially effective in damping near-surface winds, and radiative damping removes temperature anomalies that have already largely lost their associated atmospheric motions through the faster forms of damping.

Over oceans, the coupling between atmosphere and ocean does not simply remove a thermal anomaly, but mixes it into the joint system of the atmosphere and the mixed layers of the ocean. The ocean's mixed layers have a heat capacity about 20 times that of the atmosphere alone [27]. The joint disturbance is then damped further by atmospheric long-wave radiative dissipation. Consequently, its lifetime is extended from about 40 days to several years. This is one of several sources of long-term memory for the climate system. Other frequently cited examples are the presence of soil moisture and snow cover [28]. The effects of both these factors on the atmosphere can persist for several months or more.

There are some additional kinds of random disturbances to the climate system that have intrinsically long lifetimes. Those most studied are the radiative perturbations due to stratospheric aerosol from volcanic eruptions [29]. The aerosol is in the form of sulfate particles which form from sulfur dioxide (SO_2). The extent of the volcanic perturbation on climate depends on the amount of volcanic material reaching the stratosphere and on the sulfur content of its source [30]. Gases and small particles have a typical residence time in the stratosphere of about 2 years [31]. The duration of the climatic perturbation is determined by the finite lifetime of the stratospheric aerosol plus the additional memory from oceanic heat storage.

As already suggested, the oceans are the "flywheel" for storing thermal energy in the climate system. If the oceans, with an average depth of 5 km, were uniformly mixed, they would equilibrate with the atmosphere on a time scale of about 400 years. In reality, only the near-surface waters are in rapid contact with the atmosphere; this reservoir is typically about 70 m in depth, with a thermal time scale of several years. Thermal perturbations applied to the ocean over thousands of years would change ocean temperatures at all depths. External thermal perturbations occurring over a 10- to 100-year period, such as that resulting from increasing concentration of atmospheric carbon dioxide (CO_2), however, largely transfer energy to the mixed-layer reservoir and the reservoir of ocean intermediate water, which together require several decades or more to equilibrate to thermal perturbations.

Viewing the ocean as vertically stacked reservoirs is somewhat unsatisfactory, since most of the exchange to levels below the surface mixed layers occurs at high latitudes. In essence, the deeper layers outcrop at the surface in these high-latitude regions of water mass formation [32]. So far, only variations in the temperatures of ocean surface mixed layers have been related to observed climatic change.

The most well-known example of 1–2 year climatic memory provided by ocean temperature anomalies is the phenomenon referred to as "El Niño" or "Southern Oscillation" [33].

Other memory reservoirs of the climate system include the polar sea ice, with time scales of a few years, and continental ice sheets, with time scales of 10^3 to 10^4 years [34]. *Figure 9.8* illustrates the

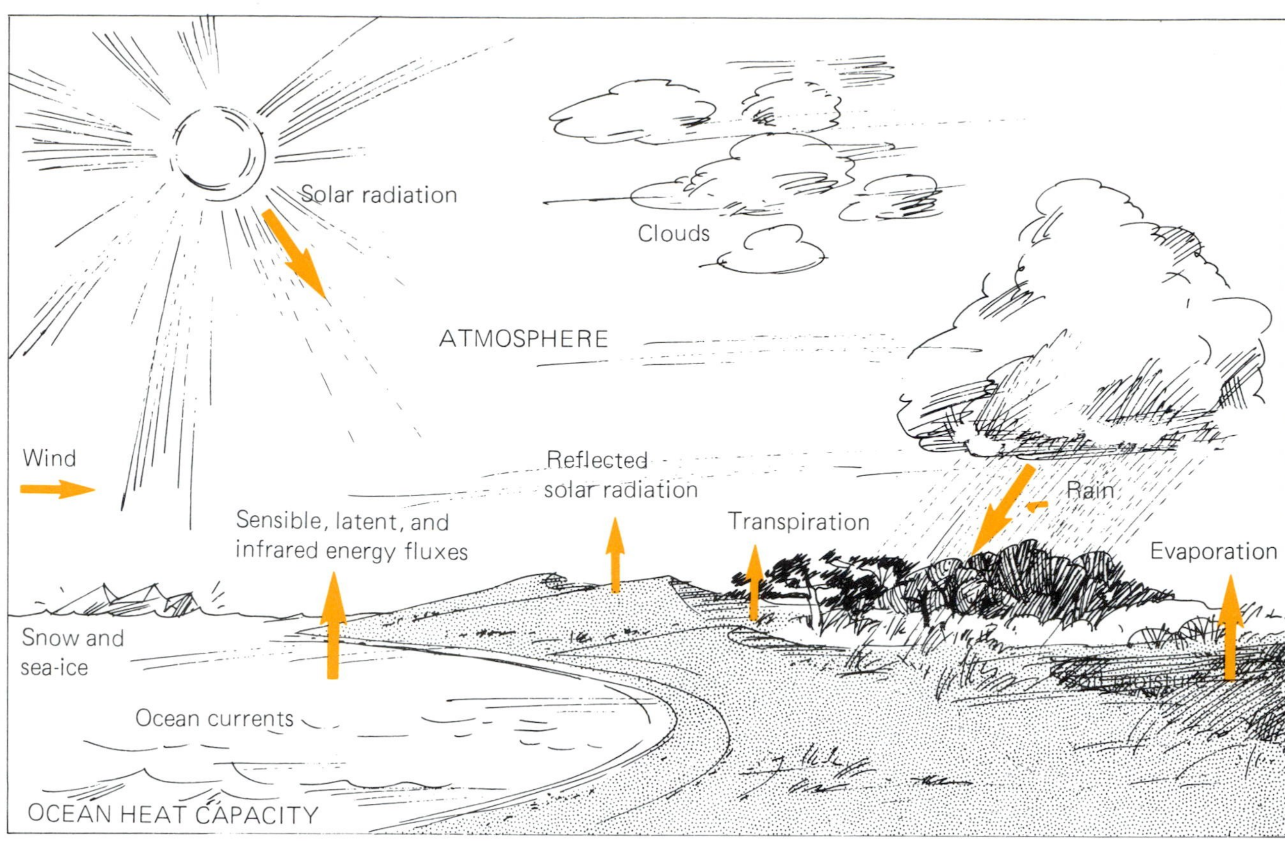

Figure 9.8 Mechanisms contributing to climatic memory on a time scale of months to decades. The oceans gain energy from solar radiation and return this energy to the atmosphere through sensible, latent, and infrared thermal energy fluxes with a time lag that ranges from a few months to many decades. Snow and sea ice reflect solar radiation and so promote cooler temperatures; hence their continued presence. Likewise, moisture in soil lasts for up to several weeks or more before it is lost, and tends through increased evapotranspiration to promote rainfall and thus maintain itself for a yet longer period.

most commonly discussed mechanisms contributing to climatic memory on a time scale of months to decades. *Figure 9.9* summarizes the range of time and space scales of the mechanisms discussed.

The average surface temperature of the Northern Hemisphere, and presumably of the whole Earth, has varied over a range of about 1 °C since the first extensive instrument records made in the 1880s, as shown in *Figure 9.4.* However, hemispheric temperature changes over a time scale of decades are generally a residual of considerably larger, up to 5 °C, regional temperature-change patterns [35].

In summary, climate-system variability is the result of disturbances that randomly originate from either internal energy sources or transient external perturbations. The subsequent decay of the original disturbance with time is governed by various climatic reservoirs and dissipative processes within the system. It is, however, not yet possible with the limited historical records of climatic change to closely associate observed climatic variability with particular perturbations and reservoirs. It is, nevertheless, widely recognized that short-term weather variability resulting from mesoscale and synoptic scale systems is a major source of noise in longer term climatic records. Much of the additional variability, at least of global average temperature, may be the result of variations in volcanic aerosol.

One further possible complexity, yet only established in very simple models for atmospheric winds, is the presence of "almost intransitive" states [36], i.e., there may be an infrequent switching from one

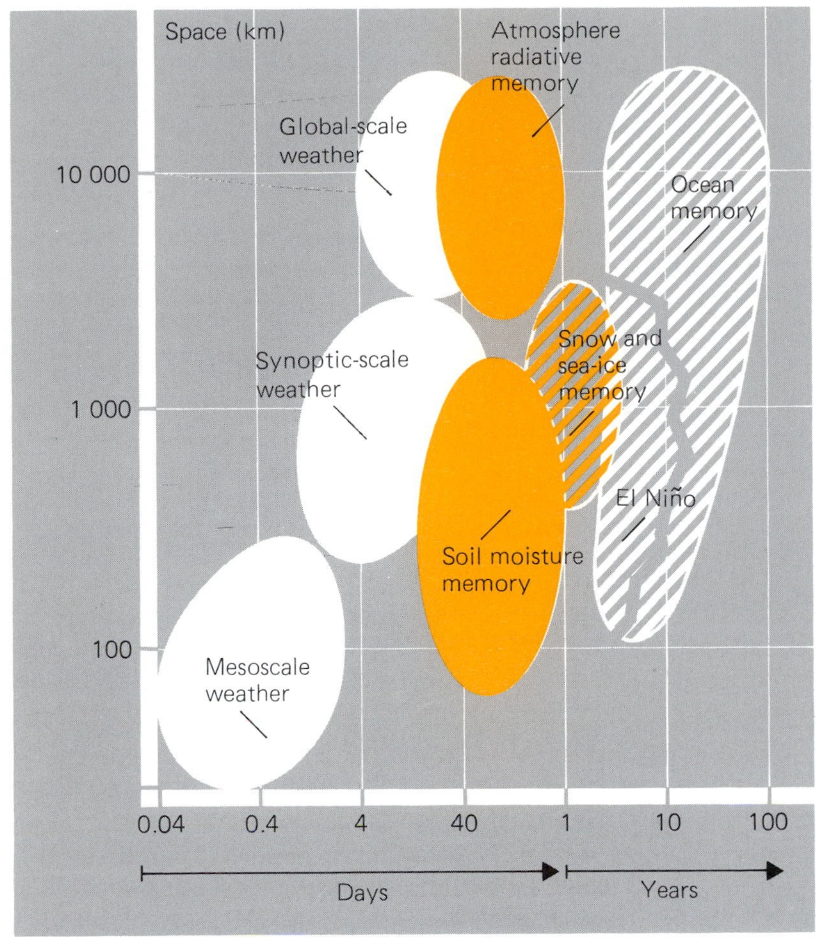

Figure 9.9 Space–time scales of the different climatic processes discussed in the text. The ordinate indicates representative space scales and the abscissa gives representative time scales.

climate regime to another. A mechanical analog would be a ball rolling around in a hole (i.e., a potential well) which, once in a while, is bumped to some other hole. This possibility of regime switching has often been suspected by practitioners of statistical climate prediction, who have found that statistical relationships established over one time period become invalid for a later time period. However, faulty technique and ignorance of slow system memories may have been the source of many such difficulties. If regime switching does occur in the climate system, it could be of major concern for evaluating the impacts of human activities on climate, as suggested by the mathematicians' use of the term "catastrophe" for such switches [37].

Regimes previously not accessible might in the future be reached by the climate system, or the rates of occurrence or lifetimes of some regimes might be altered [38]. Such possibilities are the crux of the question of resiliency of the climate system to human activities. Unfortunately, little progress has been made in studies on this aspect of climate beyond the provision of mathematical examples. Hence, most of what can be said of future climate is in a framework of linear analysis.

Fluxes of energy and water in the global climate system

Solar fluxes

The flux of solar energy in space at the Earth's average distance from the sun and integrated over all wavelengths is known as the solar constant. About 2% of this radiation occurs in the ultraviolet band at wavelengths <0.4 μm and is mostly absorbed in the stratosphere or above. Nearly half of the solar energy occurs in the visible part of the solar spectrum at wavelengths between 0.4 and 0.7 μm. The remaining solar energy occurs at near-infrared wavelengths, mostly from 0.7 to 4 μm. The solar constant has often been estimated during the last century, but with good accuracy only after the advent of rocket and satellite platforms. Recent measurements give a value of 1368 ± 7 W/m^2 [39]. Because the Earth has four times as much area as a flat disc of equivalent radius, the average solar flux incident on the top of the atmosphere is one fourth the solar constant, i.e., 342 W/m^2.

The global average of various vertical energy fluxes within the climate is illustrated in *Figure 9.10* [40]. As the solar radiation passes through the atmosphere, some is reflected and some is absorbed. Reflection includes upward Rayleigh scattering from atmospheric molecules (about 5%), and scattering from cloud particles (about 17%) and from aerosol particles (about 2%). About 24% of the incident solar radiation is absorbed by atmospheric constituents; 4% is absorbed in the visible and ultraviolet bands, mostly by O_3; and 20% is absorbed in the near-infrared, by water vapor and cloud droplets [41].

More than half of the solar energy entering the atmosphere reaches the Earth's surface, where further reflection takes place. About 6% of the original incident solar beam is reflected from the Earth's surface to space, in approximately equal proportions from the liquid oceans, from sea ice and snow cover, and from snow-free land [42]. Most of the flux incident at the surface is absorbed. On average, about 30% of the incident solar radiation is reflected from the climate system [43]. The fraction of solar radiation reflected into space is called the planetary albedo. About 240 W/m^2 of solar radiation is absorbed and available to drive climatic processes.

Thermal infrared emission and vertical energy transfer

The climate system in equilibrium must not gain or lose energy. Thus, the absorbed solar radiation is removed by other processes. On the global average, removal is by thermal infrared emission. Ocean, land, vegetative surfaces, and clouds emit radiation,

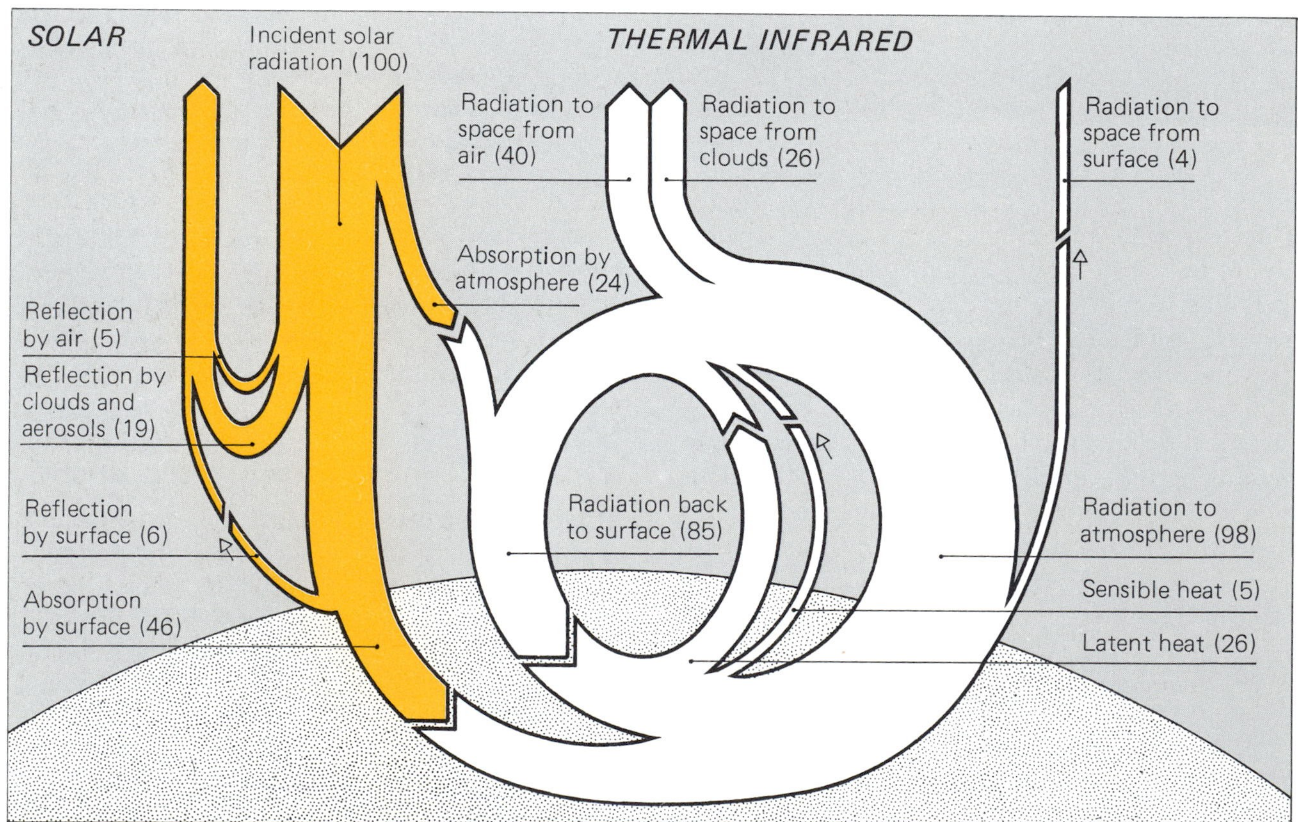

Figure 9.10 Global average of vertical energy fluxes [solar (yellow) and thermal infrared] and atmospheric absorption [40].

according to the Stefan–Boltzman law, in proportion to the fourth power of their temperature. Radiatively active molecules in the atmosphere also emit thermal radiation at a rate that increases with increasing temperature, depending on the wavelengths of their emission spectra. Water vapor, CO_2, and O_3 molecules are the dominant emitters. These, as all emitters of thermal radiation, are also absorbers. Thus, radiation emitted from the surface is absorbed in the atmosphere and a lesser amount reemitted from that same absorber. Some of this is, in turn, reabsorbed at a higher level and again a lesser amount reemitted. A fraction of the radiation emitted from the surface and from all levels in the atmosphere escapes to space. Indeed, at a high-enough level, referred to as the top of the atmosphere, there is no more absorption and all radiation passing through such a level escapes to space. Most thermal radiation reaching the top of the atmosphere has been emitted from the colder high levels in the atmosphere. Consequently, the amount of thermal radiation that leaves the atmosphere is equal to that which would have been emitted from a surface at a temperature of −18 °C, i.e., 33 °C colder than the global average surface temperature of 15 °C. A temperature of −18 °C corresponds to the average atmospheric temperature about 6 km above the Earth's surface.

Since much of the absorption of solar radiation occurs at the Earth's surface, but thermal emission is largely from high levels in the atmosphere, some mechanism for the vertical transfer of energy within the atmosphere must exist. The required upward movement of energy is provided, in part, through radiative absorption and reemission by atmospheric clouds and molecules. However, the bulk of the energy is transferred upward by small- and large-scale atmospheric motions. Some energy is carried by sensible heat fluxes, that is, upward movement of warm air. However, most is transferred as latent energy carried by water vapor. Thus, condensation of water vapor into rain and snow directly supplies most of the energy removed from the atmosphere by thermal infrared radiation.

Energy fluxes at the Earth's surface

Of the 240 W/m^2 of solar radiation absorbed in the climate system, approximately 150 W/m^2 is absorbed globally at the Earth's surface. This energy is returned to the atmosphere, about 80 W/m^2 as latent heat of water vapor, 20 W/m^2 as sensible heat, and the remainder as the difference between upward and downward thermal–infrared radiative fluxes [44]. Liquid water is turned into vapor at the surface through evaporation. The latent heat released into the atmosphere through rain and snowfall must balance the energy input to evaporation at the Earth's surface. Equivalently, the net global rainfall of about 1.0 m/yr is balanced by 1.0 m/yr of water evaporated. However, about one third of the rain and snowfall on land is not evaporated but instead is carried to the oceans by rivers. For this reason and because of the large fraction of land that is arid, the average evaporation from land per unit area is less than half that from the oceans, i.e., only about 0.5 m/yr [45]. In vegetated areas much of the evaporation occurs in the interior of leaves and the water is lost by transpiration through leaf stomata. Perhaps about half of the evaporation from land surfaces occurs in the form of transpiration. *Figure 9.11* [46] illustrates the various terms in the global hydrological cycle. In arid regions, evaporation generally cannot exceed the amounts of rainfall, which may be less than 0.3 m/yr. This relative shortage of evaporation in arid regions is compensated by excess fluxes of sensible heat, sometimes with daily averages exceeding 100 W/m^2. These relatively large amounts of heat added to the lowest atmospheric layers drive thermal convection and vigorously stir the daytime atmosphere through a thickness of at least several kilometers.

The lower atmospheric levels stirred by thermal convection and friction are known as the planetary boundary layer. During nighttime or wintertime conditions over land with no convective heating, this layer may collapse to a thickness of <0.1 km. It is the planetary boundary layer that serves as the immediate reservoir for pollutants put into the atmosphere from the Earth's surface. Thin or tightly capped (i.e., inversion layer) planetary boundary layers are thus responsible for increased severity of pollution levels.

Spatial and temporal variations

The incidence, absorption, and reflection of solar radiation and the responding fluxes of sensible and

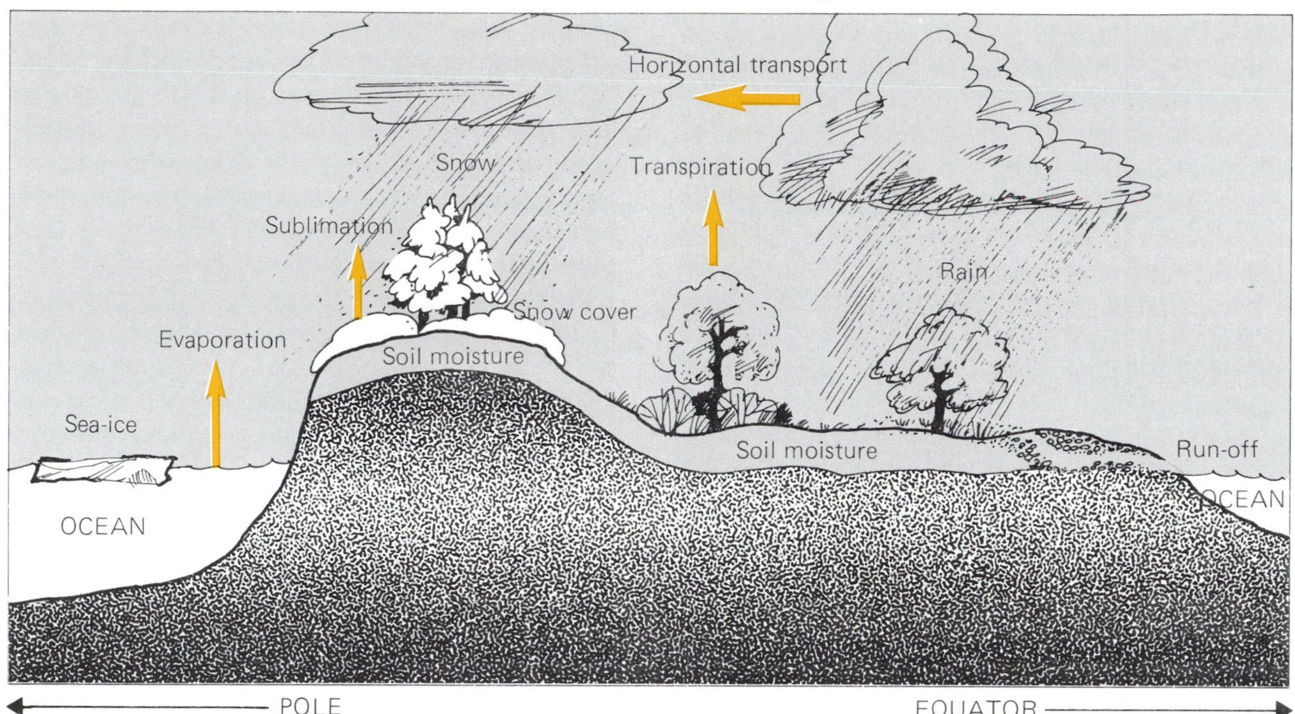

Figure 9.11 Sketch of global hydrological cycle processes.

latent heat and infrared thermal fluxes, as described above, vary with latitude, season, and other climatic conditions. The amount of absorbed solar radiation varies on an annual average from $>300\ W/m^2$ in tropical latitudes to $<100\ W/m^2$ in high latitudes, as illustrated in *Figure 9.12* [43]. The net radiative flux at the top of the atmosphere also varies widely with longitude, weather and climatic fluctuations, and seasons.

Global temperature change from disruptions of radiative energy fluxes

A framework for ranking potential global climate modifications

The atmospheric concentration of CO_2 has increased by about 25% since preindustrial times, largely because of the burning of fossil fuel [47]. Its concentration is expected to be more than double preindustrial values by the end of the twenty-first century [48]. But why are these points so remarkable? There are other trace gases in the atmosphere, albeit in more minute quantities, whose relative concentrations have been increased much more by human activities. Much of the haze over continental areas is now of human origin. There have been large decreases in recent years in the extent of tropical forests. Could these changes be yet more important for climate? Perhaps the construction of more highways or the increasingly large number of solar collectors on roofs could change climate. Some systematic approaches are obviously needed simply to compare and establish the relative importance for climate of all such changes in the environment caused by human activities. After such a comparison, those changes that appear most significant for climate can be studied further to improve quantitative projections of future effects.

The mathematical tools we use both to establish relative effects and to make detailed projections of future climate are climate models. The simplest such model is the zero-dimensional energy balance model, which is inferred from the global average balance between the absorbed solar radiation and the thermal infrared radiation leaving the atmosphere. Many more elaborate climate models have

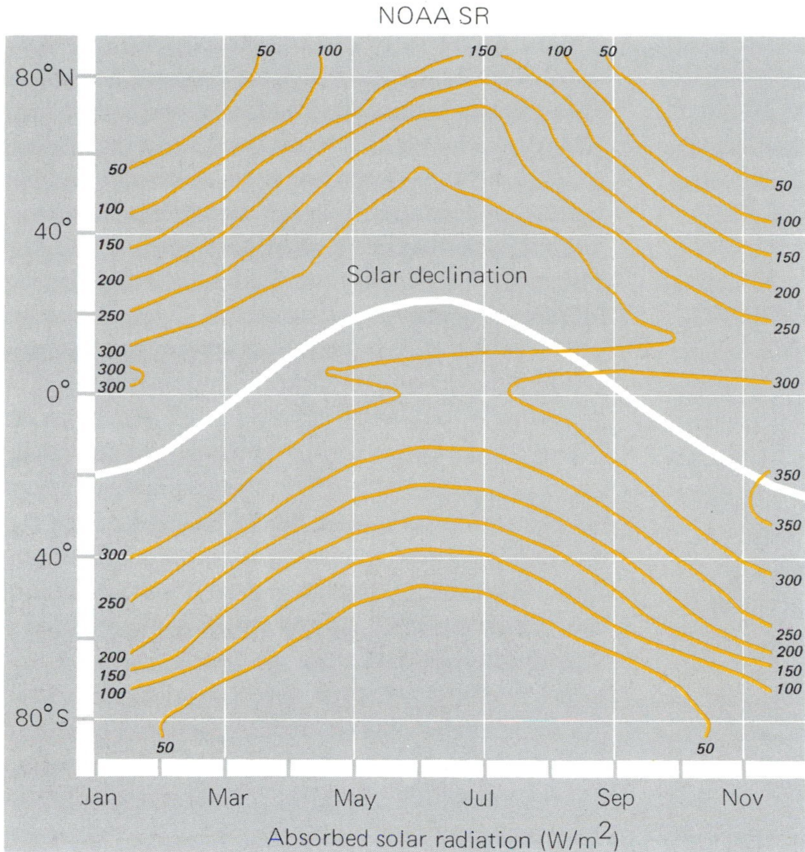

Figure 9.12 Absorbed solar radiation versus latitude and time of year [43].

been developed; the physically most complete are the general circulation models (GCMs), reviewed on pp 275–277. All such models satisfy in principle the constraint of global energy balance.

The term "external change" is used for any changes made outside the climate system that, in turn, can effect a change in climate [49]. Let ΔQ be the change in net global radiation at the top of the atmosphere resulting from some such external change, e.g., a changed solar constant or a change in concentration of radiatively active gases. For equilibrium, the climate system must adjust its loss of thermal and reflected solar radiation to balance ΔQ.

The thermal emission changes with changing temperatures of the Earth's surface, of atmospheric molecules, and of clouds. As discussed in the previous section, large-scale winds and ocean currents redistribute thermal energy horizontally within the climate system, and small-scale convective processes move energy vertically. To some extent, then, the temperature at any point in the climate system is linked to the global average surface temperature. In simple climate models, the net thermal emission leaving the atmosphere is taken to be directly proportional to surface temperature. Examples of possible failures of this assumption are discussed in the next section.

Changes in reflected solar radiation resulting from internal climate dynamics would largely be the result of changes in the extent and reflective properties of clouds or of large-scale fields of ice and snow. Sea ice in the Arctic Ocean and around Antarctica is especially significant because of its relatively large area in the spring and summer, seasons of greatest local incident solar flux [50]. The area of snow and ice is closely related to the extent of average air temperatures below 0 °C. Thus, many simple climate models have used local surface temperature to define the margin of high albedos due to snow and ice. To the extent that local surface temperature varies with global temperature, the contribution to global albedo by ice and snow is also proportional to global temperature [51].

Global feedback model

It is useful to quantify the discussion of potential changes of global radiation balance through the use of a very simple model. Global radiation balance is assumed to be approximated by

$$\lambda \Delta T = \Delta Q \qquad (9.1)$$

Here, ΔT is the change in global average surface temperature and λ is the global "feedback parameter", to be discussed later.

Equation (9.1) indicates that changes in the global fluxes of thermal or solar radiation due to human activities change global temperature in proportion to the magnitude of the flux change (*Figure 9.13*).

Application of equation (9.1) to estimate long-term climatic change simply requires values for ΔQ and for λ. However, if changes occur on a time scale of less than 100 years or so, it is necessary to consider an additional "thermal inertia" term in equation (9.1) that represents the fraction of ΔQ required to provide heat to the oceans. In this section, I am largely concerned with the various human contributions to ΔQ. However, some comments on a likely range of values for λ and ocean thermal inertia modification are necessary for perspective, though neither are well-established quantities.

The feedback term λ can be represented by the sum of individual contributions to climate feedback. Especially large and uncertain are the feedbacks associated with changes in atmospheric water vapor concentrations and in various radiative properties of cloudiness and the ice–snow albedo feedback; that is, the change of reflected solar radiation with changes in the cover of snow and sea ice resulting from temperature change (see pp 277–279).

I have developed elsewhere a detailed analysis of the most likely values of these individual contributions to the global feedback parameters and their uncertainties [52]. The most remarkable new inputs into such an analysis are the results of recent modeling studies [3, 53]. The first of these studies [53] indicates the possibility of a much larger cloud feedback than previously thought likely, whereas the second indicates the possibility of a large ice-albedo feedback. Together, these studies suggest that past estimates of the most likely values of λ may have been somewhat too large and that its precise value is considerably more uncertain than previously believed. For the purposes of this chapter, I take $\lambda = 1.5\ \mathrm{W/m^2\ °C}$ [54].

Ocean thermal inertia generally reduces the global temperature change from that implied by equation (9.1). For time scales too short for ocean surface waters to warm, i.e., a few weeks or less, the temperature change over land in response to small global radiation perturbations is only about 10% of the response implied by equation (9.1) [55]. For time scales of a few years or less, the oceanic heat uptake involves primarily the surface mixed layers. On time scales of decades, the thermal capacity of deeper layers must also be included and the transfer of energy to these layers may still reduce the global warming to not much more than half of that implied by equation (9.1).

Equation (9.1) can only be safely used without regard for ocean thermal inertia on a century time

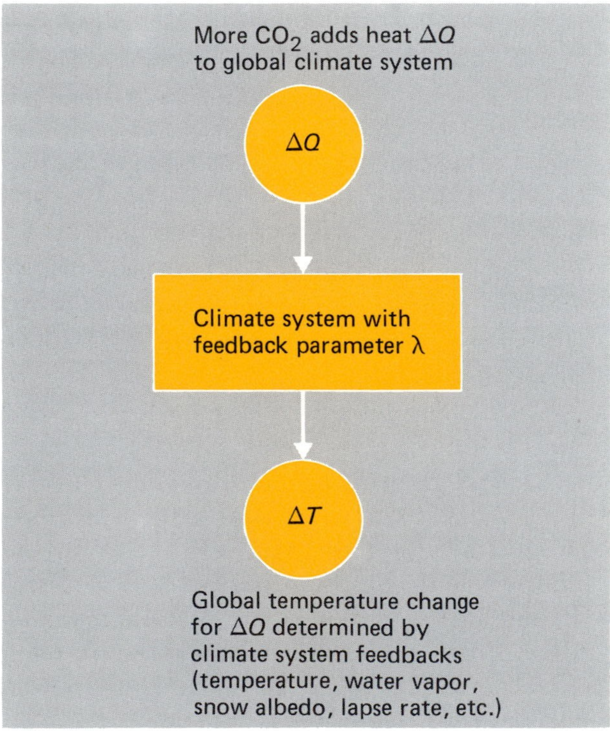

Figure 9.13 Calculation of global average surface temperature change from an energy balance climate model requires a prescription of the changed heating of the climate system ΔQ for present temperatures and the climate system feedback parameter λ.

scale. However, it still provides qualitative guidance for shorter time scales.

Comparison of natural and human contributions to variations in global radiation balances

One measure of the significance of changes in global energy balance mediated by human activities is their magnitudes relative to natural variations. Long-period natural variations are, however, still poorly characterized. Solar fluxes have been observed to vary around mean values by about 0.1% on a day-to-day basis [56]. Measured global temperatures over the last 100 years are not inconsistent with hypothesized solar output oscillations of the same magnitude with periods of 76, 22, and 12 years [57].

Volcanic aerosols in the past may have reduced absorbed global solar radiation by up to about 1%, averaged over a year [58]. However, these perturbations generally decay with the 2-year residence time of the stratospheric aerosol, so that long-term effects depend on the occurrence of multiple volcanic eruptions.

Global radiation balance may also vary with natural fluctuations in the climate system, e.g., year-to-year variations in cloud cover or snow cover. However, previous satellite measurements of global outgoing infrared radiation and reflected solar radiation have been too inaccurate to detect natural year-to-year variations at the expected levels of less than 1% of the absorbed solar radiation.

In summary, natural fluctuations in global absorbed solar radiation lie between 0.1 and 1% of the mean value of 240 W/m^2. On time scales of decades or longer, these variations are unlikely to exceed 0.4% or to be less than 0.1% of the absorbed solar radiation. Thus, external changes in global radiative fluxes as small as 0.1% are only of serious concern for future global climate, according to the above analysis, provided that many such changes of the same sign are added together.

For example, at current rates, enough additional CO_2 is added every 3 years to the atmosphere to reduce the thermal radiation leaving the troposphere by about 0.1%. Thus, the amount of CO_2 added to the atmosphere in any one year is of little concern, but the cumulative effect of a half-century's worth of additional CO_2 added to the atmosphere could easily exceed the 1% level of global radiation disruption.

Trace gases

There are several atmospheric trace species other than CO_2 whose concentrations are now increasing in the atmosphere (e.g., Crutzen and Graedel, Chapter 8, this volume) and whose potential effects on climate are known to be large and to add to the warming effect of CO_2 [59]. Gases in minute concentrations may be radiatively important if they have strong absorption bands in the 8 to 12 μm window of atmospheric thermal radiation, which contains about 30% of the upward infrared radiation emitted by the Earth's surface. At other wavelengths, strong absorption by atmospheric water vapor and CO_2 greatly weakens the effects of any other radiative absorbers. Gases in very minute concentrations absorb primarily in line cores and, thus, their effects are linear in relation to concentration, in contrast, e.g., to CO_2 whose effect varies with the logarithm of its concentration. Concentration of a trace gas must reach the order of 1 p.p.b. or larger for it to be important, given that it has strong absorption in the 8 to 12 μm window.

Extensive examination of all known atmospheric trace gases has revealed only a few that have large enough concentrations and strong enough absorption bands in the correct window to contribute significantly to future climate change. However, present indications are that practically all these gases are increasing in the troposphere and that their total future effect may be comparable to that of increasing CO_2 content.

Perhaps the most important trace gases from the viewpoint of future climate effects are the chlorofluorocarbons F-11 and F-12 ($CFCl_3$ and CF_2Cl_2, respectively). These compounds are analogous to CO_2 in that their current accumulation in the atmosphere is not only unambiguously known to result from human economic activities, but also emissions from the known sources of these compounds agree well with the measured global atmospheric increases. The compounds F-11 and F-12 are almost entirely inert in the troposphere and are only destroyed by hard ultraviolet radiation in the stratosphere. They thus have a lifetime in the climate system of 50 to 100 years [60]. Excess CO_2 has an even longer lifetime, being removed only by mixing into deep ocean waters and eventual incorporation into ocean sediments [61].

Another long-lived species of importance is nitrous oxide (N_2O), which also has a lifetime of about 150 years. Its primary natural sources are the soil and oceanic nitrogen cycles as mediated by microorganisms, but these are not well-quantified. The reasons for the current increase of 0.3% N_2O per year are not known [62], but this increase is thought to be more likely the result of fossil fuel combustion than of disruption of the soil nitrogen cycles. The implication is that any potential climatic effect due to changes in amounts of long-lived atmospheric compounds would last for at least a century, even after the sources of the additional concentrations have been removed.

Other important species for climate are methane (CH_4) [63] and tropospheric [64] and stratospheric [65] O_3. These compounds have relatively short lifetimes, but future changes in their concentrations may be the result of changes in long-lived trace atmospheric species or other, but slowly reversible, environmental changes. Thus, the climatic effects of changes in CH_4 or O_3 could be persistent. Atmospheric CH_4, the result of environmental microbiological processes, is now increasing for unknown reasons. The chemical cycles of tropospheric and stratospheric O_3 are known to be affected by other trace atmospheric gases, including those mentioned above. No significant trends in total global O_3 concentrations have yet been unambiguously detected, in part because of the relatively large natural variability of O_3 concentrations. However, O_3 in the lower troposphere above Europe and North America has increased by roughly 2% per year since the late 1960s [66].

Albedo changes from aerosol and land surface changes

Human activities can change the albedo of land surfaces or the radiative properties of the atmosphere through the introduction of aerosols. These factors are considered here only with regard to their effects on global radiation balance. Their regional climatic effects, likely to be larger, are discussed later. Past conversions of forests into agricultural lands may have increased the reflection of solar radiation by as much as 1 W/m^2 over human history, but probably by no more than 0.2 W/m^2 over the last 30 years [67]. The largest contributors to planetary albedo increase are removal of tropical forests and increasing the area of deserts. Changes likely over the next 100 years would increase planetary reflection by less than 1 W/m^2 [68] and probably not more than 0.3 W/m^2.

Humans introduce aerosols into the atmosphere through industrial activities, fossil fuel and biomass burning, and agricultural practices. Important components of these aerosols are carbon particles, sulfate particles, and natural soil particles. These aerosols have short, about 10 days, residence times in the atmosphere and their sources are localized, so their atmospheric concentrations are highly variable [69]. For this reason, and because of the wide range of radiative properties, the impact of anthropogenic aerosols on net radiative fluxes at the top of the atmosphere cannot be readily characterized, but is possibly small. The climatic effects of natural aerosols are somewhat better estimated.

The expected large increases in solar collectors by the year 2100 would decrease albedo over the area covered by the collectors, especially in desert regions, by as much as 0.1 to 0.3, but probably would decrease the global average absorbed solar radiation by less than 0.1 W/m^2. For example, an albedo decrease of 0.3 applied over 1% of the total land area would decrease the global average reflected radiation by about 0.1 W/m^2 [68]. Likewise, even a large amount of direct energy release would have only a small effect, e.g., 50 terawatts of energy release, five times the current usage, is equivalent to 0.1 W/m^2 averaged over the globe [70]. Thus, any possible global energy change due to surface albedo changes, aerosols, or direct energy release in the next 100 years is likely to be quite small compared to the expected radiative effects of trace gases by 2100, and perhaps even to natural variability.

Synthesis in terms of a surprise-free scenario for global warming

The predominant effect of future human activities on global climate is likely to be one of warming. Only modifications of surface albedo and perhaps increases in aerosol concentrations are expected to counter this tendency, and even then not by a large amount; a likely upper limit is 1 W/m^2. *Table 9.1* suggests the likely range of trace-gas warming effects by the year 2100. The value and uncertainty indicated for CO_2 are from the Monte Carlo analysis by Nordhaus and Yohe for future energy use given in the recent NAS publication *Changing Climate* [71]. Darmstadter, Chapter 5, this volume, discusses these and other energy projections and emphasizes the fragility of such projections, their great uncertainty, and strong dependence on modeling assumptions.

The concentrations of $CFCl_3$ and CF_2Cl_2, indicated in *Table 9.1*, correspond approximately to steady-state values assuming double the current emission rates. Doubling of N_2O, CH_4, and tropospheric O_3 concentrations from current concentrations has been suggested by a number of authors [72], but this is largely an educated guess inferred from current trends and concepts as to likely causes. The last four entries in *Table 9.1* are even wilder guesses [73]. They mainly illustrate the point that "all other" trace gases beyond the first seven entries in *Table 9.1* can probably be neglected in considering future radiative effects. All the 95% uncertainty limits, except that for CO_2, are my own estimates.

On the basis of the expected warming of 9 W/m^2, ΔQ in *Table 9.1*, I suggest that by the year 2100 the expected net effect of human activities since preindustrial times will be to increase global temperature by about 6 °C. According to the uncertainty range of the figures in *Table 9.1*, if there were no drastic changes from the expected levels of the world economy and development, the actual change in ΔQ would probably lie between 3 and 15 W/m^2. According to equation (9.1), with

Table 9.1 Trace gas radiative effect scenario for the year 2100

Species	*Current concentrations*	*Likely cause of increase*	*Expected concentrations in year 2100*	*Range of (95%) likely values*	*Expected ΔQ at top of troposphere* (W/m^2)	*Dependence of ΔQ on concentration*
CO_2	345 p.p.m.	Fossil fuel, soil, and biomass carbon	720 p.p.m.	500 to 1200 p.p.m.	4.5	Logarithm
CH_4	1.7 p.p.m.	Changes in OH or sources	3 p.p.m.	1 to 10 p.p.m.	0.5	Square root
N_2O	0.3 p.p.m.	Changes in nitrification and denitrification, fossil fuel	0.5 p.p.m.	0.3 to 0.8 p.p.m.	0.5	Square root
O_3 (troposphere)	Profile	Anthropogenic change in tropospheric chemistry	×2	×0.7 to ×5	1.2	Linear
O_3 (stratosphere)	Profile	Cl_x, NO_x	3% total column loss but 50% loss at 40 km, 5% increase in lower stratosphere	1% column gain to 20% column loss	0.2	Not known
$CFCl_3$	0.20 p.p.b.	Anthropogenic release	2.0 p.p.b.	1 to 5 p.p.b.	0.5	Linear
CF_2Cl_2	0.33 p.p.b.	Anthropogenic release	4.0 p.p.b.	2 to 10 p.p.b.	1.1	Linear
CCl_4	0.14 p.p.b.	Anthropogenic release	1.0 p.p.b.	0.1 to 10 p.p.b.	0.2	Linear
CH_3CCl_3	0.15 p.p.b.	Anthropogenic release	3.0 p.p.b.	0.5 to 3 p.p.b.	0.1	Linear
CH_3Cl	0.6 p.p.b.		1.0 p.p.b.	0.1 to 10 p.p.b.	0.01	Linear
All other chlorocarbons	0.1 p.p.b.		1.0 p.p.b.		0.2	Linear
				Total (rounded)	9±6	

$\lambda = 1.5\ W/m^2$, the consequence of this ΔQ change would be a global temperature increase of between 2 and 10 °C over past historical values. These results, of course, involve considerable subjective judgment [74] and are made without a thorough study of all the questions involved, ranging from those of future human economic development through biogeochemical cycles to the complex processes in the climate system. However, I believe my confidence limits and most likely estimates are defensible and could be much more thoroughly quantified without being drastically revised, provided we can assume that trace gas emissions will not be constrained to ameliorate their environmental effects. Thus, the estimates provide a useful basis for discussion of some aspects of future global climate. It is unlikely that temperatures will increase uniformly at all latitudes. However, it is not yet possible to estimate regional changes with any confidence [75].

Internal redistributions of energy in the climate system including disruptions of the hydrological cycle

In the previous section I discussed global climatic change resulting from changes in the absorption of solar radiation or in the emission of infrared radiation as viewed from the top of the atmosphere. Changes in the concentrations of atmospheric trace gases are the most likely cause of such a disruption of the global radiative budget. These gases, with the exception of tropospheric O_3, are long-lived compared to atmospheric-transport time scales of a few years or less, and so changes in their concentrations would occur nearly uniformly in all parts of the troposphere. Their radiative effects, likewise, vary but little with latitude.

In this section, by contrast, I consider possible future environment perturbations whose primary effect is that of redistribution of energy within the climate system rather than of changing energy fluxes at the top of the atmosphere. Significant redistribution effects may result in three ways:

(1) From relatively intense modifications of net incident energy fluxes.
(2) From changes in atmospheric or oceanic transport processes on a less than global scale.
(3) From a change in internal conversions between different forms of energy.

This last category involves primarily changes in the hydrological cycle. Because human activities are largely confined to land surfaces, redistribution effects are most likely to be found over land.

Intensive modifications of net incident energy flux

The most important intense modifications of net incident energy flux over large areas involve modifications of surface albedo by extensive changes in land use. The basic response of the atmosphere to concentrated changes in energy input is to redistribute that change over a broader area. The detailed climatic effects of this energy redistribution are thus a subject of inquiry, as well as the more local effect of the changed energy input. Activities of a potentially large-enough scale include: large-scale tropical deforestation, desertification of arid lands, large areal changes in land devoted to irrigated agriculture; large increases in continental aerosol; possible vast arrays of solar collectors. The first two of these activities could reduce the albedo at the top of the atmosphere over large areas, e.g., 0.1% to 1% of global land area by between 0.02 and 0.1 or so [76]. The last two activities could decrease surface albedos by comparable amounts [77]. Thus, net radiative heating of the surface–atmosphere system could be changed by as much as 5 to 25 W/m^2 over large areas with land-surface heating increased by up to twice as much. The actual size of areas and time spans involved depend on the nature and pace of future human development, which is not projected here.

A specific example that has been studied is the effect of albedo change on drought over the Sahel region of Africa. This region, on the southern margin of the Sahara Desert, has a width of about 5° latitude and extends across Africa, about 20° longitude (about $10^6\ km^2$ or 0.7% of total global land area). Analytic and numerical studies initiated by Charney *et al.* [78] support the following hypothesis. An increase of albedo over this region by 0.2 would promote a large (e.g., about 40%) decrease in rainfall during the summer rainy season directly over the modified area. Thus, if the albedo increase were due to removal of vegetation, the resultant climatic change would tend to maintain itself.

A decrease in net incident energy in an atmospheric column is necessarily compensated by an import of energy from elsewhere. The hydrodynamic response to this energy deficit is sinking motions over the region of energy deficit. These sinking motions warm the atmospheric column and suppress rainfall, increasing further the energy deficit. Cloudiness is also decreased. With less clouds, both the absorption of solar radiation and the loss to space of infrared radiation in the surface–atmosphere system are increased. The temperature at the ground may either increase or decrease depending on changes in net radiation received at the ground and changes in evapotranspiration. It is likely that these changes would be accompanied by changes in atmospheric climate of opposite sign outside the region of surface albedo modification. Teleconnections to remote regions cannot be excluded [79]. A qualitatively similar atmospheric response would be expected to occur with large surface albedo changes elsewhere; details would depend on the existing climate of the perturbed region.

Modifications of internal energy fluxes

Generation of electricity by transfer of colder, deep ocean water to the surface [80] is an example of modifications of internal energy fluxes in the climate system. The impact of such an activity on climate depends on its extent. Possible impacts of such modifications can be estimated by comparing their horizontal scale and the magnitude of their energy flux disruptions with that of natural fluxes. Ocean temperature anomalies of a degree or larger produce large local changes in the fluxes of sensible and latent heat to the atmosphere. In the tropics, these changes are primarily compensated by changes in convective rainfall. The heat released by convective rainfall drives local vertical motions. Through teleconnections, these motions can change climate elsewhere, even at great distances [81].

Rivers flowing into the Arctic Ocean help maintain the low salinity of near-surface water and hence the stable stratification of the Arctic Ocean [82]. Thus, the proposed large-scale diversion of rivers into the Caspian and possibly Azov Seas would increase vertical mixing in the Arctic Ocean and perhaps destabilize the Arctic ice pack. However, studies to date suggest that this climate change problem may not be serious [83].

The most catastrophic climate change that could be brought on by disruption of internal energy fluxes is that which would accompany a large-scale nuclear war. Large fires following a nuclear exchange could readily produce enough smoke to prevent sunlight from reaching the surface over much of the Northern Hemisphere for a period of several weeks [84]. Since the smoke would probably decrease planetary albedo, the analysis of the previous section would predict a net increase of temperatures. However, more detailed study [85] suggests that large surface cooling is by far the most likely possibility. The atmosphere near the top of the smoke layer would be strongly heated and its temperature increased by at least several tens of degrees. This change in the atmospheric temperature profile would greatly reduce convective energy exchange between atmosphere and surface. This reduced convective energy exchange, together with a large reduction of net radiation incident at the surface, could produce temperature decreases over land in the Northern Hemisphere summer of more than 10 °C. The temperature changes over land would require only a few days to be established because of the low heat capacity of land surfaces. Temperatures over oceans, on the other hand, would decrease but little in the presence of smoke clouds of only a few weeks' duration because of the large oceanic heat capacity, but oceanic changes that did occur could persist for several years or more. The short-term climatic effects expected from a nuclear war are illustrated in *Figure 9.14*.

Over longer periods, the dominant climatic effects of nuclear war could be due to changes in trace gas concentrations in the atmosphere and in land surface properties resulting directly from the large fires and indirectly from the impact of reduced sunlight on the biosphere [84].

Disruptions of the hydrological cycle

The climatic effects of nuclear war that occur for the duration of the smoke clouds would include large disruptions in atmospheric and surface hydrological processes. More benign future human activities may also change various aspects of the hydrological cycle and in this way modify climate over large areas of the Earth. The important climatic change questions include what changes of rainfall will there be, if any, in terms of, e.g., total amounts, time, and space distributions? What changes will there be in the moisture carried in the soil and in soil erosion? What changes will there be in stream flow, either as it affects water supplies or threats of flooding?

Changes in net incidence of energy at the top of the atmosphere have feedbacks with the hydrological cycle, as previously discussed. Changes in net radiative energy absorbed at the surface are balanced by changes in sensible and latent fluxes from the surface. Likewise, changes in rainfall imply modified surface evapotranspiration.

Direct changes in surface hydrology over large areas may occur as a result of changes in vegetative cover,or in the area covered by irrigated agriculture. Tropical deforestation, in particular, implies significant modifications of surface hydrology, depending on the properties of the subsequent vegetative cover. Disruptions of the land part of the hydrological cycle, *per se*, are likely to place more stress on the global biosphere than are the consequent effects on the atmosphere. Of considerable importance are the effects of disruptions in vegetation cover on water-storage systems. Crosson, Chapter 4, this volume, emphasizes the large economic damage of silt loading. Changes in water supply resulting either from changes in vegetative cover or future climatic change are also of serious concern.

The micrometeorological effects of tall vegetation differ considerably from those of short vegetation. Because of greater surface roughness, interception, i.e., reevaporation from leaves, usually removes a larger fraction of incident rainfall from tall than from short vegetation and makes the evapotranspiration more dependent on atmospheric humidity and leaf temperature and less dependent on net incident radiation [86].

To the extent that runoff is increased, evapotranspiration is decreased. A change in the rate of evapotranspiration relative to that of sensible heat fluxes could modify the time and spatial characteristics of rainfall. Larger sensible heat fluxes would

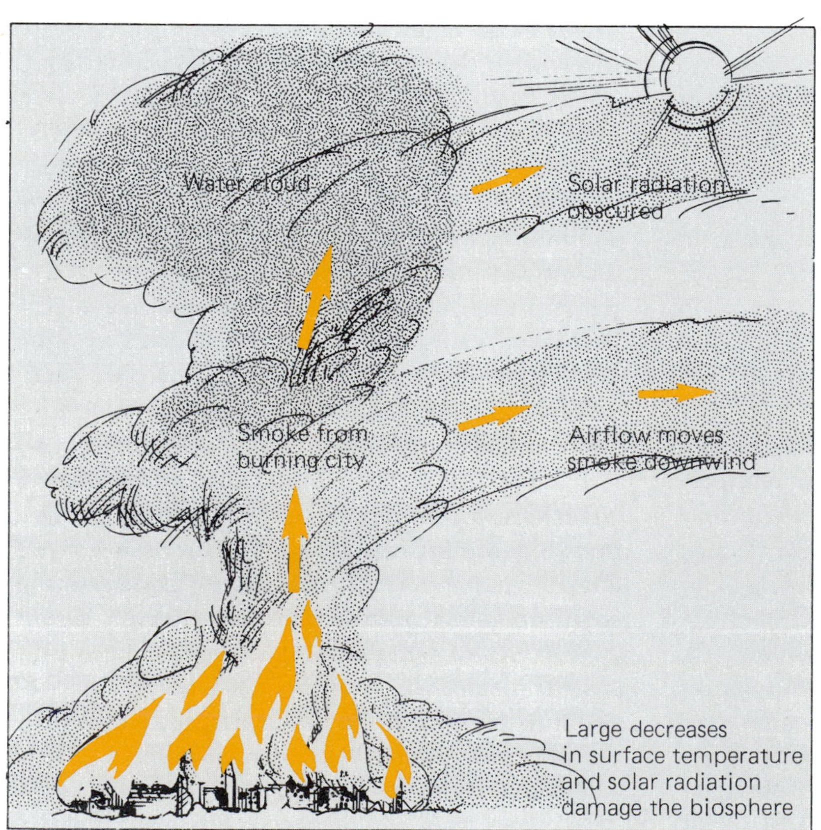

Figure 9.14 Climatic effects of nuclear war.

likely promote more intense, more local, and more sporadic convective rainfall. Such a change in the rainfall is also a likely consequence of a warmer climate, such as would result from increased concentrations of atmospheric trace gases. More intense and local rainfall would produce more soil erosion damage, more runoff, and more drought-like conditions, even if the total average rainfall remained the same.

Conversion of forest or semiarid grasslands into irrigated agriculture [87] replaces a surface that may have had considerable resistance to evapotranspiration with a surface that both freely transpires and freely evaporates water vapor. Increases of water flux to the atmosphere with a change to irrigation would thus be largest in arid lands. Detailed studies of the possible climatic consequences of large-scale conversion into irrigated agriculture have not yet been made and would require realistic descriptions of the future extent of such a conversion.

Models for simulating the climate system

Processes modeled

The climatic processes described in previous sections can be quantitatively treated only through the use of complex computer models [88]. This section is intended as a brief introduction to such models. The most comprehensive models of the atmospheric part of the climate system are called general circulation models (GCMs) [89]. These models have been developed over the last 30 years in conjunction with the development of numerical weather-prediction models [90]. The atmosphere is represented by a three-dimensional mesh of points, typically separated by about 5° latitutude or longitude and by about 1 km in the vertical dimension. The lower boundary for this system is the Earth's continents and oceans with topography smoothed to the model resolution. The models solve for atmospheric winds, on an hour-by-hour basis, using the fluid form of Newton's equations of motion. These winds are driven by pressure differences that are generated by spatial variations in temperature. Temperatures are calculated by application of the laws of thermal energy conservation at each model grid point. Clouds are formed in the models with properties that determine their effects on solar and thermal radiation.

Thus, GCMs in simulating atmospheric climate also necessarily simulate all the day-to-day weather variations whose statistics make up climate. Temperatures are changed by atmospheric motions, latent heat release, absorption of solar radiation, absorption and emission of thermal radiation, and by transfers of sensible heat from the surface. The GCMs also solve an equation for conservation of water vapor at each grid point. Water vapor is supplied to the atmosphere by the surface through evaporation. It is moved around in the atmosphere by winds and removed from the atmosphere by condensation into rainfall.

A GCM becomes a complete climate model with the addition of submodels for processes at the Earth's surface which maintain conservation of energy and water for the total climate system. These submodels include an ocean model, a model for sea ice, and models for land surface processes, including soil moisture and snow cover. With these ingredients, a GCM responds to the constraints of global energy balance as discussed on pp 263–266. The ingredients of a climate model are illustrated in *Figure 9.15.*

Some current limitations

The GCMs have tended to use submodels that are considerably oversimplified compared to state of the art descriptions for surface processes. Oceans, for example, are in reality as complex a dynamical system as the atmosphere and their surface temperatures, the condition needed for the atmospheric model, can only be properly obtained using ocean GCMs coupled to the atmosphere. However, it has not been practical to use ocean GCMs in most climatic change studies. Instead, the horizontal transport of heat by the oceans has been neglected and their temperature calculated assuming the ocean at each grid point to be a uniformly mixed reservoir for heat. Sea ice has, likewise, been represented by a one-dimensional slab controlled only by thermodynamic processes. The important effects of sea-ice movement have been ignored.

Oversimplified ocean and sea-ice models, together with questionable descriptions for ice and

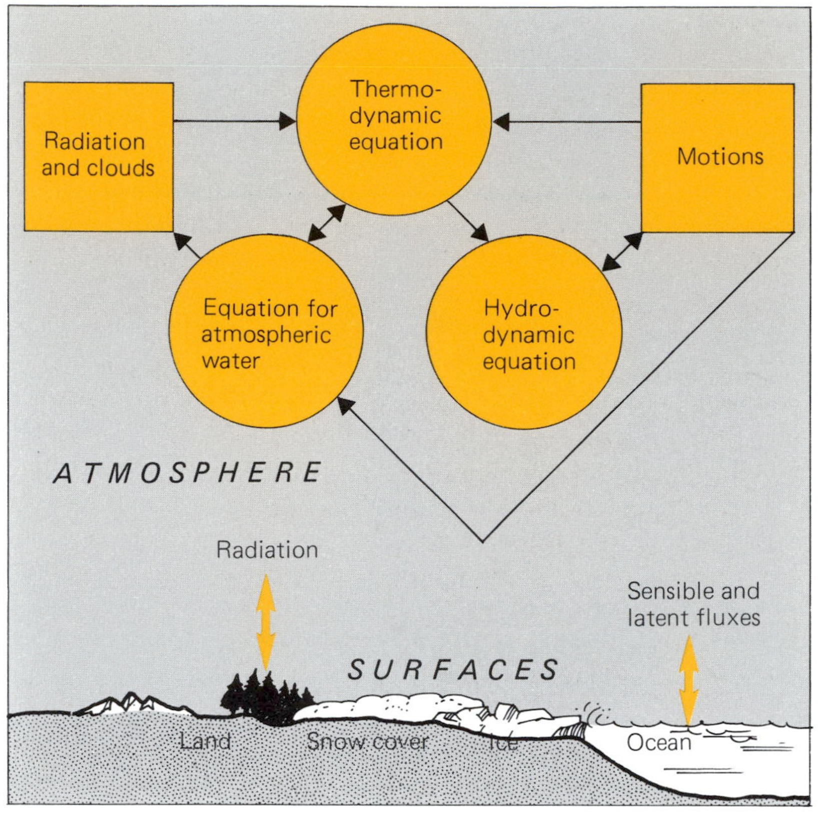

Figure 9.15 Components of three-dimensional climate models.

snow albedos, suggest that the ice albedo radiative feedbacks in GCMs are of doubtful accuracy [91].

Likewise, the models used for land surface water, snow budgets, and evapotranspiration have been too simple to allow quantitative evaluation of the effects of realistic changes in land surface processes on climate. Modelers have instead concentrated on large arbitrary changes in surface albedo or in wetness to help demonstrate the hypothesis that such changes could be of major importance for climate. It is likely that much more sophisticated and realistic descriptions of oceans, sea ice, and land surfaces will be incorporated into GCMs in the next few years [92]. One practical constraint of any global model, including GCMs, is their relatively coarse horizontal spatial resolution, i.e., a few hundred kilometers or more. This limitation is a result, in part, of lack of input data on finer scales, but more especially of the rapid increase of computational cost with higher resolution. Many of the land processes, both natural and human mediated, that interact with the climate system occur as much finer scale mosaics than can be represented in the models. Techniques for averaging over this fine structure to obtain inputs at the model resolution are still poorly developed. These difficulties limit the degree of confidence that can be placed in model-calculated responses to changes at the Earth's surface that have a detailed spatial structure. Presumably, the development and application of procedures for coupling global scale to mesoscale models could alleviate some of the resolution difficulties.

Model variability

The day-to-day weather variations that GCMs generate in the processes of simulating climate provide both an opportunity and a limitation. As discussed on pp 254–263, impacts from future climatic change will not only depend on shifts in climatic mean terms but also on both the present weather variability and possible changes in it. The questions of whether and how much weather variability would change with climatic change can be considered only with GCMs. The existence of model variability, however, implies that exact estimates of mean climatic changes would only be possible for an infinite ensemble of climate simulations. Any one simulation consists of a climatic signal with superimposed model "noise". Climatic change must be obtained from the difference between a noisy control simulation and a noisy perturbation simulation [93]. The practical effect of this limitation is that only climatic changes large compared to natural variability can be quantitatively obtained. Calculating the collective average and variances about this average of several control and several perturbation simulations allows estimates of the variability of model averages, that is, estimates of the model noise level. Such estimates can also be made by sampling over several time intervals sufficiently far apart in the simulations to be uncorrelated.

Usefulness of climate models for estimating future climatic change and its impacts

Climate models, in particular GCMs, are our only means for determining bounds for possible future climatic changes. Details as to future global changes are still rather uncertain, and possible regional variations from global changes are still largely speculative. However, even with this present state of the art, GCM climate models do produce physically consistent scenarios for future climates that can be obtained by no other means and that include regional and temporal variability. This variability, though not likely to be correct in detail, should, nevertheless, still be quite useful for developing impact study methodologies sophisticated enough to recognize the importance of such variability.

Not entirely improbable climatic changes that might result from human activities

The most extreme climatic changes that have been discussed theoretically are the ice-covered Earth and runaway greenhouse instabilities. These examples are mentioned here, first to dismiss them as too improbable to be of serious concern, even though either of them would annihilate the terrestrial biosphere. They are mentioned, second, to suggest possible paths of exploration for less catastrophic, but more probable future extreme climate changes.

On p 268, I introduced equation (9.1), $\lambda \Delta T = \Delta Q$, to compare possible perturbations to global radiative balances. Evidently, for a given ΔQ the feedback parameter λ determines the equilibrium response of global temperature. The global feedbacks that may contribute most significantly to λ are water vapor, ice albedo, and cloud radiative feedbacks. The most extreme possible future global climatic change is thus defined by combining the upper limits to possible ΔQ with a lower limit to λ. Thus, it is appropriate to discuss the processes contributing to λ and their possible ranges.

Water vapor feedback

Water vapor feedback acts to increase climatic change. That is, with a warmer climate the concentrations of atmospheric water vapor increase. Water vapor is, at present, the largest contributor to the trapping of outgoing infrared fluxes, so increased water vapor would enhance considerably an initial warming. This positive feedback is estimated to reduce λ by roughly 1.5 W/m^2 °C for present conditions [94]. The precise value depends on possible changes in the vertical and latitudinal distributions of water vapor with climatic change [53].

The action of positive water vapor feedback depends on increases in atmospheric and surface temperatures. Could water vapor feedback be increased enough to make λ negative? Any climatic state with a negative value of λ would be unstable and change to some other state [95]. In the present context, temperatures might then increase to a much hotter value, reaching or exceeding the boiling point of water. This "runaway greenhouse" effect would only be possible, if at all, for a climate many tens of degrees warmer than today's. Whether such an instability is theoretically realizable appears to depend on concomitant changes in the vertical distribution of water vapor and in the vertical temperature profile [96]. The more immediate question is whether or not less drastic and more likely warming situations would be modified by smaller increases in water vapor feedback.

Ice albedo feedbacks

Ice albedo feedback refers to change in the cover of ice and snow with changing surface temperatures. More ice and snow are expected with colder temperatures and would reflect more solar radiation and hence further reduce temperatures. Currently, ice albedo feedback is believed to decrease λ by about 0.5 W/m^2 °C, with various studies providing estimates between about 0.0 and 0.9 W/m^2 °C [97]. This positive feedback increases with colder temperatures and may have been much larger during past ice age climates. If ice albedo feedback were large enough to make λ negative, the climate system would drop to another equilibrium configuration with complete global cover of ice and snow. Climate modeling estimates of the negative ΔQ required to give this instability range between 2 and 10% of the absorbed solar radiation [98].

Recent studies with seasonal models have suggested that ice albedo feedback is dominated by the effects of changing cover of sea ice [99]. With increasing temperatures, a maximum possible change in planetary albedo from decreased snow and ice can be estimated from the albedo decrease resulting from the removal of all sea ice and seasonal snow cover. The Antarctic and Greenland continental ice sheets are presumed to be unable to change much in the next 100 years [34]. For a rough estimate, we consider only sea ice, taken to cover 5% of the globe and reflecting a seasonal average of about 60 W/m^2 more than would an ocean surface. These figures imply an upper limit of about 3 W/m^2 for the maximum possible increase in absorbed solar radiation from a decrease in sea ice due to a warmer climate.

Cloud radiative feedbacks

Cloudiness influences both the planetary albedo and the trapping of infrared radiation. Clouds with low-level tops are important primarily for their reflection of solar radiation, which depends on the area of cloud coverage, the amount of liquid water in the cloud, and the size of the cloud droplets. Clouds modify infrared fluxes by emitting radiation at colder temperatures, that is in lesser amounts than the radiation emitted from lower levels and from the surface, which is absorbed in part by the cloud bottoms [100].

High, thin cirrus clouds may be largely transparent to solar radiation, but are still important for trapping infrared radiation. Since tropospheric temperatures generally decrease upward, increases in the altitude of cloud tops reduces emission of infrared radiation to space and warms the climate system [101].

The existence and radiative properties of clouds depend on local dynamical and convectional processes and on the nature and distribution of cloud condensation nuclei. There are few obvious connections between cloudiness and global climatic changes. A warmer climate would be expected to have higher rainfall rates, thus more intense storm systems, and probably more convective cloud systems. Cloud radiative properties might conceivably provide a large positive feedback on global climatic change, but this feedback is sufficiently uncertain that even its sign is doubtful [53, 102]. Present cloud properties and possible changes over sea-ice margins are of special concern for evaluating the magnitude of ice albedo feedbacks.

Limits to potential climatic changes due to disruption of global radiative fluxes

One conclusion from the above discussion is that the precise value of the global feedback parameter λ is rather uncertain even for present conditions, but especially for large changes from present conditions. A probable lower limit to λ, excluding ice albedo feedback but otherwise for present conditions, is about 1.2 W/m^2 °C which would only be realized with a large positive cloud feedback. If we take 15 W/m^2 (*Table 9.1)* as the maximum likely increased global heating by the year 2100, add to this 3 W/m^2 for ice albedo feedback, and neglect possible effects of ocean thermal inertia, we infer a maximum possible global temperature increase of about 15 °C. Since the upper limit to heating and the inverse to λ are estimated to be reached or exceeded with a probability of 2.5%, a global temperature increase equal to or greater than 15 °C formally has a probability of 0.06%. However, a warming by 9 °C or more by the end of the twenty-first century has a 1% probability according to my analysis. Thus, it is not totally improbable that the climate system could achieve in the next 100 years conditions as warm as the Cretaceous era of 100 million years ago when polar temperatures were 10 to 20 °C warmer and tropical temperatures were perhaps 5 °C warmer than present. With such drastic temperature increases, sea levels would rise by more than 1 m as a result of ocean thermal expansion. Another 6 m of sea-level rise would be possible, if the West Antarctic ice sheet began to disintegrate rapidly [103]. A year-round ice-free Arctic Ocean would itself greatly change climatic conditions in high northern latitudes.

It is also not entirely improbable, on the other hand, that global temperatures in the next 100 years would change little from their present values. However, the chance of an occurrence of ice-age-like conditions in the next 100 years is extremely improbable, less than one chance in a million, in my opinion. Drastic climatic change resulting from the impact of a large meteorite, as believed responsible for global extinctions at the Cretaceous–Tertiary boundary, is even less likely [104].

Extreme regional changes

On the basis of the current large rates of deforestation, it is plausible that natural tropical forests will largely disappear over the next 100 years [105]. The most sensible replacement would be managed silviculture and agricultural systems of comparable or greater productivity. However, conversion into vast expanses of wastelands with little vegetative cover of economic value is not entirely improbable. Such a change would imply large changes in regional and perhaps global climate.

Also, whereas deserts are most likely to stay approximately where they are over the next 100 years, it is not totally impossible that human misuse and overuse of arid lands could accelerate natural processes of desertification to the point where most arid lands would be converted into deserts over the next 100 years. Extensive desertification would also imply quite significant modifications in regional and global climate.

As a possible trend in the other direction, the extent of irrigated agriculture in arid lands could greatly increase with concomitant implications for the climate system [106].

None of these extreme possibilities of large-scale land changes has yet been studied by climate modelers with any degree of realism. However, likely bounds on possible climatic change could be established through consideration of past modeling studies of the effects of large hypothetical changes in land surface hydrology or albedo [107].

Concluding discussion

Summary of the analytical framework for climate change

Some related themes are emphasized in this chapter. First, it is very important to recognize that the climate system cannot be simply represented by average conditions, but involves a wide range of spatial and temporal variability.

Second, human activities may change climate on a global scale through large-scale disruptions of energy fluxes. The possible disruptions are of two kinds. The first consists of global changes in the net radiative-energy absorption of the climate system. Such a change is likely to modify global temperature and produce other accompanying climatic variations. The second kind of disruption has as a primary impact a redistribution of energy within the climate system. Energy redistributions may be forced by less-than-global changes in net radiative absorption, or by disruptions in internal energy fluxes within the climate system, or by modifications of hydrological processes, especially at the surface.

The most powerful tools for quantitative evaluation of the climatic consequences of energy flux disruptions are climate models, especially GCMs as described on pp 275–277. However, in most cases considerable improvements in the models are necessary before adequate confidence can be placed in the results.

Characterizing extreme climatic change

Research into the extreme possibilities of future climatic change might be both different from and more useful than research devoted to most likely possibilities. From a rational decision-making, economic viewpoint, reducing the *extreme boundaries* of 95 or 99% probable climatic change may be much more valuable than attempting to refine the statement of the *most likely* climate change. Past modeling studies have helped to elucidate ranges of possible climatic change, but on an accidental and nonsystematic basis. That is, much of the modeling work has not yet reached a state of maturity where there are consistent and accepted treatments for cloud radiation and ice–snow albedo feedback processes and, even for the same formulations, these processes may give rather different results in different GCM models. Thus, the Hansen GCM [53] has a large positive cloud feedback unlike that found in any other model, and the Washington and Meehl [3] seasonal cycle GCM simulations have indicated a larger ice albedo feedback than found in any other model. With an excessive insistence on a scientific consensus, such "outlier" results are likely to disappear; modelers are generally quite good at tuning model parameters within the range of possible values so that their model provides what are regarded as acceptable results. Modelers could be encouraged to examine the possibilities of promoting simulations of extreme possibility by suitable selections of parameter values within their acceptable range. For example, what would result from combining Hansen's large cloud feedback with Washington and Meehl's large ice albedo feedback?

In principle, important model parameters can be identified, their values assigned a probability distribution, and these parameters varied according to Monte Carlo simulation techniques. However, besides extreme computational expense, such an approach would first require much more understanding and agreement than at present as to what are the most important model parameterizations, what are optimum formulations, and how uncertain they are. For example, in spite of intensive effort over the last decade to solve the cloud radiation problem, no solution is yet in sight.

Another approach to characterizing extreme possibilities of climatic change is through study of the past observational record. To the extent that past radiation balance changes can be identified, past climatic variations help put bounds on the sensitivity of climate models to future perturbations in radiative forcing.

Current monitoring systems, extended into the future, should attempt to detect not only shifts in mean conditions, but also shifts in the occurrence of extreme events and the occurrence of previously unrecorded extreme events.

Closer interaction between climate modeling and impact studies

Modelers sometimes marvel at the (seeming to them) relatively large efforts devoted to the study of climatic impacts. They may fail to recognize the extent of effort needed simply to develop effective methodologies in the impact area. Impact experts and other scientists whose studies relate to questions of climatic change may, on the one hand, have unrealistic expectations of the level of information that might be available from climate models while, on the other hand, actually consider in their studies much less information than could be available from climate models. Third world countries may lack the scientific expertise to evaluate their own situations in light of the conclusions provided by climate modeling and impact studies.

Closer communications between the different interests mentioned above are clearly desirable. Yet many disciplinary experts already feel resentful of the amount of their time they must devote to participation in interdisciplinary workshops and writing for interdisciplinary or nonexpert audiences. How can interdisciplinary communications be better promoted in the area of climate studies? More meetings and papers or international programs are probably not the answer. I think the solution must lie in a closer integration between the research programs of modeling centers and the research activities of the users of climate studies. This objective might be achieved, in part, through more extensive programs of visiting scientists. Perhaps electronic communication systems can be used to maintain contacts between already acquainted colleagues in cross-disciplinary areas.

A specific example of modeling information that should be (but has not yet been) exploited for impact studies is the wide range of natural weather and climate variations that are simulated by GCMs. Most of this information is lost in the process of condensing model output into a form suitable for publication. Climatic impacts that depend on frequencies of threshold events might, for example, be most effectively studied by evaluating the changes in frequencies of threshold events directly within the models rather than by using model estimates of mean changes to evaluate threshold events. Such a study would require making the desired information one of the quantities that is saved and analyzed from the ephemeral streams of millions of numbers that are spewed forth during any given model simulation.

Directions of climate model improvements

Substantial improvements in climate modeling capabilities occur on a 10-year time scale and are now actively being promoted by the World Climate Research Program. With emphasis on a continuation of current research, the climate modelers should, in a decade, be able to include, with some confidence, simulations of cloud radiative properties. The inclusion of realistic ocean models will probably still be in its infancy, but perhaps be done well enough to allow sea-ice albedo feedbacks to be predicted with some degree of confidence. These improvements may allow predictions of future global temperature change for given scenarios of radiative absorbers with confidence limits of, e.g., ±30%.

To these improvements perhaps will be added refined parameterizations of land surface processes and an improved understanding of the atmospheric and surface hydrological cycle and its interactions with internal atmospheric dynamics. The modeling advances should be sufficient to allow model predictions on the scale of a thousand kilometers or so, and for monthly averages to be made with some degree of correspondence to expected future climatic change. Besides simple climatic mean change, such quantities as possible changes in stochastic variability, diurnal amplitudes, etc., could be useful.

A decade from now, limitations of global model resolution may still preclude reliable information on a scale of a few hundred kilometers or less. This defect might be alleviated, in part, over selected areas if the technology of coupling limited-area mesoscale models to global models is adequately developed. In any case, models will never provide precise predictions. The necessary existence of natural variability in models and in reality and of remaining model limitations imply that the maximum available information on future climatic change must necessarily be a probability distribution of outcomes. Thus, the improvement of methodologies for establishing uncertainty ranges and, ideally, probability distributions for model projections should be emphasized.

Understanding climatic change as part of a system for sustainable development of the biosphere

If the modeling improvements discussed above are realized, the resultant tools should allow an evaluation of potential climatic change and its impacts to be an important aspect of rational decision making regarding the directions of future human development. Furthermore, increased capabilities for monitoring climate, as discussed by Izrael and Munn, Chapter 13, this volume, may be able to suggest mid-course corrections in developmental activities when model predictions stray from reality. Such possibilities are but part of the likely further large improvements in modeling and monitoring all the aspects of global environment as we plunge further into the present "information age" [108].

If the anticipated progress in climate modeling and monitoring occurs, serious constraints would still remain in the creation of "usable knowledge". Many of the modifications of the climate system now occurring as a result of human activity require hundreds of years to be reversed, and it is questionable whether human institutions can halt these modifications. Thus, adaptation may be the only feasible response to large-scale climatic change. A strategy of adaptation would result in the least disruption of social systems, especially with the benefit of early warning from modeling studies and monitoring systems [109].

Climate models can project future climate only with considerable uncertainties. Although there is and must be the uncertainty resulting from inaccuracies in the models themselves, greater uncertainty lies in the projection of future changes in the model inputs, e.g., the future concentrations of atmospheric CO_2 and other trace gases. Understanding better the likely changes in such inputs requires, on the one hand, an improved understanding of the natural biogeophysical system as proposed in the IGBP program (McElroy, Chapter 7, this volume) and, on the other hand, a better understanding of the human systems that may produce stress on the climate system.

Sustainability in fragile climate areas

The most significant future climatic change may occur globally. However, any such change will vary from region to region in both its manifestation and impact. Developing future sustainable biospheric systems may require that special attention be given to relatively undeveloped but ecologically fragile lands. Such lands occur in the Arctic [110] and South American tropics (e.g., as reviewed by Crosson, Chapter 4, this volume). The climate system may be especially sensitive to changes in these regions [111]. Furthermore, both the Arctic and the humid tropics are likely to be greatly affected by climatic change over the next 50 years. By then, summer Arctic sea ice may have largely disappeared and the tropics may be under increasingly severe stresses from increased flooding, soil erosion, seasonal aridity, and, for some crops, heat stress. These changes may have considerable implications for the role of these regions in the atmospheric chemical system. Thus, some emphasis in research programs involving biospheric sustainability should be given to better understanding of the human, biotic, and climatic interactions in these two regions.

Further reading

Carbon Dioxide Assessment Committee (1983), *Changing Climate*, 490 pp (National Academy of Sciences, Washington, DC).

Dickinson, R. E. (1982), Modeling climate changes due to carbon dioxide increases, in W. C. Clark (Ed), *Carbon Dioxide Review: 1982*, pp 101–133 (Oxford University Press, New York).

Houghton, J. T. (Ed) (1984) *The Global Climate*, 233 pp (Cambridge University Press, Cambridge, UK).

Ingersoll, A. P. (1983), The atmosphere, *Scientific American*, **249** (3), 162–175.

Jäger, J. (Ed), *Climate and Energy Systems. A Review of Their Interactions*, 231 pp (John Wiley and Sons, New York).

Schneider, S. H. and Londer, R. (1984), *The Co-Evolution of Climate and Life*, 563 pp (Sierra Club Books, San Francisco, CA).

Notes and references

[1] This trapping is commonly called the "greenhouse effect". It is due to atmospheric clouds and radiatively active atmospheric gases, especially water vapor and CO_2, and to a lesser extent O_3, CH_4, N_2O, and the chlorofluorocarbons. For a summary of the role of atmospheric radiative absorbers in the climate system, see Dickinson, R. E. (1982), Modeling climate changes due to carbon dioxide increases, in W. C. Clark (Ed), *Carbon Dioxide Review: 1982*, pp 101–133 (Oxford University Press, New York).

[2] For example, in the moist tropics the mean daily temperature range is typically 8 °C, the mean seasonal range about 2 °C, and periods of extreme weather give ±5 °C variations about the mean conditions. As a second example, in Siberia the mean seasonal temperature range is 50 °C, the mean daily temperature range is 15 °C in summer, half that in winter, and periods of extreme weather vary the temperature by ±25 °C about the mean conditions.

[3] Washington, W. M. and Meehl, G. A. (1984), Seasonal cycle experiment on the climate sensitivity due to a doubling of CO_2 with an atmospheric general circulation model coupled to a simple mixed layer ocean model, *Journal of Geophysical Research*, **89**, 9475–9503.

[4] Manabe, S. and Stouffer, R. J. (1980), Sensitivity of a global climate model to an increase of CO_2 concentration in the atmosphere, *Journal of Geophysical Research*, **85**, 5529–5554.

[5] Wigley, T. M. L., Angell, J., and Jones, P. D. (1985), Analysis of the temperature record, in M. C. MacCracken and F. M. Luther (Eds), *State-of-the-Art Report on the Detection of Climate Change*, Ch 4 (DOE, Washington, DC).

[6] Barron, E. J. and Washington, W. M. (1984), The role of geographic variables in explaining paleoclimates: results from Cretaceous climate model sensitivity studies, *Journal of Geophysical Research*, **89**, 1267–1279.

[7] US Committee for the Global Atmospheric Research Program (1975), *Understanding Climatic Change: A Program for Action* (National Academy of Sciences, Washington, DC).

[8] Gribbin, J. and Lamb, H. H. (1978), Climatic change in historical times, in J. Gribbin (Ed), *Climatic Change*, pp 68–82 (Cambridge University Press, Cambridge, UK); Kutzbach, J. E. and Otto-Bliesner, B. L. (1982), The sensitivity of the

African–Asian monsoonal climate to orbital parameter changes for 9000 years BP in a low-resolution general circulation model, *Journal of the Atmospheric Sciences*, **39**, 1177–1188.

[9] A recent readable review of studies of ice ages over the last million years is given in Schneider, S. H. and Londer, R. (1984), *The Co-Evolution of Climate and Life* (Sierra Club Books, San Francisco, CA).

[10] Rasmusson, E. M., Arkin, P. A., Chen, W.-Y., and Jalickee, J. B. (1981), Biennial variations in surface temperature over the United States as revealed by singular decomposition, *Monthly Weather Review*, **109**, 587–598.

[11] For a recent discussion of the difficulties of such analyses, see Pittock, A. B. (1978), A critical look at long-term sun–weather relationships, *Reviews of Geophysics and Space Physics*, **16**, 400–416.

[12] Monin, A. S. and Vulis, I. L. (1971), On the spectra of long-period oscillations of geophysical parameters, *Tellus*, **23**, 337–345; Kutzbach, J. E. and Bryson, R. A. (1974), Variance spectrum of Holocene climatic fluctuations in the North Atlantic sector, *Journal of the Atmospheric Sciences*, **31**, 1958–1962; Mitchell, J. M. (1966), Stochastic models of air–sea interaction and climatic fluctuation, in J. D. Fletcher (Ed), *Proceedings of the Symposium on the Arctic Heat Budget and Atmospheric Circulation*, pp 45–74, Rand Corporation memorandum RM-5233; and see note [7].

[13] The viewpoint of atmospheric variability in time presented here is an extension of that of Hasselman, K. (1976), Stochastic climate models, Part I. Theory, *Tellus*, **28**, 473–485, and the earlier discussion of Mitchell, note [12].

[14] Leith, C. E. (1973), The standard error of time-averaged estimates of climatic means, *Journal of Applied Meteorology*, **12**, 1066–1069.

[15] This point is simply illustrated by the requirements of agriculture in temperate latitudes. The growing season is limited by the number of consecutive frost-free days, which depends on cold extremes of temperature in fall and spring. Productivity is also limited by heat damage during hot extremes in midsummer. The cold and hot extreme temperatures depend not only on weather variations but also on the amplitude of the diurnal temperature cycle. In marginal lands, agricultural abandonment may follow small shifts in mean climate which increase the frequencies of crop failures [see Parry, M. (1981), Evaluating the impact of climate change, in C. D. Smith and M. Parry (Eds), *Climate Change* (Department of Geography, University of Nottingham, UK)]. These shifts in frequency of crop failure correspond to previously rare events becoming more common. Concerns over the possibility of increasing sea level with climate change deal in large part with the resulting increased damage done by high tides and storm surges rather than the damage done by sea level at its average level [see Hoffman, J. S., Keyes, D., and Titus, J. G. (1983), *Projecting Future Sea Level Rise*, report from the US Environment Protection Agency (Washington, DC)]. Mearns, L. O., Schneider, S. H., and Katz, R. W. (1984), Extreme high temperature events: changes in their probabilities with changes in mean temperatures, *Journal of Climate and Applied Meteorology*, **23**, 1601–1613, have recently discussed how increases in mean summer temperatures can greatly increase the frequency of extreme temperatures that cause heat damage to corn crops. Human societies can and do build infrastructures that buffer them against the shorter term expected climatic variations. For example, dams are built for flood control and protection against drought, and grains are stockpiled for protection against crop failures. However, such devices may be overwhelmed by shifts in the frequencies or magnitude of threshold events. For example, a dam may hold a one-year reserve of water, but a shift in climate means may greatly increase the probability of two consecutive drought years.

[16] For example, Koeppe, C. E. and DeLong, G. C. (1958), *Weather and Climate* (McGraw-Hill, New York).

[17] Atmospheric disturbances of less-than-global scale are mostly generated near or at the Earth's surface. Their energy density decays away from the source region with a vertical decay scale proportional to their horizontal scale. The exceptions are transient wave-like disturbances near the equator and the quasi-stationary global planetary waves at the middle latitudes.

[18] See, e.g., Lighthill, J. and Pearce, R. P. (1981), *Monsoon Dynamics* (Cambridge University Press, Cambridge, UK).

[19] For a recent review of climatological planetary waves, see Dickinson, R. E. (1980), Planetary waves: theory and observation, *Orographic Effects in Planetary Flows*, 450 pp, GARP Publications Series No. 23, Global Atmospheric Research Programme (GARP) (WMO–ISCU Joint Scientific Committee, Geneva).

[20] For a review of the dynamics of such systems, see Lilly, D. K. (1979), The dynamical structure and evolution of thunderstorms and squall lines, *Annual Review of Earth and Planetary Sciences*, **7**, 117–161.

[21] For example, Madden, R. A. and Shea, D. J. (1978), Estimates of the natural variability of time-averaged temperatures over the United States, *Monthly Weather Review*, **106**, 1695–1703; Leith, note [14].

[22] Madden, R. A. (1978), Further evidence of traveling planetary waves, *Journal of the Atmospheric Sciences*, **35**, 1608–1618.

[23] Atmospheric winds have a scale of 10 m/s. A column of air loses momentum to the surface at a rate r of about $H^{-1}C_{\mathrm{D}}u$, where $H = 10^4$ m and

$C_D u \approx 10^{-2}$ m/s, u being the surface wind and C_D the drag coefficient. Thus $r = 10^{-6}/s = 0.1/day$.

[24] This number has sometimes been assumed to be as small as 10 days. The value given here includes water vapor feedback and is obtained by assuming that a column of air has a heat capacity of 10^7 J/m^2 °C and radiative damping of 2 W/m^2 °C. Thus, the damping rate is $2\times10^{-7}/s \simeq (40\ \text{day})^{-1}$.

[25] Soil is in contact with the atmosphere through molecular diffusion and thus the soil heat storage, largely resulting from water in the soil, depends on the time scale involved. For a 1-day time scale, the soil heat capacity corresponds to that of about 0.1 m of water and for a 100-day time scale to about 1 m of water. By comparison, the atmosphere's heat capacity corresponds to that of about 3 m of water.

[26] The local rate of energy exchange between ocean and atmosphere is given by $H+LE$, where $H = \rho_a C_p(T_0 - T_a)C_D u$, and $E = \rho_a(q_0 - q_a)C_D u$, where L = latent heat constant, ρ_a = density of air, T_a = temperature of air, q_a = water vapor mixing ratio of air, q_0 = saturated water vapor mixing ratio at the surface of the ocean, and C_p = specific heat of air. The term H is the sensible heat flux and the term E is the evaporation rate. The latter is controlled, in part, by ocean temperature since q_0 is a function of ocean surface temperature.

[27] The ocean mixed layer is about 50 to 70 m in depth, whereas the heat capacity of the atmosphere corresponds to that of about 3 m of water.

[28] Namias, J. (1980), Severe drought and recent history, *Journal of Interdisciplinary History*, **4**, 697–712; see also Namias, J. (1981), Snow covers in climate and long-range forecasting, *Glaciological Data*, **11**, 13–26; Walsh, J. E. (1984), Snow cover and atmospheric variability, *American Scientist*, **72**, 50–57; Yeh, T.-C., Wetherald, R. T., and Manabe, S. (1983), A model study of the short-term climatic and hydrologic effects of sudden snow-cover removal, *Monthly Weather Review*, **111**, 1013–1024. The concept of sources of climate memory is also discussed at considerable length in *Scientific Papers Presented at WMO/ISCU Study Conference on Physical Basis for Climate Prediction on Seasonal, Annual, and Decadal Time Scales*, WCP-47 (WMO, Geneva), held at Leningrad, USSR, 13–17 September 1982. Snow and soil moisture feedbacks are also invoked by Barnett, T. P., Heinz, H. D., and Hasselmann, K. (1984), Statistical prediction of seasonal air temperature over Eurasia, *Tellus*, **36A**, 132–146.

[29] For example, Robock, A. (1983), The dust cloud of the century, *Nature*, **301**, 373–374. See also the entire November 1983 issue of *Geophysical Research Letters*, **10**.

[30] For example, see Rampino, M. R. and Self, S. (1984), The atmospheric effects of El Chichon, *Scientific American*, **250**, 48–57.

[31] For example, National Research Council (1979), *Stratospheric Ozone Depletion by Halocarbons: Chemistry and Transport* (National Academy of Sciences, Washington DC).

[32] For recent reviews of the role of oceans as part of the climate system, see Woods, J. D. (1984), The upper ocean and air–sea interaction, in J. T. Houghton (Ed), *The Global Climate*, pp 79–106 (Cambridge University Press, Cambridge, UK); and Bretherton, F. P. (1982), Ocean climate modeling, *Progress in Oceanography*, **11**, 93–129.

[33] For example, Pan, Y. H. and Oort, A. H. (1983), Global climate variations connected with sea-surface temperature anomalies in the eastern Equatorial Pacific Ocean for the 1958–73 period, *Monthly Weather Review*, **111**, 1244–1258. A popular summary is given in Brock, R. G. (1984), El Niño and world climate: piecing together the puzzle, *Environment*, **26**, 14–26, 37–39.

[34] For example, see Bentley, C. R. (1984), Some aspects of the cryosphere and its role in climatic change, in J. E. Hansen and T. Takahashi (Eds), *Climate Processes and Climate Sensitivity, Maurice Ewing Series 5*, pp 207–220 (American Geophysical Union, Washington, DC).

[35] Global surface temperature trends over the last 100 years have been discussed by Hansen, J., Johnson, D., Lacis, A., Lebedeff, S., Lee, P., Rind, D., and Russel, G. (1981), Climate impact of increasing atmospheric carbon dioxide, *Science*, **213**, 957–966. Regional 10-year trends have been analyzed by van Loon, H. and Williams, J. (1976), The connection between trends of mean temperature and circulation at the surface: Part I. Winter, *Monthly Weather Review*, **104**, 365–380.

[36] Lorenz, E. N. (1968), Climatic determinism, *Meteorological Monographs*, **5**, 1–3. Lorenz has developed a simple set of differential equations representing a baroclinic wave interacting with a zonal mean temperature gradient. Depending on choice of parameters, the set exhibits a wide range of variable behaviors, including constant solutions, simple periodic solutions, solutions with multiple periods, and solutions with various degrees of irregularity, including some that can be interpreted in terms of regime jumping. A lucid discussion is given in his Crafoord Prize Lecture, Lorenz, E. N. (1984), Irregularity: a fundamental property of the atmosphere, *Tellus*, **36A**, 98–110.

[37] Zeeman, E. C. (1976), Catastrophe theory, *Scientific American*, **234**, 65–83; Vickroy, J. G. and Dutton, J. A. (1979), Bifurcation and catastrophe in a simple, forced, dissipative quasi-geostrophic flow, *Journal of the Atmospheric Sciences*, **36**, 42–52; Arnold, V. I. (1984), *Catastrophe Theory*, translated by R. K. Thomas (Springer-Verlag, Berlin).

[38] Examples of multiple equilibria have been found in one-dimensional energy balance climate models. The most widely recognized alternative climate state is that of an ice-covered Earth, which is produced by reducing the incident solar radiation for

a long enough time to some fraction, e.g., 0.9, of current values. This state is stable, once reached, even with incident solar radiation increased back to present values, because the ice and snow-covered system has a much greater reflection of solar radiation. Other examples in simple climate models have been more subtle and less obviously related to any real physical process. The periodic occurrence of past ice ages has stimulated discussions of simple climate models as represented by nonlinear, self-excited oscillators, e.g., Ghil, M. and Tavantziz, J. (1983), Global Hopf bifurcation in a simple climate model, *SIAM Journal of Applied Mathematics*, **43**, 1019, 1041; Källen, E., Crafoord, C., and Ghil, M. (1979), Free oscillations in a climate model with ice sheet dynamics, *Journal of the Atmospheric Sciences*, **36**, 2292–2303; Nicholis, C. (1984), Self-oscillations and predictability in climate dynamics – periodic forcing and phase locking, *Tellus*, **36A**, 217–227; and Saltzman, B. and Sutera, A. (1984), A model of the internal feedback system involved in late quaternary climatic variations, *Journal of the Atmospheric Sciences*, **41**, 736–745. Oceanic scientists are currently intrigued by the possibility of two stable modes involving either strong or weak deep-water production in the North Atlantic coupled to differences in evaporation over precipitation and continental runoff, e.g., Broecker, W. S., Peteet, D. M., and Rind, D. (1985), Does the ocean–atmosphere system have more than one stable mode of operation?, *Nature*, **315**, 21–25.

[39] Willson, R. C., Gulkis, S., Janssen, M., Hudson, H. S., and Chapman, G. A. (1981), Observations of solar irradiance variability, *Science*, **21**, 700–702.

[40] The basic figure is from Clark, note [1], p 446. The numbers have been somewhat revised and are consistent with Budyko, M. I. (1982), *The Earth's Climate: Past and Future* (Academic Press, New York).

[41] See Lacis, A. A. and Hansen, J. E. (1984), A parameterization for the absorption of solar radiation in the Earth's atmosphere, *Journal of the Atmospheric Sciences*, **31**, 118–133, for a popular formula for solar radiation absorption by O_3 and water vapor.

[42] About 65% of the globe is covered by liquid water with an average surface albedo of 0.06, about 25% by snow-free land with an average albedo of 0.2, and about 10% by snow and ice with an average albedo of approximately 0.6, but with less-than-average solar incidence. As a consequence of atmospheric and cloud absorption and reflection, the albedo at the Earth's surface must be multiplied by some factor between 0.4 and 0.5 to obtain its contribution to albedo at the top of the atmosphere.

[43] For example, Ohring, G. and Gruber, A. (1983), Satellite radiation observations and climate theory, in B. Saltzman (Ed), *Advances in Geophysics, Vol. 25, Theory of Climate* (Academic Press, New York).

[44] Since these numbers are not known that well, I have a penchant for rounding them. The values I quote agree approximately with those of Budyko, note [40]. A brief discussion of global energy balance is given by Ingersoll, A. P. (1983), The atmosphere, *Scientific American*, **249**, 162–175. Budyko quotes 1.1 m/yr as the annual rate of evaporation and rainfall. I use the value of 1.0 m/yr which has been derived by several other investigators but may be near the lower end of what is likely, e.g., Jaeger, L. (1983), Monthly and areal patterns of mean global precipitation, in A. Street-Perrott, M. Beran, and R. Ratcliffe (Eds), *Variations in the Global Water Budget* (D. Reidel Publishing Co., Dordrecht).

[45] Budyko, note [40].

[46] Based on a figure in Meehl, G. (1984), Modeling the Earth's climate, *Climatic Change*, **6**, 259–286.

[47] 1985 CO_2 concentrations are about 345 p.p.m.v. (parts per million by volume). Estimates for preindustrial times have been mostly between 260 and 290 p.p.m.v., e.g., Elliott, W. P. (1984), Meeting report. The pre-1958 atmospheric concentrations of carbon dioxide, *Transactions of the American Geophysical Union*, **26**, 416–417; Pearman, G. I. (1984), Pre-industrial atmospheric carbon dioxide levels: a recent assessment, *Search*, **15**, 42–75; Wigley, T. M. (1983), The preindustrial carbon dioxide level, *Climatic Change*, **5**, 315–320.

[48] Carbon Dioxide Assessment Committee (1983), *Changing Climate* (National Academy of Sciences, Washington, DC).

[49] The use of the term external change depends on what is assumed to comprise the climate system, as discussed by Leith, note [14]; see also Schneider, S. H. and Dickinson, R. E. (1974), Climate modeling, *Reviews of Geophysics and Space Physics*, **12**, 447–493.

[50] Robock, A. (1983), Ice and snow feedbacks and the latitudinal and seasonal distribution of climate sensitivity, *Journal of the Atmospheric Sciences*, **40**, 986–997; Barry, R. G., Henderson-Sellers, A., and Shine, K. P. (1984), Climate sensitivity and the marginal cryosphere, in J. E. Hansen and T. Takahashi (Eds), *Climate Processes and Climate Sensitivity, Maurice Ewing Series 5*, pp 221–237 (American Geophysical Union, Washington, DC).

[51] Relationships between annual mean and latitudinal temperature and albedo were first derived by Budyko, M. I. (1969), The effect of solar radiation variations on the climate of the Earth, *Tellus*, **21**, 611–619; and Sellers, W. D. (1969), A global climate model based on the energy balance of the Earth–atmosphere system, *Journal of Applied Meteorology*, **8**, 392–400. North, G. R., Cahalan, R. F., and Coakley, J. A. (1981), Energy-balance climate models, *Reviews of Geophysics and Space Physics*, **19**, 91–122, give a review of energy balance climate models.

[52] Dickinson, R. E. (1986), How will climate change? The climate system and modelling of future climate, in B. Bolin, B. R. Döös, J. Jäeger, and R. A. Warrick (Eds), *The Greenhouse Effect, Climatic Change, and Ecosystems. A Synthesis of Present Knowledge*, Ch 5 (Wiley, Chichester, UK).

[53] Hansen, J., Lacis, A., Rind, D., Russell, G., Stone, P., Fung, I., Ruedy, R., and Lerner, J. (1984), Climate sensitivity: analysis of feedback mechanisms, in J. E. Hansen and T. Takahashi (Eds), *Climate Processes and Climate Sensitivity, Maurice Ewing Series 5*, pp 130–163 (American Geophysical Union, Washington, DC).

[54] In note [52] an expected value of λ, i.e. $\bar{\lambda}$, of 1.3 W/m^2 °C inferred.

[55] A result of this sort has frequently been obtained in GCM sensitivity studies with ocean temperatures held fixed, e.g., Coakley, J. A. and Cess, R. D. (1985), Response of the NCAR community climate model to the radiative forcing by the naturally occurring tropospheric aerosol, *Journal of the Atmospheric Sciences*, **42**, 1677–1692.

[56] There are no direct measurements of solar fluxes over long time periods. The satellite measurements of Willson *et al.*, note [39], showed a range of 0.2% in solar fluxes, depending on sunspot coverage, from which a likely magnitude of longer period variations may be inferred.

[57] Gilliland, R. L. (1982), Solar, volcanic, and CO_2 forcing of recent climatic events, *Climatic Change*, **4**, 111–131.

[58] For example, Hansen, J. E., Wang, W.-C., and Lacis, A. A. (1978), Mount Agung eruption provides test of a global climatic perturbation, *Science*, **199**, 1065–1068.

[59] The climatic effects of atmospheric trace gases have been discussed in WMO (1982), *Report of the Meeting of Experts on Potential Climatic Effects of Ozone and Other Trace Gases*, WMO Global Ozone Research and Monitoring Project, Report No. 14 (WMO, Geneva). The material in this report has been further developed and updated by Ramanathan, V., Singh, H. B., Cicerone, R. J., and Kiehl, J. T. (1985), Trace gas trends and their potential roles in climate change, *Journal of Geophysical Research*, **90**, 5547–5566. Another recent review is that of Dickinson, R. E. and Cicerone, R. J. (1986), Future global warming from atmospheric trace gases, *Nature*, **319**, 109.

[60] Owens, A. J., Steed, J. M., Miller, C., Filkin, D. L., and Jesson, J. P. (1982), The atmospheric lifetimes of CFC 11 and CFC 12, *Geophysical Research Letters*, **9**, 700–703; Stordal, F., Isaksen, I. S. A., and Horntveth, K. (1985), A diabatic circulation two-dimensional model with photochemistry simulations of ozone and long-lived tracers with surface sources, *Journal of Geophysical Research*, **90**, 5757–5776.

[61] The role of the oceans in controlling CO_2 concentrations is reviewed by Baes, C. F. (1982), Effects of ocean chemistry and biology on atmospheric carbon dioxide, in W. C. Clark (Ed), *Carbon Dioxide Review: 1982*, pp 187–203 (Oxford University Press, New York); and also Brewer, P. G. (1984), Carbon dioxide and the oceans, pp 186–215 in Carbon Dioxide Assessment Committee, note [48]. Related CO_2 changes on geological time scales are discussed by Knox, F. and McElroy, M. B. (1984), Changes in atmospheric CO_2: influence of the marine biota at high latitude, *Journal of Geophysical Research*, **89**, 4629–4637; Berner, R. A., Lasagna, A. C., and Garrels, R. M. (1983), The carbonate–silicate geochemical cycle and its effect on atmospheric carbon dioxide over the last 100 million years, *American Journal of Sciences*, **283**, 641–683; and see McElroy (Chapter 7, this volume).

[62] The observed trends have been discussed by Weiss, R. W. (1981), The temporal and spatial distribution of tropospheric nitrous oxide, *Journal of Geophysical Research*, **86**, 7185–7195; Khalil, M. A. K. and Rasmussen, R. A. (1983), Increase and seasonal cycles of nitrous oxide in the Earth's atmosphere, *Tellus*, **35B**, 161–169; and see McElroy (Chapter 7, this volume).

[63] Observed trends for CH_4 are discussed by Rasmussen, R. A. and Khalil, M. A. K. (1984), Atmospheric methane in the recent and ancient atmospheres: concentrations, trends, and interhemispheric gradient, *Journal of Geophysical Research*, **89**, 11 599–11 605.

[64] The environmental effects of tropospheric O_3 have been reviewed by Hov, O. (1984), Ozone in the troposphere: high level pollution, *Ambio*, **13**, 73–79.

[65] The climatic effects of stratospheric O_3 are reviewed in Ramanathan, V. and Dickinson, R. E. (1979), The role of stratospheric ozone in the zonal and seasonal radiative balance of the Earth–troposphere system, *Journal of the Atmospheric Sciences*, **36**, 1084–1104. The latest attempts to project future change are given in National Academy of Sciences (1983), *Causes and Effects of Stratospheric Ozone Reduction: An Update* (National Academy Press, Washington, DC).

[66] Angell, J. K. and Korshover, J. (1983), Global variation in total ozone and layer-mean ozone: an update through 1981, *Journal of Climate and Applied Meteorology*, **22**, 1611–1627. See also, Logan, J. A. (1985), Tropospheric ozone: seasonal behavior, trends, and anthropogenic influence, *Journal of Geophysical Research*, **90**, 10 463–10 482.

[67] Sagan, C., Toon, O. B., and Pollack, J. B. (1979), Anthropogenic albedo changes and the Earth's climate, *Science*, **206**, 1363–1368; Efimova, N. A. (1983), Effects of forest felling on the planetary albedo and air temperature, *Meteorologiya i Gidrologiya*, **5**, 20–25; Henderson-Sellers, A. and Gornitz, V. (1984), Possible climatic impacts of land cover transformations, with particular emphasis on tropical deforestation, *Climatic Change*, **6**, 231–256.

[68] Changes from tropical forest to grassland would likely increase surface albedos from 0.03 to 0.10 above the forest albedo value. Severe degradation of grassland to near-desert conditions could increase surface albedos by more than 0.10. A much smaller average albedo change (*ca.* 0.005) has been inferred for West Africa by Gornitz, V. and NASA (1985), A survey of anthropogenic vegetation changes in West Africa during the last century – climatic implications, *Climatic Change*, **7**, 285–326. After correcting for atmospheric effects, cf. note [42], a land albedo increase of 0.03, averaged over all land surfaces, would increase reflected radiation by about 1.5 W/m^2.

[69] For a recent review of atmospheric aerosols, see Prospero, J. M., Charlson, R. J., Mohnen, V., Jaenicke, R., Delany, A. C., Moyers, J., Zoller, W., and Rahn, K. (1983), The atmospheric aerosol system: an overview, *Reviews of Geophysics and Space Physics*, **21**, 1607–1629.

[70] A terawatt is 10^{12} watt. The area of the Earth is 5×10^{14} m^2. Present energy use is about 10 terawatts and some projections of future energy growth have reached a usage as high as 50 terawatts; see, e.g., Figure 3.11 of Jäger, J. (1983), *Climate and Energy Systems. A Review of Their Interactions* (John Wiley and Sons, New York).

[71] Nordhaus, W. D. and Yohe, G. W. (1984), Future paths of energy and carbon dioxide emissions, pp 87–152 in Carbon Dioxide Assessment Committee, note [48]. A more recent assessment gives somewhat lower values for future CO_2 concentrations, see Bolin *et al.*, note [52].

[72] For example, Ramanathan, V. (1980), Climatic effects of anthropogenic trace gases, in W. Bach, J. Pankrath, and J. Williams (Eds), *Interactions of Energy and Climate*, p 269 (D. Reidel Co., Dordrecht).

[73] Ramanathan *et al.*, note [59], have discussed all the radiatively active trace gases listed in *Table 9.1* and many others, including their current concentrations and present trends, and have suggested a likely range of concentrations in the year 2030. Dickinson and Cicerone, note [59], suggest a range of scenarios for the more important trace gases to the year 2050.

[74] My credentials for making such a one-man Delphi judgment include, besides participation for the last 12 years in climate modeling research and several general reviews of related subjects, my past involvement in many previous attempts at making uncertainty estimates regarding global environmental change, including the 1974 Climatic Impact Assessment Program study, membership of the first three NAS panels that evaluated the question of O_3 change because of chlorocarbons, and the 1979 NAS CO_2 climate change evaluation. I must admit that all such past attempts to establish uncertainty estimates have had serious difficulties.

[75] My statement regarding the current lack of ability to project regional climate change is based on an evaluation of the current level of understanding of the physical processes responsible for regional climate change and the fact that the most complete models, e.g., Washington and Meehl, note [3], Hansen *et al.*, note [53], and Manabe and Stouffer, note [4] all give quite different results for regional climate change.

[76] See note [68], Sagan *et al.*, note [67], and Henderson-Sellers and Gornitz, note [67].

[77] See note [68] and the discussion on p 270 for solar collectors. Irrigation would decrease surface albedos by 0.03 to 0.10. See note [87] for a discussion of the extent of irrigated agriculture.

[78] Charney, J. G., Quirk, W. J., Chow, S.-H., and Kornfield, J. (1977), A comparative study of the effects of albedo change on drought in semi-arid regions, *Journal of the Atmospheric Sciences*, **34**, 1366–1385.

[79] Horel, J. D. and Wallace, J. M. (1981), Planetary-scale atmospheric phenomena associated with the southern oscillation, *Monthly Weather Review*, **109**, 813–829; Wallace, J. M. and Gutzler, D. S. (1981), Teleconnections in the geopotential height field during the northern hemisphere winter, *Monthly Weather Review*, **109**, 784–812; Blackmon, M. L., Lee, Y.-H., and Wallace, J. M. (1984), Horizontal structure of 500 mb height fluctuations with long, intermediate, and short time scales as deduced from lag-correlation statistics, *Journal of the Atmospheric Sciences*, **41**, 961–979.

[80] This topic is reviewed by Jäger, note [70].

[81] See, for example, Webster, P. J. (1981), Mechanisms determining the atmospheric response to sea surface temperature anomalies, *Journal of the Atmospheric Sciences*, **38**, 554–571; Horel, and Wallace, note [79]; Wallace and Gutzler, note [79].

[82] For example, Aagaard, K., Coachman, L. K., and Carmack, E. (1981), On the halocline of the Arctic Ocean, *Deep-Sea Research*, **28A**, 529–545.

[83] Micklin, P. P. (1981), A preliminary systems analysis of impacts of proposed Soviet river diversions on Arctic Sea ice, *Transactions of the American Geophysical Union* **62**, 489–493; Semtner, A. J. (1984), The climatic response of the Arctic Ocean to Soviet river diversions, *Climatic Change*, **6**, 109–130; Holt, T., Kelly, P. M., and Cherry, B. S. G. (1984), Cryospheric impacts of Soviet river diversion schemes, *Annals of Glaciology*, **5**, 61–68. The last paper finds some observational support for decreased sea-ice concentration associated with reduced river flow and has an extensive list of references to previous Russian-language research papers. A summary of past environment research in the USSR on this topic has been given by Voropaev, G. (1984), Diversion of water resources in the Caspian seabed, *Options*, **2**, 6–9, from IIASA (Laxenburg, Austria).

[84] Crutzen, P. J. and Birks, J. W. (1982), The atmosphere after a nuclear war: twilight at noon, *Ambio*, **11**, 114–125.

[85] Turco, R. P., Toon, O. B., Ackerman, T., Pollack, J. B., and Sagan, C. (1983), Nuclear winter: global consequences of multiple nuclear explosions, *Science*, **222**, 1283–1292; Covey, C., Schneider, S. H., and Thompson, S. L. (1984), Global atmospheric effects of massive smoke injections from a nuclear war: results from general circulation model simulations, *Nature*, **308**, 21–25; Crutzen, P. J., Galbally, I. E., and Brühl, C. (1984), Atmospheric effects from post-nuclear fires, *Climatic Change*, **6**, 323–364; Ramaswamy, V. and Kiehl, J. T. (1985), Sensitivities of the radiative forcing due to large loadings of smoke and dust aerosols, *Journal of Geophysical Research*, **90**, 5597–5613; Aleksandrov, V. V. and Stenchikov, G. L. (1983), *On the Modelling of the Nuclear War* (Computing Center, USSR Academy of Sciences, Moscow); Thompson, S. L. (1985), Global interactive transport simulations of nuclear war smoke, *Nature*, **317**, 35–39; Malone, R. C., Auer, L. H., Glatzmaier, C. A., Wood, M. C., and Toon, O. B. (1985), Influence of solar heating and precipitation scavenging on the simulated lifetime of post-nuclear war smoke, *Science*, **230**, 317–318.

[86] For example, Shuttleworth, W. J. and Calder, I. R. (1979), Has the Priestley–Taylor equation any relevance to forest evaporation? *Journal of Applied Meteorology*, **18**, 639–646.

[87] Currently, about 0.015 of the total land area of the Earth (223×10^6 ha) is devoted to irrigated agriculture, and the area is growing at a rate of about 2% per year according to FAO data quoted by Lindh, G. (1981), Water resources and food supply, in W. Bach, J. Pankrath, and S. H. Schneider (Eds), *Food–Climate Interactions*, pp 239–260 (D. Reidel Publishing Company, Boston). The maximum limit of irrigation area is about double the present value, see Linnemann, H., de Hoogh, J., Keyzer, M. A., and Van Heemst, H. D. J. (1979), *MOIRA: Model of International Relations in Agriculture* (North-Holland, Amsterdam).

[88] A detailed summary of current capabilities of climate models has been published by the World Meteorological Organization in Gates, W. L. (Ed) (1979), *Report of the JOC Study Conference on Climate Models: Performance, Intercomparison and Sensitivity Studies*, GARP Publication Series No. 22 (WMO, Geneva). A briefer discussion occurs in Dickinson, R. E., note [1]. For a recent review of climate models, see G. Meehl, note [46].

[89] For a recent review of GCMs, see Simmons, A. J. and Bengtsson, L. (1984), Atmospheric general circulation models: their design and use for climate studies, in J. T. Houghton (Ed), *The Global Climate*, pp 37–62 (Cambridge University Press, Cambridge, UK).

[90] The first global GCM study was reported by Smagorinsky, J. (1963), General circulation experiments with the primitive equations. I. The basic experiment, *Monthly Weather Review*, **91**, 99–164.

[91] Discussed in more detail in Dickinson, R. E. (1985), Climate sensitivity, in S. Manabe, K. Miyakoda, and I. Orlanski (Eds), *Issues in Atmospheric and Oceanic Modeling*, pp 99–129 (Academic Press, New York).

[92] For one such scheme, Dickinson, R. E. (1984), Modeling evapotranspiration for three-dimensional global climate models, in J. E. Hansen and T. Takahashi (Eds), *Climate Processes and Climate Sensitivity, Maurice Ewing Series 5*, pp 58–72 (American Geophysical Union, Washington, DC).

[93] See, for example, Chervin, R. M. and Schneider, S. H. (1976), On determining the statistical significance of climate experiments with general circulation models, *Journal of the Atmospheric Sciences*, **33**, 405–412.

[94] The importance of this term was first noted in the radiative–convective model study of Manabe, S. and Wetherald, R. T. (1967), Thermal equilibrium of the atmosphere with a given distribution of relative humidity, *Journal of the Atmospheric Sciences*, **24**, 241–259. For a review of radiative–convective models, see Ramanathan, V. and Coakley, J. A., Jr. (1978), Climate modeling through radiative–convective models, *Reviews of Geophysics and Space Physics*, **16**, 465–489.

[95] This is seen most readily by noting that if a heat capacity term C is added to equation (9.1) it becomes $C\partial\Delta T/\partial t + \lambda\Delta T = \Delta Q$, from which it is seen that if λ is negative, ΔT grows exponentially away from an equilibrium solution.

[96] For recent examinations of water vapor feedback for very large warming, see Kasting, J. F., Pollack, J. B., and Ackerman, T. P. (1984), Response of Earth's atmosphere to increases in solar flux and implications for loss of water from Venus, *Icarus*, **57**, 335–355; Lal, M. and Ramanathan, V. (1984), The effects of moist convection and water vapor radiative processes on climate sensitivity, *Journal of the Atmospheric Sciences*, **41**, 2238–2249.

[97] These values may be inferred from reported climate model results by dividing reported changes in absorbed solar radiation by the model change in global average temperature. High side values were found in the original energy balance models of Budyko, note [51], and Sellers, note [51]. The simulation of Washington and Meehl, note [3], gives an ice albedo feedback decrease of λ of about 0.7 W/m^2 °C. The value of approximately 0.0 was found by Washington, W. M. and Meehl, G. A. (1983), General circulation model experiments on the climatic effects due to a doubling and quadrupling of carbon dioxide concentration, *Journal of Geophysical Research*, **88**, 6600–6610, for an annual average mixed-layer ocean model.

[98] For example, North *et al.*, note [51].

[99] See note [50] and Washington and Meehl, note [3]. These studies emphasize the importance of the seasonal cycle in defining the albedo change and the consequent temperature change. The thermal

inertia of the ocean where seasonal sea ice is absent in summer results in maximum temperature changes in winter lagging the change in reflected radiation by 6 months.

[100] For example, Ramanathan and Coakley, note [94].

[101] For example, Manabe and Wetherald, note [94].

[102] For example, Cess, R. D. (1976), Climatic change: an appraisal of atmospheric feedback mechanisms employing zonal climatology, *Journal of the Atmospheric Sciences*, **33**, 1831–1843.

[103] For example, Mercer, J. H. (1978), West Antarctic ice sheet and CO_2 greenhouse effect: a threat of disaster, *Nature*, **271**, 321–325. Past ice ages have required at least several thousands of years to develop. The possibility cannot be completely excluded that the forthcoming large anthropogenic warmings could somehow knock the climate system into a direction of continental cooling. However, the absence of any plausible mechanism makes this possibility very unlikely.

[104] The only such event documented geologically occurred 65 million years ago at the Tertiary–Cretaceous boundary, see Alvarez, W., Kauffman, E. G., Surlyk, F., Alvarez, L. W., Asaro, F., and Michel, H. V. (1984), Impact theory of mass extinctions and the invertebrate fossil record, *Science*, **223**, 1135–1141. Species extinctions have been found to have a 26 million year periodicity that may imply an astronomical event with that period, as delightfully discussed by Gould, S. J. (1984), The cosmic dance of Siva, *Natural History*, **93**(8), 14–19. On the basis of this hypothesis, the probability of collision with a large asteroid or comet would greatly increase in 13 million years. Gould, S. J. (1985), Some opinions unfit to print, *Discover*, **November**, 86–91, rebuts criticisms of this theory published as editorials in *Nature* and the *New York Times*.

[105] The past history of deforestation is discussed by Richards, Chapter 2, this volume. A review by country of the current status of tropical deforestation is given by Myers, N. (1980), *Conversion of Tropical Moist Forests* (National Academy of Sciences, Washington, DC). The largest areas of untouched tropical forest are found in the Amazon Basin. Recent reviews of the currently accelerating pace of deforestation in this region and the cultural and economic framework are given by Hecht, S. B. (1985), Environment, development and politics – capital accumulation and the livestock sector in Eastern Amazonia, *World Development*, **13**(6), 663–684; and Fearnside, P. M. (1986), Human use systems and the causes of deforestation in the Brazilian Amazon, in R. E. Dickinson (Ed), *The Geophysiology of Amazonia Vegetation and Climate Interactions* (Wiley, New York)..

[106] Linnemann *et al.*, note [87].

[107] E.g., as reviewed by Mintz, Y. (1984), The sensitivity of numerically simulated elements to land-surface boundary conditions, in J. T. Houghton (Ed), *The Global Climate*, pp 79–106 (Cambridge University Press, Cambridge, UK).

[108] Many of the future potential changes posing stresses for the climate system could have even more serious consequences for other aspects of the global environment. Perhaps the greatest value of global climate modeling studies will be as prototypes for the future development of physically based global environmental models that will provide a framework for including environmental effects in economic analyses of global development. Only by much enhancing our exploitation of environmental information can we hope to tread a sensible path to the sustainable society, e.g., in directions outlined by Brown, L. R. (1981), *Building a Sustainable Society* (W. W. Norton & Co., New York). Contemporary sociologists and popular authors emphasize that the USA, and presumably the rest of the world, is becoming an information society, see Naisbitt, J. (1984), *Megatrends* (Warner Books, New York), or Dizard, W. P. (1982), *The Coming Information Age* (Longman, New York). The present technological systems need to be extended to a global infrastructure for analysis and application of environmental information, see Ausubel, J. (1980), Economics in the air – an introduction to economic issues of the atmosphere and climate, in J. Ausubel and A. K. Biswas (Eds), *Climate Constraints and Human Activities*, pp 13–60 (Pergamon Press, Oxford, UK), and Ausubel, J. H. (1981), Can we assess the impact of climate changes?, *Climatic Change*, **5**, 7–14.

[109] Policy implications of changes in climate are discussed by, e.g., Schelling, T. C. (1983), Climate change: implications for welfare and policy, pp 449–482 in Carbon Dioxide Assessment Committee, note [48].

[110] The oil and gas deposits in the Alaskan Arctic are estimated to contain up to 40% of the remaining undiscovered crude oil or oil-equivalent in natural gas within the USA, see Weeks, W. F. and Weller, G. (1984), Offshore oil in the Alaskan Arctic, *Science*, **225**, 371–378.

[111] Salati, E. and Vose, P. B. (1984), Amazon basin: a system in equilibrium, *Science*, **225**, 129–137; Dickinson, note [105]; The Polar Group (1980), Polar atmosphere–ice–ocean processes: a review of polar problems in climate research, *Reviews of Geophysics and Space Physics*, **18**, 525–543; and NRC Committee on the Role of Polar Regions in Climatic Change (1984), *The Polar Regions and Climatic Change* (National Academy Press, Washington, DC).

Commentary

T. M. L. Wigley

Dickinson has provided an excellent (and somewhat alarming) analysis of probable and possible future climatic changes. The picture he paints, which is in accord with the majority of current scientific opinion, shows the world at a climatic watershed, between a past climate dominated by natural fluctuations and a future climate dominated by the uncontrolled effects of man. The changes in store may well be much larger and much more rapid than anything we have yet experienced. If so, they will have to become an integral part of future agricultural, hydrological, and economic planning. But to integrate climatic change into the socioeconomic fabric of our lives requires better information than just global mean temperature; scenarios (or, better still, predictions) of the regional details of future climates are required.

General circulation models are the best available tool for constructing such scenarios of future climate. But are they, as Dickinson states, the only tool available? In a decade or so, GCMs may well be sufficiently well-developed for us to place some confidence in their predictions on spatial scales down to 1000 km. At present, however, these models are extremely limited in their ability to project regional climatic change, as evidenced by the lack of regional consistency between three of the most recent GCM studies of the effects of elevated CO_2 (Dickinson, note [81]). Given this inconsistency and the other limitations that Dickinson summarizes with, at times, embarrassing honesty, is it reasonable to assert that GCMs "do produce physically consistent scenarios for future climates that can be obtained by *no other means*" (p 277, my emphasis)? In what sense are GCM results physically consistent? And what other methods are available? Are they so inferior to the GCM approach?

The model differences noted above can be traced to differences in the ways that some of the complex physical processes are parameterized. The results produced by all models must, of course, be consistent with their own internal simplifications of real-world physical processes; in that sense they are all "physically consistent". But they are clearly inconsistent with the real world. Internal consistency is an obvious prerequisite of any method of scenario development; that a method produces a good approximation to the real-world response is, however, the final test of its value. GCMs can be heavily criticized as far as their performance in the latter area is concerned. As Dickinson rightly emphasizes, the marked differences between model simulations of present-day climate and real-world data, as well as the noted inconsistencies in their predictions of the response to elevated CO_2 concentrations, demonstrates their current strictly limited (arguably negligible) value as far as regional studies are concerned.

Of course, models will be improved and, perhaps, they *will* produce regionally consistent results in ten years' time. But even then, will we be able to rely on their predictions? The model studies to which Dickinson refers are all concerned with the steady-state response of the climatic system to imposed external forcings, such as a doubling of atmospheric CO_2 concentration. He notes, however, that the regional details of the transient response to a time-dependent forcing (i.e., the real-world case) may differ considerably from the steady-state regional details. To predict the transient response we need three things: a properly validated model, the right initial conditions, and a prediction of future forcing. Can we hope to have these in ten years' time? How does one validate a time-dependent model? The signal-to-noise problems of validating steady-state results are difficult enough; for the transient response case they may be well nigh insurmountable. Will the oceanic and atmospheric observational network be good enough to be able to specify the initial conditions? The World Ocean Circulation Experiment offers some hope in this direction, but there is much yet to be learned regarding the monitoring requirements for the ocean. And how can we predict future forcing by, for example, such important factors as volcanic eruptions? At present, we do not even know the past forcing record with any confidence.

GCMs are *not* the only way to produce physically consistent scenarios of future climate. Blanket assertions, as they are, can only harm the credibility of the modeling community. We are not considering means of providing *definite* predictions (although it was once claimed that this was the role of GCMs); we are considering the estimation of the *possible*, of plausible responses of the climate system to external forcing. And the range of possible responses can be derived in a number of ways. The past climatic record contains a wealth of information that can be used (and has been used) to generate future climatic scenarios. This information can usefully complement model-generated data, either to fill in gaps where there is large uncertainty in model output or, perhaps, to provide subgrid scale detail. And, of course, as a means of validating model performance, the use of observational evidence is a fundamental aspect of modeling work.

There can be little doubt that data from the real world are physically realistic, and that recent instrumental data do not reflect any idealized steady state of the climate system, but actually follow its transient response. But are these data relevant? Future climatic forcing will probably be due largely to the effects of radiatively active gases, such as CO_2, CFMs, CH_4, etc., whereas past forcing is thought to be due to changes in CO_2, volcanic activity, and (possibly) the output of the Sun. Will the climate system respond in a similar way to dominantly radiative gas forcings, as it has responded to past forcings? Is the response critically dependent on the initial state, a state that is currently much warmer than it was, say, 100 years ago? The answers to these questions, probably respectively "no" and "yes", simply mean that the past cannot be used to *predict* the future. But past data *can* be used to produce physically realistic future scenarios. We now apparently expect no more than this from GCMs. Each method has its strengths and weaknesses.

The most detailed data-based scenarios have been those produced using the instrumental records from the twentieth century. The basic method is to compare periods when the world as a whole was warm with periods when cooler conditions prevailed, and to extrapolate the differences. In other words, a past, globally-warm period is used as an analog for the future. Three early works [1] used composites of data from individual warm or cold years, but this method has been criticized because the processes that cause interannual climatic variations may well be very different from those involved in secular climatic change. As an alternative, Lough *et al.* and Palutikof *et al.* [2] have compared data from the warmest and coldest 20-year periods (*viz.* 1934–1953 and 1901–1920) in the period since 1900. These authors also compared scenarios produced using different analog periods. They found both similarities and differences, a result which should remind us of the current state of the GCM art, where different methods (i.e., models) also produce different scenarios.

The telling distinction between data- and model-based scenarios is that we cannot hope to improve the former to the same extent that models will be improved. Perhaps the only way forward with purely data-based scenarios is to develop them as model-guided and selective extrapolative tools. The scenarios mentioned above have capitalized on the early twentieth century warming, essentially using this as an analog for future warm*ing* (as opposed to warmth). A renewed global warming began around the mid-1960s, since which time global mean temperature has increased by about 0.3 °C. We could, therefore, use this most recent warming episode and, guided by model results and other physical insights into the possible causes of climatic change, cautiously and selectively extrapolate trends into the future. Alternatively, it may be possible to synthesize model results with past data, to take those features of model output that we have most confidence in, and to use these to choose appropriate data from the past record with which to fill in the details.

As the years go by and CO_2 and other trace gas concentrations increase, we may well be exploring uncharted territory further and further, entering climatic conditions that we have no experience of. As this happens, what is now the recent past may become increasingly less relevant. In terms of its relevance, the carpet of past climates is rolling up behind us; but, at the same time, new carpet rolls out under our feet. As climate models are improved, new and relevant data that can be used for scenario development will be continually generated. Let us not lose sight of the potential value of these data.

Notes and references (Comm.)

[1] Wigley, T. M. L., Jones, P. D., and Kelly, P. M. (1980), Scenarios for a warm, high-CO_2 world, *Nature*, **283**, 17–21; Williams, J. (1980), Anomalies in temperature and rainfall during warm Arctic seasons as a guide to the formulation of climate scenarios, *Climatic Change*, **2**, 249–266; Namias, J. (1980), Some concomitant regional anomalies associated with hemispherically averaged temperature variations, *Journal of Geophysical Research*, **85**, 1580–1590.

[2] Lough, J. M., Wigley, T. M. L., and Palutikof, J. P. (1983), Climate and climate impact scenarios for Europe in a warmer world, *Journal of Climatology and Applied Meteorology*, **22**, 1673–1684; Palutikof, J. P., Wigley, T. M. L., and Lough, J. M. (1984), *Seasonal Scenarios for Europe and North America in a High-CO_2, Warmer World*, Technical Report TR012. (US Department of Energy, Carbon Dioxide Research Division, Washington, DC).

Chapter 10 The resilience of terrestrial ecosystems: local surprise and global change

C. S. Holling

Editors' Introduction: Adequate explanations of long-term global changes in the biosphere often require an understanding of how ecological systems function and of how they respond to human activities at local levels.

Outlined in this chapter is one possible approach to the essential task of linking physical, biological, and social phenomena across a wide range of spatial and temporal scales. It focuses on the dynamics of ecological systems, including processes responsible for both increasing organization and for occasional disruption. Special attention is given to the prevalence of discontinuous change in ecological systems, and to its origins in specific nonlinear processes interacting on multiple time and space scales. This ecological scale of analysis is linked "upward" to the global scale of biogeochemical relationships and the "Gaia" hypothesis (see Chapters 7, 8, and 9), and "downward" to the local scale of human activities and institutions (see Chapters 3, 11, and 14).

Introduction

Considerable understanding has been accumulated during the last decade of the way the world "ticks" in its various parts, and progress has been made in recognizing what those parts are and the need to interrelate them. Only after such developments, and therefore only recently, could we begin to address global ecological questions effectively. There are four key questions. How do the Earth's land, sea, and atmosphere interact through biological, chemical, and physical processes? How do ecosystems function and behave to absorb, buffer, or generate change? How does the development of man's economic activities, particularly in industry and agriculture, perturb the global system? How do people – as individuals, institutions, and societies – adapt to change at different scales? In this chapter I respond to the second question by exploring the way ecosystems function and behave, but with the other three problems in mind.

During much of this century global change has been slow, although some important cumulative effects have occurred. The gradual expansion, on a global scale, of economic and agricultural development is well represented by regular increases in atmospheric carbon dioxide (CO_2) of approximately one part per million per year [1]. This increase is attributable to a 2–4% annual increase in burning fossil fuels and, in part, to deforestation. The climate during this century has been benign relative to other periods. Marine fisheries stocks, although typically variable, were largely steady during the period 1920–1970, at least in relation to the apparent sharp shifts that occur among such species as Pacific sardine or Atlantic herring every 50 to 100 years [2]. Problems of environmental pollution have increased in geographical scale from the highly local to the size of air basins or watersheds, but slowly enough that the effects have been largely ameliorated [3]. The deterioration of Lake Baikal in the USSR has been slowed and that of Lake Erie in North America has been stopped. Fish have returned to the River Thames and the extreme smogs of London are now only memories. Atmosphere, oceans, and land, coupled through biological, chemical, and physical forces, have apparently been able to absorb the global changes of this century.

But now qualitative change [4], as distinct from gradual quantitative change, seems possible. Man's industrial and agricultural activities have speeded up many terrestrial and atmospheric processes, expanded them globally, and homogenized them. Four qualitative changes are suggested. First, such changes are being considered as ecological, not simply environmental. For example, pollution can no longer be viewed as inertly burdening the atmosphere. Rather, its impacts on vegetation can accelerate the consequences by impairing the regulatory processes that are mediated by vegetation. Second, the intensity of the impact of man's activities and their acceleration of the time dynamics of natural processes can influence the coupling of long-term regulatory phenomena that link atmosphere, oceans, and land. Third, some of these qualitative changes are likely to be irreversible in principle. So long as the change is local, it can be reversed because there are alternative sources both of genetic variability and species and for the renewal of air and water. But this becomes less and less an option as the change becomes more homogeneous with increasing scale, from local to continental to hemispheric, and then to global. Finally, with the option of reversibility reduced, increasing emphasis will be placed on adapting to the inevitable. But individual, institutional, and social adaptation each have their own time dynamics and histories. There has been little experience in translating the remarkable adaptive responses of individuals to local changes [5] into responses to international and global ones.

In order to analyze ecosystem function and behavior in such a way that global changes can be related to local events and action, I consider four topics. The first is a conceptual framework that can help focus treatment of the contrasts between global and local behavior and between continuous and discontinuous behavior. Since the framework describes different perceptions of regulation and stability, it provides the necessary background for the second topic: the particular causal relations and processes within ecosystems, the influence of external variation on them, and their behavior in time and space. The third topic synthesizes our present understanding of the structure and behavior of ecosystems in a way that has considerable generality, and organizational power. The fourth connects that understanding to our knowledge of global phenomena and of local perception and action.

The conceptual framework: Gaia and surprise

In this chapter I discuss ecosystems, but first the relationships between ecosystems and two other key aspects of the global puzzle must be established; namely, with global biogeophysical events, and with societal perception and management. In the former case, some image is needed of the way the global systems in the atmosphere, oceans, and land interact. That image is provided by the Gaia hypothesis [6]. And to relate our understanding of the behavior of local ecosystems to the way societies perceive and manage those systems the concept of surprise is needed: Gaia and surprise are dealt with in turn.

Gaia

Gaia is the "global biochemical homeostasis" hypothesis, proposed by Lovelock and Margulis [6, 7], that life on Earth controls atmospheric conditions optimal for the contemporary biosphere. The Gaia hypothesis presumes homeostatic regulation at a global level. An example is the maintenance of 21% oxygen in the air, a composition representing the highest possible level to maximize aerobic metabolism, but just short of the level that would make Earth's vegetation inflammable. The residence time of atmospheric oxygen is of the order of thousands of years, a time scale that renders methane (CH_4) production by anaerobic organisms an important regulator of oxygen concentration. The mechanism proposed includes the burying of a small amount of the carbonaceous material of living matter each year and the production of CH_4, which reacts with oxygen, thereby providing a negative feedback loop in the system of oxygen control. Similarly, linked biological and geological feedback mechanisms have been proposed for the regulation of global temperature. The regulation is mediated by control of CO_2 in the atmosphere at concentrations that have compensated for increasing solar radiation over geological time [8].

Even though the Gaia hypothesis is speculative, at least there is more and more evidence for the dynamic role of living systems in determining the composition of many chemicals in the air, soil, and water [9]. And at a smaller geographical scale, as discussed later, there are many ecosystem processes that cybernetically regulate conditions for life.

There are three reasons why I use the Gaia hypothesis as one of my two organizing themes. First, by being rooted in questions of regulation and stability through identifiable biological, chemical, and physical processes, it gives a direction for relevant scientific research – for disproof of the hypothesis if nothing else. Second, this is the only concept I know that can, in principle, provide a global rationale for giving priority to rehabilitation, protection of ecosystems, and land use management. If CH_4 production, for example, provides an essential negative feedback control for ozone (O_3) concentration, then the recent 1–2% annual increase in CH_4 content is important and priority should be given to considering major changes in its primary sources – i.e., wetlands, biomass burning, and ecosystems containing ruminants and termites [10]. Finally, an examination of global change concerns not only science but policy and politics. In a polarized society where certitude is lacking, Gaia has some potential for bridging extremes by providing a framework for understanding and action.

Surprise

Just as Gaia is global, the second organizing theme of surprise is, necessarily, local. Surprise concerns both the natural system and the people who seek to understand causes, to expect behaviors, and to achieve some defined purpose by action. Surprises occur when causes turn out to be sharply different than was conceived, when behaviors are profoundly unexpected, and when action produces a result opposite to that intended – in short, when perceived reality departs *qualitatively* from expectation.

Expectations develop from two interacting sources: from the metaphors and concepts we evolve to provide order and understanding and from the events we perceive and remember. Experience shapes concepts; concepts, being incomplete, eventually produce surprise; and surprise accumulates to force the development of those concepts. This sequence is qualitative and discontinuous. The longer one view is held beyond its time, the greater the surprise and the resultant adjustment. Just such a sequence of three distinct viewpoints, metaphors, or myths has dominated perceptions of ecological causation, behavior, and management [11].

Equilibrium-centered view: nature constant

This viewpoint emphasizes not only constancy in time, but also spatial homogeneity and linear causation. A familiar image is that of a landscape with a bowl-shaped valley within which a ball moves in a way determined by its own acceleration and direction and by the forces exerted by the bowl and gravity. If the bowl was infinitely large, or events beyond its rim meaningless, this would be an example of global stability. Such a viewpoint directs attention to the equilibrium and near-equilibrium conditions. It leads to equilibrium theories and to empirical measures of constancy that emphasize averaging variability in time and "graininess" in space. It represents the policy world of a benign nature where trials and mistakes of any scale can be made with recovery assured once the disturbance is removed. Since there are no penalties of size, only benefits to increasing scale, this viewpoint leads to notions of large and homogeneous economic developments that affect other biophysical systems, but are not affected by them.

Multiple equilibria states: nature engineered and nature resilient

This second viewpoint is a dynamic one that emphasizes the existence of more than one stable state. In one variant the instability is seen as maintaining the resilience of ecological systems [12]. It emphasizes variability, spatial heterogeneity, and nonlinear causation. A useful image is that of a landscape of hills and valleys with the ball journeying among them, in part because of internal processes and in part because exogenous events can flip the ball from one stability domain to another. This viewpoint emphasizes the qualitative properties of important ecological processes that determine the existence of stable regions and of boundaries separating them. Continuous behavior is expected over defined periods that end with sharp changes induced by internal dynamics or by exogenous events, at times large, at times small.

The length of the period of continuous behavior often determines the magnitude of the subsequent change and affects policy recommendations. For example, one would argue from an equilibrium-centered viewpoint that climate warming due to the accumulation of "greenhouse" gases will proceed slowly enough for ecological and social processes to adapt of their own accord. Efforts to facilitate adjustment are unnecessary because existing crop types, for example, are likely to develop and be well adapted to prevailing conditions. However, the second viewpoint of dynamic, nonlinear nature suggests just the opposite: that slow changes of the type expected might be so successfully absorbed and ignored that a sharp, discontinuous change becomes inevitable.

Similarly, spatial graininess, which is small relative to the range of movement of an organism, is presumed to be averaged out in the equilibrium-centered approach [13]. The nonlinear viewpoint, however, presents the possibility that small-scale events cascade upward, as has been described for climatic behavior [14]. But for ecological systems, Steele [15, 16] notes that widely ranging animals feed on small-scale spatial variability. For example, if fish could not discover and remain in plankton patches they could not exist.

This second viewpoint can produce two variants of policy. One assumes that the landscape is fixed or that sufficient knowledge is available to keep it fixed. It is a view of nature engineered to keep variables (the ball) away from dangerous neighboring domains. It occurs in the responsible tradition of engineering for safety, of fixed environmental and health standards, and of nuclear safeguards.

The alternative variant sees that key features of the landscape are maintained by the journeys of the ball, by variability itself testing and maintaining the configuration. This is resilient nature in which the experience of instability is used to maintain the structure and general patterns of behavior. It is assumed in the design that there is insufficient knowledge to control the landscape and hence one attempts to retain variability while producing economic and social benefit [12, 17]. In such cases variables are allowed to exceed flexible limits so long as natural and designed recovery mechanisms are encouraged. Designs have been proposed, for example, for dealing with pollution [18], environmental hazards [5], water resources [19], and pest management [20].

Organizational change: nature evolving

The final viewpoint is one of evolutionary change. Later a number of examples are presented to demonstrate that successful efforts to constrain natural variability lead to self-simplification and so to fragility of the ecosystem. A variety of genetic, competitive, and behavioral processes maintain the values of parameters that define the system. If the natural variability changes, the values shift: the landscape of hills and valleys begins to alter. Stability domains shrink, key variables become more homogeneous (e.g., species composition, age structure, spatial distribution), and perturbations that previously could be absorbed no longer can be.

The resultant surprises can be pathological if continuing control requires ever-increasing vigilance and cost. But if control is internal and self-regulated, i.e., homeostatic, then the possibility opens for organizational change because the benefits of being embedded in a larger ecological or social system significantly exceed the costs of local control.

An example from biological evolution is the remarkably constant internal temperature maintained by endothermic (warm-blooded) animals in the presence of large changes in external temperature. A large metabolic load is required to maintain a constant temperature. As expected, the range of internal temperatures that sustains life becomes narrower than for (cold-blooded) ectotherms. Moreover, the typical endotherm body temperature of around 37 °C is close to the upper lethal temperature for most living protoplasm. It does not represent a "policy" of keeping well away from a dangerous threshold.

The evolutionary significance of this internal temperature regulation is that maintenance of the highest body temperature, short of death, allows the greatest range of external activity for an animal [21]. Speed and stamina increase and activity can be maintained at high and low external temperatures, rather than forcing aestivation or hibernation. There is hence an enhanced capability to explore environments and conditions that otherwise would preclude life. The evolutionary consequence of such temperature regulation was the suddenly available opportunity for dramatic organizational change and explosive radiation of adaptive life forms. Hence the reduction of internal resilience as a consequence of effective self-regulation was more than offset by the opportunities offered by other external settings.

Hence the study of evolution requires not only concepts of function, but also concepts of organization – of the way elements are connected within subsystems and the way subsystems are embedded in larger systems. Food webs and the trophic relations that represent them are an example and have long been a part of ecology. Recently some revealing empirical analyses have demonstrated remarkable regularities in such ecosystem structures [22], with food webs of communities in fluctuating environments having a more constrained trophic structure than those in constant environments [23].

These and related developments, connected in turn to hierarchy theory [24] on the one hand, and the stability and resilience concepts described earlier, on the other, are starting to provide the framework required for comprehending organizational evolution [25]. Although not as well developed as equilibrium, engineering, and resilience concepts, such developments are an essential part of any effort to understand, guide, or adapt to global change.

These views of nature represent the different concepts people have of the way natural systems behave, are regulated, and should be managed. Surprise can occur when the real world is found to behave in a sharply different way from that conceived. The perception can be ignored, resisted, or acknowledged depending on how extreme the departure is and depending on how flexible and adaptable the observer is. Although observer and system are interlinked, I do not explore the psychology and dynamics of individual, institutional, and social adaptation in this chapter, though this is ultimately necessary if we want to understand and design sustainable systems. But in the next section I examine a number of ecological systems to determine which of the views of nature most closely matches reality.

Dynamics of ecosystems

Resilience and stability

This chapter relies heavily on the distinction between resilience and stability. Since that distinction was first emphasized [12] a significant literature has developed to test its reality in nature, to expand the theory, and to apply this to management and design. Much of what follows is drawn from the literature. The distinction relies on definitions that recognize the existence of different stability structures of the kind described in the previous section. There are four main points. First, there can be more than one stability region or domain, i.e., multiequilibrium structures are possible. Second, the behavior is discontinuous when variables (i.e., elements of an ecosystem) move from one domain to another because they become attracted to a different equilibrium condition. Third, the precise kind of equilibrium – steady state or stable oscillation – is less important than the fact of equilibrium. Finally, parameters of the system that define the existence, shape, and size of stability domains depend on a balance of forces that may shift if variability patterns in space and time change. In particular, reduced variability through management or other activities is likely to lead to smaller stability regions whose contraction can lead to sharp changes because the stability boundary crosses the variables, rather than the reverse.

This leads to the following definitions. Stability (*sensu stricto*) is the propensity of a system to attain or retain an equilibrium condition of steady state or stable oscillation. Systems of high stability resist any departure from that condition and, if perturbed, return rapidly to it with the least fluctuation. It is a classic equilibrium-centered definition.

Resilience, on the other hand, is the ability of a system to maintain its structure and patterns of behavior in the face of disturbance. The size of the stability domain of residence, the strength of the repulsive forces at the boundary, and the resistance of the domain to contraction are all distinct measures of resilience.

Stability, as here defined, emphasizes equilibrium, low variability, and resistance to and absorption of change. In sharp contrast, resilience emphasizes the boundary of a stability domain and events far from equilibrium, high variability, and adaptation to change.

However, one school of ecology so strongly emphasizes linear interactions and steady state properties [26, 27, 28] that resilience is treated in the opposite way to that described above. It is defined as "how fast the variables return towards their equilibrium following a perturbation [28]" and is measured by the characteristic return times. In terms of the definitions used in this chapter, this concerns only one facet of stability and has nothing to do with the qualitative distinctions that I believe are important.

In addition to the growing number of tests and demonstrations of the key features of resilience, there have been two major expansions of theory and example. One is Levin's excellent analysis and review of patterns in ecological communities [29]. Levin first placed experimental, functional, and behavioral descriptions within a formal mathematical frame. More important, he made two qualitative additions. One was to explore spatial patterns of multistable systems by analyzing the consequences of diffusion. The second was to make a sharp distinction between variables associated with different speeds or rates of activity, partly to facilitate analysis but more to stress the consequences of coupling subsystems whose cycles are of different lengths. The second major expansion was that of Allen and Starr who extended the analyses of ecosystem patterns for a wide range of examples [25]. Most significantly, they embedded theory, measurement, and modeling relevant to resilience and stability into a hierarchical framework. More than any recent development, this framework provides a means of studying community structure and of treating evolution or organizational change.

Ecosystem scale

These three developments in analysis – of multistable systems [12], of spatial diffusion [29], and of hierarchies [25] – concern the coupling of nonlinear subsystems of different scales in time and space. They are fundamental to understanding how predictable change is, whether or not historical accidents are important, and how to achieve a balance between anticipatory design and adaptive design. Clark [30] has provided a useful classification of the relevant scales for a wide range of geophysical, ecological, and social phenomena. The scales range from square centimeters to global and from minutes to thousands of years. In the present analysis I concentrate on ecological systems covering scales from a few square meters to a few thousand square kilometers and from a few years to a few hundred years.

These scales represent ecosystems, which are defined here as communities of organisms in which internal interactions between the organisms determine behavior more than do external biological events. External abiotic events do have a major impact on ecosystems, but are mediated through strong biological interactions within the ecosystems. It is through such external links that ecosystems become part of the global system. Hence, the spatial scale is determined by the dispersal distance of the most mobile of the key biological variables. The structure of eastern North American spruce–fir forests, for example, is profoundly affected by the spruce budworm, which periodically kills large areas of balsam fir. The modal distance of dispersal of adult budworms is of the order of 50 km [20], but movements are known to extend up to 200 km. The relevant spatial area over which internal events dominate can therefore cover a good part of east-central North America. And the minimum area for analysis has to be of the order of 70 000 km^2.

Similarly, the time span of up to a few hundred years is set by the longest-lived (slowest-acting) key biological variables. In the case of the spruce–fir forests the trees are the slowest variables with a rotation age of about 70 years. Any effective analysis therefore must consider a time span that is a small multiple of that – of the order of 200 years.

Eugene Odum [31], more than anyone else, has emphasized that such ecosystems are legitimate units of investigation, having properties of production, respiration, and exchange that are regulated by biological, chemical, and physical processes. Hence they represent distinct subsystems of the biogeochemical cycles of the Earth. They are open, since they receive energy from the sun and material and energy from larger cycles. In regulating and cycling this material through biotic and abiotic processes, outputs are discharged to larger cycles. Ecosystems hence are Gaia writ small.

Succession

One dominant theme of ecosystem study has been succession – the way complexes of plants develop after a disturbance. Clements' scheme of succession has played an important role in guiding study and theory [32]. He emphasized that succession leads to a climax community of a self-replicating assemblage of plants. The species comprising the assemblage are determined by precipitation and temperature. Plant colonization and growth are seen as proceeding to the stable climax. Initial colonization is by pioneer species that can grow rapidly and withstand physical extremes. They so ameliorate these conditions as to allow entry of less robust but more competitive species. These species in turn inhibit the pioneers but set the stage for their own replacement by still more effective competitors. Throughout this process, biomass accumulates, regulation of biological, chemical, and physical processes becomes tighter, and variability is reduced until the stable climax condition is reached and maintained. This scheme represents a powerful equilibrium-centered view in which disturbances by fire, storm, or pest are treated as exogenous (and somehow inappropriate) intrusions into a natural order. Clements gave an analogy to an organism and its ability to repair damage [32].

During the past 15 years this view has been significantly modified by a wide range of studies of ecosystems – some dominated by disturbance, some not – and by experimental manipulation of defined ecosystem units such as the classic Hubbard Brook Watershed Study [33]. Before describing those developments, however, it is useful to relate this view of succession to another powerful equilibrium-centered notion: that of r and K strategies.

MacArthur and Wilson [34] proposed this classification to distinguish between organisms selected for efficiency of food harvest in crowded environments (K-selected) and those selected simply to maximize returns without constraint (r-selected). The designations come from the terms of the logistic equation, where K defines the saturation density (stable equilibrium population) and r the instantaneous rate of increase. MacArthur [35] pointed out the contrast between "opportunist" species in unpredictable environments (r-strategists) and "equilibrium" species in predictable ones (K-strategists). Pianka [36] and Southwood *et al.* [37] have emphasized that these represent extremes of a continuum, but that a variety of life histories and biological and behavioral features correlate with the two strategies. Briefly, r-strategists have a high reproductive potential, short life, high dispersal properties, small size, and resistance to extremes. They are the pioneers of newly disturbed areas or the fugutive species that ever occupy transient habitats. K-strategists have lower reproductive potential, longer life, lower dispersal rates, large size, and effective competitive abilities. They represent, therefore, the climax species of Clements or those that occupy stable, long-lasting habitats.

There clearly are communities that have developed a climax maintained through plant-by-plant replacement in the manner proposed by Clements. Lorimer [38], for example, examined the history of presettlement forests in northeastern Maine, USA, and found that the time interval between severe disturbances was much longer than that needed to obtain a climax, all-age structure. Other examples are presented in an extensive review of forest succession edited by West *et al.* [39]. But the Clementsian view of succession as analogous to the recovery of an organism from injury, with an ordered and obligatory sequence of replacements of one species by another, is oversimplified and limiting for several reasons.

First, many communities are subjected to regular or irregular disturbances severe enough to kill established plants over areas of a few square meters (the size of a tree) to several thousand square kilometers. Traditionally these disturbances – fires, landslides, storms, floods, disease, insect pests, and herbivore grazing – have usually been viewed as external to the system. But when they occur at a frequency related to the life span of the longest-lived species, the plants themselves can become increasingly adapted to the disturbance and so make the event an internally triggered and maintained phenomenon. This is particularly well recognized for fire. Mutch [40], for example, demonstrated that vegetation of fire-adapted species was significantly more combustible than that of related species in communities not subject to fires. Similarly, Biswell [41] describes the twig development and proliferation of chaparral species that significantly increase the inflammability of plants that are 15 years and older. This coincides with a typical burn cycle of similar duration. Such "accidents designed to happen" are more common than is usually recognized and further examples involving agents other than fire are described later.

As a consequence, there are many examples of what I imagine consistent Clementsian ecologists would be forced to see as self-inflicted wounds to the ecosystem "organism". Such disturbances have a wide range of periodicities set by the dynamics of the slowest variable [42]. Fire frequency in the Pacific Northwest of North America, for example, occurs every 400–500 years, and this period is related to the potential 100-year life span of Douglas fir [43]. Eastern white pine forests experience a fire periodicity of 100–300 years in presettlement times [44, 45], while cyclic changes of 200-year periodicity are proposed for elephant populations; this period is due to the recovery time required for the tasty (to elephants) and long-lived baobab trees of East Africa [46]. The fire controlled period throughout much of the boreal forest of northern North America was between 51 and 120 years [47, 48]. Chaparral in California [49] is adapted to a more frequent cycle of 10–50 years.

The second significant departure from Clements' notions is even more fundamental. Some disturbances can carry the ecosystem into a different stability configuration or domain. At times this happens after a long period of exploitation has apparently reduced the resilience of the ecosystem. For example, fishing in the Great Lakes has been argued to have set the stage for a radical change in fish communities from a system dominated by a large species to one of smaller species [50], the overall biomass remaining constant. The resilience was reduced to a level where small stresses from the physical environment, from man, or from biological invasion triggered the new configuration. In a similar way, shifts of savannas from mixed grass–shrub systems to ones dominated by shrubs are often triggered by a modest drought after being conditioned by an extended period of cattle grazing [51]. In other instances, the magnitude of the triggering event is so great or the resilience of the ecosystem so naturally low that new configurations emerge quite independent of previous management. Bormann and Likens [33, p 189] present one such example of a burn in a spruce–hardwood forest on thin soils that transformed part of the ecosystem into a bare-rock–shrub system. Hence there is not just one climax state; there can be more than one.

Third, species that are important late in the sequence can be present together with pioneer species at the initiation of old field succession [52] or forest succession [43, 48]. The resultant successional sequence is hence much more in the form of a competitive hierarchy as described by Horn [42]. Early in a sequence, the opportunist species grow rapidly, dominate for a short time, but ultimately cannot withstand crowding and competition from other more persistent species. Marks [53] presents a particularly clear demonstration of this sequence and of the opportunist role of pin cherry in reestablishing disturbed hardwood systems in New Hampshire, USA. Late in the sequence pin cherry trees are almost totally absent, having been squeezed out by more competitive trees like beech and red maple. After disturbance, however, seeds long dormant in the soil germinate, pin cherry trees flourish, and begin to be eliminated again after about 20 years.

Finally, invasion of species after disturbance as well as during succession is highly probable, particularly in tropical lowlands [54]. This, combined with the competitive hierarchical relations mentioned above, make the tropical forest highly individual in character and very diverse. So many species are capable of filling a particular niche that succession is better described by life history traits and tree geometry. There is, moreover, considerable advantage in dealing with succession in terms of such properties, since they determine successional status [55] - whether in tropical, temperate, or arctic regions.

That is why the idea of *r*-selected and *K*-selected species was introduced earlier. Each strategy is associated with distinctive traits that, in exaggerated form, contrast the exploitative and opportunistic species that dominate early in the succession with the consolidating or conservative species that dominate later through competition. Moreover, the terms can be used to refer to two principal ecosystem functions: exploitation and conservation. Early in succession biotic and abiotic exploitative processes dominate. These lead to the organization and binding of nutrients, rapid accumulation of biomass, and modification of the environment. Eventually, conservative forces begin to dominate, with competition being the most important aspect. This leads to increased organization through trophic and competitive connections, to reduced variability, and, if not interrupted, to reduced diversity.

Ecosystems, however, are also systems of discontinuous change. In addition to the successional process leading to increasing order there are periods of disorganization. The examples mentioned earlier were of large-scale disruption that affect extensive areas. But change of this kind also occurs in the ecosystems that most closely achieve a climax condition. Individual trees senesce, creating local gaps. The only difference is that the spatial scale is small and the disruptions are not necessarily synchronous. A complete picture of the dynamics of ecosystems therefore requires additional functions to those of exploitation and conservation. Such functions relate more to the generation of change and the introduction of disorder.

Forces of change

To identify these functions and their effects, I initially analyze a small number of examples of ecosystems that demonstrate pronounced change and that have been examined in detail. They can be classified as follows: forest insect pests, forest fires, grazing of semiarid savannas, fisheries, and human disease. Many of the examples also represent systems that have been subjected to management. In a sense, the management activities can be viewed as diagnostic, for they introduced external changes that helped expose some of the internal workings of the natural, unmanaged system. In addition, a number of the management approaches were very much dominated by the goal of achieving constancy through externally imposed regulations. Hence the implicit hypothesis was an equilibrium-centered one and the experiences in managing forests, fish, and other organisms can be viewed as weak tests of that hypothesis.

To give an impression of the consequences I consider the following examples:

(1) Successful suppression of spruce budworm populations in eastern Canada using insecticides certainly preserved the pulp and paper industry and employment in the short term by partially protecting the forest. But this policy has left the forest and the economy more vulnerable to an outbreak over a larger area and of an intensity not experienced before [20].
(2) Suppression of forest fire has been remarkably successful in reducing the probability of fire in the national parks of the USA. But the consequence has been the accumulation of fuel to produce fires of an extent and cost never experienced before [56].
(3) Semiarid savanna ecosystems have been turned into productive cattle-grazing systems in the Sahel zone of Africa, southern and east Africa, the southern USA, northern India, and Australia. But because of changes in grass composition, an irreversible switch to woody vegetation is common and the systems become highly susceptible to collapse, often triggered by drought [51].
(4) Effective protection and enhancement of salmon spawning on the west coast of North America are leading to a more predictable success. But because this triggers increased fishing and investment pressure, less productive stocks become extinct, leaving the fishing industry precariously dependent on a few enhanced stocks that are vulnerable to collapse [57].
(5) Malaria eradication programs in Brazil, Egypt, Italy, and Greece have been brilliant examples of sophisticated understanding combined with a style of implementation that has all the character of a military campaign. But in other areas of the world, where malaria was neither marginal nor at low endemic levels, transient success has led to human populations with little immunity, and mosquito vectors resistant to DDT. As a consequence, during the past five years some countries have reported a 30- to 40-fold increase in malaria cases compared with 1969–1970, signaling a danger not only to the health of the population, but also to overall socioeconomic development.

In each of these examples, the policy successfully reduced the probability of an event that was perceived as socially or economically undesirable. Each was successful in its immediate objective. Each produced a system with qualitatively different properties. All of these examples, and others that fall into the five classes, represent "natural" systems that are coupled to management institutions and to the society that experiences the success or endures the failure of management. Here I focus principally on the natural system.

Despite the large number of variables in each example, the essential causal structure and behavior can be represented by interaction among three sets of variables. These represent three qualitatively different speeds, or rates of activity, corresponding to rates of growth, generation times, and life spans (*Table 10.1*). It becomes possible, as a consequence, to proceed in two directions: to achieve a qualitative understanding of the natural system and to achieve a detailed policy design. The first objective draws upon the theory of differential equations [58]. The second draws upon more recent developments in simulation modeling and optimization [20]. Both are very much connected to hierarchy theory [25].

In many of the examples, both objectives have been pursued. Here I concentrate more on the efforts to achieve a qualitative understanding in order to define research agendas. The other objective, concentrating on detail, is also useful but more in terms of defining operational management agendas. Since distinctively different speeds can be identified for the variables, four steps of analysis are possible [58]:

(1) Analyze the long-term behavior of the fast variable, while holding the slow variables constant.
(2) Define the response of the slow variables when the fast ones are held fixed.
(3) Analyze the long-term behavior of the slow variables, with the fast variables held at their corresponding equilibria.
(4) Combine the preceding steps to identify needs for extra coupling, so that, when added, the behavior of the full system is described.

I now summarize the procedure, emphasizing the main conclusions. The "fast" dynamics are determined by the way key processes affect change in the fast variable at different fixed values of the slow variables. An example is shown in *Figure 10.1*. The important point is that a long history of experimental analysis of ecological processes has led to generalization of the qualitative form of system response, the condition for each distinct form, and the features that determine where impact is greatest (see Holling and Buckingham [63] for predation and competition and Peters [64] for a variety of ecological processes related to body size). Hence, this knowledge can be applied to understanding behavior at a more aggregate level of the hierarchy. Equally important, it frees the analysis from the need for detailed quantification, setting the stage for research designs that are both more economical and more appropriate.

In continuing this qualitative emphasis, attention is then focused on conditions for increase and conditions for decrease of the fast variable. The boundary between the two represents either transient or potentially stable equilibria. The conditions for these equilibria can be organized to show the set of all equilibria, i.e., the zero isoclines for the fast variable. Four examples are shown in *Figure 10.2*: for spruce budworm (a), jack-pine sawfly (b), forest fire (c), and savanna grazing (d).

Table 10.1 Key variables and speeds in five classes of ecosystems.

The system	*The variables*			*Key references*
	Fast	*Intermediate*	*Slow*	
Forest insect pest	Insect	Foliage	Trees	[58, 59]
Forest fire	Intensity	Fuel	Trees	[60]
Savanna	Grasses	Shrubs	Herbivores	[51]
Fishery	Phytoplankton	Zooplankton	Fish	[16]
Human disease	Disease organism	Vector and susceptibles	Human population	[61, 62]

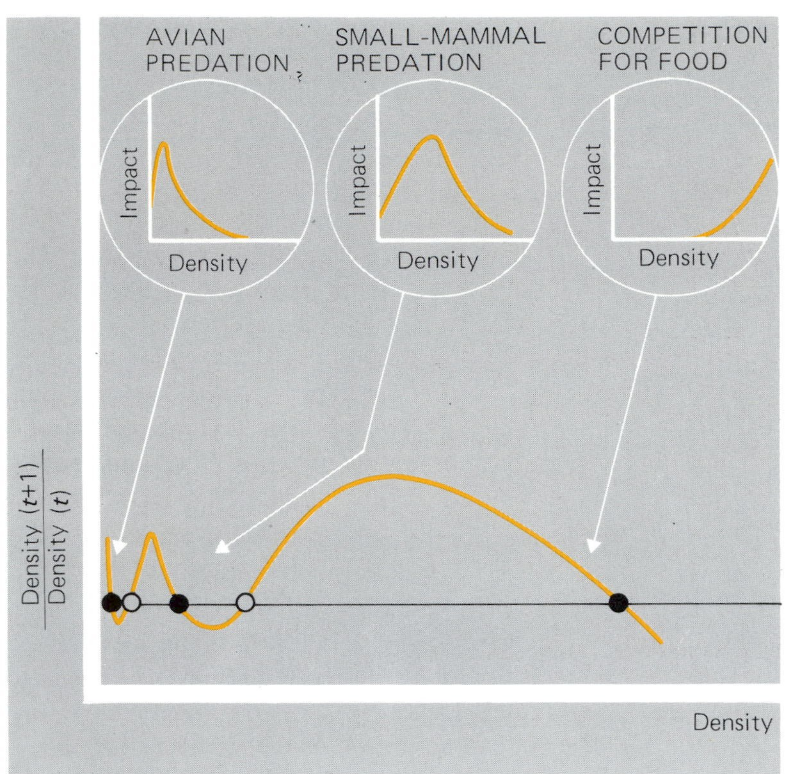

Figure 10.1 A stylized recruitment curve for jack-pine sawfly at a fixed level of slow and driving variables, showing the contributions of three of the key processes. The horizontal line represents the conditions where the population density of the next generation is the same as that of the present generation. Intersections of the recruitment curve with this line indicate potential equilibria, some potentially stable (closed circles), some unstable (open circles).

These equilibrium structures show that there are a number of stability states controlled by the slow variable. There are many other examples given in the literature: for 15 other forest insect pests [59], for other grazing systems [62, 65], for fish [17, 66], and for human host–parasite systems [62].

Two main points emerge at this stage of the analysis. First, discontinuous change occurs because there are multiple stable states. As the slow variables change (tree growth, fuel accumulation, herbivore population increase), different equilibria suddenly appear, and when other equilibria disappear, the system is suddenly impelled into rapid change after periods of gradual change. The basic timing of these events is set by the dynamics of the slow variable.

Second, external stochastic events can lead to highly repetitive consequences. All of the surfaces shown in *Figure 10.2* would be better represented as fuzzy probability bands to reflect "white noise" variability in weather conditions. But this modification typically changes the precise timing of events by a trivial amount. For long periods the systems are in a refractory state and the triggering event is totally or strongly inhibited.

In the insect pest cases [*Figures 10.2(a)* and *(b)*], for example, a variety of predators, chiefly birds, introduce such a strong "predator pit" that insect populations are either becoming extinct or being kept at very low densities. Similarly, the reflexively folded set of unstable equilibria for fire [*Figure 10.2(c)*] can turn stochastic ignition events, such as lightning strikes, into highly predictable outbreaks of fire. If the surface is low [dotted line in *Figure 10.2(c)*], then the average ignition intensity of B triggers a fire at C which consumes the fuel and hence extinguishes itself at A. This is similar to the cycles of ground fires experienced prior to fire management in the mixed-conifier forests of the Sierra Nevada in western USA [56, 60]. In several areas they occurred with a remarkably consistent interval of seven to eight years, and helped maintain conditions for tree regeneration and nutrient cycling. In addition, these light fires killed only some of the young white fir, thereby introducing and maintaining gaps in the forest canopy and, in essence, producing natural fire breaks. However, if the undersurface is raised because of increased moisture or effective fire control practices [broken

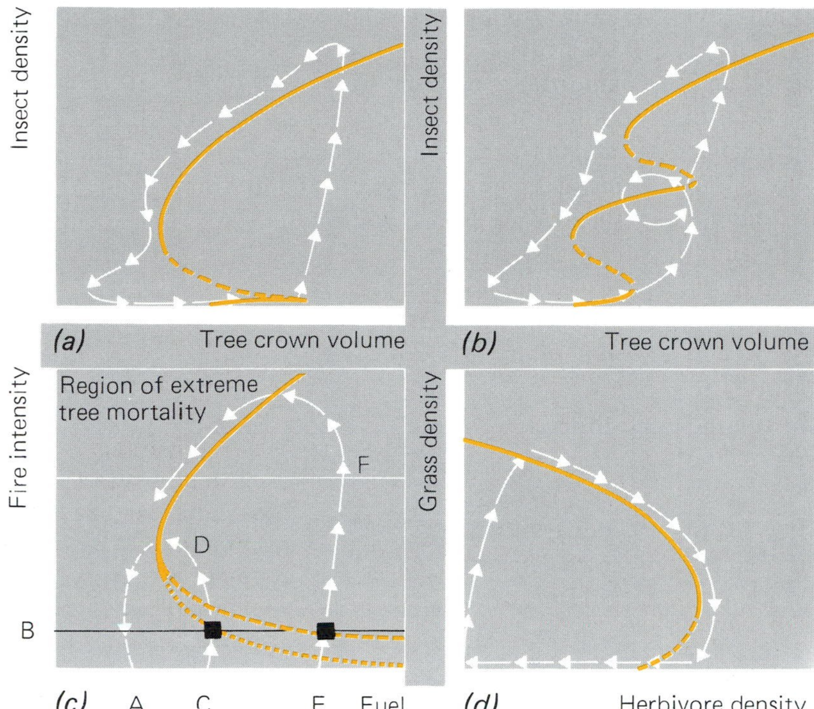

Figure 10.2 Zero-isocline surfaces showing the equilibrium values of the fast variable at different fixed levels of the slow variables. The full lines represent stable surfaces and the broken lines unstable ones. Typical trajectories are shown by the arrows. (*a*) Spruce budworm and balsam fir; (*b*) jack-pine and sawfly; (*c*) forest fire and fuel; (*d*) savanna grass and herbivore grazing.

line in *Figure 10.2(c)*], more fuel must accumulate before an average ignition event triggers a fire [point E, *Figure 10.2(c)*]. This results in a longer period before a fire, but also in a more intense fire, corresponding to a natural long-period fire cycle of the kind mentioned earlier or to the unexpected failure of a fire control policy.

Because of these properties, pulses of disturbance should not be seen as exogenous events. Insect outbreaks, forest fires, overgrazing, sudden changes in fish populations, and outbreaks of disease are determined by identifiable processes affecting the fast variable, whose impacts are modified by the magnitudes of the slow variables. As a consequence, changes in the slow variables eventually result in a condition where a sharp disturbance is inevitable.

A fuller definition of those properties requires two more steps. First, the equilibrium structure set by the fast variable is affected by both intermediate-speed and slow variables. A three-dimensional representation of the zero isocline can then be shown. An example for spruce budworm is shown in *Figure 10.3*. Finally, zero-isocline surfaces are constructed for both the intermediate and slow variables in order to explore the intersections between them. A more formal and rigorous treatment can be found in Ludwig *et al.* [58]. *Figure 10.4* shows, as an example, the isocline surface for tree crown volume laid over that for budworm. Where these two surfaces intersect (line AB) represent the only places where a stable equilibrium for both budworm and trees might be possible. But this can only be realized if the stable portion of the zero isocline for foliage, the intermediate-speed variable, also intersected the line AB. The surface for foliage is folded something like that of the budworm, with a stable surface and an unstable reflexed one. For values of foliage area below this unstable surface, foliage production cannot match natural foliage depletion, so that the foliage eventually disappears. Although it cannot be clearly shown in *Figure 10.4*, it happens that the foliage zero-isocline surface lies under the major portion of the budworm surface. As a consequence, there is no stable intersection.

Thus the unmanaged budworm system is in a state of continuous and fundamental disequilibrium. If one variable is on a stable zero isocline, the others are usually not on theirs. If two of the variables happen to be simultaneously on their stable isoclines, the third one is never on its stable isocline. It is a system under continual dynamic change, always chasing ever-receding equilibria. But control

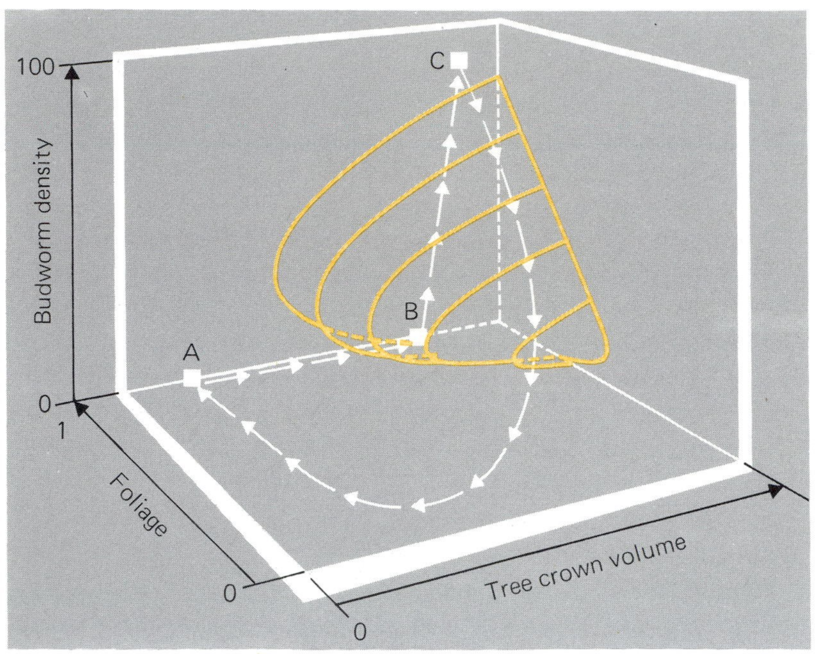

Figure 10.3 Zero-isocline surface for budworm as a function of foliage and tree crown volume. The trajectory shows a typical unmanaged outbreak sequence.

is never completely lost because of the existence of the independent single- or two-variable equilibrium states.

Only an analysis involving three variables would expose this behavior. In such cases, equilibrium-centered concepts, definitions, and measurements that require the existence of at least one non-zero equilibrium (termed here a system equilibrium) are simply irrelevant. There is no system equilibrium.

With these examples I can now add to the review of the main conclusions by summarizing the role of the three distinct speeds of variables in producing cyclic variations of different periods. Under different conditions each of these speeds can dominate the dynamics.

In the spruce–budworm system the period is set by the slowest variable, the tree, and the other variables interact with it. The same is true of unmanaged savanna-grazing systems, where intensive periods of overgrazing lead to depletion of above-ground vegetation. This is followed by emigration or high mortality of ungulates allowing early recovery of perennial grasses with underground storage. Sometime later, annual grasses begin to dominate through competition with the former. Many forest fire systems are similarly controlled, in that the slowest variable sets the cycle and the fast and intermediate variables follow.

In all these examples the variability produces diversity as a consequence of a cyclic shifting of the competitive advantage between species. Balsam fir can outgrow spruce, and would do so except that budworm preferentially attacks balsam and suddenly shifts the balance [67]. The high photosynthetic efficiency of annual grasses places them at an advantage over those perennial grasses that invest a considerable part of their biomass in underground storage. But intense overgrazing tips the balance the other way, so that both types are refused [51]. And Loucks' analysis of long-period forest fire cycles makes a similar point regarding maintenance of species diversity [44]. In all these cases the cycle length is set by the slowest variables and other variables are driven according to that cycle. In every example, the high variability encourages species diversity and spatial heterogeneity.

In a second class of cases the basic timing is set by the intermediate-speed variable. The slowest variable is largely disengaged because a stable oscillation develops around a single system equilibrium in which this variable persists. An example is the ground fire cycle described earlier for the mixed-conifer forests of the Sierra Nevada [*Figure 10.2(c)*, cycle C–A]. A number of forest insect systems show this pattern [59]. For example, the European larch–budmoth system in Alpine

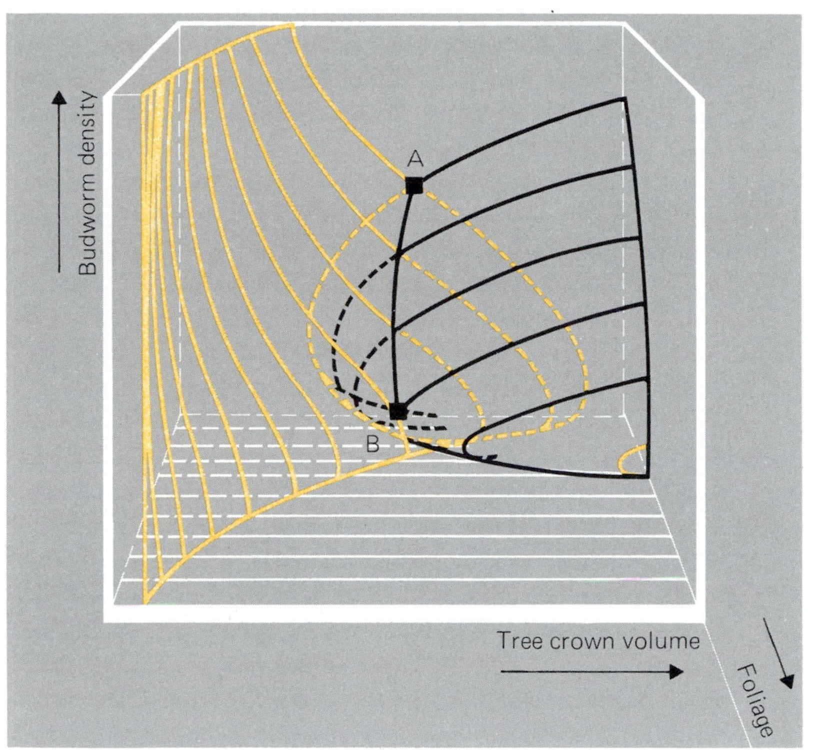

Figure 10.4 Overlay of the zero-isocline surface for tree crown volume (a measure of forest age) against the budworm surface of *Figure 10.2*. The line AB is where the two surfaces intersect.

regions of Switzerland [68] shows a remarkably persistent cycle with an 8- to 10-year period that has persisted for centuries. Both insect and foliage follow this cycle but there is little tree mortality.

Finally, there are patterns in which the fast variable dominates and the intermediate and slow variables are little affected. A particularly interesting example is the jack-pine–sawfly system, [69] and *Figure 10.2(b)*. At intermediate tree ages, predation by small mammals establishes a "predator pit" at moderate population densities of sawfly. At these densities, the small-mammal predation causes enough mortality to allow the sawfly and its parasites to establish a high-speed stable limit cycle of 3–4 years. Foliage is little affected and, as a consequence, neither is tree mortality. Such oscillations can persist for some years, but eventually the system is shifted to a different pattern by a change in climate or forest stand. Other forest insect systems show this pattern, as does endemic malaria where vectoral capacity is high [61].

In summary, the key features of this analysis of the forces of change leads to the following observations:

(1) There can be a number of locally stable equilibria and stability domains around these equilibria.

(2) Jumps between the stability domains can be triggered by exogenous events, and the size of these domains is a measure of the sensitivity to such events.

(3) The stability domains themselves expand, contract, and disappear in response to changes in slow variables. These changes are internally determined by processes that link variables and, quite independently of exogenous events, force the system to move between domains.

(4) Besides exogenous stochastic events, different classes of variability and of temporal and spatial behavior emerge from the form of equilibrium surfaces and the manner in which they interact. There can be conditions of low equilibrium with little variability. There can be stable-limit cyclic oscillations of various amplitudes and periods. And there can be dynamic disequilibrium in which there is no global equilibrium condition and the system moves in a catastrophic manner between stability domains, occasionally residing in extinction regions. There also exists the possibility of "chaotic" behavior.

The one overall conclusion is that discontinuous change is an internal property of each system. For long periods change is gradual and discontinuous behavior is inhibited. Conditions are eventually reached, however, when a jump event becomes increasingly likely and ultimately inevitable.

Synthesis of ecosystem dynamics

Ecosystem functions

It was mentioned in an earlier section that there are two aggregate functions that determine ecosystem succession: an exploitation function (related to the notion of r-strategies) that dominates early and a conservation function (related to K-strategies) that dominates late in the succession. The conclusion of the preceding analysis of forces of change is that there is a third major ecosystem function. The increasing strength of connection between variables in the maturing ecosystem eventually leads to an abrupt change. In a sense, key structural parts of the system become "accidents waiting to happen".

When the timing is set by the slowest variable, the forces of change can lead to intense, widespread mortality. When the timing is set by the faster variables the changes are less intense and the spatial impact, while synchronous over large areas, is more patchy. But even in those instances, individuals constituting the slow variables eventually senesce and die. The difference is that the impact is local and is not synchronous over space.

There is both a destructive feature to such changes and a creative one. Organisms are destroyed, but this is because of their very success in competing with other organisms and in appropriating and accumulating the prime resources of energy, space, and nutrients. The accumulated resources, normally bound tightly and unavailable, are suddenly released by the forces of change. Such forces therefore permit creative renewal of the system. I call this third ecosystem function "creative destruction", a term borrowed from Schumpeter's economic theory [70].

Although the change is triggered by such a function, the bound energy, nutrients, and biomass that accumulated during the succession are not immediately available. There is therefore a fourth and final ecosystem function. One facet of that function is the mobilization of this stored capital through processes of decomposition that lead to mineralization of nutrients and release of energy into the soil. The other facet includes biological, chemical, and physical processes that retain these released nutrients, minimizing losses from leaching. This fourth function is one of ecosystem renewal.

These processes result in a pulse of available nutrients after disturbance. In many instances surprisingly little is lost from the ecosystem through leaching. In other instances so much is lost that algal blooms may be triggered in receiving waters [33, 71]. The kinds of retention mechanisms are not well understood because of the difficulty of studying soil dynamics at an ecosystem scale. But experimental manipulation of whole watersheds through harvesting, removal of structural organic material from the soil surface, and herbicidal inhibition of vegetative regrowth has begun to allow some of the mechanisms to be identified [33, 72]. They include colloidal behavior of soil, rapid uptake by the remaining vegetation whose growth is accelerated by the disturbance, and low rates of nitrification that keep inorganic nitrogen in ammonia pools rather than as the more soluble nitrates [73]. In addition a recent experiment demonstrated that rapid uptake of nutrients by microbes during decomposition is a major process preventing nitrogen losses from areas of harvested forests [72].

Such processes of release and retention after disturbance define the renewal function. Hill [74] emphasizes their importance in reestablishing the cycle of change and hence in determining the resilience of ecosystems. Of particular importance are the processes of retention. When savannas become dominated by woody shrubs, it is because of the loss of water retention capacity of perennial plants and soil. Similarly, intensified burning of upland vegetation in Great Britain has caused the vegetation to shift irreversibly from forest cover to extensive blanket bogs [75]. On sites where soils are poor, rainfall is high, and temperatures are low, the result has been loss of nutrients and reduced transpiration and rainfall interception, leading to waterlogged soils, reduced microbial decomposition, and the development of peat. The original tree species, such as oak, cannot regenerate because of wetness, acidity, and nutrient deficiency. In a similar vein, tropical rain forests may have a low resilience to large-scale disturbance. Many of the tree species have large seeds with short dormancy periods. These features facilitate rapid germination and regrowth of vegetation in small disturbed areas, but make it impossible to recolonize extensive areas of cleared land [76]. Partly as a result, extensive land clearance in the Amazon basin has led to permanent transformation of tropical forest areas into scrub savanna [77].

The full dynamic behavior of ecosystems at an aggregate level can therefore be represented by the sequential interaction of four ecosystem functions: exploitation, conservation, creative destruction, and renewal (*Figure 10.5*). The progression of events is such that these functions dominate at different times: from exploitation, **1**, slowly to conservation, **2**, rapidly to creative destruction, **3**, rapidly to renewal **4**, and rapidly back to exploitation. Moreover, this is a process of slowly increasing organization or connectedness (**1** to **2**) accompanied by gradual accumulation of capital. Stability initially increases, but the system becomes so overconnected that rapid change is triggered (**3** to **4**). The stored capital is then released and the degree of resilience is determined by the balance between the processes of mobilization and of retention. Two properties are being controlled: the degree of organization and the amount of capital accumulation and retention. The speed and amplitude of this cycle, as indicated earlier, are determined by whether the fast, intermediate, or slow variable dominates the timing.

These patterns in time have consequences for patterns in space. Rapidly cycling systems generate ecosystems that are patchy. Tropical ecosystems are an example. Slowly cycling systems produce higher amplitude, discontinuous change that tends to occur as a wave moving across space. In the case of uncontrolled budworm outbreaks, for example, the wave

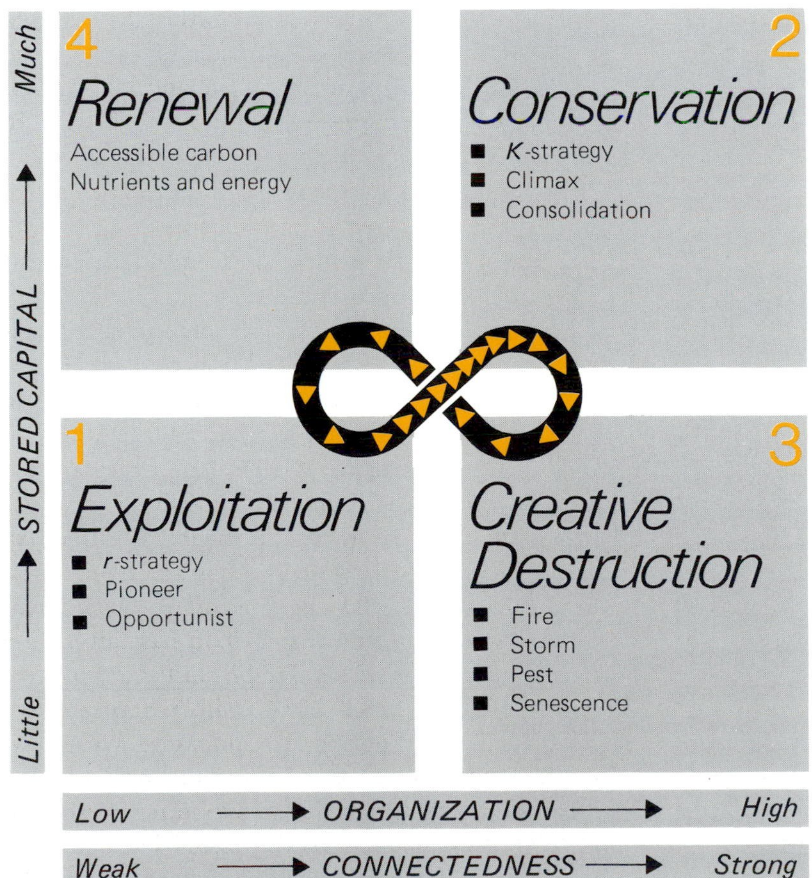

Figure 10.5 The four ecosystem functions and their relationship to the amount of stored capital and the degree of connectedness. The arrowheads show an ecosystem cycle. The interval between arrowheads indicates speed, i.e., a short interval means slow change, a long interval rapid change.

takes about 10 years to sweep across the province of New Brunswick, Canada.

The factors determining the size distribution of the areas of disturbance, however, are not well understood. If the considerable understanding of time dynamics could be connected to an equal understanding of spatial patch dynamics, then questions of global change could be better anticipated and better dealt with by local policies.

Levin [29, 78] has developed an effective framework for analysis and description of patches. There are two parts:

(1) Patch size and age distributions as related to birth and death rates of patches.
(2) The response of species to the regeneration opportunities existing in patches of different sizes and ages.

The "fast" and "slow" designations are part of the analysis, as well as diffusion and extinction rates. It is, therefore, completely compatible with the analysis presented here, and has begun to be applied to forest systems [79] together with Mandelbrot's theory of fractals to relate extinction laws and relative patchiness [80]. As a consequence Mandelbrot proposes a descriptive measure of patchiness and succession that is scale independent and has considerable value for any effort to measure patch dynamics and disturbance.

Complexity, resilience, and stability

This synthesis helps clarify the relationship between complexity and stability. It was long argued that more species and more interactions in communities conferred more stability, the intuitive notion being that the more pathways that were available for movement of energy and nutrients, the less would be the effect of removal of one. However, May's analysis of randomly connected networks showed that increased diversity, in general, lowered stability [81]. This means that ecosystems are not randomly connected. The issue has been significantly clarified by Allen and Starr [25] and the treatment here provides further support of May's observation.

First, measures of stability referred to typically did not distinguish between stability and resilience – systems with low stability can often demonstrate high resilience. Second, ecosystems have a hierarchical structure, and for this reason it has been possible to capture the essential discontinuous behaviors with three sets of variables operating at different speeds. Other species and variables are dramatically affected by that structure and the resultant behavior, but do not directly contribute to it. Hence the relevant measures of species diversity, which is one measure of complexity, should not involve all species, but only those contributing to the physical structure and dynamics.

The significant measure of complexity, therefore, concerns the degree of connectedness within ecosystems. As Allen and Starr demonstrate, the higher the connectedness, i.e., the complexity, the lower the probability of stability. They present examples both from theory and from the empirical literature to demonstrate the point. A system can also become so underconnected that critical parts go their own way independent of each other – resilience disappears. Tropical farming based upon monocultures and extensive land clearance is highly overconnected, particularly through pest loads [82]. Hence stability and ultimately resilience are lost. Introduction of patches through traditional shifting of agriculture or through the breaking of connections by interplanting different cultivars produces a farming pattern more akin to the highly patchy, less connected natural system.

Third, the pattern of connectedness and the resultant balance between stability and resilience are a consequence of the pattern of external variability that the system has experienced. Systems such as those in the tropics, which have developed in conditions of constant temperature and precipitation, therefore demonstrate high stability but low resilience. They are very sensitive to disturbances induced by man. On the other hand, temperate systems exposed to high climatic variability typically show low stability and high resilience, and are robust to disturbance by man.

The present analysis adds an important fourth ingredient. Hierarchies are not static in the kinds or strengths of connections. The degree of connectedness changes as the ecosystem is driven by the four ecosystem functions. Succession introduces more connectedness, and hence increasing likelihood of instability. An overconnected condition develops, triggering a discontinuous change. The connectedness is sharply reduced thereby, to be followed by reorganization and renewal. The destabilizing effect produced by overconnectedness generates variability, which in turn encourages the development and maintenance of processes conferring resilience, particularly during the period of low connectedness and recovery. Collapse of resilience, or escape to a different stability domain, can occur, however, if the system becomes too underconnected during the destabilized phase of this cycle. This can happen if processes of mobilization are not balanced by processes of retention. Since those processes occur dominantly in soil, any exploration of global change must place a high priority on developing a better and more extensive understanding of soil dynamics in relation to the cycles driven by the four ecosystem functions.

Connections

The previous section addressed the question posed in the introduction concerning the capacity of ecosystems to absorb, buffer, or generate change. It concentrated on the processes and functions that lead to cycles of ecosystem growth, disruption, and renewal. The periods and amplitudes of those cycles are defined by qualitatively distinct speeds of a small number of key variables. Their ability to maintain structure and patterns of behavior in the face of disturbance, i.e., their resilience, is determined by the renewal function whose properties are, in part, maintained by pulses of disturbance.

The timing and spatial extent of the pulses emerge from the interaction between external events and an internally generated rhythm of stability/instability. Industrial societies are changing the spatial and temporal patterns of those external events. Spatial impacts are more homogeneous; temporal patterns are accelerated. An understanding of impacts of global change therefore requires a framework to connect the understanding developed here for ecosystem dynamics to that developed for global biogeochemical changes on the one hand and societal developments on the other hand. There are transfers of energy, material, and information among all three, as suggested in *Figure 10.6*. Biosphere studies now concentrate on changes in the amount, speed, and scale of those transfers and what should or should not be done about them. The Gaia hypothesis, as indicated earlier, provides a focus for discussing the interaction between ecosystems and global biogeochemical cycles. Surprise provides a focus for discussing the interaction between ecosystems and society.

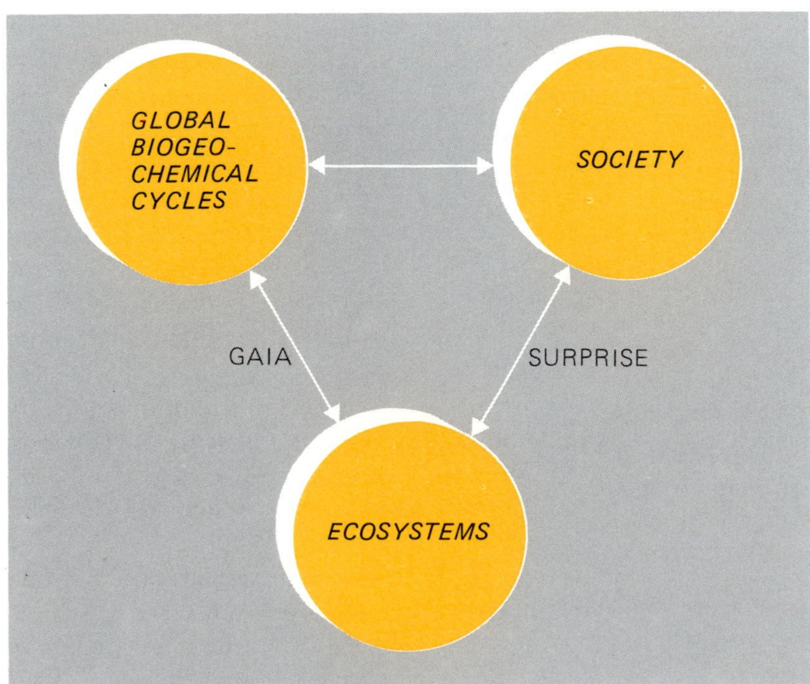

Figure 10.6 Connections.

Gaia

The spatial and temporal patterns generated by the four ecosystem functions form the qualitative structure of ecosystems. A small number of variables and species are fundamental to determining that structure. And the resultant architecture of an ecosystem offers a variety of niches which are occupied by different species that are affected by the ecosystem structure, but contribute little to it. But while contributing little to the structure, they could contribute significantly to exogenous biogeochemical cycles. This could be determined by drawing on the extensive literature that identifies the major components of geochemical exchange and regulation in plants, animals, and soils. The scheme presented here can provide a way to organize this knowledge so that the interactions of specific ecosystems with external biogeochemical cycles and their possible regulatory roles can be better understood.

Moreover, if the key processes of homeostatic regulation in atmospheric cycles could be demonstrated, such an approach could also provide a way to identify the ecosystems that contribute most to the feedback control. In order to do so, the qualitative analysis outlined here for a few systems could be expanded into a comparative study of the structure of ecosystem dynamics in each of the life zones defined by Holdridge [83] or by Soviet geographers (as reviewed in Grigor'yev [84]) on the basis of climate data.

Such a study would provide a descriptive classification for determining ecosystem responses to global environmental changes. The responses that are most critical are the qualitative patterns of behavior. These patterns are determined by the fast, intermediate, and slow variables during ecosystem growth and disruption and by the mobilization/retention processes during renewal. They also emerge from the way the resultant internal dynamics modify climatic variability. The latter is determined by global atmospheric and oceanic processes, which in turn set the variability in the physical environment.

Steele has reviewed the temporal behavior of physical variables in the ocean and atmosphere [85]. If the well defined periods of days and seasons are removed, the underlying trend for physical conditions in the oceans is for variance to increase as a function of period. The increase is close to the square of the period and occurs at all time (and space) scales. This "red noise" is in contrast to "white" noise where variance is independent of scale (see Dickinson, Chapter 9, pp 257–260, this volume).

For periods up to about 50 years, physical variation in the atmosphere, unlike in the oceans, is close to being white noise. Thereafter variation seems to follow a red spectrum, suggesting a coupling of atmospheric and oceanic processes. As described earlier, many terrestrial ecosystem cycles have a period from a few decades to one or two hundred years, driven by the slowest variables. Even if this similarity in cycle periods is a coincidence rather than due to adaptation, changes in the external forcing frequency induced by man's activities could be transmitted and transformed by the existing response times of ecosystems in unexpected ways. Now that the qualitative dynamics of a number of ecosystems are beginning to be better understood, a fruitful area of research can be developed to demonstrate, by example, how changes in the frequency pattern of external forcing can affect ecosystem stability and resilience.

Surprise

Man's efforts to manage ecosystems can be viewed as weak experiments testing a general hypothesis of stability/resilience. In many of the examples discussed earlier, the management goal was to reduce the *variability* of a target variable by applying external controls. Crudely, it represented an equilibrium-centered view of constant nature. All the cases examined were successful in achieving their short-term objectives, but as a consequence of that success, each system evolved into a qualitatively different one.

The evolution took place in three areas. First, the social and economic environment changed. More pulp mills were built to exploit the protected spruce–balsam forests; more recreational demand was developed in the parks protected from fire; more efficient and extensive fisheries were developed to exploit salmon; more land was used for cattle ranches on the savannas; and more development was possible in those areas protected from malaria.

Second, the management agencies began to evolve. Effective agencies were formed to spray insects, fight fires, operate fish hatcheries, encourage cattle ranching, and reduce mosquito populations. And the objective of these agencies naturally shifted from the original socioeconomic objective to one that emphasized operational efficiency: better and better aircraft, navigation, and delivery systems to distribute insecticide; better and better ways to detect fires and control them promptly.

These changes in the socioeconomic environments and in the management institutions were generally perceived and were rightly applauded. But evolution occurred in a third area – the biophysical – whose consequences were not generally perceived.

Because of the initial success in reducing the variability of the target variable, features of the biophysical environment which were implicitly viewed as constants began to change to produce a system that was structurally different and more fragile. Reduction of budworm populations to sustained moderate levels led to accumulation and persistence of foliage over larger and larger areas. Any relaxation of vigilance could then lead to an outbreak in a place where it could spread over enormous areas. Reduction of fire frequency led to accumulation of fuel and the closing of forest crowns so that what were once modest ground fires affecting limited areas and causing minor tree mortality became catastrophic fires covering large areas and causing massive tree mortality. And similarly, increased numbers of salmon led to increases in size and efficiency of fishing fleets and extinction of many native stocks; maintenance of moderate numbers of cattle led to changes in grass composition toward species more vulnerable to drought and errors of management; persistent reduction in mosquitoes led to gradual increases in the number of people susceptible to malaria, and to mosquitoes resistant to insecticide.

In short, the biophysical environment became more fragile and more dependent on vigilance and error-free management at a time when greater dependencies had developed in the socioeconomic and institutional environment. The ecosystems simplified into less resilient ones as a consequence of man's success in reducing variability. In these cases, connectedness increased because of spatial homogenization of key variables: foliage for budworm, fuel and canopy structure for fires, efficient but vulnerable grasses for savannas, numbers of stocks and ages of fish, and the number of people susceptible to malaria.

The hypothesis of constant nature encountered the surprising reality of resilient nature. If control falters, the magnitude and extent of the resultant disruptive phase can be great enough to overwhelm the renewal process.

Just as ecosystems have their own inherent response times, so do societal, economic, and institutional systems. How long an inappropriate policy is successful depends on how slowly the ecosystem evolves to the point when the increasing fragility is perceived as a surprise and potential crisis.

The response to such surprises is alarm, denial, or adaptation and is similarly related to the response times of different groups in society and of the management institutions. For example, forest fire policy in the national parks of the western USA has recently changed radically to reinstate fire as the natural "manager" of the forests. This rather dramatic adaptation was not made easily and rested upon the existence of an alternative policy and of technologies to implement it, on a climate of understanding, and on costs that were relatively modest compared with my other examples. But it might be equally important that the critical variables of fuel and forest tree composition changed at the slowest rate of all the examples. It was some sixty years before the change became critical. I argue that the relevant time unit of change for a management institution is of the order of 20 to 30 years, the turnover rate of employees. As a consequence, by the time the problem became critical there was a new generation of experts and policy advisors who would be more willing to recognize failures of their predecessors than of their own. In addition, the slowness of change allowed the accumulation of knowledge of the processes involved and the communication of that growing understanding to a wide range of actors.

In contrast, the changes in the budworm/forest systems proceeded faster. Insecticide spraying began on a large scale in the mid-1950s with conditions of vulnerability building to a critical point by the early 1970s. In this case, the 15–20 year period was insufficient to accumulate and, most important, disseminate an understanding of the problem. Alternative policies or technologies were not developed and the parents of the original policies were still central actors and defenders of the past. Adaptive change has been an agonizing process and is only now showing signs of occurring [86].

There are insufficient examples to make these remarks anything more than speculation: but they do identify a research priority to determine the time dynamics that lead to increasing dependencies of societies on policies that have succeeded in the past, to examine increasing rigidities of management institutions, and to increase sensitivity to surprise. The research effort should be based on case studies that cover as wide a spectrum of man's activities as possible – economic, technological, and behavioral.

Such a comparative study requires collaboration among a number of disciplines. But it is essential to involve practical experience in business, government, and international organizations as well. It is only possible now because so many place priority on understanding change. Equally important, frameworks for understanding change can be found in economics, technology, institutional behavior, and psychology that provide some possible connections to the framework presented here for ecosystems. Examples are suggested in *Table 10.2*.

The analogies suggested by *Table 10.2* might simply represent common ways for people to order their ignorance. But there are strong hints, at least from analysis of institutional organizations from the perspectives of cultural anthropology [88, 89] and of technological developments [87], that functions similar to the four ecosystem functions operate in societal settings, although the results can be very different. Some comparative studies already exist that have both predictive and descriptive power. An example is Thompson's analysis of the very different decisions that were made in the UK and California concerning the siting of liquefied gas plants [89]. And regarding technological development, consider Brooks' argument (Chapter 11, this volume) and this quote [87, p 253]:

> One reason for this situation is that, as a particular technology matures, it tends to become more homogeneous and less innovative and adaptive. Its very success tends to freeze it into a mould dictated by the fear of departing from a successful formula, and by massive commitment to capital investments, marketing strucures and supporting bureaucracies. During the early stages of a new technology many options and choices are possible, and there are typically many small competing units, each supporting a different variation of the basic technology, and each striving to dominate the field. Gradually one variation begins to win, and the economies of scale in marketing and production then begin to give it a greater and greater competitive edge over rival options. The technical options worth considering become narrower and narrower; research tends to

Table 10.2 Possible analogies between ecosystem function and functions or typologies proposed for other systems.

Subject	*Function or typology*			
Ecosystem	Exploitation (r)	Conservation (K)	Creative destruction	Renewal
Economics [e.g., 70]	Innovation, Market, Entrepreneur	Monopolism, Hierarchy, Saturation, Social rigidity	Creative destruction	Invention
Technology [e.g., Brooks 87]	Innovation	Technological monoculture, Technological stalemate	Participatory paralysis	Expert knowledge
Institutions [e.g., 88, 89]	Entrepreneurial market	Caste, Bureaucracy	Sect	Ineffectual
Psychology [e.g., Jung as in 90]	Sensation	Thinking	Intuition	Feeling

> be directed increasingly at marginal product improvements or product differentiation, and the broader consequences of application tend to be taken more and more for granted. Elsewhere I have spoken of this as a "technological monoculture". What happens is that through its very success a new technology and its supporting systems constitute a more and more self-contained social system, unable to adapt to the changes necessitated by its success.

He later adds [87, p 256]:

> The paralysis of the decision process by excessive participation will eventually result in a movement to hand the process back to élites with only broad accountability for results according to then current social values. Eventually effects on certain social expectations will become sufficiently serious so that distrust of the experts will revive and there will be a new wave of demands for participation until the frustration of more diffuse social interests will again result in reversion to experts.

This analysis and speculation is completely in harmony with the cyclic processes described here for ecosystems.

Such analogues at the least suggest that a formal comparative study of different cases could help provide an empirical basis to classify the timing of key phases of societal response to the unexpected: in detecting surprise, in understanding the source and cause of surprise, in communicating that understanding, and in responding to surprise. Such a classification can help introduce a better balance between prediction, anticipation, and adaptation to the known, the uncertain, and the unknown features of our changing world.

Recommendations

Ecosystems have a natural rhythm of change the amplitude and frequency of which is determined by the development of internal processes and structures in a response to past external variabilities. These rhythms alternate periods of increasing organization and stasis with periods of reorganization and renewal. They determine the degree of productivity and resilience of ecosystems.

Modern technological man affects these patterns and their causes in two ways. First, traditional resource-management institutions constrain the rhythms by restricting them temporally and homogenizing them spatially. Internal biophysical relationships then change, leading to systems of increasing fragility, i.e., to a reduced resilience. Moreover, modern man and his institutions operate with a different historical rhythm that can mask indications of slowly increasing fragility and can inhibit effective adaptive responses, resulting in the increased likelihood of internally generated surprises, i.e., crises. Second, the increasing extent and intensity of modern industrial and agricultural activities have modified and accelerated many global atmospheric processes, thereby changing the external variability experienced by ecosystems. This imposes another set of adaptive pressures on ecosystems when they are already subject to local ones. As a consequence, locally generated surprises can be more frequently affected by global phenomena, and in turn can affect these global phenomena in a web of global ecological interdependencies.

We now have detailed examples and analyses of ecological patterns, largely from northern temperate regions, that demonstrate the role of variables of different rates of action and reveal the importance of functions that trigger change and renewal in maintaining resilience. The resultant synthesis indicates that there is now less of a priority to develop predictive tools than to design systems with enough flexibility to allow recovery and renewal in the face of unexpected events – in short, there needs to be a better balance established between anticipation, monitoring, and adaptation [91].

The design effort would be facilitated by research to test and expand the conclusions in three ways. First, the ecosystem synthesis should be extended to further examples of four critical ecosystems: arctic, arid, humid tropical, and marine, since each has patterns and structures different from northern temperate ecosystems. Second, the analyses of time responses and rhythms of change described here should be extended more explicitly to the links between natural/societal systems, particularly regarding the history of economic, technological, and resource development. Third, there is a need and opportunity to develop a set of well replicated mesoscale experiments in order to reduce the ambiguity of problems occurring because of local surprise and global interconnection. These are given more specifically in the following sections.

A comparative study of resilience and ecosystem recovery

Purpose: to define early warning signals of pathologically destructive change and to design self-renewing resource systems.

Data are required to extend the description of time patterns to allow comparison between northern temperate, arctic, arid, humid tropical, and marine systems. Processes that trigger change and facilitate renewal should be identified and classified in terms of their effects on stability, productivity, and resilience. The former requires information as to the role of slow variables in triggering pulses of disturbance. The latter particularly requires a study of soil processes, the balance between nutrient mobilization and retention, their sensitivity to disturbance, and their rates of recovery after small- to large-scale disturbances (e.g., from natural patch formations to man-made land clearance and drainage).

A comparative study of sources and responses to surprise in natural–social systems, particularly economic technological, and resource development

Purpose: to define conditions that determine how much to invest in action (decide policy and act now), anticipation (delay and find out more), or adaptation (forget the immediate problem and invest in innovation).

The generation of sharp change, its detection, and adaptation of policy responses depend on the interaction between the response times of the managed (natural) system, of the institutions managing the systems, and of the economic and social dependencies that develop.

It now seems possible to classify resource, ecological, and environmental problems not only in terms of uncertainty of their consequences, but also in terms of uncertainty of societal response. Those requiring priority attention are not necessarily those that have the greatest impact, but those likely to generate a pathological policy response. The analysis presented here for ecosystems could be usefully applied to interactions between three components.

One of these components concerns the organization and time dynamics of management institutions. Focus and direction can be given by combining the analysis of surprise with the experience and orientation that has matured in hazards research studies [5] and in institutional analyses from the perspective of cultural anthropology [88, 89]. The second component concerns the geophysicochemistry of the atmosphere and oceans that increasingly connects regional economic development with global ecological interdependency through the ecosystems. Focus and direction can be given by the Gaia hypothesis of Lovelock [6] and the system dynamic studies of Steele [16] which view the atmosphere, oceans, and living systems as an interacting, self-regulated whole. The third and final component is society itself, particularly the historical patterns of economic and technological development that reveal how attitudes are formed, technological monocultures developed, and innovations either inhibited or enhanced. Focus and direction can be given by combining an understanding of ecosystem surprise with historical analyses of change, such as those of McNeill [92].

International mesoscale experiments

Purpose: to develop a set of internationally replicated experiments involving areas from a few square kilometers to a few thousand that can test alternative hypotheses developed to explain particular impacts of man's activities and to determine remedial policies.

Our understanding of the structure and behavior of ecosystems, and of how exploitation and pollution affect them, comes from a synthesis of knowledge of ecological, behavioral, physiological, and genetic processes. Much of that knowledge has been developed from the solid tradition of experimental, quantitative, and reductionist science which can now be generalized and synthesized to propose quantitative structures, qualitative behaviors, and qualitative consequences of impacts. Although synthetic, they essentially represent hypotheses because the arguments are based on studies that could be accommodated in the laboratory or in a few hectares. Ecosystems (as well as people's responses to them) operate on scales of a few square kilometers to several thousand square kilometers. That is where our knowledge and experience is the weakest.

The experiments would be designed to clarify alternative explanations of and policies for problems that emerge from the extension and intensification of industrial and agricultural development. Rather than discussing or investigating these endlessly, it should now be possible to design experiments that distinguish between alternatives. It is essential to concentrate on experiments in which the tests are qualitative in nature, the duration short (less than 5 years by drawing on fast/slow definitions of variables), the spatial scale in the "meso" range, and the policy consequences international. International replication and collaboration then becomes part of the design, which could ultimately contribute to institutional solutions as well as to scientific and policy understanding [93].

Notes and references

[1] See, for example, McElroy (Chapter 7, this volume) and Dickinson (Chapter 9, this volume).

[2] Steele, J. H. and Henderson, E. W. (1984), Modeling long-term fluctuations in fish stocks, *Science*, **224**, 985–987.

[3] Clark, W. C. and Holling, C. S. (1985), Sustainable development of the biosphere: human activities and global change, in T. Malone and J. Roederer (Eds), *Global Change*, pp 474–490, Proceedings of a symposium sponsored by the ICSU in Ottawa, Canada (Cambridge University Press, Cambridge, MA).

[4] Qualitative change, in the sense used here, is structural change; that is, changes in the character of relationships between variables and in the stability of parameters. Such changes challenge traditional approaches to development, as well as to control, which is in itself a qualitative change in the combined ecological–social system.

[5] Burton, I., Kates, R. W., and White, G. F. (1977), *The Environment as Hazard* (Oxford University Press, New York).

[6] Lovelock, J. E. (1979), *Gaia: A New Look at Life on Earth* (Oxford University Press, Oxford, UK).

[7] Lovelock, J. E. and Margulis, L. (1974), Atmospheric homeostasis by and for the biosphere: the Gaia hypothesis, *Tellus*, **26**, 1–10.

[8] Lovelock, J. E. and Whitfield, M. (1982), Life span of the biosphere, *Nature*, **296**, 561–563.

[9] Bolin, B. and Cook, R. B. (1983), *The Major Biogeochemical Cycles and Their Interactions* (John Wiley, Chichester, UK).

[10] Crutzen, P. J. (1983), Atmospheric interactions – homogeneous gas reactions of C, N and S containing compounds, in Bolin and Cook, note [9], pp 67–114.

[11] Holling, C. S. (1977), Myths of ecology and energy, in *Proceedings of the Symposium on Future Strategies for Energy Development*, Oak Ridge, TN, 20–21 October, 1976, pp 36–49 (Oak Ridge Associated Universities, TN). Republished in L. C. Ruedisili and M. W. Firebaugh (Eds) (1978), *Perspectives on Energy: Issues, Ideas and Environmental Dilemmas* (Oxford University Press, New York).

[12] Holling, C. S. (1973), Resilience and stability of ecological systems, *Annual Review of Ecology and Systematics*, **4**, 1–23.

[13] Levins, R. (1968), *Evolution in Changing Environments* (Princeton University Press, Princeton, NJ).

[14] Lorenz, E. N. (1964), The problem of deducing climate from the governing equations, *Tellus*, **16**, 1–11.

[15] Steele, J. H. (1974), Spatial heterogeneity and population stability, *Nature*, **248**, 83.

[16] Steele, J. H. (1974), *Structure of Marine Ecosystems* (Harvard University Press, Cambridge, MA).

[17] Peterman, R., Clark, W. C., and Holling, C. S. (1979), The dynamics of resilience: shifting stability domains in fish and insect systems, in R. M. Anderson, B. D. Turner, and L. R. Taylor (Eds), *Population Dynamics*, pp 321–341 (Blackwell Scientific, Oxford, UK).

[18] Fiering, M. B. and Holling, C. S. (1974), Management and standards for perturbed ecosystems, *Agro-Ecosystems*, **1**, 301–321.

[19] Fiering, M. B. (1982), A screening model to quantify resilience, *Water Resources Research*, **18**, 27–32; and Fiering, M. B. (1982), Alternative indices of resilience, *Water Resources Research*, **18**, 33–39.

[20] Clark, W. C., Jones, D. D., and Holling, C. S. (1979), Lessons for ecological policy design: a case study of ecosystem management, *Ecological Modelling*, **7**, 1–53.

[21] Bennett, A. F. and Ruben, J. A. (1979), Endothermy and activity in vertebrates, *Science*, **206**, 649–654.

[22] Briand, F. and Cohen, J. E. (1984), Community food webs have scale-invariant structure, *Nature*, **307**, 264–267.

[23] Cohen, J. E. and Briand, F. (1984), Trophic links of community food webs, *Proceedings of the National Academy of Sciences*, **81**, 4105.

[24] Simon, H. A. (1973), The organization of complex systems, in H. H. Pattee (Ed), *Hierarchy Theory*, pp 1–28 (George Braziller Inc., New York).

[25] Allen, T. F. H. and Starr, T. B. (1982), *Hierarchy. Perspectives for Ecological Complexity* (University of Chicago Press, Chicago, Ill, and London).

[26] Patten, B. C. (1975), Ecosystem linearization: an evolutionary design problem, *American Naturalist*, **109**, 529–539.

[27] Webster, J. R., Waide, J. B., and Patten, B. C. (1975), Nutrient recycling and the stability of ecosystems, in *Mineral Cycling in Southeastern Ecosystems*, ERDA Symposium Series, CON-740-513, pp 1–27.

[28] Pimm, S. L. (1984), The complexity and stability of ecosystems, *Nature*, **307**, 321–326.

[29] Levin, S. A. (1978), Pattern formation in ecological communities, in Steele, J. A. (Ed), *Spatial Pattern in Plankton Communities*, pp 433–470 (Plenum Press, New York).

[30] Clark, W. C. (1985), Scales of climate impacts. *Climatic Change*, **7**(1), 5–27.

[31] Odum, E. P. (1971), *Fundamentals of Ecology* (W. B. Saunders, Philadelphia, PA).

[32] Clements, F. E. (1916), Plant succession: an analysis of the development of vegetation, *Carnegie Institution of Washington Publication*, **242**, 1–512.

[33] Bormann, F. H. and Likens, G. E. (1981), *Patterns and Process in a Forested Ecosystem* (Springer, New York).

[34] MacArthur, R. H. and Wilson, E. O. (1967), *The Theory of Island Biogeography* (Princeton University Press, Princeton, NJ).

[35] MacArthur, R. H. (1960), On the relative abundance of species, *American Naturalist*, **94**, 25–36.

[36] Pianka, E. R. (1970), On *r*- and *K*-selection, *American Naturalist*, **104**, 592–597.

[37] Southwood, T. R. E., May, R. M., Hassell, M. P., and Conway, G. R. (1974), Ecological strategies and population parameters, *American Naturalist* **108**, 791–804.

[38] Lorimer, C. G. (1977), The presettlement forest and natural disturbance cycle of northeastern Maine, *Ecology*, **58**, 139–148.

[39] West, D. C., Shugart, H. H., and Botkin, D. B. (1981), *Forest Succession. Concepts and Application* (Springer, New York).

[40] Mutch, R. W. (1970), Wildland fires and ecosystems – a hypothesis, *Ecology*, **51**, 1046–1051.

[41] Biswell, H. H. (1974), Effects of fire on chaparral, in T. T. Kozlowski and C. E. Ahlgren (Eds), *Fire and Ecosystems*, pp 321–364 (Academic Press, New York).

[42] Horn, H. S. (1976), Succession, in R. M. May (Ed), *Theoretical Ecology. Principles and Application*, pp 187–204 (Blackwell Scientific, Oxford, UK).

[43] Franklin, J. F. and Hemstrom, M. A. (1981), Aspects of succession in the coniferous forests of the Pacific Northwest, in West *et al.*, note [39], pp. 212–229.

[44] Loucks, O. L. (1970), Evolution of diversity, efficiency and community stability, *American Zoologist*, **10**, 17–25.

[45] Heinselman, M. L. (1973), Fire in the virgin forests of the Boundary Waters Canoe Area, Minnesota, *Quarterly Research*, **3**, 329–382.

[46] Caughley, G. (1976), The elephant problem – an alternative hypothesis, *East African Wildlife Journal*, **14**, 265–283.

[47] Rowe, J. S. and Scotter, G. W. (1973), Fire in the boreal forest, *Quarterly Research*, **3**, 444–464.

[48] Heinselman, M. L. (1981), Fire and succession in the conifer forests of northern North America, in West *et al.*, note [39], pp 374–405.

[49] Hanes, T. L. (1971), Succession after fire in the chaparral of southern California, *Ecological Monographs*, **41**, 27–52.

[50] Regier, H. A. (1973), The sequence of exploitation of stocks in multi-species fisheries in the Laurentian Great Lakes, *Journal of the Fisheries Research Board, Canada*, **30**, 1992–1999.

[51] Walker, B. H., Ludwig, D., Holling, C. S., and Peterman, R. M. (1981), Stability of semi-arid savanna grazing systems, *Journal of Ecology*, **69**, 473–498.

[52] Pickett, S. T. A. (1982), Population patterns through twenty years of oldfield succession, *Vegetatio*, **49**, 45–59.

[53] Marks, P. L. (1974), The role of pin cherry (*Prunus pensylvanica* L.) in the maintenance of stability in northern hardwood ecosystems, *Ecological Monographs*, **44**, 73–88.

[54] Goméz-Pompa, A. and Vázquez-Yanes, C. (1981), Successional studies of a rain forest in Mexico, in West *et al.*, note [39], pp 246–266.

[55] Horn, H. S. (1981), Some causes of variety in patterns of secondary succession, in West *et al.*, note [39], pp 24–35.

[56] Kilgore, B. M. (1976), Fire management in the national parks: an overview, in *Proceedings of Tall Timbers Fire Ecology Conference*, Vol. 14, pp 45–57 (Florida State University Research Council, Tallahassee, FL).

[57] Larkin, P. A. (1979), Maybe you can't get there from here: history of research in relation to management of Pacific salmon, *Journal of the Fisheries Research Board, Canada*, **36**, 98–106.

[58] Ludwig, D., Jones, D. D., and Holling, C. S. (1978), Qualitative analysis of insect outbreak systems: the spruce budworm and the forest, *Journal of Animal Ecology*, **44**, 315–332.

[59] McNamee, P. J., McLeod, J. M., and Holling, C. S. (1981), The structure and behavior of defoliating insect/forest systems, *Research on Population Ecology*, **23**, 280–298.

[60] Holling, C. S. (1980), Forest insects, forest fires and resilience, in H. Mooney, J. M. Bonnicksen, N. L. Christensen, J. E. Latan, and W. A. Reiners (Eds) *Fire Regimes and Ecosystem Properties*, USDA Forest Service General Technical Report, pp 20–26 (USDA Forest Service, Washington, DC).

[61] Macdonald, G. (1973), *Dynamics of Tropical Disease* (Oxford University Press, London, UK).

[62] May, R. M. (1977), Thresholds and breakpoints in ecosystems with a multiplicity of stable states, *Nature*, **269**, 471–477.
[63] Holling, C. S. and Buckingham, S. (1976), A behavioral model of predator–prey functional responses, *Behavioral Science*, **3**, 183–195.
[64] Peters, R. H. (1983), *The Ecological Implication of Body Size* (Cambridge University Press, Cambridge, UK).
[65] Noy-Meir, I. (1982), Stability of plant–herbivore models and possible applications to savannah, in B. J. Huntley and B. H. Walker (Eds), *Ecology of Tropical Savannahs* (Springer, Heidelberg, FRG).
[66] Clark, C. W. (1976), *Mathematical Bioeconomics* (Wiley, New York).
[67] Baskerville, G. L. (1976), Spruce budworm: super silviculturist, *Forestry Chronicle*, **51**, 138–140.
[68] Baltensweiler, W., Benz, G., Boven, P., and Delucchi, V. (1977), Dynamics of larch budmoth populations, *Annual Review of Entymology*, **22**, 79–100.
[69] McLeod, J. M. (1979), Discontinuous stability in a sawfly life system and its relevance to pest management strategies, *Current Topics in Forest Entomology*, General Technical Report WO-8 (USDA Forest Service, Washington, DC).
[70] Schumpeter, J. A. (1950), *Capitalism, Socialism and Democracy* (Harper, New York); and Elliott, J. E. (1980), Marx and Schumpeter on capitalism's creative destruction: a comparative restatement, *Quarterly Journal of Economics*, **95**, 46–58.
[71] Vitousek, P. M. and White, P. S. (1981), Process studies in succession, in West *et al.*, note [39].
[72] Vitousek, P. M. and Matson, P. A. (1984), Mechanisms of nitrogen retention in forest ecosystems: a field experiment, *Science*, **225**, 51–52.
[73] Marks, P. L. and Bormann, F. H. (1972), Revegetation following forest cutting: mechanisms for return to steady-state nutrient cycling, *Science*, **176**, 914–915.
[74] Hill, A. R. (1975), Ecosystem stability in relation to stresses caused by human activities, *Canadian Geographer*, **19**, 206–220.
[75] Moore, P. D. (1982), Fire: catastrophic or creative force? *Impact of Science on Society*, **32**, 5–14.
[76] Goméz-Pompa, A., Vázquez-Yanes, C., and Guevara, S. (1972), The tropical rain forest: a non-renewable resource, *Science*, **177**, 762–765.
[77] Denevan, W. M. (1973), Development and imminent demise of the Amazon rain forest, *Professional Geographer*, **25**, 130–135.
[78] Levin, S. A. and Paine, R. T. (1974), Disturbance, patch formation and community structure, *Proceedings of the National Academy of Sciences*, **71**, 2744–2747, and Paine, R. T. and Levin, S. A. (1981), Intertidal sandscapes: disturbance and the dynamics of pattern, *Ecological Monographs*, **51**, 145–178.
[79] Hastings, H. M., Pekelney, R., Monticcolo, R., van Kannon, D., and Del Monte, D. (1982), Time scales, persistence and patchiness, *BioSystems*, **15**, 281–289.
[80] Mandelbrot, B. B. (1977), *Fractals: Form, Chance and Dimension* (Freeman, San Francisco, CA).
[81] May, R. M. (1971), Stability in multi-species community models, *Mathematical Biosciences*, **12**, 59–79.
[82] Janzen, D. H. (1983), Tropical agroecosystems, in P. H. Abelson (Ed), *Food: Politics, Economics, Nutrition and Research* (AAAS, Washington, DC).
[83] Holdridge, L. R. (1947), Determination of world plant formations from simple climate data, *Science*, **105**, 367–368.
[84] Grigor'yev, A. Z. (1958), The heat and moisture regime and geographic zonality, *Third Congress of the Geographical Society of the USSR*, pp 3–16.
[85] Steele, J. H. (1985), A comparison of terrestrial and marine ecological systems, *Nature*, **313**, 355–358.
[86] Baskerville, G. L. (1983), *Good Forest Management, a Commitment to Action* (Dept. Natural Resources, New Brunswick, Canada).
[87] Brooks, H. (1973), The state of the art: Technology assessment as a process, *Social Sciences Journal*, **25**, 247–256.
[88] Douglas, M. (1978), *Cultural Bias*. Occasional Paper for the Royal Anthropological Institute No. 35 (Royal Anthropological Institute, London).
[89] Thompson, M. (1983), A cultural bias for comparison, in H. C. Kunreuther and J. Linnerooth (Eds), *Risk Analysis and Decision Processes: The Siting of Liquified Energy Gas Facilities in Four Countries* (Springer, Berlin).
[90] Mann, H., Siegler, M., and Osmond, H. (1970), The many worlds of time, *Journal of Analytical Psychology*, **13**, 33–56.
[91] Walters, C. J. and Hilborn, R. (1978), Ecological optimization and adaptive management, *Annual Review of Ecology and Systematics*, **9**, 157–188.
[92] McNeill, W. H. (1982), *The Pursuit of Power* (University of Chicago Press, Chicago, Ill).
[93] An earlier version of this chapter was published in Malone and Roederer, note [3].

Commentary

F. di Castri

From my own perception, the chief aim of Holling's chapter is to provide from the start a kind of conceptual framework for the proposed program, the *Sustainable Development of the Biosphere.* There have been and will be other international endeavors such as this: worth mentioning are UNESCO's *Man and the Biosphere Programme* (MAB), which started in 1971 and is ongoing, and ICSU's *International Geosphere–Biosphere Programme* (IGBP) – also called *Global Change* – for which a feasibility study is being undertaken. For none of these has an essay of conceptualization – comparable to this chapter – been attempted; probably there were good reasons for disregarding this aspect, not the least being consideration of the difficulties involved in achieving a sensible agreement on such conceptual issues, especially when large and heterogeneous groups of countries and scientists have to work together.

In addition, the MAB Program has almost unavoidably shifted toward a loose coordination of very heterogeneous packages of national projects; a number of which certainly provide enlightening empirical insights for a theorization of some man–ecosystem interactions. Nevertheless, the local and specifically problem-oriented nature of the best MAB field projects prevents an approach – through MAB – to global concerns.

On the other hand, IGBP initially had almost exclusively global views. It is now increasingly recognized within IGBP that a large number of explanations for the functioning of the biosphere should come from local and regional studies – chiefly of a biological nature – and that the notion of spatial and temporal scale is the key for the success of the overall program. However, it is unlikely that a program of fundamental science, like IGBP, can become really involved in local societal issues.

There is, therefore, a niche for the kind of concerns addressed in Holling's chapter , where a conciliation is proposed between two outcomes, local surprise (of a biological but mostly societal type) and global change (considered chiefly from a biogeochemical point of view). There would be merit if Holling's views were also discussed within MAB and IGBP, as these two programs are seeking new avenues of research and a better definition of their framework.

Concerning specifically Holling's chapter, one can be tempted to react in two opposing ways: either to accept everything by intuition and sympathy – being attracted by the cohesion of the argument and the stimulating challenge of the ideas – or to block reject these speculations. In fact, several statements are not supported by scientific evidence (but this is to be expected in such an article), some hypotheses are not testable in an experimental and quantitative way (as the author recognizes), and one cannot easily visualize how they could be proved or disproved; there is also much use (and perhaps some abuse) of analogical reasoning (with all the charm and the risk involved in analogies). I suspect that whichever position one is inclined to – acceptance or rejection – is a matter of personal feeling and behavior rather than of a difference in scientific background. From my viewpoint, I will try to establish which are topics where a consensus might emerge, as well as which are the most controversial aspects. I focus my comments on the three pillars of Holling's conceptual building: local surprise, the four ecosystem functions, and the Gaia hypothesis.

As regards local surprise in relation to societal and institutional behavior, while admittedly similar concepts might be put forward using different terms – e.g., perception – I believe that Holling's presentation has definite advantages, even from a pedagogical viewpoint *vis-a-vis* both scientists and policy-makers. I cannot agree more that in the contemporary context – both socioeconomic and scientific – management institutions (and political establishments) should not desperately seek prediction and control, but rather should settle their organization and their policy on the acceptance of, and the adaptation to, surprise effects. Flexibility should be the *sine qua non* condition for their decisions. By the way, little prediction can be provided at present by the key science of ecology, and I suspect that the same is true for the key disciplines of economy and sociology.

In relation to the Gaia hypothesis, it seems to be at present one of the few workable concepts when addressing research on global biospheric problems, even if the end result be the disproof of it. I imagine that Gaia is also the underlying hypothesis of IGBP. In addition, the use of Gaia and of local surprise to enlighten the interfaces between biospheric and more traditional ecological issues, on the one hand, and between ecological and societal issues on the other, has the advantage of focusing on *processes* of a scientific and decision-making nature rather than on ill-defined interdisciplinary linkages. Interdisciplinarity, when it is considered and implemented as an end in itself – and not as a tool for addressing new complex problems – leads too often to sterile research and to verbose descriptions and nonexplanatory results.

Almost paradoxically, in view of Holling's background, the most controversial points may well refer to the ecosystem concept as defined by Holling's four functions – exploitation, conservation, creative destruction, and renewal. To be kindly sarcastic, and admittedly somewhat unfair, one would be tempted to think that Holling's conceptual construction is too harmonious and too beautiful to accurately equate with reality. The supporting examples given by Holling are well presented, but I am doubtful whether they should be generalized to such an extent, and anyway they may lead to different interpretations. Furthermore, I wonder whether these views on ecosystem functions can convince anyone who is not already converted or preadapted to such ideas. From my viewpoint, I believe that Holling's views on this matter represent a very stimulating framework for discussion, but I am somewhat skeptical of the possibility of his hypotheses being proved (or disproved). In fact, too many elements evoked to support these four ecosystem functions are either nonmeasurable or interpreted at present in ways too controversial for good scientific communication. No one is more aware than Holling of the many (and sometimes opposite) interpretations of the terms stability and resilience [1]; but also the adaptive strategies of species represent more a reference concept than a quantifiable parameter (and the place of so-called *r*- and *K*-species in an ecological succession is not so clearly defined or so fixed as *Figure 10.5* may suggest).

As a matter of fact, I believe that the whole ecosystem concept is in crisis at present, at least if the ecosystem is taken as a kind of well-defined supraorganism, and not as a useful methodological tool to study problems of interactions and of system behavior as regards different ecological multispecies units. A too extreme and exclusive view of ecosystem properties can alienate from ecology some of the disciplines, such as ecophysiology, population biology, and genetics, that are essential for the explanation of most of the ecological processes (and this is already happening in a number of countries [2]).

The other key aspect of the theoretical framework proposed by Holling deals with *connections*, as exemplified in *Figure 10.6.* Ecosystems clearly represent the crossroads of all the systems, being intermediate between biosphere and society. Some geographers and economists could again object to the use of ecosystems; they may argue that the exchanges of energy and products exceed the boundaries of any given ecosystem, and may prefer other units such as "human use systems" or "resource systems". Nevertheless, even if these systems help to establish interfaces with societies, they cannot facilitate linkages with global biospheric issues; furthermore, as regards ecosystems, these possible objections may imply simply a different interpretation of the definition and hierarchical scale of ecosystems. I am personally more worried, as regards connections, about the excessive use of analogies. I need evidence to be convinced that ecosystems are just "Gaia writ small"; changing of scale implies usually the emerging (and the disappearing) of new functions and properties. On the other hand, analogies, such as those presented in *Table 10.2* between ecosystems and other systems of a societal nature *sensu lato*, can perhaps improve the understanding between disciplines and their different approaches, but do not serve operational purposes for research and may be dangerous and misleading if they are taken too strictly.

After all, it is amazing to see to what extent a science in crisis, such as ecology – in crisis for several reasons, not least the confusion between ecological and environmental problems [2] – is capable of "exporting" so many concepts, through an analogical process, to other sciences that probably face a comparable crisis, such as geography, economics, or sociology. I wonder whether ecology itself is not shifting at present from a biological science to a sociological one, with all the epistemological and methodological implications involved in this move.

Finally, in relation to Holling's recommendations, they all represent essential avenues for research. However, on the basis of previous experience in international programs, such as the *International Biological Programme* and the *Man and the Biosphere Programme* [3], I suspect that it is seldom realized how high are the costs, the manpower involvement, and the intrinsic difficulties represented by these proposals. Another obstacle, which must be overcome for the development of ecology, is our lack of a theoretical basis for the comparison of ecosystems. We do not know yet how to evaluate the degree of extrapolatability and predictability of our research from one to another ecosystem, or from one to another period of time. This is true even within the same ecosystem type, e.g., comparisons between temperate forests of the northern and southern hemispheres; the difficulties in comparing more complex and less studied ecosystems, such as those of the humid tropics, or of more heterogeneous systems, such as the Mediterranean-type ones, increase exponentially. When different types of ecosystems are compared, for instance terrestrial and aquatic ones, or forests and grasslands, it becomes almost impossible to escape from the pitfall of overgeneralization, with little regard for reality.

This remark is not intended to underestimate the importance of a comparative ecology or to discourage students from undertaking this kind of research. On the contrary, I feel that these comparisons represent now the main challenge for ecology. It is, nevertheless, essential to fully understand the limits and the methodological background for such comparisons. If they are placed in the framework of strictly climatic and old-fashioned classifications of life zones, such as those of Holdridge and Grigor'yev, as suggested here, I do not hesitate in predicting many "surprise effects" when the results are compared; and these surprises may well derive from the lack of understanding or knowledge of the diverse origins of these ecosystems, on the weight of the past, including both geological and historical factors (and man's impact in fashioning ecosystem patterns is much older than usually considered). A strong injection into such research of evolutionary ecology (I mean evolution in the strict sense of biological evolution) and of explanatory biogeography – with special emphasis on the invasion of alien species – would be likely to reduce considerably these surprises.

I regret that, because of space limitations, I cannot refer to many other enlightening points of this chapter, such as continuity and discontinuity in ecological processes, linear and nonlinear interactions, etc. As with most of Holling's work one is captured by the argument and – in agreement or disagreement – cannot resist the temptation of being involved in the discussion; and this is precisely the goal that a chapter of this nature should achieve.

Notes and references (Comm.)

[1] As quoted by Holling, Pimm, S. L. (1984), The complexity and stability of ecosystems, *Nature*, **307**, 321–326.

[2] di Castri, F. (1984), *L'écologie. Les défis d'une science en temps de crise.* Rapport au Ministre de l'Industrie et de la Recherche (La Documentation Française, Paris).

[3] di Castri, F. (1985), Twenty years of international programmes on ecosystems and the biosphere: an overview of achievements, shortcomings and possible new perspectives, in T. F. Malone and J. R. Roederer (Eds), *Global Change* (Cambridge University Press, Cambridge, UK).

PART FOUR

SOCIAL RESPONSE

Social response

Contributors

About the Contributors

Harvey Brooks is a former Dean of Engineering and Applied Physics at Harvard University and currently Professor of Technology and Public Policy in Harvard's John F. Kennedy School of Government. He has held many affiliations with government, industry, and foundations and is currently Chairman of the US Committee for IIASA for the American Academy of Arts and Sciences.

Mark Cantley is responsible for the European Community's "CUBE" (Concertation Unit for Biotechnology in Europe). Educated in mathematics (B.A., Cambridge), economics (M.Sc., L.S.E.), operational research, and accounting and finance, he spent four years with the British Iron and Steel Research Association, then moved to Lancaster University, to develop teaching and research interests in OR and corporate strategic planning. He also spent two years at IIASA studying scale and learning in production systems, and strategic control in psychogeriatric care. In 1979, he joined the EEC's FAST programme (Forecasting and Assessment in Science and Technology), become responsible for the "Bio-Society" activities, and contributed to the development of the Community strategy for biotechnology in Europe, in whose implementation he is now engaged.

Michael Gwynne is a Fellow of Balliol College, Oxford University, and Director of both the Global Environment Monitoring System (GEMS) and the Environment Assessment Service, both part of the United Nations Environment Programme. Educated at the Universities of Edinburgh and Oxford, he is the author of many scientific papers and government and international reports in the fields of agriculture, range management, pastoralism, conservation, ecology, and remote sensing.

Nigel Haigh is head of the London office for the Institute for European Environmental Policy. He studied mechanical sciences at Cambridge University, UK, and for ten years worked as a patent lawyer before turning to the environmental field. In 1984 he published *EEC Environmental Policy and Britain*, the first attempt in any country to analyze the interaction between the EEC, national government, and local government in the adoption and implementation of environmental policy.

Yuri Izrael is Chairman of the USSR State Committee for Hydrometeorology and Control of Natural Environment, Head of the Hydrometeorological Department of the Council of Ministers of the USSR, and a Corresponding Member of the USSR Academy of Sciences. He graduated from the Central Asian University of Tashkent and is the author of many fundamental works on atmospheric physics and problems of natural environment control.

Giandomenico Majone is Professor of Statistics at the University of Calabria in Italy, but is currently at the John F. Kennedy School of Government, Harvard University. He was educated at the Universities of Padua and California, and at the Carnegie Institute of Technology. Majone has published papers on Bayesian statistics, decision theory, and systems and policy analysis.

Ted Munn is Leader of the Environment Program at IIASA, Laxenburg, Austria and Professor at the Department of Physics and an Associate at the

Institute of Environmental Studies, University of Toronto. Educated at McMaster University and the Universities of Toronto and Michigan, Munn is author of the textbooks *Descriptive Micrometeorology* and *Biometeorological Methods*, as well as being Editor-in-Chief of the *International Journal of Boundary-Layer Meteorology* and Editor of SCOPE publications. He designed the Global Environmental Monitoring System for UNEP, for whom he acts as senior consultant from time to time. He was recently elected Fellow of the Royal Society of Canada.

Martin Parry is University Lecturer in the Department of Geography at the University of Birmingham, UK. Educated at the Universities of Durham, the West Indies, and Edinburgh, he directed research for the Social Sciences Research Council's study of moorland change and upland management, and was instrumental in promoting amendments to the UK Wildlife and Countryside Act (1981). Parry is just completing a joint UNEP/IIASA Project on "Integrated Approaches to Climatic Impacts: The Vulnerability of Food Production in Marginal Areas", as part of the UNEP World Climate Programme.

Masatoshi Yoshino is a Professor at the University of Tsukuba, Japan, and is Chairman of the working group *Tropical Climatology and Human Settlements* of the International Geophysical Union. Educated at the University of Kyoiku, Tokyo, Yoshino has worked at the Universities of Bonn and Heidelberg, FRG, and at Hosei University, Tokyo, Japan. He is author of several books, including *Climate in a Small Area* and editor of both *Water Balance in Monsoon Asia* and *Climate and Agriculture in Monsoon Asia.*

Chapter 11 The typology of surprises in technology, institutions, and development

H. Brooks

Editors' Introduction: Most studies of the future are dominated by smooth projections that are free of significant discontinuities, random events or, more generally, of surprises. Social learning to cope with surprise is also neglected. The IIASA Biosphere Project seeks to understand the role of surprise in shaping interactions between human activities and the environment, and to design methods that can accommodate both surprise and social responses to it in assessments of future prospects for sustainable development.

Addressed in this chapter are the nature and origins of surprise in interactions among technology, institutions, and society, thus complementing the analysis in Chapter 10 of surprise in the management of ecological systems. Proposed is a typology of surprises that affect these interactions, with special attention to the ways in which certain long-term trends and scale-effects increasingly predispose systems to discontinuous and often unexpected change.

Introduction

That the world is in a state of transition toward a single interdependent system is by now conventional wisdom, but the nature and extent of this interdependence varies greatly between different parts of the system. Economic interdependence among nations has been institutionalized in many specialized agencies and forums. While this institutionalization is far from complete, especially with respect to the management of technological developments, it has progressed much further than the institutionalization of ecological interdependence. The two themes of ecological and economic interdependence are strongly linked through a third theme – the development and diffusion of technology. The most dramatic development of the last 30 years has been the growth of the technological capacity of the industrialized countries, as well as of many of the developing countries. The world population of scientists and engineers, like the volume of international trade, has grown at thrice the rate of the population and at about twice that of world GNP, and this population is generally expected to increase further. In addition, the decline in the cost of transportation, and especially of communications, relative to other economic factors has dramatically enhanced the mobility of both capital and technology, creating an acceleration of structural change in the world economy, and expanding the spatial and temporal extent of the ecological consequences of economic and technological development.

The expansion of economic, technological, and ecological interdependence has stimulated a growing volume of research on its implications and consequences. The International Institute for Applied Systems Analysis (IIASA) itself is one institutional manifestation of this expansion. Much of the work to date has been based, implicitly or explicitly, on an evolutionary paradigm – the gradual, incremental unfolding of the world system in a manner that can be described by surprise-free models, with parameters derived from a combination of time series and cross-sectional analyses of the existing system. This is also true of work with econometric models, of energy models, and of models of the global environment [1].

The focus on surprise-free models and projections is not the result of ignorance or reductionism so much as of the lack of practically usable methodologies to deal with discontinuities and random events. The multiplicity of conceivable surprises is so large and heterogeneous that the analyst despairs of deciding where to begin, and instead proceeds in the hope that in the longer sweep of history surprises and discontinuities will average out, leaving smoother long-term trends that can be identified in retrospect and can provide a basis for reasonable approximations to the future. The underlying assumption is that surprises can be isolated from the underlying trends so that their interactions can be neglected in the first approximation, or at least approximated through the parameterization of smooth models, much as the phenomena of turbulence can be parameterized in the equations of motion for fluids.

One of the purposes of the Biosphere Project at IIASA is to understand how far we can and what is required to depart from this zero-order approximation, and to establish if it is even a legitimate approximation to the real world. In this chapter I hope to deal with the issue of surprise in relation to the interaction between technology, human institutions, and social systems. Other chapters deal with surprise more specifically in terms of the biosphere itself. I fear, however, that our understanding of discontinuities and surprises in this broader sociotechnical framework is more rudimentary and poorly formulated than it is in the biosphere–development interactions analyzed so elegantly (for example) by Holling (Chapter 10, this volume).

A typology of surprise

We can divide the surprises dealt with in this chapter into three general types:

(1) Unexpected discrete events, such as the oil shocks of 1973 and 1979, the Three Mile Island (TMI) reactor accident, political coups or revolutions, major natural catastrophes, accidental wars.
(2) Discontinuities in long-term trends, such as the acceleration of USA oil imports between 1966 and 1973, the onset of the stagflation phenomenon in OECD countries in the 1970s, the decline in the ratio of energy consumption growth to GNP growth in the OECD countries after 1973.
(3) The sudden emergence into political consciousness of new information, such as the relation between fluorocarbon production and stratospheric ozone (O_3), the deterioration of central European forests apparently due to air pollution, the discovery of the recombinant DNA technique, the discovery of asbestos-related cancer of industrial workers.

These three types of surprises are, of course, interrelated. Discrete events may be the trigger for a permanent change in long-term trends. New information may be the result of discontinuities in long-term trends that had been latent but took a long time to become apparent. Much new information is such that we realize we should have known about it if we had only had the sense to look for it.

Some implicit assumptions

In order to keep the discussion of this chapter within finite bounds we make some assumptions that implicitly rule out certain types of surprises. In particular, we assume continued growth of interdependence and associated linkages in time and space, as well as among different types and classes of problems which previously appeared unrelated. Second, we rule out discussion of the possible consequences of a major resurgence of protectionism and economic nationalism, of general war, or of a sudden major shift in power alignments among nations. It is not that such events are improbable, but including them is beyond the scope of a single chapter.

Our discussion is also mainly restricted to technology and to technology–society interactions. We do not deal with the biosphere as such, since that is adequately covered in the other chapters.

Alternatives to surprise-free thinking

As hinted at above, it would be a mistake to reject surprise-free thinking out of hand. It is not, however, my contention that we must accept a fundamental paradigm change that sweeps away all that has gone before. Rather it is a question of understanding the limits of current projections and models, and especially of how long-term changes in both environmental and social parameters interact with short-term changes and random events, much as the concepts of eddy diffusion and momentum transfer are used to explain average flows in large-scale atmospheric and oceanic phenomena. On the other hand, because we are interested in the impact on human beings and their institutions, we must also gain an understanding of how long-term trends predispose systems toward certain kinds of surprises and discontinuities or toward greatly amplified responses to small, random events. Unfortunately, knowing that a system is becoming more predisposed to certain catastrophic events does not tell us when these events will occur. We might have continued for another 20 years without a TMI incident, for example, which might have produced very different outcomes in terms of worldwide energy developments.

By making nonlinearities and surprises a center of methodological concern, we do not intend to denigrate the possibility of elucidating trends of longer term. We must recognize, however, that from the viewpoint of individuals, organizations, and nations, smooth development is the exception rather than the rule. Nonlinearities, perception thresholds, and effects of scale are particularly important in determining human perceptions and reactions, and hence the feedback from external events to changes in behavior that affect the impact of environmental, technological, or economic trends on society.

In what follows I give several examples by way of illustration.

Alternation of perceptions of resource scarcity and abundance

For most of twentieth century, and indeed prior to that, there has been a regular cycle of alternating concerns about the scarcity of key natural resources followed by complacency about their abundance. The swings in these perceptions seem to lean far beyond objective reality in both directions. Often the alternations have been triggered by dramatic short-term events which raise intense public concern for a brief period, but not for long enough to alter the direction of policy permanently. What is more remarkable, however, is the degree to which generalized public concerns influence expert appraisals of the situation in any one period.

The most dramatic example of recent times is probably those events immediately surrounding the Arab oil embargo in 1973, but pessimism regarding resources, especially energy resources, had been growing for several years prior to this, after a peak of optimism in the 1960s [2]. In fact, the famous "limits-to-growth" debate had preceded the event by a year or two [3]. The environmental movement of the late 1960s had as part of its agenda a strong element of concern over the depletion of nonrenewable resources. The oil embargo was followed within a year by a worldwide shortfall in cereal grain harvests and by an explosion in raw material prices which drove the first inflationary burst of the 1970s [4]. All this contributed to a widespread perception of dwindling resources, both by the general public and by a significant proportion of experts. It was then strongly reinforced by the second oil shock of 1979.

Yet, looking at these events in retrospect, it can be argued that both the public and many experts confused short-term, "fast-cycle" incidents with more gradual long-term trends. To be sure, although the events that triggered the crisis occurred with random timing, their probability of occurrence had been greatly increased by the more gradual underlying changes in world energy trade.

For example, the peaking of the USA domestic oil and gas production in about 1970 had triggered a very rapid acceleration of USA demands on the world oil market, and this set the stage for the unexpected success of the OPEC cartel in raising oil prices. These trends, which were apparent in retrospect [5], had been slowly but surely raising the vulnerability of the world oil system to small disturbances, even though the particular events which precipitated the 1973 and 1979 crises were unpredictable as to their timing, and perhaps not even inevitable. These events also sensitized the whole international trading system to potential commodity shortages and doubtless amplified the grain and commodity price explosions that occurred in 1974. Moreover, the policy response of the USA to the sudden rise in the price of imported oil had the perverse effect of further stimulating the rise of USA oil imports, thus enabling the second oil shock of 1979, in which a far smaller supply shortfall than in 1973 caused a much more serious market disruption, in part due to self-protective inventory building. The 1979 crisis, triggered by the Iranian revolution, was itself related to the politically destabilizing effect of the dramatic rise in the inflow of oil revenues to the Persian Gulf region. What we see here is a whole series of interrelated surprises, and yet it is not clear that anything more than an increased probability of such surprises could have been predicted from the context. The exact course of the crisis appears to have been highly dependent on the exact timing of events, which was essentially random. We can also see in retrospect that many actions designed to cope with short-term problems, such as USA domestic price controls and the crude oil allocation system, had the effect of greatly exacerbating the impact of these events on a time scale of 5–10 years [6].

We can also see, however, that effects lasting for decades and triggered by crisis-generated conditions – namely dramatic, price-induced energy conservation [7] – tend to drastically mitigate the problem of apparent shortages on this time scale, and indeed sow seeds of doubt on the widespread perception of a true physical resource shortage. Today we see a fundamental split developing among the experts regarding the prospects on a 30–50 year time horizon. The events of the 1970s have generated a sharp, almost discontinuous downward trend in energy/output ratios for both industry and consumer services throughout the industrialized world, as shown in *Figure 11.1* [7] for the USA. It now seems apparent that the full effects on output/energy ratios of the oil price shocks of the 1970s will not work themselves through the system for another decade [7, 8]. According to Hogan [7], more than three quarters of the oil conservation effect has been due to improved end-use efficiency rather than to slower economic growth in the industrialized countries. The combination of increasing diversification of oil supply sources with the improved energy efficiency of durable goods has evidently reduced the sensitivity of the system, both to cartel activity and to unforeseen political events, such as the Iran–Iraq war, which has hardly affected world oil markets so far.

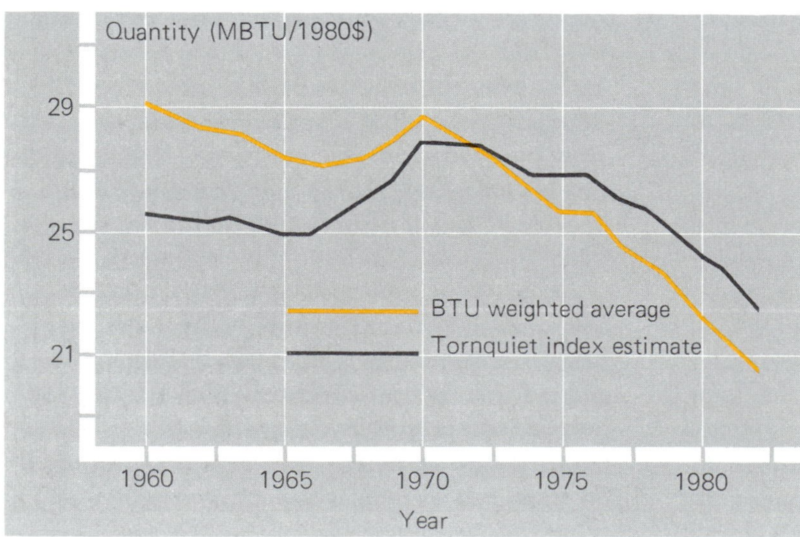

Figure 11.1 Aggregate energy indexes per $GNP for the USA.

Despite the current relaxation of the crisis, however, experts are sharply divided about trends beyond the next ten years. Many believe that the present oil glut is temporary [8], and that prices will begin their upward trend again in the late 1980s or early 1990s with the system becoming once again vulnerable to random political events in the Middle East or elsewhere. There are others who believe that the two oil shocks have set world oil consumption on a new trajectory relative to economic growth, and that the discovery of oil and gas resources outside OPEC will continue to place downward pressures on the real price level of oil, and hence of all competing energy forms, as far into the future as can be foreseen [9]. Some even fear that many present marginal, high cost sources of energy, such as Alaskan and North Sea oil and gas, may become subeconomic before they are depleted, so that they will have to be either shut down or receive massive government support, either through direct subsidies or cartel-like arrangements between consumers and low-cost producers so as to limit production from the lowest cost sources [9].

Thresholds and nonlinearities

The benefits of technology often increase in proportion to its scale of application, whereas many environmental and social phenomena resulting from the application of technology increase highly nonlinearly with increasing scale [10]. Thus a technological activity may become strongly established with influential vested interests during the linear regime, before the disbenefits that increase nonlinearly with scale begin to manifest themselves or become apparent to a wider public. The situation is illustrated in *Figure 11.2*, which is intended to suggest how the benefits and disbenefits of a technology might vary with its scale of application. The two solid black and yellow curves show the benefits (linear) and the disbenefits (nonlinear), respectively, while the broken curve shows the net marginal benefit as a function of scale (essentially the derivative of the difference between the benefit and the disbenefit curves). As the scale increases the marginal benefit of further increments in scale may become negative, as indicated. This could be viewed as a form of the "tragedy of the commons [11]," in which the result of a participant acting for his or her own benefit becomes a disaster for all. One of the classic examples is that of traffic, e.g., automobile traffic, where, particularly in urban areas at peak traffic hours, the disbenefits set in quite abruptly at a critical traffic density. The threshold density for the onset of disbenefits can be raised by increasing the peak capacity of the roads, but this requires a

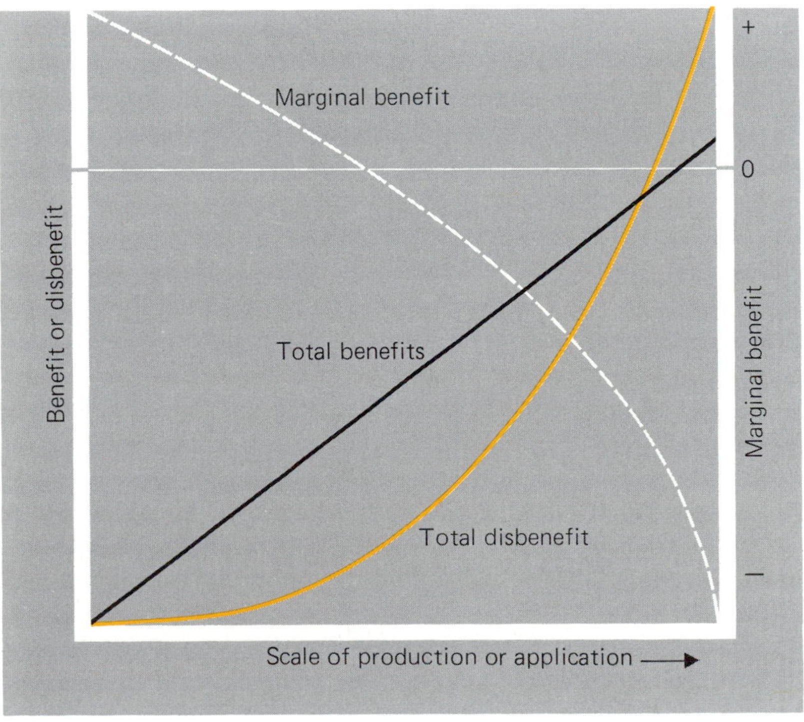

Figure 11.2 Variation of benefits and disbenefits of a technology with scale of application.

very high capital cost per additional vehicle accommodated [12]. Furthermore, the additional capacity concentrates vehicles even more in peak periods and on peak routes, with the result that there is relatively more unused capacity at other times, thus further adding to the incremental investment per vehicle accommodated. Moreover, this leaves out such additional *social costs* as the preemption for highway construction of socially valuable urban land and the displacement or disruption of neighborhoods.

Another example of such a threshold effect arising from nonlinearities also occurs in connection with automobiles, namely the smog resulting from auto emissions. Smog is an indirect product of photochemical reactions among several pollutants; these reactions are known to vary nonlinearly with pollutant concentrations. Thus smog problems appear permanent at some critical traffic density or automobile population per unit area [13]. This is illustrated, again schematically, in *Figure 11.3*,

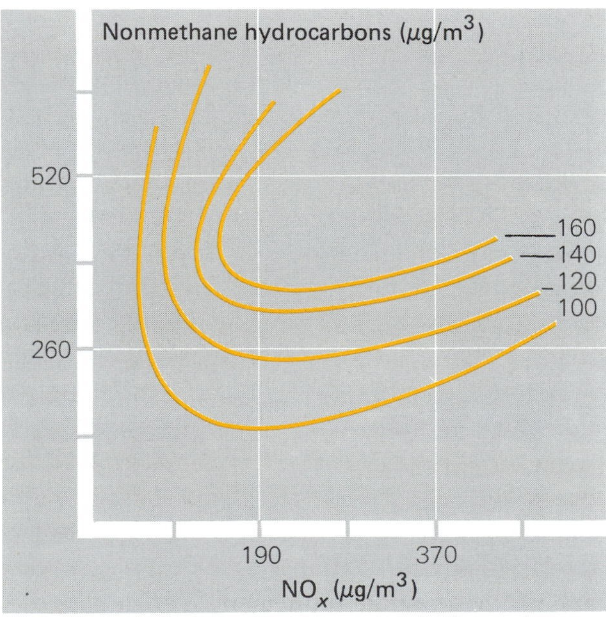

Figure 11.3 Maximum oxidant level as a function of total emission concentrations (relative scale) [13].

which shows oxidant level as a function of total emissions.

An example of a perceptual threshold occurs in connection with airport noise. It is well known from surveys of residents near airports that the complaint rate increases much more rapidly than in proportion to the noise intensity or noise perception index (NPI), when this index rises above some threshold level [14]. Similarly, the public reaction to the siting of large industrial facilities, especially power plants, seems to be more than proportional to the size or capacity of the facility. Moreover, there are bandwagon effects such that successful opposition to the siting of facilities in one community encourages opposition in other communities, so that public opposition to specific types of facilities (such as hazardous waste dumps or processing facilities) is contagious and somewhat like an epidemic, with the result that the probability of opposition to any one facility increases with the number of previous attempts to site such facilities within a given time period.

Another interesting example of a political threshold effect occurs in the displacement of labor. When a large company, such as Chrysler, Lockheed, or British Leyland, is threatened with bankruptcy due to foreign competition or other causes, political sensitivities are very high, and some form of government rescue operation is frequently forthcoming, even though many more people may be losing their jobs in the same geographical area due to bankruptcies of small businesses in a comparable period of time [15]. In other words, the size and geographical concentration of institutions changes public perceptions about their importance and their relation to the political process in a way that is not simply proportional to the number of individuals affected. Similarly, many simultaneous fatalities in an air crash attracts much more public attention and political action than the steady attrition of automobile accidents.

Another example is those fears of worker displacement resulting from the introduction of robot and other computer-based technologies in manufacturing. The metalworking occupations, which appear to be most amenable to displacement by computerized manfacturing technologies, are highly concentrated in particular geographical areas, such as the five states in the US Great Lakes region [16]. This geographical and sectoral concentration is likely to lower the threshold for political intervention in the process of conversion to computerized manufacturing. Effects that are quantitatively larger, but more dispersed either in space or over time, are likely to attract less political attention and consequent policy response than smaller effects that are more concentrated and exceed some sort of local "critical mass". This situation appears characteristic of many different kinds of "negative externalities" of technical or economic–structural changes, ranging from air pollution to employment displacement. It may generate a bias in political attention that reduces the effectiveness of social responses to external trends.

Latent surprises

A source of surprise that has come to attention with great force only recently, but is likely to grow in importance, is the appearance of evidence for delayed effects of technological activities that occurred in the relatively distant past. One example is the appearance of cancers resulting from exposure of relatively large populations to very low concentrations of carcinogenic chemicals or low levels of ionizing radiation. The most dramatic example is probably that of asbestos [17]. In many instances the latency period can extend over most of a human life span. The nightmare scenario associated with this is the appearance of a cancer "epidemic" many decades after the exposure of a large population over a long period of time with no apparent ill effects. The most well-documented and quantitatively important case of this is probably that of smoking and lung cancer [18]. Otherwise most of the nightmare scenarios have occurred only in limited, occupationally exposed populations, such as asbestos workers or coal miners with black lung. Another example of the nightmare scenario that has attracted attention is the rapid introduction of artificial sweeteners into diet soft drinks – first saccharin, and now aspartane [19]. The problem is the exposure of large and youthful populations to nonnatural substances within a time period that is short compared with the possible latency period for the appearance of secondary effects. An even more speculative, and possibly more dramatic, case might be the spread into the human population of bacteria resistant to antibiotics, which develop from feeding antibiotics to cattle or poultry [20]. The appearance of widespread forest damage as a result of air pollution in both Central Europe and North America [21] may be a similar latency phenomenon, in which the effects observed arise from causes that accumulate over long periods of time.

The potential importance of the latency phenomenon is underlined by evidence that the incidence of cancer in large populations is highly variable geographically, implying that most cancers are of environmental origin [22], though not necessarily associated with industrial products, as sometimes asserted. Indeed, the consensus among experts seems to be that these geographical variations are primarily associated with diet and lifestyle [22]. While the relative incidence of different types of cancer has been changing over time, there is no evidence to suggest the existence or prospect of a cancer epidemic associated with industrial activity or economic development, apart from the well-documented case of smoking. Because of the latency phenomenon, however, the absence of a contemporary cancer epidemic is not as strong an argument as might be desired against the possibility that such an epidemic might appear some time in the future as a result of apparently innocuous activities taking place now. The continual exposure of large numbers of people to substances not common in nature is a potential source of future surprise, which is likely to decline only gradually as our scientific understanding of the specific biological mechanisms of carcinogenesis and mutagenesis slowly improves.

Moreover, the asbestos story, the Love Canal case, and the Agent Orange affair show that large numbers of people do not have to suffer latency effects to create large societal feedbacks. The inherent surprise element in the phenomenon itself tends to be amplified by public and political responses to what would have been considered isolated events a generation ago. Increased scientific knowledge is itself a source of public reaction, enhanced by improved communication and media reporting, and by the continual refinement of epidemiological and other techniques for establishing small, but statistically significant, associations between exposures and disease incidence. While these effects are not necessarily nonlinear in the sense described earlier (although they can be nonlinear also), their appearance occurs in a way that is quite analogous to nonlinear effects; that is, an apparently beneficial activity is introduced and widely diffused long before any indication of possible disbenefits or social costs becomes apparent.

The carcinogenesis case is typical of a wide class of phenomena of latency. For example, the present acute concern over the consequences of past disposal of toxic chemical wastes is very similar in character. The social costs of activities undertaken several decades ago, largely in ignorance of potential consequences, suddenly became apparent and the source of great public anxiety and demands for action, including the search for scapegoats. The slow spread of toxic wastes in the aquifer provides the latency mechanism here. Until very recently, the planning horizon for the safe disposal of chemical wastes was quite short – a matter of decades – in contrast to that for radioactive wastes, where millenia have always been at issue. Yet, particularly in the case of toxic heavy metals, chemical wastes can be thought of as having an infinite half-life, and ought to be subject to time horizons as long as, say, the long-lived actinides in radioactive wastes [23]. The entry of wastes into groundwater is the mechanism of latent surprise for a host of industrial activities, and provides an illustration of how focusing societal planning on short-range phenomena while ignoring underlying trends in the slow variables may generate the seeds for future surprise and crisis (see Holling, Chapter 10, this volume).

A slightly different aspect of the latency phenomenon is illustrated by the use of chemical pesticides, particularly the more persistent ones. Here, the development of resistant pests or pathogens is a gradual phenomenon resulting from cumulative use, which at any point in time can be countered by increasing the intensity of application. However, the end result is the gradual oversimplification of the agricultural ecosystem, decreasing its natural resilience, and making it more vulnerable to large perturbations. The declining gains in agricultural productivity from the use of pesticides are illustrated in *Table 11.1* [24].

Table 11.1 Pesticide usage and agricultural yields in selected world areas [24].

Area or nation	*Pesticide use* (g/ha)	*Rank*	*Yield* (kg/ha)	*Rank*
Japan	10790	1	5480	1
Europe	1870	2	3430	2
USA	1490	3	2600	3
Latin America	220	4	1970	4
Oceania	198	5	1570	5
India	149	6	820	7
Africa	127	7	1210	6

A somewhat different class of latency phenomena to have received public attention recently is typified by carbon dioxide (CO_2) build-up and stratospheric O_3 depletion. In the case of toxic waste disposal the latency aspect really arose out of ignorance; if we had known more about groundwater transport of toxic chemicals we might have developed a safer disposal method earlier. In the CO_2 and stratospheric O_3 cases, by contrast, scientific understanding has enabled us to foresee the future effects of current activities long before any actual damage has occurred. The discovery itself, however, came as a surprise long after the industrial activities which would ultimately give rise to the effect had become well established and the subject of strong vested interests. Moreover, in the CO_2 case the human adaptations required may be considerable [25] and no less difficult by virtue of the fact that the need for them was discovered so far in advance. On the other hand, the slowness in the development of the problem compared to the time scale for the introduction of innovations in human societies provides an opportunity for adaptation which is absent in the other examples of latency described here.

Positive *versus* negative surprises

In the examples discussed so far we have been dealing with surprises and nonlinear effects described as disbenefits – adverse effects that became manifest after an apparently beneficial activity had become well established. In discussing projections of the future to accommodate surprise, however, it is important not to overlook positive surprises, which are equally possible, and could be more important for planning in the long run.

One of the most important sources of positive surprise is human ingenuity, which almost always tends to be underestimated. In his classic paper, *The Principle of the Hiding Hand*, Hirschman [26] has codified this notion for development projects in developing countries, but it is equally applicable to technological undertakings or even to the overcoming of environmental threats. Hirschman points out that many highly successful projects would not have been undertaken at all if their inventors or developers had foreseen all the difficulties. These difficulties are frequently surmounted as a result of unexpected innovations, which were sought only because the prior commitment to a project had become so deep that its proponents and managers were unwilling to be deterred by setbacks. Indeed, reliance on this hiding hand is probably the most distinctive characteristic of successful entrepreneurs, who frequently announce what they are going to accomplish and then challenge the engineers and managers to prevent them becoming liars [27]. One of the less fortunate characteristics of large, bureaucratic organizations is that they are more likely to know what cannot be done, since they have better access to experts thoroughly conversant with the difficulties and histories of past failures. Probably this is one of the principal reasons why small entrepreneurial businesses are frequently more successful innovators than larger organizations with superior technical resources.

One example of positive surprise can be found in environmental, health, and safety regulation. Regulations when first propounded are often technically unrealistic, but frequently technology has been more successful in finding a way to meet them than the experts in the affected industries believed possible. In the early days of automobile emission regulations in the USA, the conflict between fuel efficiency and pollution control objectives was thought to constitute an inexorable technical trade-off. Yet within a few years most of the conflict had been eliminated as a result of new technological developments, such as the three-way catalyst and microprocessor combustion controls. The average fuel efficiency of USA automobiles of models of recent years has more than doubled since 1973, and the prospects for further improvements in fuel consumption are brighter than for the much less efficient cars of the early 1970s; in other words, the more that has been achieved, the lower the apparent barriers to further achievement [28]. At the time of the first oil crisis the world automobile industry was widely described as mature, with a highly refined technology and little prospect for fundamental innovation. It was predicted that competition would be mostly in styling and accessories, with a gradual and marginal improvement in manufacturing costs. Yet now the technology of the industry has become much more intensive, in many respects beginning to be once again a "high tech" industry, with competition more on the basis of technology than at any time since the early days of the century [28]. In addition, through a series of dramatic innovations in management practices, the Japanese automobile industry has set entirely new production standards for passenger cars and light trucks, both for product quality and reliability and for cost per vehicle [28]. While it is true that innovation in the industry is still incremental and cumulative rather than revolutionary or epochal, the cumulative impact of a succession of small innovations will be much greater than anything foreseen a few years ago. The many who saw the modern automobile as a technological troglodyte [29] about to go the way of the dinosaurs, the interurban trolley, or the transatlantic ocean liner, have been proved wrong.

One can cite many other examples of possible or likely positive surprises that were barely conceivable at the time of the "limits-to-growth" debate of the early 1970s:

(1) The potential application of genetic engineering to food production and the detoxification of organic wastes [30].
(2) Developments in medium-scale, combined-cycle electric power generators fueled by natural gas or low BTU gas from an integral coal gasifier [31].
(3) Sophisticated applications of microelectronics to the control and monitoring of energy consumption, with a dramatic potential for providing energy services using less primary energy.
(4) The discovery of much greater resources of deep natural gas [32].
(5) Developments in ceramics, composites, and fiber-reinforced plastics with the possibility of substitution for metals and alloys used currently [33].
(6) Low-polluting, closed-cycle food production and industrial processes [34].
(7) Development of cost-competitive solar energy systems.

(8) The wide adoption of waste water recycling and purification systems.
(9) The discovery of natural or synthetic chemical compounds that provide protection against cancer.
(10) Development and widespread deployment of an inexpensive and reliable system of radio communication for technology transfer to people in remote areas [35].

Also, within the last few decades we have seen the discovery of North Sea, Mexican, and Arctic oil and gas fields; the development of optical communications and signal processing systems; and the use of remote-sensing satellites for resource and environmental assessment. Many of these technologies have advanced to, or close to, operational status faster than most people predicted when the first possibilities were announced.

One example of a positive nonlinearity is the effects of the learning curve and of economies of scale on the competitiveness of new technologies. As Rosenberg and Frischtak [36, pp 11–12] have emphasized, there can be highly nonlinear relationships between the rate of improvement of a new technology and its rate of adoption and diffusion. In its earliest stages the costs of new technology tend to exceed those of existing technologies that fulfil the same function, so that the rate of adoption is slow, often disappointing the expectations of enthusiastic proponents. As incremental improvements continue, however, and learning curve effects and economies of scale in production are realized, the competitive advantage of the new technology reaches a take-off point, after which it penetrates the market with unexpected rapidity.

Another source of positive nonlinearity in the adoption and diffusion of technological innovations can be described as the strength of the backward and forward linkages of a particular innovation within the economy as a whole [36, pp 16–17]. Backward linkages may occur through the new demand for expenditures on manufacturing equipment or facilities, or for input components required to implement an innovation in the economy. Forward linkages can occur either through "reduction of the price of the products into which the innovation enters as an input" or through innovation inducing "the creation and diffusion of new products and processes that, in their turn, would bring about the widespread adoption of the original innovation [36, p 17]." Examples are the microchip, which so improved the performance of many consumer durables that people could not afford to be without the new models, and many innovations in materials which enabled radical improvements in productivity or performance across a wide range of industries. The basic point is that "innovations flowing from a few industries may be responsible for generating a vastly disproportionate amount of technological change, productivity improvement, and output growth in the economy [36, p 18]." The development of these interindustry linkages frequently involves long latency periods before the cross-industry benefits can be realized, and at the same time the rapid market expansion generated by these linkages drives learning curves and economies of scale, which further accelerate the adoption of the triggering innovation. Even when the rate of technological progress appears to be more or less uniform and steady, its economic impact may be nonlinear and tend to exhibit a kind of threshold effect as the linkages act synergistically. *Figure 11.4* and *Table 11.2* illustrate how in the last decade the

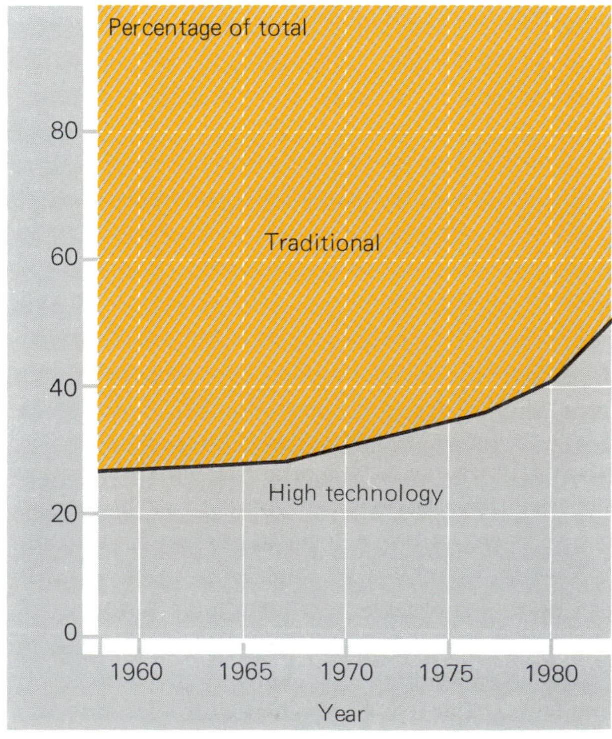

Figure 11.4 Shifts in manufacturing investment by equipment type 1958–1982 [37] and see *Table 11.2.*

Table 11.2 Shifts in manufacturing investment by equipment type (%).

Equipment	*1958–1962*	*1963–1967*	*1968–1972*	*1973–1977*	*1978–1982*
High-Tech, total	26.5	27.6	29.9	34.0	45.4
Office machinery	8.4	8.7	9.2	11.2	19.2
Communications machinery, photo, and electronic equipment	18.1	18.9	20.7	22.8	26.2
Traditional, total	73.5	72.4	70.1	66.0	54.6
Agricultural equipment	7.2	6.9	5.6	6.0	4.1
Construction equipment	4.7	5.1	4.6	4.5	3.3
Industrial machinery	25.4	24.8	22.5	20.5	17.4
Transportation equipment	25.7	26.4	27.6	26.5	22.3
autos, trucks, and buses	18.6	18.7	20.0	21.0	17.2
airplanes, ships, and railroad equipment	7.1	7.7	7.6	5.4	5.0
Other: engines, service machinery, and household equipment	10.5	9.1	9.8	8.6	7.5

capital investment in equipment embodying information technologies has accelerated due to these linkages with the whole economy [37]. One may also speculate that the increasingly global interdependencies in the world economy make such linkages more significant than in the past because markets can now be generated in a short time on the world scale.

Of course, it is somewhat misleading to describe the nonlinearities in the adoption and spread of new technologies as exclusively positive or benign. The acceleration of the adoption of a new technology frequently implies acceleration of the abandonment of an old one, or the displacement of a part of the labor force. Two recent examples are the newspaper publishing industry in the 1970s [38] and the mechanical watch industry somewhat earlier. The prospect of disruption of metal fabricating industries, mentioned earlier, or the office automation revolution [39] offer examples of possible negative effects that are still speculative. Moreover, the often sudden acceleration of adoption rates beyond a certain threshold leaves less time for assessment of longer term social costs and for planning the necessary social adaptations, such as retraining the work force, managing new types of effluents or wastes, or changing international trade relations.

In other words, the sharp separation we have drawn in order to compare negative and positive nonlinearities in technological progress may be a somewhat artificial one. On the one hand negative surprises, such as automobile emissions or the energy crisis, can give rise to a new wave of innovation, as it did in the automobile industry. The innovation wave, initially directed at a narrow problem, stimulates exploration of related innovations that would not have been thought of but for the initial stimulus of the negative externality. On the other hand positive nonlinearities increase adjustment stresses for the work force and management.

The question of cross-industry and cross-national linkages among innovations and their effect on economic performance has been little researched. There is, for example, little or no information on the quantitative importance of such linkages in relation to rates of economic growth, investment, or foreign trade.

Technological monocultures

In the early stages of the emergence of a new industry built around a fundamental innovation, such as the automobile, the airplane, the semiconductor, or the computer, the structure of the industry is very fluid, characterized by a high degree of diversity and experimentation. Frequently there are many small firms, each exploring a somewhat different technological approach. Competition among firms tends to be focused on technological innovation aimed primarily at product performance rather than price, or even on such qualities as reliability, compatibility with other products of the same genre, or service and maintenance. As competition continues one particular technological approach begins to emerge as the dominant technology [40]. Competition begins to center more on successive incremental improvements in this dominant technology and on small, cumulative manufacturing and managerial innovations that bring down production costs and improve reliability and standardization [41]. As the dominant technology emerges, its competitive position increasingly benefits from the cost advantages due to higher volume production than its competitors, in terms of both direct economies of scale and progress along the learning curve in the refinement of manufacturing, service, and marketing.

However, the very snowballing success of the dominant technology tends to steadily narrow the technical basis of competition. The search for cumulative improvements covers a smaller and smaller domain of technical possibilities, even as it becomes more intensive within that domain. In the process many technical possibilities which were deemed very promising in the early experimental stage receive declining attention from designers. In particular, options that might have been inherently superior either in cost or performance or both, but that require more development or depend upon more numerous or more problematic ancillary innovations, may simply fall by the wayside because of the growing cost advantage of the dominant technology, an advantage that arises from its head start in the market. Moreover, in a technical race becoming gradually more focused on cost reduction, technical factors that might affect higher order social impacts or risks of the technology also receive declining attention. Nevertheless, as the technology and the industry mature and the scale of application increases, they enter the regime in which the nonlinear appearance of disbenefits begins. In other words, new problems resulting from the scale of application become important just when the broad type of R & D program that might have helped anticipate such problems has been phased out, because it is no longer in the main line of development necessary to the commercial success of the dominant technology. Yet it is often just at this point of initial commercial success that the dominant technology can become vulnerable to unexpected side effects which can generate a societal reaction against it. This emergence of dominant technologies that develop unexpected side effects at a critical scale of application is what I have referred to as the problem of "technological monocultures [42]." The term derives from the obvious analogy with agricultural or forest monocultures which, because of their density, become vulnerable to insect pests, pathogens, environmental stresses, or the absence of ancillary inputs, such as water or fertilizer. Like agricultural monocultures, technological monocultures are highly successful in a stable and predictable environment (or market). Though more efficient than alternatives they are less robust when the environment becomes less predictable.

In the case of technological monocultures the problem may go even further than in the agricultural metaphor, because the organizations that have successfully commercialized the dominant technology acquire some power to influence the external environment or market in which they operate. They may thus be tempted to direct their energies toward increasing the predictability of this environment through ever more sophisticated marketing techniques, or by lobbying for protection against environmental regulations or foreign competition. J. K. Galbraith has, indeed, popularized the idea of technology controlling its own political and market environment in his book, *The New Industrial State* [43]. This book was written at a time when American corporations were at the zenith of their competitive strength and appeared invulnerable to foreign competition or societal regulation, as typified by the monolithic USA automobile industry. The decade of the 1970s, however, showed clearly the limitations of the power of USA corporations to control their environment in their own interest. In fact, any success in controlling their environment in the short term may actually be inimical to the long-range interests of corporations, because it delays the adaptive measures that they will eventually be forced to take anyway.

The automobile industry

The evolution of automobile technology, especially in the USA, where it had its first large successes, provides a good example of a technological monoculture. The USA automobile became a highly standardized product, which experienced only very slow technological evolution once the dominant technology of the gasoline internal combustion engine with electric self-starter had been established against rival technologies, such as the Stanley Steamer, the electric car, or even the diesel. The period of exuberant experimentation, with numerous firms competing with many innovations, extended through the early 1920s but gave way to the period of rationalization and standardization typified by Alfred P. Sloan's managerial and production systems at General Motors. The resultant reductions in cost and improvements in reliability of the dominant technology guaranteed that more radical competing innovations could not hope to catch up. For such technical change as continued, the locus of engineering innovation gradually shifted to Europe, which by the 1960s had already become the source of most of the more important automotive innovations. From this point of view the smaller and more highly differentiated markets in Europe were a positive advantage in stimulating bolder technological innovation. The technical issues that were about to explode in the 1970s – safety design, emissions reduction, fuel efficiency, durability – were given low priority in the USA, if not totally ignored. A new wave of innovations in these areas, when it finally came, was largely the result of a political "technology forcing" process imposed on the technology establishment of the industry from outside, first by regulation and later by foreign competition. All these issues were externalities that became significant as the scale of the industry and the automobile population grew; they were thus by-products of its commercial success, and they appeared first in the USA exactly because that was where the highest degree of market penetration of the automobile had occurred. Yet it was precisely this success that had caused these issues to be excluded from the technical search process in the industry. The same process as had raised the importance of these issues from a societal standpoint resulted in their exclusion from the technical agenda of the industry's technological establishment. It took a series of external shocks to enlarge this technological agenda – the political activism of the environmental movement which chose the automobile as its first focus; competition from newly emerged foreign competitors, principally Japan; two successive oil price shocks; and the worst economic recession since the Great Depression. It is not yet clear even today whether these shocks were sufficient to reestablish the USA technology on a long-term, viable, self-sustaining path.

Chemical technology

A somewhat analogous account can be given of USA chemical technology, another of the great technological success stories of the twentieth century. Here the slowdown in technological innovation did not occur to the same extent as in the automobile industry. Nevertheless, the scope of the innovation agenda of the chemical industry tended to narrow as its scale expanded, and the externalities of the industry were excluded from its agenda, much as for the automobile industry. Issues such as waste disposal and the management of residuals were not perceived as significant technical barriers to further market expansion or profitability. Here, again, we see the phenomenon of the innovation agenda narrowing at the very time when past technological and market successes of the industry really required that it be enlarged, not only in the broader public interest, but also in the ultimate interest of the further spread and adoption of the industry's products and technologies.

Other examples

A similar account could be given for pharmaceuticals, pesticides and herbicides, electric power generation, commercial air transport, industrialized agriculture, and many other areas in which successful innovation over an extended period has become self-limiting because of the failure to enlarge the innovation agenda sufficiently quickly, particularly in relation to the externalities of the industry, either with respect to its production processes or to its products, or to both. To be sure, in all these examples there were factors other than externalities that slowed the industry's growth. In the chemical industry, for example, the cost of energy and raw materials and the consequent price increases led to a slowing of demand growth and to a worldwide overcapacity, especially in commodity-type chemical products. Energy costs were also an important factor in slowing the growth of commercial air transport and in altering the economics of industrialized agriculture. Nevertheless, rising public awareness of the externalities of all these industries, leading to increased regulation, were also important.

Nuclear power

Civilian nuclear power is another case in point. Reactor safety and radioactive waste management were quite prominent on the research agenda in the early days, but did not expand commensurately as the industry began to experience commercial success. In the minds of many of the proponents of nuclear technology, too much emphasis on research in these areas was seen as a public admission of negative aspects of the technology, rather than as a challenge to be overcome in the interests of its long-term viability and growth. Innovative effort became focused on those aspects that most immediately related to the technology's market penetration in competition with other utility fuels, chiefly inexpensive Middle East oil prior to 1973. This resulted in a concentration on the realization of scale economies through the design of larger plants, the extension of the burn-up life of the fuel, improving neutronic efficiency, and reducing operating costs through better fuel management.

In the early days of nuclear power research, many reactor types competed for R & D funding by both government and industry. Today many experts would argue that several of these candidate types, such as the Canadian heavy water (CANDU) reactor or the high temperature, gas-cooled reactor (HTGR), had greater long-term promise in terms of efficient use of uranium, reliability, and intrinsic safety than the light water reactor (LWR) design that has become the dominant American, and eventually world, technology. But, once the LWR design began to benefit from economies of scale and descent of the learning curve through successive generations of reactors, the alternative design concepts were eliminated as they became relatively more costly and were perceived to pose more difficult technical problems that had to be overcome in the short term. The greater the market penetration of the LWR the greater its cost advantage became, and hence the less incentive for continued exploration of alternative design concepts. Effort was concentrated on incremental improvement of the LWR. Thus, a temporarily satisfactory technological alternative gradually became an insurmountable barrier to the development and assessment of designs that might have been superior in the longer run. It may even be that this premature termination of the exploration of other alternatives will prove fatal to the survival of the industry, at least in the USA where preemption by the LWR came earliest and was most complete. The continued focus on scale economies and large reactor sizes can also be regarded as an instance of the technological monoculture syndrome described earlier.

The history of nuclear power development can thus be described in terms of the emergence of a technological monoculture, the LWR, which became fatally vulnerable to a number of unforeseen shocks and surprises, ranging from the emergence of the environmental movement through the TMI episode to the drastic reduction in projected demand growth for electricity. It remains to be seen whether we will eventually face the consequences of another short-term response running counter to long-term trends. It is quite possible that the present perception, particularly in the USA, that nuclear power is not really necessary for a viable energy-supply system in the twenty-first century, is also excessively influenced by short-term considerations that arise from the current leveling-off of the demand growth for all energy and from the great excess of electrical generating capacity that was created because of the long lead times of electricity generation planning. Electricity demand appears to be much more closely related to economic growth than nonelectrical energy demand, so that if economic growth continues at current rates a shortage of electrical capacity could begin to develop a decade hence, as it did just prior to the oil embargo in 1973. If this happens it will be another example of the overpreoccupation of policy with the fast variables; that is, with temporary gluts and shortages in world fuel markets.

Unfortunately, all the preceding remarks are speculative and highly subjective. The adoption of the LWR was so complete that the hypothesis that alternative design concepts might have been superior cannot really be tested. The excellent experience of the Canadians with the CANDU design is not conclusive, because of different economic operating rules and because other countries have also had better operating experience with the USA LWR designs than has the USA itself [44].

Technological diversity

The principal lesson of the preceding account of how technological monocultures develop is that there may be an inherent value in the maintenance of technological diversity that is analogous to the value of the maintenance of genetic diversity in natural and man-made ecosystems. The existence of and the considerable depth of knowledge about many alternative technological options is a potential source of systemic self-renewal and adjustment to new circumstances. Such technological diversity has a social value which is not captured by the usual considerations of efficiency, market share, or organizational growth, which tend to drive the evolution of technological systems in industrial societies. An overall system that is less efficient or more costly because it requires the infrastructure for a diversity of technologies may nevertheless have greater viability or survival potential in an environment subject to shocks, surprises, or discontinuities in long-term trends. Yet there are few apparent rewards to organizations, or even nations, in directing efforts toward such diversity. In a competitive system they may incur loss before the advantages of diversity have a chance to be tested by events. Fortunately, international competition may supply some of this diversity. In the nuclear power field, for example, the USA may be in a position to revive its capabilities by importing technology from abroad if future necessity so dictates [45]. The Germans, for example, have maintained a viable HTGR program, while several other nations have maintained a viable program of liquid-metal cooled reactors – sufficient, at least, to provide a foundation for future growth in the next century should circumstances make this attractive.

Fast and slow variables

In part, the problem of monocultures arises from the tendency of sociotechnical systems to respond preferentially to "fast variables" in the environment, as compared to "slow variables," especially when the latter are less familiar or predictable. There is a close analogy between the fast and slow variable effects discussed by Holling (Chapter 10, this volume) and by Timmerman (Chapter 16, this volume) on natural ecosystems and such effects on the phenomena discussed in this chapter. In fact, Clark [46] has explicitly extended the Holling concept to include social systems.

The policy question is: Where are the driving forces for expansion of the innovation agenda to come from? In the examples I have cited these forces derived mostly from the political process, but usually belatedly and with what often appeared to be unnecessary conflict and organizational trauma. In many cases, notably that of oil price controls in the USA in the early 1970s, the political response exacerbated rather than alleviated the problem. Low and declining real gasoline prices after 1973, for example, provided positive disincentives to the USA automobile industry to innovate in fuel efficiency, and made it more vulnerable to competition from a country (Japan) that allowed the price mechanism to drive innovation.

A possible argument is that the technological diversity necessary to assure greater robustness of systems, such as energy supply and demand, in the face of uncertainty and surprise represents a positive externality that justifies some form of collective or public subsidy. But there is a problem of who chooses the technologies. Investment in technological diversity has an aspect of insurance, which in turn implies risk sharing, but it is difficult to see how to make this work in practice. The incentives that operate in governmental systems are usually of no longer term, and frequently of shorter term, than those that operate in the market system.

Adaptation and adjustment of individuals, groups, and communities

The adaptation and adjustment of individuals, groups, nations, and whole societies to changes in their environment or in their own internal relationships can be viewed as a group problem. In a recent paper Breznitz [47], in dealing with the psychology of individual adjustment and adaptation, puts forward several concepts that seem equally applicable to the problem-solving behavior of larger groups and communities, and have their counterpart in some of the literature of organizational behavior [48]. In what follows I make fairly extensive use of his ideas, but in a context very different from his.

Transition to more inclusive cooperation

An important idea that does not have a counterpart in individual adjustment is that successful problem solving in communities frequently requires a transition to higher degrees of cooperation among groups that previously looked upon themselves as in conflict with each other over access to the same resources. This transition to higher levels of integration among groups is often the most difficult and dangerous part of the long-term adjustment of man to his environment. Frequently, it is accompanied by a prolonged period of ambiguity or ambivalence between cooperative and conflict behavior among coalescing communities. An extreme example is the succession of increasingly destructive wars in Europe which preceded the present, apparently successful, integration of Western Europe into a single, comparatively cooperative community in which violent conflict among nations seems almost unthinkable. Conflict and competition have by no means disappeared, but they are well contained and do not break into widespread violence. The disappearance of extreme confrontations in Western Europe is now taken for granted.

One can think of the modern world as having evolved through history from a situation of fragmentation into tiny islands of cooperation, each in periodic conflict with other islands in what each regarded as a zero-sum game. The interdependency of these islands gradually increased through social complexity, shared resources, and a shared environment until a time was reached when more cooperation than conflict among the islands became natural. The positive-sum aspects of the game gradually became more apparent as the complexity of the relations between the hitherto separate communities and cultures grew. Today we are in the midst of a very painful transition to a single, interdependent global community in which it should be increasingly apparent that the benefits of cooperation to all its interdependent parts exceed the rewards of rivalry. By the same token, however, the ambivalence within each part between cooperative and conflictual behavior is at a maximum, with a resultant high degree of instability and uncertainty.

The danger of partial solutions

Breznitz divides the problem-solving processes of individuals into three stages: presolutions, partial solutions, and the best solution [47]. The presolution stage is characterized by intensive exploration of the individual's environment and of the resources and information available for the solution of his or her problem. This exploration typically results in the identification of a partial solution, but one which is far from the best available. It is analogous to a subsidiary peak in a mountain range. Nevertheless, it is a sufficiently successful adaptation to greatly reduce the individual's incentives to continue further experimentation and exploration, thus reducing the probability of going on to find a better adaptation, let alone the best available. It seems fairly evident that the same behavior characterizes organizations or communities just as much as individuals. In the literature of organizational theory this is labeled as "satisficing". In social systems, among the partial solutions found first are likely to be those involving less cooperation among previously separate groups (families, tribes, corporations, states, nations). There is thus a natural tendency for groups to settle for adaptations that could be improved upon by cooperation at a higher level of inclusiveness – higher levels of integration among groups. If individual groups find adequate partial solutions that do not necessitate cooperation with other groups, they will tend to stop there, thus failing to search for opportunities for cooperative solutions that would have been superior and satisfied the mutual interests of several groups.

In the energy crisis of 1973, for example, the first impulse of each country was to seek a solution that minimized the impact on its own economy without reference to impacts on other countries or the system of consuming nations as a whole. In the 1973 crisis this was tempered by the fact that not all the actors were nations, and the multinational oil companies, acting in *their* interests, provided an impulse toward transnational solutions, ultimately involving a remarkably equal division of the supply shortfall among all the industrialized nations [49]. In addition, with some encouragement from the oil multinationals, the principal consuming nations created the International Energy Agency (IEA), which was able to make modest plans for supply sharing in future emergencies [50]. However, the magnitude of the shortfall required for triggering the sharing agreement was greater than that of the crisis of the Iranian revolution, and the IEA was therefore ineffective.

The tendency toward fixation on the first reasonably successful partial solution is well exhibited in the phenomenon of technological monoculture described earlier. The narrowing of the innovation agenda with the advent of the first commercial success reflects exactly the kind of behavior in organizations that Breznitz describes in individuals. Initial commercial success of a product frequently represents a partial solution, which not only suppresses further efforts to find better solutions, but also reinforces types of organizational behavior that stabilize this partial solution, making the organization more resistant to change or even to the reception of information that might suggest the possible need for change. If the organization has sufficient market power or access to the political process, it may even try to alter the environment rather than adapt to it. In the early days of the environmental movement and the energy crisis in the USA automobile companies, oil companies, and utilities were all slow to accept the possibility that they were in "a new ball game", and they treated the new market and political forces buffeting them as temporary aberrations that provided no clear signal as to future trends, and consequently did not require a strategic response, that is, a major alteration in long-term plans (ten years or more).

Breznitz and the organizational theorists also show that there is an asymmetry in the effects of good news and bad news on organizational behavior in relation to partial solutions [47]. Worsening of the environment, especially when it occurs in the form of a shock, stimulates problem solving and exploratory behavior, but good news tends to relax problem solving and reduce sensitivity to the possibility of better alternatives and the desirability of searching for them. In the USA in the years following the oil embargo, price controls on gasoline and fuel were good news for the automobile companies and drivers, which canceled out most of the impulse toward adjustment that had been created by the crisis. It took almost a decade for world prices to be internalized in the USA economy.

Response pools

According to Breznitz [47] the capacity of individuals to adapt to their environment is determined by what he calls their "response pool", the repertory of behavioral responses on which they draw to meet challenges presented by their environment. Probably such response pools are even more limited in scope for organizations than for individuals. According to Breznitz "unstable environments increase, therefore, the adaptation value of response pool variability [47, p 43]," which affects the evolution of organizations and communities, much as genetic variability within populations affects biological evolution. Market systems and pluralistic political systems or organizations in general have a greater capacity for adaptation because the variability of their response pools is enhanced relative to more centralized systems of organization. On the other hand, as we have seen, the dynamics of success in market competition can also lead to the emergence of dominant organizations whose survival or viability is based on a successful partial solution – an incomplete, and ultimately inadequate, adaptation. Moreover, such organizations often have significant capacity to influence their environment temporarily, so enhancing its apparent stability and thus delaying adjustment measures. With such organizations – perhaps typified by the automobile companies and electric utilities in the USA – if a significant change in environment or period of environmental turbulence succeeds a long period of stability (or, even more fatally, stable growth), the low variability of the response pools of the dominant organizations renders them unprepared to cope. In fact, such a model probably describes the effects of the turbulent conditions of the 1970s on all of western society, following the unprecedented period of stable growth from 1952 to 1973.

Behavioral insurance

Breznitz [47] divides response pools into two classes: the "currently active response pool" (CARP), and the "general response pool" (GRP). CARP consists of those organizational responses that can be considered as part of the "standard operating procedures" of the organization, the formulas which the organization has developed for solving the familiar and recurrent problems that it encounters in its day-to-day environment. It is the set of responses that Graham Allison identifies in his "organizational process" model of decision making in organizations and bureaucracies [51]. The GRP, on the other hand, consists of those organizational responses that are not currently in active use, but are implicit in the professional training, traditions, and human resources of the organization. The GRP, in other words, consists of those latent responses that can be called forth under stress, but that are less automatic, and take a longer time to mobilize and orchestrate, than CARP responses. They are those responses more dependent on organizational leadership. Of course, the distinction between CARP and GRP is not a sharp one; rather they define ranges on a spectrum.

In order to increase the capacity of organizations or communities to adapt to future changes it is important to deliberately cultivate their response pool variability. This involves periodic testing of the environment as well as periodic retesting of previously discarded behavioral responses. The latter is particularly difficult because over time organizations develop an inventory of things that everybody knows do not work, and this inventory itself becomes a part of the organization's CARP, long after the reasons why certain responses did not work at the time they were discarded have been long forgotten; reasons that may be no longer valid, if they ever were.

The deliberate cultivation of response pool variability for the explicit purpose of promoting future organizational adaptability to a potentially turbulent environment is called by Breznitz "behavioral insurance" (BI) [47]. Emergency planning and training are one form of behavioral insurance that has often proved itself in practice. BI also tends to reduce irrational or ineffective behavior when an emergency arises and it also enhances the self-confidence of the organization in encountering surprises. Thus, BI is a way of directly addressing the issue of surprises and discontinuities in planning for the future.

Another form of BI, in both governments and corporations, is investment in research and development. This should probably be understood more broadly than conventional corporate R & D activities, including organizational and work practice experimentation, not just experimentation with hardware. Indeed, much of the current literature on industrial innovation seriously underestimates the role of social innovation, even in the introduction and diffusion of new hardware [52]. The extraordinary success of the Japanese automobile industry in penetrating European and American markets owes more to innovations in work-force management and the social organization of production than it does to specific technological innovations in manufacturing [28].

One way in which given national societies expand their behavioral response pools (and thus take out BI) may be through the activities of what James Q. Wilson [53] calls "policy entrepreneurs." Examples include successful publicists, like Rachel Carson, Ralph Nader, Amory and Hunter Lovins, and also many organized public interest groups, and such efforts as the "limits-to-growth" debate launched by the Club of Rome [54]. Such efforts might be better judged not on the accuracy or reliability of the information they disseminate or the views they espouse, but on their indirect contribution to the more general broadening of the response pools of institutions and of societies. Disruptive, and even "irresponsible", as such activities may appear in the short run to some critics, they can be regarded as a form of BI for the society, which may contribute to its future adaptability. The important point is that the notion of variability in the societal response pool is inseparable from the existence of error. One cannot experiment and explore without making errors, and the errors will probably always outnumber the appropriate responses. One must hope that natural selection through trial and error causes the correct organizational responses to dominate through a strengthening of the organizations or societies that try them.

One way in which the variability of the response pool can be promoted is by a conscious effort, when addressing a problem, to prolong the experimentation of the presolution period. The organization should continue to probe possibilities even after an apparently satisfactory first solution has been found. Failure of less promising or more difficult approaches should not be accepted as "givens" for all time. Sometimes solving a problem the hard way can be more useful to society than a more elegant solution because it leads to a deeper understanding, which, though of little short-term usefulness, opens up more options that may eventually be of use in dealing with surprises and discontinuities. This is the counterpart in behavioral terms of the technological diversity recommended in the approach to technological innovation in the preceding section.

A possible research agenda

Nonlinearities

We need to analyze many more detailed case studies to exemplify or verify the mechanisms of nonlinear appearance of secondary impacts. These studies should include examples of the nonlinear or threshold public response to externalities, as well as specific attempts to better quantify relations between causes and outcomes for the purely physical effects, such as automobile emissions and sulfate aerosols (acid rain, interaction between sulfates and oxidants, etc.). A systematic search for examples of nonlinearities in the past, which can be well documented and analyzed, might be the prelude to a search for possible future nonlinearities and for synergistic effects in existing industrial and developmental activities.

What might have been

Although the literature is replete with retrospective case histories of the development of technological impacts and the societal and human responses thereto, for the most part these histories restrict themselves to describing and interpreting what happened, to identifying the players, and to describing why they took the positions they did and what kind of influence they had on outcomes. Few attempt the much more difficult task of exploring what might have happened or what alternative choices might have been if the participants had been more sensitive to underlying, longer term trends. The actual design of policies that might have been followed had we known then what we know now, and the detailed exploration of the likely consequences of these policies, is an ambitious task that has seldom been attempted, yet I think much could be learned from it. It would be important not only to suggest which possible alternative policies or paths of development might have been, but also to search for the possible secondary consequences of these alternatives with the same care as we try to identify and describe, retrospectively, the unforeseen side effects of the policies or paths actually followed.

For example, knowing what we know today of trends in USA oil imports in the late 1960s, what policies could have been adopted that would have mitigated our vulnerability to the Arab oil embargo in 1973? What, if any, other undesirable consequences would the implementation of these policies have led to? Some of the alternative policies that might have been followed include:

(1) Gradual restructuring of the USA tax system to avoid subsidy of oil consumption, and even to encourage efficiency in energy end-use.
(2) A tariff on oil imports to subsidize the creation of a strategic petroleum reserve, perhaps concerted among all the OECD countries.
(3) A more vigorous government R & D and private incentive program to encourage development of energy technologies based on domestic resources.
(4) Retention of the oil import quota system introduced in the 1950s.

Would any of these measures have been politically feasible, given the general circumstances that existed in the mid-1960s? Would it have been possible to design a coalition of interests that would have supported any of these policies or some combination of them? Would limitations on oil imports have resulted in accelerated depletion of USA petroleum reserves, rendering the USA even more vulnerable than it actually was by the mid-1970s? Would higher energy prices in the 1960s have created greater weaknesses in the USA economy and in its competitive position relative to other countries, or would they have helped the USA to identify inherent weaknesses earlier and begin to correct them?

As another example, can one invent a detailed strategy that the management of USA automobile companies might have followed in the 1960s and 1970s that would have avoided, or at least mitigated, the disastrous deterioration that took place after 1979? Was there a better strategy and timetable for reducing automobile emissions than the tortuous course that the USA has followed since the passage of the Clear Air Amendments of 1970?

In the case of surprise, and essentially random, events such as the oil shocks, TMI, or Love Canal, it would be interesting to speculate what might have happened in terms of societal response and public policy if one or more of these systemic shocks had come earlier or later than they actually did. To what extent did the effect of any one of these events depend on other events that were happening simultaneously? Suppose the Arab oil embargo had not occurred at a time when the whole industrialized world was simultaneously following stimulative economic policies and monetary expansion in the wake of the 1970–1971 recession?

Regier and Baskerville (Chapter 3, this volume) have traced the degradation of two important ecosystems as a consequence of what they term "exploitative development." It would be an extremely interesting and informative exercise to produce alternative scenarios for the development of these two ecosystems. What would have constituted a truly "sustainable" strategy for exploitation in these two cases, and what, if anything, would have been the net economic cost of following this strategy? Would it have been inevitable that any more sustainable strategy would have foundered on political infeasibility because of local pressures for economic development, or can we see in retrospect the possibility of forging a political coalition that would have sustained this strategy? Would it be possible to develop a systematic methodology of accounting for the benefits and costs of various alternative development strategies for specific ecosystems, or for specific resource systems in association with their ecosystems? Even granting that the design of any such accounting system entails important value judgments along the way, it may be possible to develop several alternative accounting systems that involve different value judgments. These would be compared by testing them with several development scenarios for ecosystems on the historical development of which we already have detailed knowledge. One of the difficulties of such a program would be the validity of isolating the system under study. To what extent would the benefit–cost accounting system break down because of linkages and feedbacks from other ecosystems or sociotechnical systems to which the given system is linked?

A major purpose of the research strategy described above would be to help define more precisely the whole concept of "sustainable development", which is at the heart of the IIASA program, by actually reconstructing past development scenarios in alternative ways and discovering whether it is possible to evaluate these alternative scenarios on some plausible scale of "sustainability". Perhaps the concept of sustainability cannot be expressed in terms of a single objective function, but has to be treated as multidimensional. Even in this case, however, it must be possible to devise some system of valuation. This system might not be an intellectual process or methodology, but rather a social or political process that proceeds according to a set of ground rules that can be specified or agreed to in advance.

Innovation linkages

In this chapter I have suggested that one of the important sources of nonlinearities in the impact of emerging technologies may arise from the linkages among innovations, such as the effect of one particular innovation in generating demand for others required to make it economically successful. Such linkages may also include demands on resources or impacts on ecosystems, which can be either increasing or decreasing. A simple example would be the impact of information technology and communications on the demand for energy or raw materials, or the development of closed-cycle manufacturing or food production systems in which most of the material inputs are recycled [34].

Most of the research to date that points to such linkages is qualitative and impressionistic. Can the relationships be made more quantitative and precise, e.g., by actually tracing the market growth of a number of innovations hypothesized to be causally linked, or tracking actual transactions among firms? Can we develop a convincing way of delineating the macroeconomic effects of specific innovations or clusters of innovations? In the case of sustainable or nonsustainable development or exploitation scenarios, which we proposed investigating in the preceding section, it would be important to understand how alternative scenarios are linked to the economy of a whole region or country. Is there a way of using the strategy applied to a particular ecosystem or resource base to give plausible predictions about the behavior of the larger sociotechnical systems in which it is embedded?

There is even less information or research on how clustering and synergism among innovations affect the generation of externalities or lead society to neglect research directed at identifying and overcoming externalities.

Globalization of linkages

The globalization of the world economy has apparently been accompanied by globalization of the process of technological innovation. Innovations developed in one country become standard practice of the industry worldwide in a much shorter time than in the past. This leaves less time for society to assess the potential impact of new technologies and affords less opportunity for countries to control or regulate the secondary effects of technology within their own boundaries. International competition may force the rapid adoption of some technologies internationally, with little opportunity for individual societies to decide whether or not they wish to accept the social or environmental consequences that may accompany these technologies. Regulation is for the most part a national activity, decided upon through national, political mechanisms. Yet national controls on the use of technologies, particularly new or emerging technologies, are difficult or impossible to implement without a high level of cooperation and agreement among nations.

Two examples of the above problem may be worth citing in order to illustrate the kind of research that may be needed. The first relates primarily to the transfer of technologies among the highly industrialized countries and the second to the transfer of technologies from industrialized to less developed countries. Many countries are very apprehensive about the impacts of information-based technologies, both on aggregate employment and on the quality of working life and the mix of skill requirements in the labor force. Several of the Scandanavian countries, for example, have legislation that controls the manner and rate of introduction of computerized systems into various work settings and gives organized labor considerable influence in the decision process [55]. However, it is extremely difficult to implement such laws or regulations on a national basis when controls on technology introduction can strongly affect the competitiveness of national firms in the world economy. This is particularly true of the very open economies of many European nations, which are heavily dependent on exports. Regardless of the merit of such regulations, the point is that the imperative of national competitiveness tends to disenfranchise a national electorate, which elects to maintain certain social or environmental standards within its boundaries. This phenomenon, of course, is not unknown within the USA and is what led the USA to increasingly go federal in various environmental and social regulations, but in this case there is a national constituency. In the international sphere there is really no corresponding constituency, and therefore no real mechanism for reaching a social consensus on a strategy of economic development which can respect the value of sustainability.

The other example that presents even greater difficulties is that of the transfer of technologies from developed to developing countries, where there are wide differences in the level of infrastructure between the two societies. This can work to the disadvantage of both societies. For example, the transfer of labor-intensive manufacturing operations from developed to developing countries to take advantage of low labor costs may provide economic benefits to the developing country, but if the discrepancy in labor standards – not only wages but also health and safety – is too large, the effect of the competition may be simply to erode standards in the developed country, especially if the difference is so large that it cannot be offset by more sophisticated labor-saving technology and management methods. Not only high wages, but also health and safety standards in the workplace may be most easily offset by the replacement of people with machines, which can work in environments not fit for humans. In either case, competition from low-wage countries displaces labor in high-wage countries. On the side of the developing country, its workers may be considered as being exploited for the benefit of consumers in the developed country. Not only workers, but the environment of the developing country may be exploited to lower the cost of goods to consumers in developed countries. Yet the interdependencies within the global environment mean that what happens in the developing countries cannot be considered in isolation. To an increasing extent, both resources and the environment and biosphere form part of a global "commons". To the extent that the whole world has an interest in the nondegradation of key ecosystems or resource systems anywhere in the world [56], the inability to regulate exploitation to a reasonable standard of sustainability in the developing country becomes a problem for the world as a whole. On the other hand, to the extent that the shift of manufacturing occurs because the developing country environment truly has higher assimilative capacity, then it could be argued that the transfer of technology is legitimate, dictated by a genuine, natural advantage.

Further reading

Bauer, R. S., Rosenbloom, R. S., and Sharpe, L. (Eds) (1969), *Second Order Consequences: A Methodological Essay on the Impact of Technology* (MIT Press, Cambridge, MA).

Brooks, H. and Hollander, J. M. (1979), United States energy alternatives to 2010 and beyond: the CONAES study, *Annual Review of Energy*, **4**, 1–70.

Clark, W. C. (1985), Scales of climate impacts, *Climatic Change*, **7**, 5–27.

Greenberger, M. L., Crenson, M. A., and Crissey, B. L. (1976), *Models in the Policy Process: Public Decision Making in the Computer Era* (Russell Sage Foundation, New York).

OTA (1982), *Global Models, World Futures, and Public Policy: A Critique*, OTA-R-165 (Office of Technology Assessment, US Congress, Washington, DC).

Schelling, T. C. (1984), Anticipating climate change, implications for welfare and policy, *Environment*, **26**(8), 6–9.

Tukey, J. (Chairman) (1973), Panel on Chemicals and Health of the President's Science Advisory Committee, *Chemicals and Health* (Science and Technology Policy Office, National Science Foundation, Washington, DC).

UNESCO (1973), The social assessment of technology, *International Social Science Journal*, **25** (3) (United Nations Educational, Scientific, and Cultural Organization, Paris).

Notes and references

[1] Greenberger, M. L., Brewer, G. D., Hogan, W. W., and Russell, M. (1983), *Caught Unawares: The Energy Decade in Retrospect* (Ballinger, Cambridge, MA); Council on Environmental Quality and the Department of State (1980), *Entering the Twenty-First Century: The Global 2000 Report to the President*, 2 Vols (Barney, G. O., Project Director) (US Government Printing Office); Office of Technology Assessment (1982), *Global Models, World Futures, and Public Policy: A Critique*, OTA-R-165 (Mills, W., Project Director) (OTA, US Congress, Washington, DC).

[2] Brooks, H. and Hollander, J. M. (1979), United States energy alternatives to 2010 and beyond: the CONAES study, *Annual Review of Energy*, **4**, 1–70; Landsberg, H. (1973), Low-cost abundant energy: paradise lost?, *Annual Report*, pp 27–49 (Resources for the Future, Washington, DC).

[3] Meadows, D. H., Meadows, D. L., Randers, J., and Behrens, W. W., III (1972), *The Limits to Growth: A Report for the Club of Rome's Project on the Predicament of Mankind* (Universe Books, NY); Greenberger, M. L., Crenson, M. A., and Crissey, B. L. (1976), *Models in the Policy Process: Public Decision Making in the Computer Era* (Russell Sage Foundation, New York); *cf.* The world models, pp 158–182.

[4] National Commission on Supplies and Shortages (Rice, D. B., Chairman) (1976), *Government and the Nation's Resources* (US Government Printing Office), *cf.* Ch 4: The experiences of 1973–1974, pp 47–75; Tilton, J. E. (1977), *The Future of Non-Fuel Minerals*, especially Chapters 2, 5, and 6 (The Brookings Institution, Washington DC).

[5] Landsberg (1973), note [2].

[6] Alm, A. A., Colglazier, E. W., and Kates-Garnick, B. (1981), Coping with interruptions, in D. A. Deese and J. S. Nye (Eds), *Energy Security*, Ch 10, pp 303–346 (Ballinger, Cambridge, MA).

[7] Hogan, W. W. (1984), *Patterns of Energy Use*, Energy and Environmental Policy Center, Discussion Paper Series No E-84-04, p 18 (John F. Kennedy School of Government, Harvard University, Cambridge, MA).

[8] Energy Modeling Forum (1982), *World Oil*, EMF Report 6 (E. Zausner, Chairman) (EMF, Stanford University, Stanford, CA).

[9] Odell, P. (1983), IIASA International Energy Workshop, Laxenburg, Austria.

[10] Bauer, R. S., Rosenbloom, R. S., and Sharpe, L. (1969), *Second Order Consequences: A Methodological Essay on the Impact of Technology* (MIT Press, Cambridge, MA).

[11] Hardin, G. (1968), The tragedy of the commons, *Science*, **162**, 1243–1248.

[12] Webber, M. W. (1976), The BART experience – What have we learned? *The Public Interest*, No 45, (Fall), 79–108.

[13] Jacoby, H. S., Steinbruner, J. D., Weinstein, M. C., Clark, I. D., Appleman, J. M., and Ahern, W. R., Jr. (1973), *Clearing the Air: Federal Policy on Automotive Emissions Control* (Ballinger, Cambridge, MA).

[14] Environmental Studies Board (1971), *Jamaica Bay and Kennedy Airport: A Multidisciplinary Environmental Study*, 2 Vols (National Academy of Sciences, Washington, DC); *cf.* Vol. 2, pp 92–96, 118–120.

[15] Usher, D. (1981), *The Economic Prerequisites of Democracy* (Columbia University Press, New York), *cf.* footnote 5, Ch 5.

[16] Ayres, R. U. and Miller, S. M. (1982), Robotics and the conservation of human resources, *Technology in Society*, **4**(3), 181–197.

[17] Kakalik, J. S., Ebener, P. A., Felstiner, W. L. F., and Shankley, M. G. (1983), *Costs of Asbestos Litigation*, RAND/R-3042-ICJ (The Institute for Civil Justice, The RAND Corporation, Santa Monica, CA).

[18] Panel on Chemicals and Health of the President's Science Advisory Committee (Tukey, J., Chairman) (1973), *Chemicals and Health* (Science and Technology Policy Office, National Science Foundation, Washington, DC); *cf.* especially Appendix C, pp 151–166.

[19] Committee for a Study on Saccharin and Food Safety Policy (Robbins, F. C., Chairman) (1978), *Saccharin: Technical Assessment of Risks and Benefits*, Part 1 of a Two-Part Study (National Academy Press, Washington, DC); Havender, W. R. (1983), The science and politics of cyclamate, *The Public Interest*, No 71 (Spring), 17–32.

[20] Schell, O. (1984), *Modern Meat: Antibiotics, Hormones, and the Pharmaceutical Farm* (Random House, New York); *cf.* also review by Jeffrey Blumberg in *The Diet and Nutrition Newsletter*, **3**(2), Tufts University, April, 1985.

[21] Wetstone, G. S. and Rosencranz, A. (1983), *Acid Rain in Europe and North America: National Responses to an International Problem* (Environmental Law Institute, Washington, DC).

[22] Roberts, L. (Ed) (1984), *Cancer Today: Origins, Prevention, and Treatment* (IOM/NAS Press, Washington, DC); *cf.* Ch 5, The epidemiology of diet and cancer, pp 49–61; Ch 6, Dietary carcinogens and anti-carcinogens, pp 63–72; Office of Technological Assessment (1981), *Assessment of Technologies for Determining Cancer Risks from the Environment*, OTA-H-138 (Gough, M., Project Director) (OTA, US Congress, Washington, DC).

[23] Murray, J. P., Harrington, J. J., and Wilson, R. (1982), Risks of hazardous chemicals and nuclear wastes: a comparison, *CATO Journal*, *II*, 565–606.

[24] Study of Critical Environmental Problems (1970), *Man's Impact on the Global Environment: Assessment and Recommendations for Action* (MIT Press, Cambridge, MA), Table 2.5, p 118, and Table 2.6, p 119.

[25] Schelling, T. C. (1984), Anticipating climate change, implications for welfare and policy, *Environment*, **26**(8), 6–9, 28–35.

[26] Hirschman, A. (1967), The principle of the hiding hand, *The Public Interest*, No 6 (Winter), 10–25.

[27] Brooks, H. (1984), Seeking equity and efficiency: public and private roles, in H. Brooks, L. Liebman, and C. Schelling (Eds), *Public–Private Partnership: New Opportunities for Meeting Social Needs*, p 14 (Ballinger, Cambridge, MA).

[28] Roos, D. and Altshuler, A. (Codirectors) (1984), *The Future of the Automobile, Report of MIT's International Automobile Program* (MIT Press, Cambridge, MA); *cf.* Ch 4, Technological opportunities for adaptation, pp 77–105.

[29] Brown, L. R., Flavin, C., and Norman, C. (1979), *Running on Empty: The Future of the Automobile in an Oil-Short World* (W. W. Norton, New York).

[30] Office of Technology Assessment (1981), *Impacts of Applied Genetics: Micro-Organisms, Plants, and Animals*, OTA-HR-132 (Z. Harsanyi, Project Director) (OTA, US Congress, Washington, DC).

[31] Parsons Company (1978), *Preliminary Design Study of an Integrated Coal Gasification Combined Cycle Power Plant*, prepared for Southern California Edison Company (Electric Power Research Institute, Palo Alto, CA).

[32] Gold, T. and Soter, S. (1982), Abiogenic methane and the origins of petroleum, *Energy Exploration and Exploitation*, **1**(2), 89–104; Gold, T. (1982), *Testimony Before the Subcommittee on Energy Development and Applications*, Committee on Science and Technology, US House of Representatives, Roswell, New Mexico, July 30.

[33] Ashby, M. (1979), The science and engineering of materials, in P. W. Hemily and M. N. Özdas (Eds), *Science and Future Choice*, Vol I, pp 19–48 (Oxford University Press, Oxford, UK).

[34] Revelle, R. (1977), Let the waters bring forth abundantly, in Y. Mundlak and S. F. Singer (Eds), *Arid*

Zone Development: Potentialities and Problems (Ballinger, Cambridge, MA).

[35] Markhoff, J. (1984), Bulletin boards in space: amateur radio pioneering promises low-cost global communications, *Byte*, **May**, 88–91.

[36] Rosenberg, N. and Frischtak, C. R. (1984), Technological innovation and long waves, *Cambridge Journal of Economics*, **8**, 7–24.

[37] Eckstein, O. C., Caton, C., Brinner, R., and Duprey, P. (1984), *The DRI Report on Manufacturing Industries*, p 31 (Data Resources, Inc., Lexington, MA).

[38] Smith, A. (1980), *Goodbye Gutenberg: The Newspaper Revolution of the 1980s* (Oxford University Press, New York); *cf.* Ch 6: Printing unions and technological change, pp 207–237.

[39] Marschall, D. and Gregory, J. (1983), *Office Automation: Jekyll or Hyde? Highlights of the International Conference on Office Work & New Technology*, October 1982 (Working Women Education Fund, Cleveland, OH).

[40] Abernathy, W. J. (1977), *The Productivity Dilemma: Roadblock to Innovation in the Automobile Industry* (Johns Hopkins University Press, Baltimore, MD).

[41] Imai, K. and Sakuma, A. (1983), An analysis of Japan–US semiconductor friction, *Economic Eye, a Quarterly Digest of Views from Japan*, **4**, 13–18.

[42] Brooks, H. (1973), The state of the art: technology assessment as a process, *The Social Assessment of Technology, International Social Science Journal*, **25**(3), 253.

[43] Galbraith, J. K. (1978), *The New Industrial State*, 3rd edn (Houghton Mifflin, Boston, MA).

[44] Hansen, K. and Winje, D. (1984), *Disparities in Nuclear Power Plant Performance in the US and the FRG*, Report MIT-EL-84-018, *cf.* also *e-lab newsletter*, October–December, 1984 (MIT Energy Laboratory, Cambridge, MA).

[45] Lester, R. K. (1985), National policy options for advanced nuclear power plant development, in MIT Nuclear Engineering Department, *Report of Program on Nuclear Power Plant Innovation, Vol. II* (MIT, Cambridge, MA).

[46] Clark, W. C. (1985), Scales of climate impacts, *Climatic Change*, **7**, 5–27.

[47] Breznitz, S. (1985), Educating for coping with change, in M. Frankenhauser (Ed), *Ancient Humans in Tomorrow's Electronic World* [Noble Networks (Advertising and PR Communications) Ltd., London, UK]; the ideas drawn upon in the present discussion are presented more fully in an unpublished paper by Breznitz submitted for a conference of the same title, held at Aspen Institute, Wye Plantation, MD, in September 1984.

[48] Cyert, R. and March, J. (1963), *A Behavioral Theory of the Firm* (Prentice-Hall, Englewood Cliffs, NJ).

[49] Stobaugh, R. B. (1975), The oil companies in the crisis, in R. Vernon (Ed), *The Oil Crisis in Perspective, Daedalus*, **104**(4), 179–202.

[50] Deese, D. A. and Nye, J. S. (Eds), *Energy and Security*, Appendix A, pp 427–430 (Ballinger, Cambridge, MA); Keohane, R. (1978), The International Energy Agency, *International Organizations*, **Autumn**, 929.

[51] Allison, G. T. (1971), *Essence of Decision: Explaining the Cuban Missile Crisis*, pp 67–96 (Little Brown & Co., Boston, MA).

[52] Brooks, H. (1982), Social and technological innovation, in S. B. Lundstedt and E. W. Colglazier, Jr (Eds), *Managing Innovation: The Social Dimensions of Creativity, Invention and Technology*, Ch 1, pp 1–30 (Pergamon Press, Elmsford, NY).

[53] Wilson, J. Q. (1981), Policy intellectuals and public policy, *The Public Interest*, No 64 (Summer), 31–46.

[54] Forrester, J. W. (1971), *World Dynamics* (Wright-Allen Press, Inc., Cambridge, MA).

[55] Schneider, L. (1984), Technology bargaining in Norway, in H. Brooks, *et al.*, *Technology and the Need for New Labor Relations*, Discussion Paper Series No 129D (John F. Kennedy School of Government, Harvard University, Cambridge, MA).

[56] Myers, N. and Myers, D. (1983), How the global community can respond to international environmental problems, *Ambio*, **12**(1), 20–28.

Commentary

M. Cantley

How do societies learn to manage better their technologies and their ecosystems? Harvey Brooks gives some useful pointers, if not complete answers, in this exciting and important chapter. It is important because, through case examples and the identification of deep common features, he deepens our understanding and sharpens our perception of the societal processes involved in the advancement and control of innovation.

The questions addressed become more urgent as we tighten the knots of our global interdependence, both economic and ecological; the two increasingly linked by technology. Brooks highlights a common mechanism of technological success leading, through economies of scale and cumulative learning effects, toward technological monocultures. The unforeseen adverse and nonlinear effects which then emerge as a result of dominant scale may demand a broader repertoire of response capabilities than has survived the competitive battles of the earlier stages of innovation. Resilience demands diversity. Technological and social mechanisms too focused on short-term competitive success, or *ad hoc* responses to the fast variables, reduce diversity.

Brooks commends R & D as behavioral insurance, in both corporations and governments, but understood more broadly than just hardware. Certainly there are some attempts – one could cite the strategic environment units of some large multinationals, the development of business theories on the management of strategic surprise [1], or some of the governmental futures groups, such as the UK's Systems Analysis Research Unit (deceased), the US Congressional Office of Technology Assessment, or the European Community's FAST (Forecasting and Assessment in Science and Technology). Such CASSANDRA units (Centres for the Analysis of Science and Society, Advising on New Directions for Research and its Applications) face Cassandra's problem – the forecasting may be perfect, but she couldn't win political credibility (until too late) [2].

Faced with the recurrent prospect of Schumpeterian gales of creative destruction, can we improve our weather forecasting and contingency planning? Brooks's chapter is relevant in drawing our attention to the slow variables. For these, religion may be a better guide than technological economics. In primitive cultures, it is religion – or let us say the sociocultural value system – that enshrines the accumulated societal wisdom and transmits it across generations. But we have to learn which elements of locally specific taboos can be jettisoned; and which elements remain of value in a changing, technologically interdependent world.

It is striking to consider how closely the Brooks analysis matches the classic Emery and Trist typology [3] of the causal texture of organizational environments, in which 20 years ago they predicted and emphasized the emergence of the increased complexity of *turbulent fields*, "In these, dynamic processes, which create significant variances for the component organizations, arise not simply from (their) interaction, but also from the field itself," because of:

> i. The growth . . . of organizations, and linked sets of organizations, so large that their actions are both persistent and strong enough to induce autochthonous processes in the environment . . .
> ii. The deepening interdependence between the economic and the other facets of the society . . .
> iii. The increasing reliance on research and development to achieve the capacity to meet competitive challenge. This leads to a situation in which a change gradient is continuously present in the environmental field.
>
> For organizations, these trends mean a gross increase in their area of relevant uncertainty. The consequences which flow from their actions lead off in ways that become increasingly unpredictable: they do not necessarily fall off with distance, but may at any point be amplified beyond all expectation . . . [3].

In such systems, Emery and Trist suggest as a possible solution "the emergence of *values that have overriding significance for all members of the field* [3]." This again fits well with Brooks's analysis: given that we are ". . . in the midst of a very painful transition to a single, interdependent global community in which it should be increasingly apparent that the benefits of cooperation to all its interdependent parts exceed the rivalry [p 340]," then we need to emphasize cooperative values; the difficulty of this switch he acknowledges, since at the same time ". . . the ambivalence within each part between cooperative and conflictual behavior is at a maximum, with a resultant high degree of instability and uncertainty [p 340]."

To stress cooperative values in the management of our planet is not incompatible with the emphasis on diversity – for given the certainty of ignorance and error, there is all the greater need for openness, for the intellectual freedom and self-critical spirit of a Popperian "open society", even where it may appear in the short run to be at a disadvantage against purposeful targetting by more monolithic or consensual societies [4].

What underlies Brooks's chapter and, indeed, surfaces explicitly is a picture of a societal learning system. This global learning system operates under conditions of increasingly rapid knowledge acquisition and near instantaneous diffusion, through R & D and the information and communication technologies to which it has given rise (one of the examples of the positive feedback nonlinearities with which Brooks illustrates his typology of surprises). Yet this learning system, in spite of its massive accumulation of experience, has somehow to remain sufficiently self-critical to jettison or to reorganize its knowledge, its perceptions, its in-built assumptions, whenever – or, preferably, just before

– their obsolescence is made manifest [5]. It has to learn, in the context of technological development, the difference between *full speed ahead,* and *optimal speed, right direction.*

Learning should mean performance improvement – are we becoming better managers? One of the most refreshing features of the chapter is its balanced consideration not only of negative surprises, of the emergence of unforeseen and long-latent surprises, but also of positive surprises. And among these latter, "One of the most important sources . . . is human ingenuity, which almost always tends to be underestimated [p 332]." I could have cheered out loud; for here was always the weakest point of the clockwork, mechanistic "limits to growth" models: they were nonteleological, devoid of intelligent life and human purposes. Brooks cites, in the same context of positive surprises, the application of genetic engineering to food production and of information to energy control. Such triumphs as these of human ingenuity, of mind over matter, of that key concept "dematerialization" [6] – it sounds better in the original French – are the crux of the matter. For they illustrate precisely the practical significance, the declining ratio of materials to gross national product, implied by the Club of Rome's happier and more recent publication, *No Limits to Learning* [7]. The chemistry of function and targetted effect must displace the chemistry of bulk processing, with its side effects and unintended externalities.

Harvey Brooks offers an analysis, and a research agenda, of interdisciplinary, intellectual challenge, and of relevance to learning how to improve the management of our technology, our societies, and our planet. Though as to his proposed studies of "what might have been [p 343]," I confess that it brought immediately to my mind that remarkable character of Michael Frayn, Ivan Kudovbin [8]. In fact, an incisive "might-have-been" study on electric power plant designs and scale in the UK from 1950 onwards was undertaken by Abdulkarim and Lucas [9].

Notes and references

[1] See, for example, Ansoff, H. I. (1975), Managing surprise and discontinuity – strategic response to weak signals, *California Management Review,* **18**(2), 21–33.

[2] For details, see Homer, *The Iliad* (Greek), or Virgil *The Aeneid, Book IX* (Latin).

[3] Emery, F. E. and Trist, E. L. (1965), The causal texture of organizational environments, *Human Relations,* **18**, 21–32; reprinted in F. E. Emery (Ed) (1969), *Systems Thinking* (Penguin, Harmondsworth, UK).

[4] This conflict is discussed in the context of Western Europe in Cantley, M. F. and Holst, O. (1981), Europe in a changing world: open society or purposeful system?, in J. P. Braun (Ed), *Operational Research '81* (North-Holland, Amsterdam).

[5] For a similar analysis of societal learning in technological systems, see Cantley, M. and Sahal, D. (1980), *Who Learns What? A Conceptual Description of Capability and Learning in Technological Systems,* IIASA Research Report RR-80-42 (IIASA, Laxenburg, Austria).

[6] See in particular the paper by Sargeant, K., Biotechnology, connectedness and dematerialisation, presented at *Biotechnology '84,* at the Royal Irish Academy, Dublin, 1–2 May (proceedings to be published by the Society for General Microbiology).

[7] Botkin, J. W., Elmandjra, M., and Malitza, M., *No Limits to Learning,* A Report to the Club of Rome (Pergamon Press, Oxford, UK).

[8] In one of Frayn's weekly columns in *The Observer,* some 20 years ago, the precocious historical achievements, inventions, and creations of Ivan Kudovbin were described in glowing terms. He kudovbin – but he wasn't.

[9] Abdulkarim, A. J. and Lucas, N. J. D. (1977), Economies of scale in electricity generation in the United Kingdom, *Energy Research,* **1**, 223.

Chapter 12 International institutions and the environment

G. Majone

Editors' Introduction: Institutions and organizations provide a wide array of means through which societies can respond to the challenges of sustainable development. Efforts to design better social response capabilities require a better understanding of the strengths and weaknesses of alternative institutional structures in dealing with long-term, large-scale environmental issues.

In this chapter institutional options for dealing with environmental problems in the larger context of international relations are examined. Global treaties and regulations represent only one extreme form of international organization. Less formal "regimes" based on shared principles, people, and values often have a greater influence on the actual course of events. In setting the tone and agenda of such regimes, key roles are played by the international scientific community and, somewhat surprisingly, by regional and even national initiatives.

The internationalization of ecological issues

Students of international relations tend to agree that issues related to environmental and resource interdependence will become increasingly important in the coming decades; and that this trend will have to be reflected in new institutional arrangements, as well as in a greater sensitivity of trade and development policies to ecological concerns. Since in the post World War II period money and trade have ceased to be "low politics" for top policymakers and become "high politics", on the same level as the traditional concerns of diplomacy, so, it is argued, the next decades may witness policy and institutional developments in the area of ecological interdependence that are comparable to the earlier innovations in economic interdependence.

The analogy can be pushed a bit further. The international economic system has recently become the object of study for different academic disciplines. Both international economics and international politics theorize about its functioning – the roles and relationships of actors in it, and the nature and courses of change affecting it. The approach is increasingly interdisciplinary: international monetary relations, for example, are part of the international political system, since it is governments that are the main actors; but they are also part of the international economic system, since a subject of the governments' deliberation is money [1].

Similarly, issues of environmental and resource interdependence cannot be easily separated from other economic and political aspects of international relations. True, the study of ecological interdependence is still at the embryonic stage. As long as the nature of key ecological interdependencies remains poorly understood, the scope of the issues and roles of different national and international actors must remain ill-defined. It would, however, be a serious pitfall to search for solutions exclusively in the area of natural science: environmental problems need not have "environmental" solutions, especially at the international level [2]. Political and economic factors can be expected to interact with transnational environmental issues in three different ways:

(1) International environmental policies and institutions will have to rely to a great extent on the existing structure of international relations. In fact, international cooperation on ecological issues may depend on broader political agreements. Thus, the structure of détente of the 1970s and the results of the first round of the Helsinki Conference on Security and Cooperation in Europe made possible the Helsinki Convention on the Pollution of the Marine Environment in the Baltic.
(2) Environmental concerns will increasingly affect international economic and political relations as the notion of economic interdependence is expanded to include that of environmental and resource interdependence. Conversely, it has been argued that a major obstacle to the implementation of environmentally sound development strategies lies in the present structure of the international political and economic order.
(3) International environmental relations, in and of themselves, are political relations since national actors are sovereign in the international arena.

A related point of methodology is that interdependence, although it has an objective physical and economic basis, is primarily a conceptual construct, a way of interpreting the changing nature of international relations. Contrary to the assumptions of traditional theories of international relations, the interdependence model recognizes that contemporary societies are connected by multiple channels – including informal links among governmental, professional, and scientific elites, and functional relations among international organizations – and that the international agenda is no longer monopolized by questions of military and political security, but includes a variety of issues in the areas of science, technology, economics, and the environment. As Keohane and Nye have noted [3], in a world of multiple and imperfectly linked issues, in which coalitions are formed transnationally and transgovernmentally, the potential role of international organizations is vastly increased. Such organizations set the international agenda, act as catalysts for the formation of coalitions, and provide arenas for political initiatives and linking of problems. By defining salient issues, providing advance warning, and deciding which issues should be linked, international organizations, but also international bodies like scientific and professional associations, can help determine governmental priorities and even the relative importance of different departments and policy positions within governments.

In their search for improved understanding of the changing nature of international relations, scholars have used a variety of approaches. If the key concept in the 1950s was "international system", in the 1970s and 1980s the notion of "international regime" has proved to be analytically useful in explaining an increasingly complex and interdependent world [4].

International regimes – formal and informal

Regimes are defined as the principles, norms, rules, and decision-making structures on the basis of which the expectations of actors converge with respect to any given kind of issue. Such networks of rules and procedures are pervasive characteristics of the international system because no patterned behavior can sustain itself for any length of time without generating a congruent regime [5].

The management of natural resources, such as marine fisheries and deep seabed mining, has proved to be a particularly fruitful area of application of the regime paradigm. Attention to the practical problems of managing natural resources inevitably leads to the adoption of an institutional perspective: society's decisions about the use of such resources are seen to depend crucially on the social institutions, rules, and practices that serve to order and implement the decisions of those interested in the use of those resources. From this viewpoint, the essential feature of resource regimes is the conjunction of convergent expectations and patterns of behavior or practice. Such a conjunction tends to produce conventionalized behavior, i.e., behavior based on recognized social conventions [6].

According to our definition, regimes are special kinds of social institutions. Now, institutions may find explicit recognition in contracts, constitutions, treaties, or formal organizations, but this is not necessary for their emergence or effective functioning. Some of the most fundamental social institutions – the family, natural languages, exchange markets – exist almost independently of formal sanctions.

In other cases, explicit organizations may facilitate the operation of an institution (regime), or even become an essential element of it. The regional fishery management councils set up by the American Fishery Conservation and Management Act of 1976 and the Baltic Marine Environment Protection Commission created by Articles 12 and 13 of the 1974 Helsinki Convention are examples. But highly formalized institutional arrangements usually also depend on associated informal patterns. The cost of implementing even the most precisely drawn legal contract would be prohibitive in the absence of informal sanctions, tacit agreements, mutual trust, and other nonformalized relations [7]. Similarly, informal elements of regimes may serve to provide interpretations of ambiguous aspects of formal agreements, to deal with issues not explictly covered by these arrangements, to facilitate communication and information, and so on. Thus, in studying international regimes it is important to make a distinction between the regime as a social institution and the contractual and organizational elements that accompany it [6].

National and subnational regimes for natural resources and the environment can usually rely on specialized agencies, such as courts, police forces, and research organizations, for information gathering, monitoring, enforcement, and the settlement of disputes. At the international level, however, comparable institutions either do not exist or are severely underdeveloped. For example, the United Nations has no administrative apparatus for monitoring activities carried out under the regime for Antarctica or for settling disputes pertaining to deep seabed mining. Hence, even when there is a clear need for formal organization, international regimes lack the extensive institutional support enjoyed by national regimes. This is another reason for the importance of informal and semiformal arrangements with relevant public agencies in the member states and with independent professional and research organizations.

Globalism versus functional regionalism

As already noted, international regimes for natural resources and the environment can hardly operate in the absence of some explicit organization. Information about environmental damage and abatement costs must be obtained, activities monitored, and general standards interpreted and adapted to changing circumstances. In some cases, revenues – pollution taxes or royalties for deep seabed mining, for example – must be collected and disposed of; applications must be reviewed and licenses granted. As a minimum, a small secretariat is needed for keeping regular contacts with the national member organizations, scheduling meetings, and keeping records.

Recognition that formal organizations are often necessary for operating international regimes immediately raises classic issues of institutional design: staffing, financing, management controls, discretionary powers, jurisdictional boundaries. Jurisdictional problems loom particularly large in the debate on international regimes and are reviewed in this section. Questions related to the power that international organizations can realistically expect to be granted are discussed in the next section.

The current popularity of the notion of ecological interdependence reflects a growing concern of international public opinion with problems of transboundary pollution, international trade in hazardous substances, and especially with global issues, such as worldwide changes in climate or the risk of depletion of the Earth's ozone (O_3) layer by the inert chlorofluorocarbons used in a variety of domestic and industrial applications.

It is tempting to assume that regulation of such global problems requires the creation of global regimes, but this is not always, or even usually, the case. Interdependence is not synonymous with globalism, so the best approach is often that of "thinking globally while acting locally" – or regionally.

At the United Nations Conference on the Human Environment held in Stockholm in 1972, some people argued that the need for a common jurisdiction over global resources, such as the oceans or the stratosphere, can be met effectively only by transnational structures with sufficient regulatory power – for example, an international regulatory body for the stratosphere.

In theory, such a global organization would avoid the free rider problem, since all countries would be forced to regulate. Uniform regulation would provide no incentives for multinational firms to move production from heavily regulated countries to countries where regulation is weaker or nonexistent. Regulatory costs would be equitably distributed between rich and poor nations. Expensive duplications in research, testing, data collection, and monitoring would be avoided.

However, the idea of a world regulatory agency immediately runs into serious conceptual and practical problems. Observe, first, that some global problems (e.g., climatic change) have regionally different impacts and, hence, their solutions should also be regionally differentiated. In other cases the consequences are global, but the causes are localized in particular areas (e.g., the chlorofluorocarbon problem, see below).

Second, persons, firms, and even capital are not globally mobile. Economic, political, and cultural barriers all limit the actual domain of mobility of the factors of production. Hence, a global authority is not needed to prevent firms from moving production from regulated to unregulated or less regulated countries.

Nor is there a strong case for global organization on equity grounds. A large international body can take important decisions only through some process of bargaining in a highly politicized atmosphere, where all issues, including equity, are inevitably colored by the prevailing state of North–South or East–West confrontation. What is perhaps even more serious is the tendency of organizations with a large and heterogeneous membership to define equity in terms of equality of treatment. Thus, one would expect a world regulatory agency to set uniform standards that are too strict for some members (e.g., poor countries which may well prefer to accept a higher level of risk if it can give them faster economic development) and not sufficiently stringent for others.

Moreover, uniform regulation, like a worldwide ban on a new drug because of the risk of certain side effects, would make it impossible to obtain experience that would reduce uncertainty, and thus deprive the world of a potentially valuable product. Conversely, approval of a product on a worldwide scale, rather than in a limited number of countries, could have disastrous consequences if the product subsequently poses serious health hazards. If thalidomide had been approved by a world food and drug administration and had come into use simultaneously everywhere, the consequences would have been far worse [8].

Transactions costs – the costs of identifying the parties and persuading them to agree; of collecting the information necessary for maintaining and revising the initial agreement; of monitoring and policing performance by members – tend to increase rapidly with jurisdictional domain, functional scope, and size of membership. Hence, a global organization would be extremely expensive to set up and operate.

Again, the membership of such a global organization is bound to be highly heterogeneous with respect to preferences, risk perceptions, level of economic development, cultural, and legal traditions.

Last, but certainly not least, it is highly improbable that most states would be willing to set up an organization endowed with adequate power and authority to deal with environmental issues on a global scale. I conclude that creating a global organization to regulate transboundary pollution and environmental hazards is not a feasible, or even desirable, solution, though broad international conventions, in areas such as international trade in hazardous substances, remain a desirable long-term goal.

To understand why, and under what conditions, certain global problems may be solved faster and more effectively by seeking agreements between a limited number of countries rather than setting up a global regime, consider the problem of O_3 depletion due to emissions of chlorofluorocarbons into the atmosphere.

Production of chlorofluorocarbons is remarkably concentrated: 12 countries account for 80% of the total world production of the major chlorofluorocarbons. These countries are all members of OECD, utilize similar technologies in production, share the same technical perception of the problem, have similar political systems and similar preferences for environmental quality, and are extensively linked by a variety of transnational networks. Hence, it should not be too difficult to reach an effective agreement limiting production of chlorofluorocarbons in these countries. At the same time, given their share in world production and use of chlorofluorocarbons, such an agreement would go a long way toward solving the global problem of O_3 depletion.

The attractiveness of this strategy is increased because regulations would be unlikely to induce significant shifts in the location of chlorofluorocarbon production facilities to countries with less stringent regulations. There are several reasons for this, including the particular structure of the industry, the expectation of many executives that substantial convergence will occur among OECD countries with regard to chlorofluorocarbon regulation over the next 5 to 10 years, and their belief that these countries would impose countermeasures (selective import duties, border adjustments, quotas, and embargoes) to block any observed relocation to "chlorofluorocarbon pollution heavens". Hence, regulation in this area would not exert, at present, any great influence on patterns of industrial location. As the analysts of Resources for the Future conclude, "the emergence of regulation differentials among nations will represent a shift at the margin in the locational calculus of CFC-related firms [9]."

Beside the direct benefit of a significant reduction of world chlorofluorocarbon emissions, the strategy suggested here would produce important indirect consequences. Bohm [10] has usefully classified these indirect effects of "policy linkages" into push and pull effects. In the push case, the regulated countries influence the willingness of other countries to take action by such means as making further extensions of chlorofluorocarbon control policy (e.g., going beyond a ban on the use of chlorofluorocarbons in aerosols) dependent on their promise to initiate regulatory actions. Alternatively, chlorofluorocarbon regulation could be used to bargain over other regulatory issues (e.g., shipment of radioactive materials or dumping of toxic wastes in the oceans).

In addition to influencing the willingness of other countries to take regulatory action, the regulated countries can influence their ability to do so by providing cost information, technical data, know-how, and administrative experience acquired in controlling chlorofluorocarbon emissions.

Important pull effects are the moral example set by the regulated countries, policy imitation, and what may be called the "demonstration effect". Moral example can exert a powerful influence on the international community by influencing public opinion and decision makers through the media, scientific channels, and the policy debate.

Many problems in natural resource management and transnational pollution are regional rather than global, and there is no good reason for expanding the scope of management schemes beyond the boundaries of the relevant ecological and socioeconomic systems. In fact, the historical record shows that most viable international arrangements involve regional institutions having responsibility for dealing with fairly specific issues in some geographically or functionally demarcated area.

These results conform with the predictions of functional federalism. According to this theory, the trade-off between scale economies and other factors pressing for the expansion of jurisdictional boundaries, and diversity of preferences, pushing in the opposite direction, leads to many overlapping jurisdictions, each dealing with its own set of highly specialized and closely related problems.

Unlike global organizations (or purely national regimes) regional authorities can be defined so as to make their boundaries congruent with the relevant ecological and socioeconomic systems. This fact, together with the greater homogeneity of preferences, cultural traditions, and levels of development, justifies the expectation of negotiations leading to productive results. For all these reasons, regional regimes typically represent the preferred institutional alternative for the management of international environmental problems.

Nevertheless, functional regionalism has its limitations. As Cooper [11, p 51] points out:

> [A] system of functional federalism would, in the absence of a higher authority willing and able to sacrifice its vested interests in particular jurisdictions, inhibit bargaining and political compromise across functional, jurisdictional boundaries. For much of the time it is useful to have each issue operate on its own track, with its own set of conventions and sanctions to influence behavior. But from time to time the inability to bargain across issue areas would prevent communities from reaching an optimal configuration of public goods.

However, this problem may be more serious at the national than at the international level. The reason for this lies, paradoxically, in the intrinsic weakness of international organizations. As already noted, international organizations are always weaker than comparable national organizations. Lacking sufficient regulatory authority, they depend on the cooperation and support of national governments, and these are in a position to bargain across functional boundaries. In fact, it may be argued that one of the important functions of international organizations consists precisely in providing national members with a forum for policy discussion and political compromise.

International organizations: limitations and possibilities

If the view of international organizations as incipient world governments is archaic [3], it is also naive to expect that such organizations could become minigovernments with the power to define and implement optimal programs of conservation or pollution abatement, independent of the policies, attitudes, and socioeconomic structures of the member states.

Often, the responsibility for the regulation of transboundary environmental problems is assigned to organizations originally created for other purposes. A good example is the bilateral International Joint Commission (IJC) of the USA and Canada, originally established to oversee the provisions of the Boundary Waters Treaty of 1909. Although water quality was a consideration in that treaty, IJC paid relatively little attention to environmental issues until 1972, when the USA and Canada concluded an agreement on water quality in the Great Lakes, specifying quality objectives and means to achieve them, and extending the authority of the IJC to advise and make recommendations concerning a broad range of environmental problems.

But even when they are created for a specific purpose, international organizations are much more constrained than the corresponding national regimes; they have fewer and less powerful policy instruments at their disposal than functionally equivalent national agencies, and cannot rely to the same extent on a network of supporting institutions. Compare, for example, the powers usually granted to international fisheries commissions with the comprehensive management system set up by the US Fishery Conservation and Management Act to govern the harvesting of fish in the extensive area known as the fishery conservation zone.

Again, the major function of the International Baltic Sea Fishery Commission is to prepare and submit to the member states recommendations with regard to measures for management of the natural resources of the Baltic. Aside from collecting, analyzing, and disseminating statistical data, it does not carry out scientific research itself, but simply submits proposals for coordinating such research and seeking the services of the member states and of scientific organizations. The powers of the Baltic Marine Environment Protection Commission are also limited, being mostly confined to defining pollution criteria and objectives for pollution abatement. Implementation of both international regimes for the Baltic is based on national enforcement [12].

The fact is that for a number of political and economic reasons, governments are reluctant to entrust the welfare of their citizens to other governments, just as citizens generally dislike the thought that their welfare may depend on decisions taken by officials of other governments or of international organizations. Moreover, where scientific uncertainty is high, and when technical and economic conditions are rapidly changing, the risks of far-reaching international agreements are objectively great. Governments respond to these risks by reducing the scope of agreements and by surrendering as little national power as possible [13].

For all these reasons, it seems likely that future international organizations for dealing with ecological interdependencies will not be radically different from some of the international fisheries and environment commissions already in existence. The main test of their success will be their capacity to obtain information that is accepted by all interested parties as a basis for productive negotiations, and to put forth proposals that are widely accepted as being based on solid facts and reasonable assumptions. One important method of gaining acceptability comprises using experts from the national bureaucracies and research institutions in an innovative and intellectually stimulating way, so giving the experts an incentive not only to serve their own national institutions, but also to put their scientific and professional skills at the disposal of the international organization [14].

Such transgovernmental relations are not the only means by which the lack of formal regulatory powers may be compensated. The importance of informal relations for the smooth functioning of international regimes has already been noted. Informal channels of communication and persuasion are especially relevant in the present context. For issues with long time horizons, a certain degree of persuasion, education, and politicization, in the sense of the generation of controversy and debate, is probably a necessary condition for proper attention and timely decisions [15]. Science – man's most sophisticated system of communication and persuasion, and the only international institution enjoying worldwide acceptance regardless of ideological and cultural differences – can play a crucially important role in this regard, giving legitimacy as well as a factual basis to ecological concerns.

In fact, transnational groupings of individual scientists, ranging from the Scientific Committee on Problems of the Environment and the International Union for the Conservation of Nature and Natural Resources to the Baltic Marine Biologists and Baltic Oceanographers, are already investigating and politicizing different aspects of ecological interdependence. However, such efforts tend to be episodic and uncoordinated. Moreover, they are often remote from concrete management problems and oblivious to the political and economic realities of the international system.

Supranational institutions organized to deal with particular groups of problems, in a broad perspective of global interdependence, can fill the gap that exists at present between transnational scientific groups and the international system in its political, economic, and institutional components. Such institutions can mobilize and coordinate scientific knowledge, and provide mechanisms by which this knowledge may capture the attention of policymakers, while increasing the sensitivity of the scientific community to the policy implications of new knowledge about environmental and resource interdependence.

The institutional model that emerges from the discussion of the limitations and possibilities of international organizations bears little resemblance to Weberian bureaucracy. Rather, it suggests a web of intergovernmental, transgovernmental, and interorganizational relations at different levels of formalization. As Keohane and Nye [3] have argued, neither leadership nor effective power can come from international organizations; but such organizations can provide the cognitive and conceptual basis for the day-to-day policy coordination on which effective multiple leadership depends.

Notes and references

[1] Strange, S. (1976), *International Monetary Relations* (Oxford University Press, London).

[2] Schelling, T. C. (1975), Environmental concerns and international conflict, in *Organization of the Government for the Conduct of Foreign Policy*, **1**, 231–242 (US Government Printing Office, Washington, DC).

[3] Koehane, R. O. and Nye, J. S. (1977), *Power and Interdependence* (Little, Brown, and Company, Boston, MA).

[4] Puchala, D. J. and Hopkins, R. F. (1982), International regimes: lessons from inductive analysis, *International Organization*, **36**, (2), 245–276.

[5] Krasner, S. D. (1982), Structural causes and regime consequences; regimes as intervening variables, *International Organization*, **36**, (2), 185–205.

[6] Young, O. R. (1982), *Resource Regimes* (University of California Press, Berkeley, CA).

[7] Macaulay, S. (1963), Non-contractual relations in business, *American Sociological Review*, **28**, 55–70.

[8] Coleman, J. S. (1982), *The Asymmetric Society* (Syracuse University Press, New York).

[9] Gladwin, T. N., Ugelow, J. L., and Walter, I. (1982), A global view of CFC sources and policies to reduce emissions, in J. H. Cumberland, J. R. Hibbs, and I. Hoch (Eds), *The Economics of Managing Chlorofluorocarbons*, Ch 3 (Resources for the Future, Washington, DC).

[10] Bohm, P. (1982), CFC emissions control in an international perspective, in J. H. Cumberland, J. R. Hibbs, and I. Hoch (Eds), *The Economics of Managing Fluorocarbons* (Resources for the Future, Washington, DC); see also Editors' postscript, p 359.
[11] Cooper, R. N. (1976), Worldwide versus regional integration: is there an optimum size of the integrated area?, in F. Machlup (Ed), *Economic Integration* (The MacMillan Press, London).
[12] Boczek, B. A. (1980), The Baltic Sea: a study in marine regionalisms, *German Yearbook of International Law*, **23**, 196–229.
[13] Binder, R. B. (1981), *Managing the Risks of International Agreement* (The University of Wisconsin Press, Madison, WI).
[14] Scott, A. (1976), Transfrontier pollution and institutional choice, in I. Walter (Ed), *Studies in International Environmental Economics*, pp 303–318 (Wiley – Interscience, New York and London).
[15] Keohane, R. O. and Nye, J. S. (1975), Organizing for environmental and resource interdependence, in *Organization of the Government for the Conduct of Foreign Policy*, **1**, 46–63 (US Government Printing Office, Washington, DC).

Commentary

N. Haigh

The practice of international relations in the field of the environment is, as Giandomenico Majone's chapter implies, still relatively new, but very rapid developments over the last few years now provide a body of experience against which to test some of the propositions put forward in the chapter.

The UNEP *Register of International Treaties and Agreements in the Field of the Environment* [1] lists a total of 108 texts, only two of which are earlier than 1940. The first, of 1921, restricts the use of white lead in paint and arose from the work of the International Labor Organization (ILO), while the second, of 1933, is concerned with the preservation of flora and fauna, particularly in Africa. A rather more comprehensive register [2] lists 288 multilateral treaties of interest to the protection of the environment and signed before 1983, only ten of which were signed before the end of 1939. The earliest is the *Convention on the Navigation of the Rhine* of 1868 whose effect on the environment is only in establishing an international organization for a great river that carries pollutants across national frontiers. The two compilations differ in length because the second is not only more catholic in its interpretation of the 'environment', but also includes treaties or conventions that merely created international organizations that can then deal with environmental matters.

The use of the treaty or convention to regulate the environment is the natural extension of the centuries-old mechanism by which sovereigns of nation states formalized their relationships. There are, however, difficulties with treaties and conventions, whatever their subject matter. Beyond the difficulty of agreement of terms which increases with the number of parties, there are uncertainties resulting from involvement of parliaments before ratification, and there are difficulties of implementation, since there is no easy way in which one nation state can ensure action by another. In the environmental field the difficulties may be greater than in the traditional fields of international relations, since the subject matter often has to be regulated in some detail if the desired results are to be achieved. Yet regulations are often handled by local governments or agencies with quite different traditions of administration. Consequently the need has been felt for supplementary channels of communication and action.

Undoubtedly the United Nations' Conference on the Human Environment held in Stockholm in 1972 gave a major impetus for environmental activities to be undertaken by international organizations. The United Nations' Environment Programme (UNEP) is one product of that conference, and is one of the rather few international organizations that deals exclusively with broadly environmental matters. More commonly, an increasing number of previously existing international organizations have adapted themselves to the needs of international efforts at environmental protection, many on a regional basis as Majone suggests.

A compendium of *Environmental Programmes of Intergovernmental Organizations* [3] lists 17 relevant organizations, but since it was drawn up primarily for the benefit of the chemical industry of the western world it omits intergovernmental organizations of the eastern world, e.g., the Council for Mutual Economic Assistance (CMEA). The 17 organizations differ widely from one another and include the Council of Europe, the United Nations' Economic Commission for Europe (ECE), the

European Economic Community (EEC), the Food and Agriculture Organization (FAO), the Intergovernmental Maritime Consultative Organization (IMCO), the Organization for Economic Cooperation and Development (OECD), and the World Health Organization (WHO).

A category of organization not listed in this compendium is the body established by a treaty or convention to deal with one specific topic. Such a body may go beyond trying to ensure consistency of application and may itself develop new proposals. Examples are the executive body that has to meet once a year under the Convention drawn up by the UN ECE on long-range transboundary air pollution and the commissions that exist under the Paris and Oslo Conventions on pollution of the sea.

There is scope for more analysis of the effectiveness of these various international organizations and treaties. Some do little more than ensure a flow of information between countries; others have their own institutions able to generate political pressure. The reason that the Council of Europe was the first international organization to work systematically in the environmental field is, in all probability, linked to the fact that it has a Parliamentary Assembly that created a pressure long before government officials saw the environment as requiring international efforts. The lack of opportunity for public pressure, and indeed for public information, is a problem that many international organizations have still to recognize. Yet at a national level public interest has been a generator for action in the environmental field and supplies the pressure for implementation of legislation.

One particular phenomenon that deserves attention is the way a subject shifts from one international organization to another. OECD did pioneering work in the field of long-range transport of air pollutants, but activity then moved to the UN ECE following an initiative of the USSR. The EEC has since proposed much more detailed legislation, indeed, no discussion of the environmental work of international bodies can ignore the EEC, which differs from all other international organizations in having its own institutions able to adopt legislation that is binding on the Member States without further review or ratification, and that furthermore has a Court of Justice able to ensure that the legislation is applied.

Notes and references (Comm.)

[1] United Nations Environment Programme (1983), *Register of International Treaties and Other Agreements in the Field of the Environment* (UNEP, Nairobi).

[2] Burhenne, W. (Ed) (1974), *International Environmental Law – Multilateral Treaties*, four volumes, loose-leaf, Beitrage zur Umweltgestaltung B7 (Erich Schmidt Verlag, Berlin).

[3] de Reeder, P. L. (1977), *Environmental Programmes of Intergovernmental Organizations*, loose-leaf (Martinus Nijhoff, The Hague).

Editors' postscript

Since this chapter was completed, more than 20 states plus the EEC have adopted the *Vienna Convention for the Protection of the Ozone Layer*, thus adding yet another international agreement to the list discussed by Haigh. Nonetheless, the *Convention* failed to produce a common protocol for future control. Drafting work continues, but for now the regional protocols proposed by the "Toronto Group" and the EEC remain the most likely bases for actual control of O_3 damage. Majone's "functional regionalism" thus continues to provide a useful perspective on the O_3 question, despite adoption of the international *Convention*. For further information on the *Convention*, see Sand, P. H. (1985), Protecting the ozone layer: the Vienna Convention is adopted, *Environment*, **27**(5), 18–20, 40–43.

Chapter 13 Monitoring the environment and renewable resources

Yu. A. Izrael
R. E. Munn

Editors' Introduction: Reliable knowledge about how the biosphere has actually responded to human actions is a prerequisite for effective social responses to environmental problems. Lack of such knowledge is a major impediment to the design of sustainable development strategies.

In this chapter we explore how environmental monitoring systems can be made more useful for the management of long-term interactions between development and environment. We begin with a review of objectives for integrated monitoring of the physical, chemical, and biological components of the biosphere. The problems of linking local and global scales of observation are discussed, as are issues on the optimization of monitoring systems. Special attention is devoted to how monitoring systems can be designed to help address the problems of timely detection of potentially irreversible trends and of identification of unexpected environmental impacts.

Introduction

Environmental and resource management (ERM) is greatly dependent on a continuous flow of relevant information from various data-gathering systems. Without meaningful field measurements the operation of effective systems to protect human health and to manage biosphere resources wisely would be hardly possible. Yet the monitoring needs of ERM are often not given sufficient attention.

Monitoring systems are imperfect for several reasons. First, there are limitations in technology, such as the inability to measure evaporation from an open body of water during gale-force winds (when evaporation rates are greatest!). Second, there is a lack of understanding of the processes involved, e.g., of the role of microbial organisms in the soil with respect to the global sulfur cycle (but as models of the environment improve, monitoring systems become better targeted to user needs). Third, there is the high capital and operating costs of monitoring, which often impose limitations on existing and new programs.

Fourth, there is the problem that the needs of users change over the years, requiring modifications in system design to accommodate new applications. This phenomenon is illustrated by the six stages in programs to monitor urban air pollution. In Stage 1, a monitoring network is established *to describe* ambient air quality, i.e., to obtain a general overview of spatial gradients, seasonal cycles, and frequency distributions of pollutants. In Stage 2, the program objective is *to predict* air quality, requiring that a model be formulated and tested and that emission inventories and meteorological measurements be incorporated in the monitoring system. Once the model is operating satisfactorily, data needs become more sharply focused; the objective in the next stage, Stage 3, is to check that the prediction system remains in calibration.

In Stage 4 (management), which may sometimes precede Stage 2, the goal is to ensure that air quality criteria or standards are achieved. The associated monitoring system is thus designed to check "hot spots" and may be supplemental to the network required with respect to Stages 2 and 3.

The air quality criteria used with respect to Stage 4 are obtained from the published literature, as summarized by the World Health Organization (WHO), for example. Contributing to the development of dose–response relations contained in such summaries may be epidemiological studies using various health indicators and ambient air pollution data of the Stage 1 or 4 type. But urban populations spend much of their time indoors and in commuting from homes to offices, factories, and schools. So in Stage 5 (health effects research), three new sets of factors need to be monitored: human exposure to air pollution (using personal samplers); selected health indicators (e.g., asthma attacks) [1, 2]; and activity patterns (e.g., hours spent at home, outdoors, in offices, etc.).

Finally, recognizing that some pollutants, such as lead and pesticides, reach people by various pathways (air, food, drinking water), air quality monitoring has to be integrated into a larger program for monitoring food and drinking water (Stage 6).

This example (monitoring urban air pollution) is not exceptional and it is important to review periodically the ability of monitoring systems to meet changing program needs.

Historically, monitoring has been subdivided into four categories:

(1) Physical.
(2) Chemical (including environmental pollution).
(3) Biological–ecological (including renewable resources, such as forests and fisheries).
(4) Socioeconomic.

This framework is used here for convenience, although the existence of strong interlinkages is recognized, requiring an integrated approach. As an example, management of the chemical environment requires knowledge not only of the concentrations of various pollutants, but also of:

(1) The physical environment (e.g., wind fields, river flows, ocean currents).
(2) The biological environment (e.g., exchange rates to and from terrestrial and marine biota).
(3) The socioeconomic environment (e.g., production rates of anthropogenic substances).

We conclude this introduction with some definitions. An Intergovernmental Working Group defined *monitoring* as "a system of continued observation, measurement and evaluation for defined purposes [3]." The urban air pollution example given earlier is very much in the spirit of this definition, particularly with respect to the last three words: "for defined purposes." However, the definition is rather general, and another term, *integrated monitoring*, is often used (see, for example, Izrael [4, 5, 6]):

> The repeated measurement of a range of related environmental variables and indicators in the living and non-living compartments of the environment, for the purpose of studying large parts of the biosphere as a single system.

It seems clear that an *integrated approach* to monitoring is essential, in which biological–ecological indicators are some of the main building blocks in system design. To illustrate the consequences of not recognizing this principle, we cite the well-established EMEP (European Monitoring and Evaluation Programme or, more formally, the Cooperative Programme for the Monitoring and Evaluation of Long Range Transmission of Air Pollutants in Europe) system for monitoring atmospheric pollution [under the auspices of the Economic Commission for Europe (ECE)] and BAPMoN [Background Air Pollution Monitoring Network under the auspices of the World Meteorological Organization (WMO)], which failed to provide a warning of forest damage in central Europe.

Included in the monitoring process, of course, are quality control, data assimilation in user-friendly form, and system optimization, i.e., assessment to determine whether the system design could be improved to meet the prescribed objectives.

Designing or redesigning a monitoring system

Sometimes an opportunity arises to design a monitoring system from first principles. More often, however, a system is already operating and the question is whether it is able, perhaps with modification, to meet the needs of a new program. To answer this, the following factors must be considered:

(1) What questions are to be asked of the monitoring system? This has already been mentioned but here it is useful to list some of the objectives of monitoring:

   (*a*) Exploratory (curiosity monitoring).
   (*b*) Research and model development.
   (*c*) Support for operational prediction systems.
   (*d*) Regulatory.
   (*e*) Early detection of trends.

(2) With what space and time resolutions are the questions to be answered? Sometimes it is desirable to observe fine-scale detail with the monitoring system; in other cases, the preferred strategy is to remove this noise.

(3) With what accuracy are the questions to be answered? The precision with which an objective is to be met must be expressed in statistical terms, e.g., is a system designed to provide early warnings of trends of 1, 5, or 10% change per annum, with a 90, 95, or 99% confidence?

(4) What are the measurement errors? Are they likely to be random or is there bias? In the case of ground-based monitoring stations, measurement error includes that due to nonrepresentativeness of observational sites. (For a useful statement on the representativeness of meteorological stations, see Nappo [7].)

(5) What is the space variability of the field and how strong are the space correlations? Research on this topic was pioneered by Gandin [8] and his associates at the Main Geophysical Observatory in Leningrad. In ecosystem studies, there is the special problem of patchiness, i.e., areas with an abundance of a species surrounded by areas sparsely populated by that species.

(6) What are the time variabilities and autocorrelations in the data?

(7) What are the practical constraints that limit the successful operation of a system? For ground-based monitoring, these may include difficulty in finding a representative site, lack of a suitable power supply, and lack of protection from vandalism. For monitoring from space platforms, practical problems include the long lead time required for system design, and the assimilation of data that accumulate at a very rapid rate.

(8) Finally, there is the problem of designing a monitoring system sufficiently flexible to meet two or more objectives, or to be responsive to an evolving understanding of the processes being monitored [9, 10]. This is an interesting optimization problem that should be of particular interest to the International Institute for Applied Systems Analysis (IIASA).

Three further points should be made. First, a mismatch will sometimes become apparent between program objectives and system capability. The manager should then try to estimate the additional resources required to meet objectives and the degree to which tolerances (space and time resolutions, precision) must be relaxed if the budget remains fixed. Second, it should be emphasized that the framework for a monitoring system is dependent on the designer's concept of the processes being monitored, and competing models of the environment could imply different monitoring strategies. As an example, the information needs for resource management depend greatly on whether an input–output or stress–response model of the biosphere is being used. Third, there is the problem of *retrospective* or *historical monitoring*, in which the data are fragmentary or impossible to check. In order to have an historical perspective of biogeochemical cycles, for example, it would be of great value to know, over the last few centuries, at least:

(1) The atmospheric concentrations of carbon dioxide (CO_2) and other gases.

(2) Changes in land-use patterns over successive decades.

Many clues about historical trends exist in tree rings, sediments, snow and ice cores, early government and church records, and personal diaries. However, these data are not easily retrieved in a compatible form.

Monitoring the physical environment

Various geophysical organizations have long been involved in the observation of the physical environment. Hydrometeorological services of various countries possess rather rich experience in this field. The World Weather Watch (WWW) of WMO coordinates activity on weather observation, including methods, quality control, and data exchange throughout the world.

Particularly for time scales of years, and even decades, the interpretation of geophysical data often requires information on the chemical environment (see the next section) and on long-term changes in properties of the Earth's surface, including assessment of:

(1) Anthropogenic heat fluxes into the biosphere.
(2) Changes in surface albedo.
(3) Changes in the character of the Earth's surface due to human activity, including potential physical and ecological consequences (special attention should be given to urbanization).
(4) The role of agriculture in various geophysical and ecological processes (deforestation, redistribution of water resources, the change of the character of the underlying surface, mass production of monocultures, wide use of chemical protection methods).
(5) The role of exploitation and unintentional destruction of renewable resources – biological (deforestation, fires, marsh drainage, etc.) and geochemical (soil impoverishment, overconsumption of water leading to desertification, etc.). It is especially important to stress the danger of mass destruction of tropical forests.

Monitoring the chemical environment

Although it is difficult, if not impossible, to find sites on Earth where the chemical environment resembles that of preindustrial times, it is nevertheless useful to take measurements at so-called baseline or background stations. This is done, for example, at WMO BAPMoN sites.

To assess the impact of mankind on the chemical environment, the following substances should be monitored at both background and impact location levels [4]:

(1) Atmospheric inputs of sulfur dioxide (SO_2) and nitrogen dioxide (NO_2) and subsequent transformations to particulates in the environment.
(2) Biogenic elements, particularly phosphorus and nitrogen and their fluxes.
(3) Propagation of radioactive products from nuclear explosions and nuclear power plants, particularly cesium-137, strontium-90, and krypton-85.
(4) Fluxes and biospheric cycles of heavy metals, particularly mercury, cadmium, and lead (in the atmosphere, their deposition on land surfaces, in streams and other land water bodies, and in oceans).
(5) Organochlorine compounds such as pesticides, polychlorinated biphenyls (PCBs; in the atmosphere, surface waters, oceans).
(6) Petroleum products such as polychlorinated aromatic hydrocarbons (PAH) benzo[*a*]pyrene (in the ocean).
(7) Stratospheric and tropospheric aerosols, including particle size distributions. This information would be used to study aerosol budgets, the role of highly dispersed tropospheric aerosols in climatic change, and the possibility that highly dispersed aerosols from forest fires could penetrate the stratosphere.

Associated tasks are to study:

(1) The environmental behavior of various substances.
(2) The pathways.
(3) The role of atmospheric and oceanic transport.
(4) The atmospheric distribution of substances – both existing and potential.
(5) Substance fluxes, using balance approaches.
(6) Accumulation rates of pollutants in surface waters and the oceans.
(7) Major fluxes between biosphere compartments, keeping track of anthropogenic components separately.

This program of monitoring should help reveal the ecological consequences of any observed build-up.

Monitoring the biological–ecological environment

The variability of biological–ecological systems

Biological–ecological systems contain tremendous internal variability, both in space (e.g., the patchiness problem) and in time (e.g., population explosions at irregular intervals). Furthermore, the responses of these systems to external stresses may be exceedingly variable, with a broad range of responses to a given stress. Children and the very old are particularly sensitive to air pollution, for example.

Applications in environmental pollution management

For the reasons already given, monitoring strategies designed to help protect the health of people and of ecosystems often concentrate on monitoring the environment rather than the organisms being stressed. Protection of people and of ecosystems is supposed to be assured if environmental exposures do not exceed specified values obtained from published dose–response curves. Nevertheless, biological–ecological monitoring does play an important role in several key areas [11–15]:

(1) *Research.* In order to develop or improve dose–response relations, a continuing program of biological monitoring under a range of environmental conditions is required. In the case of epidemiological studies, for example, various health indicators have been used, including death rates, hospital admittances, school and office absenteeism rates, census information, and personal diaries. In most countries such data are incomplete and not very accurate. For example, death certificates and hospital admittance forms give no information on smoking habits, which could dominate air pollution exposures.

(2) *Early warning.* In this case, species that are especially sensitive to particular kinds of environmental stress are used as indicators of damage. For example, delphinium plants and tobacco leaves are sensitive to oxidants. An extreme example is the miner's canary, but the absence of certain species of fish in a river or of lichens in a city also suggests that the local environment is somewhat unhealthy.

(3) *Accumulator species.* Lichens, moss, fish, and wildlife often accumulate potentially toxic substances in measurable amounts, e.g., pesticides and metals, even though the substances cannot be detected in the surrounding medium. This is one of the justifications for the mussel watch program [11]. Also the *moss bag* [12] can be used to estimate the deposition of heavy metals.

(4) *Integration of the effects of multiple pathways or multiple pollutants.* In many cases, a pollutant reaches the target by several pathways (e.g., air, drinking water, food). Alternatively, several pollutants may be present at the same time, sometimes producing synergisms; an oxidant smog contains a multiplicity of pollutants, for example. So it is sometimes rather wistfully suggested that dose–response relations would be more useful if they could be developed for typical mixtures of substances rather than for individual pollutants. However, the practical difficulties in calibrating such a system have still to be overcome.

(5) *Indicators of spatial patterns.* Conventional networks of monitoring stations can sometimes be supplemented by biological indicators, e.g., areal extent of leaf-tip damage around a smelter; number of telephone complaints per unit area of odors from a refinery; absence or presence of particular species. A second justification for mussel watch [11] is as an indicator of spatial patterns.

Applications in resource management

Introduction

Monitoring methodologies reported in the scientific literature have, until recently, been concerned exclusively with weather–climate and pollution applications. However, this is beginning to change as the needs of resource managers are being stated more explicitly [16–21]. Below we discuss only a few examples of this type of application, but we expect considerable methodological developments in the next decade.

The local scale

Biological–ecological monitoring is increasingly required on the local scale with respect to land-use planning and assessment of environmental impacts. Before permission to proceed with a project can be obtained, many jurisdictions require so-called *baseline monitoring*, which may include, e.g., fish and wildlife censuses. Current approaches to baseline monitoring have been criticized [17, 18] on several grounds, but particularly because species abundances depend on several factors, such as the presence of predators, the availability of food, and the suitability of habitats for nesting. Unless these other factors are monitored, the ecologist has no basis for predicting the impact of development. Also monitoring the presence or absence of rare or endangered species introduces some very difficult extreme-value statistical considerations. An ecosystem approach is therefore recommended, in which predator–prey relations and life cycles of selected species are monitored [16, 17].

Postassessment appraisals of the accuracy of predictions contained in an environmental impact assessment also require biological–ecological data. The permit to construct a power station or smelter should specify the monitoring strategy to be adopted, i.e., biological indicators to be used, network design, frequency of monitoring, and action to be taken if ecosystem behavior is significantly different from that predicted in the environmental impact statement. In this connection, a precondition for receiving money as a loan or grant from a government or an international body, such as the World Bank or United Nations Development Programme (UNDP), ought to be that the developer be required to carry out postconstruction biological monitoring.

As emphasized by Beanlands and Duinker [17], a main goal of ecological monitoring should be hypothesis testing.

The intermediate scale

Various requirements exist for biological–ecological data on the national, provincial, or county scale. Here we discuss two particular examples: forest management and resource accounting. First, however, it is worth quoting a principal recommendation contained in a recent United Nations Environmental Programme (UNEP) collection of papers on ecological monitoring of arid zones [19], *viz.* that a three-pronged approach be used for biological–ecological monitoring on the intermediate scale:

(1) Ground observations from both fixed and mobile stations.
(2) Aircraft observations.
(3) Satellite observations.

Forest management As emphasized by Regier and Baskerville [20], successful resource management requires detailed biological–ecological data. Yet it is difficult in most countries to obtain reasonable estimates of even such a simple quantity as the rate at which prime agricultural land is lost to urbanization.

For forest management in New Brunswick, Canada [20, 21], the elements included in the current forest monitoring program are:

(1) Abundances of the several commercially important tree species.
(2) Age distributions of these species.
(3) Current state of susceptibility of these species to damage by the spruce budworm.
(4) Annual updates.
(5) Spatial resolution of a few hectares, i.e., the size of an individual stand.

Regier and Baskerville [20] are convinced that forest sustainability can be achieved only if management is on the mesoscale.

Resource accounting Current national accounting systems in most countries concentrate on economic indicators, such as gross national product (GNP), cost of living, trade surpluses, etc. However, there has been recent interest in establishing an equivalent system for the resource sector. So a new term, *resource accounting*, is used by economists and statisticians to denote a data gathering and retrieval system for keeping track of stocks and flows of resources [22, 23, 24]. Although this is a good idea, there are several obstacles to be overcome before a system can become operational:

(1) *Conceptual difficulties.* The following problems in definitions arise at the outset:

   (*a*) *Quantity of a resource.* A substance is a resource only if it is of use to someone and if it is economic to obtain (in comparison with the cost of a substitute). For example, should forests in remote areas be included in forestry accounts?
   (*b*) *Quality of a resource.* How should the quality of a resource, such as agricultural land, be measured?
   (*c*) *Multiple uses of a resource.* Quite often a renewable resource is quantified in different ways by different users. For example, a river may provide drinking water, commercial fishing, recreation, dilution of sewage, hydroelectric power, and industrial coolant.
   (*d*) *Wastes.* In the course of using a resource (both renewable and nonrenewable), wastes may be created, thus degrading the environment and resource base. How is this to be characterized in the national accounting system?
   (*e*) *Multiple users of the resource accounting system.* It is difficult to design a system that will meet the needs of multiple users. For example, one of the objectives for operating such a system is to popularize the idea of resource management. But this goal is in conflict, to a certain extent, with that of providing very technical data to modelers of economic systems.

(2) *Practical difficulties.* The information needed for resource management is often compiled by different levels of government and by different departments at the same level of government. Assembling data from these disparate sources is often a frustrating task. In this connection, Rapport and Friend [23, 24] have recommended that renewable resource data should be organized according to ecozones (such as the Great Lakes basin) rather than according to census tracts.

Despite these difficulties, serious attempts are being made in several countries, e.g., Norway, France, and Canada, to introduce resource accounts into the national accounting system.

The large scale

On the large scale two or more jurisdictions often have a common interest in sustaining a resource, such as fisheries. It is critical that they agree upon the integrity of the statistical data sets and monitoring systems being used. This means that data collected in different countries must be intercomparable, with regular intercomparisons of monitoring and analysis protocols. The WMO has a long tradition in this field, but only since the UN Stockholm Conference have other intergovernmental bodies attempted to standardize collection systems of biological and ecological data. UNEP has played a major role in this effort.

Information needs in the field of intergovernmental ERMs are application specific. So existing biological data banks and monitoring systems may be of high quality, but may not be appropriate for the application in mind.

One of the first steps in developing a strategy for monitoring (to meet a specific objective) is, of course, to undertake an inventory of existing data banks and monitoring systems. Independently, however, an evaluation should be made of the real needs for biological–ecological data. Matches and mismatches can then be assessed in an attempt to produce an acceptable monitoring strategy.

Biological–ecological monitoring in the context of "integrated monitoring"

An interesting practical example of the concept of *integrated monitoring* is the proposal by Brejmeyer [25] for a network of stations to monitor the pine forests of Poland, one station in each of the 49 provinces of that country. The sites are to be in protected areas (national parks, biological reserves), where the character of the surrounding area is unlikely to change. The main monitoring components are:

(1) Ecosystem functioning:

   (*a*) Production of new, green material.
   (*b*) Loss of organic material.

(2) Environmental pollution:

   (*a*) Air pollution concentrations.
   (*b*) Heavy metal concentrations in lichen, bark, and needles.
   (*c*) Sulfur and heavy metal concentrations in the litter and upper soil layers, and in earthworms.

(3) Phytosociological data:

   (*a*) Basic properties of site.
   (*b*) Species diversity and abundance.
   (*c*) Microclimate.

The objective is to develop a system that is relatively simple to operate, but that can provide meaningful inputs to ecosystem models, including pollutant pathway analyses. The data may also be useful for those who study global biogeochemical cycles.

In Sweden, a National Environmental Program (PMK) has been established to keep track of long-term changes in environmental conditions, including the fluxes of pollutants in and between various media [26]. This work is carried out in or near some 20 small watersheds, mostly in national parks or nature reserves.

Monitoring socioeconomic factors

Monitoring of socioeconomic factors is important in ERM for two reasons:

(1) Some of the driving variables in ERM models are socioeconomic. For example, market prices have a considerable effect on the rates of production of lumber and wheat. Similarly, an indicator of overcapitalization of the fishing industry might provide an early warning of economic collapse.
(2) Some of the socioeconomic variables may be useful surrogates for ERM variables. For example, the daily production of energy from a coal-fired power station can be used to infer the emissions of sulfur and CO_2 from the chimney. As a second example, farm productivity might be helpful in interpreting soil erosion rates.

Unfortunately, the space and time resolutions of conventional socioeconomic data are frequently inappropriate for ERM applications. In studies of long-range transport and acid deposition, for example, atmospheric modelers require inventories of daily emissions, but sometimes must be content with annual or seasonal estimates. So it is important that the requirements for socioeconomic data to support ERM studies should be specified in as much detail as possible.

The use of indicators

The designer of a monitoring system has a very large selection of variables from which to choose a relevant and manageable subset. It is clearly important to make a careful choice. What are the critical points in a complex system (such as the world carbon cycle) where observations will be of most use? Can several variables be combined into a single quantity (such as Reynolds number, Richardson number, or the Monin–Obukhov length), which will meaningfully represent the net effect of several processes [27].

There are many examples of such indicators, although the question has not yet been studied from an interdisciplinary perspective. Here we illustrate our ideas in terms of pollution potential, or carrying capacity, of marine ecosystems [28]. It is clearly important to estimate the maximum pollution loading that can be placed on individual ocean ecosystems, regions, and the world oceans as a whole without sustainability being impaired.

The assimilative capacity of a marine ecosystem, A_{mi}, for the given pollutant i (or sum of pollutants) for ecosystem m is a measure of the maximum amount of pollutants (in terms of the whole zone or the unit volume of the marine ecosystem) that can be accumulated, destroyed, transformed (biologically or chemically), and removed per unit time due to sedimentation, diffusion, or any other transport process out of the ecosystem that does not disturb its normal functioning. Assimilative capacity A_{mi} characterizes the ability of the marine ecosystem with respect to the dynamic accumulation of toxic substances and to the active removal of various pollutants, with the main ecosystem properties being preserved.

The numerical value of A_{mi} depends on many natural and anthropogenic factors: stream velocity, turbulence, water exchange, water temperature, and structure and functioning of biotic components, as well as the chemical and physical properties of the pollutants entering the marine ecosystem. The whole complex of natural phenomena responsible for "self-purification" of the marine environment can be reduced to a few important processes.

Three main processes that, for all practical purposes, determine the assimilative capacity of the marine environment (ecosystem) can be singled out: hydrodynamic processes ($A_{v,w}$), microbiological oxidation (A) of organic pollutants, and biosedimentation ($A_{k,v}$), i.e.,

$$A_{mi} = A_{v,w} + A + A_{k,v} \tag{13.1}$$

When assessing A_{mi} it is important to introduce critical anthropogenic loadings, critical ecosystem processes, and critical (with respect to the impact) marine species. When determining assimilative capacity A_{mi}, two indices are of great importance: maximum permissible (to provide normal functioning of the ecosystem) concentration of the given ith substance C (or sum of substances) and a multiplicity coefficient K of the amount of substance (per unit volume) that can be assimilated by the given ecosystem and removed per unit volume as a result of various processes (for a given averaging time, determined for the permissible concentration C_{pi}). Then

$$A_{mi} = C_{pi} K_i V \tag{13.2}$$

where V is the marine volume under consideration.

Criteria for oceanic ecological well-being are based on the concept of assimilative capacity. The approach described may also be applied to terrestrial ecosystems (including fresh-water ecosystems).

The above remarks indicate that there are some real opportunities in the field of monitoring optimization, in terms of the development of integrated approaches and of maximum use of indicators for ecological purposes.

Space–time optimization of monitoring systems

Optimization of a system to meet a single objective

As mentioned previously, *optimization* of monitoring systems is used in two senses: space–time network optimization, and selection and fine tuning of subsets of the most relevant variables. We do not deal with this second case except to note that it is of fundamental importance, particularly with respect to large programs, such as the Global Atmospheric Research Program (GARP) and the International Geosphere Biosphere Program (IGBP), and also to note that optimization is achieved through models and sensitivity analyses.

Turning to the first case, measurements of an environmental element or indicator are correlated in space and time, the correlations decreasing with increasing separations in space and/or time. In addition, measurements of two environmental elements or indicators are cross-correlated. For these reasons, monitoring systems contain a certain amount of duplicate information. World weather patterns, for example, are coherent and it is possible to predict synoptic developments from surface observing stations 300 km apart and upper atmosphere stations 600 km apart. These separation distances were obtained by trial and error during the last century, but they have since been derived on statistical grounds.

The problem of optimization of a monitoring system has been widely studied in meteorology and in air pollution, based largely on the correlation fields and associated structure functions [8]. These studies reveal that optimization depends on:

(1) The objective(s) of monitoring.
(2) The tolerances permitted by the user with respect to meeting the objective.
(3) The extent of measurement errors, including those due to poor siting of monitoring stations.
(4) The strength of the correlation fields.
(5) Constraints imposed by budgets.

The problem is greatly simplified if only a single objective and tolerance are to be met. Then there are three approaches that can be taken to optimization [29]:

(1) *The statistical approach.* If sufficient data have already been collected from an existing monitoring system, statistical tests can be applied to determine whether the system is meeting specified goals and to improve its performance.
(2) *The modeling approach.* If the processes being monitored are well understood (even if there are no data for the region or application of interest), mathematical models can be used to design a system that will meet designated performance standards.
(3) *The combined statistical and modeling approach.* Best of all, if monitoring data exist and the environmental processes being monitored are reasonably well understood, a combined statistical and modeling approach can be used to optimize the system. For example, root-mean-square differences between observed and predicted values could provide a basis for selecting the most important elements of a monitoring system.

Efforts should be made to try to apply these approaches to the environmental and resource management field. There is also a need to quantify the uncertainty in optimization procedures, i.e., due to the fact that data sets from different time periods yield slightly different optimization results. Recent developments in the analysis of fuzzy data sets using the maximum entropy principle [30] may be worth exploring here.

Optimization of a surface network to meet multiple objectives

A large body of information exists in the engineering literature with respect to system optimization to meet *multiple* objectives. This area of study has hardly been tapped by environmental and resource management researchers, despite its obvious importance. We recommend that IIASA serve as a focal point for such work. We also briefly sketch a recent environmental application [31] to the design of a monitoring network for urban air quality that is to be optimal for more than one type of pollutant. The objective in this study was to detect violations in air quality standards, but because the sources contributing to the concentrations of one pollutant are different from sources contributing to another, an optimal network would be pollutant specific, increasing operating costs considerably. So the authors of this study selected subsets of common sites, using the principle of Pareto optimality [31]. Each pollutant was treated as a subutility, and network optimization occurred when none of the subutility functions worsened when any other improved.

Optimization of monitoring from space to meet multiple objectives

A rather different example of optimization of a monitoring system is to be found in the study of Kondratyev and Pokrovsky [10] concerning the spectral intervals to be used for monitoring the Earth's resources from space. The "best" spectral bands depend on the application, and Kondratyev and Pokrovsky indicate that in oceanography, hydrology, geology, and forestry–agriculture they have identified 102, 33, 36, and 32 applications, respectively. Because of the high cost of satellites and of ground-based transmission and data-processing centers it is necessary to rationalize system design. Given the relative importance of each requirement, the incremental costs of making additional measurements, and the signal-to-noise ratio in each spectral interval, the authors suggest that a least-squares technique can be used to optimize the system to meet the majority of user needs and tolerances.

Monitoring to provide early detection of change

Strategies for environmental and resource management often assume that conditions prevailing over the last few years give a good indication of future states (i.e., of mean values, frequency distributions, and probabilities of occurrence of various types of rare events such as floods and droughts). But this is not likely to be so. In the first place, the historical records of many environmental variables contain low-frequency oscillations of the order of decades or longer, e.g., the Little Ice Age. This means, for example, that a water management system based on rainfall records over the last 30 years may not be optimal for the next 30 years. Second, the environmental impact of mankind, which has always been important locally and sometimes regionally, now threatens to become continental and even global, with increasing probabilities of irreversible trends (e.g., CO_2 greenhouse warming) and flip-flops (see p 371) to new steady states (e.g., large-scale desertification).

UNEP [32] has expressed the view that forward planning must reckon with considerable uncertainty and surprise, and that one of the main tasks of the next decade is to design appropriate early warning systems. Most existing monitoring systems were not created with this application in mind; in terms of early warning, they provide information only on the current noise level in the environment.

We deal with three related questions in turn.

Irreversible trends

To anticipate an irreversible trend in an environmental variable it is necessary to have a model of the phenomenon. But because the model will have been calibrated on current or historical data, predictions will be rather uncertain. Furthermore, the early stages of a trend will be difficult to detect experimentally because of great natural variability in environmental conditions. Thus, it is important to try to optimize early warning monitoring systems, e.g., by a careful selection of indicators, monitoring sites, and averaging times. For example, it is likely that if stratospheric ozone (O_3) depletion were to occur, there would be particular heights, latitudes, and seasons where the effect could first be isolated [33].

Wigley and Jones [34] studied this general question in the context of CO_2 climate warming, using signal-to-noise ratios. They obtained a signal from predictions of surface-temperature warming from a climatic model in which the atmospheric CO_2 concentration was doubled, as shown in *Figure 13.1* [35]. The noise was the computed variance of temperatures for the years 1941–1980. The resultant

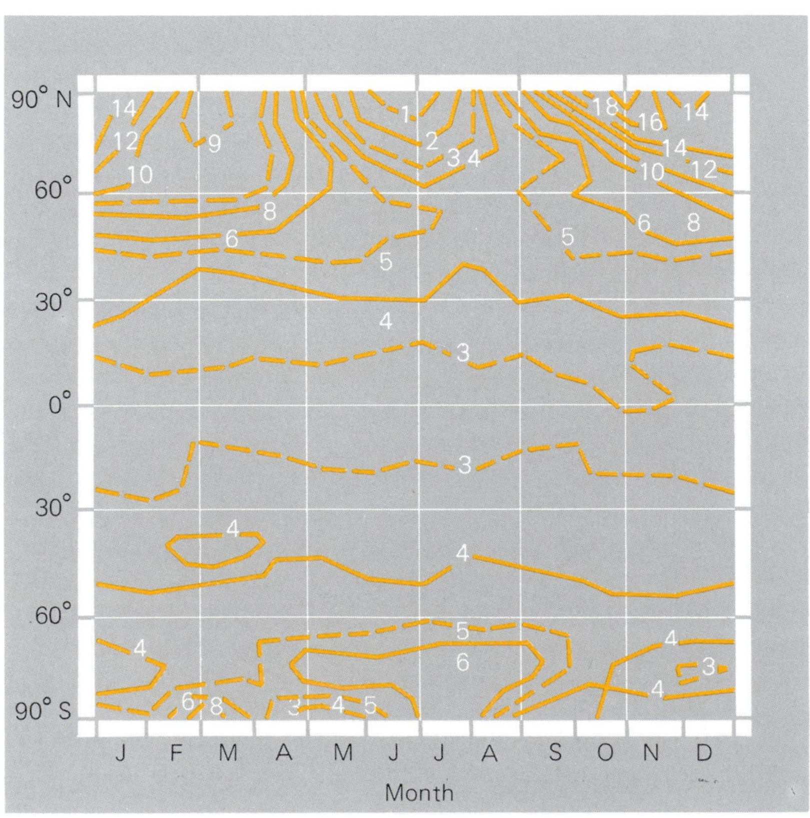

Figure 13.1 Latitude–time distribution of zonal mean difference in surface-air (70 m altitude) temperatures (K) between present and quadrupled CO_2 experiments [35].

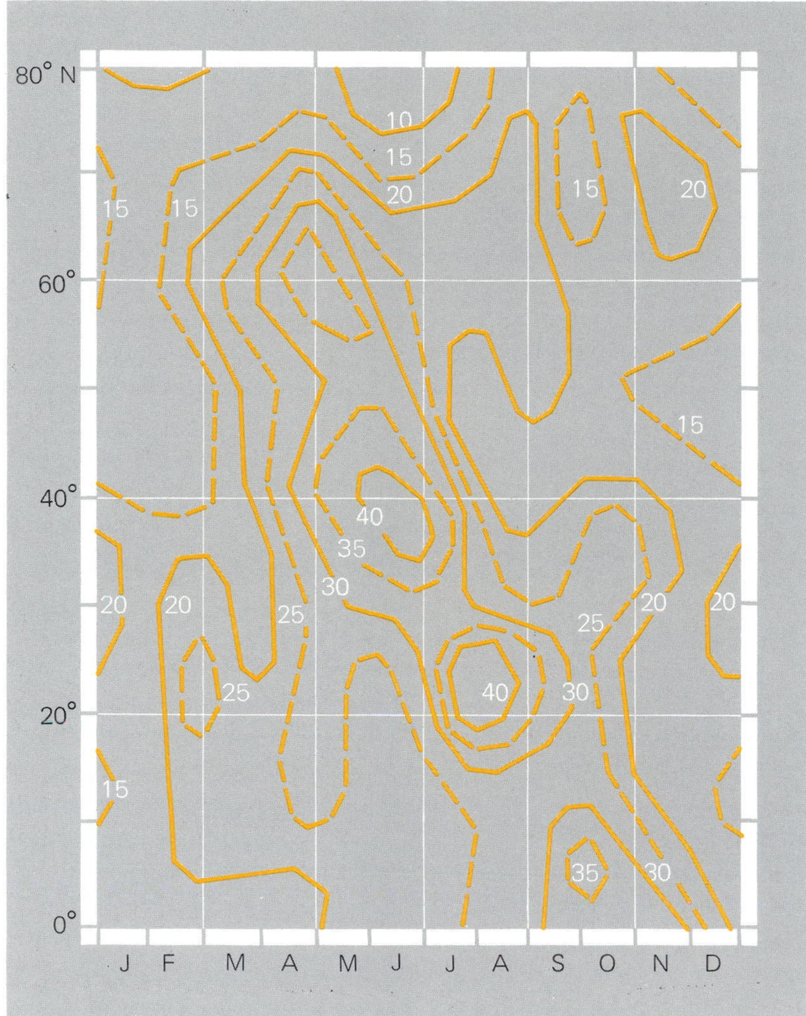

Figure 13.2 Signal-to-noise ratio for predicted CO_2-induced changes in surface-air temperature as a function of latitude and month. The signal is based on the numerical modeling results of Manabe and Stouffer [35]. The noise has been calculated from grid-point surface-temperature data. The value for month j at latitude L is the areally weighted average of grid points at $L-5$, L, and $L+5$, and the noise level is proportional to the standard deviation of month-j values over the period 1941 to 1980, corrected for autocorrelation effects [34].

isopleths of signal-to-noise ratio are shown in *Figure 13.2* [34], which suggests that CO_2 induced warming is likely to be detected first in middle latitudes in summer, even though the warming will be greatest at high latitudes in autumn and winter.

Overshoots, flip-flops, and other discontinuities

In recent years, developments in such fields as catastrophe theory have indicated that many systems have outer limits. When these limits are approached, a chance event or a slight enhancement of the current conditions may cause a system to overshoot, to flip-flop rapidly to a new and radically different steady state, or even to collapse. These ideas are well understood in engineering and the physical sciences (with respect to resonance causing, e.g., the collapse of the Tacoma Narrows bridge some years ago, feedback on microphones, critical damping of wind vanes, and the "almost intransitivity" of climate [36]); these phenomena are generally modeled with second-order differential equations having more than one solution.

Application of these ideas to ecological and even to socioeconomic [37] systems is a welcome development of the last few years. Models that include Monte Carlo simulations of randomness are now sufficiently elaborated that it should soon be possible to identify the key elements of an early warning monitoring system. For terrestrial ecosystems, Holling (Chapter 10, this volume) has suggested patchiness and soil composition as priority indicators. Field studies are also in progress or planned, e.g., with respect to the flip-flop of grassland to desert due to overgrazing (Brian Walker, personal communication).

In connection with socioeconomic applications, there is the added complexity that the resilience of society may change drastically with time, due to the introduction of new technology, for example. So it will be necessary to monitor not only the environmental stresses, but also the response characteristics of the system being stressed.

Examining the effectiveness of proposed management strategies in avoiding irreversible trends and flip-flops

In this application, it is supposed that several possible management strategies are to be compared, the objective being to select the one that minimizes the possibility of subsequent irreversible trends or flip-flops. Then certain key elements of the environment are to be monitored to provide the earliest possible indications that the strategy is working or not working. (It is conceivable that a strategy would not be acceptable due to difficulty in determining its effectiveness because of environmental variability.)

This area of interest can best be illustrated with an example. Suppose that it is predicted that acidic deposition will cause irreversible damage to lakes unless sulfur emissions from large point sources are reduced by x%. Then a number of management questions arise:

(1) Assuming that acidic deposition would decrease proportionally to a decrease in sulfur emissions, what length of record of the chemical constituents of precipitation would be required before the decrease in acidic deposition would be demonstrated?
(2) What is the minimum percentage reduction in sulfur deposition that could be detected in y years with 95% confidence?
(3) What is the optimum array of monitoring stations for answering these questions?

A statistical framework for dealing with this kind of problem was developed at a recent workshop [38]. Among the recommendations made is that the signal-to-noise approach be used.

There are many similar kinds of environmental issues, e.g., stratospheric O_3 depletion, CO_2 greenhouse warming, desertification, soil salinization, health effects of ionizing radiation. One of the questions that needs to be discussed in each case is the adequacy of existing monitoring systems for comparing proposed alternative management practices whose objective is to avoid undesirable change.

Monitoring to support studies of sustainable development: the main considerations and research needs

When designing or redesigning a monitoring system, it is essential at the outset to specify goals and tolerances. A high-quality data set that happens to be available may be largely irrelevant for the problem at hand. It is pointed out by McElroy (Chapter 7, this volume) in the context of IGBP that programs should be formulated around specific questions and hypotheses. They should not be driven by the availability of new experimental techniques, however powerful. Nevertheless, it may be illuminating to determine the questions that existing monitoring systems and data banks are capable of answering, including uncertainty bands. In any event, it is always useful to compile an inventory of such monitoring systems.

A second main conclusion is that monitoring and modeling are interdependent, interactive processes. For this reason, and also because the goals for monitoring often change over the years, an adaptive monitoring strategy is recommended. The program should not become locked into specific sampling and analysis protocols, for example. To illustrate with three competing resource models:

(1) *Input–output.* The main variables of interest are the *flows* from one sector or location to another, for example, the energy inputs to and from a city [39].

(2) *Stress–response.* Variables of interest are:

(*a*) *Stress.* Water quality, air quality, overconsumption of the resource (e.g. overfishing), etc.

(*b*) *Response.* Changes in the age distribution, size, and commercial value of the stock; indicators of ecological damage; etc.

(3) *Forest sustainability.* In the New Brunswick case, the management strategy to achieve sustainable forest yields is clearly in focus, requiring a mesoscale information network (p 365 and [20, 21]).

It is clear that the data requirements in these three cases are quite different. The possibility that the resource monitoring system may have to accommodate some other conceptual framework at some future time should therefore not be overlooked.

A third conclusion is that developing early warning indicators of irreversible changes and discontinuities is at the leading edge of research, and some real progress could be made in the next 5–10 years. Current ideas should be elaborated on several fronts:

(1) *Historical analogies.* There are many reasonably well-documented cases of discontinuities and surprises (prolonged droughts, collapses of resource-based industries, ecological population explosions, etc.). What indicators could have been used to detect these phenomena at a very early stage? As pointed out by Lovelock (personal communication) in an epidemiological context, the *annual death rate* is not a very useful early warning indicator of an epidemic. What other indicators should be used?

(2) *Ecological applications.* C. F. Holling (Chapter 10, this volume) has some interesting ideas on early warnings of ecological shifts. Initiatives such as his should be encouraged, as well as attempts to test such theoretical hypotheses with field data.

(3) *Studies of feedbacks.* One of the suggestions made at the IIASA Workshop was that the health of a biogeophysical, ecological, or socioeconomic system is maintained as a balance amongst various positive and negative feedbacks. Then an early indication that the system might go out of control would be given by a sharp increase or decrease in the strength of one of the feedbacks, with a counteracting response from one of the others. An appropriate early warning system would then require identification, quantification, and monitoring of these feedback mechanisms.

Finally, we wish to stress that if *sustainable development of the biosphere* becomes a long-term IIASA project, monitoring should become an integral component. Whether there is a separate subproject labeled *monitoring* is not important, and it might be better to have a monitoring element within each of the other subprojects. In any event, the work could not all be done at IIASA, and a recommended strategy is to involve scientists in several countries, with coordination and summer workshops at IIASA. The main monitoring themes to be considered could be:

(1) Optimization of monitoring systems to meet various objectives.

(2) Design of a conceptual framework for resource accounting.

(3) Design of monitoring systems to provide early warnings of irreversible changes and discontinuities.

(4) Conceptual studies to support the development of IGBP proposed by the International Council of Scientific Unions.

In the first two cases, IIASA has the special advantage of having system analysts and economists already working within the Institute.

Further reading

Two books that provide an excellent overview of current views on integrated monitoring systems are:

World Meteorological Society (1980), *Proceedings of the First International Symposium on Integrated Global Monitoring of Environmental Pollution,* Special Environmental Report No. 15 (WMO, Geneva). (Also published in Russian by USSR

State Committee for Hydrometeorology and Control of Natural Environment.)

Gidrometeoizdat (1983), *Proceedings of the Second International Symposium on Integrated Global Monitoring of Environmental Pollution* (Gidrometeoizdat, Moscow) (in English and Russian).

A synthesis of current practices for optimization of air quality monitoring networks is given in:

Munn, R. E. (1981), *The Design of Air Quality Monitoring Networks* (Macmillan, Basingstoke, UK).

A collection of papers that provides a useful entry into the literature on monitoring the biosphere, including natural resources, is:

United Nations Environment Programme (1980), *Selected Works on Ecological Monitoring of Arid Zones*, GEMS/PAC Info. Series No. 1 (UNEP, Nairobi, Kenya).

Notes and references

[1] World Health Organization (1983), *Estimation of Human Exposure to Air Pollution*, WHO Offset Pub. No. 69 (WHO, Geneva).

[2] Fugas, M. (1982), *Human Exposure to Carbon Monoxide and Suspended Particulate Matter in Zagreb, Yugoslavia*, WHO Offset Pub. EFF 82.33 (WHO, Geneva).

[3] Intergovernmental Working Group (1971), *Proceedings of Meeting of Intergovernmental Working Group on Environmental Monitoring* (IGWG, Geneva).

[4] Izrael, Yu. A. (1974), Global observation system. Prediction and assessment of the change in the natural environment state, principles of monitoring, *Meteorologia i hydrologia*, No. 7, 3–8.

[5] Izrael, Yu. A. (1979), *Ecology and Control of Natural Environmental State* (Gidrometeoizdat, Leningrad).

[6] Izrael, Yu. A. (1980), Main principles of the monitoring of natural environment pollution, in *Proceedings of Symposium on the Development of Multi-Media Monitoring of Environmental Pollution, Riga 1978*, Special Envir. Rep. 15 (WMO, Geneva).

[7] Nappo, C. J. (1982), The workshop on the representativeness of meteorological observations, *Bulletin of the American Meteorological Society*, **63**, 761–764.

[8] Gandin, L. S. (1963), Objective analysis of meteorological fields, *Gidriometeoizdat* [English translation (1965) by the Israel Program for Scientific Translations, Jermal, No. 242].

[9] Smith, D. E. and Egan, B. A. (1979), Design of monitoring networks to meet multiple criteria, *Journal of the Air Pollution Control Association*, **29**, 710–714.

[10] Kondratyev, K. Ya. and Pokrovsky, O. M. (1983), Techniques for the selection of spectral intervals for multispectral survey of the earth's resources from space, *Advances in Space Research*, **3**, 251–255.

[11] Goldberg, E. (1978), The mussel watch, *Environmental Conservation*, **5**, 101–125.

[12] Goodman, G. and Roberts, T. M. (1971), Plants and soils as indicators of metals in the air, *Nature*, **231**, 287–292.

[13] Goodman, G. and Roberts, T. M. (1983), *Environment Monitoring and Assessment*, **3**, 205–403 (a collection of papers on biological monitoring).

[14] Manning, W. J. and Feder, W. A. (1980), *Biomonitoring Air Pollutants with Plants* (Applied Science Publishers, London).

[15] Woodwell, G. M. (Ed) (1984), *The Role of Terrestrial Vegetation in the Global Carbon Cycle*, SCOPE 23 (John Wiley, Chichester, UK, and New York).

[16] American Association for the Advancement of Science (1983), *Resource Inventory and Baseline Study Methods for Developing Countries* (AAAS, Washington, DC).

[17] Beanlands, G. E. and Duinker, P. N. (1983), *An Ecological Framework for Environmental Impact Assessment in Canada* (Federal Environmental Assessment and Review Office, Canada).

[18] Hilborn, R. and Walters, C. J. (1981), Pitfalls of environmental baseline and process studies, *EIA Review*, **2**, 265–278.

[19] United Nations Environmental Programme (1980), *Selected Works on Ecological Monitoring of Arid Zones*, GEMS/PAC Info. Series No. 1 (UNEP, Nairobi).

[20] Regier, H. A. and Baskerville, G. J., Chapter 3, this volume.

[21] Erdle, T. and Jordan, G. (1984), *Computer-based Mapping in Forestry: A View from New Brunswick*, preprint (Department of Natural Resources, University of New Brunswick, Fredericton, Canada).

[22] Economic Commission for Europe (1982), *Comparative Analysis of Approaches Used in Natural Resource Accounting Schemes*, CES/AC. 40/18 (ECE, Geneva).

[23] Rapport, D. and Friend, A. (1979), *Towards a Comprehensive Framework for Environmental Statistics: A Stress–Response Approach* (Statistics Canada, Ottawa).

[24] Statistics Canada (1978), *Human Activity and The Environment* (Statistics Canada, Ottawa).

[25] Brejmeyer, A. (1981), Monitoring of the functioning of ecosystems, *International Journal of Environmental Monitoring and Assessment*, **1**, 175–193.

[26] Bernes, C., Giege, B., Johansson, K., and Larsson, J. E. (1985), Design of an integrated monitoring

programme in Sweden, *International Journal of Environmental Monitoring and Assessment*, **6**(2), 113–126.

[27] For example, the flow of a fluid through a pipe depends on pipe diameter, *D*, flow velocity, *U*, and fluid viscosity, *V*. These quantities are often combined to yield a dimensionless Reynolds number DU/V [cm(cm/s)/(cm^2/s)], which characterizes many different fluids, flow rates, and pipe diameters.

[28] Izrael, Yu. A. and Tsyban, A. V. (1983), On assimilation capacity of the World Ocean, *Doklady AN SSSR*, **272**(3), 702–705.

[29] Munn, R. E. (1981), *The Design of Air Quality Monitoring Networks* (Macmillan, Basingstoke, UK).

[30] Skilling, J. (1984), The maximum entropy method, *Nature*, **309**, 748–749.

[31] Modak, P. M. and Lohani, B. N. (1985), Optimization of ambient air quality monitoring networks Part III, *International Journal of Environmental Monitoring and Assessment*, **5**, 39–53.

[32] United Nations Environment Programme (1982), *The Environment in 1982: Retrospect and Prospect*, UNEP/GC(SSC)/2 (UNEP, Nairobi).

[33] Campbell, J. (Ed) (1978), *Proceedings NASA-Sponsored Symposium on Ozone Trend Detectability* (available from Dr. J. Campbell, NASA, Langley Research Center, Hampton, VA).

[34] Wigley, T. M. L. and Jones, P. D. (1981), Detecting CO_2-induced climate change, *Nature*, **292**, 205–208.

[35] Manabe, S. and Stouffer, R. J. (1979), A CO_2-climate sensitivity study with a mathematical model of the global climate, *Nature*, **282**, 491–493.

[36] See Dickinson, Chapter 9, this volume, for a discussion of the almost intransitive theory of Lorenz.

[37] Burton, I. (1963), The vulnerability of cities, in R. White and I. Burton (Eds), *Approaches to the Study of the Environmental Implications of Contemporary Urbanization*, MAB Tech. Note 14, pp 111–117 (UNESCO, Paris).

[38] Munn, R. E. (1984), *The Detection of Trends in Wet Deposition Data: Report of a Workshop*, Environment Monograph No. 4 (Institute for Environmental Studies, University of Toronto).

[39] Whitney, J. and Dufournaud, C. (1983), The impact of urbanization on the environment: an ecological input–output model, in R. White and I. Burton (Eds), *Approaches to the Study of the Environmental Implications of Contemporary Urbanization*, MAB Tech. Note 14, pp 83–95 (UNESCO, Paris).

Commentary

M. Gywnne

The links between monitoring and management whereby those who plan and those who manage environmental resources are supplied with a steady flow of appropriate, reliable field information upon which to base their decisions and actions are, as set out in this chapter, so obvious that it seems impossible for anyone to doubt the need. Yet those of us who have to design and implement such monitoring programs know that this is far from the case. Even now there are many people who say that there is no need to spend money on obtaining yet more information about what we already know. We have sufficient knowledge, these critics say, to enable us to act now. In the developing world, environment and resource monitoring are often viewed as an unnecessary ploy to delay development action, as well as a waste of good money that could be better spent on meeting more immediate practical development needs, such as direct measures to combat desertification. This attitude, however, is not the exclusive prerogative of developing nations. In a recent international forum on the effects of acidic deposition, for example, many from the Nordic countries were arguing along similar lines. In this case the cry was that we already know enough about the effects of acid rain upon lakes and forests so that there is no need to waste valuable time in seeking more information. Once again the call was for immediate action based upon what we already know.

Even among those who recognize the general need for monitoring activities, there are some who do not see why it is necessary to continue monitoring once management actions have been implemented. The need to monitor the effectiveness of management so that improvements in management strategies can be made is often not realized. Nor is the desirability fully appreciated of continuing surveillance monitoring in order to detect further unexpected changes in the environment or resource state.

That people still have these attitudes is, I think, probably the fault of those who are responsible for the monitoring programs themselves: too often monitoring is seen as an end in itself. Although the programs usually have clearly stated objectives, these objectives are often couched in terms suggesting that monitored data are only to help understand processes, to be used to develop models, or to indicate further research needs. In other words, the obvious practical benefits of monitoring programs are frequently not indicated so that those who are asked to participate in or support monitoring programs are not fully aware of the practical links to management and the resultant economic benefits and improvements in the quality of life.

The relationships between monitoring and assessment are also not fully appreciated. Yet it is when monitored data are evaluated in such a way that an assessment statement can be made on the condition and trend of particular environmental variables or natural resources that people most readily comprehend what is happening to their environment. They can thus better appreciate the value of the monitoring programs that produced this information and the practical benefits that can result. Without this understanding by both the informed public and the planners it is doubtful whether many essential monitoring programs will be able to continue.

The demonstration of the practical benefits that can arise, or are arising, from particular monitoring programs must be seen as an essential component in the design and implementation of those programs. This aspect is one that I feel is not adequately covered in this chapter though I think it is appreciated by the authors. It is an aspect that we and our colleagues in the United Nations agencies have learned the hard way, through designing and implementing the many and various worldwide monitoring networks that make up the Global Environment Monitoring System (GEMS) as we know it today. This lesson had not been learned in the early days and so the relevant elements were not given sufficient prominence in the initial GEMS networks, such as the European Monitoring and Evaluation Programme (EMEP) and the Background Air Pollution Monitoring Network (BAPMoN). All new GEMS monitoring programs have built-in periodic assessments and relatively frequent outputs of plain language statements on environmental conditions and trends. The recent WHO/UNEP report [1] on urban air pollution, based on the findings of one of the GEMS health-related monitoring networks, is an example of a plain language assessment statement that was given wide distribution in English, French, and Spanish, and led to immediate interest in further participation in the network.

The authors point out that costs may require a monitoring program to meet more than one set of objectives and consequently to monitor more than one set of variables at any given site. One-site, multiple-data-set monitoring is also at the heart of the new integrated monitoring approach as now used, for example, in some GEMS' temperate forest monitoring projects. Monitoring of this type may result in a reduction of both accuracy and precision with regard to individual variables. These reductions will have to be considered on a case-by-case basis, but they will, in general, be acceptable as long as the methods used allow the data collected from different sites in different countries to be intercomparable in a meaningful way. The importance of quality assurance procedures and regular intercomparisons of monitoring and analytical methods in this respect cannot be over emphasized. Without the consequent ability to compare monitored data in a meaningful way it is not possible to build reliable global, regional, and national pictures of the state of the variables concerned.

The environment is a complex composite of systems – physical, biological, and socioeconomic – and few management decisions can be taken without regard to them all. Traditional access to environmental data no longer meets the demands of planners faced with a world where the nature of environmental change is infinitely complex. With the development of computers that can handle and analyze the immense quantities of data that large-area environmental monitoring dictates, a comprehensive global data base is now possible.

The basis of this approach is a geographical information system in which various data planes gathered through inventory and monitoring can be related to each other and to a particular part of the Earth's surface – be it the whole globe or a local area. Data planes can include, for example, topography, soils, water resources, vegetation, fauna, land-use, climate, and atmospheric constituents, as well as various socioeconomic inputs.

Globally, such a system can be used to make statements about global trends, such as changes in climate and increasing desertification. Nationally, it can be used at any scale of planning from nationwide agricultural schemes to the siting of roads and human settlements. To make better use of the information held in its various data banks, GEMS has created such a system in GRID – the Global Resource Information Database. GRID enables various GEMS monitored data sets, combined with information from other sources, to be focused on the environmental and developmental problems of specific geographical areas, thus allowing more informed planning and management decisions to be made. This approach is a new and important aspect of environment and resource monitoring which I feel has not been given sufficient attention by the authors. It provides a major bridge between monitoring and management.

Limitations of space in this commentary prevent me from elaborating on two other points touched upon by the authors. First, historical monitoring – the obtaining of information on past environmental conditions from sources available to us today – is important for the proper understanding of what we see and measure in our current environment. I can do no more here than direct interested readers to a new GEMS report on this very neglected, but important, subject [2]. Second, economists and sociologists have not kept pace with scientists in deriving suitable and meaningful socioeconomic monitoring methods. Consequently, socioeconomic inputs to environmental monitoring and assessment procedures remain inadequate and unsatisfactory. This situation is changing, but slowly, so that it will be some years yet before the imbalance is corrected.

No single account of environment and resource monitoring can adequately deal with all aspects of the subject, so it is not surprising that this chapter does not cover some of them satisfactorily – renewable resource monitoring, for example. Nevertheless, the chapter does represent a useful attempt to put forward the principles behind the design and implementation of environment and resource monitoring programmes.

Note and references (Comm.)

[1] UNEP/WHO (1984), *Urban Air Pollution 1973–1980*, UNEP/WHO Report (World Health Organization, Geneva).

[2] MARC (1985), *Historical Monitoring*, MARC Report No 31 (Monitoring and Assessment Research Centre, London).

Chapter 14 Some implications of climatic change for human development

M. L. Parry

Editors' Introduction: Social processes of adaptation and adjustment are central to the interactions of human development and the environment. Relatively little is understood, however, about how such processes respond to long-term environmental changes.

This chapter is focused on the responses of agricultural development to long-term changes in climate. It is argued that such changes often increase societies' vulnerability to disruptions that may ultimately be precipitated by other events. (This is a specific example of the "slow variable" issue more generally discussed in Chapters 10 and 11.) In addition, however, it is shown that long-term changes in climatic means may exert some of their most significant social impacts through associated changes in the frequency of extreme events. The implications of such changes for the economic risks faced by individual farmers and for the resource opportunities available to individual regions are discussed.

Introduction

A recent report by the United Nations Environmental Programme (UNEP) on world environmental trends referred to three main current concerns with our climate: rising carbon dioxide (CO_2) concentrations, acidification of rain and snow, and stratospheric ozone (O_3) depletion [1]. These were part of a list of environmental changes which included loss of tropical forests, desertification, waterlogging, salinization, etc. It was curious to some, particularly from the so-called South, that the report and the meeting it addressed preoccupied itself primarily, at least in the field of climate, with what those of the South consider as problems of the North; that is, with anthropogenically caused changes of climate. To their minds the overriding concern should be the awesome hazard posed *now* by natural climatic variations, particularly in those countries least able to cope with them [2]. Their argument is leant some support by the continuing impact weather has on the modern (increasingly buffered and integrated) food system. The year 1972 is a favorite example: serious droughts occurred in the Sahel, USSR, India, Australia, and South America, with a million or more excess deaths in India and Bangladesh alone attributed to this bad-weather year [3]. Depleted grain supplies in several of the major food-producing areas led to increasing food prices and some famine. There was a 3% drop in total world grain production in 1972. World food stocks dropped by more than a quarter after 1971 to 134×10^6 tons in 1974 [4] and, although they recovered in the late 1970s to precrisis levels, the level today of grain security as a proportion of consumption remains below the 1969 level.

An argument could thus be adduced that the most pressing climatic issue today is existing short-term variability, not future medium- or long-term average change. This is given further weight by the fact that those regions of the world with the most variable climates are generally located near the Tropics – that is, often in countries with less developed economies which, on the face of it, are less able to deploy resources to make good any impact from unfavorable weather (*Figure 14.1*). Moreover, there are also indications that variability of production of certain food grains has increased over the last decade in many tropical countries [6]. At this point we should circumnavigate the related question (*viz.*, how far is level of development related to climatic resources), merely noting that there is a respectable amount of scientific literature on this subject and that here is yet another dimension of the climate–development issue: that climate as a resource varies greatly from one part of the globe to another and that this *climatic change over space* may be of greater consequence for human development than *climatic change over time*. In fact, in many countries and for a wide range of crops the present interannual variability of yields is greater than those predicted for a range of *future climatic scenarios* (from large cooling to large warming) in the National Defense University Study [6, 7]. Climatic hazards and climate-related agricultural pests and diseases are much more prevalent in tropical than in temperate climates. The loss of gross national product (GNP) to natural hazards is, on average, 20 times greater in less-developed than in developed countries [8]. Hence the view that the task of reducing the effects of this "great climatic anomaly" between North and South is a more pressing issue than that of preventing or reducing the threat of change to the *status quo* by (for example) an increasing atmospheric concentration of CO_2 [9]:

> While the fact and magnitude of [climatic change] are still uncertain, variability is always present and often high. Fortunately many of the policies that could prepare us better for the first contingency are also those that would provide some insurance against the second [6].

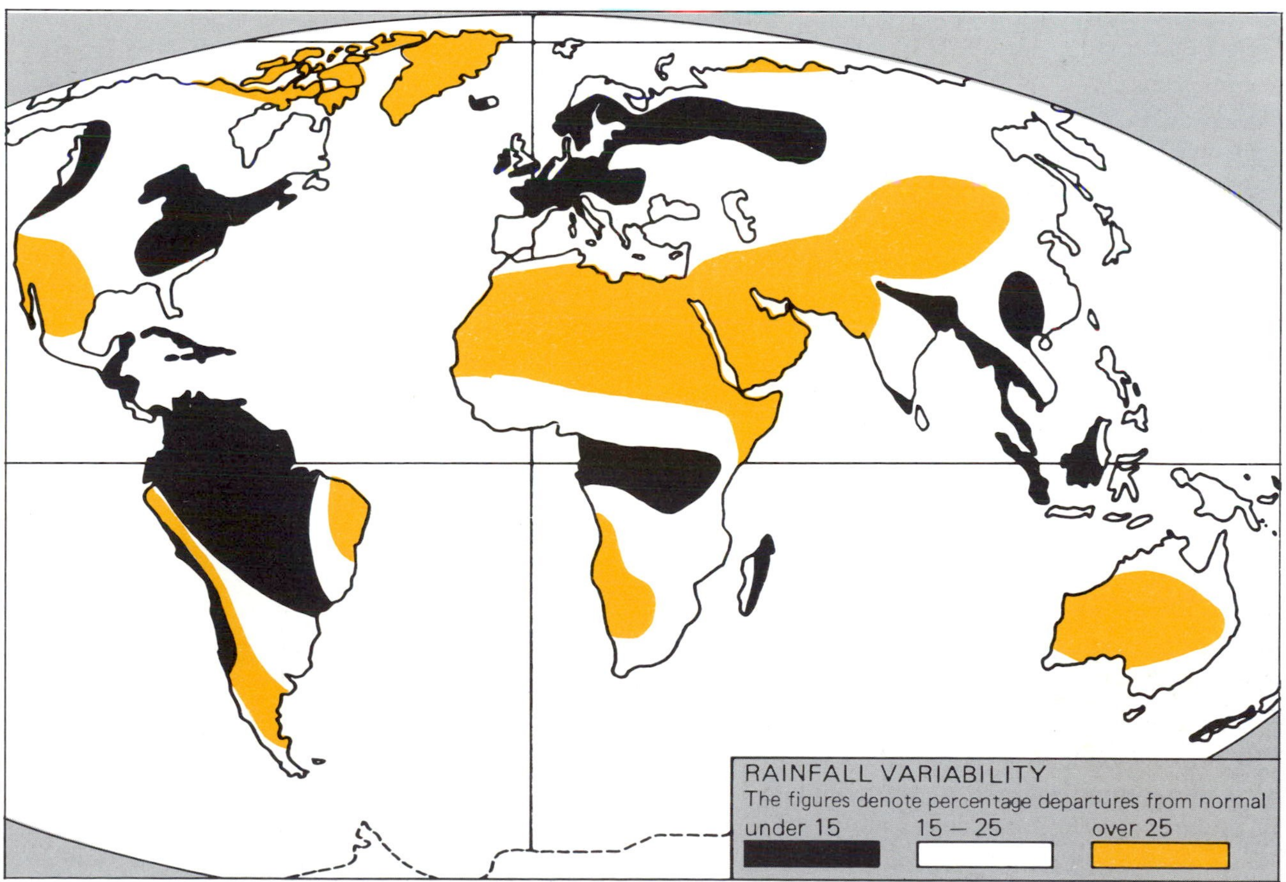

Figure 14.1 The global pattern of climatic variability, as indicated by the percentage departure of annual rainfall from the mean annual normal rainfall. Note that, in general, the most variable rainfall occurs near the Tropics (adapted from [5]).

Yet changes of climate have occurred over the longer term and are likely to occur in the future (see Dickinson, Chapter 9, this volume). I draw a distinction between medium-term climatic changes (occurring over 10 to 200 years) and long-term ones (occurring over more than 200 years). We thus have three types of climatic fluctuation: *climatic variability*, the observed year-to-year differences in climatic variables, to which economic and social systems generally *adjust*; *medium-term climatic changes*, to which society probably needs to *adapt* in order to avoid undesirable impact; and *long-term climatic changes*, which occur on time scales too large to be considered significant for the planning horizons of most societies.

It may seem a little unusual to classify climatic phenomena on the basis of their human impact, but if it is ultimately our intention to specify the role of climatic changes in human development, then it is helpful to attempt, at the outset, to integrate the approaches of social and natural science, and thus emphasize their complementary roles. In particular we should exploit the complementarity between direct (or causal) methodologies of the natural scientist and "adjoint methodologies" of the social scientist [10]. There has been a tendency to focus on the former, performing, for example, a sensitivity study by perturbing one input variable (such as CO_2 concentration) and tracing the effects (climate → agriculture → individual farmer → food prices → society). In the adjoint approach, for example, one might consider the perturbations, both climatic and nonclimatic, that influence farmers' decisions and perceptions, and trace those that have a climatic origin to a number of climatic variables and their relationship to the level of atmospheric CO_2 (society → food prices → individual farmer → agriculture → climate). Both approaches may be necessary to understand the full complexity of the interactions, but the adjoint approach has received less attention. Its advantage for those concerned with climatic impact is that, by evaluating perturbations caused by climate in relation to perturbations caused by other sources (technology, demand, etc.), the social scientist can produce outputs that can be expressed more readily in terms of policy.

In this chapter I emphasize *medium-term* changes of climate because these occur on time scales to which societies can respond. *Figure 14.2* illustrates a pronounced example of this – the dramatic medium-term change in mean annual temperature experienced in Iceland after 1920. I subsequently show that the economic effects of this were quite dramatic.

Long-term, or secular, climatic changes have occurred and no doubt have affected the resource base of societies in the past. But the task of disentangling those changes from the web of other variables, and to do so with such imprecise data, is daunting. Moreover, I am unsure what we could learn from such an exercise when we come to think about planning our own responses to climatic change. The time scales simply exceed our ability and imagination. The secular trend in *Figure 14.3* illustrates such long-term changes [12]: yet embedded within this trend are important medium-term fluctuations, such as that in the Northern

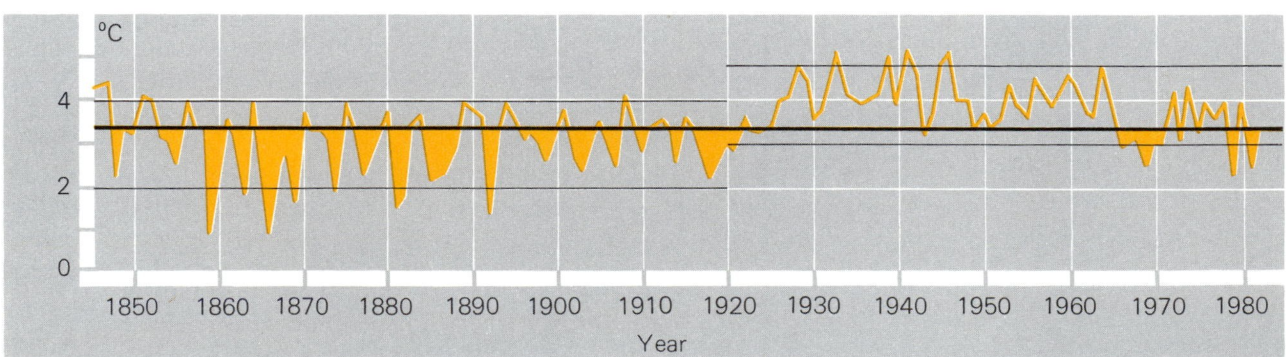

Figure 14.2 Annual mean temperature at Stykkisho'lmur, Iceland. Values below the 1846–1982 average are shaded yellow and extreme warm and cold years are delineated for two arbitrarily chosen periods. Extremes are defined as those with a return period of 10 years or more [11].

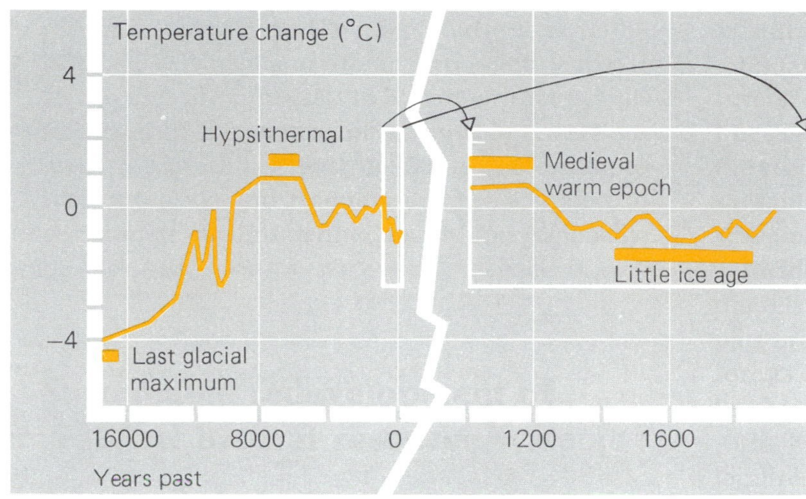

Figure 14.3 Schematic representation of temperature changes during the past 18 000 years [12].

Hemisphere between the so-called "Medieval warm epoch" and "Little ice age". In northern Europe, between these periods, mean annual temperatures were reduced by about 2 °C, with significant implications for agricultural potential [13].

Although it falls outside my remit implied by the title of this chapter, I would exclude consideration of short-term climatic variability only with reluctance. It is the major source of climatic impact today, and almost certainly will be in the future. More importantly, it is likely that *short-term impacts are the mechanism through which any longer term change is felt.* For example, the impact of a possible CO_2-induced warming on agriculture is likely to be felt as much through changes in the frequencies (or risk) of specific events (such as days of plant moisture stress or short moisture-limited growing seasons) as through the change in mean conditions [14]. Short-term impacts will thus be embedded in any long-term future change.

In general, I focus here on impacts on the biological resource base, that is on agriculture, fisheries, and forestry *potential,* primarily because such first-order effects are more readily measurable and less obscured by changes in technology, institutions, social values, etc. For the same reasons I also place as much emphasis on the possible effects of known, measurable climatic changes in the present or recent past as on those which may occur in the medium-term future, for example from an increase in atmospheric concentration of CO_2.

The questions I address are these: What assumptions underlie the study of climatic change and human development? What has been the recent progress in comprehending the array of relationships? Do we need to redefine the research problem into forms more amenable to solution and more appropriate for application in policy? What conceptual framework is best suited to future research? What are the research priorities? I try to answer these in turn, focusing largely on first-order impacts on food supply. In doing this, I do not mean to diminish the role of climatic change on the secondary and tertiary sectors (or on those direct impacts from other biophysical systems, such as from rising sea levels or from disease). No doubt these are important, but they are likely to be less so than impacts in agriculture, fisheries, and forestry.

Clarifying the assumptions

It is important to clarify some distinctions and assumptions that have tended to obfuscate the study of climatic change and human development. First, we must be clear about the time scale. Short-term climatic variations can generally be expected to have a short-term effect on society, unless they happen to trigger a response that has been preconditioned by other long-term factors.

Second, we can thus assume a distinction between (*a*) hypotheses of climatic change as proximate or precipitating causes of economic and social change and (*b*) those that are underlying or indirect causes of economic or social change. For example, we should be extremely wary of assuming a climatic "cause" when the weather may merely have *precipitated* impacts which were fundamentally economic, social, or political in origin [15]. What, for example, were the real "causes" of the Sahelian crisis in the mid-1970s – meteorological drought or enhanced vulnerability due to economic and political developments insensitive to an environment that has always been changeable?

This brings us to the third distinction: that between *contingently necessary* and *contingently sufficient conditions* for an occurrence [16]. It is probable that increased vulnerability was a necessary condition (or precondition) for the Sahelian disaster; without it the economic system might have been more resilient to the meteorological drought. It was not, however, a sufficient condition, because it required a further event (meteorological drought) to precipitate the effect. On the other hand, a broadly similar effect could have been triggered by a different short-term event, for example, a political upheaval. If this were the case, then climate was *neither* a sufficient *nor* a necessary condition for the Sahelian crisis, nor for many other events of which it may be deemed a precipitating factor [17]. It may, however, trigger a cascade of misfortunes which, in concert, have a lasting impact on the progress of human development. Such is the scenario portrayed in a recent study of the role of climate in the crisis facing Sumeria in the Tigris–Euphrates lowlands around the beginning of the first millennium BC, when drought may have decreased available water supplies and led, in turn, to increased salinization, and hence to reduced crop yields. As a result less food may have been stored, thus increasing the vulnerability of society to recurrent environmental impacts [18].

Finally, we should examine the set of four assumed relationships or functions by which we normally proceed to disaggregate the complexity of the connection between climate and society [19]:

(1) *Function 1.* Utility function (that is satisfaction in farming, fishing, etc.) that depends on:

(*a*) Personal characteristics.
(*b*) The probability distribution of financial performance.

Increasingly recognized as important here are attitudes to risk of climatic impact which may modify the traditionally assumed goal of maximizing expected profits.

(2) *Function 2.* The second general relationship is a mix of behavioral and environmental effects which are part of (1b). That is, a probability distribution of financial performance that depends on:

(*a*) Allocative decisions.
(*b*) The probability distribution of yields.
(*c*) The probability distribution of prices.
(*d*) Institutions and government interventions.

Both the probability distributions (2b) and (2c) may often be related to climate, and government interventions may be related to mitigating climatic impacts.

(3) *Function 3.* Probability distribution of yields that depends on:

(*a*) Climate.
(*b*) Allocative decisions.

The influence of climate on yields can be estimated either empirically or by simulating, for example, plant responses. Many of the allocative decisions [both in (2a) and (3b)] are influenced by climate, being intended to mitigate the effect of climatic variations.

(4) *Function 4.* Probability distribution of prices that depends on:

(*a*) Consumer demand (i.e., incomes, tastes, etc.).
(*b*) Institutions and government interventions.
(*c*) Supply and thus also climate.

Climate may act directly on demand and directly on prices.

Anderson summarizes the relationships as follows:

> Taken together, Functions 1 to 4 depict an agricultural sector directly dependent on climate .... Each of Functions 1 to 4 has a dual interpretation depending on whether it is being construed as a planning (*ex ante*) device or an observational recording (*ex post*) device. Broadly speaking the *ex ante* relationships can be characterized as risk (or uncertainty) and the *ex post* relationships as instability [20].

Thus climatic fluctuations can make themselves felt through several functions operating in both the agricultural and nonagricultural sectors. It is not simply a matter of modeling crop yields and tracing the downstream effects of their variations on the wider economy.

Preliminary models

To organize this array of relationships into an analytical framework a number of conceptual models have been developed, which have generally become more complex in recent years, allowing for an increasing variety of feedback processes to be described. In this section, I outline the recent development of these models and illustrate their use with one or two examples.

Impact models

Impact models are graphic representations of those types of study methodology which view climate as causing impacts in a more or less linear, noninteractive fashion [21]. In some cases the mechanisms are treated as a single unordered set, the impacts (effects) being immediately contingent to the climatic event (cause). Little account is taken of intervening factors.

A basic model of this kind characterizes the hypothesis of climatically induced settlement change as an explanation for the decline of the Mycenaean civilization between about 1200 and 900 BC [22]. The abandonment of Mycenaean palaces and evacuation of settlements in the south Peloponnese has generally been attributed to invasion by the Dorians from the northeast. But the Dorians might have been moving into an area that had already been depopulated by outward migration – an evacuation induced by famine caused by recurring drought; whatever the reasons, there is archaeological evidence that the Mycenaean settlement shifted to the northwest Peloponnese.

Any weakening of the polar high pressure at this time would have caused a more northerly track of the Atlantic depressions which, generally, bring spring and autumn rain to the Aegean. If this had occurred, then the only rainfall that would have continued to be reliable would have fallen on the western Peloponnese. The original Mycenaean settlements could have suffered persistent drought, and responded by migrating.

In archaeological and documentary terms, there is little evidence to support this hypothesis of climatically induced migration: and the alternative hypothesis of military overthrow is not refuted [23]. Moreover, little evidence exists either for the climatic changes *per se* or for their possible effects in changing vegetation in the region [24]. However, an analysis of patterns of winter climate in the south Aegean has revealed that a type of hemispheric circulation characterized by patterns consistent with that proposed for the Mycenaean drought occurred during the period November 1954 to March 1955, and can be used as an analogue of the dry weather postulated for the late Mycenaean period [25]. The drought pattern could, therefore, have occurred – but did it occur at that time and over several consecutive years? Can we specify in much greater detail the supposed connections between meteorological drought, changes in the agricultural resource base, and the shift of settlement?

Similar simple causal models characterized earlier excursions into the study of climate and cultural history, studies now associated with the philosophy of determinism [26]. In 1907 Ellsworth Huntington argued for the existence of climatic "pulsations", which periodically drove the nomadic peoples of central Asia to the fringes of the sedentary world in Europe and southwest Asia [27]. He also suggested that Mayan migrations in central America and the desertions of Roman settlements in Syria were in part a consequence of climatic shifts [28]. These ideas were subsequently rejected and it became fashionable in the 1940s and 1950s (the height of the possibilist movement in geography [29]) to ridicule Huntington's work. Today, particularly with the renewed interest in links between cultural history and environmental change a more reasonable judgment might be that Huntington's naive conclusions stemmed from insufficient data and inadequate analytical techniques rather than from unpromising hypotheses.

Studies of the cyclical connection between climate and history still have a toehold in science, but generally lack sufficient rigor to be convincing. An

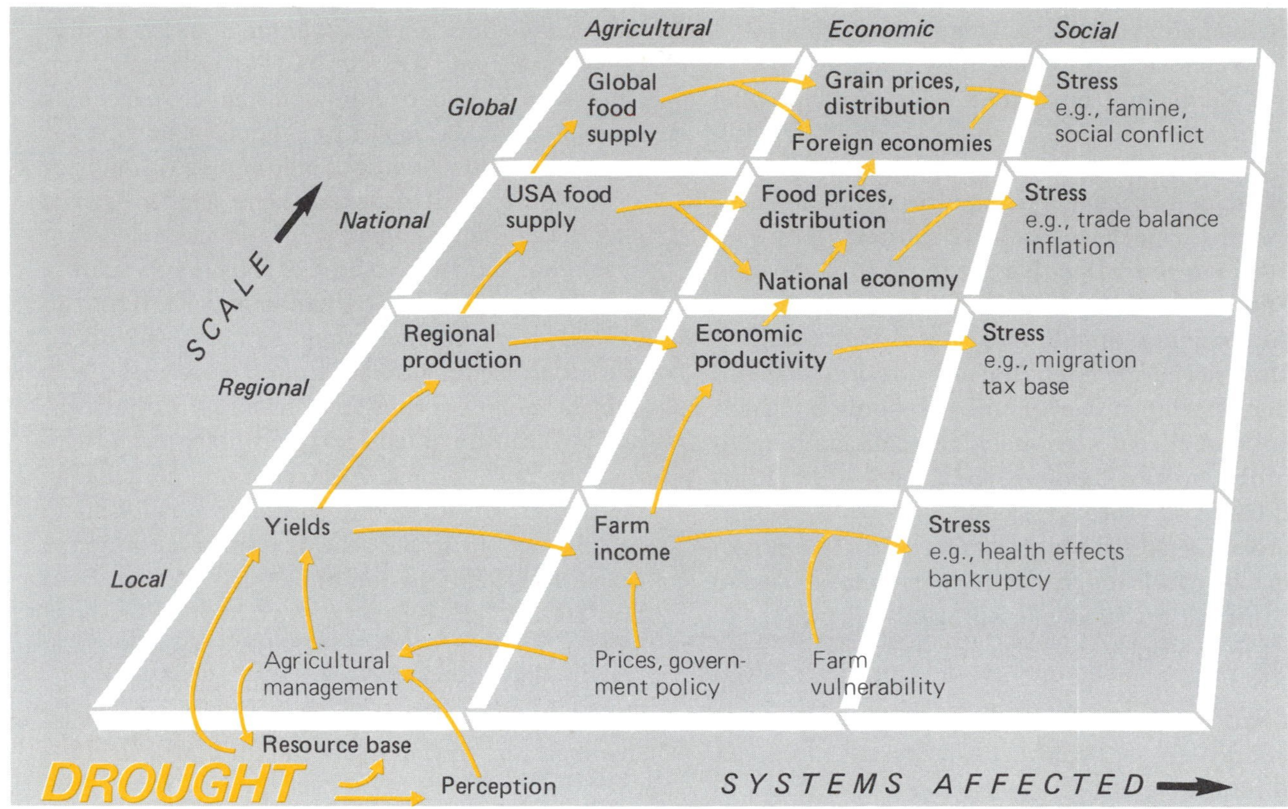

Figure 14.4 The hypothetical pathways of drought impacts on society [32].

example is the hypothesized relationship between a supposed 80-year temperature cycle, the rate of population increase, and the degree of social disturbance (measured by a weighted index of historical chronologies) [30]. More promising is a recent exploration of the connection between weather and "long swings" in the agricultural sector [31].

More detailed studies of climate–society interactions attempt to trace the effects of the climatic event as they cascade through physical and social systems, meeting other sets of intervening factors which disguise and modify the link. A specific example is the work by Warrick and Bowden, tracing the impacts of drought occurrence in the US Great Plains [32]. They traced a variety of pathways that drought impacts could take, spanning a variety of spatial scales (from local to global) and a variety of systems (from agricultural to social). From the source of impact in the lower left of *Figure 14.4* the pathway can be traced from the first-order (direct) physical impact to the higher order (less direct) effects on society. What starts as meteorological drought becomes agricultural drought, and subsequently, perhaps, a perturbation in the wider economy.

Interactive models

A greater degree of realism can be introduced by considering adapation and adjustment to climatic change. By *adjustment* we mean the short-term conscious response to a perceived impact or risk of impact (for example, building a dam to store additional water for a drought period). By *adaptation* we mean the long-term, often intuitive (subconscious) response to an environment which embodies many sets of risks and opportunities (for example, a system of slash-and-burn agriculture which, over the years, has matched farming activities in different parts of the world to the local agricultural resource base) [33]. These two types of interactions allow us to consider, first, a whole suite of other factors, such as the *vulnerability* or *resilience* of different systems to impact, which in turn affect how societies can (and do) respond to climatic change. These are considered elsewhere in this volume by Holling (Chapter 10) and Timmerman (Chapter 16). Second, these interactions introduce the issue of perception, that is, the cognition of potential or actual impact, the risks or benefits involved, and the opportunities to avoid or take advantage of them. To illustrate, faced with the task of defining a set of researchable problems, Kellog and Schware simplified the CO_2 problem into a hierarchy of issues [34]:

> a. For a given scenario of fossil fuel use, determine the future levels of CO_2 concentration in the atmosphere.
> b. For a given scenario of CO_2 concentration, determine the resulting climatic change in terms of regional and seasonal patterns of temperatures, precipitation, and soil moisture, among other variables.
> c. For a given scenario of polar-region climatic change, determine what will happen to the volume of the major ice sheets and the subsequent effect on sea level.
> d. For a given scenario of regional and seasonal climatic change, determine the effects on specific activities such as agriculture, land use, water resources, industry, transportation, and energy requirements on a country-by-country or region-by-region basis.
> e. For a given scenario of specific socioeconomic effects, determine the probable or desirable responses of societies.
> f. Along with the preceding set of considerations, determine the influences on earlier aspects of the problem of feedbacks resulting from implementing alternative strategies to mitigate the impacts or, perhaps, to avert the climatic change.

Yet by (realistically) including the last of these tasks the dimensions of the research problem are expanded enormously. A suite of *ex ante* relationships are now a part of the research problem:

(1) How to measure the risks from increasing CO_2?
(2) How to identify the alternative possible responses to these risks?
(3) How to evaluate these responses?

For a second, and somewhat contrasting, illustration of the importance of perception in the climate → society and society → climate relationship we turn back the calendar almost 1000 years – to those Norse settlements that gained a foothold along the Greenland coast in *ca.* AD 985, but were extinguished by AD 1500, probably between 1350 and 1450. In the thirteenth century the population, which then totaled about 6000 in two settlements on the south and southwestern coasts of Greenland, was subjected to a synergistic interaction of stresses from hostile Eskimos (*Inuit*), the decline of the European market for walrus ivory, and challenging climatic fluctuations, particularly during *ca.* 1270–1300, *ca.* 1320–1360, and *ca.* 1430–1460:

> Had these factors not coincided when they did, Norse Greenland very probably would have survived the fifteenth century and might well have endured to the present in some form. However, this sort of explanation treats human response to climatic stress as a minor and dependent variable . . . we must consider not only the nature of the external stresses that seemed to have killed Norse Greenland, but also the reason for that society's selection of ultimately unsuccessful response to such stress [35].

McGovern concludes that failure to adapt to the changing circumstances explains much of the Norse decline. The Norse continued to emphasize stock raising (cattle, sheep, and goats) in the face of reduced capacity of the already limited inner-fjord pastures. The options to improve their hunting skills and to exploit the rich seas around (just as the Inuit successfully did) were not taken up. One reason for this may have been simply poor management by the elite in a highly stratified religious society:

> Baldly put, a society whose administrators (as well as its peasants) believe that lighting more candles to St Nicholas will have as much (or more) impact on the spring seal hunt as more and better boats is a society in serious trouble [35].

In a successful society we would expect adaptive strategies to be in constant interchange with the environment, allowing a continuous identification of new risks and potential benefits within that environment, and incorporating these into revised strategies that preserve and propagate the more successful variables of the system. This adaptive system can be visualized as the three-dimensional intersection described by social behavior, technology, and resource opportunities (*Figure 14.5*) [36]. If McGovern is right, the Norse society had surrendered some of the adaptive abilities that it displayed 300 years earlier at the time of settlement to a hierarchical social structure that may

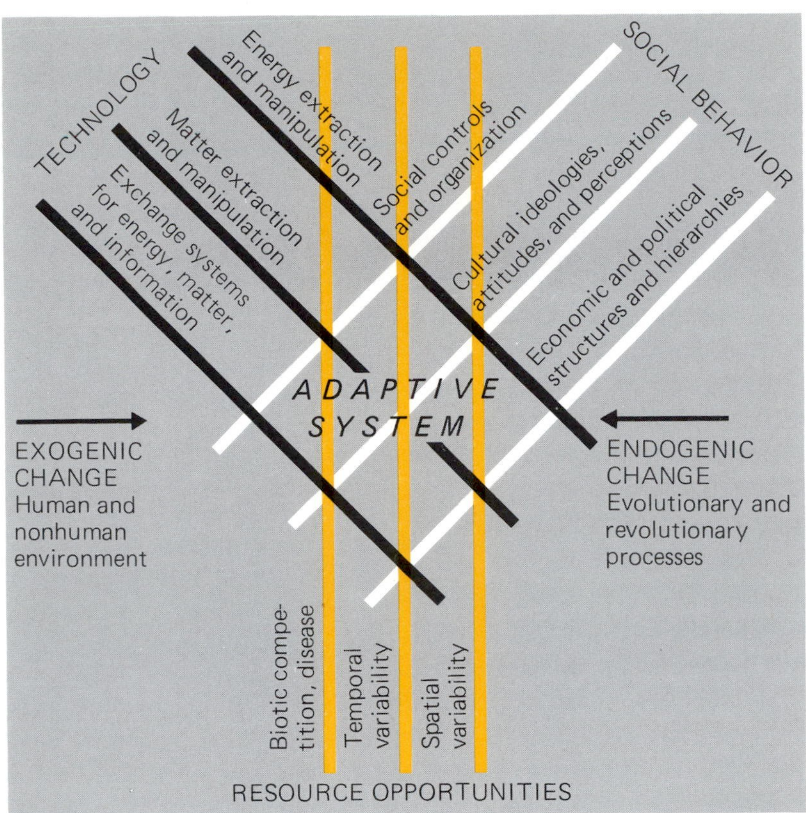

Figure 14.5 A three-dimensional model for the interactive variables of an adaptive system [36].

have been appropriate in Demark but was much less so in Greenland.

This brief examination of adaptation and adjustment points to two conclusions. First, most social and economic systems have adapted, perhaps slowly and over many years, so that they can absorb a certain range of exogenous perturbations (for example, short-term climatic variations) without a major disruption of the systems' normal behavior. These adaptations are probably based on the experience of perturbations in the past and are presumed to accommodate a certain perturbation range in the future. The entire range is unlikely to be accommodated because the cost of insuring against the least likely events is greater than the full (unmitigated) cost of the impact from those events. For example, farming systems in a drought-prone region may be adapted to accommodate droughts of a certain degree of severity and frequency. But more severe droughts are considered sufficiently infrequent to be ignored. Thus, Saarinen found that the farmers on the US Great Plains tended, in their farming strategies, to ignore droughts with a long return period [37].

It is fair to assume that agricultural and other systems of a successful society have developed to accommodate a major part of the expected range of climatic anomalies. Less rare and less extreme anomalies are absorbed without major shock to the system. However, our second conclusion is that the rare and extreme event (for example, those occurring with a probability of 0.05 or less) may exceed the resilience of the system, and may cause a major perturbation that can cascade into other subsets of the system (e.g., wider sectors of the economy), which would not normally expect to experience such a perturbation and are not adapted to absorb them. A change in the central tendency of climate or its variability by altering the frequency of the extreme events can thus upset the pattern of adjustment. For example, *Figure 14.2* illustrates for Iceland the change in frequency of extremely warm or cool years between two arbitrarily chosen periods, and the range of adjustment theoretically required to maintain resilience to extremes with a return period of not more than 10 years. To optimize its use of climatic resources the Icelandic economy (essentially limited to farming and fishing) would have had to respond over about 30 years to quite marked changes in the frequency of extremes.

Redefining climatic change

If this analysis is correct, the impact from climatic change comes not only from the average, but also from the *extreme event*. Thus, even if there is a change in the central tendency of the climate, an important medium through which this change is felt may be through a change in the *frequency of the extreme anomalies*, such as the number of years of particularly difficult weather that, for example, reduce food supplies, raise prices, and bring scarcity or (conversely) bumper years that flood the food market, drop prices, and reduce farm incomes to the point where the planting area in the following year can be affected. This has important implications for the CO_2 question, because one of the obstacles to active government interest in the impact of possible future climatic change rather than of present-day climatic variability is government's overriding concern with the short term rather than the long term. In general, the concern is with impacts from short-term anomalies such as floods, droughts, and cold spells. This suggests that a useful form in which long-term climatic change can be expressed to the policymaker is as a *change in the frequency of such anomalies*. A further advantage of this approach is that the change can be expressed as a *change in the risk of impact*. Government programs could then be devised to accommodate specified, tolerable levels of risk by adjusting activities as necessary to match the change of risk. Finally, climatic change can also be viewed as a *change in resource opportunities* where increasing risks close some doors, but reduced risks open others elsewhere. I deal with each of these approaches in turn.

Interpreting climatic change as a change in the frequency of extreme events

One of the critical issues concerning the impact of possible CO_2-induced climatic changes is how costly it would be for present economic systems to adapt. Any answer to this question presumes a knowledge of how, in general, society adapts to environmental change. At present it is fair to say that we do not know. One possibility is that society gradually adapts to slow cumulative tendencies in the perceived average. That is a comforting thought, particularly when changes in mean tendencies of climate might reasonably be expected to change slowly with steady increases in atmospheric CO_2; and this has probably encouraged thinking about matching rates of technological change (for example, of crop breeding) to rates of climatic change [38].

Our alternative hypothesis, however, is that impacts come not from the average, but from the extreme event – and that society is more likely to respond to its perception of a *change in frequency* of these extreme events [39]. To illustrate, few farmers (whether commercial or subsistence) plan activities on their expectation of the average return. They either gamble on good years or insure against bad ones [40]. More risk-prone farmers tend to tune their activities to bad years, commercial farmers (such as those on the Canadian prairies) to good years. The prairie community is thus periodically stressed when the reverse conditions prevail [41].

The impact from extremes can alter according to three types of climatic change. To illustrate this, let us assume that annual rainfall or temperature has a Gaussian (normal) distribution about the mean $\bar{x}$ and variance σ^2. We also assume that agriculture is only seriously affected if rainfall or temperature is outside the range of $(\bar{x} \pm k)$ [*Figure* 14.6(*a*)]. In the first type, $\bar{x}$ changes but σ^2 remains unchanged [*Figure 14.6*(*b*)]. The frequency of negative extremes (as defined originally by the agricultural system which we presume to be slow to adapt to the new climate regime) is increased, and that of positive extremes is reduced. If this change occurred for rainfall on the Canadian prairies we might expect the bad years to far exceed the good years.

In the second type of climatic change, σ^2 changes but $\bar{x}$ does not [*Figure 14.6*(*c*)]. There is increased risk of both deficient and excessive temperatures and/or rainfall. Both $\bar{x}$ and σ^2 change in the third type [*Figure 14.6*(*d*)] [42].

Such changes feature frequently in the instrumental records for all but the most stable climates. For example, the rainfall record for 22 stations in India (*Figure 14.7*) indicates, for 1890–1919, a

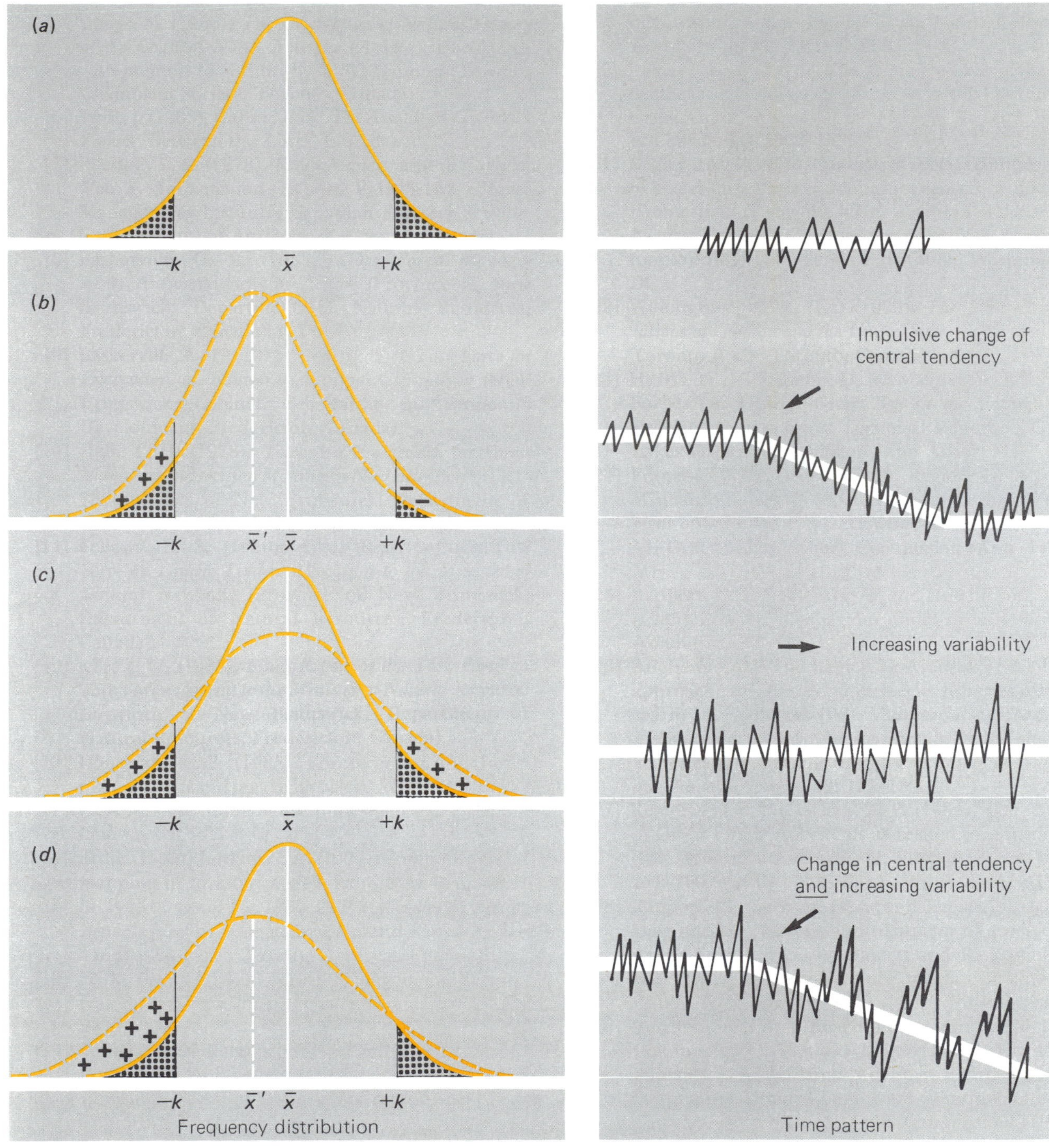

Figure 14.6 Three types of climatic changes with respect to the vulnerability of agriculture. Shaded areas show the frequency of extreme events to exceptionally high (+) or low (−) yields [42].

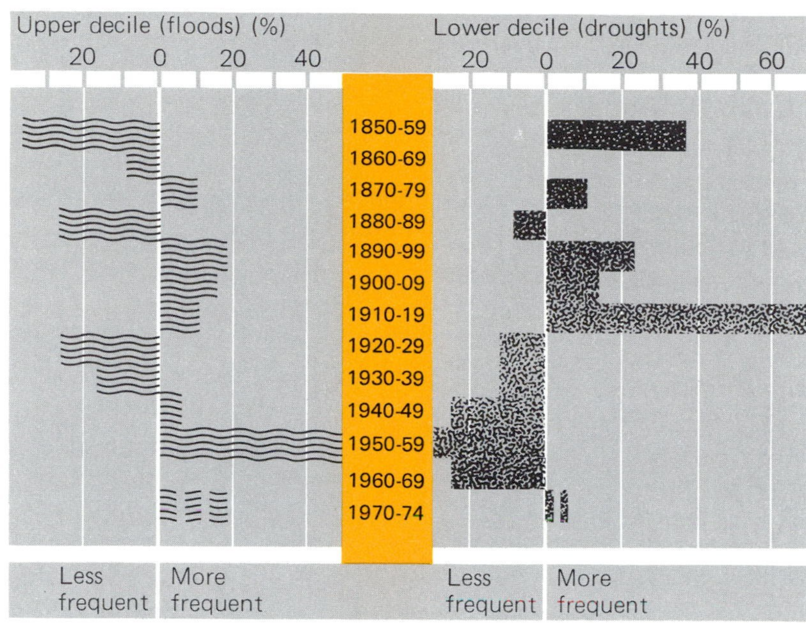

Figure 14.7 Extreme summer monsoon rainfall in India (22 stations), June–September [43].

period rich in extremes of both floods and droughts, but for 1920–1939 a period with a few extremes (i.e., the second type of climatic change). In the 1950s a high frequency of floods was accompanied by a low frequency of droughts, while the reverse was true in the 1850s (the first or third type of climatic change) [43].

Interpreting climatic change as a change in level of risk

If social and economic systems are tuned to the frequency of likely impact from extreme climatic events, it is logical to assume that human activities are being matched to a perceived level of climatic risk. One way of evaluating climatic change in human terms is therefore to consider it as a *change in level of risk*, that is, in the probability of an adverse or beneficial event, such as shortfall from some critical level of output or excess above the expected yield. In agriculture, for example, we might assume that farmers are entrepreneurs whose activities are based upon the expected return from their gambling on a certain mix of good and bad years. This has important implications for the impact of climatic change for two reasons. First, because broadly linear changes in mean temperature and precipitation (either over time or between one place and another) have strongly *nonlinear aspects* when redefined as the probability of occurrence of a certain anomaly. For example, let us suppose that extremely cold winters or dry summers occur with a probability of $p = 0.1$. The return period of a single extreme is 10 years, while the return for the occurrence of two consecutive extremes is 100 years (assuming independence and a normal distribution of frequencies). Any change in climate will lead to a change in p, either through changing variability, which will change p directly, and/or through a change in mean conditions, which must also change p if extremes are judged relative to an absolute threshold. Alternatively, p may change through changes in some critical impact threshold as a result of land-use changes, new crops or crop mixes, increasing population pressure, and so forth. If p becomes 0.2, then the return period for a single severe season is halved to 5 years. The return period for consecutive severe seasons, however, is reduced by a factor of four to only 25 years. Thus, linear changes in climate can cause quasi-exponential changes in the probability of crop failure or, conversely, bumper harvests [44].

The second reason why change of risk may be crucial is that, for many activities, these levels of risk are already critical. This is self-evident in the case of marginal farmers whose risks are already high. Amongst nonmarginal farmers, however, change of risk is also likely to have a similar quasi-exponential effect on the risk of one type of farming (crop type, mix of crops, and livestock, etc.) being more profitable than another, i.e., on the *comparative advantage* which one farming type may have over another. We may conclude that *surprisingly large changes in farming types may be necessary to accommodate apparently small changes in average climate.*

An increase in average temperature would thus cause substantial increases in the probabilities of high extremes of temperature which, in regions like the US Corn Belt where maize can be close to its maximum-temperature tolerance limits, can have significant deleterious effects on crop growth and final yield. For example, a 1.7 °C increase in the mean temperature at Des Moines (Iowa) would increase threefold the likelihood of occurrence of a run of five consecutive daily maximum temperatures of at least 35 °C [45].

One means of developing scenarios for a high-CO_2 warm world has been to base them on climatic patterns prevailing in analogous circumstances, either in the distant past (from paleoenvironmental data [34]) or in the historical period (from the instrumental climatic record [46]). Bergthorsson has suggested a similar approach in assessing the *impact* of a CO_2-inducing warming on the Icelandic economy [11]. The quite sudden warming around 1920 resulted in mean annual temperatures some 1.2 °C higher for 1931–1960 than they were for 1873–1922. What impacts would a similar increase over present temperature have on today's Icelandic economy? We could expect substantial increases in hay production on improved pastures, in carrying capacity of the unimproved rangelands, and in production of sheep meat, with reduced spending on imported feed substitutes. All this would significantly alter Iceland's balance of payments.

Translated into the frequency of extreme events, or of certain levels of production contingent upon such events, the effects of a recent warming, such as that in Iceland, can be surprising. Consider, for example, the warming in northwest Europe between about 1660 and 1980. If we compare the cool 50 years (1661–1710) and the warm 50 years (1931–1980) on the long instrument-based temperature record for central England from 1659, the difference in mean annual temperature is 0.75 °C. That difference lies well within the range of warming at 60 °N predicted by most general circulation models for a doubling of atmospheric CO_2. Changes in occurrence of extreme events between the two periods thus enable an estimate to be made of similar changes that might occur as a result of increased CO_2 (though we should not automatically assume that the same relationship between mean and variance would hold in a high-CO_2 world).

One major change between 1661–1710 and 1831–1980 has been in the probability of crop failure in northern and upland Europe due to inadequate summer temperatures. At locations near the present upper limit of cereal cropping this probability has been reduced tenfold (from 0.2 to 0.02). Thus the risk to cropping posed by crop failure has altered markedly with climatic warming in the past three centuries; and it is quite probable that a broadly similar alteration in risk levels would occur with a CO_2-induced climatic warming [47].

Interpreting climatic change as a change in resource opportunities or economic options

Implicit in the assumption that a change in risk affects economic activity is that perception of the risk change affords information about changes in resource opportunities or economic options. Since, in agriculture at least, climate can reasonably be construed as a resource, climatic change can produce benefits or disadvantages that may require an adjustment to match altered resource levels. One important path of these impacts is through range of choice:

> Changes in climate can alter the range of options that may compete for investment of time, money, and other resources. Moreover, the *perception* of these changed options is often important because the timing of investment in relation to weather can significantly influence the return on that investment. For example, the timing of farming operations (ploughing, sowing, harvesting, etc.) frequently explains much of the variation in yields from farm to farm at the local level. Changes in climate might tend to enhance the mismatch between weather and farming operations because of a lag in management response to changes in, most importantly, the "time windows" for planting and harvesting. For this reason, crop selection is probably one of the most effective means of response to an adverse climatic change, for the development of new strains or the

introduction of new crops can serve to keep open these time windows sufficiently to allow adequate yields to be maintained [47, emphasis in original].

The task of capitalizing on an altered array of options is quite complex. To simplify matters, let us assume:

(1) That all farmers wish to maximize their income (rather than some of them, say, wishing to minimize their risk).
(2) That the average climate is unchanged, only the weather varying from year to year.

Our farmer (in, say, the US midwest or USSR Ukraine) must decide whether to plant wheat or barley, the yields of which are most affected by rainfall in the growing season [48]. The situation is depicted here as a decision tableau (*Figure 14.8*). The farmer receives a return of r and u for wheat and s and t for barley under wet and dry years, respectively. His estimates of the probabilities of each of those types of year will affect his decision regarding the optimum cropping mix.

EVENTS / ACTIONS	Above average precipitation (A)	Below average precipitation ($\bar{A}$)	Expected returns (ER)
Plant wheat (W)	r	u	$ER(W) = p_0 r + (1 - p_0) u$
Plant barley ($\bar{W}$)	s	t	$ER(\bar{W}) = p_0 s + (1 - p_0) t$
Prior probabilities	p_0	$1 - p_0$	

Figure 14.8 The decision tableau, including the expected returns associated with the actions under prior information, for the agricultural decision-making situation presented in the text [48].

In reality the picture is complicated in three ways:

(1) The real (i.e., instrument observed) probabilities of types of year can change from decade to decade.
(2) Different farmers have different attitudes toward risk, thus affecting their optimum cropping mix.
(3) Different farmers have different estimates of the probabilities of types of weather-year, thus affecting their response.

Any of these three factors can change the optimum farming strategy quite markedly. To illustrate, let us assume that farmers in central India seek an optimum mix of two crops – sorghum and maize. The payoff from these crops varies, with sorghum faring better in dry years and maize in wet. A hypothetical payoff–climate function for the crops is given in *Figure 14.9.*

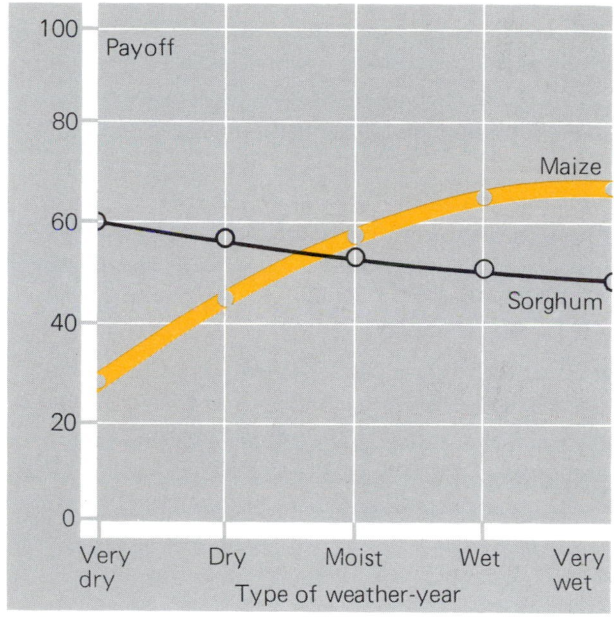

Figure 14.9 Hypothetical payoff versus climate functions for maize and sorghum; for explanation, see text.

Now let us assume that there are four types of farmer:

(1) Those wishing to maximize average payoff over the medium term (i.e., over 10 years), regardless of short-term variations in payoff.
(2) Those wishing the same, but having regard to short-term variations, ensuring that no annual payoff is less than two thirds of the maximum average.

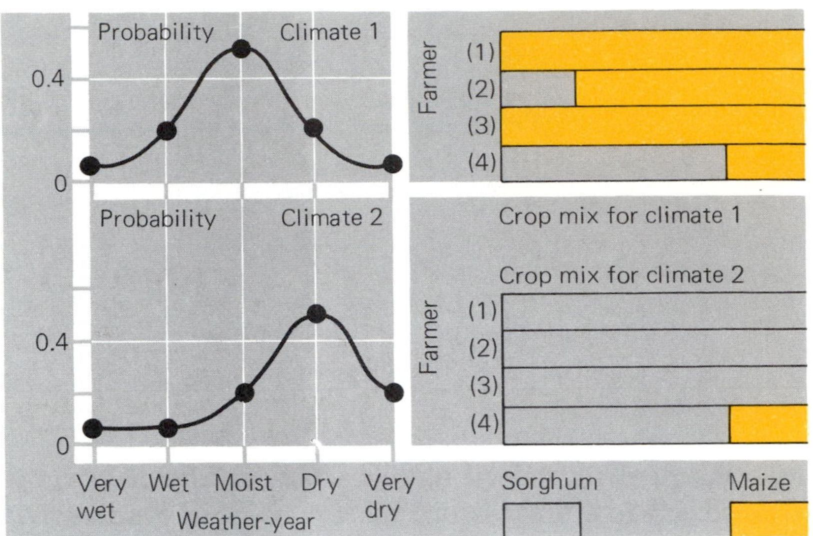

Figure 14.10 Cropping strategies and climatic change.

(3) Those acting as (2), but ignoring in their calculations low-probability extremes (i.e., 0.05 or less).
(4) Those wishing to maximize the minimum payoff in the worst possible year.

From a payoff matrix we can identify, for a given climate, the optimum crop mix for each type of farmer (*Figure 14.10* and *Table 14.1*). For climate 1, farmer (1) would choose a maize and sorghum mix of 100:0, farmer (2) 70:30, farmer (3) 100:0, and farmer (4) 20:80. Thus, for the same climate, farmers with different goals and different attitudes choose, not surprisingly, different strategies.

Now let us assume (*Figure 14.10*) a spell in which dry years are more frequent, the sort of change that might occur over a 10- or 20-year period, as for monsoon rainfall in India (*Figure 14.7*). In seeking to minimize his risks, farmer (4) has, in fact, chosen the strategy that is also most resilient to the hypothesized climatic changes: he can maintain an unchanged crop mix and still achieve his optimum payoff. In contrast, the others must change their crop mixes quite radically if they are to maintain what they regard as an optimum payoff.

To sum up: apparently small climatic changes (particularly changes in mean tendency of climate) can have marked effects on the frequency of extreme events, on levels of risk, and on the pattern of optimum response strategies. If, for example, those farmers seeking to maximize *average* payoff continued, under the drier climate we have hypothesized, to pursue their concentration on

Table 14.1 Payoff for different crop mixes in different types of year.

	Maize: sorghum crop mix										
Type of year	0:100	10:90	20:80	30:70	40:60	50:50	60:40	70:30	80:20	90:10	100:0
Very dry	61.0	57.9	54.8	51.7	48.6	45.5	42.4	39.3	36.2	33.1	30.0
Dry	58.0	56.9	55.8	54.7	53.6	52.5	51.4	50.3	49.2	48.1	47.0
Moist	55.0	55.4	55.8	56.2	56.6	57.0	57.4	57.8	58.2	58.6	59.0
Wet	52.0	53.5	55.0	56.5	58.0	59.5	61.0	62.5	64.0	65.5	67.0
Very wet	49.0	51.2	53.4	55.6	57.8	60.0	62.2	64.4	66.6	66.8	71.0
Climate 1											
Mean payoff	55.0	55.2	55.5	55.7	55.9	56.2	56.4	56.6	56.9	57.1	57.3
Climate 2											
Mean payoff	57.3	56.3	55.4	54.5	53.6	52.7	51.8	50.9	50.0	49.1	48.2

maize they would, in fact, have achieved the lowest average payoff possible. Thus, the impact of climatic change on those who either do not perceive it, or who perceive it and choose not to adjust, can be surprisingly severe.

Toward a framework for considering the impact of climatic change

Can any large-scale, long-term patterns and relationships be seen to connect climatic change and human development? For the present, at least, we can borrow a term from Scottish law which allows a verdict of *Not Proven* (i.e., although there is probably a case to answer, there is insufficient evidence to convict), and it is uncertain whether we shall ever have sufficient evidence simply because the concept of "proof" in history is not the same as in physical sciences. We can say with certainty that history is played out on a fluctuating stage, fluctuations to which the successful actors have had to be extraordinarily responsive. This is a function of the "endless variability of climate on all time scales [49]." But to be certain that long-term climatic changes have had a lasting impact on the development of human affairs, we must have an answer to the "What . . . if" question: *What* would have been the path of human development *if* the climate had not changed? Can we, with confidence, put forward one single major train of events in human affairs for which we have the answer?

Let us consider a conditional counterproposition: that if there had been no climatic change over the past millennium, then the history of modern man would not, in any general sense, have been different. Following this argument, some historians conclude that:

> There are no explanatory lacunas [*sic*] which need to be filled by reference to the presumed effects of (long-term) climatic deterioration . . . the most fruitful approach would appear to be the investigation of the impact, and short-run implications of climatic shocks experienced by the productive system [50].

This returns us to an earlier conclusion: that short-term shocks, embedded in long-term change, are an important mechanism of climatic impact. Thus, even if we are only interested (as in the present context) in long- or medium-term changes at quite large spatial scales, we need to comprehend the short-term mechanisms of impact if our policies of response are to be appropriate. In the remainder of this chapter I discuss, first, methods available for measuring sensitivity to climatic impacts and, second, the range of adaptations and adjustments that are suggested by such measurements.

Estimating sensitivity to climatic change

Two broadly different approaches have been taken. One approach is the use of detailed semidescriptive case studies, concentrating on particular periods of apparently marked climatic impact on particular economies and societies [51]. The abundance of detail and the attention often given to large numbers of variables, in addition to imperfections of the data, frequently mean that rigorous modeling is not feasible. But such studies may yield hypotheses that can later be tested more rigorously. An illustration of this approach is Garcia's case study of the 1972 Sahelian drought, which seriously questioned the premise of meteorological drought as the main cause of the crisis and which, by implication, has brought into question the array of assumptions by which we have, rather simplistically, linked climate and society [52].

Although the descriptive approach is often the best-suited to historical enquiry, some case studies of climatic impact have been based on inductive forms of reasoning and incorporate the weaknesses of inductivism [53]; that is, they lack an unbroken logic to establish the relationships between the various propositions linking climate and development and have not sought to test the propositions either empirically or theoretically. An alternative is to construct a model of the climate–development interaction and to test this model against historical actuality using time series data of chosen dependent (i.e., impact) and independent (i.e., climatic) variables [54].

If we harness to this approach the prediction of *isopleth shifts* we have a method that can help identify areas affected by climatic change or variability. It focuses on the shift of limits or margins representing boundaries between arbitrarily defined classes, the classifications being of vegetation, land use, yields, and so on. In this sense, the boundaries delimit zones on maps that can undergo a spatial shift for a given change of climate, thus defining impact areas. An example of this method is illustrated in *Figure 14.11* [47] where the impact of climatic change is described in terms of the resultant change in the probability of harvest success or failure. The weather for a number of years, described by a set of meteorological data, can be expressed as a probability of risk or reward using an appropriate model. When calculated for a number of stations this probability level can be mapped geographically as an isopleth. Scenarios of changing

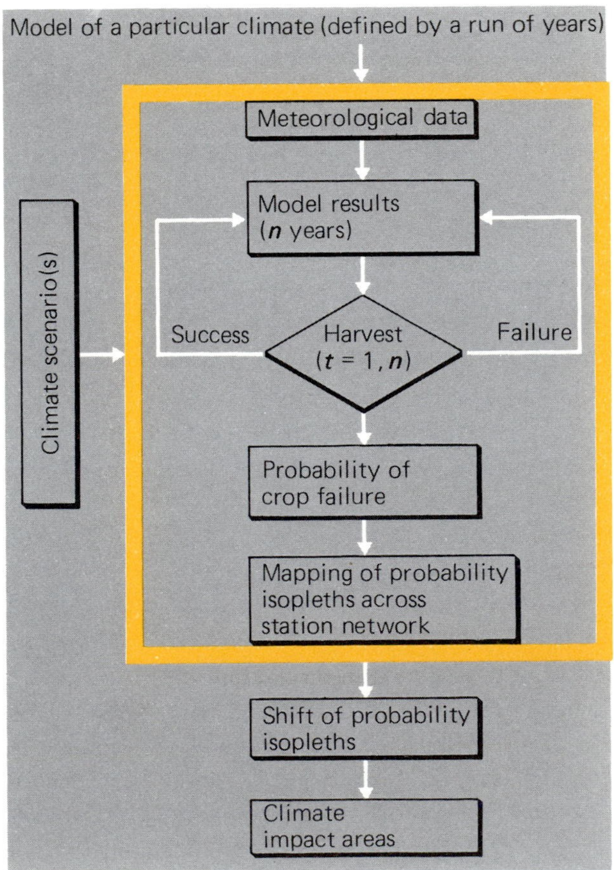

Figure 14.11 Identifying areas of climate impact, using shifts in the isopleths of probability of crop failure [47].

climates can then be used as inputs to the model to produce geographical shifts of the probability isopleth, which are then identified. The areas delimited by these shifts represent areas of specific climatic impact. We can use this approach to estimate the climatic sensitivity of numerous variables, some of which are directly affected by climate (e.g., climatic classifications, measures of agricultural potential) and some of which are more distantly related (e.g., actual agricultural yield, regional and national food production). I take these two categories in turn.

Estimating the sensitivity of agricultural potential to climatic change

One means of portraying the direct effect of a climatic change is by mapping the shift in climatic zones. Since most climatic classifications (like that of Köppen) are closely related to vegetation classes, the shift of such zones essentially reflects a shift of climatic potential for plant growth. *Figure 14.12* illustrates the sharp spatial effect in eastern Australia when mean annual precipitation significantly decreased in the period 1911–1940 as compared with the previous 30 years [55]. At Bourke in New South Wales the isohyets had to be redrawn about 100 km to the east of their original line.

More specifically related to ecosystem needs are climate–life-zone classifications. That by Holdridge attempts to represent the broad distribution of terrestrial ecosystem complexes as a function of annual temperature and precipitation [56]. A perturbation of one or both of these variables would thus alter the distribution of life zones. A preliminary experiment for a doubling of CO_2 climate using the GCM results of Manabe and Stouffer [57] indicates quite substantial changes in the distribution of ecosystems (*Figure 14.13.*) [58].

More closely related to agricultural potential are agroclimatic indices which may give, for example, in semiarid areas a quantitative evaluation of the water balance during the growing season. To illustrate, the isopleths of rainfall excess over potential evapotranspiration have been mapped for each year of the 1969–1973 Sahelian drought. In the driest year, 1972, these isopleths underwent a southward shift of about 120–200 km [59]. This sort of year-by-year analysis of the shift of agroclimatic zones may

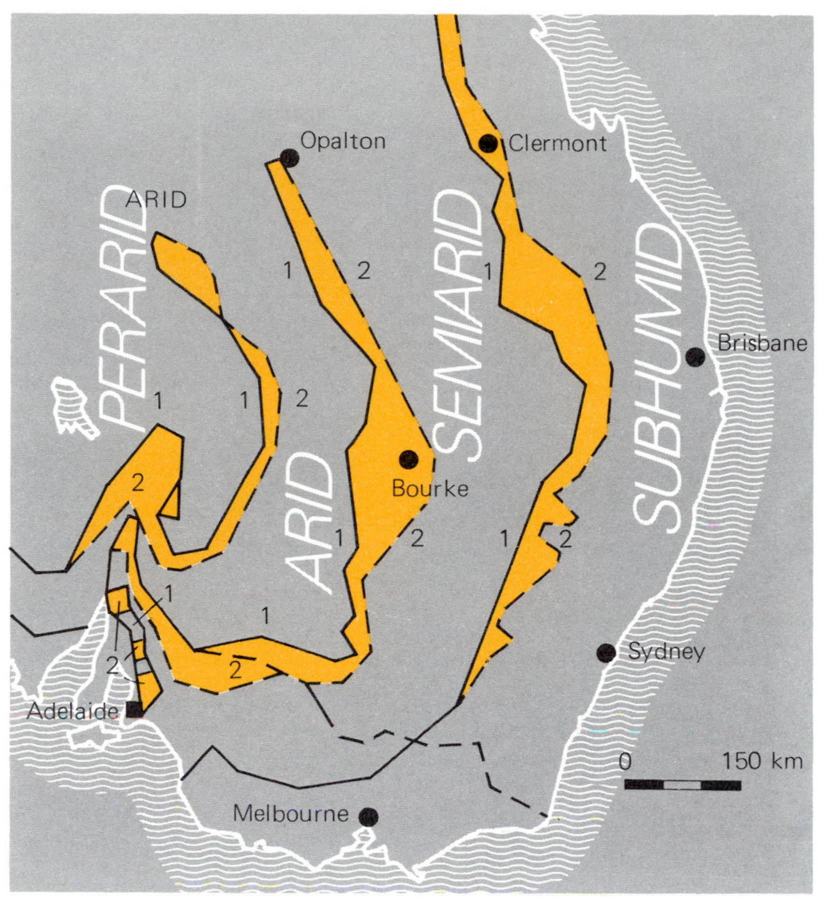

Figure 14.12 Shifts in the climatic belt boundaries for southeast Australia: 1 is 1881–1910 and 2 is 1911–1940 [55].

permit a better understanding of the effect of climatic variability on agriculture on a regional basis.

Very recently some experiments have been made to estimate the effect of a CO_2-induced climatic change on yields predicted by crop–weather simulation models. Data for changes in temperature, rainfall, and cloudiness under a doubling of atmospheric CO_2, derived from experiments with a global circulation model developed at the Goddard Institute for Space Studies (GISS), have been used as inputs for a variety of crop–weather simulation models in Canada, the USSR, Japan, Iceland, and Finland [60]. As an example, *Figure 14.14* illustrates the predicted changes in spring-wheat yields in Saskatchewan.

Estimating the sensitivity of crop yields to climatic change

We can also estimate the impact that might result today (with present-day technology) from a recurrence of a climatic event known to have had a severe impact in the past. What, for example, would be the impacts on wheat production if the US Great Plains were to suffer another severe, prolonged drought? During 1932–1937 this helped bring about *ca.* 200 000 farm bankruptcies or involuntary transfers and the migration of more than 300 000 people from the region. Assuming 1975 technology and a 1976 crop area the impact would still be sharp (*Figure 14.15*) [61]. For a recurrence of the worst weather-year, 1936, simulated production shows a drop of 25%, reducing national wheat production by about 15% (assuming average production elsewhere in the USA). The cumulative effect could be substantial: yearly yields simulated for the weather during 1932–1940 average about 9–14% below normal and amount to a cumulative loss over the decade equal to about a full year's production in the Great Plains.

Similarly, Saskatchewan wheat yields for the 1930s would, with 1978 technology, have been 20% below average, according to yield functions that relate average provincial wheat yields to corresponding May to August net evaporation data for the period 1921–1978 [62].

A conceptual framework

The most promising aspect of these approaches is that they offer the prospect of constructing a *hierarchy* of models to analyze further the cascade of climatic impacts through various elements of the biophysical, economic, and social systems. The pathways and linkages can be traced with three sets of models – of climatic changes, of climatic impacts on potential and actual yield, and of the downstream economic and social effects of these (*Figure 14.16*). Scenarios using outputs from general circulation models, data from instrumental climatic records, or a combination of the two, are used as inputs to impact models to predict potential or actual yield responses to climatic changes. To trace the downstream effect of yield changes, outputs from the impact models are used as input to economic models (farm simulations, regional input–output models, etc.). It is then possible to consider what policies best mitigate certain impacts at specified points in the system.

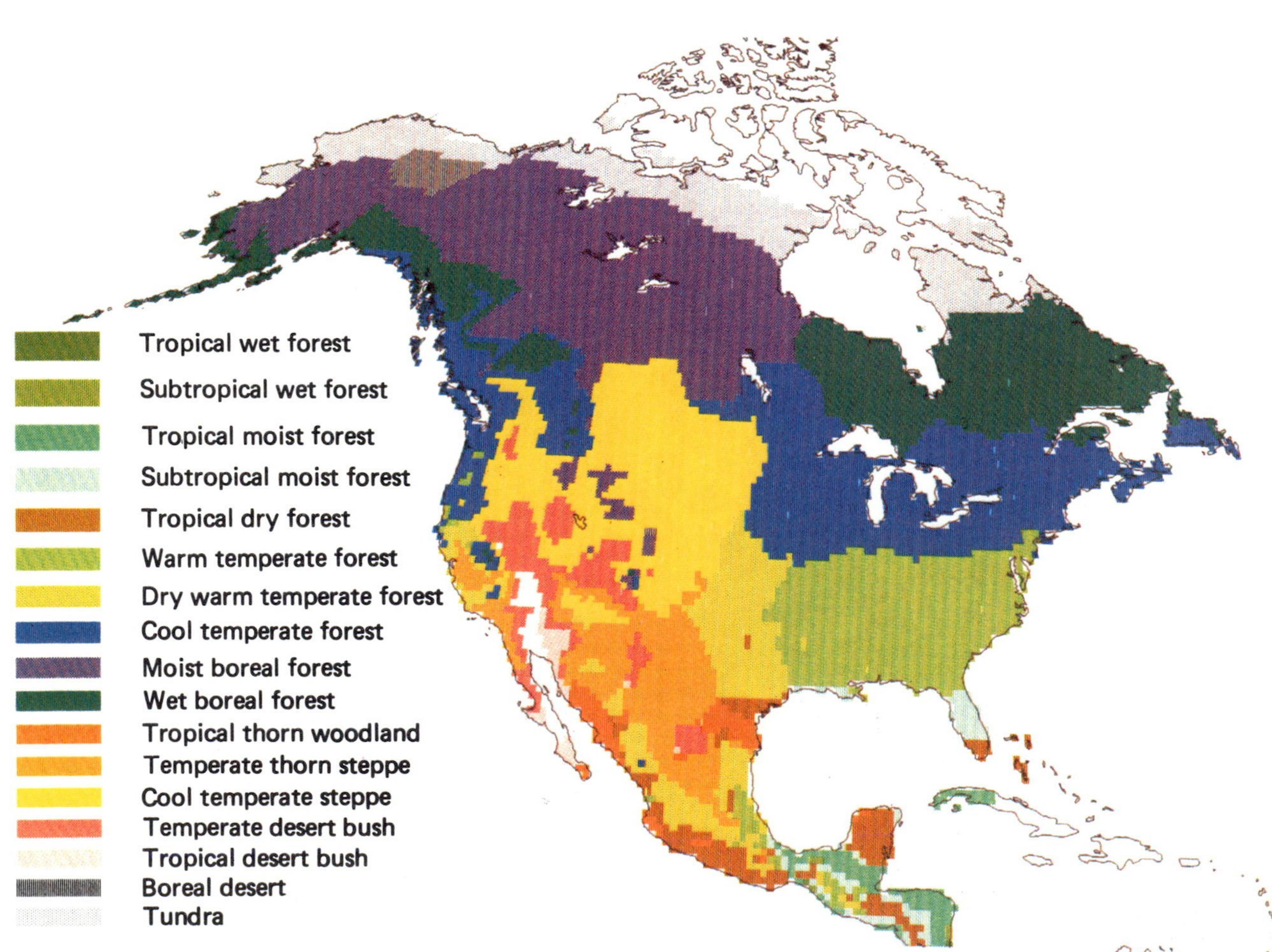

Figure 14.13 (*a*) Holdridge life-zone classification for present conditions of mean annual biotemperature and precipitation [58].

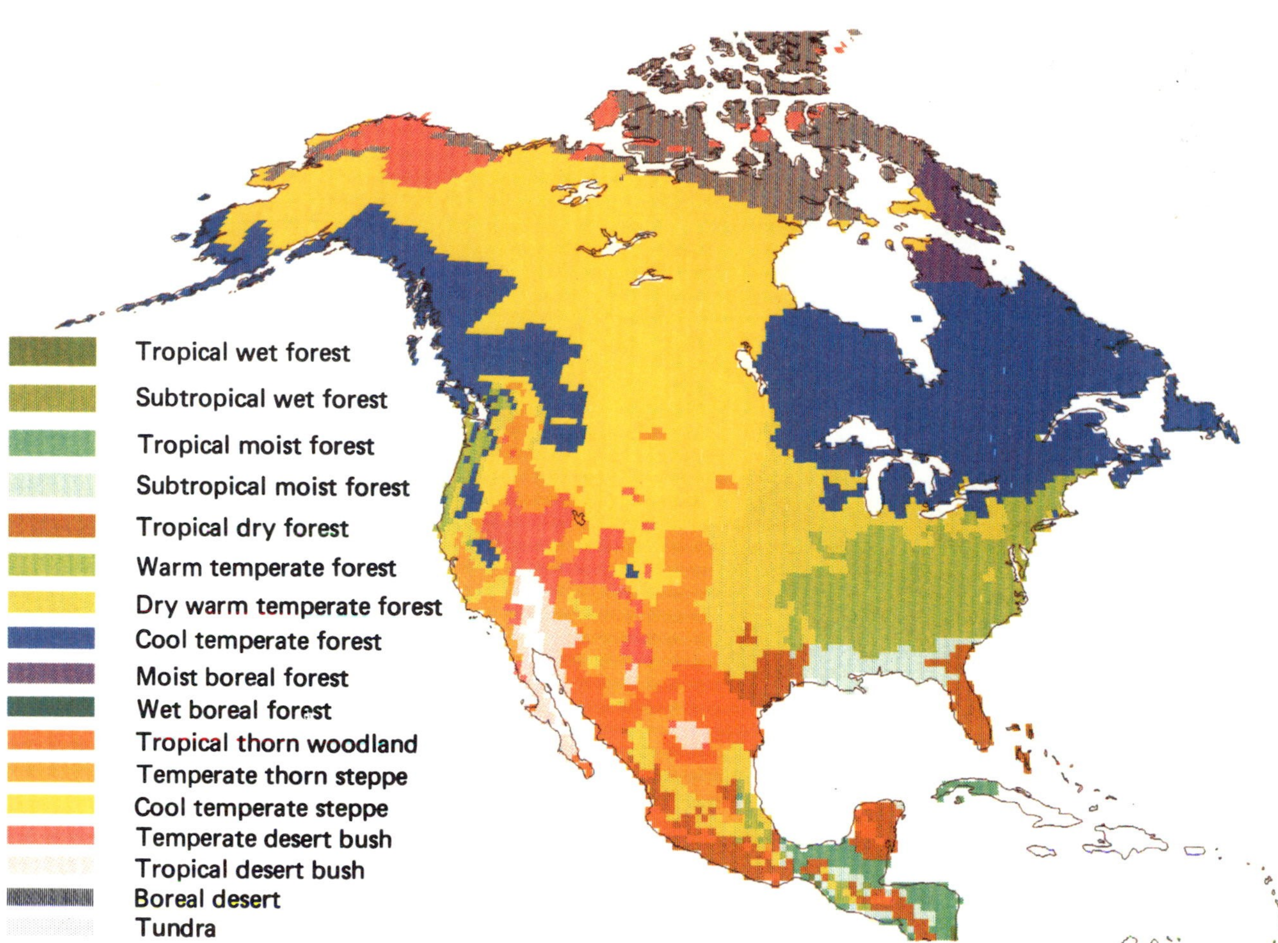

Figure 14.13 (*b*) Holdridge life-zone classification for doubling of CO_2 conditions of biotemperature (assuming unchanged precipitation) [58].

Conclusions

Options and policies

There are many potential responses to mitigate the impacts of climatic variability or change. These may range from a multitude of long-standing adaptations within the economic or social systems (for example, widespread tillage practices which reduce the likelihood of pest infestations resulting from certain weather conditions) to specific large-scale projects, such as the construction of reservoirs and irrigation systems. Considered at the global level, an expansion of the area under assured irrigation is the surest way to increase food security, particularly in the tropics and subtropics where rainfall reliability is often low. At the regional and local level a huge array of measures is both available and in use to combat short-term climatic variability. To illustrate, the options for drought mitigation in Saskatchewan range from tax adjustments to water pricing, and from snow management to stubble mulching (*Table 14.2*, p 401). During 1930–1980 the accumulated 50-year cost of these measures was $1.3 billion (in 1980 terms), compared to the direct *annual* cost of (unprevented) impact from drought to the Saskatchewan economy of $4.7 billion.

It is less easy to identify the response options to a possible future change of climate, such as that which might result from increasing atmospheric CO_2 and *much* less easy to say what would be their cost and their effectiveness. As in most issues of pollution abatement, there are four possible lines of approach:

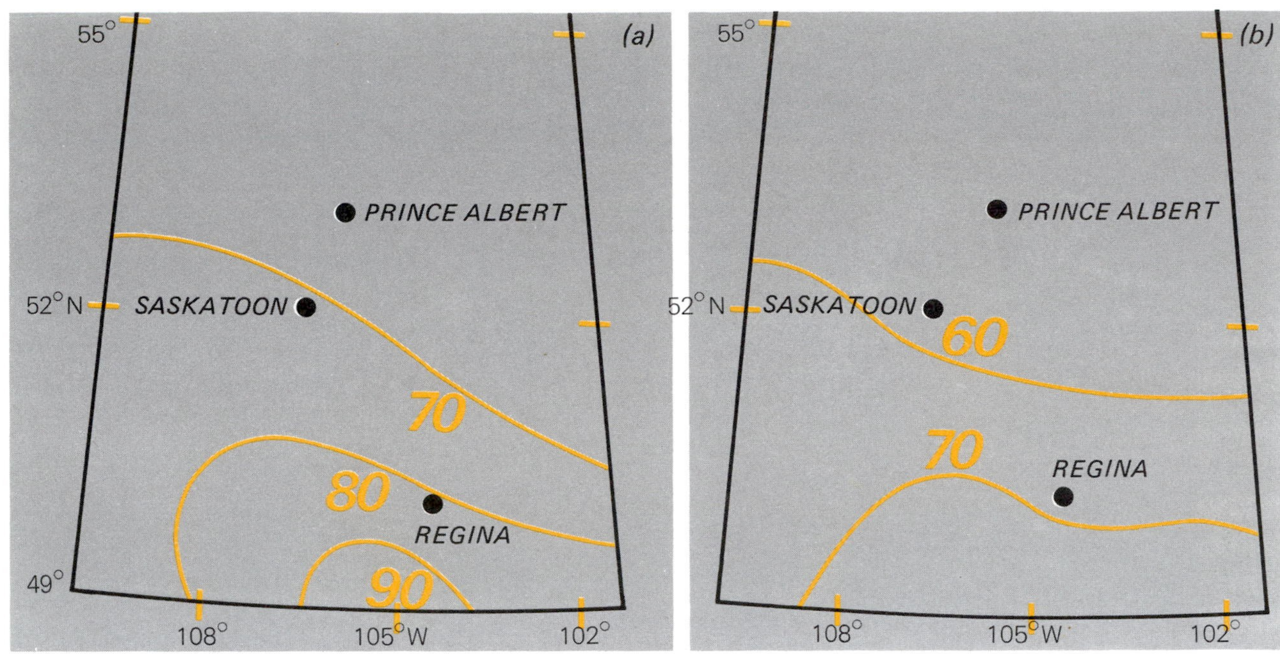

Figure 14.14 Change in spring-wheat yields (percentage of 1951–1980 normal) in Saskatchewan resulting from doubling-of-CO_2 experiments, with the GISS general circulation model: (*a*) for predicted changes in temperature and precipitation; (*b*) for predicted changes in temperature, and normal precipitation (from Stewart in [60]).

(1) Decrease the effluent output (i.e., reduce CO_2 production).
(2) Reduce polluting effects of effluent (i.e., remove CO_2 from effluents or from the atmosphere).
(3) Pay the price of making good the damage (i.e., make countervailing modifications to climate, weather, or hydrology).
(4) Suffer a reduced environmental utility (i.e., adapt to increasing CO_2 and changing climate).

Within these broad lines of response are many technological, economic, and social devices to avoid damage, remove it, or adapt to it (*Table 14.3*, p 402) [63].

Adapting to climatic change

Most of the preventive or adaptive mechanisms listed in *Table 14.3* involve the transfer of resources from other sectors or other regions. This presumes a certain degree of technological ability and a certain degree of economic integration. In general, the greater the level of technological development and social organization, the better are societies able to lessen the impact of minor climatic stresses [64]. At the same time, however, the degree of integration with other economies may affect the extent to which the impact is devolved or shared. To a partially closed economy, such as that of Norse Greenland in the fourteenth century, extreme climatic stress may cause a major shock to the system. A similar, though temporary, impact was felt in the USA Great Plains as a result of drought in the 1890s when thousands of square miles on the dry margins of the Plains were effectively depopulated. By the 1930s, however, the Plains "were a full part of the open and enlarging national economy which directly shared the drought impact [65]." Increasing economic integration may thus devolve the impact of changing climate while technology and social organization enables adaptation to it, though there are numerous intervening factors that are likely to bring very many exceptions to this rule (*Figure 14.17*).

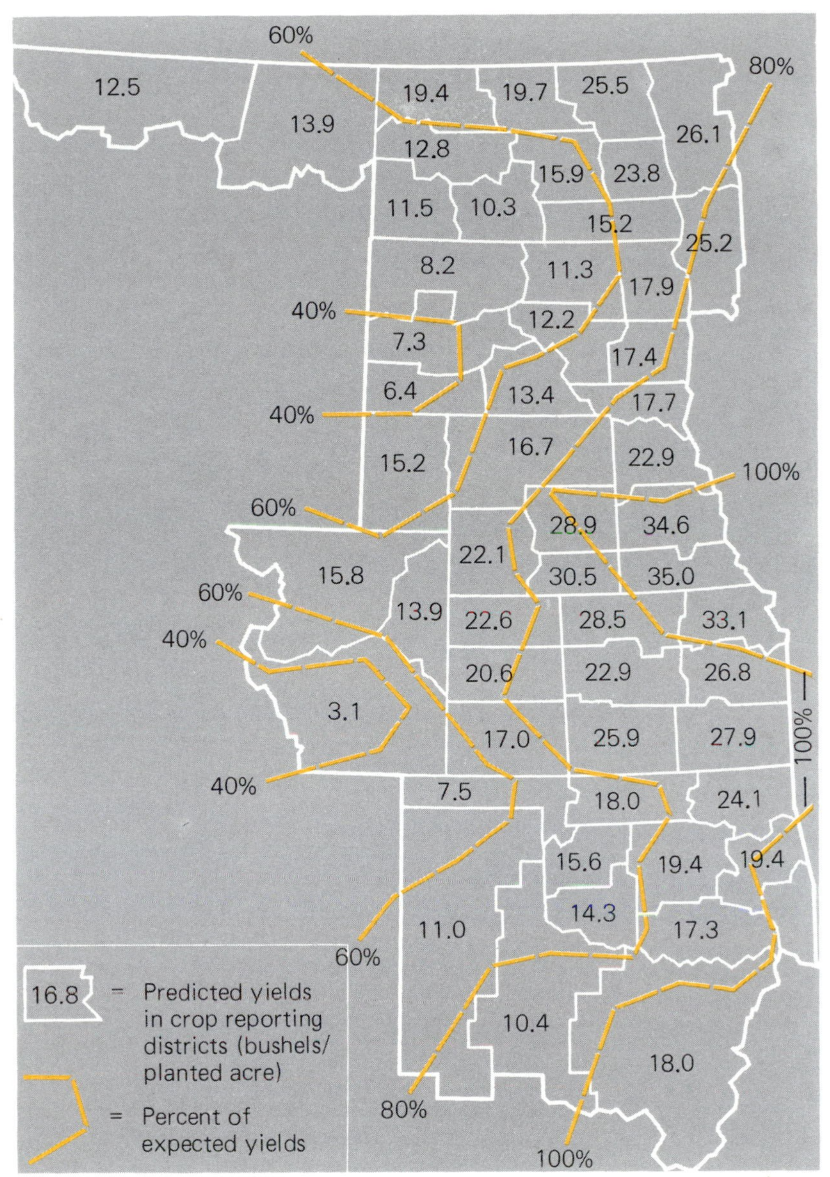

Figure 14.15 Simulated wheat yields for US Great Plains: 1936 weather, 1975 technology [61].

Research priorities

The naiveté of the foregoing discussion illustrates something of the youth of assessment of climatic impact. We have a very long way to go, and many paths to follow. Assuming limited resources, only a few of these paths can be pursued at any one time, so we have some difficult choices to make. What follows is a personal selection of these (*Figures 14.18*, p 405, and *Table 14.4*, p 404).

First, we require more specific and user-oriented information regarding climatic change (its likelihood, nature, magnitude, areal extent, duration and, most important, rate of onset). This information needs to be expressed in terms readily applicable to the user (and often as derived parameters, such as date of first and last frost, days of heat stress, etc.) which would enable agroclimatic measures, such as length of growing season, to be more readily assessed. It is also important that information on variability be available, to provide estimates of possible changes in those extreme conditions that we have seen to be important in agricultural decision-making.

Second, an important path of climatic impact on the most climate-sensitive sectors of our society (farming, fishing, and forestry) is an *indirect* one – i.e., through changes in other physical systems (soil

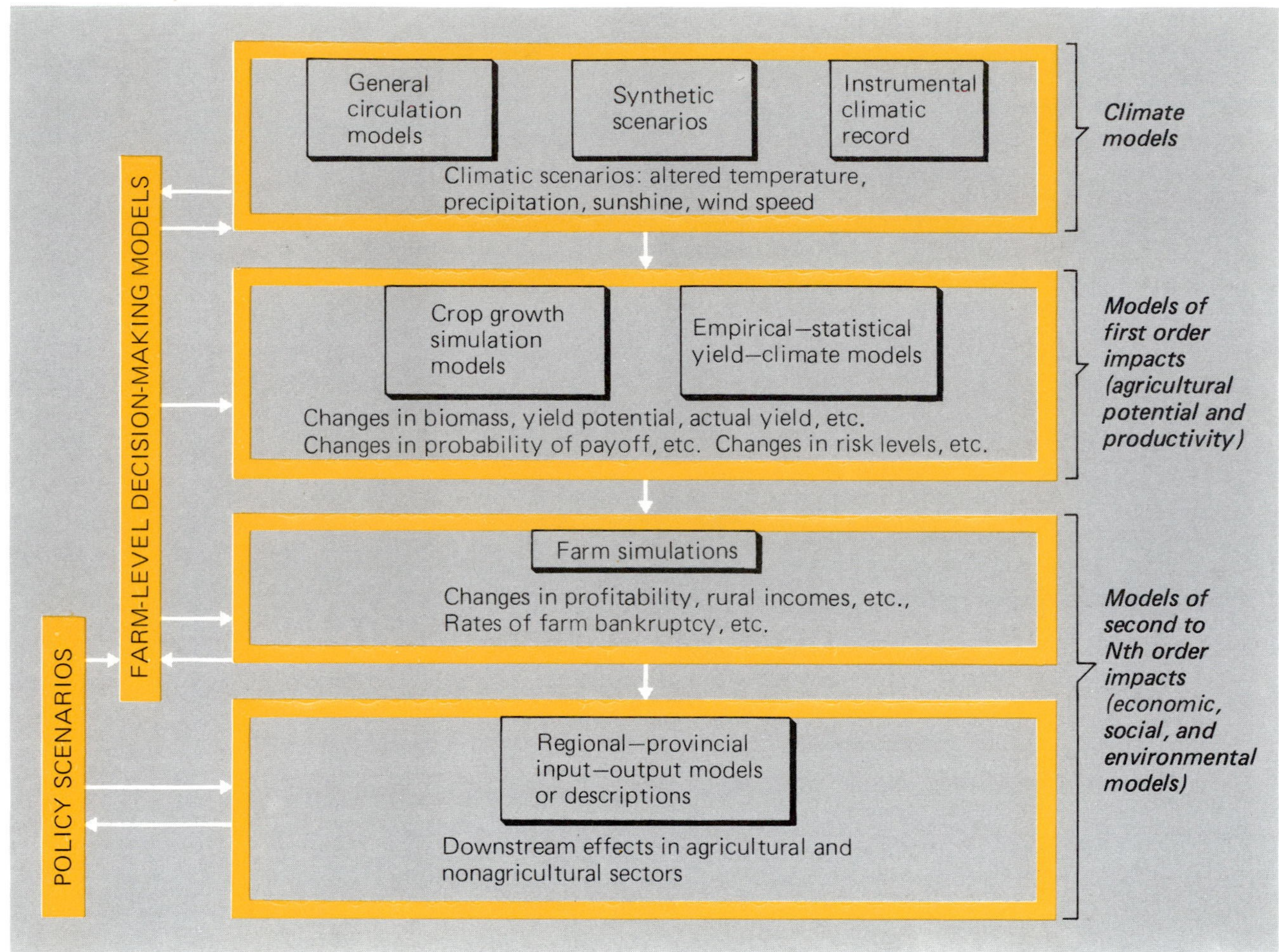

Figure 14.16 A hierarchy of models for the assessment of climate impacts and the evaluation of policy responses [60].

chemistry, ocean currents, agricultural pests and diseases, and their vectors). Although we have barely begun to grasp these interactions they clearly have a major influence on some production systems. The collapse of the Peruvian anchory fishery as a result of the 1972 El Niño is one example. Probably more widespread is the effect of climatic conditions on disease. Outbreaks of many plant diseases (such as potato blight, wheat rust, etc.) are triggered by specific weather conditions. Climatic changes that affected the frequency of these conditions could alter the incidence of such outbreaks.

Furthermore, we need to specify with greater precision the interaction between climate and other resources in the primary sector, by modeling (empirically or by simulation) crop–climate, timber–climate, and fisheries–climate relationships. It should then be possible to trace, with greater confidence, the downstream effects of these first-order impacts on other sectors of economy and society by reference to a hierarchy of the three types of models we have considered: climatic, impact, and economic.

Finally, we need to focus on the human side of the climate–human development equation: the constraints on choice, the role of decision making at the enterprise (e.g., farm) level, and the process of formulating strategies to cope with climatic change.

Specific activities

The above is a rather abstract shopping list. How can these be incorporated into concrete research activities over the next few years? Four interlocking

Table 14.2 Drought mitigation options in Saskatchewan [62].

(A) *Water augmentation*
1. Weather modification
2. Irrigation/dugouts/wells

(B) *Conservation of water*
1. Agricultural cultivation practices:
   summer fallow
   stubble mulching
   strip cropping
   contouring and terracing
   land leveling
   minimum tillage
2. Other cultural practices:
   chemical weed control
   fertilization
   snow management
   shelterbelts
3. Bioengineering
4. Water supply protection:
   irrigation canal linings
   evaporation loss retardants
   low-flow showerheads, and other residential water-saving devices
5. Waste water reuse:
   irrigation return flows/recirculation of industrial waste water

(C) *Modification of water demand*
1. Economic incentives and penalties:
   metering/pricing policies
2. Legal mechanisms:
   building code regulations
   user priorities
   water-use quotas
3. Voluntary changes in use:
   educational programs

(D) *Modification of intrasectoral characteristics*
1. Operational flexibility:
   farm size
   drought resistant crops and livestock
   diversification of production
   off-farm work
   information and management assistance
   change cropping plans
   haul water
2. Land-use management/regulation:
   zoning/land-use restrictions
   removal of marginal land from crop production
   preservation of wetlands
3. Prediction and forecasting:
   soil moisture/crop growth monitoring
   stream flow monitoring
   water supply/demand projections
4. Alternative design of urban water systems:
   interbasin transfers
   system construction/enlargement/upgrading
   deep wells
   water bank

(E) *Spread/share losses and costs*
1. Financial protection:
   drought crop insurance
   feed/fodder bank
   short-term credit
   tax adjustments
2. Relief and rehabilitation:
   low-interest loans
   direct transfers
   subsidies/grants

activities, which build on existing studies and would extend our knowledge in the desired directions, come to my mind:

(1) Studies to improve our understanding of the sensitivity of agroecological potential (both at global and regional scales) to climatic change. Agroecological zone surveys [by FAO, Office de la Recherche Scientifique et Technique Outre-Mer (ORSTROM), and national governments) have now been completed for many countries, being intended as a planning tool in agricultural development. *Prime facie*, these zones are very sensitive to certain types of climatic change, including that likely to result from increasing atmospheric CO_2. Taking a number of alternative scenarios of change (both CO_2-induced and those analogous to recently observed climatic fluctuations) it should be possible to examine the shift of such zones and the significance of these shifts for agricultural development programs. At the outset such studies could be undertaken in areas that are both ostensibly highly sensitive to climatic fluctuations and

Table 14.3 CO_2-induced climatic change: Framework for policy choices [63].

	Policy choices for response[a]			
Possibly changing background factors	*Reduce CO_2 production*	*Remove CO_2 from effluents or atmosphere*	*Countervailing modifications*	*Adapt to increasing CO_2 and changing climate*
Natural warming, cooling, variability			*Weather*: enhance precipitation; modify, steer hurricanes and tornadoes	*Environmental controls*: heating/cooling of buildings, area enclosures other adaptations: habitation, health, construction, transport, military
Population: global distribution, nation, climate zone, elevation (sea level), density				*Migrate*: intranationally; internationally
Income: global average; distribution				*Compensate losers*: intranationally; internationally
Governments				
Industrial emissions: non CO_2 greenhouse gases; particulates			*Climate*: change production of gases, particulates; change albedo of ice, land, ocean; change cloud cover	
Energy: per capita demand; fossil versus nonfossil	*Energy management*: reduce energy use; reduce role of fossil energy; increase role of low-carbon fuels	*Remove CO_2 from effluents*: dispose in ocean, land; dispose of by products in land, ocean		
Agriculture: *forestry, land use, erosion*: farming and other dust; agricultural emissions (N_2O, CH_4)	*Land use*: reduce rate of deforestation preserve undisturbed carbon-rich landscapes	*Reforest*: increase standing stock, fossilize trees		*Change agricultural practice*: cultivation; plant genetics *Change demand for agricultural products*: diet *Direct CO_2 effects*: change crop mix; alter genetics
Water: supply, demand, technology, transport, conservation, exotic sources (icebergs, desalinization)			*Hydrology*: build dams, canals; change river courses	*Improve water-use efficiency*

[a] Responses may be considered at individual, local, national, and international levels.

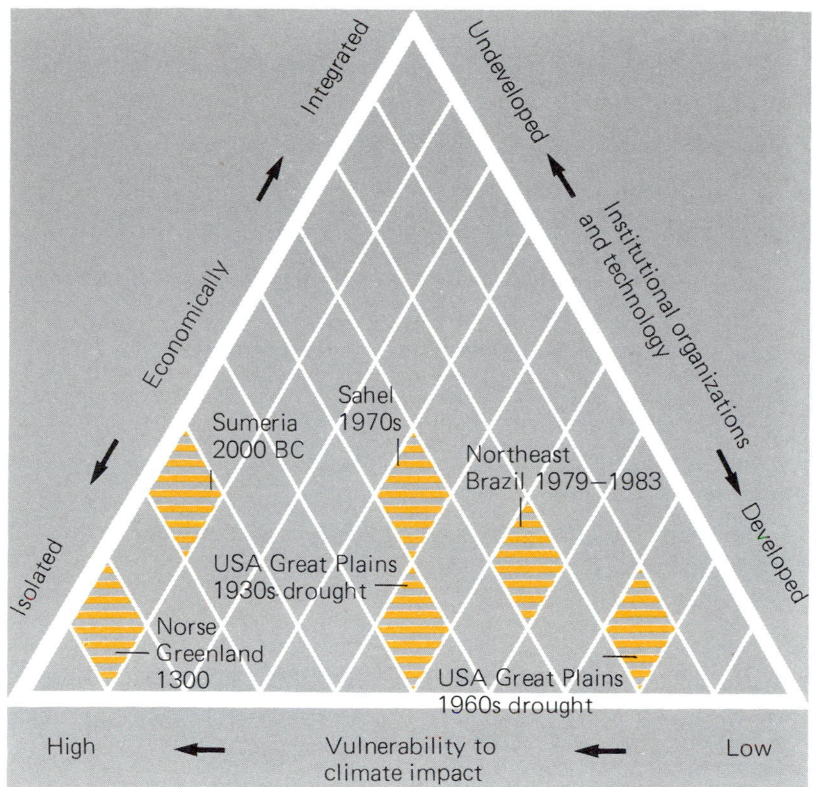

Figure 14.17 Technology, integration, and the vulnerability of society to impact from climatic change.

central to the world food system. The wheat belts in the Canadian Prairies and eastern Australia are appropriate examples. Subsequently, these could be extended to provide inputs for a global assessment of agroecological sensitivity. Current research funded by the International Institute for Applied Systems Analysis (IIASA), the United Nations Environment Programme (UNEP), and FAO would provide the starting point for this activity.

(2) Parallel to this, and using a similar agroecological data base and similar climatic scenarios, it would be logical to pursue a sensitivity study of food systems which are, on past experience, highly vulnerable to climatic change. The Sahel, northeast Brazil, and east Africa are probably the most appropriate choices, given the available data and the urgency of the problem in these regions. This activity could tie in with associated programs in which the UN University, FAO, UNEP, and IIASA are now involved.

(3) We have considered changes in agroecological potential to be the major first-order impact of climatic change. But how do these cascade through the different sectors of a society and how are they thus modified? Some answers might well be provided in a few concrete studies, focusing on regions where robust climate, impact, and economic models exist, and where these can be linked together in order to trace climatic impacts through the entire regional or national economic system. The provincial economic models developed for Manitoba and Saskatchewan are the only complete examples that I know of. There may well be others, particularly in the USSR. Experiments with hierarchies of such models for CO_2-induced and other climatic changes would provide us with information about the impact of the secondary and tertiary sectors.

(4) What sort of institutional frameworks are best suited to planning for sustained agricultural potential and production in the face of climatic change? We need some substantive research – ideally some *experiments in planning.* An appropriate example is *Projecto Nordeste,* a medium-term heavily financed federal program to "drought-proof" northeast Brazil. This project, which is now in the stage of preliminary planning, presents an opportunity to evaluate

Table 14.4 Climatic change impact analysis: some research needs.

(1) More appropriate data on climatic change for impact analysis:
- greater spatial and temporal resolution of climatic data for impact analysis
- more information on likely rate of change
- other data (e.g., wind speed), including derived parameters (e.g., growing season length, days of high temperature stress, heating degree days, etc.)

(2) Improved understanding of the relationship between changes of climate and changes in other physical systems: specifically, climate and:
- soil and soil nutrients, erosion, salinization, etc. (especially with regard to agricultural potential)
- ocean temperatures, currents (especially with regard to fisheries potential)
- pest and diseases, and their vectors (especially with regard to agriculture)

(3) Improved understanding of the relationship between climatic variations and *first-order impacts* in unmanaged ecosystems, agriculture, and energy–transport; specifically:
- climate and timber growth, fisheries and crop yields (empirically or by simulation modeling)
- climate variation and changes in the agroecological resource base
- direct impacts of climate on energy requirements (especially for space heating), and of weather on transportation costs

(4) Greater precision in tracing the downstream effects of climatic change in *secondary and tertiary* sectors (i.e., second- to *n*th-order impacts); specifically:
- by using *hierarchies of linked* climate, impact, and economic models to trace the pathways of impacts

(5) and (6) Greater precision in simulating impacts at the enterprise level (e.g., by farm simulations) and regional level (e.g., by input–output models)

(7) Improved understanding of the role of perception of climatic change (its actual or potential impacts) and the array of potential responses:
- of the appropriate measures of climatic change–impact (e.g., changes in mean, in frequency of extreme events, in risk levels)
- of the appropriate measures of costs and benefits of impacts
- of the choice range of potential or actual adjustive mechanisms; their costs and benefits

(8) Improved understanding of adaptive responses at the enterprise level, especially of:
- firm and farm-level decision-making, particularly in the face of risk, uncertainty, and constraints on choice
- the resilience–vulnerability of economic systems (from sensitivity analyses of farming types and of fishery and forestry systems)

(9) Strategy formulation at the regional, national, and international level:
- the range of possible policy responses
- the process of policy formulation
- the constraints on choice (with respect to other resource needs and problems)

the array of response strategies and the institutional frameworks most suited to their implementation. *Projecto Nordeste* aims to harness interdisciplinary research with integrated planning, bringing together climate modelers, meteorologists, agroclimatologists, agronomists, economists, planners, and politicians. Starting with FAO-type surveys of agroecological potential the program seeks to take the input of climatic impact analysis through to the level of federal and regional legislation.

Each of the four activities outlined above have a number of subareas which link to one another: they

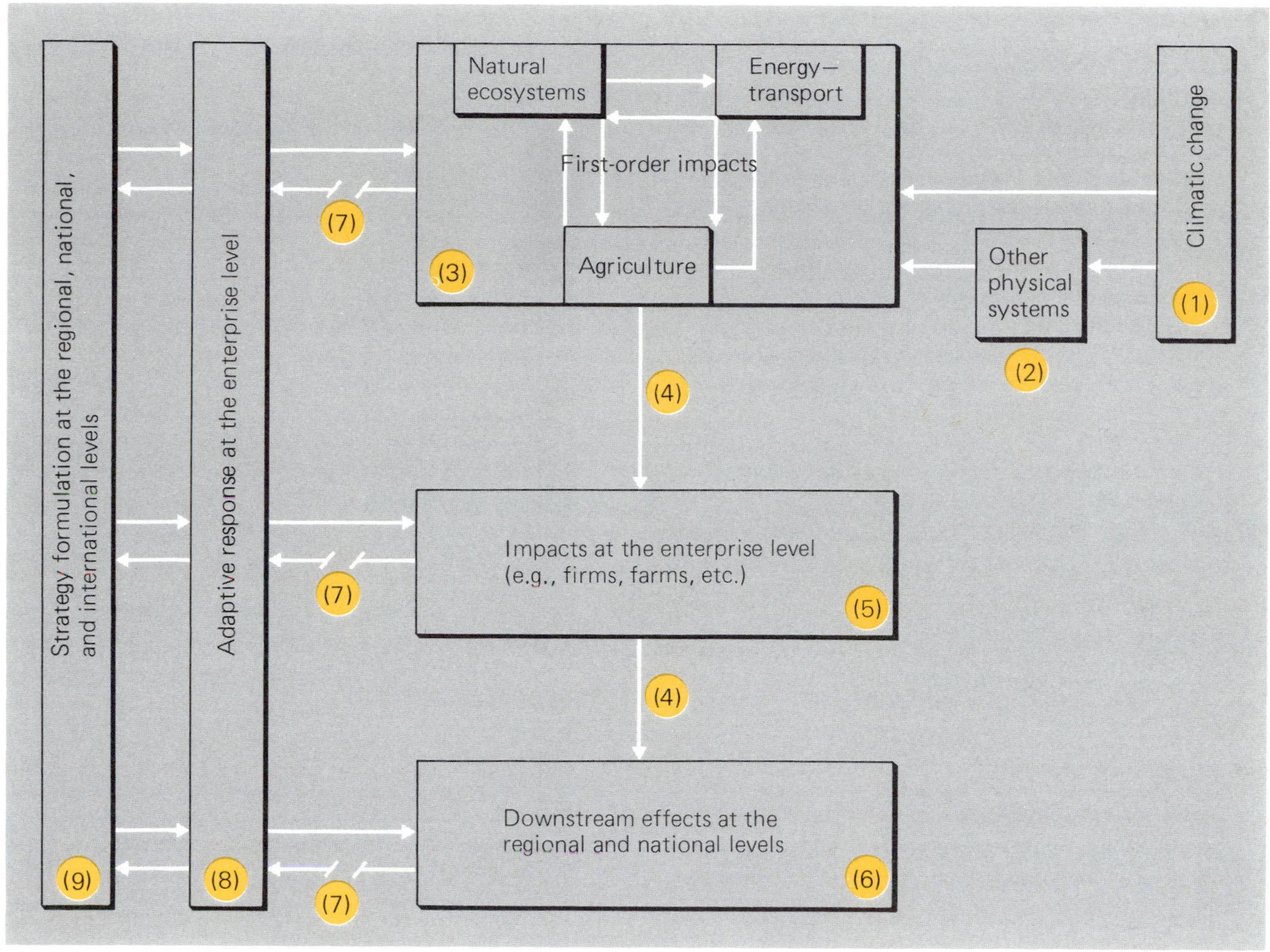

Figure 14.18 Climatic impact analysis: some research priorities. For (1)–(9) see *Table 14.4.*

are interwoven, not discrete. There are many other paths; but those outlined here are feasible, relatively inexpensive tasks from which the likely payoff is substantial; and they would draw together the threads of existing work by a number of different organizations into a more coherent program for a sustainable development of the biosphere.

Notes and references

[1] Holdgate, M. W., Kassas, M., and White, G. F. (1982), World environmental trends between 1972 and 1982, *Environmental Conservation*, **9**(1), 11–29.

[2] See, for example, a ringing criticism by Bandyopadhyaya, J. (1983), *Climate and World Order*, p 25 (South Asian Publishers, New Delhi). Although seriously flawed in its treatment of the climatic data, this treatise by a political scientist is a useful antidote to a Northern-centric view which permeates many climate impact assessments. See also Myrdal, G. (1968), *Asian Drama* (Pantheon, New York).

[3] Brown, L. R. (1978), *The Twenty-Ninth Day* (Norton, New York).

[4] US Department of Agriculture (1982), *World Grain Situation and Outlook: Foreign Agricultural Circular* (US Government Printing Office, Washington, DC). It must be emphasized that it is extraordinarily difficult to ascribe this drop, with any confidence, to specific causes – climatic or nonclimatic.

[5] Petterson, S. (1958), *Introduction to Meteorology*, 2nd edn, p 281 (McGraw-Hill, New York).

[6] Oram, P. A. (1984), Sensitivity of agricultural systems to climatic change, *Climatic Change*, **7**(1), 129–152.

[7] National Defence University (1980), *Crop Yields and Climate Change to the Year 2000* (NDU, Washington).

[8] Burton, I., Kates, R. W., and White, G. F. (1978), *The Environment as Hazard* (Oxford University Press, New York).

[9] Bandyopadhyaya, J. note [2], p 27.
[10] The term "adjoint", which normally refers to the transpose of a square matrix or a determinant, is used here to indicate that such methodologies have features in common with direct methodologies, but opposite in nature.
[11] Bergthorsson, P. (1984), A note on the climatic change in Iceland between the time periods 1873–1922 and 1931–1960, paper presented at the Study Conference on the Sensitivity of Ecosystems and Society to Climatic Change, summarized in Parry and Carter, note [39].
[12] Wigley, T. M. L. (1982), personal communication quoted by Jäger, J. (1983), *Climate and Energy Systems* (Wiley, Chichester).
[13] These effects are discussed in Parry, M. L. (1978), *Climatic Change, Agriculture and Settlement* (Dawson, Folkstone, UK).
[14] The effects of long-term changes in mean climate on the risk of extreme events are considered by Parry, M. L. and Carter, T. R. (1985), The effect of climatic variations on agricultural risk, *Climatic Change*, **7**(1), 95–110.
[15] Glantz, M. (1978), Render unto weather . . . – An editorial, *Climatic Change*, **4**, 305–306.
[16] Nagel, E. (1961), *The Structure of Science* (Routledge, London).
[17] A similar distinction is drawn, from the same example, between "facts" and "pseudo-facts" by Garcia, R. (1981), *Nature Pleads Not Guilty* (Pergamon, Oxford, UK).
[18] Johnson, D. L. and Gould, H. (1984), The effect of climate fluctuations on human populations: a case study of Mesopotamian society, in A. K. Biswas (Ed), *Climate and Development*, pp 117–138 (Tycooly International Publishing, Dublin).
[19] This section draws heavily from Anderson, J. R. (1978), Impacts of climatic variability on the Australian agriculture, in Commonwealth Scientific and Industrial Research Organization, Department of Science, *The Impact of Climate on Australian Society and Economy*, Appendix (CSIRO, Canberra).
[20] Anderson, note [19], p 4.
[21] For more detailed discussion see Kates, R. W. (1985), The interaction of climate and society, in R. W. Kates, with J. Ausubel and M. Berberian (Eds), *Climatic Impact Assessment: Studies of the Interaction of Climate and Society*, pp 3–36 (Wiley, New York).
[22] Carpenter, R. (1966), *Discontinuity in Greek Civilization* (Cambridge University Press, Cambridge, UK).
[23] Dickinson, O. (1974), Drought and the decline of the Mycenae: some comments, *Antiquity*, **48**, 228–230.
[24] Wright, H. E. (1968), Climatic change in Mycenaean Greece, *Antiquity*, **42**, 123–127.
[25] Bryson, R. A., Lamb, H. H., and Donley, D. L. (1974), Drought and the decline of the Mycenae, *Antiquity*, **48**, 46–50.
[26] Environmental determinism postulates that human actions are ultimately determined by the natural environment.
[27] Huntington, E. (1907), *The Pulse of Asia*, pp 365–367 (Yale University Press, New Haven, CT). Huntington's theory was an extension of that developed by Brückner, E. (1890), Klimaschwungen seit 1700, *Geographische Abhandlungen*, **4**(2), 261–264.
[28] Huntington, E. (1925), *Civilization and Climate*, 3rd edn (Yale University Press, New Haven, CT).
[29] Environmental possibilism postulates that the natural environment allows man certain choices over which free will can be exercised.
[30] Takahashi, K. (1981), Climatic change and social disturbance, *Geojournal*, **5**(2), 165–170.
[31] Solomon, S. (1984), The role of the agricultural sector in British long swings *ca.* 1850–1913, submitted. The "long swing" or "Kuznets swing" has a period of *ca.* 20–30 years, as distinct from the "long wave" or "Kondratieff wave" with a period of *ca.* 40–60 years.
[32] Warrick, R. and Bowden, M. (1981), Changing impacts of drought in the Great Plains, in M. Lawson and M. Baker (Eds), *The Great Plains, Perspectives and Prospects* (Center for Great Plain Studies, Lincoln, NE).
[33] For further details see notes [8] and [21].
[34] Kellogg, W. W. and Schware, R. (1981), *Climate Change and Society* (Westview, Boulder, CO).
[35] McGovern, T. H. (1981), in T. M. L. Wigley, M. J. Ingram, and G. Farmer (Eds), *Climate and History*, Ch 17 (Cambridge University Press, Cambridge, UK).
[36] Butzer, K. W. (1982), *Archaeology as Human Ecology: Method and Theory for a Contextual Approach* (Cambridge University Press, New York).
[37] Saarinen, T. F. (1966), *Perception of the Drought Hazard on the Great Plains*, Research Paper No. 106, (Chicago University Geography Department, Chicago, Ill).
[38] Waggoner, P. E. (1983), Agriculture and a climate changed by more carbon dioxide, in National Research Council, *Changing Climate*, pp 383–418 (National Academy Press, Washington, DC).
[39] Warrick, R. A. and Riebsame, W. E. (1983), Societal response to CO_2-induced climate change: opportunities for research, in R. S. Chen, E. Boulding, and S. H. Schneider (Eds), *Social Science Research and Climatic Change*, pp 44–60 (Reidel, Dordrecht, Netherlands); Parry, M. L. and Carter, T. R. (1984), *Assessing the Impact of Climate in Cold Regions*, Summary Report SR-84-1 (International Institute for Applied Systems Analysis, Laxenburg, Austria).
[40] Edwards, C. (1978), Gambling, insuring and the production function, *Agricultural Economics Research*, **30**, 25–28.
[41] McKay, G. A. and Williams, G. D. V. (1981), *Canadian Climate and Food production*, Canadian

Climate Centre Report No. 81–83, unpublished manuscript.
[42] This section draws heavily on the work of Fukui, H. (1979), Climatic variability and agriculture in tropical moist regions, in World Meteorological Organization, *Proceedings of the World Climate Conference,* pp 426–479, WMO No. 537 (WMO, Geneva).
[43] Flohn, H. (1978), Time variations of climatic variability, in K. Takahashi and M. M. Yoshino (Eds), *Climatic Change and Food Production,* pp 311–318 (University of Tokyo Press, Tokyo).
[44] Parry, M. L. (1984), The impact of climatic variations on agricultural margins, pp 351–368 in Kates *et al.,* note [21].
[45] Mearns, L. O., Katz, R. W., and Schneider, S. H. (1984), Changes in the probabilities of extreme high temperature events with changes in mean global temperature, *Journal of Climatology and Applied Meteorology,* **23**, 1601–1613.
[46] See, for example, Wigley, T. M. L., Jones, P. D., and Kelly, P. M. (1980), Scenario for a warm, high-CO_2 world, *Nature,* **283**, 17–21; and Williams, J. (1980), Anomalies in temperature and rainfall during warm Arctic seasons as a guide to the formulation of climate scenarios, *Climate Change,* **2**, 249–266.
[47] Parry and Carter, note [39].
[48] This example is taken from Murphy, A. H., Winkler, R. L., and Katz, R. W. (1983), The use of decision analysis to assess the actual and potential value of short-range climate forecasts, *International Meeting on Statistical Climatology,* September 26–30, Lisbon, preconference proceedings (Instituto Nacional de Meteoroligia e Geofisca, Lisbon).
[49] Hare, F. K. (1979), Climatic variation and variability, in World Meteorological Society, *Proceedings of the World Climate Conference,* WMO No. 537, pp 51–87 (WMO, Geneva).
[50] Anderson, J. L. (1981), History and climate: some economic models, Ch 14 in Wigley *et al.,* note [35].
[51] Post, J. D. (1977), *The Last Great Subsistence Crisis in the Western World* (Johns Hopkins University Press, Baltimore, MD); Pfister, C. (1981), An analysis of the Little Ice Age climate in Switzerland and its consequences, Ch 8 in Wigley *et al.,* note [35].
[52] Garcia, note [17].
[53] See for example the Mycenaean example, p 383.
[54] The statistical techniques applicable to climate impact studies are discussed by Wigley, T. M. L., Huckstep, N. J., Ogilvie, A. E. J., Farmer, F., Mortimer, R., and Ingram, M. J. (1984), Integrated assessment of the impact of climate in historical studies, pp 529–564 in Kates *et al.,* note [21].
[55] Gentili, J. (1971), *Climates of Australia and New Guinea* (Elsevier, Amsterdam).
[56] Holdridge, L. R. (1947), The determination of world plant formations from simple climatic data, *Science,* **105**, 367–368.
[57] Manabe, S. and Stouffer, R. J. (1980), Sensitivity of a global model to an increase of CO_2 concentration in the atmosphere, *Journal of Geophysical Research,* **85**, 5529–5554.
[58] Emanuel, W. R., Shugart, H. H., and Stevenson, M. P. (1984), Climatic change and the broad-scale distribution of terrestrial ecosystem complexes, *Climatic Change,* **7**(1), 29–43.
[59] Mattei, F. (1979), Climatic variability and agriculture in the semi-arid tropics, in World Meteorological Society, *Proceedings of the World Climate Conference,* pp 475–509, WMO No. 537 (WMO, Geneva).
[60] This work is in progress at IIASA, and is partly funded by UNEP. The results will appear in Parry, M. L., Carter, T. R., and Konijn, N. (Eds), *Assessment of the Impact of Climatic Change on Agriculture, Vol 1, In High-Latitude Regions* (Reidel, Dordrecht) (forthcoming).
[61] Warrick, R. A. (1984), The possible impacts on wheat production of a recurrence of the 1930s drought in the US Great Plains, *Climatic Change,* **6**, 27–37.
[62] Saskatchewan Drought Studies (1968), *Interim Report, Phase 1* (Saskatchewan Environment, Regina, Canada).
[63] Schelling, T. C. (1983), Climatic change: implications for welfare and policy, in National Research Council, *Changing Climate,* pp 449–482 (National Academy Press, Washington, DC).
[64] Bowden, M. J., Kates, R. W., Kay, P. A., Riebsame, W. E., Warrick, R. A., Johnson, P. L., Gould, H. A., and Weiner, D. (1981), The effect of climatic fluctuations on human populations: Two hypotheses, Ch 21 in Wigley *et al.,* note [35].
[65] Bowden *et al.,* note [64], p 509.

Commentary

M. Yoshino

Introduction

The various effects of human activities on climate occur over different time scales, of which the medium-term change of climate and its subsequent effect on socioeconomic conditions has been rather neglected by researchers. Generally, short-term implications, such as weather–crop relationships and weather–health relationships, have been physiologically measured and analyzed extensively and, on the other hand, long-term implications of climatic change, with respect to the history of civilizations, for example, have long been studied by historians and climatologists. Medium-term implications are relatively difficult to study, because of the lack of homogeneous data and because the applicable methods of study include those of both natural science and the socioeconomic sciences. Parry describes these comprehensively in this chapter.

General

In order to analyze quantitatively the implications of climatic change for human society, one must pay attention first to the whole choice of relationships; for instance, the chain climate–agriculture–individual farmer–food prices–society. This is not an easy task, even if we consider only a part of the process between climate and agriculture. Climate itself can be considered in many terms, such as frequency distribution, extreme value, standard deviation, periodicity, and secular trend, in addition to the mean, average, or normal values. In agriculture, there are aspects to be considered, such as a change of main crop, a change of cash (economic) crop, development of a land-use pattern, and a change of political system or economic structure, in addition to the generally discussed components, such as the introduction of new varieties and cultivation techniques. Also, the family construction (work force) of individual farmers and their traditional customs in cultivation change. Food prices are determined by policymakers in many countries. So the quantitative relationships between climate and society are quite obscured in this labyrinth of changing inputs.

Furthermore, it is very important that human society adapt and adjust to the new environmental conditions. The nature of such adaptation changes as society develops. Acclimation or acclimatization has been much studied from the standpoints of medical biometeorology and cultural human geography, but little from the agroclimatological viewpoint. Adaptation or adjustment of societies should be considered not only in the short term with respect to perceived impact or risk of an impact, but also in the medium term.

Problems

There are many difficulties in such studies. Here, I summarize the problems briefly:

(1) *Data.* As mentioned above, homogeneous data prior to the late nineteenth century are generally unavailable, but they are essential for statistical work. Using old diaries or chronicles, some data can be reconstructed, but this is applicable only to limited regional studies. In this historical approach, insufficient data and inadequate analytical techniques produce inconclusive results, even if built on promising hypotheses. This might be the case for some work by Huntington, as Parry mentions.

(2) *Synoptic climatological considerations.* Medium-term and long-term changes of climate are interpreted in most cases through short-term synoptic climatological conditions. For instance, weakening of polar high pressure causes a northward shift in the track of the Mediterranean depressions, which bring rain to the Aegean. This is true for the short term, but we do not know that this also occurs in the medium- and long-term sequences, which would have affected the Mycenaean settlements.

It is not clear from our present knowledge whether the synoptic climatological conditions for the medium term are affected by the intensification of extreme cases, the higher frequency of extreme cases, or situations unpredictable from the short-term conditions.

(3) *Multiple effect.* This involves the striking conclusion that surprisingly large changes in farming types may be necessary to accommodate apparently small changes in average climate. In North Japan, the difference of growing-season mean temperatures between good and bad harvest years during the last 30 years was only 1 °C [1]. Increased concentrations will not only produce the predicted temperature increase of 4–6 °C and its consequences, but also indirect effects due to resultant changes in other climatic elements. As has been pointed out, in Canada the temperature rise will increase evapotranspiration, thus reducing soil moisture and causing strong wind erosion. A similar temperature rise in Japan implies that snow in mountainous regions may not accumulate in winter, resulting in a serious water shortage for irrigation in the succeeding growing season. As has been made clear by Sri Lanka's striking drought [2], rainfall smaller than the mean value (mean minus standard deviation) in the northeast monsoon

season caused a marked decrease in the sowing acreage of paddy cultivation in the succeeding southwest monsoon season.

We should therefore study first both the direct and indirect effects of a change in climatic element, and second the complex effects during the same season and the succeeding season.

(4) *Seasonal competition.* In the middle and lower latitudes there are broad regions with double and triple croppings or even more. In these regions, croppings have been combined in a highly developed calendar. For example, shortening the winter crop season results in an extension of the summer crop season, or shortening the dry season cultivation period results in an extension of the rainy season cultivation period, causing the change of land-use pattern to be in cycles of a year or more. This means that climatic impacts producing crop failure, as discussed in *Figure 14.11*, should be studied further in connection with the seasonal competition of the crops in the regions under consideration. The effects of the medium-term change of climate on changing seasonal land-use patterns or crop calendars are of great importance, because failure of the first crop may produce, in some cases, success of the second or third crop under some new conditions.

Notes and references (Comm.)

[1] Yoshino, M. M. (1985), *Analysis of Climatic Changes and Variations from the Japanese Instrumental and Proxy Data Record,* a paper presented at the Task Force Meeting at IIASA, Laxenburg, 1–5 April 1985.

[2] Yoshino, M. M., Kayane, I., and Madduma-Bandara, C. M. (1983), Climatic fluctuation and its effects on paddy production in Sri Lanka, *Climatological Notes,* **33**, 9–32.

PART FIVE

USABLE KNOWLEDGE

Usable knowledge

Contributors

About the Contributors

Garry Brewer is currently a Professor at Yale University with appointments in the Schools of Management and Forestry; he also serves as consultant to numerous governmental and private organizations. Previously, Brewer was a member of the Senior Staff of the Rand Corporation and Editor of the journals *Policy Sciences* and *Simulation & Games.* His present interests are natural resource management and policy issues, national security problems related to communication, command, control, and intelligence, and the development of theories and methods for policy sciences.

Jerome Ravetz is Senior Fellow at the University of Leeds in England. Educated at Swarthmore College, Pennsylvania, and Trinity College, Cambridge, he has published in the fields of history and philosophy of mathematics and sciences, scientific knowledge and its social problems, risks and regulation, and methodology in policy-related research. His current interest is with the communication of inexact quantities in the policy context of technological risks.

Steve Rayner is Research Associate in the Energy Division at Oak Ridge National Laboratory and Senior Research Associate at the Centre for Occupational and Community Research, London. Educated at the Universities of Kent and London (University College), he is currently working on an analysis of institutional requirements for a future generation of energy technologies in relation to their societal acceptability, and preparing a research program on institutional factors in the perception and management of technological risk for Oak Ridge National Laboratory.

Nicholas Sonntag is Vice President and Senior Partner of ESSA Environmental and Social Systems Analysts Ltd, Vancouver, where his responsibilities and activities involve negotiations, supervision, and participation with a number of international clients (government and private) primarily in the area of environmental resource management and policy. Educated at the University of British Columbia, Sonntag's expertise is primarily in the application of systems analytic techniques to environment and resource issues.

Michael Thompson is Principal Research Fellow of the Institute for Management Research and Development at the University of Warwick. Educated at the Royal Military Academy, Sandhurst, the Royal Military College of Science, Shrivenham, and the Universities of London (University College) and Oxford (Corpus Christi College), his research interests are in the cultural, institutional, and historical forces that shape policy debates, particularly those debates characterized by high levels of uncertainty and by persistent disagreement among experts.

Peter Timmerman is Research Associate of the Institute for Environmental Studies at the University of Toronto, where he was educated. He has worked primarily on various aspects of hazards research and planning, including the Ontario Government report on the Mississauga Derailment and work on chemical accidents for the European office of the World Health Organisation. He has recently been involved in considering the ethical and political implications of hazardous waste facility siting.

Chapter 15 Usable knowledge, usable ignorance: incomplete science with policy implications

J. R. Ravetz

Editors' Introduction: In this book, we have so far reviewed a wealth of rapidly expanding scientific knowledge that may help to illuminate the interactions between development and environment. But in every chapter we have also emphasized how limited our understanding remains relative to the complexity and magnitude of the policy problems that motivate us.

In this chapter Ravetz confronts the dilemma that, despite all our research, a truthful response to questions such as "What's going to happen to the biosphere?" will most often be "We don't know, and we won't know." He goes on to explore how, in the face of overwhelming ignorance, scientific inquiries in policy-related contexts can most responsibly and effectively be conducted. Better procedures for self-criticism and quality control in science are argued to be central to the construction of "usable ignorance". A key role is also assigned to the design of approaches through which incomplete science can be better integrated into policy debates.

Introduction: the problem

All those involved in the IIASA Project on the Sustainable Development of the Biosphere will be aware that we are attempting something novel. At one level the novelty is of style of work: we hope to achieve a genuine integration of effort of scientists, scholars, and policymakers from a variety of disciplines and a variety of political cultures. At another level the novelty is methodological: we hope that colleagues from different backgrounds will engage in dialogue and constructive criticism, even (or especially!) *outside* their spheres of expertise, so that the eventual product of our labors will be a true synthesis of our work. Further, the Project itself is self-consciously exploratory, designed to foster learning about this new task while we are engaged in accomplishing it. Finally (for the project itself), our detailed goals, and also our criteria for success, will necessarily evolve in the course of our collaborative endeavors.

All these novelties, taken together, will produce a Project that is either a challenging adventure or (in terms of investment of the time, talent, and resources of each of us) a hazardous undertaking.

It is by now well appreciated in technology that innovation by qualitative jumps rather than quantitative movements is always risky; and the greater the leap forward, the more the likelihood of a fall. Yet those of us who have already invested much in bringing the Project to this point are convinced that, with all its dangers, this strongly novel approach is necessary for the task we face; and if we succeed, it will yield great benefits, both for the public and for us, the participants, who will have the satisfaction of an important job well done.

For this special sort of project is motivated by the biggest novelty of all: the new role of science in the advancement and preservation of our industrialized civilization. For centuries the dominant theme of our science has been taken from Francis Bacon's aphorism "Knowledge and power meet in one." I need not relate here the transformation of humanity's material culture that science has brought about; nor the enhancement of human life, social, moral, and spiritual, that this has enabled through the conquest of the traditional curse of poverty (in at least the more fortunate parts of the world). But now we face a new, unprecedented problem: along with its great promises, science (mainly through high technology) now presents grave threats. We all know about nuclear (and also chemical and biological) weapons; and about the menaces of acid rain, toxic wastes, the greenhouse effect, and perhaps also the reemergence of hostile species, artificially selected for virulence by our imprudent use of drugs and pesticides. It would be comforting to believe that each such problem could be solved by a combination of more scientific research of the appropriate sort together with more good will and determination in the political and technological spheres. Doubtless, these are necessary; but the question remains: Are they sufficient? The record of the first round of an engagement with these biospheric threats is not encouraging. For example, we do not yet know when, how, or even whether global temperatures will be influenced by the new substances being added to the atmosphere. This is why, we believe, a novel approach is called for, if our science-based civilization is to solve these problems that are so largely of our own making.

Indeed, we may see the issue not merely in terms of science, but of our industrialized civilization as a whole, since it has science as the basis of its definition, the science defined by the motto of Francis Bacon. And the problem that faces us is that the sum of knowledge and power is now revealed to be insufficient for the preservation of civilization. We need something else as well, perhaps best called "control". This is more than a mere union of the first two elements; for it involves goals, and hence values; and also a historical dimension, including both the remembered past and the unknowable future.

Can our civilization enrich its traditional knowledge and power with this new element of control? If not, the outlook is grim. There are always sufficient pressures that favor short-term expedients to solve this or that problem in technology or welfare, so that the evaluative concerns and long-range perspectives necessary for control, will, on their own, lose every time. That is what has been happening, almost uniformly, in our civilization until quite recently. Only in the last few decades have scientists become aware that control does not occur as an automatic by-product of knowledge and power. Our awareness has increased rapidly, but so have the problems. And we are still in the very early stages of defining the sort of science that is appropriate to this new function.

We might for a moment step back and look at this industrialized civilization of ours. It is now about half a millenium since the start of the Renaissance and the expansion of Europe. That's roughly the standard period of flourishing for previous civilizations; will ours prove more resilient to its own characteristic environmental problems? It seems likely that some of the ancient "fertile crescent" cultures declined because of excessive irrigation; and in various ways the Romans consumed great quantities of lead. What would be our auto-intoxicant of choice?

In some ways our material culture is really rather brittle; our high technology and sophisticated economies depend quite crucially on extraordinary levels of quality control in technology and on highly stable social institutions. Whether these could absorb a really massive environmental shock is open to question. The real resilience of our civilization may lie not so much in its developed hardware and institutions, as in its capacity for rapid adaptation and change. It has, after all, continued to grow and flourish through several unprecedented revolutions: one in the common-sense understanding of nature in the seventeenth century; another in the material basis of production in the eighteenth and nineteenth centuries; and yet another in the organization of society over much of the world in the twentieth century. Perhaps it could be that the latest challenge to this civilization, resulting from the environmental consequences of our science-based technology, will be met by the creation of a new, appropriate sort of science. We can only hope so, and do our best to make it happen.

What could such a new, appropriate sort of science be? Isn't science just science? In some ways, yes; but in others it is already differentiated. We are all familiar with the differences between pure or basic research on the one hand, and applied or R & D on the other. In spite of the many points of contact and overlap, they do have distinct functions, criteria of quality, social institutions, and etiquette and ethics. To try to run an industrial laboratory as if it were within the teaching and scholarship context of a university would be to invite a fiasco; and equally so in reverse. Now we face the task of creating a style of science appropriate to this novel and urgent task of coping with biosphere problems. Of course, there are many different institutions doing research with just this end in view. Sometimes they are successful; but success is more common when they have a problem where the conditions for success can be defined and met, and where the input from research is straightforward. To the extent that the problem becomes diffuse in its boundaries (geographically, or across effects and causes), entrained in crosscurrents of politics and special interests, and/or scientifically refractory, then traditional styles of research, either academic or industrial or any mix of the two, reveal their inadequacy.

This is the lesson of the great biosphere problems of the last decade. Faced with problems not of its choosing (though indirectly of its making), science, which is the driving force and ornament of our civilization, could not deliver the solutions. When asked by policymakers: "What will happen, and when?" the scientists must, in all honesty, reply in most cases: "We *don't* know, and we *won't* know, certainly not in time for your next decisions."

If this is the best that science can do, and it seems likely to be so for an increasing number of important issues, then the outlooks for effective policy-making and for the credibility of science as a cornerstone of our civilization are not good. Yet, I believe, so long as scientists try to respond as if they face simple policy questions determined by simple factual inputs, the situation cannot improve.

But what else can scientists do except provide facts for policy? I hope that we can define the task in new terms, more appropriate to our situation; and *that* is an important component of the goal of this project.

My work on this Project has already involved me in an intellectual adventure; recasting my earlier ideas about science had led me into paradox and apparent contradiction. Rather than leading colleagues into them by gentle and easy stages, I have chosen to exhibit them boldly in the title. We all know what "usable knowledge" is, although it turns out to be far from straightforward in practice [1]: but "usable ignorance"? Is this some sort of Zen riddle? I hope not. But if we are to cope successfully with the enormous problems that now confront us, some of our ideas about science and its applications will have to change. The most basic of these is the assumption that science can, indeed, be useful for policy, but if and only if it is natural and effective, and can provide "the facts" unequivocally. So long as it seemed that those facts would be always forthcoming on demand, this assumption was harmless. But now we must cope with the imperfections of science, with radical uncertainty, and even with ignorance, in forming policy decisions for the biosphere. Do we merely turn away from such problems as beneath the dignity of scientists, or do we learn somehow to make even our own ignorance usable in these new conditions? In this exploratory chapter, I hope to show how even this paradox might be resolved, and in a way that is fruitful for us all.

Images of science, old and new

If I am correct in believing that our inherited conception of science is inappropriate for the new tasks of control of these apparently intractable biospheric problems, then we shall all have to go through a learning experience, myself included. Scientists, scholars, and policymakers will need to open up, and share their genuine but limited insights of science, so that a common understanding, enriched and enhanced by dialogue, can emerge. My present task is to call attention to the problem, and to indicate my personal, rough, provisional guidelines toward a method.

Insoluble problems

I may still seem to be speaking in paradoxes, so I will suggest a question that may illuminate the problem. For background, let us start with the historical datum that in the year 1984 we could not predict when, or even whether, the Earth's mean temperature will rise by 2 °C due to an increasing CO_2 content in the atmosphere. Yet this prediction can be cast as a scientific problem, for which there are both empirical data and theoretical models. *Why* these are inadequate is a question I must defer; but we can (I hope) all agree that here is a scientific problem that cannot be solved, either now or in any planned future. And this is only an example of a class that is growing rapidly in number and in urgency.

I believe that such problems are still very unfamiliar things; for our personal training in science progressed from certainties to uncertainties without any explicit, officially recognized markers along the path. Almost all the facts learned as students were uncontested and incontestable; only during research did we discover that scientific results can vary in quality; later we may have come across scientific problems that could not be solved; and only through participation in the governing of science do we learn of choices and their criteria.

Now I can put the question, for each of us to answer for him- or her-self: When, at what stage of my career, did I become aware of the existence of scientific problems that could not be solved? *My* personal answer is not too difficult: as a philosophically minded mathematician, early in my postgraduate studies I learned of classic mathematical problems and conjectures that have defied solution for decades or even centuries. I have reason to believe that my experience was exceptional for a scientist. Certainly, I have never seen an examination in a science subject which assumed other than that every problem has one and only one correct solution. Some may well exist, but they will be a tiny minority. Similarly, research students may learn of the tentativeness of solutions, the plasticity of concepts and the unreliability of facts in the literature. But this is a form of insiders' knowledge, not purveyed to a lay public, nor even much discussed in scholarly analyses of science.

Indeed, it is scarcely a decade since insoluble scientific problems have become "news that's fit to print". Alvin Weinberg brought them into recognition with the term "trans-science" [2]. Were these a new phenomenom of the troubled 1960s, that period when environmentalists began to raise the impossible demand that science prove the impossibility of harm from any and all industrial processes and effluents? No; ever since the onset of the scientific revolution, science had been promising far more than it could deliver. Galileo's case for the Copernican Theory rested on his theory of the tides, where he contemptuously rejected the moon's influence and instead developed a mechanical model that was far beyond his powers to articulate or demonstrate. Descartes' laws of impact, fundamental for his system, were all wrong except in the trivial cases. The transformation of the techniques of manufacture, promised by every propagandist of the century, took many generations to materialize. In the applications of science, progress toward the solution of outstanding, pressing problems was leisurely; for example the break-even point for medicine, when there came to be less risk in consulting a physician than in avoidance, seems to have occurred early in the present century.

None of this is to denigrate science; however slow it was to fulfill the hopes of its early prophets, it has now done so magnificently, nearly miraculously. My aim here is to focus our attention on a certain image of science, dominant until so very recently, where the implicit rule was "all scientific problems can be discussed with students and the public, provided that they're either already solved or now being solved". Each of us (including myself) has this one-sided experience of "science as the facts" embedded deeply in our image of science. That is why I think it is a useful exercise for each of us to recall *when* we first discovered the existence of insoluble scientific problems.

"Atomic" science

If I am still struggling to find a new synthesis out of earlier ideals and recent disappointments, in spite of having earned my living on just that task for 25 years, I cannot really expect colleagues or members of the general public to provide immediate insights that will neatly solve my problems. All I can do is to offer some preliminary ideas, to share with colleagues from various fields of practice, and to hope that out of the resulting dialogue we may achieve a better understanding of the practice and accomplishments of science, as a mixture of success and failure, and of our achieved knowledge and continuing ignorance.

It appears to me that we must now begin to transcend an image of science that may be called "atomic", for "atoms" are central to it in several ways. The conception of matter itself; the style of framing problems; and the organization of knowledge as a social possession; all may be considered atomic. I believe that such an image inhibits our grasping the new aspects of science, such as quality control, unsolvable problems, and policy choices, that are essential for an effective science of the biosphere.

The idea of atomism was at the heart of the new metaphysics of nature conceived in the seventeenth century, the basis of the achievements of Galileo, Descartes, and Newton. The particular properties of the atoms were always contested, and are not crucial. What counts is the commitment to nature being composed of isolated bits of reality, possessing only mathematical properties, and devoid of sensuous qualities, to say nothing of higher faculties of cognition or feeling. Such a basis for experimental natural science was quite unique in the history of human civilizations; and on that metaphysical foundation has been built our practice and our understanding of science.

That practice is best described as analytical or reductionist. It is really impossible to imagine laboratory work being done on any other basis. But we can now begin to see its inadequacy for some fields of practice that are largely based on science, such as medicine. To the extent that illness is caused by social or psychological factors, or indeed by mere aging, atomic style of therapy through microbe hunting is becoming recognized as inadequate or even misdirected.

With the atomism of the physical reality goes an atomism of our knowledge of it. Thus it has been highly effective to teach science as a collection of simple hard facts. Any given fact will be related to prior ones whose mastery is necessary for the understanding of it; but to relate forward and outward to the meaning and functioning of a fact in its context, be it technical, environmental, or philosophical, is normally considered a luxury, regularly crowded out of the syllabus by the demands of more important material. This is not just yet another deficiency to be blamed on teachers. In his important analysis of "normal science" T. S. Kuhn imagines an essentially myopic and anticritical activity, "a strenuous and devoted attempt to force nature into the conceptual boxes provided by professional education [3]."

Our conception of the power based on scientific knowledge is similarly atomic. Engineers are trained to solve problems within what we can now see to be exceedingly narrow constraints: operational feasibility within commercially viable costings. The environment hit engineering practice with a sudden impact in the 1970s because of protective legislation, generally first in the USA and then elsewhere. It is understandable that engineers should find it inappropriate for the fate of important dams to depend on the breeding habits of a local fish; but it does reflect on their training and outlook when they repeatedly plan for nuclear power stations in the State of California without first checking for local earthquake faults. To be sure, the calculation of all environmental variables, including the cultural and psychological health of affected local residents, does seem to take engineering far from its original and primary concerns; but the demand for such extreme measures arises from a public reaction to a perceived gross insensitivity by engineers and their employing organizations to anything other than the simplest aspects of the power over nature that they wield.

Now we have learned that power, even based on knowledge, is not a simple thing. It is relatively easy to build a dam to hold back river water; there is power. But to predict and eventually manage the manifold environmental changes *initiated* by that intrusion is another matter. The flows and cycles of energy and materials that are disrupted by the dam will, all unknown to us, take new patterns and then eventually present us with new, unexpected problems. The dam, strong, silent, and simple, simple engineering at its most classic, may disrupt agriculture downstream (Aswan, the Nile), create hydrological imbalances (Volga), or even be interpreted as imperialism (Wales)! Hence the constant need for continuous, iterative *control,* lest an atomized knowledge, applied through myopic power, set off reactions that bring harm to us all.

We may say that a sort of atomism persists in the social practice of science, where the unit of production is the paper, embodying the intellectual property of a new result. This extends to the social organization of science in the erection of specialties and subspecialties, each striving for independence and autonomy. The obstacles to genuinely interdisciplinary research in the academic context, hitherto well-nigh insuperable, point up the disadvantages of this style for the sorts of problems we now confront. It is significant that when scientists are operating in a command economy, being employees on mission-oriented research or R & D, and not in a position to seek individual advancement as subject specialists, an effective exchange of skills *is* possible. Thus the atomic ideal of knowledge is not an absolute constraint; it can be suspended in the pursuit of knowledge as power; our present task is to see whether it can be transcended in the attempt to apply knowledge, produced by independent scientists and scholars, to the new tasks of control.

Quality control in science

We may now begin to move outward from this previous atomism, to enrich our understanding of the scientific process. Here I am trying only to make explicit what every good scientist has known all along. I may put another question concerning the personal development of each of us: When did I become aware of degrees of quality in scientific materials presented ostensibly as complete, uncontestable facts? I know that for some, either exceptionally independent or having a gifted teacher, the awareness came very early, even at school. For me, the moment was in my final year at College, when I studied a table of basic physical constants. There I saw alternative values for a single constant that lay outside each other's confidence limits. I realized then that the value of a physical constant could be quite other than an atomic fact. Among the discordant set not all could be right. Was there necessarily one correct value there; or was it a matter of judgment which cited value was the best?

The issue of quality is at the heart of the special methodological problems of biospheric science. For hard facts are few and far between; in many areas (such as rate constants for atmospheric chemical reactions) today's educated guesses are likely to appear tomorrow as ignorant speculations. The problem of achieving quality control in this field is too complex to be resolved by goodwill and redoubled efforts. Later I build on Bill Clark's ideas on making a first analysis of the task.

The problem of quality control in traditional science has quite recently achieved prominence, but still mainly in connection with the extreme and unrepresentative cases of outright fraud. The enormous quantity of patient, unrewarded work of peer review and refereeing, where (in my opinion) the moral commitment of scientists is more crucial, and more openly tested, than in research itself, has received scant attention from the scholars who analyze science. Yet quality control is not merely essential to the vitality and health of normal science. It becomes a task requiring a clear and principled understanding, if the new sciences of the biosphere are to have any hope of success. The inherited, unreflected folkways and craft skills of compartmentalized academic research are inadequate here; and here we lack the ultimate quality test of practice, realized mainly through the marketplace of industrial research and R & D.

I envisage a major effort in our project being devoted to the creation of appropriate methods and styles of quality control. I hope that this will emerge naturally from reflection on their own experience by scientists who have already been engaged in such work; but it cannot be expected to form itself automatically, without explicit attention and investment of the resources of all of us. I return to this theme in the final section.

Choice in science

My next theme is that of "choice"; here too it was Alvin Weinberg who first raised the issue, early in the 1960s [4]. Previous to that, the ruling assumption, one might almost say ideology, had been that real science required an autonomy that included choice of problems and the setting of criteria for that choice. But with the advent of "big science", the public that supported the effort through a significant burden on state expenditure was inevitably going to demand some voice in the disposition of its largesse. This is not the place to discuss the detailed arrangements, or the deeper problems of that new "social contract of science". For anyone involved with the IIASA Biosphere Project is fully aware that biosphere problems are not to be solved without massive investment of funds, in which public and private corporate agencies are inevitably, and quite legitimately, involved.

All this may seem so natural, that we must remind ourselves how new it is, and also how little impact it has made on the philosophical accounts of science to which we all go for enlightenment and guidance. There is a real gap between conceptions here; if science consists of true atomic facts, whose value lies in themselves, then what possible genuine criterion of choice can there be for research? Of course, the experience of research science is that not all facts are of equal value; they vary in their interest and fruitfulness, as well as in their internal strength and robustness. Hence policy decisions on research are possible, however difficult it is to quantify or even to justify them with conclusive arguments.

When we consider the criteria for choice governing mission-oriented projects, we find some components that are more or less internal to the process and others that are not. In the former category are feasibility and cost (this latter being measured against the demands of competing projects within some preassigned limited budget). For this we must take into account the aims or objectives of the project, which are necessarily exterior to it and different in kind from the research itself, for they employ values.

In considering these external values, I make a distinction between functions and purposes; the former refers to the sort of job done by a particular device; and the latter to the interests or purposes served, or the values realized, by the job being done. Functions are still in the technical realm, while purposes belong to people and to politics. It is at the intersection of these two sorts of effects that policymaking for sciences and technology is done.

The question of feasibility, while mainly technical, is not entirely straightforward. For the assessment of feasibility depends on a prediction of the behavior of a device or system when it is eventually created and in operation. To the extent that the proposal involves significant novelty or complexity, that prediction of the future will inevitably be less than certain. Indeed, it is now clear in retrospect that the great technological developments of recent decades were made under conditions of severe ignorance concerning not merely their social and environmental effects, but even their costs of construction, maintenance, and operation. There is an old and well-justified joke that if a cost–benefit analysis had been made at the crucial time, then sail would never have given way to steam. But many American utility companies might now reply that a proper analysis, made on their behalf, of nuclear power might have protected them from the financial disasters that now threaten to engulf them.

This point is not made by way of apportioning blame for the troubles of that once supremely optimistic industry. It can be argued that, say, 15 years ago it was impossible to predict which of the possible mishaps would afflict the industry, and how serious they would be. But in that event, we should recognize the ineradicable component of ignorance, not merely uncertainty, in forecasting the prospects for any radically new technology.

Ignorance

The pervasiveness of ignorance concerning the interactions of our technology with its environment, natural and social, is a very new theme. "Scientific ignorance" is paradoxical in itself and directly contradictory to the image and sensibility of our inherited style of science and its associated technology. Coping with ignorance in the formation of policy for science, technology, and environment is an art which we have barely begun to recognize, let alone master. Yet ignorance dominates the sciences of the biosphere, the focus of our Project.

The problems of applying science to policy purposes in general have been given a handy title, usable knowledge. For those problems of the imminent future, we would do well to remind ourselves of their nature by using a title like usable ignorance. Its paradoxical quality points up the distance we must travel from our inherited image of science as atomic facts, if we are to grapple successfully with these new problems. How we might begin to do so is the theme of my discussion here.

Elements of a new understanding

To some extent the preceding conceptual analysis follows the path of the maturing understanding of many scientists of the present generation. First, as students we mastered our standard facts; then in research we became aware of quality; as we became involved in the government of science we recognized the necessity for choice; involvement in environmental problems brought us up against functions of devices and of systems, and the frequently confused and conflicting purposes expressed through politics. Still, we could imagine that there was a hard core to the whole affair, in the sort of basic, incontestable facts that every schoolboy knows. Hence the intrusion of ignorance into our problem-situation did not immediately raise the specter of the severe incompetence of science in the face of the challenges – or threats – produced by the environmental consequences of the science-based technology on which our civilization rests.

Science in the policy process

This rather comfortable picture is analogous to the traditional model of science in the policy process. We may imagine this as a meeting of two sides. The public, through some political machinery, expresses a concern that some particular purposes are being frustrated or endangered, say through the lack of clean water. Administrators then devise or promote devices and systems, physical technology, or administrative agencies to perform particular functions whereby those purposes may once again be protected. For this they need information about the natural process involved in the problem; for which they turn to the scientists. The scientists provide the necessary facts (either from the literature, or produced by research to order) which either determine the appropriate solution, or at least set boundaries within which the normal processes of political bargaining can take place. In that way, the problem is solved or, at least, effectively resolved in political terms.

However well such a model has fitted practice in the past, it no longer captures the complexity and inconclusiveness of the process of policy-related science in the case of biospheric problems [5]. Indeed, we may define this new sort of policy-related science as one in which facts are uncertain; values in dispute; stakes high; decisions urgent; and where no single one of these dimensions can be managed in isolation from the rest. Acid rain may serve as the present paradigm example of such science. This model may seem to transform the image of science from that of a stately edifice to that of a can of worms. Whether this be so, the unaesthetic quality is there in the real world we confront and with which we must learn to cope somehow.

It may help if we employ another model: how problems come to be chosen for investigation. In the world of pure or academic science, problems are *selected* by the research community. If a particular area is not yet ripe for study, available techniques being insufficiently powerful, it is simply left to wait, with no particular loss. (The adventurous or foolhardy may, of course, try their luck there.) In the case of mission-oriented work, they are *presented* by managerial superiors; though these are expected to have some competence in assessing feasibility and costs of the research in relation to the goals of the enterprise. But in policy-related science, the problems are *thrust upon* the relevant researchers, by political forces which take scant heed of the feasibility of the solutions they demand. Indeed, it will be common for such problems not to be feasible in the ordinary sense. Drawing on low prestige and immature fields, requiring data bases that simply do not exist, being required to produce answers in a hurry, they are not the sort of enquiry where success of any sort can be reasonably expected.

It may be that our traditional lack of awareness of the interaction of ignorance with scientific knowledge has been maintained because science could proclaim its genuine successes and remain at a safe distance from its likely failures. Through all the centuries when progress became an increasingly strong theme of educated common sense, science could be seen as steadily advancing the boundaries of knowledge. There seemed no limit in principle to the extent of this conquest; and so the areas of ignorance remaining at any time were not held against science; they too would fall under the sway of human knowledge at the appropriate time.

Now we face the paradox that while our knowledge continues to increase exponentially, our relevant ignorance does so, even more rapidly. And this is ignorance generated by science! An example will explain this paradox. The Victorians were totally ignorant of the problem of disposal of long-lived radioactive wastes. They had no such things, nor could they imagine their existence. But now we have made them, by science; and the problem of guaranteeing a secure storage for some quarter of a million years is one where ignorance, rather than mere uncertainty, is the state of affairs. Thus we have conquered a former ignorance, in our knowledge of radioactivity, but in the process created a new ignorance, of how to manage it in all its dangerous manifestations.

Interpenetrating opposites in science

Science in the policy process is thus a very different thing from the serene accumulation of positive and ultimately useful factual knowledge, as portrayed in our inherited image. Indeed, given the intrusion of subjective elements of judgment and choice into a sphere of practice traditionally *defined* by its objectivity, we may wonder whether there *can* be any endeavor describable as science in such circumstances. To this problem I can only begin to sketch a solution, by giving two analyses, one static and the other dynamic. The former elucidates the paradoxical, or contradictory, nature of our situation, and the latter indicates paths to resolution of the paradox.

To begin with it is necessary for us to transcend the simplistic picture of science that has been dominant for so very long. For generations we have been taught of a difference in kind between facts and values. The latter were seen to be subjective, uncertain, perhaps even basically irrational in origin. Fortunately, science supplied facts, objective and independent of value judgments, whereby we could attain genuine knowledge and also order our affairs in a proper manner. Those who protested that such a sharp dichotomy was destructive of human concerns were usually on the romantic or mystical fringe, and could be ignored in the framing of curricula and in the propaganda for science.

Similarly, the opposition between knowledge and ignorance was absolute. A scientific fact could be known, simply and finally. It could, of course, be improved upon by the further growth of science; but *error* in science was nearly a contradiction in terms. The boundary between knowledge and ignorance was not permeable; it simply advanced with each increment of science, bringing light to where darkness had hitherto reigned. Of course, there have been many disclaimers and qualifications tacked onto this simple model; we all know that science is tentative, corrigible, open-ended, and all the rest. But the idea that a fact could be understood imperfectly or confusedly, or that a great scientific discovery could be mixed with error, has been brought into play only very recently by historians of science.

Hence we are really unprepared by our culture to cope with the new phenomenon of the interpenetration of these contradictory opposites. The impossibility of separating facts from values in such a critical area as the toxicity of environmental pollutants is a discovery of recent years [6]. And the creation of relevant ignorance by the inadequately controlled progress of technology is still in the process of being articulated by philosophers [7].

An immediate reaction to these disturbing phenomena can be despair or cynicism. Some scholars have elaborated on the theme that pollution is in the nose of the beholder, and reduce *all* environmental concern to the social–psychological drives of extremist sects [8]. Politicians and administrators can take the easy way out and treat scientists as so many hired guns, engaging those who are certain to employ technical rhetoric on behalf of their particular faction. Such solutions as these, if considered as cures, are really far worse than the disease. If dialogue on these urgent scientific issues of the biosphere is degraded to thinly veiled power politics, then only a congenital optimist can continue to hope for their genuine resolution.

Viewed statically, these oppositions or contradictions show no way through. But the situation is not desperate once we appreciate that decision making is not at all a unique event requiring perfect inputs if it is to be rational. Rather it is a complex process, interactive and iterative; the logical model for it is perhaps less demonstration than dialogue. Seeing decision making (or policy formation; I use the two terms indifferently) as a sort of dialectical process, we may imagine those central contradictions of usable knowledge and usable ignorance being transcended, or synthesized, through the working of the dialectical process.

Varieties of policy-related research

First, I show how these problems of policy-related research may be differentiated, and in such a way that the natural tendency of their dynamics is toward a resolution. Drawing on recent work by myself and my colleague, S. O. Funtowicz [9], I distinguish two dimensions of such problems: systems uncertainties and decision stakes. The former refers to the complex system under consideration, including aspects that are technical, scientific, administrative, and managerial; the uncertainties are the ranges of possible outcomes, corresponding to each set of plausible inputs and decisions. The decision stakes are the costs and benefits to all concerned parties, including regulators (both field employees and administrators) and representatives of various interests, that correspond to each decision. In each case, we have complex sets of ill-defined variables for aggregation into a single index, hence each of the dimensions is only very loosely quantitative. We distinguish only the values low, medium, and high (*Figure 15.1*). When both dimensions (systems uncertainties and decision stakes) are low, we have what we may call applied science; straightforward research will produce a practical band of values of critical variables within which the ordinary political processes can operate to produce a consensus.

When either dimension becomes moderately large, a new situation emerges; we call it technical consultancy. This is easiest to see in the case of system uncertainty; the consultant is employed precisely because his or her unspecifiable skills, and his or her professional integrity and judgment, are required for the provision of usable knowledge for the policy process. It is less obvious that even when uncertainties are low, moderate decision stakes can take the problems out of the realm of the routine. But on reflection, this is the way things happen in practice. If some institution sees its interests seriously threatened by an issue, then no matter how nearly conclusive the science, it will fight back with every means at its disposal, until such time as further resistance would cause a serious loss of credibility in itself as a competent institution, and a damaging loss of power as a result. The public sees such struggles most clearly in notorious cases of pollution, when a beleaguered institution persists in harmful policies (such as poisoning its workforce or the local environment), to the point of being irresponsible, immoral, or perhaps even culpable (industrial asbestosis is a notorious recent example). The outrage in such cases is fully justified, of course; but it is an error to believe either that those particular firms are uniquely malevolent, or that all firms casually and habitually behave in such a way. No, it is just when caught in such a trap, however

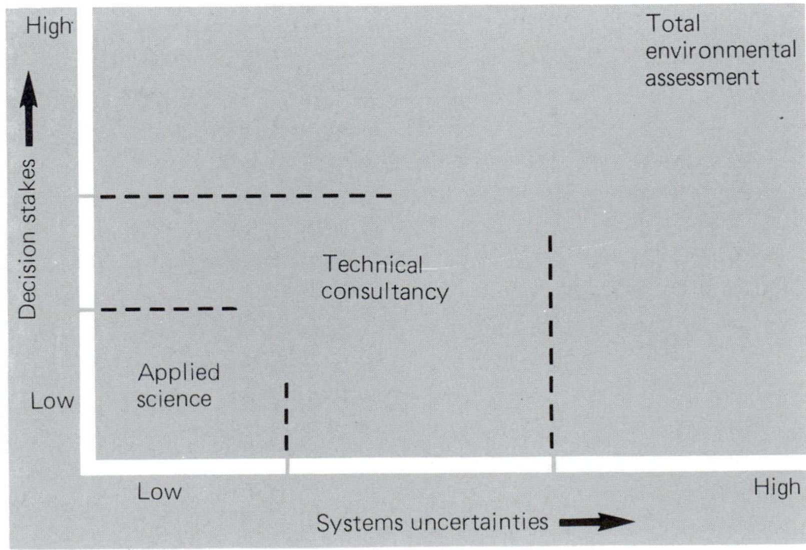

Figure 15.1 Interaction of decision stakes and systems uncertainties.

much of their own making, that institutions, like people, will fight for survival.

Such cases are fortunately the exception. It is more common for both systems uncertainties and decision stakes to be moderate. Funtowicz and I have been able to articulate a model of consultancy practice, wherein the traditional scientist's ideal of consensual knowledge is sacrificed on behalf of a more robust sort of knowledge appropriate to the problem. We call it clinical, from the field of practice in which such a style has been developed successfully. In it we eliminate safety as an attribute (the term now has a largely rhetorical meaning anyway) and substitute good performance (which may include the possibility of failures and accidents). In the same vein, we generalize probability (with its mathematical connotations) to propensity, and measure to gauge; and for prediction we substitute prognosis. In this way, we hope to express the degree to which nonquantifiable and even nonspecifiable expert judgments enter into an assessment. The outcome of the process (which is conceived as continuously iterating) is not a general theory to be tested against particular facts, but rather a provisional assessment of the health of a particular system together with the relevant aspects of its environment. I hope that this model will be useful in the Biosphere Project as it develops.

Passing to the more intractable case, where either dimension is very large, we have what we call a total environmental assessment. For here, nothing is certain, there are no boundaries or accepted methods for solving problems; the problem is total in extent, involving facts, interests, values, and even lifestyles; and total in its mixture of dimensions and components. Even here a review of history shows that in such cases a resolution can emerge. For a debate ensues, once an issue is salient; and while at first the debate may be totally polarized and adversarial in style, it may evolve fairly quickly. For both sides are attempting to gain legitimacy with the various foci of opinion, which ultimately represent power: special-interest groups, administrators, politicians, the media, respondents to opinion polls, voters. They therefore necessarily invoke the symbols of universality and rationality whereby uncommitted observers can be won over; and in however oblique and implicit a fashion, a genuine dialogue emerges. Most important in this process, new relevant knowledge is created by the requirements of the various disputants; so that the issue is brought in the direction of technical consultancy if not yet science. For example, issue-generated research can eventually transform the terms of a debate, as in the case of lead in motor fuel in the UK and Europe during the early 1980s. Events that previously had not been significant news suddenly became so; thus the various nuclear accidents of the 1950s and 1960s were of no great moment for policy purposes, while Three Mile Island was a mortal blow to the American nuclear power construction industry. Hence a problem does evolve; a dominant consensus can emerge; and then the losing side is forced into a retreat, saying what it can while the facts as they emerge tip the balance ever more decisively against it.

There is, of course, no guarantee that any particular total environmental assessment will move down scale in this way, or will do so quickly enough for its resolution to prevent irreparable harm. But at least we have here a model of a process whereby a solution can happen, analogously to the way in which great political and social issues can be (but, of course, need not be) resolved peacefully and transformed.

Debates on such issues are usually very different to those within a scientific community. They cannot presuppose a shared underlying commitment to the advance of knowledge nor presuppose bounds to the tactics employed by the antagonists. In form they are largely political, while in substance ostensibly technical or scientific. Confusion and rancour of all sorts abound. Yet, I argue, such apparently unedifying features are as consistent with effective policies for science and technology as they are for political affairs in general. And they *must* be; for the great issues of the biosphere will necessarily be aired in just such forums; there are no other forums to render these unnecessary.

The policy process and usable knowledge

Now I discuss the policy process itself, in relation to these phenomena of the interpenetration of facts and values and of knowledge and ignorance. This is not the place to develop schematic models of that process, so I will content myself with a few observations. The first is that no decision is atomic. Even if an issue is novel, even if its sponsoring agency is freshly created, there will always exist a background, in explicit law, codes of practice, folkways, and expectations, in which it necessarily operates even while reacting on the background. And once an issue exists, it is rare indeed for it to fade away. It may become less salient for policy and be relegated to a routine monitoring activity; but it can erupt at anytime should something extraordinary occur. In an earlier IIASA study [10] we saw how the concept of acid rain went through a cycle of evolution of some 15 years from its announcement by an eccentric scientist in Sweden, to its adoption as an international policy problem. Another decade may well pass before there is any effective international agreement; and yet another before the problem may be effectively brought under control. Nearly a lifetime – can we speak of a decision or the facts in such a case?

Indeed, when we look at the duration and complexity of those dialectical processes whereby a total environmental assessment problem (its common initial form) is gradually tamed, we see the necessity for a differentiation among the functions performed by the facts – or better, the inputs of technical information. Here I can do no better than to use materials recently developed by Bill Clark [11]. He starts with authoritative knowledge – the traditional ideal of science, still applicable in the case of applied science issues. This is supplemented by reporting – not in newspapers, but in the accumulation of relatively reliable, uncontroversial information on a variety of phenomena , of no immediate salience, but crucial when a crisis emerges. This is the descendant of natural history, popular in past epochs when clergymen and other gentlemen of leisure could gain satisfaction and prestige through their mastery of some great mass of material, perhaps of a locality, perhaps of a special branch of nature. The decline of this style of science, under the pressure of changing institutions and the dominant criteria of quality, is a clear example of what I have called the social construction of ignorance. Harvey Brooks has recently shown what a price we now pay for our ignorance, in the impotence of what I call the clean-up or garbage sciences in the face of our various pollution problems [12].

When science is involved in the policy process, particularly in the technical consultancy mode, then impersonal demonstrations give way to committed dialogue, and no facts are hard, massy, and impenetrable. They are used as evidence in arguments, necessarily inconclusive and debatable. In this case we invoke metaphors to describe their nature and functions; Steven Toulmin has suggested the term "maps" (not pictures, or we might say dogmas, but rather guides to action) [13]. I have developed the idea of a tool, something that derives its objectivity not so much through a correspondence with external reality as through its effectiveness in operating on reality in a variety of functions and contexts [14].

Passing to the more contested issues, we mention enlightenment, which might involve enhancing awareness or changing common sense. Perhaps the most notable example of this sort of product in recent times is *Silent Spring* by Rachel Carson [15]. Through it, the environment and its problems suddenly came into existence for the public in the USA and elsewhere. We note that this function is performed partly through the mass media; the role of investigative journalism in the press, and especially television, in enhancing the awareness of the non-scientific public (and perhaps of scientists too) should be more appreciated.

Once an issue has been made salient for the political process, then science can be a complement to interaction – that is, not being decisive in itself in any unreflective way, but correcting common-sense views, and providing crucial inputs when a debate is sharpened. To take an example from another field, the regulation of planned interference with the life cycle of embryo and fetus will not be reduced to the scientific determination of the onset of life and individuality. But, just as technical progress creates new problems of decision and regulation, scientific information can provide channels and critical points for the ethical and ideological debates on such issues.

Finally, Bill Clark mentions ritual and process; since science is the central symbolic structure of modern industrialized society, the invocation of science to solve a problem has a political power of its own. But such an action, if abused or even abortive, may lead to a wider disillusionment with the secularly sacred symbols themselves, with consequent harm to the social fabric. W. D. Ruckelshaus, sometime Administrator of the Environmental Protection Agency, has identified this danger clearly, in his warning of chaos if the agency is perceived as not doing its job [16]. Analogously, we may say that the best thing to happen to the American nuclear power industry was the outstandingly independent and critical Kemeny report on Three Mile Island [17]. If such a report had been widely and effectively denounced as a whitewash operation, the loss of credibility of the industry and of its governmental regulatory agencies could have been catastrophic.

With this spectrum of different sorts of usable knowledge, and their corresponding variety of institutions and publics, we begin to see a practical resolution of the abstract dichotomies of fact and value, knowledge and ignorance. Of course, the system as a whole is complicated, underdetermined, and inconclusive. But that means it's like social life itself, where we have many failures but also many successes. The only thing lost, through this analysis, is the illusion that the scientist is a sort of privileged being who can dispense nuggets of truth to a needy populace. Seeing the scientist as a participant, certainly of a special sort, in this complex process of achieving usable knowledge provides us with some insights on how to make his or her contribution most effective.

Toward a practical approach

Here I hope to be constructive; and I can start my argument with a topic mentioned early in my analysis of the enriched understanding of science which every researcher develops: the assessment of quality. This is frequently the first exposure of a scientist to the essential incompleteness of any scientific knowledge; not merely that there are things left to be discovered, but that the border between our knowledge and our ignorance is not perfectly defined. Even when scientific statements turn out to mean not quite what they say, they are not necessarily the product of incompetence or malevolence; rather they reflect the essential incompleteness of the evidence and the argument supporting any scientific result. In a matured field, the assessment of quality is a craft skill that may be so well established as to be nearly tacit and unselfconscious: we *know* that a piece of work is really good (or not), without being easily able to specify fully why. By contrast, one sign of the immaturity of a field is the lack of consensus on quality, so that every ambitious researcher must become an amateur methodologist in order to defend his or her results against critics.

Scientific quality – a many-splendored thing

When we come to policy-related science, that simple dichotomy of the presence or the absence of maturity is totally inadequate to convey the richness of criteria of quality, with their associated complexity and opportunity for confusion. Here I can only refer to the deep insights of Bill Clark and Nino Majone, in their taxonomy of criteria of quality among the various legitimate actors in a policy process involving science. In their table of critical criteria, they list the following actors: scientist, peer group, program manager or sponsor, policymaker, and public interest group. For each of these, there are three critical modes: input, output, and process. Mastery of that table, reproduced here (*Table 15.1*), would, I think, make an excellent introduction to the methodological problems of policy-related science.

It may well be that as this project develops, we will need to go through that exercise, if only to the extent of appreciating that the research scientist's criteria of quality are not the only legitimate ones in the process.

However different or conflicting may be the other criteria of quality they must be taken into account, not only in the reporting of research but even in its planning and execution. Now, any one of the actors in such a process must, if she or he is to be really effective in a cooperative endeavor, undertake a task that is not traditionally associated with science: to appreciate another person's point of view. This need not extend to abandoning conflicting interpretation of facts (for a fruitful debate is a genuine one) or to empathy for another's lifestyle or world view. But for strictly practical purposes each participant must appreciate what it is that another is invoking, explicitly or implicitly, when making points about the quality of contested materials.

Table 15.1 Critical criteria [18].

Critical role	*Critical Mode* Input	Output	Process
Scientist	Resource and time constraints; available theory; institutional support; assumptions; quality of available data; state of the art	Validation; sensitivity analyses; technical sophistication; degree of acceptance of conclusions; impact on policy debate; imitation; professional recognition	Choice of methodology (e.g., estimation procedures); communication; implementation; promotion; degree of formalization of analytic activities within the organization
Peer group	Quality of data; model and/or theory used; adequacy of tools; problem formulation; input variables well chosen? Measure of success specified in advance?	Purpose of the study; conclusions supported by evidence? Does model offend common sense? robustness of conclusions; adequate coverage of issues	Standards of scientific and professional practice; documentation; review of validation techniques; style; interdisciplinarity
Program manager or sponser	Cost; institutional support within user organization; quality of analytic team; type of financing (e.g., grant versus contract)	Rate of use; type of use (general education, program evaluation, decision making, etc.); contribution to methodology and state of the art; prestige; can results be generalized, applied elsewhere?	Dissemination; collaboration with users; has study been reviewed?
Policy-maker	Quality of analysts; cost of study; technical tools used (hardware and software); does problem formulation make sense?	Is output familiar and intelligible? Did study generate new ideas? Are policy indications conclusive? Are they consonant with accepted ethical standards?	Ease of use; documentation; are analysts helping with implementation? Did they interact with agency personnel? With interest groups?
Public interest groups	Competence and intellectual integrity of analysts; are value systems compatible? Problem formulation acceptable? Normative implications of technical choices (e.g., choices of data)	Nature of conclusions; equity; analysis used as rationalization or to postpone decision? All viewpoints taken into consideration? Value issues	Participation; communication of data and other information; adherence to strict rules of procedure

This important skill has been called (by Clark and Majone) "a critical connoisseurship of quality in science." One does not merely apply one's own specialist criteria blindly or unselfconsciously, however excellent or valid they may be for one's own scientific expertise or role. One must be able to assess productions from several points of view in succession, by means of an imaginative sympathy that involves seeing one's own role, one's own self, from a slight distance. It may be that I am here calling for the cultivation of attitudes proper to literary criticism, a prospect that to some may be even more alien than Zen riddles. But given the complexity of policy-related science, in response to the complexity of biosphere problems, I can envisage no easier alternative.

Usable ignorance

The preceding analysis has, I hope, made us familiar with the richness of the concept of usable knowledge in the context of incomplete science with policy implications. Now I can attempt to make sense of that paradoxical category, usable ignorance; for in many respects this defines our present task as one that is qualitatively different from the sorts of science with which we have hitherto been familiar.

First, I have indicated one approach to taming ignorance, by focusing on its border with knowledge. This should be easily grasped with the experience of research. Indeed, the art of choosing research problems can be described as sensing where that border can be penetrated and to what depth. Similarly, the art of monitoring for possible accidents or realized hazards, be they in industrial plant or environmental disruption, consists in having a border with ignorance that is permeable to signals coming from the other side; signs of incipient harmful processes or events that should be identified and controlled. Thus the technical consultancy problem is one where ignorance is managed, through expert skill, in just this way.

Where ignorance is really severe, as in total environmental assessment, then it is involved in the problem in ways that are both more intimate and more complex. For if ignorance is recognized to be severe, then no amount of sophisticated calculation with uncertainties in a decision algorithm can be adequate for a decision. Nonquantifiable, perhaps nonspecifiable, considerations of prudence must be included in any argument. Further, the nature and distribution of a wider range of possible benefits and costs, even including hypothetical items, must be made explicit. Since there can be no conclusive or universally acceptable weighting of these, the values implicit in any such weighting must be made explicit. In terms of a dialogue between opposed interests, this effectively takes the form of a burden of proof: in the absence of strong evidence on either side do we deem a system safe or do we deem it dangerous?

By such means we do not conquer ignorance directly, for that can be done only by replacing it with knowledge. But we cope with it and we ensure that by being aware of our ignorance we do not encounter disastrous pitfalls in our supposedly secure knowledge or supposedly effective technique.

The preceding account is prescriptive for future practice rather than descriptive of the past. Had ignorance been recognized as a factor in technology policy, then, for example, the nuclear power industry would today be in a far healthier state. The easy assumption that all technical problems could be solved when the time came has left that industry, and the rest of us on this planet, with such problems as the disposal of long-lived radioactive wastes. In this case we must somehow manage our ignorance of the state of human society some tens of thousands of years into the future. How many professional engineers have been prepared by their professional training for such a problem?

Coping with ignorance demands a more articulated policy process and a greater awareness of how that process operates. Great leaps forward in technology require continuous monitoring to pick up the signals of trouble as they begin to arrive; and both physical symptoms and their institutions should be designed with the ignorance factor in mind, so that they can respond and adapt in good time. (This point has been amply developed by David Collingridge in his *Critical Decision Making* [7] and other writings.)

Recognition of the need for monitoring entails that the decision process be iterative, responding in a feedback loop to signals from the total environment of the operating system. Also, the inclusion of ignorance in decision making, via the explicit assignment of burden of proof, involves a self-conscious operation of dialogue at several levels, the methodological and regulative simultaneously with the substantive. All this is very complicated, of course, and the transaction costs of running such a system might appear to be very high, not least in the absorption of time and energy of highly qualified people. But if these costs become a recognized element of the feasibility of a project, let it be so; better to anticipate that aspect of coping with ignorance than either to become bogged down in endless regulator games, or to regress to a simplistic fantasy of heroic-scale technological innovation, thereby inviting a debacle sooner or later.

Coming now to an idea about the Biosphere Project itself, I find the category of usable ignorance influencing it in several ways. First, it should condition the way we go about our work; for we will be aware that just another program of research and recommendations is not adequate to the solution of biosphere problems. Also, the concept of usable ignorance may provide topics for a special research effort within the Project. What I have described above is only a rudimentary sketch of some of the elements of a large, important, and inherently complex phenomenon. With colleagues at Leeds University, UK, I have begun to articulate themes for a coordinated research effort, involving the logic of ignorance, studies of how some institutions cope with the ignorance that affects their practice, as it reveals itself in error and failure; and more studies of how institutions cope with the threats posed by their ignorance when their monopoly of practice, or their legitimacy, is threatened.

More directly relevant to the immediate concerns of colleagues on the Biosphere Project is the way in which we will need to make our own ignorance usable. For we are, after all, inventing a new scientific style to respond to the new scientific problems of the biopshere, simultaneously with the special researches that are at its basis. We have various precedents to remind us what is *not* likely to work. The simplest is a scattered set of groups of experts, each doing their own thing and meeting occasionally to exhibit their wares. Synthesis of the efforts is then left to the organizers of the meeting and the editors of their proceedings. At a higher level, we have the experience of multidisciplinary teams, where each member must protect his or her own private professional future by extracting and cultivating research problems that will bring rewards by the special criteria of quality of his or her subject subspeciality. Here, too, the whole of the nominally collaborative effort is only rarely greater than the sum of its parts. Nor can we turn with much hope to the task force model, which does bring results in technology, for that depends critically on the simplicity of the defining problem, and on an authoritarian structure of decision and control. For our problems are multidimensionally complex by their very nature; and transnational cooperation is achieved more by cajoling than by command. Hence none of the existing styles of making knowledge usable are appropriate for ignorance.

Conditions for success

It appears, then, that we need some sort of dialectical resolution of the contradiction between the autarchy of academic-style research and the dictatorship of industrial-style development. There seem to be two elements necessary to make such a new venture a success. One is motivation: enough of us on the Biosphere Project must see it as a professional job, developing a new sort of scientific expertise in which we can continue to do satisfying work after the completion of this Project. I have no doubt that if this Project succeeds, it will become a model for many others, enough to keep all of us busy for a long time. The other element is technique: devising means whereby the genuine mutual enhancement of ideas and perspectives can be accomplished. I indicated some of these at the very beginning of this chapter, in describing some ways in which the Biosphere Project will be novel.

We may well find ourselves experimenting with techniques of personal interaction that have been developed for policy formation, but that have hitherto been considered as irrelevant to the austere task of producing new knowledge. But since we, even in our science, are trying to make ignorance usable, we should not be too proud to learn about learning, even in the research process.

The crucial element here may lie in quality assessment and the mutual criticism that makes it possible. Can we learn, sufficiently well for the task, to have imaginative sympathy with the roles and associated criteria of quality of others in different corners of this complex edifice? We will need to comprehend variety in scientific expertise, in methodological reflection, in organizational tasks, and in policy formation. If so, then we can hope to have what Bill Clark has called a "fair dialogue," in which we are each an amateur, in the best sense of the term, with regard to most of the problem on which we are engaged.

I believe that such a process is possible and that it is certainly worth a try. The environmental problems that now confront us, as residents of this planet, are now global and total. We in this group cannot hope to legislate for all of humanity over all the salient issues [19]. But we can at least indicate a way forward, showing that our civilization is genuinely resilient in meeting this supreme challenge.

Conclusion and perspective

As an historian, I like to find support and understanding in the pattern of the past as it may be extended into the future. In this connection, I can do no better than to quote from an early prophetic writing of Karl Marx. In the Preface to his *Critique of Political Economy* [20], he gave an intensely concentrated summary of past human history as he understood it, in terms of class structures and class struggles. His concluding motto was, "Mankind only sets those problems that it can solve." We must try to justify his optimism in the case of this present challenge. For we may understand it as our civilization's characteristic contradiction: the intensified exploitation of nature through the application of knowledge to power, which threatens to become self-destructive unless brought under control.

For my historical perspective on this, I would like to review the evolution of science as a social practice, as it has developed to create new powers and respond to new challenges. In the seventeenth century, the scientific revolution had two related elements: the disenchantment of nature; and the articulation of the ideal of a cumulative, cooperative, public endeavor for the advancement of knowledge. With the decay of the ancient belief in secrets too powerful to be revealed came a commitment to a new style of social relations in the production of knowledge. This was promoted as both practically necessary and morally superior. From this came the first scientific societies, and their journals provided a new means of achieving novelty while protecting intellectual property.

As this system matured in the nineteenth century, with the creation of complex social structures for the organization and support of research and researchers, the early dream of power through secular, disenchanted knowledge took on reality. For this there were developed the industrial laboratories and applied-research institutes, first in Germany, but eventually elsewhere. From these came the high technology of the present century, on which the prosperity and even survival of our cililization now depends.

The idea of using such applicable science as a significant contribution to the planned development of the means of production was first articulated in the socialist nations, and popularized everywhere by the prophetic writings of J. D. Bernal. It lost its ideological overtones during World War II; and now that planning is an essential tool even in the market-economy nations, science as "the second derivative of production" (in Bernal's phrase) is a commonplace [19]. Even academic research is now strongly guided by priorities, set in the political process, and related to the requirements of the development of the means of production and of destruction. Boris Hessen's classic thesis on *The Social and Economic Roots of Newton's Principia* may have been crude and oversimple for the seventeenth century: but for the twentieth it is a truism. There still remains a difference in slogans; in the socialist countries it is "the scientific–technological revolution"; in the others it is "don't come last in the microelectronics race"; and only time will tell how these will work out in practice.

Our present concerns are centered on the new problems of the biosphere, involving an ecological vision that ran counter to that of Bernal, and the tradition to which he was heir. The "domination of nature", the driving vision of our science-based civilization, may turn out in retrospect to have been just a disenchanted variety of magic [21]. The recently discovered fact that we cannot dominate, though we can destroy, may be the decisive challenge to our civilization. For the solution of the problem of worldwide poverty through the development of material production in imitation of the West, even if possible in the social sphere, could become ecologically devastating. Can the biosphere provide the sources and sinks for a worldwide population of a billion private automobiles? Hence, I believe the new task for science is a total one, requiring new concepts of its goals in human welfare as well as new methods of achieving knowledge and wielding power over nature under appropriate control.

This Project is necessarily restricted to our part of the problem, but it lies at the heart of it: can technological development be ecologically sustainable? If not, the prospects for our civilization resolving or transcending its characteristic contradictions are not good. Like Rome, or classical Islam, we will have had our own half a millenium of glory, and then begin a decline, enfeebled and enmeshed by problems that multiply faster than their solutions. But if we have that resilience which has characterized Europe in the past, we may yet respond to the challenge in time, and open the way to a new future for humanity.

I hope that this somewhat apocalyptic perspective does not seem too enthusiastic or grandiose. I certainly do not expect it to be shared by all of us, or even by any one in particular. But I believe that for such a task of innovation as we are engaged on, we need a perspective to give our efforts shape and direction. Indeed, we can do with several complementary perspectives out of which a synthesis, at that level, can emerge. It is in this spirit that I offer these views for your consideration and criticism.

Notes and references

[1] Lindblom, C. E. and Cohen, D. K. (1979), *Usable Knowledge: Social Science and Social Problem Solving* (Yale University Press, New Haven, CT).

[2] Weinberg, A. M. (1972), Science and trans-science, *Minerva*, **10**, 209–222.

[3] Kuhn, T. S. (1962), *The Structure of Scientific Revolutions*, p 5 (Chicago University Press, Chicago, Ill).

[4] Weinberg, A. M. (1963), Criteria for scientific choice, *Minerva*, **1**, 159–171; Weinberg, A. M. (1964), Criteria for scientific choice II. The two cultures, *Minerva*, **3**, 3–14.

[5] Otway, H. and Ravetz, J. R. (1984), On the regulation of technology 3. Examining the linear model, *Futures*, **16**, 217–232.

[6] Whittemore, A. S. (1983), Facts and values in risk analysis for environmental pollutants, *Risk Analysis*, **3**, 23–33.

[7] Collingridge, D. (1982), *Critical Decision Making* (Frances Pinter, London).

[8] Douglas, M. and Wildavsky, A. (1982), *Risk and Culture* (University of California Press, Berkeley, CA).

[9] Funtowicz, S. O. and Ravetz, J. R. (1985), Three types of risk assessment: a methodological analysis, in C. Whipple and V. T. Covello (Eds), *Risk Analysis in the Private Sector*, pp 217–232 (Plenum, New York).

[10] Whetstone G. S. (1984), *Scientific Information and Government Policies: A History of the Acid Rain Issue*, presented at the IIASA International Forum on Science for Public Policy (unpublished).

[11] Clark, W. C. (forthcoming), *Conflict and Ignorance in Scientific Inquiries with Policy Implications* (International Institute for Applied Systems Analysis, Laxenburg, Austria).

[12] Brooks, H. (1982), Science indicators and science priorities, in M. C. La Follette (Ed), *Quality in Science*, pp 1–32 (MIT Press, Cambridge, MA).

[13] Toulmin, S. (1972), *Human Understanding: The Collective Use and Evolution of Concepts* (Princeton University Press, Princeton, NJ).

[14] Ravetz, J. R. (1984), *Uncertainty, Ignorance, and Policy*, presented at the IIASA International Forum for Science and Public Policy (unpublished); an abridged version appears as Scientific uncertainty, in the US German Marshall Fund, *Transatlantic Perspectives*, No 11, April 1975, pp 10–12.

[15] Carson, R. (1962), *Silent Spring* (Houghton Mifflin, Boston, MA).

[16] Ruckelshaus, W. D. (1984), Risk in a free society, *Risk Analysis*, **4**, 157–162.

[17] Kemeny, J. G. (1979), *Report of the Presidents Commission on the Accident at Three Mile Island: The Need for Change: The Legacy of TMI* (Pergamon, New York).

[18] Clark, W. C. and Majone, G. (1985), The critical appraisal of scientific inquiries with policy implications, *Science, Technology and Human Values*, **10**(3), 6–19.

[19] Ravetz, J. R. (1974), Science, history of, *Encylopaedia Britannica*, **16**, 366–375.

[20] Marx, K. (1869), *A Contribution to the Critique of Political Economy*, p 21 (reprinted, 1971, by Lawrence & Wishart, London).

[21] Leiss, W. (1972), *The Domination of Nature* (Braziller, New York).

Commentary

S. Rayner

Philosopher of science Jerome Ravetz makes an eloquent case for the explicit recognition of the borders of scientific knowledge in the science-for-policy process. The call is far-reaching, not least because the concept of usable ignorance that Ravetz elaborates stretches the scientific imagination beyond the mere recognition of uncertainty; we may know much of which we are uncertain, little of which we are assured, and vice versa. Ravetz would have us pay attention to what is or is not known – not merely to what is uncertain about which of a known range of probable outcomes will actually occur. He is also concerned with the resilience of our institutions – those of modern civilization, not just of the scientific academy or industrial laboratory.

His proposal for resolving the dilemma of ignorance in science for policy relies on the reflexive capacity of scientific and policymaking institutions to recognize and employ different types, as well as degrees, of knowledge and nonknowledge, and to incorporate the pedigrees of scientific facts in the formal notations adopted for their expression:

> By such means we do not conquer ignorance directly, for this can be done only by replacing it with knowledge. But we cope with it and we ensure that by being aware of our ignorance we do not encounter disastrous pitfalls in our supposedly secure knowledge or supposedly effective technique [p 429].

Ravetz is arguing for a permeable boundary between ignorance and knowledge that allows us to review the antecedents and mode of gestation of a scientific fact – not just the atomized commodity that emerges from the knowledge-producing process. Usable, as opposed to banal or trivial, knowledge is in constant interaction with a state of human ignorance.

The argument is powerful, and expressed with convincing clarity. However, there is a dimension to the problem of usable ignorance that Ravetz points to but leaves unexplained. Yet this dimension is crucial to the success of his prescription.

The incorporation of usable ignorance into science for policy depends not only on the goodwill of individuals, but on the permeability of the intellectual boundaries maintained by disciplinary and political institutions. One means of boundary maintenance, long recognized by anthropologists and science-fiction writers alike, is the cultivation, within a population, of ignorance about what lies outside. In other words, we need to extend Ravetz's argument from the realm of not knowing to that of choosing (consciously or unconsciously) not to know. We must address the process of social construction and maintenance of ignorance.

Ravetz quotes Marx's *Introduction to the Critique of Political Economy*, "Mankind only sets those problems that it can solve." But, mankind's inability to solve problems does not necessarily stem from the absence of, or a limitation on, the capacity to reason; neither is it an inability to obtain data or construct facts with which to reason. It is axiomatic in cultural anthropology that problems are selected by mankind, in whole communities or specialist subsections, such as religion, science, technology, and politics, in accordance with an editing process that excludes some kinds of questions because of their societal implications. This is partly the basis of the cultural theory of risk perception.

The publication of Douglas and Wildavsky's *Risk and Culture* in 1982 [1] called the attention of risk analysts in the science-for-policy process to the argument that risks are defined and perceived differently by people in different cultural contexts. Hence, in complex societies, such as contemporary USA, it is to be expected that there will be considerable disagreement between members of various constituencies as to what constitutes a technological risk, as well as about how such risks should be prioritized for regulatory attention.

In other words, different scientific and political institutions single out for attention technological problems that they recognize as threatening the most cherished values of their preferred social order. They ignore two other kinds of problems. Most obviously, they ignore those problems for which they are confident that they have adequate institutional forms to deal with in a routine fashion. This is the case where societal trust in those institutions is maintained at a high level. The second kind of problem that both scientific and political institutions ignore includes those problems that they cannot recognize, because to do so would not merely threaten their most cherished institutional values, but would actually deconstruct the institutions themselves.

This phenomenon is known as *structural amnesia* to anthropologists, who first observed it in connection with the genealogical basis of certain East African political and judicial institutions [2]. To those who think that structural forgetfulness is a phenomenon exclusive to exotic societies, I would point out that the concept of structural amnesia has since been employed to account for the socially constructed ignorance of historical processes and geographical constraints on political action that is commonplace in certain kinds of political and religious organizations in contemporary Western societies [3].

The time has clearly arrived for the recognition of structural amnesia in science. At the same time that Ravetz was presenting the concept of usable ignorance to the Biosphere Project at IIASA, Mary Douglas [4] was addressing the annual meeting of the American Sociological Association on the topic of the scientific community's resistance to Robert Merton's account of simultaneous discoveries. During this address, Douglas implicitly pointed out that the boundaries of disciplines are particularly fertile locations for the social construction of ignorance. Leading psychologists periodically remind their disciplinary colleagues that they tend to systematically ignore the social context of individual thought and behavior, but heed to the warning is invariably short-lived. This is inevitable, given that the paradigmatic centerpiece of psychological theory is the individual psyche. To admit the sociology of knowledge into psychology would be to risk losing the disciplinary identity that distinguishes psychology from sociology. Psychology characteristically constructs its problematics in a highly individualistic framework and is therefore obliged to remain ignorant of the social context of human behavior and is consequently under strong pressure to forget it.

My argument thus far brings us directly to Ravetz's exciting contention that successful execution of the Biosphere Project requires "inventing a new scientific style," and his critique of "multidisciplinary teams, where each member must protect his or her own private professional future by extracting and cultivating research problems that will bring rewards by the special criteria of quality of his or her subject subspecialty [p 430]."

Most so-called *interdisciplinary* research is *multidisciplinary*. It consists of different perspectives being brought to bear on a problem that is defined by one of the participating disciplines – Ravetz's autarchic approach to problems. Explicit recognition that some kinds of ignorance are socially constructed may facilitate the construction of truly interdisciplinary approaches to science for policy. This will require what I understand the Biosphere Project intends, that is the joint formulation of problematics by the whole team that is to follow them up. This requires not only the good faith of those involved in monitoring their own participation, but the creation of an institutional framework in which the participants are able and prepared to call foul, and be called for fouls when disciplinary boundaries are being protected.

Notes and references (Comm.)

[1] Douglas, M. and Wildavsky, A. (1982), *Risk and Culture* (University of California Press, Berkely, CA).

[2] Evans-Pritchard, E. E. (1940), *The Nuer* (Oxford University Press, London).

[3] Rayner, S. (1979), *The Classification and Dynamics of Sectarian Forms of Organization: Grid/Group Perspectives on the Far-Left in Britain*, PhD Thesis (University of London, England); Rayner, S. (1982), The perception of time and space in egalitarian sects: a millenarian cosmology, in M. Douglas (Ed), *Essays in the Sociology of Perception* (Routledge and Kegan Paul, London).

[4] Douglas, M. (1984), *A Backdoor Approach to Thinking About the Social Order*, lecture at the American Sociological Association Annual Meeting, San Antonio, Texas.

Chapter 16 Mythology and surprise in the sustainable development of the biosphere

P. Timmerman

Editors' Introduction: The need for self-criticism and reflection on the part of those who would understand interactions between development and environment is nowhere more pressing than when dealing with our central theme of "surprise" (see Chapters 10 and 11).

Presented in this chapter is the view that our perceptions of what is surprising and what is to be expected are strongly conditioned by our necessary, but largely unexamined, myths about "how the world is". Alternative myths that emphasize equilibria or variability, limits or adaptibility, are in turn intimately bound to the cultural and institutional settings in which we find ourselves. The practical consequences of these various myths for (mis)management of the biosphere are demonstrated through a number of historical case examples. A synoptic perspective that explicitly recognizes the pluralism of social perspectives on nature and surprise is argued to be an essential foundation for usable studies of sustainable development.

Introduction

History is full of examples of managed systems that have come to grief through an inability to understand or respond appropriately to surprises. Yet failures of this kind are now more dangerous than ever before, since the margin for acceptable error in managing the Earth and its resources is rapidly disappearing.

The rise in various population and demand curves worldwide is already "bumping the ceiling" of a number of resource industries, of which the fishing industry is the most obvious. In addition, the need to maintain high, reliable outputs of goods and services from what is becoming a global network of technical monocultures [1] means that less and less of the surrounding contexts can be left to chance: vertical and horizontal integration is becoming commonplace. Little room is therefore being left to maneuver, to experiment, and to learn through trial and error. As well as on our other resources, critical pressure is being placed on the resources of time and space – time in which to recover, rehabilitate, or rethink; space in which to regroup, restock, or reinvade.

It therefore becomes necessary, in however paradoxical a form, to consider the question of possible future uncertainties, crises, catastrophes – surprises – to come seriously to terms with truly sustainable development. The paradox is, of course, that these particular threats to sustainability are largely due to the very uncertainty or improbable nature of what might happen; and that is – by definition – difficult to conceptualize.

In recent years, and perhaps in part as a cultural response to the increased concern with the future trajectories of various natural and social systems, "the expectation of the unexpected" has become a major theme of research in various areas, including mathematics, risk assessment, economic theory, systems ecology, and anthropology [2]. New theories and studies focus on the laws of chaos, on discontinuity and on disequilibrium; and this seems to be symptomatic of the beginning of a paradigm shift [3] from the realm of the normal to that of the abnormal or anomalous [4].

Paradigms and paradigm shifts are easy to proclaim and notoriously difficult to articulate. In the case of a new paradigm deliberately focusing on anomalies, the creative chaos of the shift becomes as important as the coherent pattern of any ultimate order [5].

During paradigm shifts (the world between paradigms, so to speak), the barriers between disciplines drop, just as a reversal of poles temporarily eliminates magnetic field strength. In this interim period, new combinations of sources, ideas, and conceptual models may flood in – some of which are bogus, but others of which will persist to form the groundwork for better ways of explaining and exploiting hitherto anomalous material. Of particular interest are those new combinations that recur in a wide array of subjects, indicating that similar sorts of conceptual crises are occurring across a broad front. This is what seems to be happening today with the concepts of the unexpected, uncertainty, disequilibrium, and surprise. These concepts are of particular concern for the development of plans for long-term biospheric management, both because of their obvious relevance to attempts to manage systems of bewildering complexity and because the proposal to consider the biosphere as a sustainable system provides a framework within which these concepts can begin to be examined, compared, and linked together in real-world applications.

To put all this into some tentative ordering for discussion, it has recently been suggested [6] that one approach to the management of biospheric uncertainty might be to examine existing and proposed conceptual combinations as forms of mythology. One version of this approach is provided by Holling (Chapter 10, this volume) in his discussion of ecological systems. What follows is another elaboration of the same general approach; but it is also an exploration of some other equivalent combinations of myths and myth analyses that are available in other areas subject to the same conceptual pressures.

Why myths and myth analyses? One simple answer is that many past failures of management have been due to the misapplication of various myths, especially myths about nature. A more complex justification is that, like recent terms such as vulnerability or resilience [7], surprise (which we use as the generic term) has come into use to deal with the qualitative character of quantitative events. It is an attempt to bridge (or perhaps paper over) the gap between some event or functional response of a system and the interpretation of that event or system by observers or managers.

As soon as this type of concept is employed, we enter a world where interpretation is unavoidable, and where the system being interpreted is subject to historical, irreversible time. Time, that is, wherein learning and adaptation, success and failure, can make sense. When we employ concepts like surprise, vulnerability, and resilience to evaluate the long-term performance of the biosphere or its subsystems, we are projecting possible interpretations of perceived reality over a time frame that is presumed infinite (sustainable) unless proved otherwise. This is what Thomas Schelling, in reviewing options for dealing with the carbon dioxide (CO_2) issue, calls "the problem of managing an indefinite succession of future times [8]."

There is one major constraint on this projection into infinitely unfolding historical time: the bounded space of the planet. This is the tension at the heart of the phrase "sustainable development of the biosphere" – sustainable development has time as its primary dimension, while the biosphere is the present physical and biological space wrapped around the globe. The boundedness of the biosphere has always been there implicitly, but the rise of the theme "spaceship Earth" [9] and the powerful images of Earth recently projected back to us from space have crystallized the concept of the limited sphere into a cultural artifact of profound significance. This was particularly well articulated by H. Marshall McLuhan, who wrote the following in 1970:

> Whereas the planet had been the *ground* for the human population as *figure*, since Sputnik, the planet became *figure* and the satellite surround has become the new *ground*.... Once it is contained within a human environment, Nature yields its primacy to art, and experience yields precedence to knowledge. The consequent upgrading of man's responsibilities for the planet as environment would seem rather obvious [10; emphases in original].

Part of this new ground is expressed in the efforts described elsewhere in this book – consideration of what would be entailed in upgrading our responsibility for the planet as environment (i.e., nature yielding its primacy to human art). Another part is expressed in what could be called an implosion of sensibility: once one draws a boundary around a field of activity, the internal connectivity of the parts becomes a central concern, as does the threat of internally generated surprises. Less and less can be taken for granted, since proliferating internal disruption is now conceptually *the* source of potential catastrophe. This has been summarized eloquently by Kenneth Boulding:

> The most worrying thing about the earth is that there seems to be no way of preventing it from becoming one world. If there is only one world, then if anything goes wrong, everything goes wrong. And by the generalized Murphy's Law, every system has some positive probability, however low, of irretrievable catastrophe. . . . Ultimately, of course, we must face the spaceship earth on earth. Uncertainty, however, is the principal property of the future, and time horizons themselves have an irreducible uncertainty about them [11].

To summarize then, in Holling's terminology, we have entered into a new set of mythological assumptions. Sustainable development itself can be considered as a modern form of one traditional myth, usually known as "utopia creation". This type of projection carries the potential for a number of mythical variants, ranging from the apocalypse of nuclear holocaust to the creation of earthly paradise.

What makes these particular myth projections so interesting is both that they are now capable of being modeled in great detail by computers and other modeling tools and that in their detail they make a claim on the actual future of our planet. Because of this, it is worth exploring the matrix from which these conceptual structures arose, and the previous myths that they seek to supplant.

Equilibrium myths

The myths that have held sway until recently in modern western civilization have been myths of equilibrium. These myths tend to downplay questions of instability (or the unpredictable) in favor of stability, control, and idealization against the backdrop of absolute time and space. The rise of these myths owes much to the context supplied by Newton and Laplace, whose physical theories specifically denied intrinsic unpredictability – Laplace assumed that given enough information, it would be possible to predict the position of every particle in the universe – and implied that the passage of a system through historical time was only an interesting variation on the true use of time as an absolute and therefore reversible measure [12].

More than that, however, equilibrium myths owed their power to the need to create a series of models or methods of explaining the world without constant referral to an external carrier of meaning. This need occurred with the gradual withering away of belief in a correspondence between the world and the designs of a God "out there". Newtonian mechanics pushed God as far off the stage as possible, but was unable to cope with the question of complex design – hence the eighteenth century's rhetorical gestures toward "the great chain of being" and "the absent clockmaker", not one of which was wholly convincing.

The need to explain complexity and diversity is often mediated in myth through the exploration of analogies in nature. The natural world presents the spectacle of a very complex, diverse system, apparently operating without interference. This was one of the strongest arguments throughout history for the existence of a God or gods; and mythmaking has often taken its cue from cultural attitudes toward nature. In the case of the modern west, however, the key myth creation began elsewhere, and was then transferred back to the natural world.

The key equilibrium myth for our society derives from the internally generated, self-regulating, and self-sustaining world of economics. This myth began in the confluence of the philosophy of natural sciences and the study of social systems that characterized the French and Scottish Enlightenments, and culminated in the work of Adam Smith [13]. These philosophies and studies attempted for social systems what Newton had done for physics: i.e., to discover absolute laws based on simple, active principles.

The simple active principle of the social myth was discovered by an early eighteenth century moral philosopher named Mandeville, who wrote a satirical work entitled *The Fable of the Bees* [14]. In this satire, Mandeville put forward an analysis of social interactions that made the paradoxical claim that the private vices of fraud and greed necessarily promoted the public virtues of peacefulness and wealth. Mandeville argued that seemingly chaotic or counterproductive behavior at the individual level could generate order at the societal or macrocosmic level. This model was seized on by a number of moral political philosophers, of whom Adam Smith is the best known [15].

Smith's version, immortalized in *The Wealth of Nations* [16], used the notion of the invisible hand to finesse metaphorically the question of design in the system. Then, as classical economics developed, the explanatory power of this microcosmic competition–macrocosmic order model spread into other areas. The strength of the model lay in its elegant mechanical purity and its ability to bring order to a vast array of hitherto unrelated data.

The two major weaknesses were, first, that the need for an explanation of order led economics into a myth of ideal historical equilibrium, i.e., that the system is always at or approaching a perfect stability; and, second, that the mythical model was intolerant of any externalities or alternatives that could not be captured by the mechanism. These weaknesses (which some would call strengths) have been the source of some of the main challenges to the modern order based upon this basic myth:

(1) That the system is not tending toward equilibrium, but crisis (e.g., Marxism).
(2) That nature and nonquantitative aspects of the world should not be treated as externals or pure input resources for some homogenizing market (e.g., environmentalism).

The foundations of the crisis approach, exploited by both crisis-seekers and the environmentalists, were provided by Malthus [17]. Smith's economic scheme, deriving as it did from natural law theory and with its roots in classicism, was not necessarily devoted to the infinite expansion of the economic model; rather, Smith's world is one characterized by balance and controlled equilibrium, where supply and demand eventually converge on the real natural price. Within one or two generations, however, Smith's model was rapidly harnessed to arguments for the infinite possibilities of progress. Equilibrium – without the constraint provided by natural law (or God) – became the market without any natural prices, achieving pseudoequilibrium in the short term as it moved toward long-term prosperity. True stability was shunted aside.

Malthus' proposal that the natural tendency of the economic system and the actors within the market was, in the long term, to crash against a population limit, cut – right at the outset of the refashioning of the model – to the heart of equilibrium as progress. Hence, the revulsion against Malthusianism, and the long-standing refusal to come to terms with it.

Malthus' argument was defeated, not – as often argued – by the wealth generated in the nineteenth century capitalist boom; but by the redirection of economics away from classical economic theory with its fundamental concern for the elements of production – land, labor, and potentially scarce resources – toward the utilitarian calculations of neoclassical economic theory, which put Smith's balanced economic model at the service of a new infinity: the infinite capacity of subjective desires, as expressed through the auction of the marketplace. The market was the arbiter of whether or not limits and scarcities were being approached. The underlying processes that produced those limits and scarcities were no longer relevant, unless they could be expressed in the market.

Malthus was not refuted; just rendered irrelevant. And, in this context, it should also be recalled that it was John Stuart Mill, utilitarian and economist, who in his *Principles of Political Economy* both described the transformation of economics that was taking place and also repudiated it by being the first to mention the need for an ultimately stationary economy – the forerunner of sustainable development [18].

The bypassing of Malthus helps to explain why equilibrium-as-progress also interpreted Darwin's theory of evolution in the way that it did. Darwin's discovery of the principle of evolution through natural selection following upon his reading of Malthus is legendary, and has often been cited as an example of a theory that reflected the times in which it was discovered [19]. What has not been so often noted, however, is that the myth of equilibrium in its new progressive form was so strong that it ensured that much of the subsequent history of Darwin's theory would be devoted to evolution and competition, rather than to the vagaries of natural selection.

That survival of the fittest, for example, might mean fittest for any and all environments, however harsh, was unacceptable, since certain environments – the present, for example – are valued higher than others. This meant that the historical graininess of time and space was downplayed [20]; and, furthermore, the present became a stable moral model to which nature had been tending all along.

This, then, is the context in which equilibrium myths arose and developed. They have been characterized in many forms, of which the simplest is perhaps the graphical representation of a point at the bottom of a valley or cup-shaped curve, like a pebble in a bowl (*Figure 16.1*). This represents the basic equilibrium myth of stability.

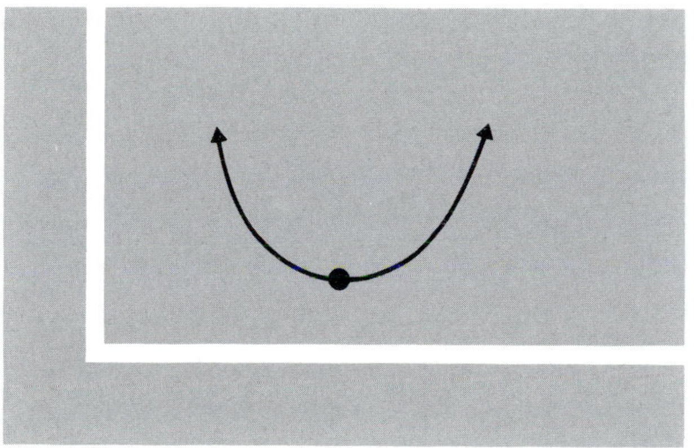

Figure 16.1 Stability.

As the previous discussion has perhaps made obvious, the myth of stability concentrates on the bottom of the bowl, with all perturbations only temporarily knocking the point away from that center. Since the point is always defined by its position relative to its resting place, there is, in principle, no outside to the bowl – indeed, any perturbation so strong as to cause the point to slip over the edge of the bowl is both physically and theoretically catastrophic, since it destroys the principles by which the myth is constructed. Any management strategy based on the myth of stability is haunted by this kind of vulnerability, and responds to it by making at least two interlocking presumptions. The first is that the best defence is to keep raising the sides of the bowl, preferably to infinity (for, as noted, the tendency to project this equilibrium system to infinity is always present). This is either done by building a lot of excess capacity or redundancy into the system, or by damping down all oscillations to consistently predictable levels.

A second presumption of the myth of stability – especially when it is used to manage natural systems – is that nature is infinitely reliable or benign. Indeed, the natural myth associated with the myth of stability is the myth of nature benign [21].

This secondary myth is necessary to the first myth: when one fish stock is exploited, go on to the next, and the first one will recover; if we could only remove this particular stressor, the ecosystem would respond favorably and predictably. Any unexpected natural event is either treated with bewilderment or shunted into the category of "Acts of God" (a revealing phrase). It is essential to the overall myth of stability that consistency and continuity of control be maintained; and it is this consistency and continuity that serves as justification for many of the management schemes in operation: surprises are threats [22].

The obverse of the myth of stability is the *myth of instability*. It, too, concentrates on the center of the bowl, but, in this case, the bowl is inverted, with the pebble on the outside. Any perturbation and the pebble rolls off down the slope (*Figure 16.2*). In natural systems, the best example of this may be a simple closed predator–prey relationship – a proliferation of prey leads to a proliferation of predators who eventually consume all the prey and then starve to death. Alternatively, one might have a system so fragile that a single, or very small, event

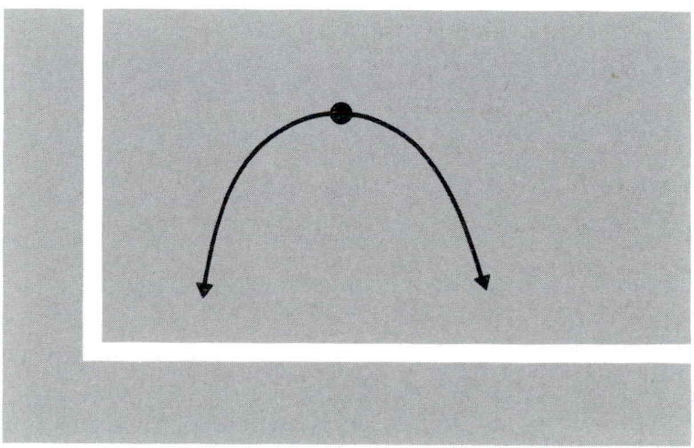

Figure 16.2 Instability.

or impact cascades through it, shattering it swiftly. Complex systems sometimes have this property, such as rockets that explode on launch due to a misplaced wire; classic examples may be found in systems of speculation, such as the South Sea Bubble of the 1720s, where a single default can quickly collapse an array of interconnected rumor.

As with the myth of stability, the myth of instability is monolithic, since everything that happens tends toward the system's destruction. As a result, its secondary nature myth is that of nature malign or intolerant. Management of the system – if there can be any management of such a system – is paranoic: the whole system is vulnerable and everything is a threat. This is the reason why the myth of instability is the simple obverse of the myth of stability, and why it has been categorized as an "equilibrium" myth – it is the demonic version of the myth of stability, obsessed in a completely uncreative way with the absolute maintenance of the *status quo*.

The third equilibrium myth, which introduces somewhat more complexity to the process of categorization, is the *myth of cyclical renewal.* It is obviously a variation on the main myth of stability, but it brings into play a sense of rhythm or pulsation. nature in this myth is benign overall; but it has regular local malignities – e.g., winter – and while the regularity of the seasons is predictable, the size of the oscillations may not be predictable year by year. This cyclical process brings into play for the first time the question of historical time; and yet neutralizes it by placing it in a broader perspective (*Figure 16.3*).

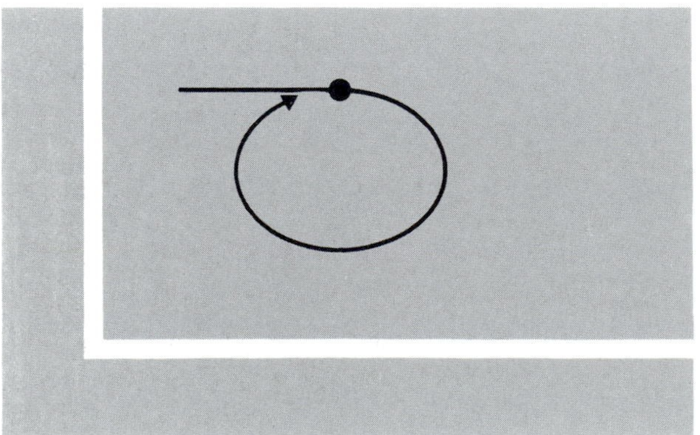

Figure 16.3 Cyclical renewal.

We are, however, introduced to the issue of scales of time [23], which begin to play a greater role as the myths complexify. It is all very well to assume that the system will eventually and regularly return to where it began, but the length of the phases of the cycle could be of importance; after all, as Keynes remarked, in the long run we are all dead. More to the point, cyclical renewal hints at the possibility that the system might be traversing a number of unstable positions that make the system appear to be in disequilibrium when looked at in a restricted or short time-frame. The only way to be assured that one is in long-term equilibrium is either to obtain some external perspective on the trajectory of the system or to operate through a complete cycle to the point where those inside the system recognize that repetition is occurring.

In addition, one wants to know if the cycle is truly cyclic, or if it is following a spiral trajectory. The incremental alternation in the trajectory may be masked by the greater yearly fluctuations or noise in the system. A now familiar example of this is provided by the Mauna Loa data on the carbon dioxide (CO_2) build-up in the atmosphere (*Figure 16.4*), where the breathing of the Earth fluctuates according to the seasonal generation of CO_2, and yet the overall trend is upward.

Variable myths

With the introduction of cycles and spirals, equilibrium myths begin to depart from their original simplicity. Eventually, perhaps as short-term disequilibria multiply, or as the dissonance between the myths and reality strengthens, alternative mythical frameworks develop as possible replacements or modifications.

The process takes two forms: one conservative and one radical. The conservative replacement still focuses on the concept of equilibrium, but the number of possible equilibria is multiplied. Using a landscape metaphor, the topography is seen as a number of peaks and valleys, not just one peak or one valley. The radical replacement, however, considers not just the points moving up and down the peaks and valleys, but also the possible movements of the

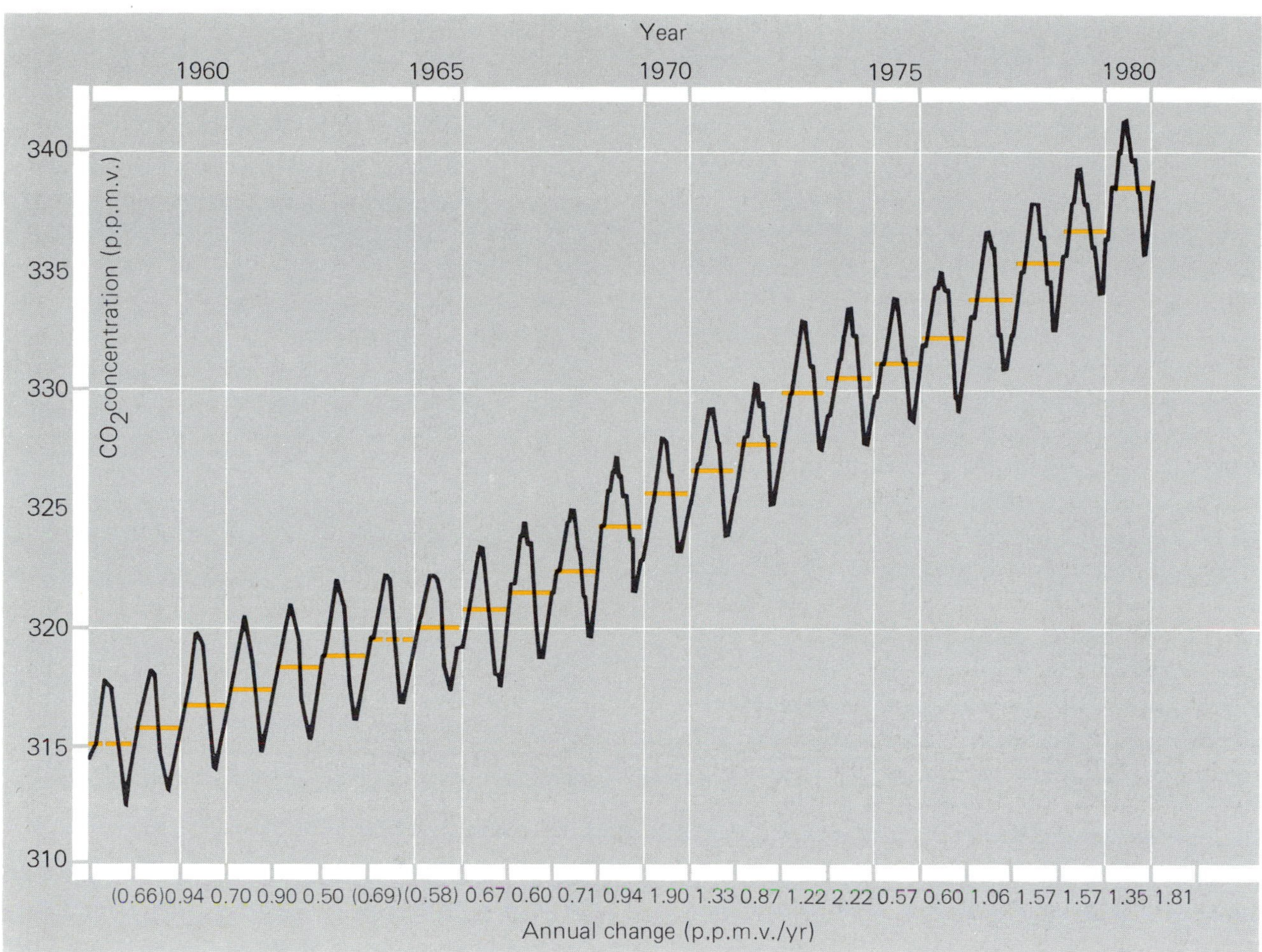

Figure 16.4 CO_2 oscillations: mean monthly concentrations of atmospheric CO_2 at Mauna Loa. The yearly oscillation is explained mainly by the annual cycle of photosynthesis and respiration of plants in the northern hemisphere [24].

peaks and valleys themselves. The first could be called the *myth of multiple stability*; the second the *myth of resilience.*

The most important early version of the myth of multiple stability – the version that historically seems to have been the first to displace the equilibrium myth from its commanding position – was, appropriately enough for this discussion, Keynes' attack on neoclassical economics in the 1920s and 1930s. The heart of Keynes' approach was the argument that, on their own terms, neoclassical economists could be faced with the possibility that the free market might well achieve equilibrium, but that equilibrium would be stuck at a point too low to achieve the benefits it was supposed to produce. However, if additional investment was created or empowered by government, the economy would be "kick-started", possibly up to a new, higher full equilibrium. Keynes contended that the evils of the depression were the result of a low equilibrium and that government management could eliminate them [25].

Apart from justifying government involvement in the marketplace, Keynes' theory brought into the open the necessity of incorporating real time into economic theory and also the necessity of considering the operation of uncertainty in the minds of investors during that time. The owning of money (in, for example, the form of savings) was not just for convenience; it was also a way of keeping one's decision-making power free or liquid in the face of an uncertain future.

The incorporation of learning, anticipatory behavior, and uncertain future into economic theory has been the springboard for much of the recent theoretical controversy in that discipline. The first generation of post-Keynesians, including Kaldor, Shackle, Sraffa, and others, explored the ramifications of what an economic theory plunged into historical time would look like, including the need to examine surprises and other forces that might thwart the perfect market [26]. And today, post-Keynesians and neoclassical economists are locked in controversy over the issue of "rational expectations", an issue explicitly relating surprises and historical time to the adequacy of the market model and, by extension, to the orthodox myth of stability [27].

According to rational expectations theory, people make mistakes in predicting the future, but they do not make the same mistakes over and over again: they begin to anticipate and they also learn how government monetary and fiscal policies reverberate through the economy. So, for example, inflationary expectations begin to be incorporated into wage settlements. For neoclassical economists (now styling themselves New Classical Economists) these expectations are instantly captured by the market, which is only seriously knocked aside by completely unpredictable surprises. This, from our perspective, is clearly another version of the myth of stability. For post-Keynesians, however, the focus is not on the market, but on the ways in which producers and consumers circumvent the market, as expressed in the aphorism: "The free market is a wonderful thing – for everybody else." Independent price-setting based on markups over costs of production, long-term wage contracts, and oligopolies are only some of the underlying structural inflexibilities veiled by the market. These underlying inflexibilities help to explain why the economic system goes into periods of involuntary destabilization and recession.

The post-Keynesians subscribe to the myth of multiple stability and their interpretation of how an economic system can suddenly produce surprises and descents to new, unfavorable, equilibrium points is practically identical to the myth of multiple stability now envisaged by those trying to manage natural systems. Many of the unexpected consequences of natural-system management are due to the basic assumption that there is one equilibrium point for the system, while, in fact, there may be a number of alternative points (*Figure 16.5*). Sudden perturbations or a long-term policy that ignores the underlying dynamics of the system can cause it to fluctuate wildly and unpredictably. In other words,

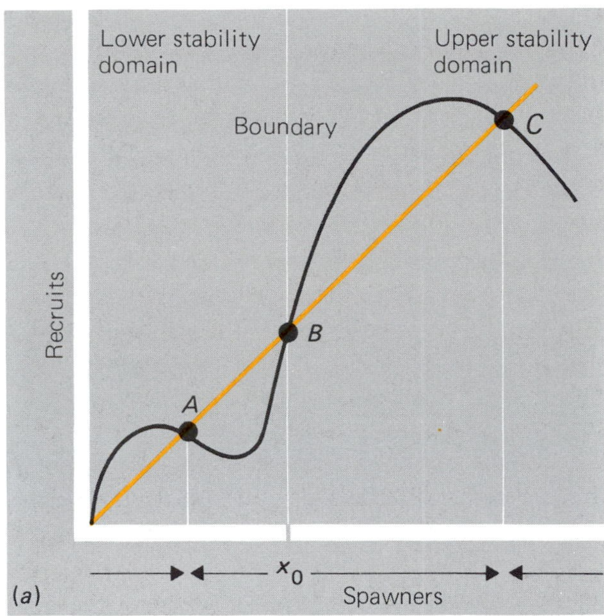

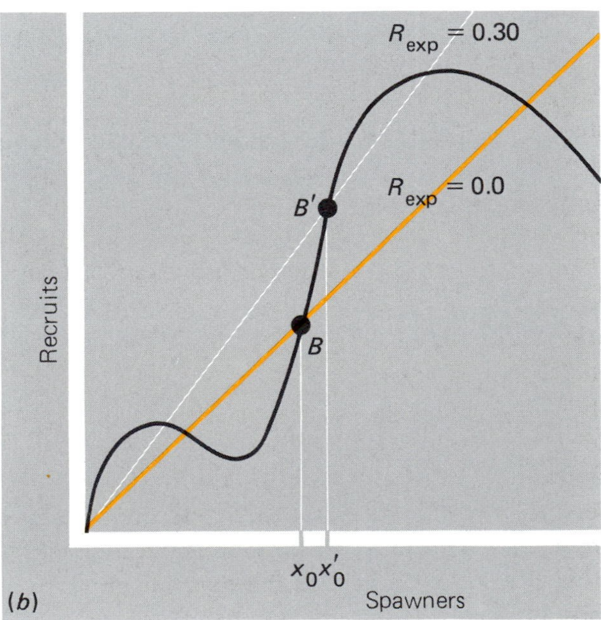

Figure 16.5 Two stability domains. The effect of predation mortality on recruitment of salmon population. (*a*) The size of the unstable boundary population under severe predation: there are two stability domains. (*b*) The replacement line (where recruits equal spawners) can shift, given changes in the exploitation rate R_{exp}. The size of the boundary population then shifts from x_0 to x'_0 [28].

the real danger is in trying to apply a myth of stability – where nature is perceived to be infinitely benign and homogenous – to a natural situation where (at the very least) a myth of multiple stability may prove more conducive to understanding.

Holling *et al.* [29] have collected a number of examples in which resource management has been counterproductive because it was based on a hope that the myth of stability was present, along with the secondary nature myth, that nature is essentially cornucopian. A more sophisticated version of this management myth recognizes that going from stock to stock, replacing one species with another as the first becomes depleted, is inadequate; but by concentrating on trying to guarantee that one particular species will not be depleted, it is hoped that one can manage the stocks of that one species while ignoring all the rest [30].

Unless this management strategy is carried out with a great deal of knowledge and understanding, it is prone to producing a myriad of surprises, which perhaps give rise to a new natural myth – nature as extraordinarily capricious, risky, or devious – requiring more and more top-heavy absolute control solutions.

A policy of maximum sustainable yield for one species may eliminate some other species or alter the underlying dynamics of the ecosystem. The system is then prone to producing unwanted surprises. These surprises can be economically damaging, since, in many cases, harvesting industries develop on the prospects of stability, and become locked into the highest supposedly sustainable yield due to their massive investments in transport, equipment, and processing facilities.

This locking-in phenomenon, the result of taking a short-term equilibrium as a consistently sustainable equilibrium, is perhaps the single most important example of the failure to shift mythologies when trying to achieve sustainable development. It stems from the power of the original myth of stability which still hovers behind the myth of multiple stability, since the heart of the modern technocratic apparatus depends upon the capacity of the original myth to flatten all externalities into market factors – all inputs are infinitely substitutable; all perturbations are temporary; the short term is indistinguishable from the long term. Only above a certain threshold are events or commodities important; below that, neither the mechanics nor the long-term state of the system is of detailed concern.

A now-famous example of this unconcern was provided by the collapse of the Peruvian anchoveta harvest after the 1972–1973 El Niño in the southeastern Pacific. Despite quasi-periodic dramatic oscillations in harvesting possibilities and repeated warnings and regulations concerning overfishing, the anchoveta industry grew. The resultant collapse meant that, with a 30% reduction in the fishing fleet, the catch in the mid-1970s was stuck at a level approximately one fifth of what it had been [31].

More complex versions of multistable phenomena are provided by predator–prey interactions with various forms of density dependence, foraging tactics, and refuges. In work, for example, on spruce budworm infestations in New Brunswick [32] it has been shown that a whole range of alternative equilibrium points (and zero equilibrium points) can be generated by different mixes of the parameters of the system and by various management practices that incidentally create an emphasis on – for instance – one predation type as opposed to another.

Furthermore, Holling and others have explored many ecological systems in which the jump from equilibrium point to equilibrium point is prompted by major phase changes in the system itself. In these situations the system can be driven into irreversible changes, often by means of a sudden "flip-flop" from one stability domain to another [33].

Another example is provided by the recent work of Steele and Henderson [34] on modeling pelagic fish populations that have interesting fluctuation patterns over the long term (e.g., Pacific sardines and Atlantic herring with 50 to 100 year peaks). These species, and others, are now being heavily fished and the fluctuations are becoming more rapid. The authors suggest that these fluctuations are jumps from one to another of three equilibrium states:

(1) A state in which population density is high in the absence of predators and is based on resource levels.
(2) A state in which population density is low in the presence of high predation levels, and is regulated by the predators.
(3) An intermediate state where the system oscillates between (1) and (2).

This is the now familiar multiequilibria world,

except that the Steele–Henderson model deliberately incorporates environmental fluctuations over time (through the use of red noise spectra of fluctuations appropriate to the sea). The model emphasizes that increases in fishing pressure or environmental change accelerate the flip-flops of the system.

In other words, operating the system under the assumption of natural persistence is not enough: the nature of that persistence needs to be examined; and the speed and quality of the oscillations of this persistence can be more important than that they keep recurring.

Of even more importance, however, is the recognition in Holling's work and in other recent models (such as that provided by Steele and Henderson) that the myth of multiple stability is not good enough to cope with a system that is being transformed by external management or that is learning or adapting to various cycles of perturbation. Assuming that there are a number of resting points in a system is only slightly more dynamic than assuming there is only one. The real paradigmatic breakthrough is toward handling a situation where everything may be changing at once.

It is here that the myth of resilience begins to take shape. Equilibrium myths are ways of picturing nature as *natura naturata* – i.e., nature as object, fixed or fixable. The myth of resilience, on the other hand, sees nature as *natura naturans* – nature naturing – i.e., nature actively altering and responding in various ways to predictable or unpredictable stresses. This means that not only must one account for the internal structure of a system and its potentialities, but also the external context of chance and unpredictable impacts must be incorporated. This is because the system is presumed capable of some sort of adaptive memory, i.e., learning through historical time.

What makes the myth of resilience so difficult to describe is that it is trying to accomplish so many things at once, against an existing frame of stable conceptual reference. Resilience quite deliberately and radically embraces just those heresies expelled by stability: ecological systems are neither in equilibrium nor disequilibrium; they are not idealizations, but realizations; chance is not only not expelled, but is often a necessary condition for system maintenance; diversity is welcomed, not steamrollered; internal dynamics are ruled not just by the competitive microcosm, but by intermediate and whole system factors – and by chance. Furthermore, a system may have a trajectory that is best observed without consideration of ideal equilibrium or climax; but as of interest in itself.

This kind of conceptualization might be illuminated by looking at recent developments in evolutionary theory and paleontology. The original attack on equilibrium Darwinism, launched by Stephen J. Gould and others, argued that evolution was in fact operating by punctuated equilibrium; that is, that species often formed quite rapidly and then exhibited stability for millions of years [35]. This suggested that high rates of speciation or spectacularly effective survival strategies might be as important as gradual honing on nature's constantly turning grindstone.

However, hardly had this theory achieved prominence when theorists were faced with the possibility that mass extinctions in the past might have been the result of sudden chance events such as the postulated impact of an asteroid on Earth at the end of the Cretaceous period [36]. This new catastrophism implies that often it may be the norm which is anomalous and that the best survival strategy may be helpless against unplanned perturbations on a large scale. Evolution thus becomes a process of surprises, mistakes, good luck, substantial failures, dead ends, and useless victories on many scales: and the world as we see it becomes a unique historical artifact.

This example may help to explain why much of recent ecological discussion has used language with a qualitative or interpretative bias – the new myth requires a new language for coping with the complex relationship between quantitative events and their qualitative expression through systems that are learning as they go along. This language must also describe how systems internalize externally generated predictable and unpredictable events. Hence, the concentration on adaptation and resilience, and the continuing controversy in ecology over whether or not diversity increases or decreases stability. These can be seen as attempts to capture conceptually the ways in which systems react to uncertainty through historical time. Two of the best known of the resilient approaches to ecological management – stress ecology and adaptive environmental management – are consistently groping with this issue [37].

Vulnerability and resilience

The controversy in ecology of diversity/vulnerability versus diversity/stability provides a good example of this situation. At its most basic, the issue depends on the ambiguities inherent in the word stability, and that depends – as we have seen – on whether the stability in question is local, global, single equilibrium, multistable, or even just the persistence through time of the shell of what was formerly a complex, thriving ecosystem.

The original conceptual model that related complexity to stability was intuitive (as much as anything), taking something from the myth of stability, as expressed in the "climax" equilibrium ecosystem model, and something from the vulnerability of simple predator–prey oscillations without time or refuge resources [38]. We have already mentioned the possible dangers of simplifying an ecosystem by single-minded management and the equation *simple equals unstable* is superficially plausible. Models have shown, however, that – to take only one example – adding more trophic levels to a system may make it less resilient. A single system such as a desert might be more persistent (and therefore resilient?) than a tropical forest which has a multitude of species, but which is essentially fragile. Yet if the forest covered a very large area, it might be difficult to change or destroy, except locally.

Instances such as this could be multiplied, given a refusal to specify what kind of systems are being compared, what kind of goals are best for the system, or what kind of perturbations it is likely to face. For example, on this last question the work of Connell, Grime, Tilman, and others indicates that a certain amount and type of stress promotes diversity in an ecosystem, by analogy with the way that the presence of a predator may keep a larger number of prey species in lower equilibrium than if the predator were removed, thereby allowing one prey species to dominate [39].

The fact is that essential to diagnosing the stability or the vulnerability of a particular system is a continuing reference to the historical and possible future contexts within which it operates. Not only has the system adapted to that context in most cases, it has also adapted the context to its own ends. The resilience of a system is due, in part, to the sum of past vulnerabilities that it has overcome and to the relevance of that learning experience to future events. It may only appear to have been resilient in retrospect, if it has been lucky enough to survive, lucky enough perhaps to meet a crisis for which it had been able to prepare. In this sense, a resilient system has as its main defense the law of probabilities which it has drawn up for itself and stored as information in the form of species, overall structure, or functional availability.

The myth of stability, as has been argued, operates on the principle of damping oscillation and maximizing available resources in defense. It is a static response to a kaleidoscopic world, and is a *survival* rather than an *adaptive* strategy.

The myth of resilience, however, assumes that all things are vulnerable and that it is essential for long-term survival (which means adaptation) to learn from events and to be lucky enough to have encountered a sequence of appropriate events. Since very often the learning process is dangerous, a resilient system treads a very fine line between learning and collapse, i.e., it often deliberately puts itself into hazardous situations – in other words, it courts surprise.

Surprise

The heterogeneity of time and the complex physical relations between a system and its context necessarily involve the formation of surprises. The attitude of a system and its managers to such surprises is, as has been (we hope) amply stated, part and parcel of its primary myth. Again, as was stated in the Introduction, surprise is an elusive concept, fraught with paradoxical uncertainty.

There are a number of ways of exploring surprise, but here it may be wise to emphasize the interactions between an event, the perceptions of that event, and the basic frame of interpretative reference which may accept or reject the implications of any particular surprise. To invoke interpretation is once again to invoke myth; and in trying to develop surprise into a working concept, it happens that actual myth criticism and anthropology have a particularly interesting role. This could be coincidental, but it is more likely to be due to simultaneous conceptual developments across a broad intellectual front.

The myth framework proposed by Northrop Frye, enlisted here because of his concentration on the transformation of myth in the modern western world from religious to secular patterns, provides a good starting point. In works such as *An Anatomy of Criticism* and *The Critical Path*, Frye [40] suggests that, until the rise of secularism, the western world operated according to the following fundamentally Christian worldview. At the top was heaven, the ideal paradise, perfection: at the bottom was hell, complete imperfection. Between these two poles was the world of nature, through which human beings wandered. Because of sin, men would be led to take the world of nature as the only real world and would eventually be dragged down to hell. If, however, they were able to perceive that their humanity partook of something higher than nature, they could become pilgrims, and would be able to climb up to God. Each of these worlds or options had its own set of metaphorical or image clusters (e.g., paradise was up toward heaven on a hill) which reinforced the structure.

Interestingly enough, Frye's pattern resembles very closely the one we have been outlining, so we may be involved in a very odd process of recapitulating aspects of previous theological myths on the secular level. The parallels are compelling:

(1) Heaven – perfect, stable, self-contained, internally defined, infinitely timeless (myth of stability).
(2) Pilgrimage nature – *Natura naturans.* Nature as a learning experience, altered by the perception of the pilgrim and the course of the journey (myth of resilience).
(3) Cyclical nature – the possibility of renewal in nature (myth of cyclical renewal).
(4) Wandering nature – *Natura naturata.* Sometimes stable, sometimes unstable, nature is unpredictable, and appears to need controlling (myth of multiple stability).
(5) Hell – imperfect, drastic, catastrophic, eternally static (myth of instability).

Essential to these mythical patterns is the use or rejection of surprise. The myth of instability (hell) produces no surprises because there is only one result of the system. The myth of stability is equivocal, but cannot theoretically cope with major surprises, because everything is theoretically knowable or under control. It proposes itself as infinitely all-knowing and timeless. For the myth of multiple stability surprises are expected, but are unpredictable or have no wider context within which to function. For the myth of resilience, however, not only is there surprise, but it is revelatory surprise. The pilgrim is constantly tested, surprised, and educated as he or she proceeds [41]. Indeed, in Christianity, such surprises are essential eruptions of meaning into hitherto inexplicable reality. These are called *epiphanies* and their function is suddenly to illuminate the underlying meaning or trajectory of the system (i.e., there is a redemptive heaven).

We can bring these speculations down to Earth by comparing such a myth system to the work of anthropologists, such as Mary Douglas [42] and Michael Thompson [43], on the different types of rationality expressed by different types of social groups. The basic rationality (or background order) of each group is seen as a function of, first, the classification by these social groupings of those anomalies or borderline cases that challenge the accepted structure of reality, and, second, each group's perception of the real risks and hazards. The social context may be illustrated by a graph (*Figure 16.6*), on which the horizontal axis represents group pressures (pressures to join and pressures to exclude others) from low to high; and the vertical line represents grid pressures (pressures to provide a framework for reality by means of rules or taboos).

The social contexts represented in *Figure 16.6* are:

(1) High grid–high group, e.g., a caste system or strong bureaucracy.
(2) Low grid–high group, e.g., a communal group with few internal rules, but strong exclusive ones.
(3) Low grid–low group, e.g., the entrepreneurial capitalist.
(4) High grid–low group, e.g., the alienated, befuddled citizen.
(5) The autonomous hermit.

Each of these five social contexts operates within a particular worldview and each has a particular relationship to those anomalies or surprises that

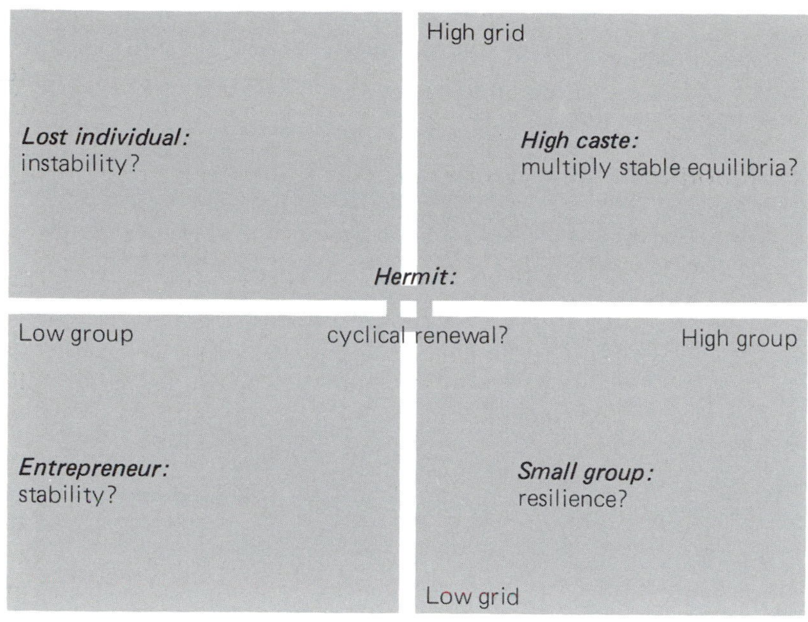

Figure 16.6 The grid–group graph.

either do not fit into the operational categories of the context or challenge the foundations of that context. This presents us with yet another matching version of our original set of myths. From *Figure 16.6*, the low grid–low group world is that of the entrepreneur, the microcosmic operator of Adam Smith's world, the myth of stability; all anomalies or surprises are either ignored or treated as gambles or as indications for the need to shift to a new set of market inputs. The high grid–low group world is the realm of the alienated citizen (or the failed entrepreneur), to whom all anomalies and surprises are indicators that "no one is really in control", i.e., the myth of instability. The high grid–high group world is that of the myth of multiple stability – there are thresholds to be maintained and surprises are indications of areas where new control strategies need to be considered. The low grid–high group world is that of the myth of resilience; here the world is seen to be discontinuous, with both short- and long-term surprises in the offing. And last, the center contains the potential for controlled cycling through each of the other zones – cyclic renewal.

Frye's myth criticism reminds us that, in the modern world, these social groupings are intertwined with the overall process of secular myth formation. This is of cardinal importance when one recognizes that since the Renaissance we have been largely concerned with the left-hand side of *Figure 16.6*, i.e., the fate of the individual. Only in the last century has the high grid–high group organization of a worldwide bureaucracy and administration come to dominate our lives. To expand on this slightly, the elimination of the theological explanation for all events, surprising or otherwise, meant that the individual now became subject only to fortune. This is the world described by Machiavelli [44] and central to the doctrine of the Renaissance humanists. In this world, the genius or strong man is master of his fate; or, if not altogether master, is possessed of such virtues that fortune would smile upon him and make him prosper. Unfortunately, this pattern of rewarded virtue is subject to surprises and uncertainties which undermine the doctrine of the self-made man. Even a Leonardo da Vinci might not be immune from a falling roof tile or a sudden unexpected fatal illness.

One of the strongest drives of modernism is the drive to remove this possibility of humiliation as much as is feasible, as well as to hide from us the ultimate undermining absurdity of death. This chapter has nothing to say about the latter; the former, however, had been transformed by the growing power of science and bureaucratic management to control much of the previously wayward world and to reduce the role of risk in our lives. This, we have argued, is the high grid–high group quadrant of the social context graph, to which the neutral, but systematic, rules of bureaucratic rationality belong. Thompson proposes that this bureaucratic group has a dyadic relationship with the entrepreneurial, individualistic group in the stability quadrant. This is easy to understand: the semiautonomous individual is still the fundamental underpinning of western society; while, at the same time, entrepreneurial capitalism remains (in the west) the engine of prosperity. The bureaucracy is thus committed to flattening out anomalies and surprises from the perspective of the myth of stability.

By virtue of its position and ethos, however, the bureaucracy is also committed to rules of systematic rationality (as famously described by Max Weber [45]). Its justification for existence is ultimately its ability to manage rationally a world of increasing complexity, since only the bureaucracy is supposedly capable of acquiring all the information necessary to calculate what would constitute "the greatest good for the greatest possible number," or provide for new Pareto optimalities (where further improvements would leave no one worse off than before). This means that the bureaucracy cannot be – or be perceived to be – a reckless gambler. Rationally, the bureaucracy is supposed to be considering the long-term good of its masters, the public.

Much of the time, however, the bureaucracy can make forgiveable mistakes, because of the ability of nature benign or cornucopian to be harnessed by the entrepreneur. In any serious conflict over the distribution of goods, the issue can be sidestepped in rich societies by giving each participant in the conflict a part or all of the ever-enlarging pie (or a promissory note that part of the pie will come to the participant in the near future). This gives the bureaucracy an even bigger stake in nature benign or cornucopian, because, when the possibility of limits or shrinkage occurs, serious political decisions based on conflicting moral values and preferred social goals may have to be made – a prospect the broadly utilitarian system is designed to avoid.

As a result, the bureaucratic management structure oscillates between nature cornucopian and nature susceptible to rational management. Its sophisticated analytic power can often present to the manager the data indicating that the system is, in fact, based on multiple stability points; but the social context continues to drive the bureaucracy toward managing the system as if there were only one such point. The examples provided in this chapter and elsewhere in this book show that this drive toward the myth of stability may cause the managed system to flip-flop instead into instability and collapse.

What the myth of resilience (and the attendant concept of surprise) proposes is that the manager of a system should try to move toward the lower right side of the graph in *Figure 16.6*. Understandably, however, the manager can be easily frightened by the short-term loss of control that such a move entails. Resilience seems to imply the loss of all control, just as if one were already in the domain of instability, but this time the manager is expected to "sit back and enjoy the ride". This flies in the face of all notions of ordinary management practice. To put it briefly, the loss of short-term control is unacceptable; with the result that long-term understanding is impossible.

It is here that the concept of surprise comes into its own and may prove to be crucial in convincing managers and policymakers that resilient systems-management is a viable proposition. First of all, surprises should be reevaluated as indispensable elements or tools for the understanding of how systems operate. If they are handled properly, they could even be seen as a valuable resource – in Nietszche's phrase: "Whatever does not kill me, strengthens me [46]."

For this reorientation to take place, however, we need to have a clearer view of what kinds of surprise we might have to handle; and, indeed, a clearer view of surprise in general. To this end, and to join together the disparate threads of our wide-ranging discussion, we make a brief attempt to produce a taxonomy of surprise, based on the concerns already outlined (see Brooks, Chapter 11, this volume).

In the context of economic theory, Shackle [47] some years ago proposed the following definition of surprise:

> ... surprise is that dislocation and subversion of received thoughts, which springs from an actual experience outside of what has been judged fully possible, or else an experience of a character which has never been imagined and thus never assessed as either possible or impossible: a *counter-expected* or else an *unexpected* event ...

Using this definition as a starting point, we can formulate a taxonomy (which would serve as the basis for the research agenda outlined in the following section). As in Popperian [48] theory, and as in the myth of resilience, surprises and uncertainties in relevant contexts can be seen as valuable sources of information.

Surprises can appear in four basic ways: they can erupt from a system, irrupt (into) a system from outside, they can bypass a system, or they can result from the system and its external context mutually creating an interactive surprise. The ability of these trajectories of surprise events to inform the observer might be ranked on this scale of increasing intensity:

(1) Anomalies – surprises that are marginal, puzzling, but not enough to alter perception.
(2) Shocks – surprises that are extensive or intensive, but which freeze the system or cause it to act inappropriately.
(3) Epiphanies – surprises that are central and reveal essential characteristics of the system dynamics in a useful way.
(4) Catastrophies – surprises that destroy a system before it can make any use of the event.

The myth of resilience is in the business of surprises that are epiphanic without being catastrophic. Uncertainty is harnessed in the service of understanding.

A research agenda

Can surprise thus described be put into the framework of the sustainable development of the biosphere?

Perhaps the most interesting suggestion prompted by the taxonomy is that surprise as a diagnostic tool may provide us with ways of coping with the paradoxes implied by the management of a bounded biosphere. For example, we might consider engineering preemptive surprises, i.e., to experiment with actual systems to see how they will respond, but at a level of surprise below the catastrophic. We could – as in stress ecology – consider how actual, inadvertent experiments-in-progress are provoking responses out of substantially unexplored complex systems. In these and other ways, we can learn to maneuver and adapt constructively to the inevitable – but unpredictable – surprises that are to come.

In this chapter I have argued, by analogy, implication, and example, that the sustainable development of the biosphere is both indicated by and subject to a parallel conceptual development, a development that could best be described as a passage through various mythologies. This conceptual development would be of only marginal real-world importance if there was no way of coming to grips with it except in literary or symbolic contexts.

It happens, however, that in economic systems, in the management of ecological systems, and in the elusive formation of policy, some of these approaches to coping with the uncertainties of the present and the even greater uncertainties of the future are already operational. And some of them are also pathological. If surprises and uncertainties are as potentially perilous for the sustainable development of the biosphere as they appear to be, then the incorporation of coping with surprise into biospheric management – and into management at other levels as well – is a priority; and it is amenable to useful research programs.

Key elements of a useful research program, which could be undertaken by the International Institute for Applied Systems Analysis (IIASA) and other suitable organizations, would include:

(1) The proposed taxonomy of surprise needs to be elaborated and, if possible, linked to other, appropriate, adaptive management techniques and concepts.
(2) Historical studies of ways in which natural and social systems have responded to past surprises and how they have incorporated the lessons learned into their warning, responding, and adapting mechanisms.
(3) Analysis of ways in which flexible, adaptive systems provoke small-scale surprises, continually testing and upgrading their responsiveness to environmental changes.
(4) Analysis of the scale of surprises, with a view to fitting their time and space characteristics into the scale change concepts now being developed in ecological and historical research [49].
(5) Perhaps most urgent is the application of the insights of the concepts of surprise and the myth of resilience to the pathology of technical

monocultures and the inappropriate management of natural systems into self-simplification, instability, and collapse. Clear, documented examples of these phenomena, with the appropriate conclusions drawn from them, may pave the way toward a reorientation of policies, thus bridging the gap between our present short-term needs and the long-term goal of a sustainable future.

Further reading

Simon, H. A. (1983), *Reason in Human Affairs* (Stanford University Press, Stanford, CA). Simon covers much of the territory of this chapter, using the opposition between bounded rationality and optimization to argue similar points. He believes – contrary to the position described in the chapter – that the world is still loosely coupled enough so as to give us leeway in preparing for surprises.

Earl, P. E. (1984), *The Economic Imagination* (Cambridge University Press, Cambridge, UK). A wide-ranging exploration and attempt at a synthesis of the behavioral psychology literature, post-Keynesian economics, and expectations theory into a new model of how the economic system works when faced with future uncertainty.

Sheffrin, S. M. (1983), *Rational Expectations* (Cambridge University Press, Cambridge, UK). A sympathetic treatment of rational expectations with only some of the mathematical models.

Rappaport, R. A. (1984), *Pigs for Ancestors* (Yale University Press, New Haven, CT). This revised and enlarged edition of Rappaport's 1968 anthropological study of the ecology of the Tsembaga of New Guinea remains the richest and most compelling attempt at a theory of cultural philosophy grounded in ecological detail.

Shrader-Frechette, K. S. (1985), *Science Policy, Ethics, and Economic Methodology* (D. Reidel, Dordrecht, The Netherlands). The current methodologies for "coping with the future" rely heavily on cost–benefit analysis, risk–benefit analysis, and environmental impact analysis for the more benevolent decision-making processes. This is the first book that goes to the heart of their philosophical assumptions about man and the world.

Cotgrove, S. (1982), *Catastrophe or Cornucopia?* (John Wiley & Sons, Chichester, UK). It is important that the sustainable development of the biosphere as a working concept evolves with considerable self-knowledge about the myths being harnessed to the projection of the various possible futures. This book provides, among other things, one way of using the anthropological insights of Mary Douglas, Michael Thompson, and others.

Notes and references

[1] The concept of "technological monocultures" was developed by Harvey Brooks (see Chapter 11, this volume).

[2] A highly selective list might include: in mathematics (and now physics) the "laws of chaos" refer to the descriptions of the quasi-periodic transition to chaos now being developed out of the work of Feigenbaum and others: see Robinson, A. L. (1982), Physicists try to find order in chaos, *Science*, **218**, 554–557; in risk assessment and anthropology, Douglas, M. and Wildavsky, A. (1982), *Risk and Culture* (University of California Press, Berkeley, CA); in economic theory, Rizzo, M. J. (1979), *Time, Uncertainty, and Disequilibrium* (D. C. Heath and Co., New York); in systems ecology, see C. S. Holling (*passim* and Chapter 10, this volume); and, for a summary volume, perhaps Prigogine, I. and Stengers, I. (1984), *Order Out of Chaos* (Bantam Books, New York).

[3] The standard reference is Kuhn, T. (1970), *The Structure of Scientific Revolutions* (University of Chicago Press, Chicago, MI). Hacking, I. (1981), *Scientific Revolutions* (Oxford University Press, Oxford, UK) contains notable articles by many of the participants in the continuing debate on paradigms and paradigm shifts.

[4] Douglas, M. (1975), *Implicit Meanings: Essays in Anthropology* (Routledge and Kegan Paul, London), especially Self-evidence, pp 276–318.

[5] Creative chaos in one form or another is emphasized by: Feyerabend, P. (1975), *Against Method: Outline of an Anarchistic Theory of Knowledge* (New Left Books, London); Lakatos, I. and Musgrave, A. (1970), *Criticism and the Growth of Knowledge* (Cambridge University Press, Cambridge, UK); Hesse, M. (1980), *Revolutions and Reconstructions in the Philosophy of Science* (The Harvester Press, Brighton, UK); Krige, J. (1980), *Science, Revolution, and Discontinuity* (The Harvester Press, Brighton, UK).

[6] This is a theme of C. S. Holling. Other than in Chapter 10, this volume, it was discussed particularly in Holling, C. S. (Ed) (1978), *Adaptive Environmental Assessment and Management* (Wiley, Chichester, UK).

[7] Timmerman, P. (1981), *Vulnerability, Resilience and the Collapse of Society*, Institute Monograph No. 1 (Institute for Environmental Studies, University of Toronto, Toronto).

[8] Schelling, T. C. (1983), Climatic change: implications for welfare and policy, in National Research Council, *Changing Climate* (National Academy Press, Washington, DC).

[9] Boulding, K. E. (1966), The economics of the coming spaceship Earth, in H. Jarrett (Ed), *Environmental Quality in a Growing Economy* (The Johns Hopkins Press, Baltimore, MD).

[10] McLuhan, H. M. (1970), Editorial viewpoint, *International Journal of Environmental Studies*, **1**, 3.

[11] Boulding, K. E. (1980), Spaceship Earth revisited, in H. E. Daly (Ed), *Economy, Ecology, Ethics: Essays Toward a Steady-State Economy* (W. H. Freeman and Co., San Francisco, CA).

[12] Nicolis, G. and Prigogine, I. (1977), *Self-Organisation in Non-Equilibrium Systems* (John Wiley and Sons, Chichester, UK).

[13] Raphael, D. D. (1969), *British Moralists 1650–1800* (Oxford University Press, Oxford, UK) provides the background; Noxon, J. (1973), *Hume's Philosophical Development* (Oxford University Press, Oxford, UK) describes the pivotal role of Newton for the research program of the social philosophers of the enlightenment.

[14] An easily obtainable edition is Mandeville, B. (1714), *The Fable of the Bees*, edited by P. Harth (1970) (Penguin Books, Harmondsworth, UK).

[15] Monro, D. H. (1972), *A Guide to the British Moralists* (Fontana Books, London) describes and selects from Francis Hutcheson, John Brown, Shaftesbury, Bishop Berkeley, etc.

[16] Various editions of the 1776 text, including *The Wealth of Nations* (J. M. Dent and Sons in the Everyman's Library Series, London).

[17] Malthus, T. R. (1798), and final version (1830), *An Essay on the Principle of Population*, edited by A. Flew (1970) (Penguin Books, Harmondsworth, UK). Malthusian scarcity – the run up to absolute limits – can be compared with Ricardian scarcity – David Ricardo's model of the differentials associated with the use of land of better or worse quality. Ricardian scarcity was essential to Marx, who, along with Engels, hated Malthus approximately as much as Malthus was ignored by the new neoclassical economists.

[18] Mill, J. S. (1857), *Principles of Political Economy*, Vol. II, p 320. There is a continuing debate over the implications of the terms stationary state, steady state, etc. See, for example, Georgescu-Rogen, N. (1977), The steady-state and ecological salvation: a thermodynamic analysis, *Bioscience*, **27**, 266–270.

[19] Darwin, C. (1859), *The Origin of Species* (Modern Library, New York). The Victorian context is described in Eiseley, L. (1957), *Darwin's Century* (Gollancz, London). Flew, A. (1984), *Darwinian Evolution* (Paladin Books, London) explores the philosophical implications of Darwin's use of Malthus.

[20] The graininess of time and the patchiness of space have become topics of considerable interest. See Clark, W. C., Jones, D. D., and Holling, C. S. (1978), Patches, movements, and population dynamics in ecological systems: a terrestrial perspective, in J. H. Steele (Ed), *Spatial Patterns in Plankton Communities* (Plenum Press, New York).

[21] See C. S. Holling, Chapter 10, this volume.

[22] Disasters comment on the society that seemed to spawn them, or that disowns them; see Hewitt, K. (1983), The idea of calamity in a technocratic age, in K. Hewitt (Ed), *Interpretations of Calamity* (Allen and Unwin, London).

[23] For which, now see Clark, W. C. (1985), Scales of climatic impacts, *Climatic Change*, **7**(1), 5–28.

[24] B. A. Bodhaire and J. M. Harris (Eds) (1981), *Summary Report 1981, Geophysical Monitoring for Climatic Change No. 10* (and previous numbers) (US Department of Commerce, National Oceanic and Atmospheric Administration, and Environmental Research Laboratories, Boulder, CO).

[25] Lekachman, R. (1966), *The Age of Keynes* (Vintage Books, New York); also Eichner, A. S. (Ed) (1979), *A Guide to Post-Keynesian Economics* (M. E. Sharpe, Inc., New York).

[26] See Eichner, note [25] for a general survey; also Bell, D. and Kristol, I. (1981), *The Crisis in Economic Theory* (Basic Books, New York). Sraffa appears more and more to be the pivotal figure for much of post-Keynesian economics. His work on imperfect competition and reswitching, and especially Sraffa, P. (1960), *The Production of Commodities by Means of Commodities* (Cambridge University Press, Cambridge, UK) presents a serious challenge to neoclassical *and* Marxian economics. In the context of the discussion of this chapter, see Carvalho, F. (1984), On the concept of time in Shacklean and Sraffian economics, *Journal of Post-Keynesian Economics*, **6**(2), 265–280.

[27] Economists currently refer to extrapolative expectations, adaptive expectations, and rational expectations. The first is based on simple extrapolation from current trends; the second is based on the assumption that people learn from the past and adapt according to their past errors. Rational expectations suggests that people do both of these things, but that they also try and use present-day information to understand how the economic system functions, and might, thereby, be more predictable. See Lipsey, R. G., Steiner, P. O., and Purvis, D. D. (1984),

Economics (Harper and Row, New York) for an overview.

[28] Peterman, R. M. (1977), A simple mechanism that causes collapsing stability regions in exploited salmonid populations, *Journal of the Fisheries Research Board Canada*, **34**(8), 1130–1142.

[29] Holling, C. S., Walters, C. J., and Ludwig, D. (unpublished), *Myths, Time Scales, and Surprise in Ecological Management* (Department of Zoology, University of British Columbia, Vancouver).

[30] Beddington, J. R. and May, R. M. (1982), The harvesting of interacting species in a natural ecosystem, *Scientific American*, **247**, 62–69, describe this phenomenon with regard to the harvesting of krill near Antarctica.

[31] Glantz, M. H. and Thompson, J. D. (1981), *Resource Management and Environmental Uncertainty: Lessons From Coastal Upwelling Fisheries* (John Wiley and Sons, Toronto).

[32] A recent summary is Cuff, W. and Baskerville, G. (1983), Ecological modelling and management of spruce budworm infested fir–spruce forests of New Brunswick, Canada, in W. K. Lauenroth, G. V. Skogerboe, and M. Flug (Eds), *Analysis of Ecological Systems: State-of-the-Art in Ecological Modelling* (Elsevier Scientific Publishing Co., Amsterdam, and the International Society for Ecological Modelling).

[33] Holling, C. S. (1973), Resilience and stability of ecological systems, *Annual Review of Ecology and Systematics*, **4**, 1–23; also Chapter 10, this volume.

[34] Steele, J. H. and Henderson, E. W. (1984), Modelling long-term fluctuations in fish stocks, *Science*, **224**, 985–987.

[35] Gould, S. J. (1984), The Ediacarian experiment, *Natural History*, **2**, 14–23.

[36] Maddox, J. (1984), Extinction by catastrophe, *Nature*, **308**, 685; and also Lewin, R. (1983), Extinctions and the history of life, *Science*, **221**, 935–937. If, as some evidence now appears to indicate, there is a 30 million year pattern to extraordinary extraterrestrial events, "chance" may be too strong a word. However, it is difficult to see how any natural system could adapt to a 30 million year cycle.

[37] Aspects of stress ecology are explored in: Rapport, D. J., Regier, H. A., and Thorpe, C. (1981), Diagnosis, prognosis, and treatment of ecosystems under stress, in G. W. Barrett and R. Rosenberg (Eds), *Stress Effects on Natural Ecosystems* (John Wiley and Sons, New York). A pioneering work in adaptive environmental management was Clark, W. C., Jones, D. D., and Holling, C. S. (1979), Lessons for ecological policy design: a case study in ecosystem management, *Ecological Modelling*, **7**, 1–53.

[38] Goodman, D. (1975), The theory of diversity–stability relationships in ecology, *The Quarterly Review of Biology*, **50**(3), 237–266.

[39] Moore, P. D. (1983), Ecological diversity and stress, *Nature*, **306**, 17, reviews this research. Pimm, S. L. (1984), The complexity and stability of ecosystems, *Science*, **307**, 321–325 proposes a taxonomy which copes with some of the ambiguities in the word *stability*, but ignores others.

[40] Frye, N. (1957), *An Anatomy of Criticism* (University Press, Princeton, NJ); Frye, N. (1971), *The Critical Path* (University of Indiana Press, Bloomington, IN); also Frye, N. (1983), *The Great Code: The Bible and Literature* (Academic Press Canada, Toronto).

[41] Bunyan, J. (1684), *The Pilgrim's Progress*, edited by N. H. Keeble (1984) (Oxford University Press, Oxford, UK).

[42] The original grid–group concept derives from the work of Basil Bernstein, as described in Douglas, M. (1973), *Natural Symbols* (Penguin Books, Harmondsworth, UK), and developed in Douglas, M. (1975), *Implicit Meanings* (Routledge and Kegan Paul, London) especially Chapter 14.

[43] Thompson, M. and Wildavsky, A. (1982), A proposal to create a cultural theory of risk, in H. C. Kunreuther and E. V. Ley (Eds), *The Risk Analysis Controversy: An Institutional Perspective* (Springer Verlag, Berlin); Thompson, M. (1984), Among the energy tribes: a cultural framework for the analysis and design of energy policy, *Policy Sciences*, **17**(3), 321–339.

[44] Skinner, Q. (1978), *The Foundations of Modern Political Thought* (Cambridge University Press, New York) describes Machiavelli's role in the rise of Renaissance humanism and the concept of *Fortuna.* The diminishing rewards of virtue are described in MacIntyre, A. (1981), *After Virtue* (University of Notre Dame Press, Notre Dame, IN).

[45] Aron, R. (1967), Max Weber, in *Main Currents in Sociological Thought, Vol. 2* (translated by R. Howard and H. Weaver) (Penguin Books, Harmondsworth, UK) is a good survey.

[46] Kaufmann, W. (translator) (1954), *The Portable Nietzsche* (Viking Press, New York), see *Twilight of the Idols, Maxims and Arrows* (first published in 1889).

[47] Shackle, G. L. S. (1972), *Imagination and the Nature of Choice* (Edinburgh University Press, Edinburgh). Shackle's earlier work, note [26], was on economic expectation: e.g., Shackle, G. L. S. (1949), *Expectation in Economics* (Cambridge University Press, Cambridge, UK), and Shackle, G. L. S. (1955), *Uncertainty in Economics and Other Reflections* (Cambridge University Press, Cambridge, UK).

[48] See Magee, B. (1973), *Karl Popper* (Fontana Books, London) for a good discussion of the implications of falsifiability.

[49] For ecological theory, see note [23]. The discussion in ecological theory of fast, medium, and slow variables in an ecosystem parallels the historical theory of Fernand Braudel, who sees history as the reflection of the interaction of three variables: *la longue durée* (geographical or large ecosystemic time); *l'histoire sociale* (economic and social history); and *l'histoire événementielle* (the history of events). For Braudel, the great events are those where all three

variables come into *conjoncture*, or phase; but most of the really important factors in the long view of history are the slower variables, while most historical writing is about the fast, spectacular, events of the lives of famous people. This is an area of potentially fruitful collaborative research. See Braudel, F. (1980), *On History* (translated by S. Reynolds) (University of Chicago Press, Chicago, Ill).

Commentary

M. Thompson

My fear is that Timmerman's important message may not cut across all the disciplinary boundaries as strongly as it should. He is too polite. My comments, therefore, are aimed solely at remedying this fault. I will try to say, as rudely as possible, what it is that I think he is saying.

Making things subservient to ideas is seldom a popular move, as Keynes discovered when he pointed out that the practical man who claims to have no use for theory is invariably the slave of some defunct economist. Hard scientists, we might expect, are scarcely likely to thank Timmerman as he deftly sorts them into Malthus-slaves and Smith-slaves on the basis of the myths of nature to which they unwittingly subscribe. But personal popularity is not what is at stake here nor, to nail another myth, are all hard scientists resistant to this seemingly threatening and insulting interpretation of their endeavors. The "invisible college" that Timmerman invokes in support of his thesis is largely made up of hard scientists, and even harder mathematicians: an elite and anarchistic band of high-achievers who are only too happy to find themselves rubbing shoulders with some of our more adventurous and hard-thinking anthropologists, theologians, literary critics, and historians. The new paradigm that Timmerman calls for is, I suggest, all around us. To ask when it will appear is like asking "When will the Mafia get into politics?"

Of course we define problems and select facts so as to defend and rationalize preferred patterns of social relations; *of course* an event is not surprising or unsurprising in itself, but only in relation to a particular set of beliefs about how the world is; *of course* a surprise is only a surprise if it is noticed by the holder of the beliefs that it contradicts. In policy debates as transscientific as the sustainable development of the biosphere, to deny the existence and centrality of such knowledge-shaping and attention-directing processes is simply to reveal that you are no longer in touch with the state of the art.

Myths of nature, as Holling has pointed out, are not lies; they are partial representations of reality. Such myths are the cultural devices by which we can capture, in elegant and simple form, some essence of experience and wisdom. Myths, in the face of our inevitably incomplete knowledge, guide our actions and moderate our fear of the unknown. Without myths we would be in a bad way. Since each is replete with accumulated wisdom, the question is not whether we should have myths or not, but, rather, which myths should we have. The obvious answer is to have them all – to pool all that accumulated wisdom and experience – but this option, it would seem, is not available. Each myth contradicts all the others, and to embrace one we must first reject the rest. Timmerman's historical survey, and a host of case studies in the area of environmental policy, show us that we do this not with reluctance but with alacrity. Different management institutions home-in onto the different myths and, once there, cling onto them as if their very lives depended on them. Cultural theory shows us that they do.

Preferred patterns of social relations – the ego-focused network of the entrepreneur, the hierarchically nested group of the bureaucrat . . . the egalitarian bounded group of the coercive utopian – are inevitably in contention with one another. They have to be fought for, they have to be defended, and each strives to ensure its own viability by inculcating in its constitutive individuals a distinctive organizational culture – a view of the world, a set of moral principles, a shared understanding of the problems it faces, and a unified blanket of ignorance – that, above all else, is designed to ensure the continued existence of that particular pattern of social relations. The bureaucrat (Kuhn's normal scientist) can no more come to terms with his structural ignorance than the entrepreneur (the interdisciplinarian) can abandon her faith in the market place of ideas, or the coercive utopian (the limits-to-growth man) his conviction that there must be radical change now, before it is too late. Each arrives on the latest, and most presumptuous, battleground – the biosphere – fully accoutred. Each sees in it what he wants to see; what she has to see: the facts that will justify the policies that will usher in the sort of planet he craves. "What a wonderful place the world would be," cry the members of each organizational culture, "if only everyone was like us," conveniently ignoring the fact that it is only the presence in the world of people who are not like them that enables them to be the way they are.

We cannot synthesize the myths and the systems of knowledge that they sustain; their fundamental contradictions preclude this. Nor can we declare a cultural truce and each bring our sustaining ignorances out into the open. There is no bias-free position; there is no knowledge untainted by the institutions that promote it; there is no cosmic exile. But what we can do is concede that this is how things are, and that the inevitable contradictions and contentions in this state of affairs lead not to chaos but to an *essential pluralism.* Nor are the cultural constraints that ever shape and update the myths of nature the only constraints on this pluralism. There are the natural constraints as well; that is where the surprises come from. So the new paradigm, though it rejects the traditional separation of facts and values that has for so long kept the sciences and the humanities away from each other, is founded not on old-style relativism but on a pattern-creating dynamic – a system – of *constrained relativism.* Cultural constraints and natural constraints are both at work and have to be explored together if we are to do the best we can for the biosphere which, like it or not, is now at our mercy. To ask "Can we manage the biosphere?" is to ask "Can we manage our institutions?"

The new institutional economics provides us with a handy guide for exploring these two important questions. It is the ever-increasing scale and complexity of our social transactions – our economic system – that has propelled us from spectator to lead actor on the biospheric stage. This means that such ability as we have to manage the biosphere is one and the same as the ability we have to manage our social transactions. Markets, hierarchies, and sects (egalitarian bounded groups) – the basic components of our institutions – are best understood as competing, yet complementary, arrangements for the handling of social transactions. Each has its advantages and its disadvantages and our concern should not be with which one is right – that is a meaningless question – but with the question of appropriateness: "Which kinds of transactions are best handled by which kind of institutional arrangement?" There is no final solution; there is no complete knowledge. All we can do is learn and tinker, tinker and learn. Diversity, contradiction, contention, and criticism are our finest resources. We must learn to husband them and make the most of them. Divided we stand; united we fall.

Chapter 17 Methods for synthesis: policy exercises

G. D. Brewer

Editors' Introduction: The ultimate challenge for the Biosphere Program is to think creatively about exceedingly complex social and environmental phenomena that interact and evolve over large time and space scales. The problems of analyzing, interpreting, and ultimately synthesizing the vast quantities of scientific and human knowledge relevant to the sustainable development of the biosphere are daunting.

Reviewed in this chapter are the alternative methods – models, simulations, and games – that have been used to address these crucial tasks. It is argued that the experience of the last decade shows large computer models to be more often obstacles than aids to creative, exploratory thinking. It is recommended that, instead, a series of new experimental "policy exercises" be undertaken, based on military experiences of free-form, manual games combined with carefully crafted strategic scenarios. Specific procedures for the policy exercises are described.

The problem

Information is the universal resource, the fundamental means by which human problems are understood and resolved. It connects our past to the present and helps us confront the future. Data we collect and analyze and decisions based on them determine how we and future generations live.

But there are difficulties. Although there are now vastly improved means of observation and measurement, particularly in the scientific realm, our capacity to analyze, interpret, and use data for social and policy purposes has failed to keep pace. Insufficient effort is not the fault. Quite the contrary, vast experience exists in the research and policy communities to create and use information; but results have been mixed.

Many possible reasons for lack of success exist. Not as common are efforts to clarify these reasons, so to make constructive and lasting improvements [1]. The following discussion is so intended. It is also wide ranging and provocative, with experiences from a variety of substantive fields and disciplines being recounted. Specific efforts to solve the energy problem, the pollution problem, the extended jurisdictions problem, or dozens of other problems subsumed within the human and biosphere rubric all offer potential insights and lessons for the future – but only if we view such experiences as observation points, not as personal investments to be both attacked and defended. Likewise, no particular intellectual discipline or professional group has been conspicuously successful in its grasp and resolution of some of mankind's most complex and consequential problems. Which is not to say that such disciplines and groups have not tried.

Many difficulties impede better creation and use of policy-relevant knowledge. The complexity of the systems involved, human perception and value, limited theories and weak methodological tools, and profound uncertainties about the future all figure prominently. These vast topics matter and merit serious, open, and relentless consideration.

Time is short; the needs are enormous. One may be optimistic or pessimistic about the prospects. The optimists, while acknowledging the difficulties, typically call attention to the propitious moment at hand in which many substantial and fundamental changes must occur [2]. The pessimists, as typically, counter with gloomy forecasts of mounting catastrophes as resources are consumed and overexploited, conflicts mount, and societies are tested as never before [3].

I prefer to side with the optimists. The alternative is simply too depressing. But casting one's lot with the hopeful is not to underestimate the difficulties nor to deny the realities. Consider for a moment the kinds of problems we face.

Beyond rationality's reach

Take the site selection and construction of a nuclear power plant. As long as the objectives are defined in technical terms, such as devising an efficient plant design or determining the amount of energy to be produced, the problem may be difficult, but it is still amenable to analysis. But when concern for the environment, safety, and other competing and value-laden objectives enter the picture, the problem passes beyond some threshold of difficulty and overwhelms attempts at rational analysis. The simplicity and clarity of technical facts yield to controversy and subjectivity [4]. Facts, as commonly understood, provide little definitive help in resolving such questions. As competing interests and goals become more insistent, the problem becomes quite insoluble. And as a consequence, it becomes nearly impossible to produce a logically consistent set of policies or programs. Attention to the "big picture" gives way to "fragmented and personalized interests and values," which are "difficult to reconcile with integrated and rational planning and foresight [2, p 15]."

Or take the problem of managing a specific fishery, especially as concerns the potential contributions science and analysis might make to it. There is no theory capable of predicting the biological, economic, political, and social consequences of a natural event such as El Niño off the Peruvian coast that resulted in the collapse of one of the largest fisheries in the world [5, 6]. There is even confusion and dispute among scientists about the underlying mechanisms of the event itself [7], whether the anchoveta will ever return to abundance [8], when this may happen, or whether anyone ought to be planning or counting on anything connected with the fishery in the future [9]. But these are precisely the kinds of questions that must be confronted, somehow.

Some focus on the economic dimensions of the fisheries problem and none would dispute the importance of doing so. The disputes arise when the economic perspective is uninformed and untempered by many other perspectives. A pure economic solution that advocates efficiency through open markets with centralized ownership of the world's fish by nonexistent international authorities is illustrative [10]. But despite its theoretical elegance, this solution is simply not politically or practically feasible. And so, it is no solution at all.

Well-meaning agents in development agencies and banks naturally fix on the investment potential of fisheries in the developing world. To many of them the fisheries problem is conceived as providing capital to increase efficient harvesting, primarily by upgrading boats and gear. Around the world, but in southeast Asia particularly, the intended efficiencies have been achieved, but at other unaccounted for and substantial costs:

> Most of the coastal fisheries of Southeast Asia are approaching or have exceeded maximally sustainable yield, because of the tremendous increases in fishing efforts over the last two decades. Ever increasing numbers of small-scale fishermen and the introduction of highly effective gear types such as the trawl have combined to pose a serious threat to these valuable but vulnerable resources [11].

The biological threat has distributional consequences, too:

> Rapid growth of trawler fleets which directly compete with small-scale fishermen has increased the pressure

and introduced a serious problem of distribution of benefits from the fisheries [11].

Technological and economic efficiencies are realized in the short term, while the resource remains viable, at the expense of social goals of more equal access to the resource:

> Small-scale production units employ substantially more people per unit of investment, result in wider and more equitable distribution of income, and spread ownership of capital over a substantial number of households [11].

The basic theme here is that investment and development decisions must be taken with serious regard for whole contexts, including important cultural, social, and political dimensions. This has not happened. The consequences are manifest and the costs are rising dangerously.

A secondary theme is biological and relates to resource vitality and the intelligence necessary to keep it so. It also calls attention to the limited tools and power of marine resource stewards. "All or nothing", high technology *versus* artisanal, are the gross choices facing the stewards. And these choices may be about the best possible under prevalent constraints – facts of life neither biologists, economists, investment bankers, nor politicians want to deal with, or are equipped to deal with.

Science and the disciplined, rational principles on which it is based are scarcely able to explain these events, account for the practical realities, or even begin to sort out or foretell their various consequences. Prevalent analytic paradigms and particular intellectual perspectives are extremely limited as a means to comprehend and inform realistic choices. And even in those instances where choices have been made, where policies and institutions exist, they "All too often are devised on the basis of ridiculously simplistic views of the resource system [5, p 237]." A very different way of thinking seems to be called for.

The clash of perspectives

Our problems are intricate and changeable and often mean very different things to those involved. Neither the problems and the settings in which they occur, nor the possible solutions stand still. They evolve naturally and in response to efforts to understand and resolve them. Furthermore, despite the considerable temptations of the optimal or best solutions that enthusiasts seek and the individual disciplines seem to provide, what is thought to be optimal from one perspective often turns out to be dreadful from another.

The clash of perspectives quickly leads to a clash of interests and to conflicts at the extreme. Little is accomplished by presuming that everyone involved appreciates specific problems the same way. Even less is accomplished by ignoring legitimate differences in view and preference, whether one agrees with or even understands them or not [12]. Individuals exhibit highly different capacities to see, comprehend, and value events, circumstances, and environments [13, 14, 15]. What people see and what it means to them is not the simple matter it appears, but it is an essential starting point on the road to improved policy knowledge.

So, for example, when the engineer says that a specified nuclear power plant design will produce a given number of kilowatts for a specific price, he or she is placing a different emphasis and value on the future than the environmentalist who worries about safety and reliability. When an economist talks about efficient markets and assumes that institutions are readily created to fit economic visions of the problem, the equally important equity, distributional, and feasibility goals others value and seek are being slighted. And so it proceeds on through a long list of people and perspectives.

So whose perception or grasp of problems is better? No one's and everyone's: all relevant perspectives matter, each illuminates problems differently, in both angle and intensity. When taken together in composite, moreover, the various views begin to give one a better sense of the complexity of the whole [16].

Perspectives assume special importance as they allow individuals to link the past with the present and to imagine the future. In its most abstract and aggregate form, the collective image has been refered to as policy – a concrete manifestation of the collective perception of the future [17, pp 25–26]. The future is filled with numerous unrealized potentialities, which individuals may imagine and appraise according to their own experiences, identities, and expectations. Hopes, fears, and surprises understandably suffuse the entire enterprise.

Surprises: wishful and fearful thinking

The future is both contingent and intentional, conditions seldom well accounted for in ordinary policy research. The key point is that data are not available about the future. Data represent an historical accounting whose persistence into the future is always problematic. Inconstancy we call "surprise."

Past patterns may change in response to systemic evolutions or because of direct intervention, i.e., policy choices. Heavy reliance on data from the past imparts a conservative bias on one's reckoning of the future. And the older the data used and/or the farther into the future one pushes analyses based on them, the more likely are surprises.

Relying on old assumptions about the world can be as misinforming as depending on old data. Ascher refers to this as "assumption drag," or "the reliance on old core assumptions" with the consequence of creating "some of the most drastic errors in forecasting [18]."

Fearful and wishful thinking about the future have also had drastic consequences, especially in military matters, but not only there [19]. Assuming absolutely the worst about potential adversaries, and then preparing accordingly, has given rise to worst-case analyses among warfare specialists and self-fulfillment by their policymaking clientele. Wishful thinking follows from selected discernment and optimistic biasing of facts about a likely opponent compared with one's own capabilities. These common and distinctive styles of viewing the future are evident in recent alarming analyses and rebuttals over the prospects of nuclear winter [20], in many of the analyses of the 1970s on the energy crisis [21], and, of course, in the Star Wars debates of the 1980s.

So what does this add up to? Among other things one needs to be as self-conscious as possible about the inherent biases found in any policy analysis and to make these available for all to see and consider; to treat all assumptions, beliefs, and perceptions as transparently as possible; and to view the problem at hand from as many different perspectives as time and talent allow [22]. These matters have deep intellectual roots and have stimulated controversies over the years. Many of the problems revolve about the methods and tools one uses in policy work, a topic to which I turn next [23].

A reconsideration of methods

There are many different categories of methods and almost an equal number of reasons why each might be used for policy purposes. This rich diversity is not well appreciated, at least as one surveys the methods commonly used and their usual limitations. In this section I set forth a range of methods and potential applications and conclude by critically listing basic problems encountered by those relying on large-scale, machine (computer) models. In subsequent sections of the chapter I build upon this critique and elaborate different methods and procedures thought to be in better keeping with the aspirations of the collective Man and Biosphere Project.

Methods: models, simulations, and games

For a variety of reasons it is difficult to define satisfactorily what a model is or to state precisely what gaming and simulation mean. However, four useful *methodological categories* are distinguishable: analytic models, machine simulations, man–machine games, and free-form games [24]. And at least a working definition of the term *model* is possible: it is a representation of an entity or situation by something else having the relevant features or properties of the original [25, p 10].

Analytic models are usually quite abstract, poor in the number of variables explicitly considered, but rich in the ease of manipulation and clarity of insight. They may describe a phenomenon or situation with a series of equations or they may represent a set of logical relationships manipulable according to formal rules.

Numerically defined variables must be well understood and their unit of measurement known. A barrel of oil, in this sense, satisfies the requirements; distrust in one's government probably does not. Despite the importance distrust may have in given realistic settings, we have very little hard knowledge about it. Rather than excluding or including it in spurious mathematical form, it may be necessary to treat distrust verbally.

Game theory, as a distinct specialty field, contains many excellent examples of analytic models [26].

Excellence here commonly derives from the analyst's clear understanding of the severe limitations of his or her model, most particularly the simplifying assumptions made to secure representation. Another aspect is clarity about the differences between the model and the part of reality it stands for. In other words, little effort is expended trying to capture everything in one model; alternative formulations, emphasizing different features and pursuing different questions, are accepted.

Simulation, a second category of methods, usually means representation of a system or organism by another, simpler, more abstract system or model having relevant behavioral similarity to the original. A machine simulation uses either digital or analog equipment to achieve representation.

In contrast to analytic models, machine simulations frequently involve many variables; many seem to make a fetish of realism. This is often misplaced and demonstrates a logical fallacy of pictorial realism: a supposition that a model's usefulness is proportional to its size and degree of detail.

Machine simulations are valuable under certain conditions, but not others. In general, they are a method of last resort to be used when:

(1) It is impossible or too costly to observe processes or events in the real world.
(2) The observed system is beyond the reach of other methods.

Machine simulations should not be used, or only used circumspectly, when:

(1) Simpler techniques exist.
(2) Data are inadequate.
(3) Objectives are not clear.
(4) Short-term deadlines must be met.
(5) The problems are minor.

The superficial attractiveness of machine models has led to their widespread use; however, such use has been highly uneven in quality, despite the considerable attention paid. Machine models have proved very hard to control. Not only are scientific and other appraisal standards poorly developed – and applied – but there is little consensus among professionals as to what are proper standards. Advocacy and adversarial uses of these tools are common.

The contrast between the abstraction of analytic models and the spurious realism of many machine simulations suggests we consider options in the middle of these.

Man–machine games place humans in the loop. People may be used either as elements of the system or as objects of study. In the former case, such as a systems engineering or an operations research study, humans may be used because they are cheaper than software; in the latter, they may be used because human factors, such as judgment, are important for the problem being analyzed.

Free-form, manual gaming involves teams and a referee group operating within the framework of a scenario. If computational equipment is used, it typically serves as an adjunct to ongoing activities. Such games are designed to conform to a plausible future situation and, in their most common, military, application, these situations are conflictful. A scenario describes key events from the past, provides a context for the present, and serves as a common pool of information during play. Actions and reactions of numerous players are examined and woven together by the referee (control) team during game play. Moves result from consideration of options, objectives, and constraints by team members, as mediated by control.

Free-form games do not prove anything in the scientific sense. They do help portray the complexities of the situations represented; their role playing aspects aid understanding of different perspectives, including contradictions and inconsistencies between them. Discovery is highly valued. Typically, positions, expectations, perceptions, facts, and procedures are revealed and subjected to challenge. Thus, imagination and innovation play central roles in the drama of the free-form game. The game allows players to challenge, and improve, the initiating scenario, including its explicit and implicit assumptions.

Purposes

In characterizing models, simulations, and games several distinctive purposes for their existence and use have been mentioned. Explicit treatment of the main possibilities is needed, however. Each of our four categories has been used for *operational, teaching and training, experimental,* and *entertainment* reasons, although not in equal measure for each kind. Of these, the operational rationale is most interesting. Operational models, simulations, and games have many different characteristic applications (*Table 17.1*).

Big models: part of the problem

The evidence mounts that large-scale modeling activities meant to inform decision and policy processes are not fulfilling their promise [27]. The problem is deeply embedded and persistent. Some of its deeper roots can be traced to basic confusions that appear among those who commission, build, and use these tools [28]; others are more fundamental and reflect a low-level understanding of differences between policy analysis and scientific work [29].

Any model, no matter what its size, embodies a single perspective on a problem. That perspective is sensitive to and qualified by the steps taken to formulate it. The significance of the model is

Table 17.1 Applications of operational models, simulations, and games.

Application	*Characteristics*
Exploration	Stimulation of constructive explorations of problems that are either not well understood or are misunderstood. Especially in free-form, scenario-based versions, discovery and realization of unimagined difficulties are opportunities that occur.
Planning	(Usually linked with evaluation). Technical, doctrinal, and procedural inquiries meant to prepare for or assess operational systems, e.g., weapons systems, logistics systems, organizations, information systems, economic systems.
Cross-check	A back-up procedure to provide additional insight and confidence to recommendations devised with other means. For example, expert opinion or consultation – primarily based on experience – may be examined with games or simulations to discover flaws or inconsistencies not reported or overlooked.
Forecasting	Making predictions, especially about poorly understood problems, is far less interesting an application than several of the others here characterized. Users must know what they want to forecast, be able to judge the value to be gained from additional forecast accuracy, and have confidence that the builders of the forecasting device possess a good abstraction of the system being studied.
Group opinion	Most realistic policy decisions are based largely on expert opinion and judgment. While little explored or used, games and simulations have operational potential for eliciting, clarifying, and improving expert opinion, considered individually or in groups.
Advocacy	A competent modeler can build just about any bias imaginable into a game or simulation. A one-sided case can be presented, unintentionally too, in support of a partisan policy or position. In a bureaucratic context, the use of models, particularly large-scale machine-based ones, has led to considerable confusion about the differences between political processes and scientific ones. Advocacy need not be pernicious, especially if its existence is openly admitted and its benefits are consciously sought.

Of all the possible links between categories of method and application, one that frequently appears is machine simulation used for forecasting and/or predictive purposes, at least this is stated as the aspiration. Other types of methods and purposes are far less evident, although the difficulties commonly encountered by those producing and using computer-based models, especially the large-scale ones, are well enough known that other approaches seem warranted. Several of these difficulties are summarized next.

furthermore sensitive to and qualified by the judgment exercised in interpreting it with respect to its surrounding context. Once a problem's logic is set out, alternative perspectives should be generated and examined for their theoretical and technical content. Thus different groups of individuals need to work on complementary analyses to illuminate a problem according to their own distinctive capabilities, biases, perspectives, and expectations [30].

Big models are seductive. The time, talent, and resources necessary to produce them represent large sunk and opportunity costs. Sunk costs, the economic paradigm informs us, are meant to be ignored; unfortunately, for many, the sunk costs of large-model projects have to be rationalized so as to justify them. Opportunity costs are seldom considered. But they may be considerable, especially in settings where research and analysis monies are finite and in demand. The seduction occurs as sunk and opportunity costs converge. The models take on lives of their own, i.e., become surrogates for the reality they mean to mimic. The opportunities lost are those related to the different perspectives and paradigms, hopes and fears, and possible insights that simply never surface to be examined, compared, challenged, and considered.

There are other costs worth mentioning, too. And these are denominated in measures of lost credibility, popular suspicion and mistrust, and heightened tensions among professionals whose perspectives and expectations about the future are sharply at odds [29].

For discussion purposes consider the following assertions about large-scale, operational, machine models as a general class of endeavor:

(1) *Such work is more adversarial than most realize.* As *de facto* operational paradigms of analysis both technocratic and partisan styles dominate, despite disclaimers and avowals of adherence to conventional scholarly or textbook modes.
(2) *The work is one-sided; it presents but one perspective on a future rich in potentialities.* Competing groups of experts hardly communicate and countervailing technical expertise is rarely called upon to strengthen a technical analysis by playing devil's advocate.
(3) *Quality control and professional standards are nil.* Were there adequate standards and expectations about their enforcement, the considerable controversies attending many big model projects would likely not occur.
(4) *Data inputs have obscure, unknown, or unknowable empirical foundations, and the relevance of much data, even if valid, is unknown.* "Hard" numbers seldom are and, in any event, represent only one perspective on problems that surpass measurement's reach. Furthermore, the future is not encompassed by the past; there are no data for the future.
(5) *The most interesting aspect of any analysis is the set of assumptions used to fashion it.* Since there are no answers, as such, for the kinds of policy problems we face, far better to concentrate on the founding premises and preferences of an analysis than on their extensions as results.

All of these assertions are meant to cause honest reflection on the prevalent choice of method. And from the forecasting and planning shortcomings experienced with big model projects [18, 31] it is both timely and appropriate to do so. For instance, it appears that if one seeks forecast accuracy, then multiple-expert opinion studies tend to be preferable to large-scale, model-based studies, because the experts pay more attention to the timeliness and relevance of their core assumptions and because more of these assumptions, or perspectives, have a chance to surface and be considered. "When the choice is between fewer expensive studies and more frequent inexpensive ones, these considerations call for the latter [18, pp 202–203]."

Free-form, manual games

The basic task is to think creatively about exceedingly complex physical and social phenomena, interacting and evolving over periods of time unlike those that usually hold the attentions of the public and its leadership. Component subtasks are to identify the relevant knowledge about such phenomena, to assemble it into a relatively coherent whole, to discover missing pieces so that research priority and responsibility may be assigned, and to stimulate thought about what, if any, human interventions to set in motion – to avert unwanted outcomes or to secure human benefits otherwise foregone.

These basics must be kept well in mind. Not to do so risks diverting intellectual attention. The applications and purposes for our analyses are these: *exploration*, *expert intragroup communication*, and *group knowledge and opinion elucidation*. Judgment and opinion will certainly matter, sufficiently so as to become integral objectives of study. Advocacy, too, will play a distinctive role.

A strong argument has been made previously for free-form, manual games to pursue these purposes. More needs to be said about the historical roots, intellectual rationales, and comparative strengths and weaknesses of this analytic form.

Historical roots

Free-form, manual games are probably as old as man in conflict [32]. Their military genesis in no way detracts from their appeal and appropriateness in other circumstances [33]. Indeed, the complexity, uncertainty, and high stakes of warfare parallel those we face in learning to sustain the biosphere. In the military realm, the difficulty of the substantive problems pushed analysts to discover thoughtful techniques in tune with the challenges [34].

One such challenge was to properly account for nonquantifiable but consequential factors, such as political ones, that somehow had to be integrated into both thinking and analyses. And, historically, the political–military exercise was created and used for just this reason [35].

Political–military exercises explored some difficult questions that eluded or exceeded the capacity of alternative analytic tools:

(1) What political options could be imagined in light of the conflict situations portrayed? What likely consequences would each have?
(2) Could political inventiveness be fostered by having those actually responsible assume their roles in a controlled, gamed environment? Would the quality of political ideas stimulated be as good or better than those obtained conventionally?
(3) Could the game identify particularly important, but poorly understood, topics and questions for further study and resolution? What discoveries flow from this type of analysis that do not from others?
(4) Could the game sensitize responsible officials to make potential decisions more realistic, especially with respect to likely political and policy consequences?

These questions are quite like those facing analysts concerned with the sustainable development of the biosphere. Think momentarily about the typical problems they encounter: messy, hard to specify (save selected parts of them), and interlaced with natural, social, and political elements and processes.

Manual, military games, such as SAFE (Strategy and Force Evaluation) have been used in comparable settings [36]. And while its substantial details differ, its general purposes parallel our own.

> SAFE provides a conceptual framework within which to generate and examine new ideas. It helps to systematize issues and supply gross evaluations of alternative strategies. It keeps the outlines of all components in two interacting "big pictures" in balance and in focus [36, p 17].

And later, these comments about simplification and abstraction.

> Games are frequently censured on the grounds that they involve useless detail, serving only to cloud the true issues, and also on the opposite grounds that in some major respect they are overaggregated.... [But each attempt to simplify] fails to include a key feature and so requires successive expansion of the structure [36, p 19].

Some problems are just difficult, in other words. A brief review of the kinds of knowledge manual, free-form games produce helps one appreciate their role.

Rationales: strengths and weaknesses

This analytic form offers comparative advantages in four broad areas:

(1) To study poorly understood dynamic processes.
(2) To study poorly understood institutional interactions.

(3) To improve the game environment by opening participation to many with different perspectives and special competences, on a continuing basis over time.
(4) To prepare players for future research, analysis, and even operational responsibility.

The many pockets of scientific knowledge about the biosphere lack, for various reasons, coherence and integration. Much of this information is dynamic, where matters of rates of change of time figure prominently. Many other aspects of it are not, especially when one seeks relationships between physical and social phenomena and processes. What, for instance, might it mean for economic and political institutions to sustain major ecological disruptions? No one knows much about this. Genuine knowledge about institutions under intense external stress is far from complete or uncontested and typically gives heavy weight to historical over potential patterns and practices.

The institutional issue also surfaces in terms of the currently weak or nonexistent global authorities or arrangements that biospherical sustenance may necessitate. There is hardly a blueprint to organize mankind to cope with or reverse tropical deforestation or genetic and species depletion. It remains to be drawn, its institutional implications explored, and its possible concrete forms tried out.

The better manual games lasted for periods of time. They were not played once and for all, nor did they produce a single solution to the formidable problems they addressed. The point that the game provided an environment and framework within which ideas could be tried out, as Paxson notes, is critical. DeWeerd takes this even farther in his definition of the "contextual approach" in manual gaming:

> The utility of embedding complex problems in a clearly defined context has long been recognized by the research community.... A contextual framework helps one to exclude irrelevant materials and permits a concentration on the central problem under analysis. One needs a context to avoid wasting time in reaching a common approach to the subject [37, p 403].

Many technical issues receive attention in this mode of analysis, primarily because the scenarios used are accessible and relatively transparent. Are participants using the same time and spatial scales and references? Are the motivating scenarios interesting and plausible – or at least interesting? Is something important left out and, if so, are efforts being made to find out what and put it in?

Participation of individuals from a wide variety of backgrounds is enhanced because proceedings are conducted in plain language. Besides providing improved access, verbal descriptions allow phenomena too complex to be represented precisely still to be considered. Depending on the professionalism of those involved, a free-form game may attain quite adequate standards of rigor and respectability, albeit of a very different nature to those sought in conventional scientific work.

The manual game environment is conducive to learning as it allows those holding specific information to share it within an understandable framework. Points of agreement and disagreement are very quickly and cheaply exposed. Questionable matters of fact may be established; matters of taste and preference can be identified as such. The power of initial assumptions comes clearer to all as well, and with self-consciousness comes the prospect of thoughtful change.

The more general issue suggested here is that players are involved, continuously and closely, so that the elements of the game (and problem) become both familiar and also open to critical exploration. Play of the game, in short, improves the game. A very common outcome of a three- or four-day manual gaming experience (the culmination of many man-hours of preparatory effort) is detailed criticism of the scenarios, the dynamics, and the institutions and individuals involved. This kind of caring professional criticism is extraordinarily uncommon in, for example, a computer-based study or model.

One needs to exercise care not to oversell this approach. It is limited, as all techniques are. Some argue that results of free-form games are unscientific because they cannot be reproduced. Replication is a fundamental experimental requirement, to be sure, but it is far less urgent when one hardly understands a problem's structure in the first place.

Dominant personalities, particularly in the control role, can be troublesome, especially in the hierarchical military setting. An antidote here emphasizes the collegial and peer aspects of the enterprise: everyone involved has something to contribute and to gain. There are no answers.

Cost is often thought to be a problem, but the issue is cloudy. When senior government officials play a manual political–military exercise for a week, the price tag can be high. But, as we are coming to learn, the expense of more conventional analytic forms can be significant too. Free-form games are relatively inexpensive, especially as compared with large-scale machine models. Also, they seldom run the risk of substituting for the reality they mean to portray.

Nor do they cut-off or restrict different or dissenting views and preferences, another important cost. A key need here is the generation of many potential pathways into the future by running frequent, inexpensive, and expert-based studies. Exploration, group opinion, shared expertise, and the clarification of various individual and institutional preferences all require such operational styles and procedures.

Scenarios

Frequent and inexpensive studies come in many shapes and sizes. However, a common feature is the scenario, about which insufficient serious inquiry exists [38]. In the following discussion the nature and role of scenarios is sketched and woven together with suggestions to consider operational modeling as applied to exploration, group opinion, and advocacy goals.

Ubiquitous, essential, and misunderstood

The scenario is the fundamental building block of *all* modeling and analysis.

> After all, it is from our anticipations of environments in which our systems are to operate (the state-of-the-world, the conflict situations, and the tasks these systems are expected to accomplish) that many of our criteria for *evaluating* the *performance* of a given system emerge. Thus, having a casual attitude toward the scenario is often tantamount to having a casual attitude toward the selection criteria. If we accept the proposition that our analyses can be no better than the criteria we employ, then we must accept the corollary proposition that our analyses can be no better than our scenarios [39, p 300, emphasis in original].

The scenario is the crux of the modeling process; it is the basis for bounding and structuring a model and it contains the criteria to appraise the work.

Scenarios have other favorable characteristics. Since they usually rely on a verbal depiction they are accessible; they are easily altered because they are tentative and contingent; and they are future-oriented as they depict past and present with both likely and desired future possibilities [40, 41].

Defining characteristics

But what are scenarios? DeWeerd sums matters up: "The scenario tells what happened and describes the environment in which it happened [37, p 2]." Brown extends this: a scenario is "a statement of assumptions about the operating environment of a particular system being analyzed [39, p 300]." Bell clarifies further: scenarios are the representations of "alternative futures," by which analysts "sketch a paradigm (an explicitly structured set of assumptions, definitions, typologies, conjectures, analyses, and questions) and then construct a number of explicitly alternative futures which might come into being under the stated conditions [42, pp 865–866]." As a general matter, the scenario is the analyst's image or conception of the process or system being represented. It is not strictly speaking a prediction, rather it is an explication of the possibilities [42, p 866].

In contrast, imagine a negative or antiscenario: an explicit statement of those features of reality the analyst is *not* attempting to represent. The employment of such would go a long way toward exposing and clarifying, in plain language, those features being emphasized and those being left out of any given analysis, as there inevitably must be. Ideally, the construction of a scenario and its antithesis would encompass all knowledge relevant to a problem, system, or process. The ideal is unattainable as a practical matter, although paying attention to important aspects of reality left out is not.

A scenario sets up the approximate initial conditions of some setting or situation and provides time-based clues as to the most likely or probable sequencing and interactions of events. Using the concept this way, and applying it to free-form or manual games, reveals the importance of the game as an exploratory device. We play the game using the scenario to examine alternative developmental pathways and the options that might be taken at various times and that could affect outcomes. Open is the creative possibility of learning something that might not have occurred independently, either to the scenarist or to other participants.

Machine models are scenario dependent, too, a point not well made in the existing literature and practice. The scenario basis for a machine model merely sets up the initial conditions through the specification of included variables and their interrelationships and the assignment of initial conditions to the parameters. The time series the model produces represents one developmental pathway. Frequently, no others are considered once a model is tuned to produce plausible values. Foregone in the process is the opportunity to create and explore alternative pathways, specified in different ways and employing different initial conditions than the one set devised [43].

Methods, applications, and scenario forms

The nature of scenarios contained within each category of method (e.g., analytic models, machine simulations) will differ according to the intended applications and technical requirements of each. This point is portrayed in *Table 17.2*.

Note that methods are ordered in *Table 17.2* from "hard" to "soft" as one moves down the column; priority of use is signified by the ordering within "*Uses*". Two additional types of methods are introduced as well, thus to enrich the spectrum of possibilities.

Analytic models typically have no explicit verbal scenario. It resides in the mind of the analyst responsible, but it exists nonetheless. In principle, the guiding scenario could be externalized and one probably should do so if the insights that analysis yields are attractive enough to consider their pursuit and adoption [44].

Table 17.2 Method types and uses: scenario forms.

Methods	*Uses*
Analytic models	Research: exploration, planning, cross-checking
Machine simulations	Research: planning, forecasting, advocacy
Man–machine	Teaching–training primarily; research: planning
Free-form	Teaching; intragroup communication; research: exploration, group opinion
Seminars	Teaching; intragroup communication; research
Group studies	Intragroup communication; research

Machine simulations typically fix the context, because of technical constraints imposed to give specification and because of omissions, and they focus on the calculable physical causes and consequences of interest. The simulation characteristically has most of its scenario embedded within the computer routine, where it is quite hard for most, other than the model builder, to see or understand it [45].

Of the very few man–machine activities still in existence, most are used for teaching and training, where the scenario is both buried in the machine and rigidly held [46]. Training air traffic controllers, for instance, can be done in a man–machine setting, the main object of which is to indoctrinate players according to rigid and well understood procedures and practices. As a research device, the man–machine game has been fruitfully employed to study and plan for human-factor implications in complex organizations and systems, e.g., for determining the optimum number, form, and location of indicator instruments in a nuclear power plant control facility. But, once more, the organizations and systems are generally specifiable and rather well understood.

Free-form, manual games are most appropriate when specification and understanding are not possible or where disagreements about them abound. But, depending on the overall purpose a free-form game serves, the form its scenario takes will vary.

When used for teaching purposes, for example, an initiating scenario would provide the student players information needed to confront a problem and would engage them actively with it. Almost by definition (and in practice) this means that the scenario will describe major moves the represented actors have made as a necessary precursor to the problem at hand. Teams are told, in effect, "You made these decisions in the past and their consequence is a nasty problem. Now deal with it."

When intended to foster intragroup communication, this scenario style won't work. In this case, the initiating scenario is drawn from a realistic situation of interest and importance to the players. No goading is required to engage them. Rather, the objective is to lay the groundwork for discussion of a real problem, typically proximate in space and time.

Research applications are of most interest to us [47]. Scenarios in this instance are clearly treated as tentative, exploratory, and incomplete, in accord with the level of incomprehension of the problem being explored [40]. As a practical matter, the scenario writer, control (game director), and all team members are peers, any or all of whom have equal right to question, modify, or improve the initial scenario.

The manning of a free-form research game is critical. A diverse array of skills and backgrounds (relevant to the problem) must be represented on the teams. The collective endeavor is intended to share these differences in systematic and insight-yielding ways. Variety in the opinion, attitudinal, and belief premises and structures represented must be sought, clarified, and accepted as legitimate by all participants. Mutual professional respect, including tolerance for differences in discipline, style, method, and central concerns, is equally essential.

What is a "good" scenario?

This is the wrong question. It should be: How well does a given scenario meet the game or research purposes for which it was intended? A good scenario should be problem oriented, as determined by its success in focusing attention and opening up unknown possibilities. It should lay out biases for collective scrutiny and consideration, not bury them under a bushel. It should not unduly violate common sense, in the senses of being honest, credible, consistent, and intellectually satisfying. It should not be expected to cover all situations and all problems; for that, many other scenarios will be needed.

A good scenario is explicit; it exposes its terms for easy reference by all interested in the problem. It is only when a scenario can be communicated to others that it can be evaluated or acted on. Implicit scenarios can hide biases, conceal assumptions, and are capable of being reshuffled to meet objections or unexpected developments. To the extent they do, they are bad.

Goodness and badness are sometimes equated with a scenario's credibility or acceptability to a user. *Realistic* in this case is used to describe a good scenario and *unrealistic* serves the bad: this can be misleading. For situations close at hand and near in time, the best approximation to conditions generally presumed to be existing by informed persons yields greatest acceptance and inspires greatest interest and confidence. But for situations more remote and further ahead in time, credibility and acceptance themselves may be lacking because of failed consensus [48].

Since no one can predict the long-term future in detailed and specific terms, any scenario set there is liable to be rejected by some as implausible, incredible, or slanted. The near term is easier and allows present trends to be extended and accepted. The longer term portends *real* uncertainties, manifest as surprises, discontinuities from past trends and practices. Studies of the long-term future will always provoke rejection of selected parts. But the point, after all, is to avoid surprises, and "to avoid surprises one cannot take for granted that the future is merely the scaled-up version of the present [49, p 3]." The problem needs to be frankly recognized by pointing out when and why a discontinuity has been introduced in the scenario and analysis. Credibility in this case cannot be expected to develop from a consensus of views, certainly not when such is lacking.

A good scenario concentrates attentions by providing a common framework and reference. It delineates boundaries of a problem clearly to satisfy demands for honesty, credibility, and relevance. It introduces as little unintentional bias into collective proceedings as possible, but it also allows intentional bias to surface and be explored and challenged. A bad scenario does none of these things. It features great unexplained leaps from one point or event to another; it uses gimmicks to achieve its ends; or it sweeps problems under the rug by attributing them to a series of miscalculations.

Policy exercises

This chapter has covered a great deal of territory, but then the aspirations of this volume (and the project it portends) invite wide scope. Despite this, several themes guided the journey, and these require summary. Afterward, a proposal based on them is offered for consideration.

Major themes

Dazzlingly large amounts of information relevant to man and the biosphere exist or can be identified as important and relevant. Much of this is rightly *scientific* in character in the sense that it concerns natural phenomena, the description and specification of which produces cumulative and shared knowledge. It is critical that this chapter's intentionally strong argument is not misunderstood so as to deny the essential role scientific information must play in our mutual concerns and deliberations.

However, there is another very broad class of information that is not, strictly speaking, scientific nearly so much as it is *human*. And lest the appropriateness of this distinction escape one, recall our given terms of reference: the discovery of constructive and lasting improvements in our understanding of the biosphere *and* man. The implications are plain enough despite the incredible difficulty of seeing them through.

Very high on the list of difficulties is coming to terms with, if not to a full understanding of, the ineluctable fact that human perceptions of the world – past, present, and particularly in the future – are highly variant. Furthermore, such diversity need not be considered evil or irrational. Indeed, from it springs the creative inspiration that marks man as such. A fundamental requirement is thus to release for the common good the creative energies of ourselves and others.

As somewhat harshly argued here, prevailing attitudes and styles of knowledge creation and use have too often done precisely the opposite – by denying the legitimacy of different perspectives and preferences, by adhering narrowly to intellectual paradigms ill-suited to the challenges all confront (and then dissolving into brittle squabbles when the limitations of each are exposed), and by favoring tools and methods used to solve problems only remotely like those facing us (and continuing to use them despite lack of success).

As a practical matter the work to be done can be identified. Realistic problems must be characterized, in their relevant contexts, according to as many points of view as possible, thus to expose differences so as to resolve them, if possible, or to live with them, if not. Such characterizations must proceed openly and with utmost respect for matters important, but unknown or even unknowable. Utmost respect means not only the assembly of technical and factual information, including contentious elements, but also subjective and preferential information which, almost by definition, will mirror wide differences and inconsensus existing in the world at large. There are no answers, only imaginings and possibilities, no one of which is necessarily best for everyone for all time.

As an intellectual matter the work ahead can also be identified, despite the enormous difficulties honest engagement with it creates. Five broad tasks, outlined below, come to mind.

Goals must be clarified. Put another way, preferred events to be sought over the long haul must be explicitly considered. The requirement extends to cover both individual and institutional preferences, where institution is taken in its broadest sense.

Past trends, giving particular emphasis to man's *interactions* with the environment, must be carefully delineated. Among other reasons for doing so is the need for clarity about the sources of problems currently in the fore. They came from somewhere and have been caused by something, and by someone.

The intellectual and scientific task to improve comprehension of what is causing what and why needs less emphasis, despite its essentiality. The difficulty comes as we try to open our minds to the possibilities that many scientific paradigms have something to contribute and that no one or group merits special treatment and regard. There are no sure bets, no royal roads.

The projective task has been emphasized throughout this chapter to include gaining as clear a vision as possible of where we are heading and why. Obviously this cannot be done or done well without an understanding of what we wish to accomplish, how well we have succeded in the past, and why. The five broad intellectual tasks are, in other words, mutually dependent and supporting. Emphasis on only one at the expense of the others will not work.

And, finally, comes the creative task of discovering and inventing new means to avert futures imaginable (but unwanted) and to secure those more preferred. At the very least, one wishes these means to succeed in time by leaving mankind no worse off. At the very best, possibilities are only limited by our own imaginations.

Basic requirements

Practical and intellectual requirements must be joined in systematic fashion, not treated piecemeal or in isolation. This is a very tall order. It also runs contrary to most current thinking and practice, which implies that extraordinary efforts will be required for its accomplishment.

The tall order I have in mind can be sketched; its development will, however, be both risky and problematic. The risk derives from many sources, of which the main ones have been previously identified. The problems are not so easily foretold, although they are to be anticipated.

A *policy exercise* is a deliberate procedure in which goals and objectives are systematically clarified and strategic alternatives invented and evaluated in terms of the values at stake. The exercise is a preparatory activity for effective participation in official decision processes; its outcomes are *not* official decisions. Those engaged in policy exercises may on occasion include those with decision authority, primarily as a means to elicit information from this point of view. However, the core analytic group responsible for the policy exercise must be ever mindful of the nonbinding, unofficial nature of their shared work and its outcomes.

The active analytic paradigm guiding overall work, I would suggest, should be a *consensual* one. This does not rule out other types, however, but only if participants self-consciously declare their intentions to pursue advocacy or technocratic or partisan ends as a means to clarify specific options or possible actions of mutual concern and interest.

Participants must, in time, learn restraint: by relaxing customary identifications with paradigm, position, and place. But at the same time, conscious efforts need to be made to include an unusual assortment of highly talented and variously trained individuals. Some sense of this follows from thinking through the five intellectual tasks previously identified. Specification of trends, for instance, is a function historians are trained to perform. But how often are historians included as peers within normal policy research efforts? Honest confrontation with matters of goals and preferences implicates humanists, philosophers, psychologists, and even literary and artistic talents. It does more than just implicate; it requires their active participation in policy exercises under conditions of respect equal to those accorded scientific specialists.

Respect is the absolutely essential ingredient, albeit an elusive one attained only through uncommon tolerance for perspectives and preferences not understood or shared. As a procedural matter, all participants in policy exercises become *objects of study*, not merely its agents. The quest of such study is better comprehension and appreciation of existing and legitimate differences, including their consequences for collective problem characterizing and clarifying work.

Questions of the following sort could guide such effort (all taken with respect to participation in the policy exercise itself):

(1) Are the participants more or less sensitized to the interdisciplinary dimensions of a problem?
(2) Are they more or less disposed to ask questions that call for information not previously considered (or appreciated)?
(3) Do they improve in their abilities to reduce tendencies to exaggerate the "hardness" of some data (and derivative conclusions); to increase their acceptance and understanding of "softer" forms?
(4) How does the policy exercise affect conceptions and images of past trends, if at all?
(5) How are conceptions of the future influenced, if at all? Are participants more or less free of negative feelings about speculating as to the future, particularly about challenging

conventional limits of thought on the part of salient groups or individuals? Are they more sanguine or pessimistic about the future? Is the future envisioned as culminating in catastrophe (if so, when?) or in revolutionary reconstructions of institutions in desirable ways? Or in fluctuating cycles without trend? Are these perspectives new or are they continuations of previous assumptions?

The point of asking and pursuing these questions explicitly recognizes that knowledge creation and use are far from neutral activities but, rather, are integral with the whole of policy processes.

Note that I have consciously avoided using the phrase problem solving, which has been prematurely and unconstructively applied in cases where even a rudimentary grasp of important matters has not yet been attained. The main intents, thus, of the policy exercise are to *discover* as much about given problems as time and talent allow; to *clarify* as openly as possible differences of opinion affecting specific and common perceptions of the problem; and to *integrate* all pertinent knowledge, from a variety of sources using a range of methods, for general edification and consideration.

The matter of methods itself deserves investigation and improvement. Previous classification efforts were intended to at least identify some overlooked and underused possibilities, several of which conform, at least in principle, to the general requirements of the policy exercise. Much more is needed here, including the possibility of devising entirely new techniques and procedures to advance the common enterprise.

The practicality of creating new methods when faced with difficult problems unyielding to the old has historical precedent: witness the bursts of creative energies attending operational and systems analyses at certain points in their evolution. More, I believe, can be recovered from these experiences if appropriate questions are asked of them. And the agenda for the policy exercise would, indeed, include such retrospective stock takings.

Procedures

The policy exercise finds its procedural roots in scenario-based, free-form games. It is as much artistic as it is scientific in its style and means, a characteristic that in no way denigrates the activity. Or, as Webster's defines art:

> **1**. creativeness. **2**. skill. **3**. any specific skill or its application. **4**. a making or doing of things that have form and beauty . . . . **8**. any craft, or its principles. **9**. cunning (Webster's, 1970, p 41).

Substantive knowledge, insight, an ability to abstract, flexibility, and a willingness to build and rebuild many representations of interesting phenomena are additional ingredients that one can imagine adding to the success of a policy exercise. The "form and beauty" in this case relate to the images and words used to portray settings, interrelationships, and events – past, present, and future. A scenario is the portrait, the verbal model [50].

Other artistic tools of the trade come to mind. Maps, drawings, graphs, and pictures – both still and dynamic – play important roles in melding imaginings with measurements and providing participants with a vivid sense of the whole [51]. Furthermore, the combination of the verbal with the diagrammatic ought to allow many different versions of problems to be created, tried out inexpensively, explained to those involved, and thrown away if faulty or no longer interesting. The emphasis here is on making many different renditions of a complex setting, not on concentrating intellectual energies on just one.

At the heart of the policy exercise is a core group of some 10 to 15 individuals, each of whom contributes a distinctive set of skills, perspectives, and concerns about the general problem. Commitment and continuity of this group are absolutely essential. Occasional outsider participation is to be encouraged, especially if important questions members of the core group pose can be answered this way. What is to be avoided is the devolution of the policy exercise into an ordinary conference or seminar format.

As a practical matter, it is not physically possible to have the core group together most of the time. However, computer teleconferencing technologies are sufficiently developed to allow far more intellectual contact, at reasonable costs, than ever before. For a policy exercise on important global issues to operate, computer teleconferencing is an essential ingredient [52]. The large-scale, once-every-great-while, global conference serves one class of purposes – calling attention to issues and sketching out an agenda are two important ones. The policy exercise, as envisioned here, is quite differently intended, as its special procedural characteristics suggest.

Reconsider the previously mentioned five intellectual tasks. Each member of the core will be asked to present, for collective appraisal, verbal and/or written assessments of the underlying goals of relevant groups and individuals in the context of past trends, factors conditioning these trends, probable future outcomes, and possible methods to deflect, encourage, or otherwise alter these outcomes. These assessments are *not* intended for public dissemination. They are for internal use only, as it were. Furthermore, these assessments are not randomly generated, but follow in response to an evolving agenda, set through mutual consultation and agreement.

The "closed shop" aspect of the core group's activities, at least as it conducts its basic work, cannot be stressed enough. Since the policy exercise is not interested in positions already adopted nearly as much as it is in creating new ones based on richer understanding, intellectual maneuvering room must be established and honored. The problem of engendering personal and professional respect has already been emphasized.

Policy is choice about the future. Participants will thus be encouraged to make independent estimates about probable outcomes for problem elements. Such independent judgments may be summarized, compared, discrepancies considered, and common expectations about the future clarified [53]. As events unfold and new information is brought to bear, the initial estimates can be reconsidered. Over time, such estimates may themselves be appraised, not to assign personal credit or blame but to determine what assumptions, scenarios, and techniques worked and why?

The policy exercise must employ many different methods. A basic reason for much of this chapter is to remind others of several promising, but underused possibilities. More will probably be needed, including the creation of entirely new techniques, methods, and procedures. But keeping the effort simple is probably a good idea at first. This belief naturally leads one to represent our problems in simple, abstract, and comprehensible ways. "Modeling simple, but thinking complex", as the old modeler's dictum holds, is the order of the day. Verbal and visual representations need particular emphasis and development . . . a main reason why scenarios have been heavily stressed here .

To store, recall, and share information – especially in a multinational, multilingual setting – audiovisual aids must be devised and used to a far great extent than conventional academic conferences and seminars do. In general, the purposes to be served are collecting, evaluating, presenting, and disseminating pertinent information about the problem context. The created environment of the policy exercise is, in effect, the collective institutional memory.

The general attitude sought for members of the policy exercise is one of *critical imagination*. Speculation based on one's exposure to the evolving collection of facts and understandings about the world is encouraged. Ego and public reputation are protected by mutual agreement and out of respect for one another and for the importance of the collective endeavor. However, wild speculation untempered with critical judgment is disallowed. The judgment sought here comes from within one's self as well as from colleagues similarly engaged.

If there is a prevalent procedural or operating style, then it would be *purposively eclectic*. Given the problem at hand, no piece of information, no means of communicating or representing it, and no means to probe and discover are to be excluded.

The roles of judgment and opinion

The problems of interest to us here are not readily dealt with technically. A considerable amount of judgment must be exercised. This does not mean a mysterious, intuitive insight that cannot be described and justified, however. It does mean taking greater care than usual to discern their sources and consequences.

Many judgments are hard to discern and analyze. An adequate account of them likely involves much discursive recitation, of at best passing interest to busy decision makers and many others. Nonetheless, working to know and perceive the judgmental bases underlying any analysis must be just as important for a decision maker as the conclusions set forth. Or, to put it otherwise: What are the assumptions and how and why were they made? What technical judgments were made and what are their consequences? These apparently simple questions are essential procedural concerns of those engaged in policy exercises.

Discovering attitudes, based on reactions to opinion statements, is possible [21, 54]. And gaining an understanding about the attitudinal and belief bases of those engaged in this collaborative endeavor provides an opportunity to clarify inevitable sources of difference and disagreement – so as to anticipate and perhaps resolve them before they become sticky and intractible.

Even more interesting possibilities could be explored. For instance, in the setting of the scenario-based, free-form game, what differences occur as a consequence of matching teams of disparate attitudes and beliefs to work a specific problem? It could be extremely interesting, as well as creating an opportunity to examine, in a controlled and collegial setting, likely future conflicts – and then working to overcome them [55].

Various group judgment techniques meant to elicit expert opinion in systematic ways have been developed over the years, although many analysts have objected to them on narrow scientific grounds [56]. Other problems have been noted as well: determining whose judgment to seek; problems of pooling ignorance (and then believing it); interpreting results and making use of them.

The general challenge that pursuit of expert opinion raises can be posed as a paradox. In trying to enumerate judgmental and preferential elements within an analysis, thus to reduce complexity somewhat, one risks adding to shared uncertainty by imparting a false sense that more is known than is realistically possible.

Advocacy

The same can be said for advocacy. It is pervasive and probably inevitable. If so, rather than futilely stemming the tide, it may be more sensible to move with its flow, so to take advantage of the powerful forces advocacy represents.

A case can be made for advocacy. Actually, there is nothing intrinsically wrong with individuals promoting a perspective, disciplinary paradigm, or a derivative policy preference. But only so long as other advocates have their day in court and the rules of the adversarial game are clearly spelled out beforehand. Perversion of the intellectual and policy processes occurs when advocacy is one-sided and when no one bothers to describe the rules . . . or play by them.

There is nothing intellectually or morally wrong for one to make the best possible case for a given line of behavior. Such may present an interesting hypothesis, the full consequences of which can be grasped only through aggressive promotion. Theological debate, in an earlier time, was explicitly adversarial. It also had the intellectual advantage of organizing questions and permitting the jury or auditors to confront more adequately different positions. And as judicial and legal practices demonstrate, when the rules are spelled out and honored, loss of a debate or case need not imply personal condemnation.

What is crucial is whether the best case is made in an intellectually honest way, and whether opportunities exist to criticize and make other "best possible" cases. But, as before, the real trick is remembering that the case made is only one of numerous possibilities, no one of which necessarily fits the expectations, hopes, and fears of all those involved.

Notes and references

[1] A recent exception is Resources for the Future's Policy Education Program, Castle, E. (1983), *Information, Communication and the Policy Process* (Resources for the Future, Washington, DC).

[2] Brooks, H. (1979), Technology: hope or catastrophe? *Technology In Society*, **1**, 3–17.

[3] Ophuls, W. (1977), *Ecology and the Politics of Scarcity* (W. H. Freeman, San Francisco, CA).

[4] This illustration is based on Rein, M. (1976), *Social Science and Public Policy*, pp 98–101 (Penguin, Baltimore, MD).

[5] Clark, C. (1981), Bioeconomics of the ocean, *BioScience*, **31**(2), 231–237.

[6] Glantz, M. H. (1979), Science, politics, and the economics of the Peruvian anchoveta, *Marine Policy*, **3**, 201–210.

[7] Cane, M. A. (1983), Oceanographic events during El Niño, *Science*, **222**, 1189–1195; Rasmussen, E. M. and Wallace, J. M. (1983), Meteorological aspects of the El Niño/Southern Oscillation, *Science*, **222**, 1195–1202.

[8] Barber, R. T. and Chavez, F. P. (1983), Biological consequences of El Niño, *Science*, **222**, 1203–1210.

[9] From a 1970 reported high of 12×10^6 tons the Peruvian anchoveta fishery plummeted to less than 0.5×10^6 tons in 1978 [6].

[10] Cooper, R. N. (1974), An economist's view of the ocean, in R. E. Osgood (Ed), *Perspectives on Ocean Policy*, pp 145–165 (John Hopkins University Press, Baltimore, MD).

[11] Bailey, C. (1984), Managing an open access resource: the case of coastal fisheries, in D. Korten and R. Klaus (Eds), *People-Centered Development*, Ch. 9 (Kumarian Press, West Hartford, CT).

[12] Treating deviations from theory as being caused by imperfect people, not deficient theory, is illustrative.

[13] Barker, R. G. (1968), *Ecological Psychology: Concepts and Methods for Studying the Environment of Human Behavior* (Stanford University Press, Stanford, CA).

[14] Alexander, C. (1967), *Notes on the Synthesis of Form* (Harvard University Press, Cambridge, MA).

[15] Dubos, R. (1968), Environmental determinants of human life, in D. C. Glass (Ed), *Environmental Influences*, pp 138–154 (Rockefeller University Press and Russell Sage Foundation, New York).

[16] The theoretical issues are treated in Simon, H. A. (1969), *The Sciences of the Artificial* (MIT Press, Cambridge, MA); and the practicalities are covered in La Porte, T. R. (Ed) (1975), *Organized Social Complexity: Challenge to Politics and Policy* (Princeton University Press, Princeton, NJ).

[17] Boulding, K. (1961), *The Image* (Ann Arbor Paperbacks, Ann Arbor, MI).

[18] Ascher's dispassionate stock-taking is as eye opening as it is devastating; Ascher, W. (1978), *Forecasting: An Appraisal for Policy-Makers and Planners* (Johns Hopkins University Press, Baltimore, MD).

[19] Stech, F. (1979), *Political and Military Intention Estimation* (MATHTECH, Bethesda, MD).

[20] Turco, R. P. *et al.* (1983), Nuclear winter: global consequences of multiple nuclear explosions, *Science*, **222**, 1283–1292; Singer, S. F. (1984), The big chill? Challenging a nuclear scenario, *The Wall Street Journal*, February 3, 22.

[21] Greenberger, M., Brewer, G. D., Hogan, W., and Russell, M. (1983), *Caught Unawares: The Energy Decade in Retrospect* (Ballinger, Cambridge, MA).

[22] A parallel list was formulated by Jervis, R. (1963), Hypotheses on misperception, *World Politics*, **20**, (4), 440–449.

[23] I will address the intellectual issues in a book-length project, under IIASA auspices, with the working title *Genesis and Synthesis*, and a targeted completion date of 1986.

[24] Shubik, M. and Brewer, G. D. (1972), *Models, Simulations, and Games: A Survey*, pp 1–10, R–1060 (The Rand Corporation, Santa Monica, CA).

[25] Brewer, G. D. and Shubik, M. (1979), *The War Game: A Critique of Military Problem Solving*, p 19 (Harvard University Press, Cambridge, MA).

[26] Shubik, M. (1982), *Game Theory in the Social Sciences* (MIT Press, Cambridge, MA).

[27] Brewer discusses this at length and provides the supporting documentation, see Brewer, G. D. (1983), Some costs and consequences of large-scale social systems modeling, *Behavioral Science*, **28**, 166–185.

[28] Szanton, P. (1981), *Not Well Advised* (Russell Sage and Ford Foundations, New York).

[29] This is a major theme of a special issue of *Policy Sciences* (1984), Special issue: the IIASA energy study, *Policy Sciences*, **17**(3).

[30] Janis, I. L. and Mann, L. (1977), *Decision Making* (Free Press, New York).

[31] Ascher, W. and Overholt, W. H. (1983), *Strategic Planning and Forecasting* (John Wiley & Sons, New York).

[32] Young, J. P. (1959), *A Survey of Historical Developments in War Games*, Staff Paper no. 98 (Operations Research Office, Johns Hopkins University, Bethesda, MD).

[33] Hausrath, A. H. (1971), *Venture Simulation in War, Business, and Politics* (McGraw-Hill, New York).

[34] McHugh, F. (1966), *Fundamentals of War Gaming* (US Naval War College, Newport, RI).

[35] Goldhamer, H. and Speier, H. (1959), Some observations on political gaming, *World Politics*, **12** (1), 71–83.

[36] Paxson, E. W. (1963), *War Gaming*, RM-3489 (The Rand Corporation, Santa Monica, CA).

[37] DeWeerd, H. (1967), *Political Military Scenarios*, P-3535 (The Rand Corporation, Santa Monica, CA).

[38] My interest in this topic derives from close professional association with Harvey DeWeerd, William Jones, and Herbert Goldhamer, all exemplary exponents of free-form gaming. (DeWeerd and Goldhamer are both deceased; Jones continues work at the Rand Corporation.)

[39] Brown, S. (1968), Scenarios in systems analysis, in E. S. Quade and W. I. Boucher (Eds), *Systems Analysis and Policy Planning* (American Elsevier, New York).

[40] DeWeerd, H. (1974), A contextual approach to scenario construction, *Simulation & Games*, **5**, 403–414.

[41] de Leon, P. (1975), Scenario designs, an overview, *Simulation & Games*, **6**, 39–60.

[42] Bell, D. (1964), Twelve modes of prediction: a preliminary sorting of approaches in the social sciences, *Daedalus*, **93**, 865–878.

[43] This limitation is only slightly compensated for by running sensitivity analyses which, in usual circumstances, are minor modifications of the underlying scenario, if sensitivity analyses are done at all.

[44] Kennedy is extremely helpful about this in Kennedy, J. L. (1952), *The Uses and Limitations of Mathematical Models, Game Theory, and Systems Analysis in Planning and Problem Solving*, P-266 (The Rand Corporation, Santa Monica, CA).

[45] Kahn, H. and Mann, I. (1957), *Ten Common Pitfalls*, RM-1937 (The Rand Corporation, Santa Monica, CA); Brewer, note [27].

[46] In its heyday, the System Development Corporation in Santa Monica, California, was perhaps the most notable institution devoted to this type of activity.

[47] Brewer and Shubik, note [25, Ch 6], provide historical examples that emphasize military research and operational applications.

[48] This is the key difference between intragroup communication and research uses of free-form gaming.

[49] Morris, D. N. (1972), *Future Energy Demand and Its Effect on the Environment*, R-1098 (The Rand Corporation, Santa Monica, CA).

[50] Little or no systematic research exists against which any given scenario could be readily appraised. Indeed, one of the key purposes of the Biosphere Project is to contribute to general understanding about these and related matters.

[51] "Participants" refers to more than those directly involved and includes outside groups, citizens, and decision makers.

[52] It might be noted that Walter Orr Roberts, among others, is actively pursuing Soviet and other participation in a computer teleconference devoted to carbon dioxide issues. James Lovelock is involved with Roberts, too. Doubtless many other serendipitous linkages exist, need to be identified, and then exploited with whatever technological means available.

[53] Lest one suspect that this cannot be done, it needs be noted that a most insightful and provocative afternoon was spent by a small group of conferees puzzling over the following procedural question: What will the biosphere be like in the year 2100?

[54] Brown, S. R. (1980), *Political Subjectivity* (Yale University Press, New Haven, CT).

[55] Don Straus, formerly of the American Arbitration Association, has made a similar appeal in other forums, including IIASA.

[56] One method is presented in Dalkey, N. C. (1969), *The Delphi Method*, RM-5888 (The Rand Corporation, Santa Monica, CA).

Commentary

N. C. Sonntag

Since the early 1960s, there has been an ever increasing number of experiments attempting to use the tools of operations research to somehow optimize man's exploitation of the biosphere. In more recent years there has emerged a recognition (albeit sometimes reluctant) that the resources we wish to exploit are part of a very complex, dynamic system full of uncertainties and surprises that defy the principals of optimality and force us to be content with improvements, however modest they be. In truth, this "recognition" was a quantum leap in our level of understanding of the biophysical systems to which we are inextricably linked. As part of this quantum leap, it became increasingly obvious that the methods and tools we hoped would be the panacea for all that ails us with respect to the biosphere were not what we wanted them to be. The answers were not there for the taking. In our desperation for solutions we focused almost exclusively on product (e.g., the best decision, the best plan) and attempted to fit all of our elegant tools into each and every problem to obtain that product. The consequences were failures: some of them catastrophic, many of them costly, and most of them embarrassing. The common methods for synthesis in resource management (e.g., analytic models, simulation, man–machine games) were put on the defensive, and rightly so. It forced the practitioners to rethink the purpose and utility of their tools and thereby to learn from the failures.

In Garry Brewer's chapter the reader is first led through an eloquent and insightful synthesis of these painful lessons and then introduced to the concept of free-form, manual games. A strong argument is presented for the need to develop a free-form approach to man's development and management of the biosphere. Before presenting a specific comment I have on the free-form idea, I would like to first describe what I see as the major theme in Brewer's recommendations.

As mentioned above, until recently the use of models in resource management focused on the development of product. What Brewer is calling for, in my opinion, is a shift to a focus on process. Notwithstanding the need to develop a good product (i.e., solution), the process by which the product is developed is crucial to its acceptance and eventual implementation. The designers and implementers of policy must be responsive and responsible to a large and diverse set of interests and so demand that the synthesis process account for and integrate these interests in the analysis. Understanding exactly what this process is can be best accomplished by reviewing the chief characteristics of the process, as described in Brewer's chapter.

(1) It develops a contextual framework (through consensus) at the beginning to facilitate concentration on the central problem under analysis.
(2) It includes in the analysis all those with interests relevant to the problem and gives them every opportunity to critique and contribute from the outset.
(3) It develops and applies creative methods of communication (e.g., audiovisual aids, various levels of documentation, workshops, and other forms of collective proceedings) extensively throughout the analysis.
(4) It has a strong focus on the development and maintenance of consensus. A successful outcome is more likely if an analysis builds on a continuing set of agreements.
(5) It uses tools and methods that promote discovery, clarify key issues and differences, and integrate across disciplines.
(6) It explicitly recognizes the pervasive nature of uncertainty in biophysical and social systems. Therefore, the process is adaptive and flexible in both design and implementation.
(7) It involves an unusual assortment of highly talented and variously trained individuals at an equal level of respect and integration.

Implementation of a process with these characteristics is certainly a challenge, but not one from which we should shy away. Brewer's enthusiasm for a process like the free-form, manual games developed by the military is not only well argued, but also provides some application experience that policy analysts concerned with the development of the biosphere can turn to for some "hard knocks" insights and lessons. However, I would like to draw the reader's attention to a process very similar to the free-form approach that was developed explicitly to deal with resource management problems and now has over a 10-year track record of applications to "real world" problems. The method is called Adaptive Environmental Assessment and Management (AEAM), is described in a book of the same name [1], and is further reviewed in Sonntag *et al.* [2].

AEAM, with various levels of success, incorporates each of the above process characteristics and has, over the last decade, been applied to over 100 resource management problems in North and South America, Europe, and Southeast Asia. Of particular interest to us here is one of AEAM's key tools for synthesis and communication: development of a simulation model of the system of interest. In a typical application of AEAM, a simulation model of the biophysical–social system is built in an interactive workshop attended by 30 to 40 people representing all the key interest groups (e.g., policy, management, scientific, public, special interest) and facilitated by 4 or 5 highly experienced modelers also trained in group dynamics and facilitation. Through a series of structured exercises, the workshop participants build the model framework in a highly transparent environment and the modelers translate the framework into computer code to allow for eventual scenario gaming at the end of the workshop and after.

In direct contrast to Brewer's experience, these models have been developed successfully in situations where data were inadequate and short-term deadlines had to be met. Further, the models were designed and implemented in a style that stressed consensus and avoided any particular individual's bias (albeit recognizing the workshop has a bias). So why the difference? After some thought one realizes that philosophically the difference is not as great as it first appears. Rather, it is a difference in how a strong focus is provided to the process. In AEAM, the model serves as the framework around which to develop the policy analysis. The model's predictions (i.e., the obvious products) are useful in aiding the analyst to evaluate the options, but the more important products are the resultant consensus on structure and the sense of cooperation and communication that are established through the group process. The interesting feature is that the building of the computer model provides a very effective way of forcing participants to articulate their biases (i.e., individual mental models), thereby allowing all the participants the opportunity to review, critique, and understand each other's point of view. What is critical to appreciate is that the model (as it is manifested in computer code) usually does not play a central role in the eventual development of products. Rather, it serves a supportive function as a "state-of-the-knowledge" group hypothesis that is set up to be invalidated and then suitably revised at future workshops.

The possible entry of a free-form, manual gaming process into the realm of resource management is an exciting prospect and should be given every opportunity to contribute to our desperate need to come to grips with the issues of sustainable development of the biosphere. Where I see it playing a key role is in our attempts to evaluate policy for interactions between socioeconomic development and the natural environment that operate over large spatial and temporal scales (e.g,, acid rain). At the hemispheric to global scale, an AEAM modeling approach (as it is currently practised) would likely place unreasonable constraints on the analysis. Therefore, to follow Brewer's advice, our prevalent procedural or operating style must be purposively eclectic ("... no piece of information, no means of communicating or representing it, and no means to probe and discover are to be excluded [p 470]"), and the next step is to take the best features of the two approaches and develop a hybrid. In doing so, however, we must ensure that the people we hope to serve (i.e., policymakers and management) are involved as much in the design as in the eventual application of this new process. A failure to do so would be ignoring one of the major lessons of the previous decade.

Notes and references (Comm.)

[1] Holling, C. S. (Ed) (1978), *Adaptive Envrionmental Assessment and Management* (Wiley, New York).

[2] Sonntag, N. C., Bunnell, P., Everitt, R. R., Staley, M. J., and McNamee, P. J. (1982), *Review and Evaluation of Adaptive Environmental Assessment and Management,* Environment Canada Report (Environment Canada, Vancouver).

Author Index

The author index gives only the names that appear in the notes and references section of each chapter and discussion. Also the pages given for each entry are taken from these sections only – the reader should use the note number to find the relevant text.

Subject Index

Numbers in italics indicate a table or figure in the text. Elements and simple chemical compounds are listed by formula under Chemicals (but are also cross-referenced by full name). Compounds with more complex formula and groups are also listed under Chemicals (after the formulae) and cross-referenced by full name.